P9-DDN-504

The WORLD of PHYSICS

A Small Library of the Literature of Physics from Antiquity to the Present

by JEFFERSON HANE WEAVER

WITH ADDITIONAL COMMENTARIES AND NOTES BY
Lloyd Motz, Ph.D., and Dale McAdoo

SIMON AND SCHUSTER  *New York*

Published by Simon and Schuster
A Division of Simon & Schuster, Inc.
Simon & Schuster Building
Rockefeller Center
1230 Avenue of the Americas
New York, New York 10020
SIMON AND SCHUSTER and colophon are registered trademarks
of Simon & Schuster, Inc.
Designed by Irving Perkins Associates
Manufactured in the United States of America

10 9 8 7 6 5 4 3 2 1

Library of Congress Cataloging in Publication Data

Weaver, Jefferson.
The world of physics.

Bibliography: p.
Includes index.
Contents: v. 1. The Aristotelian cosmos and the
Newtonian system—v. 2. The Einstein universe and
the Bohr atom—v. 3. The Evolutionary cosmos and
the limits of science.
1. Physics—History—Addresses, essays, lectures—
Collected works. I. Motz, Lloyd, date.
II. McAdoo, Dale. III. Title.
QC7.5.W43 1986 530 86-1903

ISBN: 0-671-49926-2 (v. 1)
ISBN: 0-671-49930-0 (v. 2)
ISBN: 0-671-49931-9 (v. 3)
ISBN: 0-671-64216-2 (set)

Acknowledgments

THIS BOOK OWES a great deal to Professor David Finkelstein, who suggested a number of ideas and articles that were eventually incorporated into the text. Many other physicists graciously offered their time and thoughts about the contents of this anthology, and their contributions are greatly appreciated. Next, I should like to thank my editor, Bob Bender, who patiently dealt with the trials and tribulations of this project, as well as my agent, Richard Balkin, who promoted this book when it was little more than an idea. A special debt is owed to my first physics teacher, Stanley Hinden, who has always taught the subject in a straightforward and enjoyable manner. Finally, I must express my gratitude to my wife, Shelley, who endured this undertaking with good humor and patience.

This book is dedicated to my mother and father,

Helen and George Weaver

Contents

VOLUME II

The Einstein Universe and the Bohr Atom

Q Life 595

R Particles 709

K

Radioactivity

INTRODUCTION

In 1896, Wilhelm C. Röntgen, who eventually became the winner of the first Nobel Prize in Physics, made an accidental discovery in his laboratory at Würzburg, Germany, that would revolutionize existing atomic theory and lay the groundwork for future breakthroughs in the subject. These breakthroughs ranged from Max Planck's quantum theory and Niels Bohr's "shell" atom to the work in atomic disintegration and nuclear fission done by Enrico Fermi and Otto Hahn.

While using a gas-evacuated tube enclosed in cardboard, Röntgen saw on his workbench a sheet of paper coated with platinocyanide that began to glow even though the laboratory was darkened. Röntgen reasoned that penetrating rays must have passed from the tube to the paper. They could not be particles because neither an electric nor a magnetic field diverted them from their paths. But Röntgen's inability to deflect these rays with a lens caused him to conclude that they were not normal rays but ones with very short wavelengths, which he called *x-rays*.*

---

* Colin A. Ronan, *The Cambridge Illustrated History of the World's Science.* New York: Cambridge University Press, 1983, p. 493.

Simultaneously, J. J. Thomson of Cambridge University, with his graduate research assistant Ernest Rutherford, discovered that the x-rays were produced by particles emitted by the cathode (negative electrode), which then struck the anode. Thomson turned his attention to these *cathode rays.* Unlike the German physicist Philipp Lenard, who in 1894 had shown that cathode rays could pass through metal foil and, therefore, had assumed them to be a form of electromagnetic radiation, Thomson used a series of experiments to determine that the cathode rays are actually particles each with a mass some 1,800 times smaller than that of the smallest atom. Thomson's experiments proved the existence of the first subatomic particle, which he named *electron.*

While Thomson was working on cathode rays, the French physicist Henri Becquerel discovered that a uranium compound exposed to the rays of the sun gave off fluorescent emanations that fogged some photographic paper although the paper was protected by a black wrapper. A few days later, Becquerel discovered that the compound had the same effect on the paper even when the laboratory was darkened.

The brilliant Polish-French chemist Marie Sklodowska Curie invented the term *radioactivité* to describe the phenomenon observed by Becquerel. Subsequently, she and her husband, Pierre, went on to conduct extensive experiments with radioactive substances in their dilapidated Paris laboratory. The Curies demonstrated that all uranium compounds are radioactive, with an intensity proportional to the uranium content of the molecule. In experiments with thorium compounds they also found evidence of radioactivity, which was soon discovered to manifest itself in at least three ways: subjected to an electromagnetic field, some rays were deflected to one side, some were deflected to the opposite side, and others were not deflected at all. Ernest Rutherford, who had since begun working in his own laboratory, named the three types of deflected rays *alpha rays, beta rays,* and *gamma rays.* These three types differed greatly in their ability to penetrate materials.* Rutherford found that alpha and beta rays are positive and negative particles, respectively, and gamma rays are photons emitted from the excited nucleus of an atom.

Radioactivity is generated when the unstable atomic nucleus (an ex-

---

* If a penetrability index of 1 is assigned to alpha rays, then beta and gamma rays would have indexes of 100 and 10,000 respectively, because the penetrability of beta and gamma rays is 100 and 10,000 times greater than that of alpha rays.

cited nucleus) settles down to a stable state. The careful experiments done by the Curies led to the discovery of the first new radioactive element, which was named *polonium* in honor of Mme Curie's native Poland. Her samples of polonium disintegrated rapidly, having what Rutherford was later to call a *half-life* of six months. A second new element was discovered by the Curies in the same year and named *radium.* Official announcement of both discoveries was made in 1898.

Rutherford's analysis of radioactivity began in 1899 and continued for twenty years. He and his coworkers found that alpha rays travel at a speed of from five to seven percent of the speed of light (15,000 to 21,000 km/sec), and they were eventually identified with the nuclei of helium. When the helium nuclei were kept in a vessel from which they could not diffuse, they were found after a period of sixty days to be normal helium—the first example of a known element being produced during radioactive transformation. Rutherford's work at McGill University with his colleague Frederick Soddy led them to conclude in 1903 that these rays were actually particles caused by the disintegration of atoms in radioactive substances.

It fell to William Bragg of the University of Leeds to explain in 1910 that the uncertainty as to whether scientists were dealing with particles or rays was due to the fact that "when the rays struck atoms in a rarefied gas, they caused the atoms to emit high-speed electrons which, in their turn, 'knocked off' electrons from other atoms."* This process caused the gas to become electrically conducting.

Because of the negative charge of the electron and the electrical neutrality of atoms themselves, Rutherford reasoned that all atoms must have an offsetting positively charged core. Although Rutherford's model was brilliant, it was soon replaced with a "shell model" offered by the Danish physicist Niels Bohr, in which the electrons orbited the nucleus but could move only in certain discrete orbital paths. As these paths represented different energy levels, an atom emitted energy when an electron dropped from a higher to a lower orbit and absorbed energy when an electron jumped from a lower to a higher orbit. Bohr's work was especially indebted to Max Planck's quantum theory, which said that radiation appears only in discrete packets of energy, giving a discontinous view

---

* Ronan, op. cit., p. 495.

of nature that was revolutionary at that time. The Bohr model also helped explain the structure of Mendeleev's Periodic Table of the Elements by showing that "chemical properties were due to the number of orbiting electrons and these, in their turn, depended on the electric charge of the nucleus . . . [which] itself depended on the mass of the nucleus and so on the atomic weight."*

In 1919, Rutherford bombarded nitrogen with alpha particles, causing the nitrogen nucleus to disintegrate. The result was positively charged particles which Rutherford identified as *protons.* Although Rutherford's techniques could not reveal whether there were other constituent particles (such as the uncharged neutron particle, which was discovered by James Chadwick in 1932), his work showed how the nuclei of other atoms could be broken down. Rutherford's techniques would also guide the construction of the massive atom smashers such as the CERN (European Commission for Nuclear Research) facility at Geneva, which was built after World War II and used to break open the nuclei of atoms. Rutherford's work also provided much of the impetus for research in atomic fission, which led to the harnessing of nuclear power and the development of the atomic bomb.

* Ibid., p. 513.

K.1, K.2

Radioactive Substances

PIERRE CURIE

Radium and the New Concepts in Chemistry

MARIE CURIE

ON 7 NOVEMBER 1867, a baby girl, eventually to be known to the world as Marie, was born in Warsaw, Poland, into the Sklodowsky family to a physics teacher and his wife, the director of a girls' academy. Poland at that time was under the iron fist of the Tsarist Russian government. As recently as 1863, Polish revolutionary activity had quickly brought down harsh punishment from the Russian overlords. Marie's mother died of tuberculosis, and not long thereafter, Marie's father lost his post. Although she had acquired an interest in science from her father, Marie worked as a governess for several years to aid the family finances.

Eventually three of the five Sklodowsky children—Marie and an older brother and sister—managed to get to Paris, although Marie sent the other two on to France first, remaining behind to work and help them pay for their education. Finally, she arrived in Paris where she enrolled in the Sorbonne in 1891. She graduated first in her class. As a doctoral candidate, she undertook an investigation into whether rays such as those discovered by Becquerel were also emitted by elements other than uranium. She completed her thesis, *Recherches sur les substances radioactives,* in 1903; it was an extraordinary work and contained her discoveries of both polonium and radium as well as her measurement of the latter's atomic weight.

In 1895 she had married a French chemist, Pierre Curie, who was already well known in the scientific world for his discovery of piezoelectricity. Both Marie and Pierre were strongly anticlerical; their marriage took place in a Paris municipal office.

From then on, their careers were intimately linked in a remarkably successful scientific collaboration. In 1903, they shared the Nobel Prize in Physics with Henri Becquerel for the discovery of radioactivity. When her husband, Pierre, was killed in a street accident in 1906, Mme Curie took over his post as professor at the Sorbonne—the first woman ever to be so honored. (This triumph for the cause of equal rights for women was tarnished when she was nominated for membership in the Académie Française, the "Forty Immortals," and lost by a single vote cast against her because of her gender.) She was later awarded a second Nobel Prize, this time in Chemistry, in 1911 for her discovery of the elements polonium and radium. During World War I, Mme Curie equipped ambulances with portable x-ray machines and organized radiology courses for medics in the French Army. She continued her medical research as a member of the Council of the Radium Institute, which had been formed in 1914. She was elected in 1922 to the Académie de Médecine, where she was active in promoting the establishment of worker safety standards.

Marie Curie singlemindedly pursued difficult and tedious experiments, such as the extraction of minute amounts of radium from a uranium ore known as pitchblende. From several tons of pitchblende obtained in Bohemia, she patiently dissolved, filtered and crystallized, over a period of months, literally hundreds of samples of pitchblende until she had obtained about a tenth of a gram of radium. From this sample, the atomic weight was determined and its spectroscopic measurement was taken by Eugène Demarcay.

Marie Curie died tragically in 1934 of leukemia, a form of blood cancer, caused by an overexposure to radiation. She so exposed herself and her equipment to radiation (its effects were unknown then) that her notebooks are still too dangerous to handle to this day.

Pierre Curie was born in 1859 and entered the Sorbonne at the age of sixteen. Two years later he obtained his physics degree and went to work there as a laboratory assistant. In 1882 he was appointed director of the laboratory at the École Municipale de Physique et Chimie Industrielle, a post he held until 1904. Pierre and his brother Jacques discovered piezoelectricity, which is the alternate expansion and contraction of a property in response to an alternating electric field. In 1895 Pierre Curie formulated a thesis expressing the relationship between temperature and magnetism. The term *Curie point,* used in studying magnetism and temperature, honored his discoveries.

Pierre was both a theoretical and an experimental physicist who joined his wife to help isolate polonium and radium in 1898 using tedious analysis and chemical separation techniques. Marie's discovery of radium enabled Pierre to show that radioactivity diminishes at an exponential rate. Pierre's discovery paved the way for the development of archeological and geological dating methods.

The Curies' first paper was originally published in *Comptes rendus,* in 1898 and is the earliest description extant of the discovery of a new, radioactive substance in pitchblende. It was soon followed by a paper on experiments done with the help of an assistant, G. Belmon. This paper was the Curies' first published account of the discovery of polonium (already hinted at in the earlier paper) and radium. These two papers are models of the splendid French style of scientific reporting as embodied in the monumental series of *Comptes rendus* (the equivalent of the British *Philosophical Transactions,* the German *Annalen der Physik* and the American *Physical Review*).

Our selections include the Nobel lecture given by Pierre Curie in 1903 (Marie Curie declined to address the Swedish Academy) and the Nobel lecture given by Marie Curie in 1911, both of which survey the work that culminated in the discoveries of polonium and radium.

Radioactive Substances

Pierre Curie

In this life we want nothing but facts, Sir; nothing but facts.
—Charles Dickens, *Hard Times*

Although this may seem a paradox; all exact science is dominated by the idea of approximation.
—Bertrand Russell

What are the sciences but maps of universal laws; and universal laws but the channels of universal power; and universal power but the outgoings of a supreme universal mind?
—Edward Thomson

Science is always wrong. It never solves a problem without creating ten more.
—George Bernard Shaw

. . . I HAVE TO speak to you today on the properties of the *radioactive substances,* and in particular of those of *radium.* I shall not be able to mention exclusively our own investigations. At the beginning of our studies on this subject in 1898 we were the only ones, together with Becquerel, interested in this question; but since then much more work has been done, and today it is no longer possible to speak of radioactivity without quoting the results of investigations by a large number of physicists such as Rutherford, Debierne, Elster and Geitel, Giesel, Kauffmann, Crookes,

Pierre Curie, Nobel Prize in Physics Award Address, 1903. Reprinted with permission of Elsevier Publishing Co., and the Nobel Foundation.

Ramsay and Soddy, to mention only a few of those who have made important progress in our knowledge of radioactive properties.

I shall give you only a rapid account of the discovery of radium and a brief summary of its properties, and then I shall speak to you of the consequences of the new knowledge which radioactivity gives us in the various branches of science.

Becquerel discovered in 1896 the special radiating properties of *uranium* and its compounds. Uranium emits very weak rays which leave an impression on photographic plates. These rays pass through black paper and metals; they make air electrically conductive. The radiation does not vary with time, and the cause of its production is unknown.

Mme. Curie in France and Schmidt in Germany have shown that thorium and its compounds possess the same properties. Mme. Curie also showed in 1898 that of all the chemical substances prepared or used in the laboratory, only those containing uranium or thorium were capable of emitting a substantial amount of the Becquerel rays. We have called such substances *radioactive.*

Radioactivity, therefore, presented itself as an atomic property of uranium and thorium, a substance being all the more radioactive as it was richer in uranium or thorium.

Mme. Curie has studied the minerals containing uranium or thorium, and in accordance with the views just stated, these minerals are all radioactive. But in making the measurements, she found that certain of these were more active than they should have been according to the content of uranium or thorium. Mme. Curie then made the assumption that these substances contained radioactive chemical elements which were as yet unknown. We, Mme. Curie and I, have sought to find these new hypothetical substances in a uranium ore, *pitchblende.* By carrying out the chemical analysis of this mineral and assaying the radioactivity of each batch separated in the treatment, we have, first of all, found a highly radioactive substance with chemical properties close to bismuth which we have called *polonium,* and then (in collaboration with Bémont) a second highly radioactive substance close to barium which we called *radium.* Finally, Debierne has since separated a third radioactive substance belonging to the group of the rare earths, *actinium.*

These substances exist in pitchblende only in the form of traces, but they have an enormous radioactivity of an order of magnitude 2 million times greater than that of uranium. After treating an enormous amount of material, we succeeded in obtaining a sufficient quantity of radiferous barium salt to be able to extract from it radium in the form of a pure salt by a method of fractionation. Radium is the higher homologue of barium in the series of the alkaline earth metals. Its atomic weight as determined by Mme. Curie is 225. Radium is characterized by a distinct spectrum which was first discovered and studied by Demarçay, and then by Crookes and Runge and Precht, Exner, and Haschek. The spectrum reaction of radium is very sensitive, but it is much less sensitive than radioactivity for revealing the presence of traces of radium.

The general effects of the radiations from radium are intense and very varied.

VARIOUS EXPERIMENTS

Discharge of an electroscope.—The rays pass through several centimetres of lead.—A spark induced by the presence of radium.—Excitation of the phosphorescence of barium platinocyanide, willemite and kunzite.—Coloration of glass by the rays.—Thermoluminescence of fluorine and ultramarine after the action of radiation from radium on these substances.—Radiographs obtained with radium.

A radioactive substance such as radium constitutes a continuous source of energy. This energy is manifested by the emission of the radiation. I have also shown in collaboration with Laborde that radium releases heat continuously to the extent of approximately 100 calories per gram of radium and per hour. Rutherford and Soddy, Runge and Precht, and Knut Angström have also measured the release of heat by radium; this release seems to be constant after several years, and the total energy released by radium in this way is considerable.

The work of a large number of physicists (Meyer and Schweidler, Giesel, Becquerel, P. Curie, Mme. Curie, Rutherford, Villard, etc.) shows that the radioactive substances can emit rays of the three different varieties designated by Rutherford as α-, β- and γ-rays. They differ from one another by the action of a magnetic field and of an electric field which modify the trajectory of the α- and β-rays.

The β-rays, similar to cathode rays, behave like negatively charged projectiles of a mass 2000 times smaller than that of a hydrogen atom (electron). We have verified, Mme. Curie and I, that the β-rays carry with them negative electricity. The $\varkappa$-rays, similar to the Goldstein's canal rays, behave like projectiles 1,000 times heavier and charged with positive electricity. The γ-rays are similar to Röntgen rays.

Several radioactive substances such as radium, actinium, and thorium also act otherwise than through their direct radiation; the surrounding air becomes radioactive, and Rutherford assumes that each of these substances emits an unstable radioactive gas which he calls *emanation* and which spreads in the air surrounding the radioactive substance.

The activity of the gases which are thus made radioactive disappears spontaneously according to an exponential law with a time constant which is characteristic for each active substance. It can, therefore, be stated that the emanation from radium diminishes by one-half every 4 days, that from thorium by one-half every 55 seconds, and that from actinium by one-half every 3 seconds.

Solid substances which are placed in the presence of the active air surrounding the radioactive substances themselves become temporarily radioactive. This is the phenomenon of *induced radioactivity* which Mme. Curie and I have discovered. The induced radioactivities, like the emanations, are equally unstable and are destroyed spontaneously according to exponential laws characteristic of each of them.

EXPERIMENTS

A glass tube filled with emanation from radium which was brought from Paris.—Discharge of an electroscope by the rays from the induced radioactivity. Phosphorence of zinc sulphide under the action of the emanation.

Finally, according to Ramsay and Soddy, radium is the seat of a continuous and spontaneous production of helium.

The radioactivity of uranium, thorium, radium and actinium seems to be invariable over a period of several years; on the other hand, that of polonium diminishes according to an exponential law; it diminishes by one-half in 140 days, and after several years it has almost completely disappeared.

These are all the most important facts which have been established by the efforts of a large number of physicists. Several phenomena have already been extensively studied by them.

The consequences of these facts are making themselves felt in all branches of science:

The importance of these phenomena for *physics* is evident. Radium constitutes in laboratories a new research tool, a source of new radiations. The study of the β-rays has already been very fruitful. It has been found that this study confirms the theory of J. J. Thomson and Heaviside on the mass of particles in motion, charged with electricity; according to this theory, part of the mass results from the electromagnetic reactions of the ether of the vacuum. The experiments of Kauffmann on the β-rays of radium lead to the assumption that certain particles have a velocity very slightly below that of light, that according to the theory the mass of the particle increases with the velocity for velocities close to that of light, and that the whole mass of the particle is of an electromagnetic nature. If the hypothesis is also made that material substances are constituted by an agglomeration of electrified particles, it is seen that the *fundamental principles of mechanics* will have to be profoundly modified.

The consequences for *chemistry* of our knowledge of the properties of the radioactive substances are perhaps even more important. And this leads us to speak of the source of energy which maintains the radioactive phenomena.

At the beginning of our investigations we stated, Mme. Curie and I, that the phenomena could be explained by two distinct and very general hypotheses which were described by Mme. Curie in 1899 and 1900 (*Revue Générale des Sciences,* January 10, 1899, and *Revue Scientifique,* July 21, 1900).

1. In the first hypothesis it can be supposed that the radioactive substances borrow from an external radiation the energy which they release, and their radiation would then be a secondary radiation. It is not absurd to suppose that space is constantly traversed by very penetrating radiations which certain substances would be capable of capturing in flight. According to the recent work of Rutherford, Cooke and McLennan, this hypothesis seems to be useful for explaining part of the extremely weak radiation which emanates from most of the substances.

2. In the second hypothesis it can be supposed that the radioactive substances draw from themselves the energy which they release. The radioactive substances would then be in course of evolution, and they would be transforming themselves progressively and slowly in spite of the apparent invariability of the state of certain of them. The quantity of heat released by radium in several years is enormous if it is compared with the heat released in any chemical reaction with the same weight of matter. This released heat would only represent, however, the energy involved in a transformation of a quantity of radium so small that it cannot be appreciated even after years. This leads to the supposition that the transformation is more far-reaching than the ordinary chemical transformations, that the existence of the atom is even at stake, and that one is in the presence of a transformation of the elements.

The second hypothesis has shown itself the more fertile in explaining the properties of the radioactive substances properly so called. It gives, in particular, an immediate explanation for the spontaneous disappearance of polonium and the production of helium by radium. This theory of the transformation of the elements has been developed and formulated with great boldness by Rutherford and Soddy who state that there is a continuous and irreversible disaggregation of the atoms of the radioactive elements. In the theory of Rutherford the disaggregation products would be, on the one hand, the projectile rays, and on the other hand, the emanations and the induced radioactivities. The latter would be new gaseous or solid radioactive substances frequently with a rapid evolution and having atomic weights lower than that of the original element from which they are derived. Seen in this way the life of radium would be necessarily limited when this substance is separated from the other elements. In Nature radium is always found in association with uranium, and it can be assumed that it is produced by the latter.

This, therefore, is a veritable theory of the transmutation of elements, although not as the alchemists understood it. The inorganic matter would necessarily evolve through the ages and in accordance with immutable laws.

Through an unexpected consequence, the radioactive phenomena can be important in *geology*. It has been found, for example, that radium always accompanies uranium in minerals. And it has even been found that the ratio of radium to uranium is constant in all minerals (Boltwood). This confirms the idea of the creation of radium from uranium. This theory can be extended to try to explain also other associations of elements which occur so frequently in minerals. It can be imagined that certain elements have been formed on the spot on the surface of the Earth or that they stem from other elements in a time which may be of the order of magnitude of geological periods. This is a new point of view which the geologists will have to take into account.

Elster and Geitel have shown that the emanation of radium is very widespread in Nature and that radioactivity probably plays an important part in *meteorology*, with the ionization of the air provoking the condensation of water vapour.

Finally, in the *biological sciences* the rays of radium and its emanation produce interesting effects which are being studied at present. Radium rays have been used in the treatment of certain diseases (lupus, cancer, nervous diseases). In certain cases their action may become dangerous. If one leaves a wooden or cardboard box containing a small glass ampulla with several centigrams of a radium salt in one's pocket for a few hours, one will feel absolutely nothing. But 15 days afterwards a redness will appear on the epidermis, and then a sore which will be very difficult to heal. A more prolonged action could lead to paralysis and death. Radium must be transported in a thick box of lead.

It can even be thought that radium could become very dangerous in criminal hands, and here the question can be raised whether mankind benefits from knowing the secrets of Nature, whether it is ready to profit from it or whether this knowledge will not be harmful for it. The example of the discoveries of Nobel is characteristic, as powerful explosives have enabled man to do wonderful work. They are also a terrible means of destruction in the hands of great criminals who are leading the peoples towards war. I am one of those who believe with Nobel that mankind will derive more good than harm from the new discoveries.

Radium and the New Concepts in Chemistry

MARIE CURIE

[Marie Curie's] strength, her purity of will, her austerity toward herself, her objectivity, her incorruptible judgment—all these were of a kind seldom found joined in a single individual . . . The greatest scientific deed of her life—proving the existence of radioactive elements and isolating them—owes its accomplishment not merely to bold intuition but to a devotion and tenacity in execution under the most extreme hardships imaginable, such as the history of experimental science has not often witnessed.

—ALBERT EINSTEIN, *Out of My Later Years*

SOME 15 YEARS ago the radiation of uranium was discovered by Henri Becquerel and two years later the study of this phenomenon was extended to other substances, first by me, and then by Pierre Curie and myself. This study rapidly led us to the discovery of new elements, the radiation of which, while being analogous with that of uranium, was far more intense. All the elements emitting such radiation I have termed *radioactive,* and the new property of matter revealed in this emission has thus received the name *radioactivity.* Thanks to this discovery of new, very powerful radioactive substances, particularly radium, the study of

Marie Curie, Nobel Prize in Chemistry Award Address, 1911. Reprinted with permission of Elsevier Publishing Co., and the Nobel Foundation.

radioactivity progressed with marvellous rapidity. Discoveries followed each other in rapid succession, and it was obvious that a new science was in course of development. The Swedish Academy of Sciences was kind enough to celebrate the birth of this science by awarding the Nobel Prize for Physics to the first workers in the field, Henri Becquerel, Pierre Curie and Marie Curie (1903).

From that time onward, numerous scientists devoted themselves to the study of radioactivity. Allow me to recall to you one of them who, by the certainty of his judgement, and the boldness of his hypotheses and through the many investigations carried out by him and his pupils, has succeeded not only in increasing our knowledge but also in classifying it with great clarity; he has provided a backbone for the new science, in the form of a very precise theory admirably suited to the study of the phenomena. I am happy to recall that Rutherford came to Stockholm in 1908 to receive the Nobel Prize as a well-deserved reward for his work.

Far from halting, the development of the new science has constantly continued to follow an upward course. And now, only 15 years after Becquerel's discovery, we are face to face with a whole world of new phenomena belonging to a field which, despite its close connexion with the fields of physics and chemistry, is particularly well-defined. In this field the importance of radium from the viewpoint of general theories has been decisive. The history of the discovery and the isolation of this substance has furnished proof of my hypothesis that *radioactivity is an atomic property of matter and can provide a means of seeking new elements.* This hypothesis has led to present-day theories of radioactivity, according to which we can predict with certainty the existence of about 30 new elements which we cannot generally either isolate or characterize by chemical methods. We also assume that these elements undergo atomic transformations, and the most direct proof in favour of this theory is provided by the experimental fact of the formation of the chemically-defined element *helium* starting from the chemically-defined element *radium.*

Viewing the subject from this angle, it can be said that the task of isolating radium is the corner-stone of the edifice of the science of radioactivity. Moreover, radium remains the most useful and powerful tool in radioactivity laboratories. I believe that it is because of these considerations that the Swedish Academy of Sciences has done me the very great honour of awarding me this year's Nobel Prize for Chemistry.

It is therefore my task to present to you radium in particular as a new chemical element, and to leave aside the description of the many radioactive phenomena which have already been described in the Nobel Lectures of H. Becquerel, P. Curie and E. Rutherford.

Before broaching the subject of this lecture, I should like to recall that the discoveries of radium and of polonium were made by Pierre Curie in collaboration with me. We are also indebted to Pierre Curie for basic research in the field of radioactivity, which has been carried out either alone, or in collaboration with his pupils.

The chemical work aimed at isolating radium in the state of the pure salt, and at characterizing it as a new element, was carried out specially by me, but it is intimately connected with our common work. I thus feel that I interpret correctly the intention of the Academy of Sciences in assuming that the award of this high distinction to me is motivated by this common work and thus pays homage to the memory of Pierre Curie.

I will remind you at the outset that one of the most important properties of the radioactive elements is that of ionizing the air in their vicinity (Becquerel). When a uranium compound is placed on a metal plate A situated opposite another plate B and a difference in potential is maintained between the plates A and B, an electric current is set up between these plates; this current can be measured with accuracy under suitable conditions and will serve as a measure of the activity of the substance. The conductivity imparted to the air can be ascribed to ionization produced by the rays emitted by the uranium compounds.

In 1897, using this method of measurement, I undertook a study of the radiation of uranium compounds, and soon extended this study to other substances, with the aim of finding out whether radiation of this type occurs in other elements. I found in this way that of the other elements known, only the compounds of thorium behave like the compounds of uranium.

I was struck by the fact that the activity of uranium and thorium compounds appears to be *an atomic property of the element uranium and of the element thorium.* Chemical compounds and mixtures containing uranium and thorium are active in direct proportion to the amount of these metals contained in them. The activity is not destroyed by either physical changes of state or chemical transformations.

I measured the activity of a number of minerals; all of them that appear to be radioactive always contain uranium or thorium. But an unexpected fact was noted: certain minerals (pitchblende, chalcolite, autunite) had a greater activity than might be expected on the basis of their uranium or thorium content. Thus, certain pitchblendes containing 75% of uranium oxide are about four times as radioactive as this oxide. Chalcolite (crystallized phosphate of copper and uranium) is about twice as radioactive as uranium. This conflicted with views which held that no mineral should be more radioactive than metallic uranium. To explain this point I prepared synthetic chalcolite from pure products, and obtained crystals, whose activity was completely consistent with their uranium content; this activity is about half that of uranium.

I then thought that the greater activity of the natural minerals might be determined by the presence of a small quantity of a highly-radioactive material, different from uranium, thorium and the elements known at present. It also occurred to me that if this was the case I might be able to extract this substance from the mineral by the ordinary methods of chemical analysis. Pierre Curie and I at once carried out this research, hoping that the proportion of the new element might reach several percent. In reality the proportion of the hypothetical element was far lower, and it took several years to show unequivocally that pitchblende contains at least one highly-radioactive material which is a new element in the sense that chemistry attaches to the term.

We were thus led to create a new method of searching for new elements, *a method based on radioactivity considered as an atomic property of matter.* Each chemical separation is followed by a measurement of the activity of the products obtained, and in this way it is possible to determine how the active substance behaves from the chemical viewpoint. This method has come into general application, and is similar in some ways to spectral analysis. Because of the wide variety of radiation emitted, the method could be perfected and extended, so that it makes it possible, not only to discover radioactive materials, but also to distinguish them from each other with certainty.

It was also found, in using the method being considered, that it was in fact possible to concentrate the activity by chemical methods. We found that pitchblende contains at least two radioactive materials, one of which, accompanying bismuth, has been given the name *polonium,* while the other, paired with barium, has been called *radium.*

Other radioactive elements have been discovered since: actinium (Debierne), radiothorium and mesothorium (Hahn), ionium (Boltwood), etc.

We were convinced that the materials we had discovered were new chemical elements. This conviction was based solely on the atomic nature of radioactivity. But at first, from the chemical viewpoint, it was as if our substances had been, the one pure bismuth, and the other pure barium. It was vital to show that the radioactive property was connected with traces of elements that were neither bismuth nor barium. To do that the hypothetical elements had to be isolated. In the case of radium isolation was completely successful but required several years of unremitting effort.

Radium in the pure salt form is a substance the manufacture of which has now been industrialized; for no other new radioactive substance have such positive results been obtained.

The radiferous minerals are being subjected to very keen study because the presence of radium lends them considerable value. They are identifiable either by the electrometric method, or very simply by the impression they produce on a photographic plate. The best radium mineral is the pitchblende from St. Joachimsthal (Austria), which has for a long time been processed to yield uranium salts. After extraction of the latter, the mineral leaves a residue which contains radium and polonium. We have normally used this residue as our raw material.

The first treatment consists in extracting the radiferous barium and the bismuth containing the polonium. This treatment, which was first performed in the laboratory on several kilograms of raw material (as many as 20 kg), had then to be undertaken in a factory owing to the need to process thousands of kilograms. Actually, we gradually learned from experience that the radium is contained in the raw material in the proportion of a few decigrams per ton. About 10 to 20 kg crude barium sulphate containing radium are extracted from one ton of residue. The activity of these sulphates is even then 30 to 60 times greater than that of uranium. These sulphates are purified and converted to chlorides. In the mixture of barium and radium chlorides the radium is present only in the proportion of about 3 parts per 100,000. In the radium industry in France a much lower-grade mineral is most often used and the proportion indicated is far lower still. To separate the radium from the barium I have used a method of fractional crystallization of the chloride (the bromide can also be used). The radium salt, less soluble than the barium salt, becomes concentrated in the crystals. Fractionation is a lengthy, methodical operation which gradually eliminates the barium. To obtain a very pure salt I have had to perform several thousands of crystallizations. The progress of the fractionation is monitored by activity measurements.

A first proof that the element radium existed was furnished by spectral analysis. The spectrum of a chloride enriched by crystallization exhibited a new line which Demarçay attributed to the new element. As the activity became more concentrated, the new line increased in intensity and other lines appeared while the barium spectrum became at the same time less pronounced. When the purity is very high the barium spectrum is scarcely visible.

I have repeatedly determined the average atomic weight of the metal in the salt subjected to spectral analysis. The method used was the one consisting in determining the chlorine content in the form of silver chloride in a known amount of the anhydrous chloride. I have found that this method gives very good results even with quite small amounts of substance (0.1 to 0.5 g), provided a very fast balance is used to avoid the absorption of water by the alkaline-earth salt during the weighings. The atomic weight increases with the enrichment of the radium as indicated by the spectrum. The successive atomic weights obtained were: 138; 146; 174; 225; 226.45. This last value was determined in 1907 with 0.4 g of very pure radium salt. The results of a number of determinations are, 226.62; 226.31; 226.42. These have been confirmed by more recent experiments.

The preparation of pure radium salts and the determination of the atomic weight of radium have proved positively that radium is a new element and have enabled a definite position to be assigned to it. Radium is the higher homologue of barium in the family of alkaline-earth metals; it has been entered in Mendeleev's table in the corresponding column, on the row containing uranium and thorium. The radium spectrum is very precisely known. These very clear-cut results for radium have convinced chemists and justified the establishment of the new science of radioactive substances.

In chemical terms radium differs little from barium; the salts of these two elements are isomorphic, while those of radium are usually less soluble than the barium salts. It is very interesting to note that the strong radioactivity of radium involves no chemical anomalies and that the chemical properties are actually those which correspond to the position in the Periodic System indicated by its atomic weight. The radioactivity of radium in solid salts is ca. 5 million times greater than that of an equal weight of uranium. Owing to this activity its salts are spontaneously luminous. I also wish to recall that radium gives rise to a continuous liberation of energy which can be measured as heat, being about 118 calories per gram of radium per hour.

Radium has been isolated in the metallic state (M. Curie and A. Debierne, 1910). The method used consisted in distilling under very pure hydrogen the amalgam of radium formed by the electrolysis of a chloride solution using a mercury cathode. One decigram only of salt was treated, and consequently considerable difficulties were involved. The metal obtained melts at about 700° C., above which temperature it starts to volatilize. Is it very unstable in the air and decomposes water vigorously.

The radioactive properties of the metal are exactly the ones that can be forecast on the assumption that the radioactivity of the salts is an atomic property of the radium which is unaffected by the state of combination. It was of real importance to corroborate this point, as misgivings had been voiced by those to whom the atomic hypothesis of radioactivity was still not evident.

Although radium has so far only been obtained in very small amounts, it is nevertheless true to say, in conclusion, that it is a perfectly defined and already well-studied chemical element.

Unfortunately, the same cannot be stated for polonium, for which nevertheless considerable effort has already been spent. The stumbling block here is the fact that the proportion of polonium in the mineral is about 5,000 times smaller than that of radium.

Before theoretical evidence was available from which to forecast this proportion, I had conducted several extremely laborious operations to concentrate polonium and in this way had secured products with very high activity without being able to arrive at definite results as in the case of radium. The difficulty is heightened by the fact that polonium disintegrates spontaneously, disappearing by half in a period of 140 days. We now know that radium has not an infinite life, either, but the rate of disappearance is far less (it disappears by half in 2,000 years). With our facilities we can scarcely hope to determine the atomic weight of polonium because theory foresees that a rich mineral can contain only a few hundredths of a milligram per ton, but we can hope to observe its spectrum. The operation of concentrating polonium, as I shall point out later, is, moreover, a problem of great theoretical interest.

Recently, in collaboration with Debierne, I undertook to treat several tons of residues from uranium mineral with a view of preparing polonium. Initially conducted in the factory, then in the laboratory, this treatment finally yielded a few milligrams of substance about 50 times more active than an equal weight of pure radium. In the spectrum of the substance some new lines could be observed which appear attributable to polonium and of which the most important has the wavelength 4170.5 Å. According to the atomic hypothesis of radioactivity, the polonium spectrum should disappear at the same time as the activity, and this fact can be confirmed experimentally.

I have so far considered radium and polonium only as chemical substances. I have shown how the fundamental hypothesis which states that radioactivity is an atomic property of the substance has led to the discovery of new chemical elements. I shall now describe how the scope of this hypothesis has been greatly enlarged by the considerations and experimental facts which resulted in establishing the theory of atomic radioactive transformations.

The starting-point of this theory must be sought in the considerations of the source of the energy involved in the phenomena of radioactivity. This energy becomes manifest as an emission of rays which produce thermal, electrical and light phenomena. As the emission occurs spontaneously without any known cause of excitation, various hypotheses have been advanced to account for the liberation of energy. One of the hypotheses put forward at the beginning of our research by Pierre Curie and myself consisted in assuming that the radiation is an emission of matter accompanied by a loss in weight of the active substances and that the energy is taken from the substance itself whose evolution is not yet complete and which undergoes an atomic transformation. This hypothesis, which at first could only be enunciated together with other equally valid theories, has attained dominant importance and finally asserted itself in our minds owing to a body of experimental evidence which substantiated it. This evidence is essentially the following: A series of radioactive phenomena exists in which radioactivity appears to be tied up to matter in an imponderable quantity, the radiation moreover not being permanent but disappearing more or less rapidly with time. Such are polonium, radioactive emanations and deposits of induced radioactivity.

It has been established, moreover, in certain cases that the radioactivity observed increases with time. This is what happens in the case of freshly prepared radium, of the emanation freshly introduced into the measuring apparatus, of thorium deprived of thorium X, etc.

A careful study of these phenomena has shown that a very satisfactory general explanation can be given by assuming that each time a decrease of radioactivity is observed there is a destruction of radioactive matter, and that each time an increase of activity is observed, there is a production of radioactive matter. The radiations which disappear and appear are, besides, of very varied nature, and it is admitted that every kind of rays determined can serve to characterize a substance which is its source, and appears and disappears with it.

As radioactivity is in addition a property which is essentially atomic, the production or the destruction of a distinct type of radioactivity corresponds to a production or a destruction of atoms of a radioactive substance.

Finally, if it is supposed that radioactive energy is a phenomenon which is borrowed from atomic transformation, it can be deduced from this that every radioactive substance undergoes such a transformation, even though it appears to us to be invariable. Transformation in this case is only very slow, and this is what takes place in the case of radium or uranium.

* * *

The theory I have just summarized is the work of Rutherford and Soddy, which they have called *theory of atomic disintegration.* By applying this theory it can be concluded that a primary radioactive substance such as radium undergoes a series of atomic transmutations by virtue of which the atom of radium gives birth to a train of atoms of smaller and smaller weights, since a stable state cannot be attained as long as the atom formed is radioactive. Stability can only be attained by inactive matter.

From this point of view, one of the most brilliant triumphs of the theory is the prediction that the gas helium, always present in radioactive minerals, can represent one of the end-products of the evolution of radium, and that it is in the form of alpha rays that the helium atoms which are formed when radium atoms disintegrate are discharged. Now, the production of helium by radium has been proved by the experiments of Ramsay and Soddy, and it cannot now be contested that the perfectly defined chemical element, radium, gives rise to the formation of another equally defined chemical element—helium. Moreover, the investigations done by Rutherford and his students have proved that the alpha particles emitted by radium with an electric charge are also to be found in the form of helium gas in the space where they have been recovered.

I must remark here that the bold interpretation of the relationship existing between radium and helium rests entirely upon the certitude that radium has the same claim to be a chemical element as have all the other known elements, and that there can be no question of regarding it to be a molecular combination of helium with another element. This shows how fundamental in these circumstances has been the work carried out to prove the chemical individuality of radium, and it can also be seen in what way the hypothesis of the atomic nature of radioactivity and the theory of radioactive transformations have led to the experimental discovery of a first clearly-established example of atomic transmutation. This is a fact the significance of which cannot escape anyone, and one which incontestably marks an epoch from the point of view of chemists.

Considerable work, guided by the theory of radioactive transformations, has led to approximately 30 new radioactive elements being envisaged, classified in 4 series according to the primary substance; these series are uranium, radium, thorium and actinium. The uranium and radium series can, in fact, be combined, for it seems to be proved that radium is a derivative of uranium. In the radium series the last known radioactive body is polonium, the production of which by radium is now a proven fact. It is likely that the actinium series is related to that of radium.

We have seen that helium gas is one of the products of radium disintegration. The helium atoms are detached from those of radium and its derivatives during the course of the transformation. It is supposed that after the departure of four atoms of helium, the radium atom yields one atom of polonium; the departure of a fifth helium atom determines the formation of an inactive body with an atomic weight believed to be equal to 206 (20 units below that of radium). According to Rutherford, this final element is nothing more than lead, and this supposition is now being subjected to experimental verification in my laboratory. The production of helium from polonium has been directly proved by Debierne.

The relatively large amount of polonium prepared by Curie and Debierne has allowed an important study to be undertaken. This consists in counting a large number of alpha particles emitted by polonium and in collecting and measuring the corresponding volume of helium. Since each particle is a helium atom, the number of helium atoms is thus found which occupy a given volume and have a given weight. It can therefore allow us to deduce, in a general way, the number of molecules in a gram-molecule. This number, known as Avogadro's constant, is of great importance. Experiments conducted on polonium have supplied a first value for this number, which is in good agreement with the values obtained by other methods. The enumeration of alpha particles is done by an electrometric method due to Rutherford; this method has been brought to perfection by means of a photographic recording apparatus.

Recent investigations have shown that potassium and rubidium emit a very feeble radiation, similar to the beta radiation of uranium and radium. We do not yet know whether we should regard these substances as true radioactive bodies, i.e. bodies in the process of transformation.

To conclude, I should like to emphasize the nature of the new chemistry of radioactive bodies. Tons of material have to be treated in order to extract radium from the ore. The quantities of radium available in a laboratory are of the order of one milligram, or of a gram at the very most, this substance being worth 400,000 francs per gram. Very often material has been handled in which the presence of radium could not be detected by the balance, nor even by the spectroscope. And yet we have methods of measuring so perfect and so sensitive that we are able to know very exactly the small quantities of radium we are using. Radioactive analysis by electrometric methods allows us to calculate to within 1% a thousandth of a milligram of radium, and to detect the presence of 10^{-10} grams of radium diluted in a few grams of material. This method is the only one which could have led to the discovery of radium in view of the dilution of this substance in the ore. The sensitivity of the methods is still more striking in the case of radium emanation, which can be detected when the quantity present amounts, for example, to only 10^{-10} mm^3. As the specific activity of a substance is, in the case of analogous radiations, approximately in inverse proportion to the average life, the result is that if the average life is very brief, the radioactive reaction can attain an unprecedented sensitivity. We are also accustomed to deal currently in the laboratory with substances the presence of which is only shown to us by their radioactive properties, but which nevertheless we can determine, dissolve, reprecipitate from their solutions and deposit electrolytically. This means that we have here an entirely separate kind of chemistry for which the current tool we use is the electrometer, not the balance, and which we might well call the chemistry of the imponderable.

K.3

The Chemical Nature of the Alpha Particles from Radioactive Substances

ERNEST RUTHERFORD

ERNEST RUTHERFORD, the first Baron Rutherford of Nelson, was born in New Zealand in 1871, the son of a farmer. Throughout his career, he was a singularly picturesque, plain-speaking, hardworking, and extraordinarily deft experimental and theoretical physicist. After his death in 1937, he was honored by being buried in Westminster Abbey near Isaac Newton and Lord Kelvin.

In 1895, he left New Zealand for Cambridge, where he worked with J. J. Thomson, who was at that time the leading authority in electromagnetism. In 1897, Thomson and Rutherford collaborated on a paper that offered an original theory of *ionization* (the process by which atoms or molecules, or groups of atoms and molecules, become electrically charged), which helped lead Thomson to the discovery of the electron. Rutherford then turned his attention to the radiation emitted by uranium.

After accepting a professorship of physics at McGill University in Montreal in 1898, Rutherford began his experiments with radioactive elements. An eighteen-month collaboration with the Oxford-trained chemist Frederick Soddy from 1901 to 1903 produced a series of papers which laid the groundwork for the modern study of radioactivity. Rutherford and Soddy demonstrated that uranium and thorium break down by radioactivity into a series of transitional elements. At this point, Rutherford discovered that after a certain period of time half the original element remained. Rutherford coined the term *half-life* for the phenomenon. He also invented the terms *alpha, beta,* and *gamma rays.* They proved to be durable terms, although today physicists speak of alpha and beta particles and gamma rays.

In 1904, Rutherford published the first textbook on radioactivity, which was quickly hailed as a classic. He accepted the chair in physics vacated by Arthur Schuster at the University of Manchester in England in 1907 and remained there until 1919. While at Manchester, Rutherford collaborated with Hans Geiger in experiments that measured the number of alpha particles emitted by radioactive elements, such as radium. Their experiments were essential to Geiger's later efforts to build electrical and electronic particle-counting devices—Geiger counters. In 1911, Rutherford published a model of the atom in which most of its mass was packed into a dense center surrounded by empty space. Although rather simple by modern standards, Rutherford's atom was the basis for Niels Bohr's model of atomic structure. In large part, the still widely accepted Bohr atom is a more sophisticated version of Rutherford's original model.

Rutherford became director of the Cavendish laboratory at Cambridge University in 1919. In the 1920s, he and James Chadwick studied the ways in which alpha particles could be used to disintegrate some of the lighter elements. By showing that the nucleus of an atom can be altered by firing alpha particles into it, Rutherford almost single-handedly created the field of experimental nuclear physics.

The following selection by Rutherford is his Nobel address, delivered in 1908. His award was in chemistry—rather to his annoyance—because Rutherford regarded himself as a physicist. His lecture describes the critical role of alpha-particle expulsion in many radioactive phenomena. It is an extraordinary, lucid historical description of the development of alpha-particle research, concluding with Rutherford's comments on his own remarkable (and to a certain extent, to do the Nobel committee justice, *chemical*) discovery "that alpha particles, so freely expelled from the great majority of radioactive substances, are identical in mass and constitution and must consist of atoms of helium." Rutherford's Nobel address was illustrated by lantern slides and experiments on the emanation of radium.

It is far from easy to determine whether she [Nature] has proved a kind parent to man or a merciless step-mother.

—PLINY THE ELDER, *Natural History*

Here and elsewhere we shall not obtain the best insight into things until we actually see them growing from the beginning.

—ARISTOTLE, *Politics*

THE STUDY OF the properties of the α-rays has played a notable part in the development of radioactivity and has been instrumental in bringing to light a number of facts and relationships of the first importance. With increase of experimental knowledge there has been a growing recognition that a large part of radioactive phenomena is intimately connected with the expulsion of the α-particles. In this lecture an attempt will be made to give a brief historical account of the development of our knowledge of the α-rays and to trace the long and arduous path trodden by the experimenter in the attempts to solve the difficult question of the chemical nature of the α-particles. The α-rays were first observed in 1899 as a special type of radiation and during the last six years there has been a persistent attack on this great problem, which has finally yielded to the assault when the resources of the attack seemed almost exhausted.

Shortly after his discovery of the radiating power of uranium by the photographic method, Becquerel showed that the radiation from uranium like the Röntgen-rays possessed the property of discharging an electrified body. In a detailed investigation of this property, I examined the effect on the rate of discharge by placing successive layers of thin aluminum foil over the surface of a layer of uranium oxide and was led to the conclusion that two types of radiation of very different penetrating power were present. The conclusions at that period were summed up as follows:

"These experiments show that the uranium radiation is complex and that there are present at least two distinct types of radiation—one that is very readily absorbed, which will be termed for convenience the α-radiation, and the other of a more penetrative character, which will be termed the β-radiation." When other radioactive substances were discovered, it was seen that the types of radiation present were analogous to the β- and α-rays of uranium and when a still more penetrating type of radiation

---

Ernest Rutherford, Nobel Prize in Chemistry Award Address, 1908. Reprinted with permission of Elsevier Publishing Co., and the Nobel Foundation.

from radium was discovered by Villard, the term γ-rays was applied to them. The names thus given soon came into general use as a convenient nomenclature for the three distinct types of radiation emitted from uranium, radium, thorium, and actinium. On account of their insignificant penetrating power, the α-rays were at first considered of little importance and attention was mainly directed to the more penetrating β-rays. With the advent of active preparations of radium, Giesel in 1899 showed that the β-rays from this substance were easily deflected by a magnetic field in the same direction as a stream of cathode rays and consequently appeared to be a stream of projected particles carrying a negative charge. The proof of the identity of the β-particles with the electrons constituting the cathode rays was completed in 1900 by Becquerel, who showed that the β-particles from radium had about the same small mass as the electrons and were projected at a speed comparable with the velocity of light. Time does not allow me to enter into the later work of Kauffmann and others on this subject, which has greatly extended our knowledge of the constitution and mass of electrons.

In the meantime, further investigation had disclosed that the α-particles produced most of the ionization observed in the neighbourhood of an unscreened radioactive substance, and that most of the energy radiated was in the form of α-rays. It was calculated by Rutherford and McClung in 1901 that one gram of radium radiated a large amount of energy in the form of α-rays.

The increasing recognition of the importance of the α-rays in radioactive phenomena led to attempts to determine the nature of this easily absorbed type of radiation. Strutt (Lord Rayleigh) in 1901 and Sir William Crookes in 1902 suggested that they might possibly prove to be projected particles carrying a positive charge. I independently arrived at the same conclusion from consideration of a variety of evidence. If this were the case, the α-rays should be deflected by a magnetic field. Preliminary work showed that the deflection was very slight if it occurred at all. Experiments were continued at intervals over a period of two years and it was not until 1902, when a preparation of radium of activity 19,000 was available, that I was able to show conclusively that the particles were deflected by a magnetic field, though in a very minute degree compared with the β-rays. This showed that the α-rays consisted of projected charged particles while the direction of deflection indicated that each particle carried a positive charge. The α-particles were shown to be deflected also by an electric field and from the magnitude of the deflection, it was deduced that the velocity of the swiftest particles was about

2.5×10^9 cm per second, or one-twelfth the velocity of light, while the value of e/m—the ratio of the charge carried by the particle to its mass—was found to be 5,000 electromagnetic units. Now it is known from the data of the electrolysis of water that the value of e/m for the hydrogen atom is 9,650. If the α-particle carried the same positive charge as the unit fundamental charge of the hydrogen atom, it was seen that the mass of the α-particle was about twice that of the hydrogen atom. On account of the complexity of the rays it was recognized that the results were only approximate, but the experiments indicated clearly that the α-particle was atomic in mass and might prove ultimately to be either a hydrogen or a helium atom or the atom of some unknown element of light atomic weight. . . .

Consider for a moment the explanation of the changes in radium. A minute fraction of the radium atoms is supposed each second to become unstable, breaking up with explosive violence. A fragment of the atom—an α-particle—is ejected at a high speed, and the residue of the atom, which has a lighter weight than before, becomes an atom of a new substance, the radium emanation. The atoms of this substance are far more unstable than those of radium and explode again with the expulsion of an α-particle. As a result the atom of radium A makes its appearance and the process of disintegration thus started continues through a long series of stages. . . .

Another striking property of radium was soon seen to be connected with the expulsion of α-particles. In 1903 P. Curie and Laborde showed that radium was a self-heating substance and was always above the temperature of the surrounding air. It seemed probable from the beginning that the effect must be the result of the heating effect due to the impact of the α-particles on the radium. Consider for a moment a pellet of radium enclosed in a tube. The α-particles are shot out in great numbers equally from all parts of the radium and in consequence of their slight penetrating power are all stopped in the radium itself or by the walls of the tube. The energy of motion of the α-particles is converted into heat. On this view the radium is subject to a fierce and unceasing bombardment by its own particles and is heated by its own radiation. This was confirmed by the work of Rutherford and Barnes in 1903, who showed that three-quarters of the heating effect of radium was not directly due to the radium but to its product, the emanation, and that each of the different substances produced in radium gave out heat in proportion to the energy of the α-particles expelled from it. These experiments brought clearly to light the enormous energy, compared with the weight of matter involved, which was emitted during the transformation of the emanation. It can readily be calculated that one kilogram of the radium-emanation and its products would initially emit energy at the rate of 14,000 horse-power, and during its life would give off energy corresponding to about 80,000 horse-power for one day. . . .

While the evidence as a whole strongly supported the view that the α-particle was a helium atom, it was found exceedingly difficult to obtain a decisive experimental proof of the relation. If it could be shown experimentally that the α-particle did in reality carry two unit charges, the proof of the relation would be greatly strengthened. For this purpose an electrical method was devised by Rutherford and Geiger for counting directly the α-particles expelled from a radioactive substance. The ionization produced in a gas by a single α-particle is exceedingly small and would be difficult to detect electrically except by a very refined method. Recourse was had to an automatic method of magnifying the ionization produced by an α-particle. For this purpose it was arranged that the α-particles should be fired through a small opening into a vessel containing air or other gas at a low pressure, exposed to an electric field near the sparking value. Under these conditions the ions produced by the passage of the α-particle through the gas generate a large number of fresh ions by collision. In this way it was found possible to magnify the electrical effect due to an α-particle several thousand times. The entrance of an α-particle into the testing vessel was then indicated by a sudden deflection of the electrometer needle. This method was developed into an accurate method of counting the number of α-particles fired in a known time through the small aperture of the testing vessel. From this was deduced the total number of α-particles expelled per second from any thin film of radioactive matter. In this way it was shown that 3.4×10^{10} α-particles are expelled per second from one gram of radium itself and from each of its α-ray products in equilibrium with it.

The correctness of this method was indicated by another, quite distinct method of counting. Sir William Crookes and Elster and Geitel had shown that the α-particles falling on a screen of phosphorescent zinc sulphide produced a number of scintillations. Using specially prepared screens, Rutherford and Geiger counted the number of these scintillations per second with the aid of a microscope. It was found that, within the limit of experimental error, the number of scintillations per second on a screen agreed with the number of α-particles impinging on it, counted by the electrical method. It was thus clear that each α-particle produced a visible scintillation on the screen, and that either the electrical or the optical method could be used for counting the α-particles. Apart from the purpose for which these experiments were made, the results are of great interest and importance, for it is the first time that it has been found possible to detect a single atom of matter by its electrical and optical effect. This is, of course, only possible because of the great velocity of the α-particle.

Knowing the number of α-particles expelled from the radium from the counting experiment, the charge carried by each α-particle was determined by measuring the total positive charge carried by all the α-particles expelled. It was found that each α-particle carried a positive charge of 9.3×10^{-10} electrostatic units. From a consideration of the experimental evidence of the charge carried by the ions in gases, it was concluded that the α-particle did carry two unit charges, and that the unit charge carried by the hydrogen atom was equal to 4.65×10^{-10} units. From a comparison of the known value of e/m for the α-particle with that of the hydrogen atom, it follows that an α-particle is a projected atom of helium carrying two charges, or, to express it in another way, the α-particle, after its charge is neutralized, is a helium atom.

We have seen that there is every reason to believe that the α-particles, so freely expelled from the great majority of radioactive substances, are identical in mass and constitution and must consist of atoms of helium. We are consequently driven to the conclusion that the atoms of the primary radioactive elements like uranium and thorium must be built up in part at least of atoms of helium. These atoms are released at definite stages of the transformations at a rate independent of control by laboratory forces. There is good reason to believe that in the majority of cases, a single helium atom is expelled during the atomic explosion. . . .

It is very remarkable that a chemically inert element like helium should play such a prominent part in the constitution of the atomic systems of uranium and thorium and radium. It may well be that this property of helium of forming complex atoms is in some way connected with its inability to enter into ordinary chemical combinations. It must not be forgotten that uranium and thorium and each of their transformation products must be regarded as distinct chemical elements in the ordinary sense. They differ from ordinary elements in the comparative instability of their atomic systems. The atoms break up spontaneously with great violence, expelling in many cases an atom of helium at a high speed. All the evidence is against the view that uranium or thorium or radium can be regarded as an ordinary molecular compound of helium with some known or unknown element, which breaks up into helium. The character of the radioactive transformations and their independence of temperature and other agencies have no analogy in ordinary chemical changes.

Apart from their radioactivity and high atomic weight, uranium, thorium, and radium show no specially distinctive chemical behaviour. Radium for example is closely allied in general chemical properties to barium. It is consequently not unreasonable to suppose that other elements may be built up in part of helium, although the absence of radioactivity may prevent us from obtaining any definite proof. On this view, it may prove significant that the atomic weights of many elements differ by four—the atomic weight of helium—or a multiple of four. . . .

K.4

Origins of the Conceptions of Isotopes

FREDERICK SODDY

FREDERICK SODDY was born in Sussex, England, in 1877 and died in 1956 in Brighton, England. The son of a merchant, he earned his degree in 1898 at Oxford, where he was first in his class in chemistry. After spending two years doing independent research in chemistry, Soddy then went to Canada, where he had the good fortune to study with Rutherford during the latter's brief stay at McGill University. Together with Rutherford, he developed an ingenious theory of radioactive disintegration in which each radioactive element breaks down to form another element by emitting either an alpha or a beta particle. Subsequent emissions of the same type continue to the final product, the element lead. Three series of such disintegrative sequences exist in nature—the uranium series, the thorium series, and the actinium series. A fourth theoretical series, the neptunium series, was created in the laboratory in recent years. All four series have lead as their end product.

Progress in the study of end products of radioactive decay led to an abundance of subelements that could not be precisely placed in the Periodic Table; thus, the accuracy of Mendeleev's Table was questioned. Soddy boldly suggested that "it would not be surprising if the elements [resulting from the decay of uranium, thorium and radium] . . . are mixtures of several homogeneous elements of similar but not completely identical weights." He argued "that the expulsion of an alpha particle would result in a lighter element chemically inseparable from those occupying the 'next but one' position in the Periodic Table."* It was Soddy's

*Thaddeus J. Trenn, "Frederick Soddy," in *Dictionary of Scientific Biography*, Vol. 12. New York: Charles Scribner's Sons, 1975, p. 505.

solution to that problem—the identification and naming of *isotopes*—that won him the 1921 Nobel Prize in chemistry. His discovery enabled him to show how subelements of radium and thorium are actually isotopes of lead owing to their common chemical composition. "The isotopes differed in the mass of the nucleus and therefore had different radioactive characteristics (since these depended on the nature of the nucleus). On the other hand, all isotopes of a particular element had the same number of electrons in the outer regions of the atom and so had the same chemical properties (since these depended on the number and distribution of the electrons of the atom)."* How his discovery came about and how Soddy chose the word *isotopism* to describe the phenomenon is revealed in his Nobel lecture, which he delivered in 1922.

Soddy began a seventeen-year tenure at Oxford University in 1919 where he devoted himself to teaching duties and rehabilitating the laboratories there. The destruction wrought by World War I in Europe had greatly depressed him and his increasing disillusionment with the destructive ends to which science and technology had been utilized caused him to abandon active research after 1919. He became increasingly interested in economics and social issues and wrote several books in favor of "the free development of science" which he believed would lead to the rise of a "scientific civilization," that would come about only when existing decadent economic systems were abolished. Although Soddy's technocratic vision was not widely embraced, he did make a perceptive observation about energy in 1912: "[T]he still unrecognized 'energy problem' . . . awaits the future" and the world's heavy reliance on hydrocarbons for fuel is "a legacy of the past." Soddy pinned his hopes on atomic energy being able one day to "provide anyone who wanted it with a private sun of his own."

---

*Isaac Asimov, *Asimov's Biographical Encyclopedia of Science and Technology* (2nd rev'd ed.). New York: Doubleday & Co., 1982, p. 665.

The history of science is strewn with wrecks and ruins of theory—essences and principles, fluids and forces—once fondly clung to, but found to hang together with no facts of sense. And exceptional phenomena solicit our belief in vain until such time as we chance to conceive them as of kinds already admitted to exist. What science means by "verification" is no more than this, that no object of conception shall be believed which sooner or later has not some permanent and vivid object of sensation for its term.

—WILLIAM JAMES, *Psychology*

INTRODUCTION

. . . THE INTERPRETATION OF radioactivity which was published in 1903 by Sir Ernest Rutherford and myself ascribed the phenomena to the spontaneous disintegration of the atoms of the radio-element, whereby a part of the original atom was violently ejected as a radiant particle, and the remainder formed a totally new kind of atom with a distinct chemical and physical character. These disintegrations proceed successively a large number of times, so that there results a series of more or less unstable new elements between the original parent element and the ultimate unknown final product. This was a theory sufficiently challenging to the accepted doctrines of chemistry. The further detailed study of the chemical character of these successive unstable elements, produced in radioactive changes, introduced an idea which was even more subversive of the fundamental doctrines of chemistry. That idea was that the chemical elements are not really homogeneous, but merely chemically homogeneous. In some cases they are mixtures of different constituents which are only identical in their chemical character. Put colloquially, their atoms have identical outsides but different insides. Chemical analysis classifies according to the external systems of electrons which surround a small, massive internal nucleus, whereas radioactive changes, which are of the character of veritable transmutations, concern the internal constitution of this inner nucleus. They showed that the same exterior may conceal very different interiors in the atomic structure. These elements, which are

Frederick Soddy, Nobel Prize in Chemistry Award Address, 1921. Reprinted with permission of Elsevier Publishing Co., and the Nobel Foundation.

identical in their whole chemical character and are not separable by any method of chemical analysis, are now called isotopes.

I may begin with a brief statement of the earlier researches into the nature of radioactivity. The power of spontaneously emitting rays of a new kind was discovered in 1896 by Henri Becquerel for the compounds of the element uranium. The physicists sorted the rays emitted into three types, the α-, β- and γ-rays, and their real nature was quickly elucidated. The β-rays were shown to be due to the expulsion of negative electrons traveling at a far higher speed than any that can be artificially produced in the vacuum tube as cathode-rays. The γ-rays were correctly regarded as X-rays, but of a greater penetrating power. In due course the X-rays were shown to be waves of light of extraordinary short wavelength and high frequency, whilst the γ-rays are also of the nature of light, but of even shorter wavelength and higher frequency. The α-rays were first proved to be due to radiant atoms of matter carrying a positive charge and were identified, after many years of continuous work by Sir Ernest Rutherford, with helium atoms carrying two atomic charges of positive electricity. Sir William Ramsay and I had shown in 1903 that helium was being continuously generated from radium in a spectroscopically detectable quantity, and subsequent work showed that helium is generated, as α-particles, in all the radioactive changes where α-rays are expelled.

The nature of the ionization produced in gases by all these new types of radiation was quickly elucidated, and as a consequence highly sensitive and accurate methods of measurement were evolved, which, more than any other single factor, contributed to the rapid development of the subject.

On the chemical side, the work of M. and Mme. Curie had shown radioactivity to be a specific atomic property, definitely restricted to the last two of the then known elements, uranium and thorium, though the elements potassium and rubidium emit β-rays, exhibiting in this respect evidence of true radioactivity. Using the new property much as the pioneers with the spectroscope used the spectra in the discovery of new elements, they discovered the powerfully radioactive new elements, polonium and radium, and, M. Debierne, actinium, in the uranium minerals. Of the many similar new radio-elements now known, radium is still the only one the compounds of which have been prepared in a pure state, and for which the spectrum and atomic weight have been determined. As regards its whole chemical character radium is absolutely normal and its properties could have been predicted accurately for an element occupying its place in the Periodic Table. It is the last member of the family of alkaline earths and stands in relation to thorium as thorium does to uranium in the periodic system. But superimposed on this normal chemical character it possesses a radioactive character truly astounding. The energy it evolves in the form of the new radiations, measured as heat, amounts to 133 calories per gram per hour. The theory of atomic disintegration shows however that the uniqueness of radium is due mainly to the fact that its average life, 2440 years, is sufficiently long to allow the element to accumulate in the minerals in which it is formed to a ponderable quantity, and yet short enough for the rate at which the energy of disintegration is liberated to be truly surprising. In the elucidation of the various disintegration series of uranium, thorium and actinium, every degree of atomic instability is encountered. The average life varies from the order of 10^{10} years, for the primary radio-elements uranium and thorium, down to periods of the order of a minute, which marks the limit beyond which the chemical character cannot be investigated. But by physical methods members down to a period of 1/350 second have been put in evidence, and two, of the order of 10^{-6} and 10^{-11} second, are indirectly inferred.

But for all these the theory indicates, as for radium, a perfectly normal and definite chemical character as well as the superimposed radioactive character. There is no progressive change in the nature of the atoms. As they were when produced, by the explosion of the atom from which they originate, so they remain till they in turn explode. Otherwise the law of change found to hold could not apply. The chemical and spectroscopic character is that of the atom during its normal and uneventful life, whilst its radioactive character is that produced by its sudden death. But for this very definite and precise implication from the theory of atomic disintegration it might have been supposed that the chemical character of such extraordinary substances would also be extraordinary. . . .

THE EXPERIMENTAL METHOD THAT FIRST REVEALED ISOTOPES

The history of isotopes fittingly commences with the discovery of radiothorium, a new product in the thorium disintegration series, by Sir William Ramsay and Otto Hahn in 1905, which is intermediate between thorium and thorium X and generates the latter in the course of an α-ray change. A considerable quantity of thorianite, a newly discovered mineral with 60 to 70% of ThO_2 and 10 to 20% of U_3O_8, was worked up for radium by the known methods. The new substance was discovered during the fractional crystallization of the barium-radium chloride and was separated from the radium along with the inactive barium in the course of the fractionation.

In 1905 also the discovery by Godlewski of actinium X, an intermediate product between actinium and its emanation, in every way analogous to thorium X, was the commencement of the recognition of the general analogies that existed in the three series. Even so early as 1905 I said:

"The general resemblance in character and relative period of the succeeding members of the various disintegration series on the one hand, and their distinct individual pecularities on the other, are most fascinating and mysterious. We seem to have here an extension of similar relations expressed for elementary matter generally by the periodic law, and there is no doubt that any explanation of the one will have an application to the other."

Reverting now to radiothorium, all attempts to separate it from thorium compounds either failed completely or seemed successful to only a minute extent. In 1906 Elster and Geitel and also G. A. Blanc, after having discovered radiothorium as a constituent responsible for part of the radioactivity of the sediments from certain hot springs, used the experience they had gained in its separation in attempts to separate radiothorium from thorium compounds, and were, apparently, to a very slight extent, successful. Dadourian, Boltwood and McCoy and Ross found that commercial thorium salts contained only one-half as much radiothorium as the mineral from which they extracted and as the thorium they themselves separated from minerals in the laboratory. This seemed to indicate that the secret operations by which thorium is commercially extracted were much more effective in separating radiothorium than the known laboratory methods, but a search of the by-products of the thorium manufacture failed to locate the missing radiothorium.

In 1907 Hahn discovered mesothorium, a product intermediate between thorium and radiothorium and generating the latter. In 1908 he showed that mesothorium itself consisted of two successive products. The first, mesothorium 1, is produced directly from thorium and has an average life of 9.67 years. It is one of the two radio-elements, actinium being the other, which disintegrates without the expulsion of any detectable radiation, though in all probability a β-particle is expelled, but its velocity is too low for it to be detectable. Its product, mesothorium 2, gives powerful β- and γ-rays and has an average life of 8.9 hours.

In the meantime it was found that the radioactivity of the thorium compounds prepared in the laboratory from the mineral decayed during two years from preparation to one-half and became the same as that of the commercial preparations, whereas the latter increased in activity with age and the older they were the more nearly their activity became equal to that when freshly prepared from the mineral. In the process of separating the thorium, mesothorium, but not radiothorium, is separated. The latter, pending the regeneration of the former, at first decays, and subsequently, as its parent, mesothorium, begins to reaccumulate, is regenerated until finally it reaches its original equilibrium value.

"The effect then had nothing to do with the method of preparation of the thorium salt, but is purely a question of age."

It was the first effect to be recognized of many similar ones subsequently. For example in 1911 Antonoff discovered uranium Y, and attributed his success in its separation to the particular chemical processes adopted, whereas it was in reality due to the lapse of a suitable period of time between successive separations.

Boltwood concluded from his experiments that mesothorium must be removed from thorium along with thorium X when thorium is precipitated by ammonia, the original method by which thorium X was discovered.

This year, 1907, marks the first definite statement of the doctrine of the complete chemical non-separability of what are now called isotopes. McCoy and Ross, after describing many careful and prolonged attempts, all of them completely without success, to separate radiothorium from thorium, said: "Our experiments strongly indicate that radiothorium is entirely inseparable from thorium by chemical process"; and they drew from this the correct inference with regard to the radiothorium separated by Ramsay and Hahn and others: "The isolation of radiothorium from thorianite and from pure thorium nitrate . . . may have been accomplished by the separation of mesothorium, which in time changed spontaneously into radiothorium."

Thus, already in 1907 the experimental method that first revealed the existence of isotopes among the successive products of radioactive change had been applied and its implications well understood.

"Although the separation of mesothorium from thorium and of thorium X from radiothorium is easily accomplished, there is no known method of separating by chemical means either radiothorium from thorium or thorium X from mesothorium. But owing to these last-mentioned pairs' alternating in the disintegration series, each of them can readily be prepared by itself. The preparations of radiothorium obtained by Hahn and others are probably not the radiothorium existing in the original mineral, but regenerated radiothorium subsequently produced, after the separation, from the easily separated mesothorium."

The first part of the thorium disintegration series is shown below with the average life of each member and the rays expelled in its change.

$$\underset{2.10^{10}\text{ years}}{\text{Thorium}} \xrightarrow{\alpha} \underset{9.67\text{ years}}{\text{Mesothorium 1}} \xrightarrow{\beta} \underset{8.9\text{ hours}}{\text{Mesothorium 2}} \xrightarrow{\beta}:$$

$$\underset{2.91\text{ years}}{\text{Radiothorium}} \xrightarrow{\alpha} \underset{5.25\text{ days}}{\text{Thorium X}} \xrightarrow{\alpha} \underset{78\text{ seconds}}{\text{Emanation, etc.}}$$

The separation of thorium and radiothorium and of mesothorium 1 and thorium X by chemical analysis, now as then, are completely impossible. But each can be obtained alone by a suitable combination of chemical processes at suitable intervals of time. The exact procedure is dictated by the relative periods of the substances to be separated. In this connection it must be remembered, since the methods of ascertaining the nature of the products separated are purely radioactive, that it is the relative intensity of the radioactivity, or the relative number of α- or β-particles emitted per second, and not the relative weights, which is of importance. For different radioactive elements, the radioactivity is proportional to the weight divided by the period. A rapidly changing substance reforms and attains its equilibrium value correspondingly quickly, a slowly changing one correspondingly slowly.

The ease with which the mesothorium is separated, and the very long time required for it to reform in substantial amount, is the reason why it was not discovered in 1902 when thorium X was discovered. It would normally be present only in thorium compounds that have remained undisturbed a long time since preparation, or purification. This fact had later an important sequel, as we shall see.

The question whether these non-separable pairs of radio-elements are really chemically identical or not is not of importance in the argument. What is important is that 15 years ago pairs of radio-elements, actually not separable by chemical analysis, were in fact separated by successive chemical analyses at suitable intervals. If these pairs had been consecutive in the series, instead of being separated by intermediate products of different chemical character, they must have remained unresolved.

THE THEORETICAL INTERPRETATION OF ISOTOPES

In 1911 Rutherford put forward, in a somewhat tentative form, his now well-known "nuclear theory" of atomic structure, to account for the scattering of α- and β-particles in their passage through matter. To account especially for the large angles through which occasional α-particles are deflected it was necessary to suppose that these were the result of single encounters between the α-particle and an atom struck. For this to be possible, there must exist within the atom a much more powerful electric field than, for example, in the Thomson atom. Rutherford supposed that, at the centre of the atom, there existed a nucleus of very minute dimensions relatively to the atomic volume, upon which was concentrated a large charge of one sign, the rest of the atom being occupied by a number of single charges of the opposite sign which neutralized the central, or nuclear, charge. At first nothing was postulated in the theory as to the sign of this central charge, whether positive or negative, or as to the constitution of the nucleus. But it was natural to regard the central charge as positive, and the rest of the atom to be occupied by a system of negative electrons. In the next two years, much further experimental evidence on scattering accumulated, and the theory began to assume a more definite form.

"Single scattering," on this theory, should be proportional to the square of the nuclear charge. The experimental results indicated that it was approximately proportional to the square of the atomic weight of the scattering atom, and that there must be one unit of nuclear charge for two units of atomic weight, approximately. In 1911, Barkla arrived at the same conclusion as to the number of electrons in the atom from the scattering of X-rays.

In 1911, Van den Broek conceived the idea that "to each possible intra-atomic charge there corresponds a possible element" (*possible* here meant *integral*). This necessitates that successive elements in the Periodic Table should differ by one unit of nuclear charge and by one electron in the outer shell, as was subsequently practically directly established by the periodic-law generalization already considered.

To account for the supposed experimental result that there was one unit of charge for two units of atomic weight, Van den Broek at first attempted to revive the old and discarded "cubic" Periodic System of Mendeleeff, because it accommodated 120 elements. Since the last, uranium, has the atomic weight 240 it satisfies this requirement of giving on the average two units of mass per "place" i.e., per unit of charge. His attempts to fit the radioactive elements into this scheme appeared purely fanciful. But he then found that the actual experimental results for scattering were in entire accord with his own idea on the accepted Periodic Table, which accommodates some 90 elements. Thus uranium, in the last place with an intra-atomic charge of about 90, must have between 2 and 3 units of mass per unit of charge. So that, if its nucleus be imagined to be composed of 60 α-particles with charge 120, there must be present also 30 electrons to give the nuclear charge 90.

This suggestion of Van den Broek was adopted by Bohr in his theoretical researches on the structure of the atom. Bohr's views required that the electronic system is stable, so that to remove an electron involves the expenditure of energy. Hence it followed that the β-particles expelled in radioactive change must come from the nucleus and not from the external electronic system.

I had arrived at the same conclusion from totally different evidence. If, for example, the two electrons that are expelled as β-rays, when uranium X_I changes into uranium II, come from the same region of the atom as the two that are lost when U^{IV} in uranous salts is oxidized to U^{VI} in uranic salts, then the latter ought to be chemically non-separable from thorium, just as uranium X_I is. Fleck, trying this, found that, whereas there is a very close resemblance between uranium in the uranous salts and thorium, yet the two may be separated chemically without difficulty.

The expulsion of two + charges as an α-particle and of two electrons as β-particles from the nucleus causes the element to come back to the original place that it occupied in the Periodic Table. It followed therefore that the place in the Periodic Table is an expression of the *net* nuclear charge—i.e., of the difference between the numbers of positive and negative charges in the nucleus. Thus the chemically identical elements—or isotopes, as I called them for the first time in this letter to *Nature,* because they occupy *the same place* in the Periodic Table—are elements with the same algebraic or net nuclear charge, but with different numbers of + and − charges in the nucleus. On the view that the concentrated positive charge is the massive particle in the atomic structure, since positive electricity has never been observed free possessing less than the mass of an atom, the atomic weight of the isotope is a function of the *total* number of positive charges in the nucleus and the chemical character a function of the *net* number.

Though the nucleus possesses electrons, there can be no in- or outgoing of electrons between the nucleus and the external electronic system. Thus Rutherford's atom affords for the first time a clear picture of the difference between a transmutational (radioactive) change and chemical one. Changes of the number of electrons in the external system are chemical in character and produce changes in the valency of the element. These are reversible and have no effect at all on the central nucleus. Whereas changes of the nucleus are transmutational and irreversible, and they instantly impress changes upon the external electronic system to make it conform to the new nucleus. . . .

SUMMARY

We may now sum up the various distinct steps in this long and tangled story of the origins of the conception and discovery of isotopes.

1. Experimental methods are available, uniquely for the radio-elements, which enable isotopes to be severally recognized, by a suitable combination of chemical analyses at appropriate intervals of time, whereby, owing to the successive changes of the constituents, they may be separated, although chemical analysis alone is quite unable to effect this separation. This dates from 1905.

2. The complete chemical identity of isotopes, as distinct from close chemical similarity, came gradually to be recognized. McCoy and Ross were the first to express a definite opinion in this sense (1907).

3. The existence of chemical identities among the radio-elements led to the deduction that they might exist among the common elements and be responsible for the exceptions in the Periodic classification, and for the fact that the atomic weights in some cases depart widely from integral values. Strömholm and Svedberg first made this deduction (1909).

4. The recognition of the effect of the expulsion of, first, the α- and then, the β-particle (1911 and 1913) led to the correct placing of all three disintegration series from end to end in the Periodic Table. On the experimental side the name of A. Fleck, and on the theoretical side, the names of G. von Hevesy and A. S. Russell, but pre-eminently that of Kasimir Fajans, are associated with this advance.

5. The identity of isotopes was extended to include their electrochemistry (Paneth and Hevesy) and their spectra (Russell and Rossi), though here infinitesimal differences were subsequently found (Harkins and Aronberg, Merton).

6. Isotopes, on Rutherford's theory of atomic structure, are elements with identical external electronic systems, with identical net positive

charge on the nucleus, but with nuclei in which the total number of positive and negative charges and therefore the mass is different. The originator of the view that the places in the Periodic Table correspond with unit difference of intra-atomic charge is Van den Broek.

7. Moseley extended this view to the non-radioactive elements, and ultimately for the whole Periodic Table, and the definite determination of the number and sequence of the places in it became possible.

8. The chemistry of the radioactive elements and the lacunae previously existing in the radioactive series, especially in connection with the origin of actinium, have been cleared up, and this led to the discovery of a new element, proto-actinium (eka-tantalum) in uranium minerals, occupying the place between uranium and thorium and existing in sufficient quantity for the compounds of the element to be prepared in a pure state, and its spectrum and atomic weight ultimately to be determined, as Mme. Curie did for radium. Cranston and I share with Hahn and Meitner the original discovery, but the subsequent developments are due to the latter.

9. The preparation from radioactive minerals of different isotopes of lead followed, and the determination of their atomic weight, spectrum, density and other properties established that the same chemical character and atomic volume can coexist with differences of atomic weight. The work on the ionium-thorium mixture from pitchblende is a second example. . . .

K.5, K.6

Artificial Production of Radioactive Elements

IRÈNE JOLIOT-CURIE

Chemical Evidence of the Transmutation of Elements

FRÉDÉRIC JOLIOT-CURIE

IRÈNE CURIE was the elder daughter of Pierre and Marie Curie. She was brought up as an atheist, which was not unusual even in traditionally Roman Catholic France, where abuses by the clergy had engendered a strong current of popular anticlericalism that endures to this day.

Tutored privately at home by her mother and her mother's professional colleagues Paul Langevin and Jean Perrin, Irène earned her undergraduate degree at Collège Sévigné in 1914. During World War I, she served as an army nurse, assisting her mother in the operation of radiography equipment. Irène later studied mathematics and physics at the Sorbonne. Her doctoral thesis concerned variations in the range of alpha rays. She soon went to work as an assistant to her still-active mother, a job in which she met another assistant, Frédéric Joliot, a fellow atheist. Their mutual scientific interests and social attitudes led to their marriage in 1926, and they combined their surnames to give continuity to the highly distinguished Curie name.

Marie Curie developed leukemia as a result of overexposure to radioactive materials and was forced into retirement. Irène Joliot-Curie began to work together with Frédéric as a team, concentrating on radioactivity studies as well as the transmutation of elements. They shared the Nobel Prize in chemistry in 1935, the year after Marie's death. In 1937, Irène was elected professor at the Sorbonne, where she studied the neutron bombardment of uranium with P. P. Slavic; their experiments led to Otto Hahn's discovery that a neutron can split a uranium atom. Irène might have discovered the process of fission herself, but like most of her colleagues she could not accept the hypothesis that heavy elements like uranium could be split into lighter elements by neutron bombardment. Instead, Irène searched for the heavier elements she expected would be produced by the decay of uranium. She was named director of the Radium Institute in 1946 and, like her mother, died of leukemia due to overexposure to radioactive substances.

Frédéric Joliot-Curie was the son of a tradesman and was trained as an engineer at the École Supérieure de Physique et de Chimie Industrielle in Paris. In 1925, he became Marie Curie's assistant at the Radium Institute. He earned his doctorate in 1930 and began his collaboration with Irène the following year. Through the innovative use of the cloud chamber, they discovered artificial, or induced, radioactivity in 1934.* Three years later, Frédéric became Professor of Nuclear Physics at the Collège de France. His work with H. von Halban and L. Kowarski provided the first evidence that the fission of uranium could create a chain reaction. After the death of Irène in 1956, Frédéric succeeded her as head of the Radium Institute, where he worked until his own death two years later.

Both Joliot-Curies were involved in left-wing French politics. Their political activities caused the American Chemical Society to deny Irène's application for membership on political grounds. Frédéric was appointed to, and accepted, the job of director of the French Atomic Energy Program by President Charles de Gaulle. French engineers, unaided by American or British technology, completed a reactor in 1948. Joliot-Curie, who had become an active Communist Party member during World War II, was dismissed from his position in 1950, and his job was given to an old friend, neighbor, and fellow Nobel laureate, Jean Perrin. In 1951, Joliot-Curie accepted the Stalin Prize—at a time when Trofim

---

* Unlike natural radioactivity which is emitted when naturally occurring radioisotopes disintegrate, the bombardment of atomic nuclei by particles such as protons and neutrons produces artificial radioisotopes, which decay by the same process as natural isotopes.

Lysenko was prostituting Soviet biology to political ends and Stalin was insisting that Norbert Wiener's cybernetics was "reactionary Jewish pseudoscience"!

The work done by the Joliot-Curies, particularly on artificial radioactivity, is of the highest importance in pure and applied science and medicine. The first examples of transmuted elements had been produced by Rutherford and Soddy (although Rutherford disliked the term *transmutation:* "By God, Soddy," he is reported to have roared in his New Zealand accent, "they'll have us out as alchemists!"). Rutherford's oxygen-17 and Cockcroft and Walton's helium-4 are stable isotopes that are found in nature, but the Joliot-Curies' potassium-30 is the first isotope to be produced in the laboratory that has no natural counterpart on earth. It is also the first example of artificial radioactivity and the forerunner of thousands of artificially radioactive substances that have changed the quality of human life.

The Joliot-Curies' 1935 Nobel lectures, which follow, are models of clarity, and except for chronological reasons, Mme Joliot-Curie's piece could well have served as a general introduction to this entire section on radioactivity. It is lucid, comprehensive, selective; it covers a vast field with elegant economy of language. Frédéric's share of the Nobel lectures, while technically more complicated, is equally clear; it seems to end with a prophecy of the cataclysm inherent in the unwise use of atomic fission by anyone but the most ethical of scientists or, as Joliot-Curie calls them, "investigators."

Artificial Production of Radioactive Elements

IRÈNE JOLIOT-CURIE

Basic research is when I'm doing what I don't know I'm doing.

—WERNHER VON BRAUN

Research means going out into the unknown with the hope of finding something new to bring home. If you know what you are going to do, or even to find there, then it is not research at all, then it is only a kind of honourable occupation.

—ALBERT SZENT-GYORGYI

Research is the process of going up alleys to see if they are blind.

—MARSTON BATES

. . . I WOULD LIKE here to recall the extraordinary development of radioactivity, this new science which had its origin, less than forty years ago, in the work of Henri Becquerel and of Pierre and Marie Curie.

It is known that the efforts of chemists of the last century established as a fundamental fact the extreme solidity of the atomic structures which go to make up the ninety-two known chemical species. With the discovery of the radio-elements, physicists found themselves for the first time confronted with strange substances, minute generators of radiation endowed with an enormous concentration of energy; alpha rays, positively charged

Irène Joliot-Curie, Nobel Prize in Chemistry Award Address, 1935. Reprinted with permission of Elsevier Publishing Co., and the Nobel Foundation.

helium atoms, beta rays, negatively charged electrons, both possessed of a kinetic energy which it would be impossible to communicate to them by human agency, and finally, gamma rays, akin to very penetrating X-rays. Chemists had no less astonishment as they recognized in these radioactive bodies, elements which had undergone modifications of the atomic structure which had been thought unalterable.

Each emission of an alpha or beta ray accompanies the transmutation of an atom; the energy communicated to these rays comes from inside the atom. As long as they continue to exist, radio-elements have well-defined chemical properties, like those of ordinary elements. These unstable atoms disintegrate spontaneously—some very quickly, others very slowly, but in accordance with unchanging laws which it has never been possible to interfere with. The time necessary for the disappearance of half the atoms, called the half-life, is a fundamental characteristic of each radio-element; according to the substance, the value of the half-life varies between a fraction of a second and millions of years.

The discovery of radio-elements has had immense consequences in the knowledge of the structure of matter; the study of the materials themselves, and the study of the powerful effects produced on atoms by the rays they emit occupy scientific workers of numerous great research institutes in all countries.

Nevertheless, radioactivity remained a property exclusively associated with some thirty substances existing naturally. The artificial creation of radio-elements opens a new field to the science of radioactivity and so provides an extension of the work of Pierre and Marie Curie.

After the discovery of the spontaneous transmutations of radio-elements, the achievement of the first artificial transmutations is due to Lord Rutherford. Fifteen years later, by bombarding with alpha rays certain of the lighter atoms, nitrogen and aluminium for example, Lord Rutherford demonstrated the ejection of protons, or positively charged hydrogen nuclei; this hydrogen came from the bombarded atoms themselves: it was the result of a transmutation. The nature of the nuclear transformation could be firmly established: the aluminium atom, for example, captures the alpha particle and is transformed, after expelling the proton, into an atom of silicon. The amount of matter transformed cannot be weighed, and the study of radiation alone has led to these conclusions.

In the course of recent years various artificial transmutations of different types have been discovered: some are produced by alpha rays; others by protons or deuterons, hydrogen nuclei of weight 1 or 2; others by neutrons—neutral particles of weight 1, about which Professor Chadwick has just spoken. The particles expelled when the atom explodes are protons, alpha rays or neutrons.

These transformations constitute true chemical reactions which act upon the innermost structure of the atom, the nucleus. They can be represented by simple formulae as Monsieur Joliot will be telling you in a moment.

I shall now speak to you of the experiments which have led us to obtain by transmutation new radioactive elements. These experiments have been made together by Monsieur Joliot and me, and the way in which we have divided this lecture between us is a matter of pure convenience.

In our study of the transmutations with emission of neutrons produced in the light elements irradiated with alpha rays, we have noticed some difficulties in interpretation in the emission of neutrons by fluorine, sodium, and aluminium. Aluminium can be transformed, by the capture of an alpha particle and the emission of a proton, into a stable silicon atom. On the other hand, if a neutron is emitted the product of the reaction is not a known atom.

Later on, we observed that aluminium and boron, when irradiated by alpha rays do not emit protons and neutrons alone; there is also an emission of positive electrons. We have assumed in this case that the emission of the neutron and the positive electron occurs simultaneously, instead of the emission of a proton; the atom remaining must be the same in the two cases.

It was at the beginning of 1934, while working on the emission of these positive electrons, that we noticed a fundamental difference between that transmutation and all the others so far produced; all the reactions of nuclear chemistry induced were instantaneous phenomena, explosions. But the positive electrons produced by aluminium under the action of a source of alpha rays continue to be emitted for some time after removal of the source. The number of electrons emitted decreases by half in three minutes.

Here, therefore, we have a true radioactivity which is made evident by the emission of positive electrons.

We have shown that it is possible to create a radioactivity characterized by the emission of positive or negative electrons in boron and magnesium, by bombardment with alpha rays. These artificial radio-elements behave in all respects like the natural radio-elements.

Returning to our hypothesis concerning the transformation of the aluminium nucleus into a silicon nucleus, we have supposed that the phenomenon takes place in two stages: first there is the capture of the alpha particle and the instantaneous expulsion of the neutron, with the formation of a radioactive atom which is an isotope of phosphorus of atomic weight 30, while the stable phosphorus atom has an atomic weight of 31. Next, this unstable atom, this new radio-element which we have called "radio-phosphorus," decomposes exponentially with a half-life of three minutes.

We have interpreted in the same way the production of radioactive elements in boron and magnesium; in the first an unstable nitrogen with a half-life of 11 minutes is produced, and in the second, unstable isotopes of silicon and aluminium.

Chemical Evidence of the Transmutation of Elements

FRÉDÉRIC JOLIOT-CURIE

Truth in science can be defined as the working hypothesis best suited to open the way to the next better one.

—KONRAD LORENZ

The most important thing in science is not so much to obtain new facts as to discover new ways of thinking about them.

—SIR WILLIAM BRAGG

Science is not to be regarded merely as a storehouse of facts to be used for material purposes, but as one of the great human endeavors to be ranked with the arts and religion as the guide and expression of man's fearless quest for truth.

—SIR RICHARD GREGORY

THE INTERPRETATION OF our first experiments was founded, as Madame Joliot-Curie has just explained, on facts of a purely physical order. We have considered the possibility, by using the methods of radiochemistry, of substantiating our hypotheses of providing a chemical proof of the reality of the transmutations brought about.

The first unquestionable proofs of the transformation of elements into different chemical elements have been provided by the study of the phe-

Frédéric Joliot, Nobel Prize in Chemistry Award Address, 1935. Reprinted with permission of Elsevier Publishing Co., and the Nobel Foundation.

nomena of radioactivity. There is no doubt that radium is transformed spontaneously into an active gas, radon, emitting at the same time alpha particles, or helions. We may write with certainty the corresponding nuclear reaction (the atomic weight is placed above and to the left of the chemical symbol, and the atomic number to the left and below)

$$^{226}_{88}\text{Ra} \rightarrow {}^{222}_{86}\text{Rn} + {}^{4}_{2}\text{He}$$

for the quantities of the various elements taking part in this reaction can be considered sufficient for their successful identification chemically and spectroscopically.

The succession of radioactive transformations provides numerous examples in which the quantities of radio-elements are extremely small and not capable of being weighed, yet nevertheless, by the methods of radiochemistry it has been possible to examine correctly their chemical properties, and identify some of them as being isotopes of elements, active or inactive, available in large quantities.

This special kind of chemistry in which one handles unweighable quantities, sometimes of the order of 10^{-16} g, is made possible thanks to the fact that one can determine and follow by measuring the radiation emitted, infinitesimal traces of radioactive matter dispersed in the midst of other matter.

But while it is possible to write with certainty the nuclear reactions corresponding to most of the spontaneous transmutations, this is not the case for those artificially brought about.

The yield of these transmutations is very small, and the weights of elements formed by using the most intense sources of projectiles which we are able to produce at the present time are less than 10^{-15} g, representing at the most a few million atoms. It is, however, possible to deduce the nature of the atoms formed with sufficient confidence, by assuming, in order to write the corresponding nuclear reaction, that there is on the one hand conservation of the atomic weight and on the other conservation of atomic number between the reacting elements and those formed.

Thus the aluminium nucleus capturing a helion must be transformed into silicon when a proton is emitted:

$$^{27}_{13}\text{Al} + {}^{4}_{2}\text{He} = {}^{30}_{14}\text{Si} + {}^{1}_{1}\text{H} \qquad \begin{matrix} 27 + 4 = 30 + 1 \\ 13 + 2 = 14 + 1 \end{matrix}$$

The atom formed is very probably silicon, but being present in infinitesimal quantity, it is not possible to identify it chemically. On the other hand, when the atom formed is radioactive, it can be identified by applying radiochemical methods. For example, in the case where aluminium irradiated by alpha rays emits neutrons, the rule already mentioned enables us to write the following transmutation reaction:

$$^{27}_{13}Al + ^{4}_{2}He = ^{30}_{15}P + ^{1}_{0}n$$

The atom formed being radioactive, it is possible to verify that it possesses the chemical properties of phosphorus.

A piece of thin aluminium sheet, irradiated beforehand by alpha rays is attacked and dissolved in a solution of hydrochloric acid. The chemical reaction produces nascent hydrogen, which carries over the radioactive element into a thin-walled tube where it is collected over water. This separation clearly demonstrates that some element other than aluminium has been formed on irradiation by helions. It furnishes an indisputable proof of the transmutation achieved; also, traces of phosphorus would be separated from the aluminium in the same experiment.

Finally, activated aluminium is dissolved in a mixture of acid and oxidant. To the solution is added a small quantity of sodium phosphate and a zirconium salt, and it is found that the zirconium phosphate as it precipitates carries with it the radioactive element. For aluminium these experiments are delicate, for they must be made in about six minutes, the average life of the radioactive atoms which are formed being less than five minutes. Chemical tests of the same kind have shown us that the radio-element formed in boron under the action of alpha rays is an isotope of nitrogen.

We have proposed that these new radio-elements (isotopes, not found in nature, of known elements) be called radio-nitrogen, radio-phosphorus, radio-aluminium (in the case of magnesium irradiated by alpha rays) and designated by the symbols: $R^{13}N$, $R^{30}P$, $R^{28}Al$.

Immediately after these first researches, we suggested that the same phenomenon might occur for the kinds of transmutations brought about by collisions with other particles than alpha rays, for example with protons, deuterons, and neutrons.

These experiments were taken up and further developed in several countries. In England and the United States, where physicists have at their disposal equipment of very high voltages, several new elements were prepared using protons and deuterons as projectiles. In Italy first, and then in other countries, research workers, in particular Fermi and his co-workers, used neutrons, projectiles which are outstandingly suitable, to cause transmutations. A large number of new elements were created in this way, among which were radio-phosphorus $R^{32}P$, and radio-hafnium, with periods of 17.5 days and several months respectively. At the present time we know how to synthesize, often by several methods (radio-aluminium $R^{28}Al$ can be made by five different kinds of transmutation), more than fifty new radio-elements, a number already greater than that of the natural radio-elements found in the earth's crust. It was indeed a great source of satisfaction for our lamented teacher Marie Curie to have witnessed this lengthening of the list of radio-elements which she had had the glory, in company with Pierre Curie, of beginning.

The diversity of the chemical properties and the diversity of the average lives of these synthetic radio-elements will without doubt enable further advances in research in biology and in physical chemistry to be made. In order to undertake this work properly, fairly large quantities will be required. This will be achieved by the use of artificially accelerated projectiles. Equipment suitable for this is already in existence in several countries. In France, we have built two installations with which we have recently been able to obtain radio-elements in quantities a hundred times greater than those which we obtained in our first experiments. This ratio will shortly be greatly exceeded.

The method of radioactive tracers, until now confined to elements of large atomic weights, can be extended to a large number of elements distributed over the full extent of the periodic classification. In biology, for example, the tracer method, making use of synthetic radio-elements, will simplify the problems of the location and elimination of the various elements introduced into living organisms. In this case, radioactivity serves only to determine the presence of an element in a particular region of the organism. There is no point, in research of this kind, in introducing large quantities of the radioactive tracer. The quantities are determined by the sensitivity of the radiation-detecting apparatus and the size of the vegetable or animal organism. In places, which will be more readily known by the use of this method, the radiation emitted will produce its effect on the adjacent cells. For this second mode of employment, large quantities of radio-elements will be required. This will probably become a practical application in medicine.

From this overall mass of facts now evident we can realize that the few hundreds of atoms of different species which form our planet must not be considered as having been created all at one time and to last for ever. We are aware of them because they have survived. Others less stable have disappeared. It is probably some of these vanished atoms which have been recreated in our laboratories. Up to now, it has only been possible to obtain elements with a relatively brief life, extending from a fraction of a second to several months. In order to create an appreciable quantity of a much longer-lived element, an enormously intense source of projectiles would be required. Is there no hope at all of realizing this new dream?

If, turning towards the past, we cast a glance at the progress achieved by science at an ever-increasing pace, we are entitled to think that scientists, building up or shattering elements at will, will be able to bring about transmutations of an explosive type, true chemical chain reactions.

If such transmutations do succeed in spreading in matter, the enormous liberation of usable energy can be imagined. But, unfortunately, if the contagion spreads to all the elements of our planet, the consequences of unloosing such a cataclysm can only be viewed with apprehension. Astronomers sometimes observe that a star of medium magnitude increases suddenly in size; a star invisible to the naked eye may become very brilliant and visible without any telescope—the appearance of a Nova. This sudden flaring up of the star is perhaps due to transmutations of an explosive character like those which our wandering imagination is perceiving now—a process that the investigators will no doubt attempt to realize while taking, we hope, the necessary precautions.

Artificial Radioactivity Produced by Neutron Bombardment

ENRICO FERMI

ENRICO FERMI was born in Italy in 1901, the son of an inspector at the Ministry of Communications. His aptitude for mathematics and physics was rewarded by a fellowship to the University of Pisa, where he earned his doctorate in physics in 1922. After a brief collaboration with Max Born in Göttingen, Fermi became a lecturer in mathematical physics at the University of Florence. In 1926, Fermi discovered the statistical laws that now bear his name and that govern the particles subject to Pauli's exclusion principle, which holds that no two elementary particles can exist in identical quantum states. In 1927, he became Professor of Theoretical Physics at the University of Rome, where he remained until he was awarded the Nobel Prize in 1938 and subsequently emigrated to the United States.

While at Rome, Fermi developed the beta-decay theory (which explained the spontaneous transformation of a nucleus to a different state having the same mass number but a different atomic number), which combined previous work on radiation with Pauli's hypothesized neutrino. The neutrino is a particle found in weak interactions that has no mass and travels at the speed of light. It was first postulated to explain an apparent violation in the laws of conservation of energy and angular momentum in beta decay. Fermi's detailed analysis of this process led him to introduce into physics the concept of the "weak" force, which is a short-range interaction between elementary particles believed to cause the decay of many long-lived particles. The first account of his discovery was published in Italian in 1933, as the British science journal *Nature* had rejected an English version of his work as too speculative. The 1934 discovery by the Joliot-Curies of artificial radioactivity was followed by Fermi's demonstration that almost every element subject to neutron bombardment undergoes nuclear transformation. His work led to the discovery of slow neutrons (having kinetic energies of only a few electron volts), which in turn led to the discovery of nuclear fission and the production of synthetic elements.

Fermi was Professor of Physics at Columbia University from 1939 until 1942. The discovery of fission by Hahn, Strassmann and Meitner in 1939 inspired Fermi to begin work on the atomic pile, or first nuclear reactor, at Columbia; subsequently he went to the University of Chicago to continue this work to its successful conclusion. The result was the first controlled nuclear chain reaction. This triumph, which occurred beneath Chicago's stadium, reaffirmed Fermi's preeminence in nuclear and atomic physics. He was one of the chief physicists on the Manhattan Project, which successfully harnessed nuclear energy through the development of the atomic bomb.

Fermi became an American citizen in 1944 and spent the remaining ten years of his life doing high-energy physics research at the Institute for Nuclear Studies at Chicago. He also devoted considerable attention to the origin of cosmic rays, speculating that these high-energy particles might be accelerated by a universal magnetic field.

Fermi's brilliant experimental techniques as well as his inventiveness and intuition made it possible for him to demonstrate the existence of two new radioactive elements in addition to the ninety-two naturally occurring elements in the Periodic Table. These two elements were created by the neutron bombardment of uranium nuclei. He named these two new elements *ausenium* and *hesperium.* His discovery completed the work begun by Becquerel of uranium disintegration and forever ended the notion that the ninety-two elements are immutable and indestructible units of matter.

Fermi received many honors during his lifetime. Element 100 was named *fermium,* and the unit of length of 10^{-13} cm was named the *fermi.* The National Accelerator Laboratory near Chicago was also named *Fermilab* in his honor.

The progress of science consists in observing these interconnections and in showing with a patient ingenuity that the events of this evershifting world are but examples of a few general connections or relations called laws. To see what is general in what is particular and what is permanent in what is transitory is the aim of scientific thought. In the eye of science, the fall of an apple, the motion of a planet round a sun, and the clinging of the atmosphere to the earth are all seen as examples of the law of gravity. This possibility of disentangling the most complex evanescent circumstances into various examples of permanent laws is the controlling idea of modern thought.

—ALFRED NORTH WHITEHEAD, *Introduction to Mathematics*

The most useful investigator, because the most sensitive observer, is always he whose eager interest in one side of the question is balanced by an equally keen nervousness lest he become deceived. Science has organized this nervousness into a regular technique, *her so-called method of verification; and she has fallen so deeply in love with the method that one may even say she has ceased to care for truth by itself at all. It is only truth as technically verified that interests her. The truth of truths might come in merely affirmative form, and she would decline to touch it.*

—WILLIAM JAMES, *Will to Believe*

ALTHOUGH THE PROBLEM of transmuting chemical elements into each other is much older than a satisfactory definition of the very concept of chemical element, it is well known that the first and most important step towards its solution was made only nineteen years ago by the late Lord Rutherford, who started the method of the nuclear bombardments. He showed on a few examples that, when the nucleus of a light element is struck by a fast α-particle, some disintegration process of the struck nucleus occurs, as a consequence of which the α-particle remains captured inside the nucleus and a different particle, in many cases a proton, is emitted in its place. What remains at the end of the process is a nucleus

Enrico Fermi, Nobel Prize in Physics Award Address, 1938. Reprinted with permission of Elsevier Publishing Co., and the Nobel Foundation.

different from the original one—different in general both in electric charge and in atomic weight.

The nucleus that remains as disintegration product coincides sometimes with one of the stable nuclei, known from the isotopic analysis; very often, however, this is not the case. The product nucleus is then different from all "natural" nuclei, the reason being that the product nucleus is not stable. It disintegrates further, with a mean life characteristic of the nucleus, by emission of an electric charge (positive or negative), until it finally reaches a stable form. The emission of electrons that follows, with a lag in time, the first practically instantaneous disintegration, is the so-called artificial radioactivity, and was discovered by Joliot and Irène Curie at the end of the year 1933.

These authors obtained the first cases of artificial radioactivity by bombarding boron, magnesium, and aluminium with α-particles from a polonium source. They produced thus three radioactive isotopes of nitrogen, silicon and phosphorus, and succeeded also in separating chemically the activity from the bulk of the unmodified atoms of the bombarded substance.

THE NEUTRON BOMBARDMENT

Immediately after these discoveries, it appeared that α-particles very likely did not represent the only type of bombarding projectiles for producing artificial radioactivity. I decided therefore to investigate from this point of view the effects of the bombardment with neutrons.

Compared with α-particles, the neutrons have the obvious drawback that the available neutron sources emit only a comparatively small number of neutrons. Indeed, neutrons are emitted as products of nuclear reactions, whose yield is only seldom larger than 10^{-4}. This drawback is, however, compensated by the fact that neutrons, having no electric charge, can reach the nuclei of all atoms, without having to overcome the potential barrier, due to the Coulomb field that surrounds the nucleus. Furthermore, since neutrons practically do not interact with electrons, their range is very long, and the probability of a nuclear collision is correspondingly larger than in the case of the α-particle or the proton bombardment. As a matter of fact, neutrons were already known to be an efficient agent for producing some nuclear disintegrations.

As source of neutrons in these researches I used a small glass bulb containing beryllium powder and radon. With amounts of radon up to 800 millicuries, such a source emits about 2×10^7 neutrons per second. This number is, of course, very small compared to the yield of neutrons that can be obtained from cyclotrons or from high-voltage tubes. The small dimensions, the perfect steadiness and the utmost simplicity are, however, sometimes very useful features of the radon + beryllium sources.

NUCLEAR REACTIONS PRODUCED BY NEUTRONS

Since the first experiments, I could prove that the majority of the elements tested became active under the effect of the neutron bombardment. In some cases the decay of the activity with time corresponded to a single mean life; in others to the superposition of more than one exponential decay curve.

A systematic investigation of the behaviour of the elements throughout the Periodic Table was carried out by myself, with the help of several collaborators, namely Amaldi, d'Agostino, Pontecorvo, Rasetti, and Segrè. In most cases we performed also a chemical analysis, in order to identify the chemical element that was the carrier of the activity. For short living substances, such an analysis must be performed very quickly, in a time of the order of one minute.

The results of this first survey of the radioactivities produced by neutrons can be summarized as follows: Out of 63 elements investigated, 37 showed an easily detectable activity; the percentage of the activatable elements did not show any marked dependence on the atomic weight of the element. Chemical analysis and other considerations, mainly based on the distribution of the isotopes, permitted further to identify the following three types of nuclear reactions giving rise to artificial radioactivity.

$$^{M}_{Z}A + ^{1}_{0}n = ^{M-3}_{Z-2}A + ^{4}_{2}He \qquad (1)$$

$$^{M}_{Z}A + ^{1}_{0}n = {}_{Z-1}^{M}A + ^{1}_{1}H \qquad (2)$$

$$^{M}_{Z}A + ^{1}_{0}n = ^{M+1}_{Z}A \qquad (3)$$

where $^{M}_{Z}A$ is the symbol for an element with atomic number Z and mass number M; n is the symbol of the neutron.

The reactions of the types (1) and (2) occur chiefly among the light elements, while those of the type (3) are found very often also for heavy elements. In many cases the three processes are found at the same time in a single element. For instance, neutron bombardment of aluminium that has a single isotope, ^{27}Al, gives rise to three radioactive products: ^{24}Na, with a half-period of 15 hours by process (1); ^{27}Mg, with a period of 10 minutes by process (2); and ^{28}Al, with a period of 2 to 3 minutes by process (3).

As mentioned before, the heavy elements usually react only according to process (3) and therefore, but for certain complications to be discussed later, and for the case in which the original element has more than one stable isotope, they give rise to an exponentially decaying activity. A very striking exception to this behaviour is found for the activities induced by neutrons in the naturally active elements thorium and uranium. For the investigation of these elements it is necessary to purify first the element as thoroughly as possible from the daughter substances that emit β-particles. When thus purified, both thorium and uranium emit spontaneously only α-particles, that can be immediately distinguished, by absorption, from the β-activity induced by the neutrons.

Both elements show a rather strong induced activity when bombarded with neutrons; and in both cases the decay curve of the induced activity shows that several active bodies with different mean lives are produced. We attempted, since the spring of 1934, to isolate chemically the carriers of these activities, with the result that the carriers of some of the activities of uranium are neither isotopes of uranium itself nor of the elements lighter than uranium down to the atomic number 86. We concluded that the carriers were one or more elements of atomic number larger than 92; we, in Rome, use to call the elements 93 and 94 Ausenium and Hesperium respectively. It is known that O. Hahn and L. Meitner have investigated very carefully and extensively the decay products of irradiated uranium, and were able to trace among them elements up to the atomic number 96.*

It should be noticed here, that besides processes (1), (2), and (3) for the production of artificial radioactivity with neutrons, neutrons of sufficiently high energy can react also as follows, as was first shown by Heyn: The primary neutron does not remain bound in the nucleus, but knocks off, instead, one of the nuclear neutrons out of the nucleus; the result is a new nucleus, that is isotopic with the original one and has an atomic weight less by one unit. The final result is therefore identical with the

*The discovery by Hahn and Strassmann of barium among the disintegration products of bombarded uranium, as a consequence of a process in which uranium splits into two approximately equal parts, makes it necessary to reexamine all the problems of the transuranic elements, as many of them might be found to be products of a splitting of uranium.

products obtained by means of the nuclear photoeffect (Bothe), or by bombardment with fast deuterons. One of the most important results of the comparison of the active products obtained by these processes, is the proof, first given by Bothe, of the existence of isomeric nuclei, analogous to the isomers UX_2 and UZ, recognized long since by O. Hahn in his researches on the uranium family. The number of well-established cases of isomerism appears to increase rather rapidly, as investigation goes on, and represents an attractive field of research.

THE SLOW NEUTRONS

The intensity of the activation as a function of the distance from the neutron source shows in some cases anomalies apparently dependent on the objects that surround the source. A careful investigation of these effects led to the unexpected result that surrounding both source and body to be activated with masses of paraffin, increases in some cases the intensity of activation by a very large factor (up to 100). A similar effect is produced by water, and in general by substances containing a large concentration of hydrogen. Substances not containing hydrogen show sometimes similar features, though extremely less pronounced.

The interpretation of those results was the following. The neutron and the proton having approximately the same mass, any elastic impact of a fast neutron against a proton initially at rest, gives rise to a distribution of the available kinetic energy between neutron and proton; it can be shown that a neutron having an initial energy of 10^6 volts, after about 20 impacts against hydrogen atoms, has its energy already reduced to a value close to that corresponding to thermal agitation. It follows that, when neutrons of high energy are shot by a source inside a large mass of paraffin or water, they very rapidly lose most of their energy and are transformed into "slow neutrons." Both theory and experiment show that certain types of neutron reactions, and especially those of type (3), occur with a much larger cross-section for slow neutrons than for fast neutrons, thus accounting for the larger intensities of activation observed when irradiation is performed inside a large mass of paraffin or water.

It should be remarked furthermore that the mean free path for the elastic collisions of neutrons against hydrogen atoms in paraffin decreases rather pronouncedly with the energy. When, therefore, after three or four impacts, the energy of the neutron is already considerably reduced, its probability of diffusing outside of the paraffin, before the process of slowing down is completed, becomes very small.

To the large cross-section for the capture of slow neutrons by several atoms, there must obviously correspond a very strong absorption of these atoms for the slow neutrons. We investigated systematically such absorptions, and found that the behaviour of different elements in this respect is widely different; the cross-section for the capture of slow neutrons varies, with no apparent regularity for different elements, from about 10^{-24} cm^2 or less, to about a thousand times as much. Before discussing this point, as well as the dependence of the capture cross-section on the energy of the neutrons, we shall first consider how far down the energy of the primary neutrons can be reduced by the collisions against the protons.

THE THERMAL NEUTRONS

If the neutrons could go on indefinitely diffusing inside the paraffin, their energy would evidently reach finally a mean value equal to that of thermal agitation. It is possible, however, that, before the neutrons have reached this lowest limit of energy, either they escape by diffusion out of the paraffin, or are captured by some nucleus. If the neutron energy reaches the thermal value, one should expect the intensity of the activation by slow neutrons to depend upon the temperature of the paraffin.

Soon after the discovery of the slow neutrons, we attempted to find a temperature dependence of the activation, but, owing to insufficient accuracy, we did not succeed. That the activation intensities depend upon the temperature was proved some months later by Moon and Tillman in London; as they showed, there is a considerable increase in the activation of several detectors, when the paraffin, in which the neutrons are slowed down, is cooled from room temperature to liquid-air temperature. This experiment definitely proves that a considerable percentage of the neutrons actually reaches the energy of thermal agitation. Another consequence is that the diffusion process must go on inside the paraffin for a relatively long time.

In order to measure, directly at least, the order of magnitude of this time, an experiment was attempted by myself and my collaborators. The source of neutrons was fastened at the edge of a rotating wheel, and two identical detectors were placed on the same edge, at equal distances from the source, one in front and one behind with respect to the sense of rotation. The wheel was then spun at a very high speed inside a fissure in a large paraffin block. We found that, while, with the wheel at rest, the two detectors became equally active, when the wheel was in motion during the activation, the detector that was behind the source became considerably more active than the one in front. From a discussion of this experiment was deduced, that the neutrons remain inside the paraffin for a time of the order of 10^{-4} seconds.

Other mechanical experiments with different arrangements were performed in several laboratories. For instance, Dunning, Fink, Mitchell, Pegram, and Segrè, in New York, built a mechanical velocity selector, and proved by direct measurement, that a large amount of the neutrons diffusing outside of a block of paraffin, have actually a velocity corresponding to thermal agitation.

After their energy is reduced to a value corresponding to a thermal agitation, the neutrons go on diffusing without further change of their average energy. The investigation of this diffusion process, by Amaldi and myself, showed that thermal neutrons in paraffin or water can diffuse for a number of paths of the order of 100 before being captured. Since, however, the mean free path of the thermal neutrons in paraffin is very short (about 0.3 cm), the total displacement of the thermal neutrons during this diffusion process is rather small (of the order of 2 or 3 cm). The diffusion ends when the thermal neutron is captured, generally by one of the protons, with production of a deuteron. The order of magnitude for this capture probability can be calculated, in good agreement with the experimental value, on the assumption that the transition from a free-neutron state to the state in which the neutron is bound in the deuteron is due to the magnetic dipole moments of the proton and the neutron. The binding energy set free in this process, is emitted in the form of γ-rays, as first observed by Lea.

All the processes of capture of slow neutrons by any nucleus are generally accompanied by the emission of γ-rays: Immediately after the capture of the neutron, the nucleus remains in a state of high excitation and emits one or more γ-quanta, before reaching the ground state. The γ-rays emitted by this process were investigated by Rasetti and by Fleischmann.

ABSORPTION ANOMALIES

A theoretical discussion of the probability of capture of a neutron by a nucleus, under the assumption that the energy of the neutron is small compared with the differences between neighbouring energy levels in the nucleus, leads to the result that the cross-section for the capture process should be inversely proportional to the velocity of the neutron. While this result is in qualitative agreement with the high efficiency of the slow-neutron bombardment observed experimentally, it fails on the other hand to account for several features of the absorption process, that we are now going to discuss.

If the capture probability of a neutron were inversely proportional to its velocity, one would expect two different elements to behave in exactly the same way as absorbers of the slow neutrons, provided the thicknesses of the two absorbers were conveniently chosen, so as to have equal absorption for neutrons of a given energy. That the absorption obeys instead more complicated laws, was soon observed by Moon and Tillman and other authors who showed that the absorption by a given element appears, as a rule, to be larger when the slow neutrons are detected by means of the activity induced in the same element. That the simple law of inverse proportionality does not hold, was also proved by a direct mechanical experiment by Dunning, Pegram, Rasetti, and others in New York.

In the winter of 1935–1936 a systematic investigation of these phenomena was carried out by Amaldi and myself. The result was, that each absorber of the slow neutrons has one or more characteristic absorption bands, usually for energies below 100 volts. Besides this or these absorption bands, the absorption coefficient is always large also for neutrons of thermal energy. Some elements, especially cadmium, have their characteristic absorption band overlapping with the absorption in the thermal region. This element absorbs therefore very strongly the thermal neutrons, while it is almost transparent to neutrons of higher energies. A thin cadmium sheet is therefore used for filtering the thermal neutrons out of the complex radiation that comes out of a paraffin block containing a neutron source inside.

Bohr and Breit and Wigner proposed independently to explain the above anomalies, as due to resonance with a virtual energy level of the compound nucleus (i.e., the nucleus composed of the bombarded nucleus and the neutron). Bohr went much farther in giving also a qualitative explanation of the large probability for the existence of at least one such level, within an energy interval of the order of magnitude of 100 volts, corresponding to the energy band of the slow neutrons. This band corresponds, however, to an excitation energy of the compound nucleus of many million volts, representing the binding energy of the neutron. Bohr could show that, since nuclei, and especially heavy nuclei, are systems with a very large number of degrees of freedom, the spacing between neighbouring energy levels decreases very rapidly with increasing excitation energy. An evaluation of this spacing shows that whereas for low excitation energies the spacing is of the order of magnitude of 10^5 volts, for high excitation energies, of the order of ten million volts, it is reduced (for elements of mean atomic weight) to less than one volt. It is therefore a very plausible assumption that one (or more) such level lies within the slow-neutron band, thus explaining the large frequency of the cases in which absorption anomalies are observed. . . .

K.8

From the Natural Transmutations of Uranium to Its Artificial Fission

OTTO HAHN

OTTO HAHN, one of the foremost chemists of the twentieth century, was born in 1879. His father was a merchant who wanted his son to become an architect. Despite his family's opposition, Otto studied chemistry at the University of Marburg. After obtaining his doctorate in 1901, he studied with William Ramsay in London and then traveled to Canada to work for Ernest Rutherford at McGill University. He accepted an appointment to the University of Berlin in 1903 and became Professor of Chemistry there in 1910. He joined the Kaiser Wilhelm Institute of Chemistry in 1912 and served as its director from 1928 to 1945. Although Hahn had originally been interested in organic chemistry, while working for Ramsay, he was diverted into physical chemistry when he discovered a highly active form of thorium, which he called *radiothorium.* Hahn soon began to collaborate in radioactive studies with Lise Meitner; they discovered a new element, protactinium, in 1917.

Although Hahn was still a relatively unknown physical chemist when Hitler came to power in 1933, he and Meitner recognized the importance of Enrico Fermi's research on the bombardment of uranium with slow neutrons. Hahn, Meitner, and others worked on Fermi's results, and certain facts led Meitner and Hahn to conclude that Fermi had actually succeeded in breaking up—in *splitting*—the uranium nucleus. Hahn was reluctant to publish his conclusion as such, but in early 1939 he published his results without suggesting that fission of the nucleus was involved.

In the meantime, Meitner had left Germany because of the racial laws and felt free to publish Hahn's hypothesis that fission might be involved in Fermi's work. Hahn remained in Germany during the war but carefully avoided all weapons research. When World War II came to an end, Hahn, together with other leading German scientists, was arrested by American soldiers. While in detention, he received the news of the Hiroshima and Nagasaki bombings, for which he felt a great personal responsibility. It is said that he even contemplated suicide. Hahn was the only foreigner to be granted a share of the money paid by the U. S. Atomic Energy Commission following a suit brought by Fermi and other holders of the patent for the slow-neutron fission of uranium. After Hahn's death in 1968, the transuranic element 105 was named *hahnium* in his honor.

Our selection, excerpted from Hahn's Nobel lecture given in 1946 (his award was announced in 1944), is an historical account beginning with the discoveries of Becquerel and concluding with a polemical remark in the postscript that both uranium-235 and plutonium were eventually made in the U. S. and the result was the bombing of Hiroshima and Nagasaki.

Gods are born and die, but the atom endures.

—ALEXANDER CHASE, *Perspectives*

The discovery of the nuclear chain reaction need not bring about the destruction of mankind any more than did the discovery of matches.

—ALBERT EINSTEIN

There is no evil in the atom—only in men's souls.

—ADLAI STEVENSON

People must understand that science is inherently neither a potential for good nor for evil. It is a potential to be harnessed by man to do his bidding.

—GLENN T. SEABORG

The best defence against the atom bomb is not to be there when it goes off.

—THE BRITISH ARMY JOURNAL, 1949

THE YEAR 1946 marked a jubilee in the history of the chemical element uranium. Fifty years earlier, in the spring of 1896, Henri Becquerel had discovered the remarkable radiation phenomena of this element, which were at that time grouped together under the name of radioactivity.

For more than 100 years, uranium, discovered by W. H. Klaproth in 1789, had had a quiet existence as a somewhat rare but not particularly interesting element. After its inclusion in the Periodic System by D. Mendeleev and Lothar Meyer, it was distinguished from all the other elements in one particular respect: it occupied the highest place in the table of elements. As yet, however, that did not have any particular significance.

We know today that it is just this position of uranium at the highest place of the then known chemical elements which gives it the important properties by which it is distinguished from all other elements.

The echo of Becquerel's fundamental observations on the radioactivity of uranium in scientific circles was at first fairly weak. Two years later,

---

Otto Hahn, Nobel Prize in Chemistry Award Address, 1946. Reprinted with permission of Elsevier Publishing Co., and the Nobel Foundation.

however, they acquired an exceptional importance when the Curies succeeded in separating from uranium minerals two active substances, polonium and radium, of which the latter appeared to be several million times stronger than the same weight of uranium.

It was only a few years before the first surprising property of this "radiating" substance was explained. The radioactive elements decompose according to definite rules into other active elements with different chemical and physical properties, simultaneously emitting corpuscular particles, the α- and β-rays. The α-rays were characterized by Rutherford as positively charged helium nuclei, after it had been shown earlier that the β-rays consisted of negative elementary particles, the so-called electrons.

With the hypothesis, suggested by Rutherford and Soddy in 1902, of the disintegration of the atoms, the postulate of the indivisibility of chemical elements had to be abandoned.

Parallel with the investigation of the radioactive disintegration processes of uranium, which produce radium, radium emanation, the so-called active deposits, and polonium, research was carried out on thorium, at that time the second highest element of the Periodic Table. From thorium were obtained, in an exactly similar manner, strongly active disintegration products: radiothorium, mesothorium, thorium X, etc. There was also a third series of disintegration products, the actinium series, which however also started from uranium.

By the systematic study of the penetration of α-particles through thin layers of material, Rutherford in 1911 was able to propose his model of the nuclei of chemical atoms. According to this view, the atoms of the elements consist of the positively charged nucleus, representing practically the whole mass of the atom, around which move, at a relatively great distance, electrons which neutralize the nuclear charge. The charge of the nucleus undoubtedly represents the position of the element in the Periodic System.

The radioactive disintegration reactions are thus nuclear changes which are not affected by the usual processes of physics or chemistry.

We are so certain of the unchangeability of these natural laws of radioactive breakdown, that from the speed of the changes we can construct a sort of "geological clock." Actually the final substance which is produced from uranium by way of these many active intermediate products is stable lead; from the quantity of the latter formed in the originally pure uranium mineral, it is possible to calculate the age of the mineral and therefore also of the geological deposit in which it has crystallized out.

The same holds for thorium, which also produces as final product a kind of lead, although different from the "uranium lead."

Investigation proceeded further. Again the α-particles of radioactive substances formed the means for the solution of problems of nuclear physics. Following Rutherford's nuclear model, these particles were recognized as free helium nuclei. With their relatively large mass of 4 (compared with the small electron mass of about 1/1800) and their initial velocity of up to 15,000 km/sec, they formed the bullets with which it might be possible to blast a way to the nuclei of the atoms, inaccessible to other means. And again it was the genius Rutherford who made the next discovery, which was to have extremely important results. In 1919 he was able to show that by bombardment of nitrogen with the energetic α-particles of radium C, an α-particle, that is a helium nucleus, was captured by the nitrogen nucleus, and a hydrogen nucleus—a proton—left the newly formed nucleus.

The process is represented by the equation:

$$^{14}_{7}N + ^{4}_{2}He \rightarrow ^{17}_{8}O + ^{1}_{1}H + \text{energy}$$

In this equation the lower indices represent the nuclear charges, the upper ones the nuclear masses (atomic weights). The remaining energy is given up in the form of kinetic energy to the released proton, which shows that the latter is derived from the nucleus.

Here for the first time was an artificial transmutation of atoms, actually a building-up of atoms, since an oxygen of mass 17 was produced from a nitrogen of mass 14.

In the following years a large number of reactions following the same pattern were discovered. On account of the positive charge of the α-particle this type of reaction did not succeed in bringing about nuclear transmutations of the heavy elements, since their high nuclear charge of positive sign repelled the α-particle sufficiently to prevent its penetration into the nucleus.

The year 1932 brought further important discoveries: the positron, heavy hydrogen, and—last but not least—the neutron. The discovery of the neutron was the result of detective work carried out in three countries. The initial impulse was given by Bothe and Becker in Germany, who were able to detect an exceptionally penetrating radiation, believed by them to consist of γ-rays, on exposing beryllium to α-particles. Then the Joliot-Curies in France proved that the "γ-rays" which appeared in the experiments of Bothe and Becker set free hydrogen nuclei (protons) with great energy from hydrogen-containing compounds, which is impossible for γ-rays. Finally Chadwick in England gave the final explanation of these results: in addition to γ-rays there occurred uncharged neutral atomic nuclei with a mass of 1, which he called neutrons. The course of the reaction is:

$$^{9}_{4}Be + ^{4}_{2}He \rightarrow ^{12}_{6}C + ^{1}_{0}n + \gamma$$

On the discovery of neutrons, the phenomenon of isotopy, known for a number of years, could readily be explained. The chemical elements are composed of protons and neutrons. The number of charged protons determines the chemical nature of the element: the sum of protons and neutrons determines the atomic weight. Thus a larger or smaller number of neutrons does not affect the chemical nature of the element, but merely produces isotopic kinds of atoms such as had been known for a long time in the different isotopes of lead, many mercury isotopes, two of chlorine, etc. Heavy hydrogen, discovered in 1932, is similarly an isotope of ordinary hydrogen. It has in its nucleus both a proton and a neutron, and thus has a mass of 2 instead of 1. The highest chemical element known at that time, uranium, is also not a simple element, but contains isotopes of which ^{238}U is the parent of the radium series and ^{235}U that of the protactinium and actinium series.

Soon was it seen that neutrons would form projectiles specially suited for the transmutation of atoms: they have no charge and are thus not repelled by the positive nuclei of the elements. While however the (α, p) decompositions of Rutherford—(α, p) is the abbreviated formula for a reaction in which an α-particle is taken up and a proton split off—nearly always produce stable kinds of atoms which occur naturally, in the (α, n) changes—the taking up of an α-particle and emission of a neutron—the Joliot-Curies had observed an entirely new phenomenon. In 1934 they found that, on irradiation of certain elements with α-particles, not only neutrons but also positrons, that is positive electrons, appeared, even after the irradiation with α-particles had ceased. The emission of positrons fell off in the same way as had already been observed in the decay of natural radioactive elements. Artificial active elements had been produced. The first elements with which this "artificial" activity on irradiation had been observed were boron and aluminium.

From boron was produced an active nitrogen; from aluminium an active phosphorus; and in their turn these were transformed into carbon and silicon respectively, with emission of positrons, according to the reactions

$$^{10}_{5}B + ^{4}_{2}He \rightarrow ^{13}_{7}N^{*} + ^{1}_{0}n; \quad ^{13}_{7}N^{*} \xrightarrow[10\ min]{\beta+} ^{13}_{6}C$$

$$^{27}_{13}Al + ^{4}_{2}He \rightarrow ^{30}_{15}P^{*} + ^{1}_{0}n; \quad ^{30}_{15}P^{*} \xrightarrow[2.2\ min]{\beta+} ^{30}_{14}Si$$

(* indicates a radioactive isotope)

A new and large field for further investigation was hereby opened. At the same time the experimental possibilities for carrying out nuclear physical investigations were fundamentally expanded.

While so far the particles from natural radioactive elements had been the sole means of producing artificial nuclear reactions, high-voltage installations—van de Graaff generators and the cyclotron—could now be used in addition. More and more intensive irradiations could be carried out, and more and more new reactions could be realized.

For the time being, however, radium-beryllium mixtures continued to be used, since they were very convenient, though perhaps not very intensive, sources of neutrons for further investigations, especially of the newly discovered artificial radioactivity. Such sources of neutrons were prepared by thorough mixing of a well-dried, finely-powdered radium salt with the finest possible beryllium powder, and sealing the mixture into airtight glass or metal tubes. Uranium itself was only of importance because radium was obtained from it.

It was especially the Italian scientist Fermi who first realized the great importance of neutrons for the production of nuclear reactions: Fermi and his co-workers irradiated practically all of the elements of the Periodic System with neutrons, and made numerous artificial radioactive elements.

Generally, the course of the reaction was such that, especially when working with slow neutrons,* the neutrons were captured by the nucleus. Thereby in many cases an unstable isotope of the irradiated element was produced, passing into the next higher element with emission of β-rays. This latter was generally a stable atom occurring among the usual chemical elements, e.g.

$$^{127}_{53}I + ^{1}_{0}n \rightarrow ^{128}_{53}I^{*} + \gamma; \quad ^{128}_{53}I^{*} \xrightarrow[26\ min]{\beta} ^{128}_{54}Xe$$

Fermi and his co-workers continued their tests through the whole of the Periodic System up to uranium. Here also they discovered many transmutations produced by neutrons, including some very rapid ones. They proceeded from the obvious assumption that initially there are produced artificial, active, short-living uranium isotopes; as these emit β-rays Fermi inferred the production of so-called "transuraniums," representatives of the element 93 which is not known naturally, and possibly even of the still higher element 94.

Fermi's proofs were not accepted everywhere. It was pointed out that for example in the case of the so-called 13-minute element—that detected with the greatest certainty—the possibility of its being an isotope of element 91, i.e. protactinium, could not be ruled out.

At this point Lise Meitner and I decided to repeat Fermi's experiments in order to decide whether the 13-minute element was a protactinium isotope or not. This decision was taken the more readily since, by the discovery of protactinium (1917), we were familiar with its chemical properties. Moreover, a β-radiating isotope of element 91 was well known to us in the form of uranium Z, discovered by myself, which had the favourable half-life of 6.7 hours, and was available from uranium salts.

With the help of the "indicator method" we were able to prove without doubt that the 13-minute element of Fermi was neither a protactinium isotope, nor a uranium, actinium, or thorium. In accordance with the position of science at the time, Fermi's assertion should be correct, and the 13-minute element a representative of the element 93, that is a "transuranium."

We should point out here that other possibilities did not occur to anyone at that time. Since the discovery of the neutron and the application of artificial sources of radiation, a large number of most unusual nuclear reactions had been discovered; the products were always either isotopes of the irradiated substances, or their next, or at most next-but-one, neighbours in the Periodic System; the possibility of a breakdown of heavy atomic nuclei into various light ones was considered as completely excluded.

With the tests on Fermi's 13-minute element and the checking of other, rather less certain, results of Fermi, we found (later in co-operation with F. Strassmann) that the phenomena associated with the irradiation of the highest element of the Periodic System were much more complicated than had originally been supposed. Fermi and his co-workers had already, in their first communication, described two short-life β-radiating kinds of atoms (half-life 10 sec and 40 sec), which they naturally considered to be artificial isotopes of uranium produced from the original uranium by the capture of neutrons. Lise Meitner and I found, in addition, a substance with a half-life of 23 minutes, which we conclusively identified as an artificial radioactive uranium isotope. With Fermi's substances of short life, the isotopy with uranium can only be assumed, but not proved. The 23-minute element occurred without any radiation conditions in a so-called "resonance process."

As the result of many years of work, we (Hahn, Meitner, and Strassmann) had finally obtained a great number of artificial active kinds of atoms, which all appeared to be formed directly or indirectly by β-radiation from the supposed short-living uranium isotopes, and which therefore must all represent so-called transuraniums—elements higher than uranium.

According to their chemical behaviour, these could be classified into various groups, and, since in many cases the gradual production from β-radiating parent substances could be directly observed, decay schemes were drawn up extending to elements 95 and 96. In so far as the work was repeated by others, the results were always confirmed.

Independently of the transuranium investigations of Hahn, Meitner, and Strassmann just mentioned, Curie and Savitch described in 1937 and 1938 a so-called 3.5-hour substance which they had obtained by irradiation of uranium with neutrons, and of which the chemical properties could not readily be determined. According to Curie and Savitch, the substance appeared to be a rare earth, but was not actinium; it had more resemblance to lanthanum, and could only be separated from the latter by "fractional crystallization". With some hesitation Curie and Savitch decided to include the substance in the transuranium series, but the possibilities put forward by them appeared difficult to understand and unsatisfactory.

As this 3.5-hour element had been included with the transuraniums, I, together with Strassmann, tried to obtain it. After careful experiments we arrived at remarkable results, which may be formulated approximately as follows: "In addition to the transuraniums described by Hahn, Meitner, and Strassmann, there are produced by two successive α-emissions three artificial, β-active radium isotopes with different half-life times, which in their turn change into artificial β-active actinium isotopes". The conclusion that radium isotopes had been produced was the only one possible since, according to the chemical properties, only barium and radium could be considered. Barium was, according to the physical viewpoint of the time, impossible, and thus only radium was left. . . .

At the same time the production of radium under these conditions of radiation was very remarkable: α-decompositions had never been observed with neutrons low in energy, and yet here, as with the transuraniums, a number of isotopes appeared simultaneously.

The experiments were continued in various directions. The preparations were, however, always very weak and the α-rays of the most stable of the new isotopes were so strongly absorbed that thicker layers could only be investigated with poor yields of radiation. An attempt was therefore made to separate the artificial "radium" as far as possible from the barium added as carrier, in order to obtain coatings permitting easier measurement. This was done by fractional crystallization using the method of Madame Curie, a method with which we had been thoroughly familiar over a number of years. About 30 years previously I, together with Lise Meitner, had separated the radium isotope mesothorium from barium by fractional crystallization. More recently, with the assistance of a number of co-workers, the laws governing the formation of mixed crystals between radium and barium salts had been systematically investigated.

The attempts to separate our artificial "radium isotopes" from barium in this way were unsuccessful; no enrichment of the "radium" was obtained. It was natural to ascribe this lack of success to the exceptionally low intensity of our preparations. It was always a question of merely a few thousands of atoms, which could only be detected as individual particles by the Geiger-Müller counter. Such a small number of atoms could be carried away by the great excess of inactive barium without any increase or decrease being perceptible, even if the barium was precipitated in the form of barium chloride, which precipitates in a very pure form.

In order to check this, we repeated the same tests with a weak intensity of the natural radium isotopes mesothorium and thorium X. These substances were freed from every trace of their parent substance and decay products with the greatest care and, by systematic dilution, preparations were made which were only just detectable with the Geiger-Müller counter. Crystallizations were carried out with the chlorides, bromides and chromates, always with the corresponding barium salt as carrier.

The result was, as was to be expected for radium, that mesothorium and thorium X were concentrated in the first fractions of the salts named, and in fact in quantities such as we should expect from our previous experience. This proved that the few atoms of natural radium isotopes also behaved in exactly the same manner as strong preparations.

Finally we proceeded to direct "indicator tests." We mixed the pure natural radium isotopes with our artificial "radium" isotopes, also previously freed from their decay products, and fractionated the mixture in the same way as before. The result was that the natural radium isotopes could be separated from barium, but the artificial ones could not.

We checked the results in still another way. If the artificial alkaline earth isotopes were radium, then the decay products produced directly through β-emission should consist of actinium: from the element 88 should be produced the element 89. If on the other hand it was barium, then lanthanum should be formed: from element 56 the next higher element 57. With the aid of the pure actinium isotope mesothorium-2 we carried out an "indicator test" by mixing mesothorium-2 with one of the known primary decay products of artificial radium isotopes, and then carrying out the chemical separation of actinium and lanthanum by the method of Madame Curie. During the fractionation of lanthanum oxalate with actinium, the latter accumulates in the final fractions. This actually occurred with the actinium isotope mesothorium-2. The decay product of our so-called "radium isotope" however remained with the lanthanum. The artificial rare earth, which had been considered to be actinium, was really lanthanum. Thus it was established that the alkaline earth isotope, which we had believed to be radium, was in fact an artificial active barium; the lanthanum could have been produced only from barium and not from radium.

In order to make quite certain, we carried out a so-called "cycle" with barium. The most stable of the active isotopes, now identified as barium, was freed from active decay products and other impurities by recrystallization with inactive barium; one quarter of the total quantity was kept for comparison, and three quarters were subjected to the following cycle of barium precipitations: Ba-chloride→ Ba-succinate→ Ba-nitrate→ Ba-carbonate→ Ba-ferrimannite→ Ba-chloride. After passing through this series of compounds, many of which crystallized beautifully, the resulting barium chloride and the recrystallized comparison preparation were measured alternately using the same counter, with equal weights and equal thicknesses of layers. The initial activity and the increase as the result of further formation of the active lanthanum were the same for both preparations, within the limits of error: the crystallization of so many and such different salts had produced no separation of the active barium from the carrier. It could only be concluded that the active product and the carrier were chemically identical, that is, barium.

In the first communication on these tests, which "were in opposition to all the phenomena observed up to the present in nuclear physics" (January 6th, 1939), the indicator tests mentioned had not been entirely completed, and we had therefore expressed ourselves cautiously. As a second partner in the new process we assumed an element with an atomic weight of about 100, as in that case the combined atomic weights would be that of uranium, "for example 138 + 101 (e.g. element 43) gives 239!"

After the completion of the measurements in hand, and of the "cycle," the possibility of error was still further excluded.

This completion of the tests and the above-mentioned "cycle" appeared in a second communication (February 10th, 1939). This also described the splitting of the element thorium and its confirmation with the aid of indicator tests analogous to those described above. Here also reference was made to the detection of an inert gas and an alkali metal derived from it; the nature of the gas was recognized, and its separation from uranium accomplished by means of a current of air passed over the uranium during the irradiation. An active strontium and an active yttrium were identified in the uranium itself. . . .

Thus the process proceeds in such a way that the nucleus of the uranium with a charge of 92 is split into two nuclei of moderate size. If one of these is barium, which has a nuclear charge of 56, there must be produced at the same time a krypton with a nuclear charge of 36. Together these nuclei add up to 92. Both have however, as may easily be seen from the masses of uranium and of the stable isotopes of barium and krypton which occur naturally, too great a mass, and thus an excess of neutrons. They should therefore pass over into stable elements with higher nuclear charges, with emission of β-rays; and in fact, as our later experiments showed, sometimes achieve stability by way of a great number of unstable intermediate decay products.

The highest stable krypton isotope has a mass of 86. In uranium fission there is produced, among other atoms, an unstable krypton with mass 88. Uranium 235 is responsible for the fission induced by thermal neutrons, as Bohr was the first to see; this fission forms by far the larger part. If there are no side reactions then the mass of the other fission product belonging to the krypton 88, that is of the barium, should be $236 - 88 = 148$. As the highest stable barium isotope has a mass of 138, the first-mentioned product is not less than 10 units heavier. Strassmann and myself had already noted, in our second communication, the possibility that neutrons were set free in the fission process. That this was in fact the case was first established experimentally by F. Joliot. . . .

Since by the action of neutrons on uranium, fresh neutrons are liberated, the latter, if they meet uranium atoms, produce further fissions, in their turn. If more than one fresh neutron is produced, and the process is so arranged that all the fresh neutrons strike uranium atoms, then we have a chain of continuously renewing fission reactions which, like an avalanche started by a snowball, can attain enormous dimensions. Thereby the practical application of atomic energy first came into the range of possibility. S. Flügge, then attached to the Kaiser Wilhelm Institute for Chemistry, was the first to refer to this.

About ten years ago, Joliot concluded his Nobel Lecture with the following words: "If, turning to the past, we cast a glance at the progress achieved by Science at an ever-increasing pace, we are entitled to think that scientists, building up or shattering elements at will, will be able to bring about transmutations of an explosive type, true chemical chain reactions. If such transmutations do succeed in spreading in matter, the enormous liberation of usable energy can be imagined. But, unfortunately, if the contagion spreads to all the elements of our planet, the consequences of unloosing such a cataclysm can only be viewed with apprehension. Astronomers sometimes observe that a star of medium magnitude increases suddenly in size; a star invisible to the naked eye may become very brilliant and visible without any telescope—the appearance of a Nova. This sudden flaring up of the star is perhaps due to transmutations of an explosive character like those which our wandering imagination is perceiving now—a process that the investigators will no doubt attempt to realize while taking, we hope, the necessary precautions!"

What was ten years ago only a figment of our "wandering imagination", has already become to some extent a threatening reality. The energy of nuclear physical reactions has been given into men's hands. Shall it be used for the assistance of free scientific thought, for social improvement and the betterment of the living conditions of mankind? Or will it be misused to destroy what mankind has built up in thousands of years? The answer must be given without hesitation, and undoubtedly the scientists of the world will strive towards the first alternative.

POSTSCRIPT

In the preceding paragraphs we have given a general outline of how the investigation of the natural radioactivity of uranium led to the artificial splitting of that element. But this does not exhaust the possibilities of uranium. Uranium consists mainly of two isotopes with atomic weights 235 and 238. The first is present in the mixture only to the extent of $1/140$ part; nevertheless the fission process described above, which is caused with special violence by "slow" neutrons, is to be ascribed chiefly to its rare isotope. . . . Lise Meitner and the present writer were able to identify, after the irradiation of uranium with neutrons, a substance having a half-life of 23 minutes as undoubtedly a uranium isotope. (The short-lived isotopes, thought first by Fermi and later by ourselves to be artificial uranium isotopes, were actually fission products.) The 23-minute uranium was produced in a so-called resonance process by neutrons of a certain velocity. Since this substance emits β-rays, a representative of element 93, that is an actual transuranium, *must* be produced. Although we sought for it, we were unable to detect it in the very weak preparations which we had at the time. Later it was identified in the United States as a β-emitter with a half-life of 2.3 days. Thus it has an atomic weight of 239. Since this true transuranium 93 emits β-rays in turn, it must be concluded that an element 94, a further transuranium, is formed. This is a relatively long-living α-emitting substance (half-life 24,000 years). The American scientist Seaborg gave it the name plutonium, calling element 93 neptunium. A second isotope of element 93 also occurs, it is produced in another way from uranium 238, namely a so-called (n, 2n) process with the aid of neutrons of great energy: one neutron, so to speak, in passing through the uranium nucleus knocks out another one. The result is a β-emitting uranium isotope of mass 237, which thus transforms into a second isotope of which the half-life is some millions of years.

Thus the behaviour of uranium on irradiation with neutrons of different velocities, both fast and slow, is very complicated. In addition to the natural splitting process, which continues during the irradiation at a speed independent of all the other reactions, the following occur:

(1) Nuclear fission with formation of numerous artificial atoms of all elements between 30 and 64.
(2) Emission of surplus neutrons during this fission process, making a chain reaction possible.
(3) The resonance capture of a neutron with a definite energy by uranium 238, with formation of uranium 239, which in its turn is transformed into the elements neptunium and plutonium.
(4) The giving up of a surplus neutron by the ^{238}U with formation of a ^{237}U, which also forms a neptunium isotope.

As, in process (1), slow (thermal) neutrons mainly attack the rare isotope ^{235}U; process (3) is a kind of competitor of process (1), since in the resonance process (3) the neutrons are captured before they have reached thermal velocities. It is a matter of experimental technique to prevent this capture as much as possible, in order to facilitate the chain reaction. On the other hand, the extra neutrons produced in the fission of process (2) give rise to the resonance process, and thereby to the formation of plutonium. If this latter element is obtained in sufficient quantity by means of a slow controlled chain reaction carried out on a large scale (in a so-called "pile"), and is separated from uranium, then it can also act as carrier of a chain reaction. The same holds for pure ^{235}U after it has been separated from the isotope 238, since with ^{235}U the resonance process, which interrupts the chain reaction, does not occur.

Both uranium 235 and plutonium are made in the United States. The result was the bombing of Hiroshima and Nagasaki.

L

Special Relativity

INTRODUCTION

WHEN EINSTEIN'S special theory of relativity was first published in 1905, it was received with great skepticism even from well-known physicists; for years, many refused to consider the theory seriously even though elaborate equipment is not needed to see how the principle of relativity works. The sensation of relative motion, one of the basic elements of Einstein's theory, is familiar to anyone who has ever sat aboard a smooth-wheeled train standing in a darkened railway station. If that train or a train on the opposite track starts to move slowly and smoothly, it is quite impossible to tell at first glance whether the one train or the other is in motion until the appearance of a third, stationary object—a sign, a fixed light, a fixture—provides a point of reference.

As for the thought experiment known as the "twin paradox," which shows why a person traveling at or near the speed of light ages less than a twin remaining stationary at home, this principle has been proved by the observation that certain short-lived particles, traveling at close to the speed of light, endure past their expected lives when they enter the earth's atmosphere: When orbiting atomic clocks return to earth, they have been found to lag behind identical, synchronized clocks that have remained behind, indicating that they had slowed down while in motion.

A good way to examine Einstein's special theory is to learn his two fundamental postulates: (1) There is no way to tell whether an object is at rest or in uniform motion relative to a fixed ether* (see the paper by Michelson and Morley on the "luminiferous ether" below); and (2) regardless of the speed of an observer, light always moves through empty space with the same constant speed relative to the observer. These two postulates enabled Einstein to bring space and time together into a four-dimensional continuum called space-time which, unlike the absolute space and absolute time axes of the Newtonian Universe, assumed that there is no single temporal or spatial reference system in the Universe.

Einstein's special theory of relativity published along with his articles on the nature of light (which revived the corpuscular theory of light by introducing the idea of a photon—or atom of light—to explain the photoelectric effect phenomena of radiation and thus went beyond Planck's original and other quantum theory to show that radiation is always emitted in discrete packets called quanta) and Brownian motion (which proved mathematically that the behavior of gas molecules is analogous to the constant zig-zag movements of insoluble particles suspended in a liquid medium). With its publication, Einstein took his place among the greatest scientific thinkers the world has known. Yet the initial controversy surrounding his theory of relativity delayed official recognition of his achievements by the Nobel Committee for more than sixteen years. Even then his award was for his work on the photoelectric effect, rather than for his theory of relativity.

At the time of its introduction, the response to the theory of relativity was varied. Germany was then the leading country in theoretical physics, and so it was only natural that reception of the theory (especially since it was published in German) should have been more widespread than in other nations. That landmark issue of *Annalen der Physik* was barely off the presses when a physicist named Walter Kaufmann began a series of experiments on beta particles (electrons) and then declared, in effect, that

---

**The Penguin Dictionary of Physics* defines *ether* as a "now discarded hypothetical medium once thought to fill all space and to be responsible for carrying light waves and other electromagnetic waves." The ether "was assumed to be extremely elastic yet extremely light, to transmit transverse waves with the velocity of light, and to have a greater density in matter than in free space." *The Penguin Dictionary of Physics,* ed. Valerie H. Pitt. New York: Penguin Books, 1977, p. 140.

his work had demonstrated that the predictions made by Einstein in his theory were incorrect and untenable in the natural world. But Max Planck was an enthusiastic supporter of the theory. Other early backers included Wilhelm Wien (who had studied with Helmholtz and who was one of the most eminent physicists of his day), his assistant J. J. Laub, and Arnold Sommerfeld, as well as Max von Laue. Discussion was widespread and vigorous, with Einstein gaining sufficient support to enter the inner circle of the German physics community—an astounding achievement for a young man of limited fame and prestige who was employed at the time in the Swiss Patent Office (although he had already published five papers on thermodynamics).

In France, Einstein's work was almost wholly ignored—a strange circumstance when one considers the prestige of Henri Poincaré, the French authority on the pioneering work of Lorentz and a scholar often credited with having the insights and data needed to have beaten Einstein to the development of the theory. In 1899, Poincaré had given a series of lectures at the Sorbonne on electricity and magnetism, in which he discussed the principle of relativity. From the time of the appearance of Einstein's *Annalen* articles until Poincaré's death in 1912, however, Poincaré apparently never mentioned the subject, and most French physicists followed suit. Even as late as 1922, when a French scientist named Lémeray published a book supporting the Einstein theory and dared to deny the existence of the ether, he was attacked by an eminent engineer, Léon-François Le Cornu. Cartesian France had accepted Newton (Voltaire was one of the principal advocates of Newtonian philosophy), but it took an extremely long time for conservative French scholars to accept Einstein. A tradition of excessive intellectual conservatism seems to thrive in France hand in hand with its tradition of avant-garde acceptance of even the most outrageous efforts in the arts. In Pasteur's day, there was a similar reluctance to accept the idea of microbes as the source of infection. "I'll believe in microbes," a character in a contemporary French farce declared, "when I can see them with the naked eye!"

In Britain, too, the concept of relativity penetrated the national intellect at considerably less than the speed of light. Even as late as 1925, Sir Oliver Lodge was enthusiastically arguing the cause of the ether (seldom has anything so nonexistent proved as durable as the British ether). Norman R. Campbell, an early supporter of Einstein, was later joined by

Bertrand Russell and Sir Arthur Eddington, both of whom became among the most enthusiastic champions and popularizers of the new theories.

In America, reception of the new theory varied. In the 1920s, when Einstein was beginning to become something of a cult figure, newspaper columns succeeded in making a comic figure of him. His unconventional dress, the rumor that only six, eight, ten, or twelve persons in the world could understand his theory, the mysterious "fourth dimension"—all combined to provide the most confused idea of the matter and the man.

American scientists, on the other hand, were quick to grasp the principles of the new physics, and the subsequent history of both relativity and its sibling, quantum mechanics, is a triumph of international collaboration. When Einstein finally arrived for his first personal appearance in the United States, much of the country, especially its intelligentsia, took him to its heart. He made America his home after 1933 and finished his long career at the Princeton, New Jersey Institute for Advanced Study.

L.1

Space and Geometry

HENRI POINCARÉ

THE BRITISH PHYSICIST Paul Dirac once remarked that if Einstein had not proposed his special theory of relativity in 1905, someone else would have proposed it by 1910. Among those usually mentioned as being close to the discovery of relativistic principles at the time is Henri Poincaré, a French mathematician of great talent in all fields of mathematics that existed in his day. Today, mathematics is so vast a field, and so many new branches have sprung up, that it is impossible for one person to master each. Poincaré, on the other hand, made contributions to virtually every branch of his subject.

Poincaré was born in 1854 in Nancy, in northeastern France. He was not obviously gifted, since he suffered from poor vision and lack of coordination, but he was able to do well in school thanks to a prodigious memory. Henri earned his doctorate in mathematics in 1879 and became a professor at the University of Paris. His specialty was celestial mechanics, and he is remembered for his contributions to what is known as the "three-body problem," which was concerned with determining the positions and motions of three bodies in a mutual gravitational field. He also found time to study tides and their cause in the mechanics of the moon—a subject studied in greatest depth by Sir George Darwin, the son of Charles Darwin.

Aside from his specialized work in mathematics, Poincaré wrote widely on the philosophy of mathematics and the importance of mathematical creativity. Our selection, a chapter on the relation of space and geometry, is taken from the book, *The Foundations of Science*, which is subtitled *Science and Hypothesis, the Value of Science and Method.* Poincaré begins by defining geometric space as infinite, three-dimensional, homogeneous, isotropic space, and then goes on to compare it to what he calls "perceptual space," that is, the space in which our sense perceptions operate —visual, tactile and motor space. In his discussion, Poincaré reveals his extraordinary talents as a thinker and teacher. To read him is to listen to a brilliant teacher leading his pupils convincingly through a maze of thoughts to lucid conclusions.

A field of force represents the discrepancy between the natural geometry of a co-ordinate system and the abstract geometry arbitrarily ascribed to it.

—ARTHUR STANLEY EDDINGTON

The central recognition of the theory of relativity is that geometry . . . is a construct of the intellect. Only when this discovery is accepted can the mind feel free to tamper with the time-honored notions of space and time, to survey the range of possibilities available for defining them, and to select that formulation which agrees with observation.

—HENRY MARGENAU in P. A. Schilpp (ed.), *Albert Einstein: Philosopher-Scientist*

LET US BEGIN by a little paradox.

Beings with minds like ours, and having the same senses as we, but without previous education, would receive from a suitably chosen external world impressions such that they would be led to construct a geometry other than that of Euclid and to localize the phenomena of that external world in a non-Euclidean space, or even in a space of four dimensions.

As for us, whose education has been accomplished by our actual world, if we were suddenly transported into this new world, we should have no difficulty in referring its phenomena to our Euclidean space. Conversely, if these beings were transported into our environment, they would be led to relate our phenomena to non-Euclidean space.

Nay more; with a little effort we likewise could do it. A person who should devote his existence to it might perhaps attain to a realization of the fourth dimension.

GEOMETRIC SPACE AND PERCEPTUAL SPACE

It is often said the images of external objects are localized in space, even that they can not be formed except on this condition. It is also said

---

Henri Poincaré, "Space and Geometry," in *The Foundations of Science*. New York: The Science Press, 1913, pp 66–80. Reprinted by permission of Dover Publications, Inc., New York.

that this space, which serves thus as a ready prepared *frame* for our sensations and our representations, is identical with that of the geometers, of which it possesses all the properties.

To all the good minds who think thus, the preceding statement must have appeared quite extraordinary. But let us see whether they are not subject to an illusion that a more profound analysis would dissipate.

What, first of all, are the properties of space, properly so called? I mean of that space which is the object of geometry and which I shall call *geometric space.*

The following are some of the most essential:

1. It is continuous;
2. It is infinite;
3. It has three dimensions;
4. It is homogeneous, that is to say, all its points are identical one with another;
5. It is isotropic, that is to say, all the straights which pass through the same point are identical one with another.

Compare it now to the frame of our representations and our sensations, which I may call *perceptual space.*

VISUAL SPACE

Consider first a purely visual impression, due to an image formed on the bottom of the retina.

A cursory analysis shows us this image as continuous, but as possessing only two dimensions; this already distinguishes from geometric space what we may call *pure visual space.*

Besides, this image is enclosed in a limited frame.

Finally, there is another difference not less important: *this pure visual space is not homogeneous.* All the points of the retina, aside from the images which may there be formed, do not play the same rôle. The yellow spot can in no way be regarded as identical with a point on the border of the retina. In fact, not only does the same object produce there much more vivid impressions, but in every *limited* frame the point occupying the center of the frame will never appear as equivalent to a point near one of the borders.

No doubt a more profound analysis would show us that this continuity of visual space and its two dimensions are only an illusion; it would separate it therefore still more from geometric space, but we shall not dwell on this remark.

Sight, however, enables us to judge of distances and consequently to perceive a third dimension. But every one knows that this perception of the third dimension reduces itself to the sensation of the effort at accommodation it is necessary to make, and to that of the convergence which must be given to the two eyes, to perceive an object distinctly.

These are muscular sensations altogether different from the visual sensations which have given us the notion of the first two dimensions. The third dimension therefore will not appear to us as playing the same rôle as the other two. What may be called *complete visual space* is therefore not an isotropic space.

It has, it is true, precisely three dimensions, which means that the elements of our visual sensations (those at least which combine to form the notion of extension) will be completely defined when three of them are known; to use the language of mathematics, they will be functions of three independent variables.

But examine the matter a little more closely. The third dimension is revealed to us in two different ways: by the effort of accommodation and by the convergence of the eyes.

No doubt these two indications are always concordant, there is a constant relation between them, or, in mathematical terms, the two variables which measure these two muscular sensations do not appear to us as independent; or again, to avoid an appeal to mathematical notions already rather refined, we may go back to the language of the preceding chapter and enunciate the same fact as follows: If two sensations of convergence, A and B, are indistinguishable, the two sensations of accommodation, A' and B', which respectively accompany them, will be equally indistinguishable.

But here we have, so to speak, an experimental fact: *a priori* nothing prevents our supposing the contrary, and if the contrary takes place, if these two muscular sensations vary independently of one another, we shall have to take account of one more independent variable, and "complete visual space" will appear to us as a physical continuum of four dimensions.

We have here even, I will add, a fact of *external* experience. Nothing prevents our supposing that a being with a mind like ours, having the same sense organs that we have, may be placed in a world where light would only reach him after having traversed reflecting media of complicated form. The two indications which serve us in judging distances would cease to be connected by a constant relation. A being who should achieve in such a world the education of his senses would no doubt attribute four dimensions to complete visual space.

TACTILE SPACE AND MOTOR SPACE

"Tactile space" is still more complicated than visual space and farther removed from geometric space. It is superfluous to repeat for touch the discussion I have given for sight.

But apart from the data of sight and touch, there are other sensations which contribute as much and more than they to the genesis of the notion of space. These are known to every one; they accompany all our movements, and are usually called muscular sensations.

The corresponding frame constitutes what may be called *motor space.*

Each muscle gives rise to a special sensation capable of augmenting or of diminishing, so that the totality of our muscular sensations will depend upon as many variables as we have muscles. From this point of view, *motor space would have as many dimensions as we have muscles....*

I know it will be said that if the muscular sensations contribute to form the notion of space, it is because we have the sense of the *direction* of each movement and that it makes an integrant part of the sensation. If this were so, if a muscular sensation could not arise except accompanied by this geometric sense of direction, geometric space would indeed be a form imposed upon our sensibility.

But I perceive nothing at all of this when I analyze my sensations.

What I do see is that the sensations which correspond to movements in the same direction are connected in my mind by a mere *association of ideas.* It is to this association that what we call 'the sense of direction' is reducible. This feeling therefore can not be found in a single sensation.

This association is extremely complex, for the contraction of the same muscle may correspond, according to the position of the limbs, to movements of very different direction.

Besides, it is evidently acquired; it is, like all associations of ideas, the result of a *habit;* this habit itself results from very numerous *experiences*; without any doubt, if the education of our senses had been accomplished in a different environment, where we should have been subjected to different impressions, contrary habits would have arisen and our muscular sensations would have been associated according to other laws.

CHARACTERISTICS OF PERCEPTUAL SPACE

Thus perceptual space, under its triple form, visual, tactile and motor, is essentially different from geometric space.

It is neither homogeneous, nor isotropic; one can not even say that it has three dimensions.

It is often said that we "project" into geometric space the objects of our external perception; that we "localize" them.

Has this a meaning, and if so what?

Does it mean that we *represent* to ourselves external objects in geometric space?

Our representations are only the reproduction of our sensations; they can therefore be ranged only in the same frame as these, that is to say, in perceptual space.

It is as impossible for us to represent to ourselves external bodies in geometric space, as it is for a painter to paint on a plane canvas objects with their three dimensions.

Perceptual space is only an image of geometric space, an image altered in shape by a sort of perspective, and we can represent to ourselves objects by bringing them under the laws of this perspective.

Therefore we do not *represent* to ourselves external bodies in geometric space, but we *reason* on these bodies as if they were situated in geometric space.

When it is said then that we "localize" such and such an object at such and such a point of space, what does it mean?

It simply means that we represent to ourselves the movements it would be necessary to make to reach that object; and one may not say that to represent to oneself these movements, it is necessary to project the movements themselves in space and that the notion of space must, consequently, preexist.

When I say that we represent to ourselves these movements, I mean only that we represent to ourselves the muscular sensations which accompany them and which have no geometric character whatever, which consequently do not at all imply the pre-existence of the notion of space.

CHANGE OF STATE AND CHANGE OF POSITION

But, it will be said, if the idea of geometric space is not imposed upon our mind, and if, on the other hand, none of our sensations can furnish it, how could it have come into existence? . . .

None of our sensations, isolated, could have conducted us to the idea of space; we are led to it only in studying the laws, according to which these sensations succeed each other.

We see first that our impressions are subject to change; but among the changes we ascertain we are soon led to make a distinction.

At one time we say that the objects which cause these impressions have changed state, at another time that they have changed position, that they have only been displaced.

Whether an object changes its state or merely its position, this is always translated for us in the same manner: *by a modification in an aggregate of impressions.*

How then could we have been led to distinguish between the two? It is easy to account for. If there has only been a change of position, we can restore the primitive aggregate of impressions by making movements which replace us opposite the mobile object in the same *relative* situation. We thus *correct* the modification that happened and we reestablish the initial state by an inverse modification.

If it is a question of sight, for example, and if an object changes its place before our eye, we can "follow it with the eye" and maintain its image on the same point of the retina by appropriate movements of the eyeball.

These movements we are conscious of because they are voluntary and because they are accompanied by muscular sensations, but that does not mean that we represent them to ourselves in geometric space.

So what characterizes change of position, what distinguishes it from change of state, is that it can always be corrected in this way.

It may therefore happen that we pass from the totality of impressions *A* to the totality *B* in two different ways:

1. Involuntarily and without experiencing muscular sensations; this happens when it is the object which changes place;

2. Voluntarily and with muscular sensations; this happens when the object is motionless, but we move so that the object has relative motion with reference to us.

If this be so, the passage from the totality *A* to the totality *B* is only a change of position.

It follows from this that sight and touch could not have given us the notion of space without the aid of the "muscular sense."

Not only could this notion not be derived from a single sensation or even *from a series of sensations,* but what is more, an *immobile* being could never have acquired it, since, not being able to *correct* by his movements the effects of the changes of position of exterior objects, he would have had no reason whatever to distinguish them from changes of state. Just as little could he have acquired it if his motions had not been voluntary or were unaccompanied by any sensations.

CONDITIONS OF COMPENSATION

How is a like compensation possible, of such sort that two changes, otherwise independent of each other, reciprocally correct each other?

A mind already familiar with geometry would reason as follows: Evidently, if there is to be compensation, the various parts of the external object, on the one hand, and the various sense organs, on the other hand, must be in the same *relative* position after the double change. And, for that to be the case, the various parts of the external object must likewise have retained in reference to each other the same relative position, and the same must be true of the various parts of our body in regard to each other.

In other words, the external object, in the first change, must be displaced as is a rigid solid, and so must it be with the whole of our body in the second change which corrects the first.

Under these conditions, compensation may take place.

But we who as yet know nothing of geometry, since for us the notion of space is not yet formed, we can not reason thus, we can not foresee *a priori* whether compensation is possible. But experience teaches us that it sometimes happens, and it is from this experimental fact that we start to distinguish changes of state from changes of position.

SOLID BODIES AND GEOMETRY

Among surrounding objects there are some which frequently undergo displacements susceptible of being thus corrected by a correlative movement of our own body; these are the *solid bodies.* The other objects, whose form is variable, only exceptionally undergo like displacements (change of position without change of form). When a body changes its place *and its shape*, we can no longer, by appropriate movements, bring back our sense-organs into the same *relative* situation with regard to this body; consequently we can no longer reestablish the primitive totality of impressions.

It is only later, and as a consequence of new experiences, that we learn how to decompose the bodies of variable form into smaller elements, such that each is displaced almost in accordance with the same laws as solid bodies. Thus we distinguish "deformations" from other changes of state; in these deformations, each element undergoes a mere change of position, which can be corrected, but the modification undergone by the aggregate is more profound and is no longer susceptible of correction by a correlative movement.

Such a notion is already very complex and must have been relatively late in appearing; moreover it could not have arisen if the observation of solid bodies had not already taught us to distinguish changes of position.

Therefore, if there were no solid bodies in nature, there would be no geometry. . . .

Another remark also deserves a moment's attention. Suppose a solid body to occupy successively the positions α and β; in its first position, it will produce on us the totality of impressions A, and in its second position the totality of impressions B. Let there be now a second solid body, having qualities entirely different from the first, for example, a different color. Suppose it to pass from the position α, where it give us the totality of impressions A', to the position β, where it gives the totality of impressions B'.

In general, the totality A will have nothing in common with the totality A', nor the totality B with the totality B'. The transition from the totality A to the totality B and that from the totality A' to the totality B' are therefore two changes which *in themselves* have in general nothing in common.

And yet we regard these two changes both as displacements and, furthermore, we consider them as the *same* displacement. How can that be?

It is simply because they can both be corrected by the *same* correlative movement of our body.

"Correlative movement" therefore constitutes the *sole connection* between two phenomena which otherwise we never should have dreamt of likening.

On the other hand, our body, thanks to the number of its articulations and muscles, may make a multitude of different movements; but all are not capable of "correcting" a modification of external objects; only those will be capable of it in which our whole body, or at least all those of our sense-organs which come into play, are displaced as a whole, that is, without their relative positions varying, or in the fashion of a solid body.

To summarize:

1. We are led at first to distinguish two categories of phenomena:

Some, involuntary, unaccompanied by muscular sensations, are attributed by us to external objects; these are external changes;

Others, opposite in character and attributed by us to the movements of our own body, are internal changes;

2. We notice that certain changes of each of these categories may be corrected by a correlative change of the other category;

3. We distinguish among external changes those which have thus a correlative in the other category; these we call displacements; and just so among the internal changes, we distinguish those which have a correlative in the first category.

Thus are defined, thanks to this reciprocity, a particular class of phenomena which we call displacements.

The laws of these phenomena constitute the object of geometry.

LAW OF HOMOGENEITY

The first of these laws is the law of homogeneity.

Suppose that, by an external change α, we pass from the totality of impressions A to the totality B, then that this change α is corrected by a correlative voluntary movement β, so that we are brought back to the totality A.

Suppose now that another external change α' makes us pass anew from the totality A to the totality B.

Experience teaches us that this change α' is, like α, susceptible of being corrected by a correlative voluntary movement β' and that this movement β' corresponds to the same muscular sensations as the movement β which corrected α.

This fact is usually enunciated by saying that *space is homogeneous and isotropic*.

It may also be said that a movement which has once been produced may be repeated a second and a third time, and so on, without its properties varying.

In the first chapter, where we discussed the nature of mathematical reasoning, we saw the importance which must be attributed to the possibility of repeating indefinitely the same operation.

It is from this repetition that mathematical reasoning gets its power; it is, therefore, thanks to the law of homogeneity, that it has a hold on the geometric facts.

For completeness, to the law of homogeneity should be added a multitude of other analogous laws, into the details of which I do not wish to enter, but which mathematicians sum up in a word by saying that displacements form "a group."

THE NON-EUCLIDEAN WORLD

If geometric space were a frame imposed on *each* of our representations, considered individually, it would be impossible to represent to ourselves an image stripped of this frame, and we could change nothing of our geometry.

But this is not the case; geometry is only the résumé of the laws according to which these images succeed each other. Nothing then prevents us from imagining a series of representations, similar in all points to our ordinary representations, but succeeding one another according to laws different from those to which we are accustomed.

We can conceive then that beings who received their education in an environment where these laws were thus upset might have a geometry very different from ours.

Suppose, for example, a world enclosed in a great sphere and subject to the following laws:

The temperature is not uniform; it is greatest at the center, and diminishes in proportion to the distance from the center, to sink to absolute zero when the sphere is reached in which this world is enclosed.

To specify still more precisely the law in accordance with which this temperature varies: Let R be the radius of the limiting sphere; let r be the distance of the point considered from the center of this sphere. The absolute temperature shall be proportional to $R^2—r^2$.

I shall further suppose that, in this world, all bodies have the same coefficient of dilatation, so that the length of any rule is proportional to its absolute temperature.

Finally, I shall suppose that a body transported from one point to another of different temperature is put immediately into thermal equilibrium with its new environment.

Nothing in these hypotheses is contradictory or unimaginable.

A movable object will then become smaller and smaller in proportion as it approaches the limit-sphere.

Note first that, though this world is limited from the point of view of our ordinary geometry, it will appear infinite to its inhabitants.

In fact, when these try to approach the limit-sphere, they cool off and become smaller and smaller. Therefore the steps they take are also smaller and smaller, so that they can never reach the limiting sphere.

If, for us, geometry is only the study of the laws according to which rigid solids move, for these imaginary beings it will be the study of the laws of motion of solids *distorted by the differences of temperature* just spoken of.

No doubt, in our world, natural solids likewise undergo variations of form and volume due to warming and cooling. But we neglect these variations in laying the foundations of geometry, because, besides their being very slight, they are irregular and consequently seem to us accidental.

In our hypothetical world, this would no longer be the case, and these variations would follow regular and very simple laws.

Moreover, the various solid pieces of which the bodies of its inhabitants would be composed would undergo the same variations of form and volume.

I will make still another hypothesis; I will suppose light traverses media diversely refractive and such that the index of refraction is inversely proportional to $R^2—r^2$. It is easy to see that, under these conditions, the rays of light would not be rectilinear, but circular.

To justify what precedes, it remains for me to show that certain changes in the position of external objects can be *corrected* by correlative movements of the sentient beings inhabiting this imaginary world, and that in such a way as to restore the primitive aggregate of impressions experienced by these sentient beings.

Suppose in fact that an object is displaced, undergoing deformation, not as a rigid solid, but as a solid subjected to unequal dilatations in exact conformity to the law of temperature above supposed. Permit me for brevity to call such a movement a *non-Euclidean displacement*.

If a sentient being happens to be in the neighborhood, his impressions will be modified by the displacement of the object, but he can reestablish them by moving in a suitable manner. It suffices if finally the aggregate of the object and the sentient being, considered as forming a single body, has undergone one of those particular displacements I have just called non-Euclidean. This is possible if it be supposed that the limbs of these beings dilate according to the same law as the other bodies of the world they inhabit.

Although from the point of view of our ordinary geometry there is a deformation of the bodies in this displacement and their various parts are no longer in the same relative position, nevertheless we shall see that the impressions of the sentient being have once more become the same.

In fact, though the mutual distances of the various parts may have varied, yet the parts originally in contact are again in contact. Therefore the tactile impressions have not changed.

On the other hand, taking into account the hypothesis made above in regard to the refraction and the curvature of the rays of light, the visual impressions will also have remained the same.

These imaginary beings will therefore like ourselves be led to classify the phenomena they witness and to distinguish among them the 'changes of position' susceptible of correction by a correlative voluntary movement.

If they construct a geometry, it will not be, as ours is, the study of the movements of our rigid solids; it will be the study of the changes of position which they will thus have distinguished and which are none other than the "non-Euclidean displacements"; *it will be non-Euclidean geometry.*

Thus beings like ourselves, educated in such a world, would not have the same geometry as ours.

THE WORLD OF FOUR DIMENSIONS

We can represent to ourselves a four-dimensional world just as well as a non-Euclidean.

The sense of sight, even with a single eye, together with the muscular sensations relative to the movements of the eyeball, would suffice to teach us space of three dimensions.

The images of external objects are painted on the retina, which is a two-dimensional canvas; they are *perspectives.*

But, as eye and objects are movable, we see in succession various perspectives of the same body, taken from different points of view.

At the same time, we find that the transition from one perspective to another is often accompanied by muscular sensations.

If the transition from the perspective A to the perspective B, and that from the perspective A' to the perspective B' are accompanied by the same muscular sensations, we liken them one to the other as operations of the same nature.

Studying then the laws according to which these operations combine, we recognize that they form a group, which has the same structure as that of the movements of rigid solids.

Now, we have seen that it is from the properties of this group we have derived the notion of geometric space and that of three dimensions.

We understand thus how the idea of a space of three dimensions could take birth from the pageant of these perspectives, though each of them is of only two dimensions, since *they follow one another according to certain laws.*

Well, just as the perspective of a three-dimensional figure can be made on a plane, we can make that of a four-dimensional figure on a picture of three (or of two) dimensions. To a geometer this is only child's play.

We can even take of the same figure several perspectives from several different points of view.

We can easily represent to ourselves these perspectives, since they are of only three dimensions.

Imagine that the various perspectives of the same object succeed one another, and that the transition from one to the other is accompanied by muscular sensations.

We shall of course consider two of these transitions as two operations of the same nature when they are associated with the same muscular sensations.

Nothing then prevents us from imagining that these operations combine according to any law we choose, for example, so as to form a group with the same structure as that of the movements of a rigid solid of four dimensions.

Here there is nothing unpicturable, and yet these sensations are precisely those which would be felt by a being possessed of a two-dimensional retina who could move in space of four dimensions. In this sense we may say the fourth dimension is imaginable.

CONCLUSIONS

We see that experience plays an indispensable rôle in the genesis of geometry; but it would be an error thence to conclude that geometry is, even in part, an experimental science.

If it were experimental, it would be only approximate and provisional. And what rough approximation!

Geometry would be only the study of the movements of solids; but in reality it is not occupied with natural solids, it has for object certain ideal solids, absolutely rigid, which are only a simplified and very remote image of natural solids.

The notion of these ideal solids is drawn from all parts of our mind, and experience is only an occasion which induces us to bring it forth from them.

The object of geometry is the study of a particular "group"; but the general group concept pre-exists, at least potentially, in our minds. It is imposed on us, not as form of our sense, but as form of our understanding.

Only, from among all the possible groups, that must be chosen which will be, so to speak, the *standard* to which we shall refer natural phenomena.

Experience guides us in this choice without forcing it upon us; it tells us not which is the truest geometry, but which is the most *convenient*.

Notice that I have been able to describe the fantastic worlds above imagined *without ceasing to employ the language of ordinary geometry*.

And, in fact, we should not have to change it if transported thither.

Beings educated there would doubtless find it more convenient to create a geometry different from ours, and better adapted to their impressions. As for us, in face of the *same* impressions, it is certain we should find it more convenient not to change our habits.

On the Relative Motion of the Earth and the Luminiferous Ether

ALBERT MICHELSON AND EDWARD MORLEY

THE NAMES OF Albert Michelson and Edward Morley are linked forever because of one famous experiment that proved to be one of the most crucial in the entire history of experimental science because it disproved the existence of the ether medium. Their results were published in the November 1887 issue of the *American Journal of Science.*

Michelson was born in 1852 in the Prussian town of Strelno (now part of Poland). He came to the United States at the age of four, when his parents decided to seek their fortunes in the California Gold Rush. His father became a businessman, and young Albert grew up as an American. He was admitted to the United States Naval Academy and graduated with the class of 1873. He taught briefly at the Academy in the late 1870s. In 1878, he began his experiments in a subject that was to become his life-long interest: the accurate determination of the velocity of light.

To prepare himself for this difficult task he studied in France and Germany, returning to the United States to resign from the Naval Academy and to accept a job teaching physics at what was then known as the Case School of Applied Science in Cleveland, Ohio. Before going abroad he had come up with an approximate figure for the speed of light, but at Case he arrived at the elusive statistic of 186,320 miles per second (299,853 kilometers per second), only one-fiftieth of one percent off the current figure of 186,287.51 miles per second (299,792.5 kilometers per second). Michelson's 1882 figure was the most accurate until he himself refined it some twenty-five years later.

Edward Morley (1838–1923) was a professor of chemistry at Western Reserve University (Michelson's college, Case, and Western Reserve have since merged into Case Western Reserve University in Cleveland). Morley had made his reputation by his work in determining the relative atomic weights of hydrogen and oxygen.

In the 1880s, the prevailing opinion among scientists was that light waves had to be transmitted by a medium, the "luminiferous," or light-bearing ether, assumed to be invisible and all-pervasive throughout the Universe. Since all components of the Universe were known to be in motion, physicists wondered what might be the fixed reference point for this ubiquitous motion. Newton had previously assigned this function to the all-pervasive ether. In 1887, Michelson teamed up with his colleague Morley to design their famous "Michelson-Morley experiment," one of the cleverest and best known of all scientific experiments. They devised an interferometer, an instrument that divides a beam of light into two perpendicular halves. If the light had been traveling by means of ether, the two beams would return to their source, reflected by the mirrors, out of phase, thus enabling the experimenters to detect the "ether wind" caused by the motion of the earth in its orbit, if one of the beams moved parallel to the earth's velocity, and the other one perpendicular to it.*

Michelson and Morley, of course, found no trace of the ether wind, and years later Einstein, aware that there was no evidence of the ether, ignored it in his theories. The Michelson-Morley experiment is known to every physics student because it eliminated the last vestige of Newton's absolute space. The results of the experiments were first explained by the Lorentz-Fitzgerald contraction hypothesis, which proposed that the length of an object at rest in one frame of reference will appear to an observer in another frame of reference to be reduced in length in the direction of its motion. This suggestion, which was no more than an ad hoc

---

*"If the speed of light depended on the motion of the earth through the ether, then the two beams of light should not return at the same time. If they did not return at the same time, then the wave forms of the returning beams would not fit exactly, i.e., be in the same phase, and this lack of phase would show in a displacement of the interference or fringe pattern in the field of view of the interferometer." Henry A. Boorse and Lloyd Motz, *The World of the Atom*. New York: Basic Books, 1966, p. 372.

hypothesis presented by the Dutch physicist H. A. Lorentz, was later shown by Einstein to be a direct consequence of his special theory of relativity.*

Although the effects of this contradiction are quite negligible at low velocities, the theory leading to this relationship was of crucial importance to the later formulation of Einstein's theory of special relativity. The general formulas that lead to this length contraction and other relativistic results are known as the Einstein-Lorentz transformations.

---

*"It was the negative result of the experiment that in part led Einstein to one of the most fundamental ideas upon which the theory of relativity rests, namely, that the speed of light is the same for all observers regardless of how they may be moving. With this assumption, it is easy to see why the two beams return to their starting point at the same time; moreover, it is no longer necessary to posit the existence of a troublesome ether. However, before the theory of relativity was introduced, with its revolutionary concept of a constant speed of light, another attempt was made to explain the Michelson and Morley experiment. This ad hoc explanation was that the distance between the two mirrors [of the interferometer] on the line parallel to the earth's motion had decreased (because of the earth's motion) by an amount that was just big enough to enable the parallel moving beam to return at exactly the same time as the beam moving transverse to the earth's motion. This is the famous Lorentz-Fitzgerald contraction hypothesis, which Lorentz derived by picturing matter as consisting of small charged spheres (electrons) that contract in the direction of motion because of the electrical forces acting on them. This theory was not satisfactory because it tried to explain observable phenomena in terms of invisible forces and ad hoc hypotheses, which could not be tested experimentally. Einstein's theory removed this difficulty." Henry A. Boorse and Lloyd Motz, op. cit., p. 373.

Ah, but a man's reach should exceed his grasp,/ Or what's a heaven for?

—ROBERT BROWNING, *Andrea del Sarto*, 97

Science is the attempt to make the chaotic diversity of our sense-experience correspond to a logically-uniform system of thought.

—ALBERT EINSTEIN, *Out of My Later Years*

THE DISCOVERY OF the aberration of light was soon followed by an explanation according to the emission theory. The effect was attributed to a simple composition of the velocity of light with the velocity of the earth in its orbit. The difficulties in this apparently sufficient explanation were overlooked until after an explanation on the undulatory theory of light was proposed. This new explanation was at first almost as simple as the former. But it failed to account for the fact proved by experiment that the aberration was unchanged when observations were made with a telescope filled with water. For if the tangent of the angle of aberration is the ratio of the velocity of the earth to the velocity of light, then, since the latter velocity in water is three-fourths its velocity in a vacuum, the aberration observed with a water telescope should be four-thirds of its true value.

On the undulatory theory, according to Fresnel, first, the ether is supposed to be at rest except in the interior of transparent media, in which secondly, it is supposed to move with a velocity less than the velocity of the medium in the ratio $\frac{n^2 - 1}{n^2}$, where n is the index of refraction. These two hypotheses give a complete and satisfactory explanation of aberration. The second hypothesis, notwithstanding its seeming improbability, must be considered as fully proved, first, by the celebrated experiment of Fizeau, and secondly, by the ample confirmation of our own work. The experimental trial of the first hypothesis forms the subject of the present paper.

If the earth were a transparent body, it might perhaps be conceded, in view of the experiments just cited, that the intermolecular ether was at

---

Albert A. Michelson and Edward W. Morley, "On the Relative Motion of the Earth and the Luminiferous Ether," *The American Journal of Science*, Vol. XXXIV (November 1887), pp. 333–341.

rest in space, notwithstanding the motion of the earth in its orbit; but we have no right to extend the conclusion from these experiments to opaque bodies. But there can hardly be question that the ether can and does pass through metals. Lorentz cites the illustration of a metallic barometer tube. When the tube is inclined the ether in the space above the mercury is certainly forced out, for it is incompressible. But again we have no right to assume that it makes its escape with perfect freedom, and if there be any resistance, however slight, we certainly could not assume an opaque body such as the whole earth to offer free passage through its entire mass. But as Lorentz aptly remarks: "quoi qui'l en soit, on fera bien, à mon avis, de ne pas se laisser guider, dans une question aussi importante, par des considérations sur le degré de probabilité ou de simplicité de l'une ou de l'autre hypothèse, mais de s'adresser a l'expérience pour apprendre à connaitre l'état, de repos ou de mouvement, dans lequel se trouve l'éther à la surface terrestre."

In April, 1881, a method was proposed and carried out for testing the question experimentally.

In deducing the formula for the quantity to be measured, the effect of the motion of the earth through the ether on the path of the ray at right angles to this motion was overlooked. The discussion of this oversight and of the entire experiment forms the subject of a very searching analysis by H. A. Lorentz, who finds that this effect can by no means be disregarded. In consequence, the quantity to be measured had in fact but one-half the value supposed, and as it was already barely beyond the limits of errors of experiment, the conclusion drawn from the result of the experiment might well be questioned; since, however, the main portion of the theory remains unquestioned, it was decided to repeat the experiment with such modifications as would insure a theoretical result much too large to be masked by experimental errors. The theory of the method may be briefly stated as follows:

Let *sa*, fig. 1, be a ray of light which is partly reflected in *ab*, and partly transmitted in *ac*, being returned by the mirrors *b* and *c*, along *ba* and *ca*. *ba* is partly transmitted along *ad*, and *ca* is partly reflected along *ad*. If then the paths *ab* and *ac* are equal, the two rays interfere along *ad*. Suppose now, the ether being at rest, that the whole apparatus moves in the direction *sc*, with the velocity of the earth in its orbit, the directions and distances traversed by the rays will be altered thus:—The ray *sa* is reflected along *ab*, fig. 2; the angle bab_1 being equal to the aberration = *a*, is returned along ba_1, ($aba_1 = 2a$), and goes to the focus of the telescope, whose direction is unaltered. The transmitted ray goes along *ac*, is returned along ca_1, and is reflected at a_1, making ca_1e equal $90-a$, and

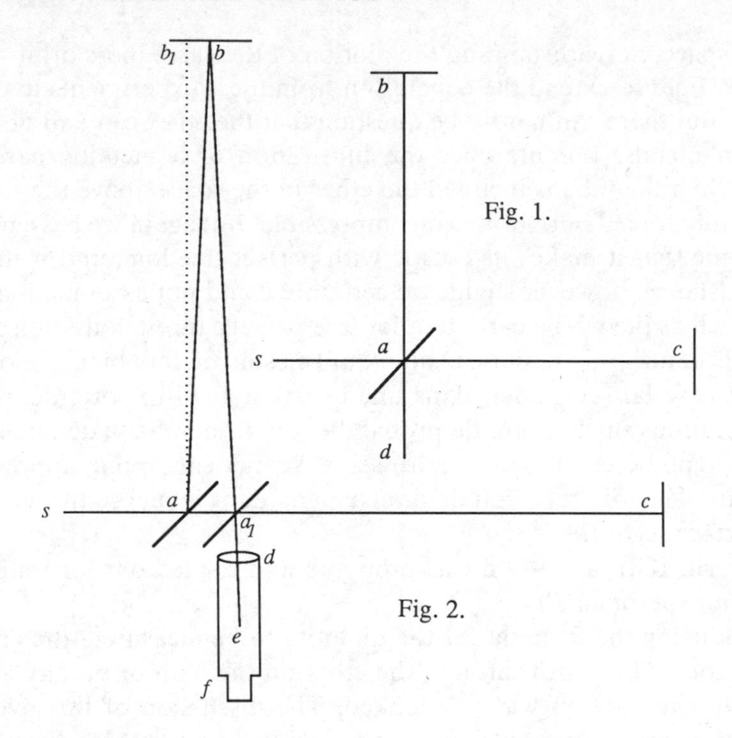

therefore still coinciding with the first ray. It may be remarked that the rays ba_1 and ca_1, do not now meet exactly in the same point a_1, though the difference is of the second order; this does not affect the validity of the reasoning. Let it now be required to find the difference in the two paths aba_1, and aca_1.

Let V = velocity of light.
  v = velocity of the earth in its orbit.
  D = distance *ab* or *ac*, fig. 1.
  T = time light occupies to pass from *a* to *c*.
  T_1 = time light occupies to return from *c* to a_1, (fig. 2.)

Then $T = \frac{D}{V - v}$, $T_1 = \frac{D}{V + v}$. The whole time of going and coming is $T + T_1 = 2D \frac{V}{V^2 - v^2}$, and the distance traveled in this time is $2D \frac{V^2}{V^2 - v^2} = 2D\left(1 + \frac{v^2}{V^2}\right)$, neglecting terms of the fourth order. The length of the other path is evidently $2D \sqrt{1 + \frac{v^2}{V^2}}$, or to the same degree of accuracy, $2D\left(1+\frac{v^2}{2V^2}\right)$. The difference is therefore $D\frac{v^2}{V^2}$. If now the whole apparatus be turned through 90°, the difference will be in the opposite direction, hence the displacement of the interference fringes should be $2D\frac{v^2}{V^2}$. Considering only the velocity of the earth in its orbit, this would be $2D \times 10^{-8}$. If, as was the case in the first experiment, $D = 2 \times 10^6$ waves of yellow light, the displacement to be expected would be 0.04 of the distance between the interference fringes.

In the first experiment one of the principal difficulties encountered was that of revolving the apparatus without producing distortion; and another was its extreme sensitiveness to vibration. This was so great that it was impossible to see the interference fringes except at brief intervals when working in the city, even at two o'clock in the morning. Finally, as before remarked, the quantity to be observed, namely, a displacement of something less than a twentieth of the distance between the interference fringes may have been too small to be detected when masked by experimental errors.

The first named difficulties were entirely overcome by mounting the apparatus on a massive stone floating on mercury; and the second by increasing, by repeated reflection, the path of the light to about ten times its former value.

The apparatus is represented in perspective in fig. 3, in plan in fig. 4,

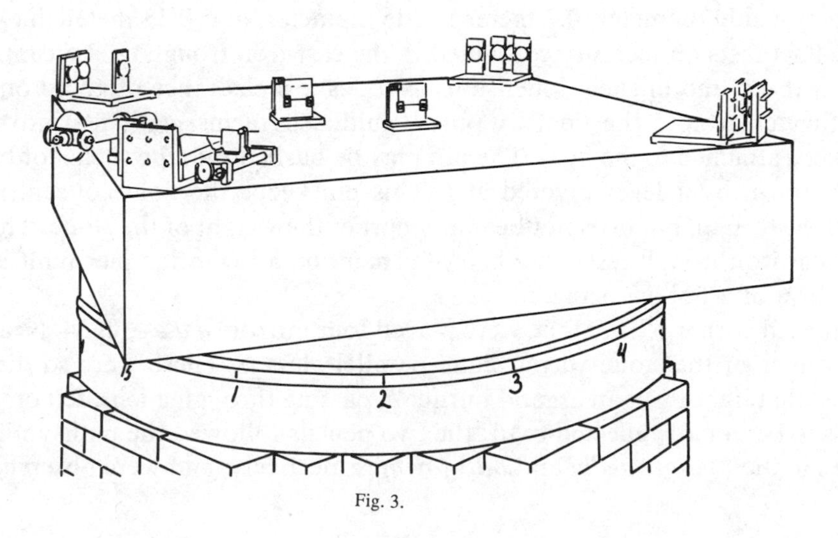

Fig. 3.

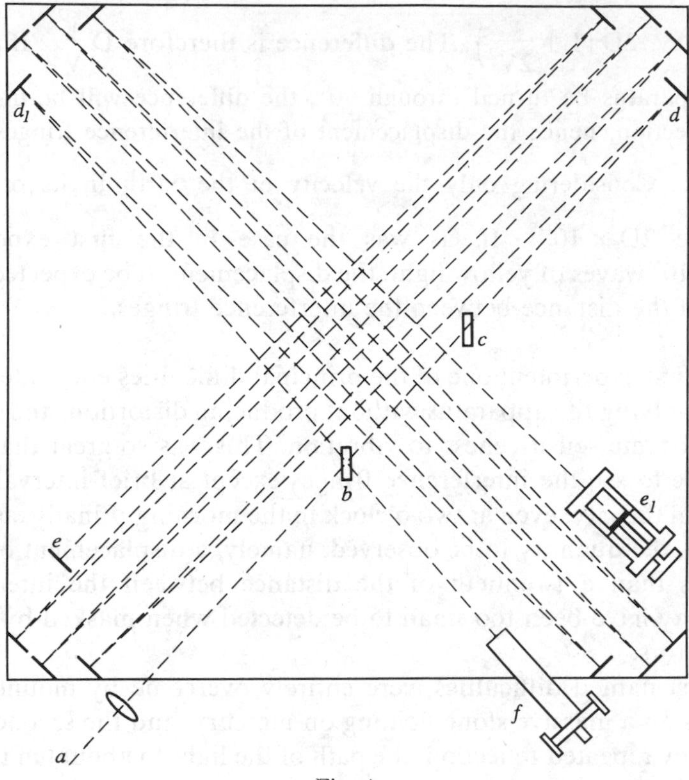

Fig. 4.

and in vertical section in fig. 5. The stone *a* (fig. 5) is about 1.5 meter square and 0.3 meter thick. It rests on an annular wooden float *bb,* 1.5 meter outside diameter, 0.7 meter inside diameter, and 0.25 meter thick. The float rests on mercury contained in the cast-iron trough *cc,* 1.5 centimeter thick, and of such dimensions as to leave a clearance of about one centimeter around the float. A pin *d,* guided by arms *gggg,* fits into a socket *e* attached to the float. The pin may be pushed into the socket or be withdrawn, by a lever pivoted at *f.* This pin keeps the float concentric with the trough, but does not bear any part of the weight of the stone. The annular iron trough rests on a bed of cement on a low brick pier built in the form of a hollow octagon.

At each corner of the stone were placed four mirrors *d d e e,* fig. 4. Near the center of the stone was a plane-parallel glass *b*. These were so disposed that light from an argand burner *a,* passing through a lens, fell on *b* so as to be in part reflected to d_1; the two pencils followed the paths indicated in the figure, *bdedbf* and $bd_1 e_1\ d_1 bf$ respectively, and were observed

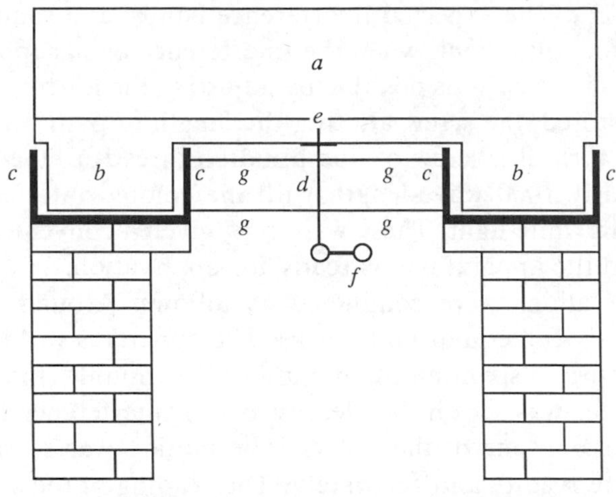

Fig. 5.

by the telescope *f*. Both *f* and *a* revolved with the stone. The mirrors were of speculum metal carefully worked to optically plane surfaces five centimeters in diameter, and the glasses *b* and *c* were plane-parallel and of the same thickness, 1.25 centimeter; their surfaces measured 5.0 by 7.5 centimeters. The second of these was placed in the path of one of the pencils to compensate for the passage of the other through the same thickness of glass. The whole of the optical portion of the apparatus was kept covered with a wooden cover to prevent air currents and rapid changes of temperature.

The adjustment was effected as follows: The mirrors having been adjusted by screws in the castings which held the mirrors, against which they were pressed by springs, till light from both pencils could be seen in the telescope, the lengths of the two paths were measured by a light wooden rod reaching diagonally from mirror to mirror, the distance being read from a small steel scale to tenths of millimeters. The difference in the lengths of the two paths was then annulled by moving the mirror e_1. This mirror had three adjustments; it had an adjustment in altitude and one in azimuth, like all the other mirrors, but finer; it also had an adjustment in the direction of the incident ray, sliding forward or backward, but keeping very accurately parallel to its former plane. The three adjustments of this mirror could be made with the wooden cover in position.

The paths being now approximately equal, the two images of the source of light or of some well-defined object placed in front of the condensing lens, were made to coincide, the telescope was now adjusted for distinct vision of the expected interference bands, and sodium light was substituted for white light, when the interference bands appeared. These were now made as clear as possible by adjusting the mirror e_1; then white light was restored, the screw altering the length of path was very slowly moved (one turn of a screw of one hundred threads to the inch altering the path nearly 1000 wave-lengths) till the colored interference fringes reappeared in white light. These were now given a convenient width and position, and the apparatus was ready for observation.

The observations were conducted as follows: Around the cast-iron trough were sixteen equidistant marks. The apparatus was revolved very slowly (one turn in six minutes) and after a few minutes the cross wire of the micrometer was set on the clearest of the interference fringes at the instant of passing one of the marks. The motion was so slow that this could be done readily and accurately. The reading of the screw-head on the micrometer was noted, and a very slight and gradual impulse was given to keep up the motion of the stone; on passing the second mark, the same process was repeated, and this was continued till the apparatus had completed six revolutions. It was found that by keeping the apparatus in slow uniform motion, the results were much more uniform and consistent than when the stone was brought to rest for every observation; for the effects of strains could be noted for at least half a minute after the stone came to rest, and during this time effects of change of temperature came into action.

The following tables give the means of the six readings; the first, for observations made near noon, the second, those near six o'clock in the evening. The readings are divisions of the screw-heads. The width of the fringes varied from 40 to 60 divisions, the mean value being near 50, so that one division means 0.02 wave-length. The rotation in the observation at noon was contrary to, and in the evening observations, with, that of the hands of a watch.

The results of the observations are expressed graphically in fig. 6. The upper is the curve for the observations at noon, and the lower that for the evening observations. The dotted curves represent *one-eighth* of the theoretical displacements. It seems fair to conclude from the figure that if there is any displacement due to the relative motion of the earth and the luminiferous ether, this cannot be much greater than 0.01 of the distance between the fringes.

Considering the motion of the earth in its orbit only, this displacement should be $2D\frac{v^2}{V^2} = 2D \times 10^{-8}$. The distance D was about eleven meters,

NOON OBSERVATIONS.

	16.	1.	2.	3.	4.	5.	6.	7.	8.	9.	10.	11.	12.	13.	14.	15.	16.
July 8	44·7	44·0	43·5	39·7	35·2	34·7	34·3	32·5	28·2	26·2	23·8	23·2	20·3	18·7	17·5	16·8	13·7
July 9	57·4	57·3	58·2	59·2	58·7	60·2	60·8	62·0	61·5	63·3	65·8	67·3	69·7	70·7	73·0	70·2	72·2
July 11	27·3	23·5	22·0	19·3	19·2	19·3	18·7	18·8	16·2	14·3	13·3	12·8	13·3	12·3	10·2	7·3	6·5
Mean	43·1	41·6	41·2	39·4	37·7	38·1	37·9	37·8	35·3	34·6	34·3	34·4	34·4	33·9	33·6	31·4	30·8
Mean in w. l.	·862	·832	·824	·788	·754	·762	·758	·756	·706	·692	·686	·688	·688	·678	·672	·628	·616
	·706	·692	·686	·688	·688	·678	·672	·628	·616								
Final mean.	·784	·762	·755	·738	·721	·720	·715	·692	·661								

P. M. OBSERVATIONS.

	16.	1.	2.	3.	4.	5.	6.	7.	8.	9.	10.	11.	12.	13.	14.	15.	16.
July 8	61·2	63·3	63·3	68·2	67·7	69·3	70·3	69·8	69·0	71·3	71·3	70·5	71·2	71·2	70·5	72·5	75·7
July 9	26·0	26·0	28·2	29·2	31·5	32·0	31·3	31·7	33·0	35·8	36·5	37·3	38·8	41·0	42·7	43·7	44·0
July 12	66·8	66·5	66·0	64·3	62·2	61·0	61·3	59·7	58·2	55·7	53·7	54·7	55·0	58·2	58·5	57·0	56·0
Mean	51·3	51·9	52·5	53·9	53·8	54·1	54·3	53·7	53·4	54·3	53·8	54·2	55·0	56·8	57·2	57·7	58·6
Mean in w. l.	1·026	1·038	1·050	1·078	1·076	1·082	1·086	1·074	1·068	1·086	1·076	1·084	1·100	1·136	1·144	1·154	1·172
	1·068	1·086	1·076	1·084	1·100	1·136	1·144	1·154	1·172								
Final mean.	1·047	1·062	1·063	1·081	1·088	1·109	1·115	1·114	1·120								

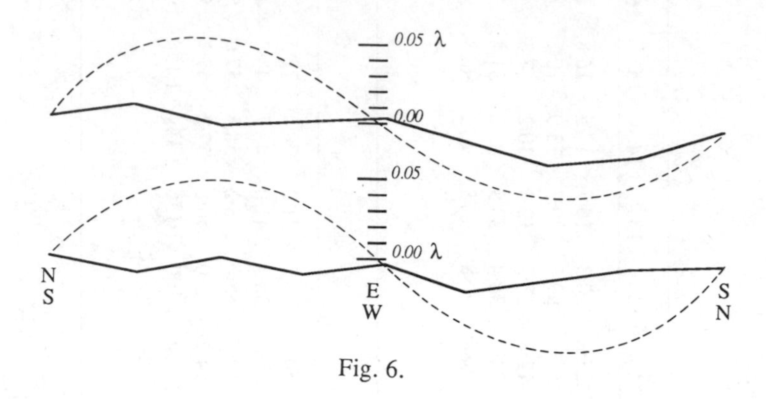

Fig. 6.

or 2×10^7 wave-lengths of yellow light; hence the displacement to be expected was 0.4 fringe. The actual displacement was certainly less than the twentieth part of this, and probably less than the fortieth part. But since the displacement is proportional to the square of the velocity, the relative velocity of the earth and the ether is probably less than one-sixth the earth's orbital velocity, and certainly less than one-fourth.

In what precedes, only the orbital motion of the earth is considered. If this is combined with the motion of the solar system, concerning which but little is known with certainty, the result would have to be modified; and it is just possible that the resultant velocity at the time of the observations was small though the chances are much against it. The experiment will therefore be repeated at intervals of three months, and thus all uncertainty will be avoided.

It appears, from all that precedes, reasonably certain that if there be any relative motion between the earth and the luminiferous ether, it must be small; quite small enough entirely to refute Fresnel's explanation of aberration. Stokes has given a theory of aberration which assumes the ether at the earth's surface to be at rest with regard to the latter, and only requires in addition that the relative velocity have a potential; but Lorentz shows that these conditions are incompatible. Lorentz then proposes a modification which combines some ideas of Stokes and Fresnel, and assumes the existence of a potential, together with Fresnel's coefficient. If now it were legitimate to conclude from the present work that the ether is at rest with regard to the earth's surface, according to Lorentz there could not be a velocity potential, and his own theory also fails.

How I Created the Theory of Relativity

ALBERT EINSTEIN

Ether and Motion

ALBERT EINSTEIN AND
LEOPOLD INFELD

ALBERT EINSTEIN wrote various versions and explanations, both popular and technical, of his special and general theories of relativity. When he was finally awarded a Nobel Prize in physics in 1921 (for his work on the photoelectric effect, not for his theories of relativity, which were still controversial in the early 1920s), he was unable to go to Sweden for the awards ceremony because of an earlier commitment to visit Japan. On 14 December 1922 (the anniversary, incidentally, of Planck's first public announcement of his theory of quanta), Einstein gave an impromptu speech at Kyoto University entitled "How I Created the Theory of Relativity" at the request of a Kyoto professor of philosophy. Einstein used no written notes, but spoke extemporaneously in German; a running translation was provided by J. Ishiwara, a professor at Tohuku University, who had studied a decade earlier with Einstein and Arnold Sommerfeld. Ishiwara kept detailed notes of the lecture and published them, in Japanese, in a local monthly *Kaizo*. In 1982, *Physics Today*, published by the American Institute of Physics, printed an English translation by Yoshimasa A. Ono, who commented, "It is clear that this account of Einstein's [formulation of his theories of relativity] throws some light on the current controversy as to whether or not he was aware of the Michelson-Morley experiment . . . in 1905."

This brief document reports in a conscientious effort at accuracy what Einstein might have answered to questions about the origins of his theory. It also provides a simple, anecdotal, nontechnical introduction to his two monumental theories.

The second selection, "Ether and Motion," is taken from a book entitled *The Evolution of Physics,* which was actually written by the Polish physicist Leopold Infeld. Although his own contribution was negligible, Einstein agreed to lend his name in order to promote the book and to assist Infeld's efforts to remain in the United States. *The Evolution of Physics* has a number of "thought experiments" which illustrate principles of classical and modern physics. This selection contains discussions of ether (a now-discarded hypothesis about the medium once believed to fill all space and through which electromagnetic waves were propagated) and the phenomenon of time dilation. The latter led to the postulation of the "twin paradox" which showed that one twin, when accelerated to a speed approaching that of light, would find that his passage of time would slow down relative to the passage of time observed by the twin who remained behind. Einstein's equations concluded that if a spaceship could reach the speed of light, the passage of time would cease altogether. The only drawback is that an infinite amount of energy would be needed to accelerate the spaceship to the velocity of light.

How I Created the Theory of Relativity

ALBERT EINSTEIN

He hath shook hands with time.

—JOHN FORD, *The Broken Heart*

Energy has mass and mass represents energy.

—ALBERT EINSTEIN, *The Evolution of Physics*

Margaret Fuller: I accept the Universe.
Thomas Carlyle: Gad! she'd better!

—(Attributed to Thomas Carlyle)

Creating a new theory is not like destroying an old barn and erecting a skyscraper in its place. It is rather like climbing a mountain, gaining new and wider views, discovering unexpected connections between our starting point and its rich environment. But the point from which we started out still exists and can be seen, although it appears smaller and forms a tiny part of our broad view gained by the mastery of the obstacles on our adventurous way up.

—ALBERT EINSTEIN, *The Evolution of Physics*

IT IS NOT easy to talk about how I reached the idea of the theory of relativity; there were so many hidden complexities to motivate my thought, and the impact of each thought was different at different stages in the de-

Albert Einstein, "How I Created the Theory of Relativity." Reprinted with permission from *Physics Today*, Vol. 35, No. 8, pp. 45–47 (1982). © 1982 American Institute of Physics.

velopment of the idea. I will not mention them all here. Nor will I count the papers I have written on this subject. Instead I will briefly describe the development of my thought directly connected with this problem.

It was more than seventeen years ago that I had an idea of developing the theory of relativity for the first time. While I cannot say exactly where that thought came from, I am certain that it was contained in the problem of the optical properties of moving bodies. Light propagates through the sea of ether, in which the Earth is moving. In other words, the ether is moving with respect to the Earth. I tried to find clear experimental evidence for the flow of the ether in the literature of physics, but in vain.

Then I myself wanted to verify the flow of the ether with respect to the Earth, in other words, the motion of the Earth. When I first thought about this problem, I did not doubt the existence of the ether or the motion of the Earth through it. I thought of the following experiment using two thermocouples: Set up mirrors so that the light from a single source is to be reflected in two different directions, one parallel to the motion of the Earth and the other antiparallel. If we assume that there is an energy difference between the two reflected beams, we can measure the difference in the generated heat using two thermocouples. Although the idea of this experiment is very similar to that of Michelson, I did not put this experiment to the test.

While I was thinking of this problem in my student years, I came to know the strange result of Michelson's experiment. Soon I came to the conclusion that our idea about the motion of the earth with respect to the ether is incorrect, if we admit Michelson's null result as a fact. This was the first path which led me to the special theory of relativity. Since then I have come to believe that the motion of the Earth cannot be detected by any optical experiment, though the Earth is revolving around the Sun.

I had a chance to read Lorentz's monograph of 1895. He discussed and solved completely the problem of electrodynamics within the first approximation, namely neglecting terms of order higher than v/c, where v is the velocity of a moving body and c is the velocity of light. Then I tried to discuss the Fizeau experiment on the assumption that the Lorentz equations for electrons should hold in the frame of reference of the moving body as well as in the frame of reference of the vacuum as originally discussed by Lorentz. At that time I firmly believed that the electrodynamic equations of Maxwell and Lorentz were correct. Furthermore, the assumption that these equations should hold in the reference frame of the moving body leads to the concept of the invariance of the velocity of light, which, however, contradicts the addition rule of velocities used in mechanics.

Why do these two concepts contradict each other? I realized that this difficulty was really hard to resolve. I spent almost a year in vain trying to modify the idea of Lorentz in the hope of resolving this problem.

By chance a friend of mind in Bern (Michele Besso) helped me out. It was a beautiful day when I visited him with this problem. I started the conversation with him in the following way: "Recently I have been working on a difficult problem. Today I come here to battle against that problem with you." We discussed every aspect of this problem. Then suddenly I understood where the key to this problem lay. Next day I came back to him again and said to him, without even saying hello, "Thank you. I've completely solved the problem." An analysis of the concept of time was my solution. Time cannot be absolutely defined, and there is an inseparable relation between time and signal velocity. With this new concept, I could resolve all the difficulties completely for the first time.

Within five weeks the special theory of relativity was completed. I did not doubt that the new theory was reasonable from a philosophical point of view. I also found that the new theory was in agreement with Mach's argument. Contrary to the case of the general theory of relativity in which Mach's argument was incorporated in the theory, Mach's analysis had [only] indirect implication in the special theory of relativity.

This is the way the special theory of relativity was created.

My first thought on the general theory of relativity was conceived two years later, in 1907. The idea occurred suddenly. I was dissatisfied with the special theory of relativity, since the theory was restricted to frames of reference moving with constant velocity relative to each other and could not be applied to the general motion of a reference frame. I struggled to remove this restriction and wanted to formulate the problem in the general case.

In 1907 Johannes Stark asked me to write a monograph on the special theory of relativity in the journal *Jahrbuch der Radioaktivität*. While I was writing this, I came to realize that all the natural laws except the law of gravity could be discussed within the framework of the special theory of relativity. I wanted to find out the reason for this, but I could not attain this goal easily.

The most unsatisfactory point was the following: Although the relationship between inertia and energy was explicitly given by the special theory of relativity, the relationship between inertia and weight, or the energy of the gravitational field, was not clearly elucidated. I felt that this problem could not be resolved within the framework of the special theory of relativity.

The breakthrough came suddenly one day. I was sitting on a chair in my patent office in Bern. Suddenly a thought struck me: If a man falls freely, he would not feel his weight. I was taken aback. This simple thought experiment made a deep impression on me. This led to the theory of gravity. I continued my thought: A falling man is accelerated. Then what he feels and judges is happening in the accelerated frame of reference. I decided to extend the theory of relativity to the reference frame with acceleration. I felt that in doing so I could solve the problem of gravity at the same time. A falling man does not feel his weight because in his reference frame there is a new gravitational field which cancels the gravitational field due to the Earth. In the accelerated frame of reference, we need a new gravitational field.

I could not solve this problem completely at that time. It took me eight more years until I finally obtained the complete solution. During these years I obtained partial answers to this problem.

Ernest Mach was a person who insisted on the idea that systems that have acceleration with respect to each other are equivalent. This idea contradicts Euclidean geometry, since in the frame of reference with acceleration Euclidean geometry cannot be applied. Describing the physical laws without reference to geometry is similar to describing our thought without words. We need words in order to express ourselves. What should we look for to describe our problem? This problem was unsolved until 1912, when I hit upon the idea that the surface theory of Karl Friedrich Gauss might be the key to this mystery. I found that Gauss's surface coordinates were very meaningful for understanding this problem. Until then I did not know that Bernhard Riemann [who was a student of Gauss's] had discussed the foundation of geometry deeply. I happened to remember the lecture on geometry in my student years [in Zurich] by Carl Friedrich Geiser who discussed the Gauss theory. I found that the foundations of geometry had deep physical meaning in this problem.

When I came back to Zurich from Prague, my friend the mathematician Marcel Grossman was waiting for me. He had helped me before in supplying me with mathematical literature when I was working at the patent office in Bern and had some difficulties in obtaining mathematical articles. First he taught me the work of Curbastro Gregorio Ricci and later the work of Riemann. I discussed with him whether the problem could be solved using Riemann theory, in other words, by using the concept of the invariance of line elements. We wrote a paper on this subject in 1913, although we could not obtain the correct equations for gravity. I studied Riemann's equations further only to find many reasons why the desired results could not be attained in this way.

After two years of struggle, I found that I had made mistakes in my calculations. I went back to the original equation using the invariance theory and tried to construct the correct equations. In two weeks the correct equations appeared in front of me!

Concerning my work after 1915, I would like to mention only the problem of cosmology. This problem is related to the geometry of the universe and to time. The foundation of this problem comes from the boundary conditions of the general theory of relativity and the discussion of the problem of inertia by Mach. Although I did not exactly understand Mach's idea about inertia, his influence on my thought was enormous.

I solved the problem of cosmology by imposing invariance on the boundary condition for the gravitational equations. I finally eliminated the boundary by considering the Universe to be a closed system. As a result, inertia emerges as a property of interacting matter and it should vanish if there were no other matter to interact with. I believe that with this result the general theory of relativity can be satisfactorily understood epistemologically.

This is a short historical survey of my thoughts in creating the theory of relativity.

Ether and Motion

ALBERT EINSTEIN
AND LEOPOLD INFELD

Ye Gods! annihilate but space and time,/ And make two lovers happy.

—ALEXANDER POPE, *The Art of Sinking in Poetry,* 11

Science has, as its whole purpose, the rendering of the physical world understandable and beautiful. Without this you only have fables and statistics. The measure of our success is our ability to live with this knowledge effectively, actively and with delight. If we succeed we will be able to cope with our knowledge and not create despair. But this also means appreciation of the plurality of knowledge. Order is not monolithic, it is plural.

—J. ROBERT OPPENHEIMER

However successful the theory of a four-dimensional world may be, it is difficult to ignore a voice inside us which whispers: "At the back of your mind, you know that a fourth dimension is all nonsense." . . . Let us not be beguiled by this voice. It is discredited. . . .

—ARTHUR STANLEY EDDINGTON

THE GALILEAN RELATIVITY principle is valid for mechanical phenomena. The same laws of mechanics apply to all inertial systems moving relative to each other. Is this principle also valid for nonmechanical

Albert Einstein and Leopold Infeld, "Ether and Motion," in *The Evolution of Physics.* New York: Simon and Schuster, 1961, pp. 164–187. Reprinted with permission of the Hebrew University of Jerusalem, Israel.

phenomena, especially for those for which the field concepts proved so very important? All problems concentrated around this question immediately bring us to the starting point of the relativity theory.

We remember that the velocity of light *in vacuo*, or in other words, in ether, is 186,000 miles per second and that light is an electromagnetic wave spreading through the ether. The electromagnetic field carries energy which, once emitted from its source, leads an independent existence. For the time being, we shall continue to believe that the ether is a medium through which electromagnetic waves, and thus also light waves, are propagated, even though we are fully aware of the many difficulties connected with its mechanical structure.

We are sitting in a closed room so isolated from the external world that no air can enter or escape. If we sit still and talk we are, from the physical point of view, creating sound waves, which spread from their resting source with the velocity of sound in air. If there were no air or other material medium between the mouth and the ear, we could not detect a sound. Experiment has shown that the velocity of sound in air is the same in all directions, if there is no wind and the air is at rest in the chosen Coordinate System (CS).

Let us now imagine that our room moves uniformly through space. A man outside sees, through the glass walls of the moving room (or train if you prefer) everything which is going on inside. From the measurements of the inside observer he can deduce the velocity of sound relative to his CS connected with his surroundings, relative to which the room moves. Here again is the old, much discussed, problem of determining the velocity in one CS if it is already known in another.

The observer in the room claims: the velocity of sound is, for me, the same in all directions.

The outside observer claims: the velocity of sound, spreading in the moving room and determined in my CS, is not the same in all directions. It is greater than the standard velocity of sound in the direction of the motion of the room and smaller in the opposite direction.

These conclusions are drawn from the classical transformation and can be confirmed by experiment. The room carries within it the material medium, the air through which sound waves are propagated, and the velocities of sound will, therefore, be different for the inside and outside observer.

We can draw some further conclusions from the theory of sound as a wave propagated through a material medium. One way, though by no means the simplest, of not hearing what someone is saying, is to run, with a velocity greater than that of sound, relative to the air surrounding the speaker. The sound waves produced will then never be able to reach our ears. On the other hand, if we missed an important word which will never be repeated, we must run with a speed greater than that of sound to reach the produced wave and to catch the word. There is nothing irrational in either of these examples except that in both cases we should have to run with a speed of about four hundred yards per second, and we can very well imagine that further technical development will make such speeds possible. A bullet fired from a gun actually moves with a speed greater than that of sound and a man placed on such a bullet would never hear the sound of the shot.

All these examples are of a purely mechanical character and we can now formulate the important questions: could we repeat what has just been said of a sound wave, in the case of a light wave? Do the Galilean relativity principle and the classical transformation apply to mechanical as well as to optical and electrical phenomena? It would be risky to answer these questions by a simple "yes" or "no" without going more deeply into their meaning.

In the case of the sound wave in the room moving uniformly, relative to the outside observer, the following intermediate steps are very essential for our conclusion:

The moving room carries the air in which the sound wave is propagated.
The velocities observed in two CS moving uniformly, relative to each other, are connected by the classical transformation.

The corresponding problem for light must be formulated a little differently. The observers in the room are no longer talking, but are sending light signals, or light waves in every direction. Let us further assume that the sources emitting the light signals are permanently resting in the room. The light waves move through the ether just as the sound waves moved through the air.

Is the ether carried with the room as the air was? Since we have no mechanical picture of the ether it is extremely difficult to answer this question. If the room is closed, the air inside is forced to move with it. There is obviously no sense in thinking of ether in this way, since all matter is immersed in it and it penetrates everywhere. No doors are closed to ether. The "moving room," now means only a moving CS to which the source of light is rigidly connected. It is, however, not beyond us to imagine that the room moving with its light source carries the ether along with it just as the sound source and air were carried along in the closed room. But we can equally well imagine the opposite: that the room travels through the ether as a ship through a perfectly smooth sea, not carrying any part of the medium along but moving through it. In our first picture, the room moving with its light source carries the ether. An analogy with a sound wave is possible and quite similar conclusions can be drawn. In the second, the room moving with its light source does not carry the ether. No analogy with a sound wave is possible and the conclusions drawn in the case of a sound wave do not hold for a light wave. These are the two limiting possibilities. We could imagine the still more complicated possibility that the ether is only partially carried by the room moving with its light source. But there is no reason to discuss the more complicated assumptions before finding out which of the two simpler limiting cases experiment favors.

We shall begin with our first picture and assume, for the present: the ether is carried along by the room moving with its rigidly-connected light source. If we believe in the simple transformation principle for the velocities of sound waves, we can now apply our conclusions to light waves as well. There is no reason for doubting the simple mechanical transformation law which only states that the velocities have to be added in certain cases and subtracted in others. For the moment, therefore, we shall assume both the carrying of the ether by the room moving with its light source and the classical transformation.

If I turn on the light and its source is rigidly connected with my room, then the velocity of the light signal has the well-known experimental value 186,000 miles per second. But the outside observer will notice the motion of the room, and, therefore, that of the source and, since the ether is carried along, his conclusion must be: the velocity of light in my outside CS is different in different directions. It is greater than the standard velocity of light in the direction of the motion of the room and smaller in the opposite direction. Our conclusion is: if ether is carried with the room moving with its light source and if the mechanical laws are valid, then the velocity of light must depend on the velocity of the light source. Light reaching our eyes from a moving light source would have a greater velocity if the motion is toward us and smaller if it is away from us.

If our speed were greater than that of light we should be able to run away from a light signal. We could see occurrences from the past by reaching previously sent light waves. We should catch them in a reverse order to that in which they were sent, and the train of happenings on our earth would appear like a film shown backward, beginning with a happy ending. These conclusions all follow from the assumption that the moving CS carries along the ether and the mechanical transformation laws are valid. If this is so, the analogy between light and sound is perfect.

But there is no indication as to the truth of these conclusions. On the contrary, they are contradicted by all observations made with the intention of proving them. There is not the slightest doubt as to the clarity of this verdict, although it is obtained through rather indirect experiments in view of the great technical difficulties caused by the enormous value of the velocity of light. *The velocity of light is always the same in all CS independent of whether or not the emitting source moves, or how it moves.*

We shall not go into detailed description of the many experiments from which this important conclusion can be drawn. We can, however, use some very simple arguments which, though they do not prove that the velocity of light is independent of the motion of the source, nevertheless make this fact convincing and understandable.

In our planetary system the earth and other planets move around the sun. We do not know of the existence of other planetary systems, similar to ours. There are, however, very many double-star systems, consisting of two stars moving around a point, called their center of gravity. Observation of the motion of these double stars reveals the validity of Newton's gravitational law. Now suppose that the speed of light depends on the velocity of the emitting body. Then the message, that is, the light ray from the star, will travel more quickly or more slowly, according to the velocity of the star at the moment the ray is emitted. In this case the whole motion would be muddled and it would be impossible to confirm, in the case of distant double stars, the validity of the same gravitational law which rules over our planetary system.

Let us consider another experiment based upon a very simple idea. Imagine a wheel rotating very quickly. According to our assumption, the ether is carried by the motion and takes a part in it. A light wave passing near the wheel would have a different speed when the wheel is at rest than when it is in motion. The velocity of light in ether at rest should differ from that in ether which is being quickly dragged round by the motion of the wheel, just as the velocity of a sound wave varies on calm and windy days. But no such difference is detected! No matter from which angle we approach the subject, or what crucial experiment we may devise, the verdict is always against the assumption of the ether carried by motion. Thus, the result of our considerations, supported by more detailed and technical argument, is:

The velocity of light does not depend on the motion of the emitting source.

It must not be assumed that the moving body carries the surrounding ether along.

We must, therefore, give up the analogy between sound and light waves and turn to the second possibility: that all matter moves through the ether, which takes no part whatever in the motion. This means that we assume the existence of a sea of ether with all CS resting in it, or moving relative to it. Suppose we leave, for a while, the question as to whether experiment proved or disproved this theory. It will be better to become more familiar with the meaning of this new assumption and with the conclusions which can be drawn from it.

There exists a CS resting relative to the ether-sea. In mechanics, not one of the many CS moving uniformly, relative to each other, could be distinguished. All such CS were equally "good" or "bad." If we have two CS moving uniformly, relative to each other, it is meaningless, in mechanics, to ask which of them is in motion and which at rest. Only relative uniform motion can be observed. We cannot talk about absolute uniform motion because of the Galilean relativity principle. What is meant by the statement that *absolute* and not only *relative* uniform motion exists? Simply that there exists one CS in which some of the laws of nature are different from those in all others. Also that every observer can detect whether his CS is at rest or in motion by comparing the laws valid in it with those valid in the only one which has the absolute monopoly of serving as the standard CS. Here is a different state of affairs from classical mechanics, where absolutc uniform motion is quite meaningless because of Galileo's law of inertia.

What conclusions can be drawn in the domain of field phenomena if motion through ether is assumed? This would mean that there exists one CS distinct from all others, at rest relative to the ether-sea. It is quite clear that some of the laws of nature must be different in this CS, otherwise the phrase, "motion through ether," would be meaningless. If the Galilean relativity principle is valid then motion through ether makes no sense at all. It is impossible to reconcile these two ideas. If, however, there exists one special CS fixed by the ether, then to speak of "absolute motion" or "absolute rest," has a definite meaning.

We really have no choice. We tried to save the Galilean relativity principle by assuming that systems carry the ether along in their motion, but this led to a contradiction with experiment. The only way out is to abandon the Galilean relativity principle and try out the assumption that all bodies move through the calm ether-sea.

The next step is to consider some conclusions contradicting the Galilean relativity principle and supporting the view of motion through ether, and to put them to the test of an experiment. Such experiments are easy enough to imagine, but very difficult to perform. As we are concerned here only with ideas, we need not bother with technical difficulties.

Again we return to our moving room with two observers, one inside and one outside. The outside observer will represent the standard CS, designated by the ether-sea. It is the distinguished CS in which the velocity of light always has the same standard value. All light sources, whether moving or at rest in the calm ether-sea, propagate light with the same velocity. The room and its observer move through the ether. Imagine that a light in the center of the room is flashed on and off and, furthermore, that the walls of the room are transparent so that the observers, both inside and outside, can measure the velocity of the light. If we ask the two observers what results they expect to obtain, their answers would run something like this:

The outside observer: My CS is designated by the ether-sea. Light in my CS always has the standard value. I need not care whether or not the source of light or other bodies are moving, for they never carry my ether-sea with them. My CS is distinguished from all others and the velocity of light must have its standard value in this CS, independent of the direction of the light beam or the motion of its source.

The inside observer: My room moves through the ether-sea. One of the walls runs away from the light and the other approaches it. If my room traveled, relative to the ether-sea, with the velocity of light, then the light emitted from the center of the room would never reach the wall running away with the velocity of light. If the room traveled with a velocity smaller than that of light, then a wave sent from the center of the room would reach one of the walls before the other. The wall moving toward the light wave would be reached before the one retreating from the light wave. Therefore, although the source of light is rigidly connected with my CS, the velocity of light will not be the same in all directions. It will be smaller in the direction of the motion relative to the ether-sea as the wall runs away, and greater in the opposite direction as the wall moves toward the wave and tries to meet it sooner.

Thus, only in the one CS distinguished by the ether-sea should the velocity of light be equal in all directions. For other CS moving relatively to the ether-sea it should depend on the direction in which we are measuring.

The crucial experiment just considered enables us to test the theory of motion through the ether-sea. Nature, in fact, places at our disposal a system moving with a fairly high velocity: the earth in its yearly motion around the sun. If our assumption is correct, then the velocity of light in the direction of the motion of the earth should differ from the velocity of light in an opposite direction. The differences can be calculated and a suitable experimental test devised. In view of the small time-differences following from the theory, very ingenious experimental arrangements have to be thought out. This was done in the famous Michelson-Morley experiment. The result was a verdict of "death" to the theory of a calm ether-sea through which all matter moves. No dependence of the speed of light upon direction could be found. Not only the speed of light, but also other field phenomena would show a dependence on the direction in the moving CS, if the theory of the ether-sea were assumed. Every experiment has given the same negative result as the Michelson-Morley one, and never revealed any dependence upon the direction of the motion of the earth.

The situation grows more and more serious. Two assumptions have been tried. The first, that moving bodies carry ether along. The fact that the velocity of light does not depend on the motion of the source contradicts this assumption. The second, that there exists one distinguished CS and that moving bodies do not carry the ether but travel through an ever calm ether-sea. If this is so, then the Galilean relativity principle is not valid and the speed of light cannot be the same in every CS. Again we are in contradiction with experiment.

More artificial theories have been tried out, assuming that the real truth lies somewhere between these two limiting cases: that the ether is only partially carried by the moving bodies. But they all failed! Every attempt to explain the electromagnetic phenomena in moving CS with the help of the motion of the ether, motion through the ether, or both these motions, proved unsuccessful.

Thus arose one of the most dramatic situations in the history of science. All assumptions concerning ether led nowhere! The experimental verdict was always negative. Looking back over the development of physics we see that the ether, soon after its birth, became the "*enfant terrible*" of the family of physical substances. First, the construction of a simple mechanical picture of the ether proved to be impossible and was discarded. This caused, to a great extent, the breakdown of the mechanical point of view. Second, we had to give up hope that through the presence of the ether-sea one CS would be distinguished and lead to the recognition of absolute, and not only relative, motion. This would have been the only way, besides carrying the waves, in which ether could mark and justify its existence. All our attempts to make ether real failed. It revealed neither its mechanical construction nor absolute motion. Nothing remained of all the properties of the ether except that for which it was invented, i.e., its ability to transmit electromagnetic waves. Our attempts to discover the properties of the ether led to difficulties and contradictions. After such bad experiences, this is the moment to forget the ether completely and to try never to mention its name. We shall say: our space has the physical property of transmitting waves, and so omit the use of a word we have decided to avoid.

The omission of a word from our vocabulary is, of course, no remedy. Our troubles are indeed much too profound to be solved in this way!

Let us now write down the facts which have been sufficiently confirmed by experiment without bothering any more about the "e—r" problem.

1. The velocity of light in empty space always has its standard value, independent of the motion of the source or receiver of light.
2. In two CS moving uniformly, relative to each other, all laws of nature are exactly identical and there is no way of distinguishing absolute uniform motion.

There are many experiments to confirm these two statements and not a single one to contradict either of them. The first statement expresses the constant character of the velocity of light, the second generalizes the Galilean relativity principle, formulated for mechanical phenomena, to all happenings in nature.

In mechanics, we have seen: If the velocity of a material point is so and so, relative to one CS, then it will be different in another CS moving uniformly, relative to the first. This follows from the simple mechanical transformation principles. They are immediately given by our intuition (man moving relative to ship and shore) and apparently nothing can be wrong here! But this transformation law is in contradiction to the constant character of the velocity of light. Or, in other words, we add a third principle:

3. Positions and velocities are transformed from one inertial system to another according to the classical transformation.

The contradiction is then evident. We cannot combine (1), (2), and (3).

The classical transformation seems too obvious and simple for any attempt to change it. We have already tried to change (1) and (2) and came to a disagreement with experiment. All theories concerning the motion of "e—r" required an alteration of (1) and (2). This was no good. Once more we realize the serious character of our difficulties. A new clue is needed. It is supplied by *accepting the fundamental assumptions* (1) and (2), and, strange though it seems, *giving up* (3). The new clue starts from an analysis of the most fundamental and primitive concepts; we shall show how this analysis forces us to change our old views and removes all our difficulties.

TIME, DISTANCE, RELATIVITY

Our new assumptions are:

1. *The velocity of light* in vacuo *is the same in all CS moving uniformly, relative to each other.*
2. *All laws of nature are the same in all CS moving uniformly, relative to each other.*

The *relativity theory* begins with these two assumptions. From now on we shall not use the classical transformation because we know that it contradicts our assumptions.

It is essential here, as always in science, to rid ourselves of deep-rooted, often uncritically repeated, prejudices. Since we have seen that changes in (1) and (2) lead to contradiction with experiment, we must have the courage to state their validity clearly and to attack the one possibly weak point, the way in which positions and velocities are transformed from one CS to another. It is our intention to draw conclusions from (1) and (2), see where and how these assumptions contradict the classical transformation, and find the physical meaning of the results obtained.

Once more, the example of the moving room with outside and inside observers will be used. Again a light signal is emitted from the center of the room and again we ask the two men what they expect to observe, assuming only our two principles and forgetting what was previously said concerning the medium through which the light travels. We quote their answers:

The inside observer: The light signal traveling from the center of the room will reach the walls *simultaneously*, since all the walls are equally distant from the light source and the velocity of light is the same in all directions.

The outside observer: In my system, the velocity of light is exactly the same as in that of the observer moving with the room. It does not matter to me whether or not the light source moves in my CS since its motion does not influence the velocity of light. What I see is a light signal traveling with a standard speed, the same in all directions. One of the walls is trying to escape from and the opposite wall to approach the light signal. Therefore, the escaping wall will be met by the signal a little later than the approaching one. Although the difference will be very slight if the velocity of the room is small compared with that of light, the light signal will nevertheless not meet these two opposite walls, which are perpendicular to the direction of the motion, quite simultaneously.

Comparing the predictions of our two observers we find a most astonishing result which flatly contradicts the apparently well-founded concepts of classical physics. Two events, i.e., the two light beams reaching the two walls, are simultaneous for the observer on the inside, but not for the observer on the outside. In classical physics, we had one clock, one time flow, for all observers in all CS. Time, and therefore such words as "simultaneously," "sooner," "later," had an absolute meaning independent of any CS. Two events happening at the same time in one CS happened necessarily simultaneously in all other CS.

Assumptions (1) and (2), i.e., the relativity theory, force us to give up this view. We have described two events happening at the same time in one CS, but at different times in another CS. Our task is to understand this consequence, to understand the meaning of the sentence: "Two events which are simultaneous in one CS, may not be simultaneous in another CS."

What do we mean by "two simultaneous events in one CS"? Intuitively everyone seems to know the meaning of this sentence. But let us make up our minds to be cautious and try to give rigorous definitions, as we know how dangerous it is to overestimate intuition. Let us first answer a simple question.

What is a clock?

The primitive subjective feeling of time flow enables us to order our impressions, to judge that one event takes place earlier, another later. But to show that the time interval between two events is 10 seconds, a clock is needed. By the use of a clock the time concept becomes objective. Any physical phenomenon may be used as a clock, provided it can be exactly repeated as many times as desired. Taking the interval between the beginning and the end of such an event as one unit of time, arbitrary time-intervals may be measured by repetition of this physical process. All clocks, from the simple hourglass to the most refined instruments, are based on this idea. With the hourglass the unit of time is the interval the sand takes to flow from the upper to the lower glass. The same physical process can be repeated by inverting the glass.

At two distant points we have two perfect clocks, showing exactly the same time. This statement should be true regardless of the care with which we verify it. But what does it really mean? How can we make sure that distant clocks always show exactly the same time? One possible method would be to use television. It should be understood that television is used only as an example and is not essential to our argument. I could stand near one of the clocks and look at a televised picture of the other. I could then judge whether or not they showed the same time simultaneously. But this would not be a good proof. The televised picture is transmitted through electromagnetic waves and thus travels with the speed of light. Through television I see a picture which was sent some very short time before, whereas on the real clock I see what is taking place at the present moment. This difficulty can easily be avoided. I must take television pictures of the two clocks at a point equally distant from each of them and observe them from this center point. Then, if the signals are sent out simultaneously, they will all reach me at the same instant. If two good clocks observed from the mid-point of the distance between them always show the same time, then they are well suited for designating the time of events at two distant points.

In mechanics we used only one clock. But this was not very convenient, because we had to take all measurements in the vicinity of this one clock. Looking at the clock from a distance, for example by television, we have always to remember that what we see now really happened earlier, just as we receive light from the sun eight minutes after it was emitted. We should have to make corrections, according to our distance from the clock, in all our time readings.

It is, therefore, inconvenient to have only one clock. Now, however, as we know how to judge whether two, or more, clocks show the same time simultaneously and run in the same way, we can very well imagine as many clocks as we like in a given CS. Each of them will help us to determine the time of the events happening in its immediate vicinity. The clocks are all at rest relative to the CS. They are "good" clocks and are *synchronized*, which means that they show the same time simultaneously.

There is nothing especially striking or strange about the arrangements of our clocks. We are now using many synchronized clocks instead of only one and can, therefore, easily judge whether or not two distant events are simultaneous in a given CS. They are if the synchronized clocks in their vicinity show the same time at the instant the events happen. To say that one of the distant events happens before the other has now a definite meaning. All this can be judged by the help of the synchronized clocks at rest in our CS.

This is in agreement with classical physics, and not one contradiction to the classical transformation has yet appeared.

For the definition of simultaneous events, the clocks are synchronized by the help of signals. It is essential in our arrangement that these signals travel with the velocity of light, the velocity which plays such a fundamental role in the theory of relativity.

Since we wish to deal with the important problem of two CS moving uniformly, relative to each other, we must consider two rods, each provided with clocks. The observer in each of the two CS moving relative to each other now has his own rod with his own set of clocks rigidly attached.

When discussing measurements in classical mechanics we used one clock for all CS. Here we have many clocks in each CS. This difference is unimportant. One clock was sufficient, but nobody could object to the use of many, so long as they behave as decent synchronized clocks should.

Now we are approaching the essential point showing where the classical transformation contradicts the theory of relativity. What happens when two sets of clocks are moving uniformly, relative to each other? The classical physicist would answer: Nothing; they still have the same rhythm, and we can use moving as well as resting clocks to indicate time. According to classical physics, two events simultaneous in one CS will also be simultaneous in any other CS.

But this is not the only possible answer. We can equally well imagine a moving clock having a different rhythm from one at rest. Let us now discuss this possibility without deciding, for the moment, whether or not clocks really change their rhythm in motion. What is meant by the statement that a moving clock changes its rhythm? Let us assume, for the sake of simplicity, that we have only one clock in the upper CS and many in the lower. All the clocks have the same mechanism, and the lower ones are synchronized, that is, they show the same time simultaneously. We have drawn three subsequent positions of the two CS moving relative to each other. In the first drawing the positions of the hands of the upper and lower clocks are, by convention, the same because we arranged them so. All the clocks show the same time. In the second drawing, we see the relative positions of the two CS some time later. All the clocks in the lower CS show the same time, but the clock in the upper CS is out of rhythm. The rhythm is changed and the time differs because the clock is moving relative to the lower CS. In the third drawing we see the difference in the positions of the hands increased with time.

An observer at rest in the lower CS would find that a moving clock changes its rhythm. Certainly the same result could be found if the clock moved relative to an observer at rest in the upper CS; in this case there would have to be many clocks in the upper CS and only one in the lower. The laws of nature must be the same in both CS moving relative to each other.

In classical mechanics it was tacitly assumed that a moving clock does not change its rhythm. This seemed so obvious that it was hardly worth mentioning. But nothing should be too obvious; if we wish to be really careful, we should analyze the assumptions, so far taken for granted, in physics.

An assumption should not be regarded as unreasonable simply because

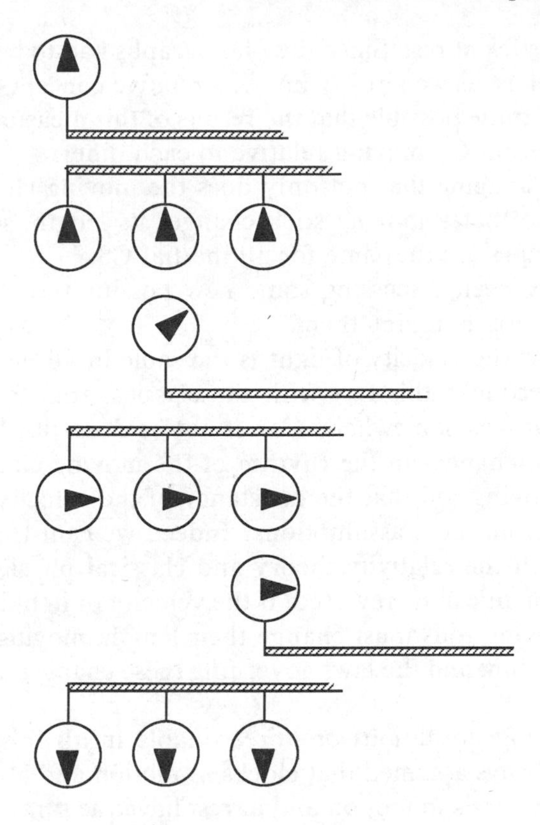

it differs from that of classical physics. We can well imagine that a moving clock changes its rhythm, so long as the law of this change is the same for all inertial CS.

Yet another example. Take a yardstick; this means that a stick is a yard in length as long as it is at rest in a CS. Now it moves uniformly, sliding along the rod representing the CS. Will its length still appear to be one yard? We must know beforehand how to determine its length. As long as the stick was at rest its ends coincided with markings one yard apart on the CS. From this we concluded: the length of the resting stick is one yard. How are we to measure this stick during motion? It could be done as follows. At a given moment two observers simultaneously take snapshots, one of the origin of the stick and the other of the end. Since the pictures are taken simultaneously we can compare the marks on the CS rod with which the origin and the end of the moving stick coincide. In this way we determine its length. There must be two observers to take note of simultaneous events in different parts of the given CS. There is no reason to believe that the result of such measurements will be the same as in the case of a stick at rest. Since the photographs had to be taken simultaneously, which is, as we already know, a relative concept depending on the CS, it seems quite possible that the results of this measurement will be different in different CS moving relative to each other.

We can well imagine that not only does the moving clock change its rhythm, but also that a moving stick changes its length, so long as the laws of the changes are the same for all inertial CS.

We have only been discussing some new possibilities without giving any justification for assuming them.

We remember: the velocity of light is the same in all inertial CS. It is impossible to reconcile this fact with the classical transformation. The circle must be broken somewhere. Can it not be done just here? Can we not assume such changes in the rhythm of the moving clock and in the length of the moving rod that the constancy of the velocity of light will follow directly from these assumptions? Indeed we can! Here is the first instance in which the relativity theory and classical physics differ radically. Our argument can be reversed: if the velocity of light is the same in all CS, then moving rods must change their length, moving clocks must change their rhythm, and the laws governing these changes are rigorously determined.

There is nothing mysterious or unreasonable in all this. In classical physics it was always assumed that clocks in motion and at rest have the same rhythm, that rods in motion and at rest have the same length. If the velocity of light is the same in all CS, if the relativity theory is valid, then we must sacrifice this assumption. It is difficult to get rid of deep-rooted prejudices, but there is no other way. From the point of view of the relativity theory the old concepts seem arbitrary. Why believe, as we did some pages ago, in absolute time flowing in the same way for all observers in all CS? Why believe in unchangeable distance? Time is determined by clocks, space co-ordinates by rods, and the result of their determination may depend on the behavior of these clocks and rods when in motion. There is no reason to believe that they will behave in the way we should like them to. Observation shows, indirectly, through the phenomena of electromagnetic fields, that a moving clock changes its rhythm, a rod its length, whereas on the basis of mechanical phenomena we did not think this happened. We must accept the concept of relative time in every CS, because it is the best way out of our difficulties. Further scientific advance, developing from the theory of relativity, shows that this new aspect should not be regarded as a *malum necessarium*, for the merits of the theory are much too marked.

L.5

Space and Time

HERMANN MINKOWSKI

A STATEMENT THAT deserves a place alongside Galileo's celebrated "*Eppur si muove!*" (which he actually said, but at a different time and place than is usually reported) comes in the first paragraph of this selection by the Russo-German mathematician Hermann Minkowski: "The views of space and time which I wish to lay before you have sprung from the soil of experimental physics, and therein lies their strength. . . . Henceforth space by itself, and time by itself, are doomed to fade away into mere shadows, and *only a kind of union of the two will preserve an independent reality*." Herein lies Minkowski's almost proverbial expression so dear to the hearts of science fiction authors: the space-time continuum.

Minkowski, born in Russia in 1864 of Jewish parents, was taken to Germany at the age of eight and was raised there. He received his degree in mathematics at the University of Königsberg in 1885. He later taught at the Swiss Federal Polytechnic, where Einstein was one of his pupils. Minkowski was one of the earliest physicists to appreciate Einstein's special theory, and his work on space and time was the early rigorous geometrical approach to relativity. He added time, designated by the variable t, as a fourth coordinate to the normal geometrical coordinates of length, width and height, usually indicated by x, y, and z, pointing out that our perception of reality is invariably in these four dimensions, with place and time considered in combination: "Nobody has ever noticed a place except at a time, or a time except at a place," he declared, in a typically epigrammatic fashion. He called the point in space indicated by the coordinates x, y, z, and t a *world-point*, adding that "the multiplicity of all thinkable x, y, z, t systems of values we shall christen the *world line*. . . . Let the variations dx, dy, dz of the space coordinates of this particle point correspond to a time element *dt*. Then we obtain, as an image, so to speak, of the eternal career of the particle point, a curve in the world, a *world-line*, the points of which can be referred unequivocally to the parameter *t* from minus infinity to plus infinity." By using his world-lines, Minkowski, with Einstein, can describe the entire Universe.

Aware of the tremendous debt he owed to his former teacher, who had helped to popularize the theory of relativity among physicists, Einstein described Minkowski's work in the following way:

> Minkowski's important contribution to the theory lies in the following: Before Minkowski's investigation, it was necessary to carry out a Lorentz transformation on a law in order to test its invariance under such transformations; he, on the other hand, succeeded in introducing a formalism such that the mathematical form of the law itself guarantees its invariance under Lorentz transformations.

Minkowski died in the winter of 1909 after having invited Max Born to collaborate with him at Göttingen; it fell to Born to go through Minkowski's papers and publish several of Minkowski's unfinished physics articles.

A good notation has a subtlety and suggestiveness which at times make it seem almost like a live teacher.

—BERTRAND RUSSELL

Nothing puzzles me more than time and space; and yet nothing troubles me less as I never think about them.

—CHARLES LAMB

Differing from Newton and Schopenhauer . . . Ts'ui Pen did not think of time as absolute or uniform. He believed in an infinite series of times, in a dizzily growing, ever spreading network of diverging, converging and parallel times. This web of time—the strands of which approach one another, bifurcate, intersect or ignore each other through the centuries—embraces every possibility.

—JORGE LUIS BORGES, "The Garden of the Forking Path"

THE VIEWS OF space and time which I wish to lay before you have sprung from the soil of experimental physics, and therein lies their strength. They are radical. Henceforth space by itself, and time by itself, are doomed to fade away into mere shadows, and only a kind of union of the two will preserve an independent reality.

I

First of all I should like to show how it might be possible, setting out from the accepted mechanics of the present day, along a purely mathematical line of thought, to arrive at changed ideas of space and time. The equations of Newton's mechanics exhibit a two-fold invariance. Their form remains unaltered, firstly, if we subject the underlying system of spatial co-ordinates to any arbitrary *change of position;* secondly, if we change its state of motion, namely, by imparting to it any *uniform translatory motion;* furthermore, the zero point of time is given no part to play. We are accustomed to look upon the axioms of geometry as finished with,

---

Hermann Minkowski, "Space and Time," in *The Principles of Relativity*. New York: Dover Publications, Inc. 1923. Reprinted with permission of Methuen & Co., London.

when we feel ripe for the axioms of mechanics, and for that reason the two invariances are probably rarely mentioned in the same breath. Each of them by itself signifies, for the differential equations of mechanics, a certain group of transformations. The existence of the first group is looked upon as a fundamental characteristic of space. The second group is preferably treated with disdain, so that we with untroubled minds may overcome the difficulty of never being able to decide, from physical phenomena, whether space, which is supposed to be stationary, may not be after all in a state of uniform translation. Thus the two groups, side by side, lead their lives entirely apart. Their utterly heterogeneous character may have discouraged any attempt to compound them. But it is precisely when they are compounded that the complete group, as a whole, gives us to think.

We will try to visualize the state of things by the graphic method. Let x, y, z be rectangular co-ordinates for space, and let t denote time. The objects of our perception invariably include places and times in combination. Nobody has ever noticed a place except at a time, or a time except at a place. But I still respect the dogma that both space and time have independent significance. A point of space at a point of time, that is, a system of values x, y, z, t, I will call a *world-point*. The multiplicity of all thinkable x, y, z, t systems of values we will christen the *world*. With this most valiant piece of chalk I might project upon the blackboard four world-axes. Since merely one chalky axis as it is, consists of molecules all a-thrill, and moreover is taking part in the earth's travels in the universe, it already affords us ample scope for abstraction; the somewhat greater abstraction associated with the number four is for the mathematician no infliction. Not to leave a yawning void anywhere, we will imagine that everywhere and everywhen there is something perceptible. To avoid saying "matter" or "electricity" I will use for this something the word "substance." We fix our attention on the substantial point which is at the world-point x, y, z, t, and imagine that we are able to recognize this substantial point at any other time. Let the variations dx, dy, dz of the space co-ordinates of this substantial point correspond to a time element dt. Then we obtain, as an image, so to speak, of the everlasting career of the substantial point, a curve in the world, a *world-line*, the points of which can be referred unequivocally to the parameter t from $-\infty$ to $+\infty$. The whole universe is seen to resolve itself into similar world-lines, and I would fain anticipate myself by saying that in my opinion physical laws might find their most perfect expression as reciprocal relations between these world-lines.

The concepts, space and time, cause the x, y, z-manifold $t = 0$ and its two sides $t > 0$ and $t < 0$ to fall asunder. If, for simplicity, we retain the same zero point of space and time, the first-mentioned group signifies in mechanics that we may subject the axes of x, y, z at $t = 0$ to any rotation we choose about the origin, corresponding to the homogeneous linear transformations of the expression

$$x^2 + y^2 + z^2.$$

But the second group means that we may—also without changing the expressions of the laws of mechanics—replace x, y, z, t by $x - \alpha t, y - \beta t, z - \gamma t, t$ with any constant values of α, β, γ. Hence we may give to the time axis whatever direction we choose towards the upper half of the world, $t > 0$. Now what has the requirement of orthogonality in space to do with this perfect freedom of the time axis in an upward direction?

To establish the connexion, let us take a positive parameter c, and consider the graphical representation of

$$c^2t^2 - x^2 - y^2 - z^2 = 1.$$

It consists of two surfaces separated by $t = 0$, on the analogy of a hyperboloid of two sheets. We consider the sheet in the region $t > 0$, and now take those homogeneous linear transformations of x, y, z, t into four new variables x', y', z', t', for which the expression for this sheet in the new variables is of the same form. It is evident that the rotations of space about the origin pertain to these transformations. Thus we gain full comprehension of the rest of the transformations simply by taking into consideration one among them, such that y and z remain unchanged. We draw (Fig. 1) the section of this sheet by the plane of the axes of x and t—the upper branch of the hyperbola $c^2t^2 - x^2 = 1$, with its asymptotes. From the origin O we draw any radius vector OA' of this branch of the hyperbola; draw the tangent to the hyperbola at A' to cut the asymptote on the right at B'; complete the parallelogram $OA'B'C'$; and finally, for subsequent use, produce $B'C'$ to cut the axis of x at D'. Now if we take OC' and OA' as axes of oblique co-ordinates x', t', with the measures $OC' = 1$, $OA' = 1/c$, then that branch of the hyperbola again acquires the expression $c^2t'^2 - x'^2 = 1$, $t' > 0$, and the transition from x, y, z, t to x', y', z', t' is one of the transformations in question. With these transformations we now associate the arbitrary displacements of the zero point of space and time, and thereby constitute a group of transformations, which is also, evidently, dependent on the parameter c. This group I denote by G_c.

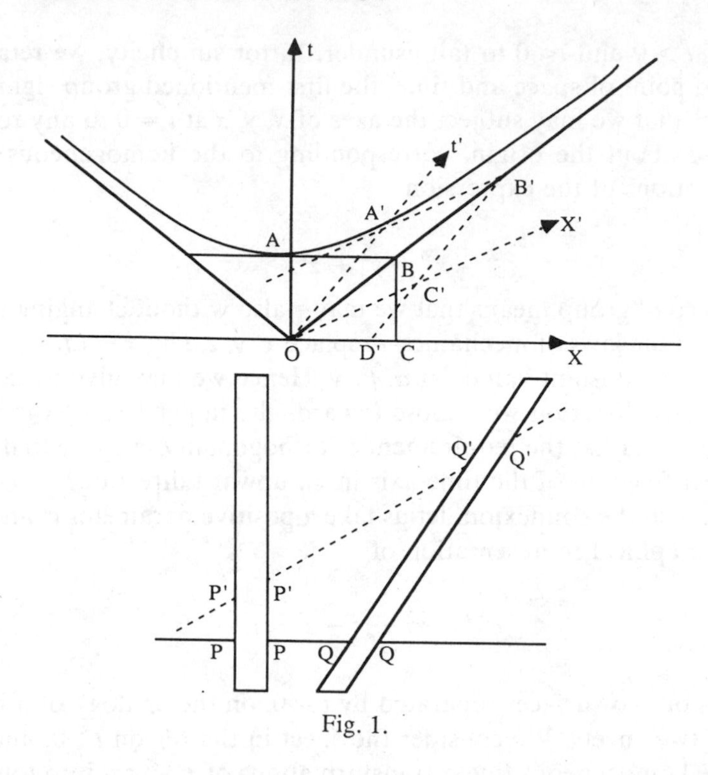

Fig. 1.

If we now allow c to increase to infinity, and $1/c$ therefore to converge towards zero, we see from the figure that the branch of the hyperbola bends more and more towards the axis of x, the angle of the asymptotes becomes more and more obtuse, and that in the limit this special transformation changes into one in which the axis of t' may have any upward direction whatever, while x' approaches more and more exactly to x. In view of this it is clear that group G_c in the limit when $c = \infty$, that is the group $G\infty$, becomes no other than that complete group which is appropriate to Newtonian mechanics. This being so, and since G_c is mathematically more intelligible than $G\infty$, it looks as though the thought might have struck some mathematician, fancy-free, that after all, as a matter of fact, natural phenomena do not possess an invariance with the group $G\infty$, but rather with a group G_c, being finite and determinate, but in ordinary units of measure, *extremely great*. Such a premonition would have been an extraordinary triumph for pure mathematics. Well, mathematics, though it now can display only staircase-wit, has the satisfaction of being wise after the event, and is able, thanks to its happy antecedents, with its senses sharpened by an unhampered outlook to far horizons, to grasp forthwith the far-reaching consequences of such a metamorphosis of our concept of nature.

I will state at once what is the value of *c,* with which we shall finally be dealing. It is the velocity of the propagation of light in empty space. To avoid speaking either of space or of emptiness, we may define this magnitude in another way, as the ratio of the electromagnetic to the electrostatic unit of electricity.

The existence of the invariance of natural laws for the relevant group G_c, would have to be taken, then, in this way:

From the totality of natural phenomena it is possible, by successively enhanced approximations to derive more and more exactly a system of reference $x, y, z, t,$ space and time, by means of which these phenomena then present themselves in agreement with definite laws. But when this is done, this system of reference is by no means unequivocally determined by the phenomena. It is still possible to make any change in the system of reference that is in conformity with the transformations of the group G_c and leave the expression of the laws of nature unaltered.

For example, in correspondence with the figure described above, we may also designate time t', but then must of necessity, in connexion therewith, define space by the manifold of the three parameters $x', y, z,$ in which case physical laws would be expressed in exactly the same way by means of x', y, z, t' as by means of x, y, z, t. We should then have in the world no longer *space,* but an infinite number of spaces, analogously as there are in three-dimensional space an infinite number of planes. Three-dimensional geometry becomes a chapter in four-dimensional physics. Now you know why I said at the outset that space and time are to fade away into shadows, and only a world in itself will subsist.

II

The question now is, what are the circumstances which force this changed conception of space and time upon us? Does it actually never contradict experience? And finally, is it advantageous for describing phenomena?

Before going into these questions, I must make an important remark. If we have in any way individualized space and time, we have, as a world-line corresponding to a stationary substantial point, a straight line parallel to the axis of *t;* corresponding to a substantial point in uniform motion, a straight line at an angle to the axis of *t;* to a substantial point in varying motion, a world-line in some form of a curve. If at any world-point x, y, z, t we take the world-line passing through that point, and find it parallel to any radius vector OA' of the above-mentioned hyperboloidal sheet, we can introduce OA' as a new axis of time, and with the new concepts of space and time thus given, the substance at the world-point concerned appears as at rest. We will now introduce this fundamental axiom:

The substance at any world-point may always, with the appropriate determination of space and time, be looked upon as at rest.

The axiom signifies that at any world-point the expression

$$c^2dt^2 - dx^2 - dy^2 - dz^2$$

always has a positive value, or, what comes to the same thing, that any velocity v always proves less than c. Accordingly c would stand as the upper limit for all substantial velocities, and that is precisely what would reveal the deeper significance of the magnitude c. In this second form the first impression made by the axiom is not altogether pleasing. But we must bear in mind that a modified form of mechanics, in which the square root of this quadratic differential expression appears, will now make its way, so that cases with a velocity greater than that of light will henceforward play only some such part as that of figures with imaginary co-ordinates in geometry.

Now the impulse and true motive for assuming the group G_c came from the fact that the differential equation for the propagation of light in empty space possesses that group G_c. On the other hand, the concept of rigid bodies has meaning only in mechanics satisfying the group $G\infty$. If we have a theory of optics with G_c, and if on the other hand there were rigid bodies, it is easy to see that one and the same direction of t would be distinguished by the two hyperbolodial sheets appropriate to G_c and $G\infty$, and this would have the further consequence, that we should be able, by employing suitable rigid optical instruments in the laboratory, to perceive some alteration in the phenomena when the orientation with respect to the direction of the earth's motion is changed. But all efforts directed towards this goal, in particular the famous interference experiment of Michelson, have had a negative result. To explain this failure, H. A. Lorentz set up an hypothesis, the success of which lies in this very invariance in optics for the group G_c. According to Lorentz any moving body must have undergone a contraction in the direction of its motion, and in fact with a velocity v, a contraction in the ratio

$$1 : \sqrt{1 - v^2/c^2}$$

This hypothesis sounds extremely fantastical, for the contraction is not to be looked upon as a consequence of resistances in the ether, or anything of that kind, but simply as a gift from above—as an accompanying circumstance of the circumstance of motion.

I will now show by our figure that the Lorentzian hypothesis is completely equivalent to the new conception of space and time, which, indeed, makes the hypothesis much more intelligible. If for simplicity we disregard y and z, and imagine a world of one spatial dimension, then a parallel band, upright like the axis of t, and another inclining to the axis of t (see Fig. 1) represent, respectively, the career of a body at rest or in uniform motion, preserving in each case a constant spatial extent. If OA' is parallel to the second band, we can introduce t' as the time, and x' as the space co-ordinate, and then the second body appears at rest, the first in uniform motion. We now assume that the first body, envisaged as at rest, has the length l, that is, the cross section PP of the first band on the axis of x is equal to $l.\ OC$, where OC denotes the unit of measure on the axis of x; and on the other hand, that the second body, envisaged as at rest, has the same length l, which then means that the cross section $Q'Q'$ of the second band, measured parallel to the axis of x', is equal to $l.OC'$. We now have in these two bodies images of two equal Lorentzian electrons, one at rest and one in uniform motion. But if we retain the original co-ordinates x, t, we must give as the extent of the second electron the cross section of its appropriate band parallel to the axis of x. Now since $Q'Q' = l.OC'$, it is evident that $QQ = l.OD'$. If dx/dt for the second band is equal to v, an easy calculation gives

$$OD' = OC\ \sqrt{1 - v^2/c^2}\,,$$

therefore also $PP{:}QQ = 1 : \sqrt{1 - v^2/c^2}$. But this is the meaning of Lorentz's hypothesis of the contraction of electrons in motion. If on the other hand we envisage the second electron as at rest, and therefore adopt the system of reference $x'\ t'$, the length of the first must be denoted by the cross section $P'P'$ of its band parallel to OC', and we should find the first electron in comparison with the second to be contracted in exactly the same proportion; for in the figure

$$P'P' : Q'Q' = OD : OC' = OD' : OC = QQ : PP.$$

Lorentz called the t' combination of x and t the local time of the electron in uniform motion, and applied a physical construction of this concept, for the better understanding of the hypothesis of contraction. But the credit of first recognizing clearly that the time of the one electron is just as good as that of the other, that is to say, that t and t' are to be treated identically, belongs to A. Einstein. Thus time, as a concept unequivocally determined by phenomena, was first deposed from its high seat. Neither Einstein nor Lorentz made any attack on the concept of space, perhaps because in the above-mentioned special transformation, where the plane of x', t' coincides with the plane of x, t, an interpretation is possible by saying that the x-axis of space maintains its position. One may expect to find a corresponding violation of the concept of space appraised as another act of audacity on the part of the higher mathematics. Nevertheless, this further step is indispensable for the true understanding of the group G_c, and when it has been taken, the word *relativity-postulate* for the requirement of an invariance with the group G_c seems to me very feeble. Since the postulate comes to mean that only the four-dimensional world in space and time is given by phenomena, but that the projection in space and in time may still be undertaken with a certain degree of freedom, I prefer to call it the *postulate of the absolute world* (or briefly, the world-postulate).

III

The world-postulate permits identical treatment of the four coordinates x, y, z, t. By this means, as I shall now show, the forms in which the laws of physics are displayed gain in intelligibility. In particular the idea of acceleration acquires a clear-cut character.

I will use a geometrical manner of expression, which suggests itself at once if we tacitly disregard z in the triplex x, y, z. I take any world-point O as the zero-point of space-time. The cone $c^2t^2 - x^2 - z^2 = 0$ with apex 0 (Fig. 2) consists of two parts, one with values $t < 0$, the other with values $t > 0$. The former, the front cone of O, consists, let us say, of all the world-points which "send light to O," the latter, the back cone of O, of all the world-points which "receive light from O." The territory bounded by the front cone alone, we may call "before" O, that which is bounded by the back cone alone, "after" O. The hyperboloidal sheet already discussed

$$F = c^2t^2 - x^2 - y^2 - z^2 = 1, t > 0$$

lies after O. The territory between the cones is filled by the one-sheeted hyperboloidal figures

$$-F = x^2 + y^2 + z^2 - c^2t^2 = k^2$$

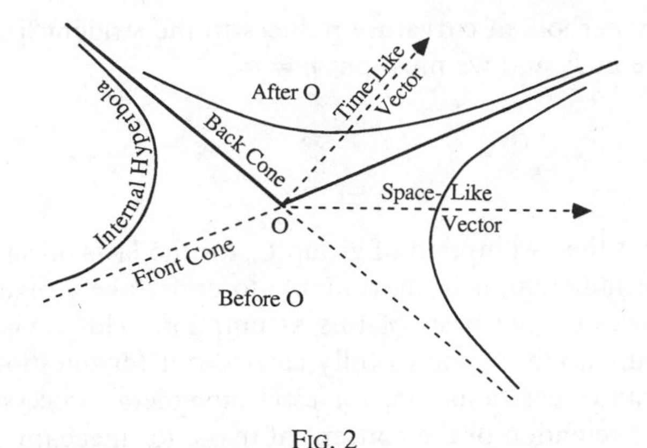

FIG. 2

for all constant positive values of k. We are specially interested in the hyperbolas with O as centre, lying on the latter figures. The single branches of these hyperbolas may be called briefly the internal hyperbolas with centre O. One of these branches, regarded as a world-line, would represent a motion which, for $t = -\infty$ and $t = +\infty$, rises asymptotically to the velocity of light, c.

If we now, on the analogy of vectors in space, call a directed length in the manifold of x, y, z, t a vector, we have to distinguish between the time-like vectors with directions from O to the sheet $+F=1, t > 0$, and the space-like vectors with directions from O to $-F = 1$. The time axis may run parallel to any vector of the former kind. Any world-point between the front and back cones of O can be arranged by means of the system of reference so as to be simultaneous with O, but also just as well so as to be earlier than O or later than O. Any world-point within the front cone of O is necessarily always before O; any world-point within the back cone of O necessarily always after O. Corresponding to passing to the limit, $c = \infty$, there would be a complete flattening out of the wedge-shaped segment between the cones into the plane manifold $t = 0$. In the figures this segment is intentionally drawn with different widths. . . .

Now, as is readily seen, there is a definite hyperbola which has three infinitely proximate points in common with the world-line at P, and whose asymptotes are generators of a "front cone" and a "back cone." Let this hyperbola be called the hyperbola of curvature at P. If M is the centre of this hyperbola, we here have to do with an internal hyperbola with centre M. Let ρ be the magnitude of the vector MP; then we recognize the acceleration vector at P as the vector in the direction MP of magnitude c^2/ρ.

If . . . the hyperbola of curvature reduces to the straight line touching the world-line in P, and we must put $\rho = \infty$.

IV

To show that the assumption of group G_c for the laws of physics never leads to a contradiction, it is unavoidable to undertake a revision of the whole of physics on the basis of this assumption. This revision has to some extent already been successfully carried out for questions of thermodynamics and heat radiation, for electromagnetic processes, and finally, with the retention of the concept of mass, for mechanics. . . .

We can determine the ratio of the units of length and time beforehand in such a way that the natural limit of velocity becomes $c = 1$. If we then introduce, further, $\sqrt{-1}\,t = s$ in place of t, the quadratic differential expression

$$d\tau^2 = -dx^2 - dy^2 - dz^2 - ds^2$$

thus becomes perfectly symmetrical in x, y, z, s; and this symmetry is communicated to any law which does not contradict the world-postulate. Thus the essence of this postulate may be clothed mathematically in a very pregnant manner in the mystic formula

$$3.10^5\ km = \sqrt{-1}\ \text{secs.}$$

V

The advantages afforded by the world-postulate will perhaps be most strikingly exemplified by indicating the effects proceeding from a point charge in any kind of motion according to the Maxwell-Lorentz theory. Let us imagine the world-line of such a point electron with the charge e, and introduce upon it the proper time τ from any initial point. In order to find the field caused by the electron at any world-point P_1, we construct the front cone belonging to P_1. The cone evidently meets the world-line, since the directions of the line are everywhere those of time-like vectors, at the single point P. We draw the tangent to the world-line at P, and construct through P_1 the normal P_1Q to this tangent. Let the length of P_1Q be r. Then, by the definition of a front cone, the length of PQ must be r/c. Now the vector in the direction PQ of magnitude e/r represents by its components along the axes of x, y, z, the vector potential multiplied by c, and by the component along the axis of t, the scalar potential of the field excited by e at the world-point P. Herein lie the elementary laws formulated by A. Liénard and E. Wiechert.

Then in the description of the field produced by the electron we see that the separation of the field into electric and magnetic force is a relative one with regard to the underlying time axis; the most perspicuous way of describing the two forces together is on a certain analogy with the wrench in mechanics, though the analogy is not complete. . . .

The validity without exception of the world-postulate, I like to think, is the true nucleus of an electromagnetic image of the world, which, discovered by Lorentz, and further revealed by Einstein, now lies open in the full light of day. In the development of its mathematical consequences there will be ample suggestions for experimental verifications of the postulate, which will suffice to conciliate even those to whom the abandonment of old-established views is unsympathetic or painful, by the idea of a pre-established harmony between pure mathematics and physics. . . .

L.6

Relativity

RUDOLF PEIERLS

SIR RUDOLF PEIERLS was born in Berlin in 1907. His extensive education took him to the universities of Berlin, Munich, and Leipzig in Germany and Einstein's old school, the Federal Institute of Technology in Zurich. He taught briefly in Italy, Denmark, and England, where he finally settled after Hitler came to power. He was Professor of Mathematical Physics at Birmingham University from 1937 to 1963, then moved on to Oxford, where he was Professor of Physics until his retirement in 1974. His chief research was in solid-state physics, where he made substantial contributions in the fields of conductivity, lattice theory, and magnetism. He also did significant research in quantum theory and nuclear reactions. In 1940, together with Otto Frisch, he wrote a brief paper, "On the construction of a 'super-bomb' based on a nuclear chain reaction in uranium." Peierls was among the first to suggest that a chain reaction in a relatively small quantity of uranium-235 would release the enormous energies of an "atom bomb." This seminal paper resulted in the organization of a British committee on atomic research chaired by George Thomson.

After the war, Peierls published a book entitled *The Laws of Nature,* which included the chapter on relativity that follows. Peierls's contribution to the vast literature aimed at providing the nonscientist with some understanding of relativity is impressive. In simple, clear language, peppered with some mathematics, he takes the tyro by the hand and gently introduces him to the doctrine that rocked the scientific world in the early decades of the twentieth century. In his introduction, he proposes a triple test which must be met by any new principle in science, and explains why the concept of relativity seems strange and difficult: it is the physics of both the infinitely vast and the infinitely small—of the cosmos and the subatomic world of particles. Few other explanations of relativity are so accessible to the general reader.

The real revolution that came with Einstein's theory . . . was the abandonment of the idea that the space-time coordinate system has objective significance as a separate physical entity. Instead of this idea, relativity theory implies that the space and time coordinates are only the elements of a language that is used by an observer to describe his environment.

—MENDEL SACHS, "Space-Time and Elementary Interactions in Relativity," *Physics Today.*

THE IDEAS OF relativity at first met with rather strong opposition both amongst physicists and amongst philosophers. The physicists met the new hypothesis in the critical spirit in which it is their business to regard any new idea, until it has passed the triple test which physics requires. It must firstly leave undisturbed the successes of earlier work and not upset the explanations of observations that had been used in support of earlier ideas. Secondly it must explain in a reasonable manner the new evidence which brought the previous ideas into doubt and which suggested the new hypothesis. And thirdly it must predict new phenomena or new relationships between different phenomena, which were not known or not clearly understood at the time when it was invented. This process took some time because relativity is important only for objects moving with a speed comparable to that of light. As such objects were not readily available the opportunities for tests were few, and in many cases the tests required very difficult observations of high precision. Since then particles moving with high velocity have become commonplace in any physics laboratory. We are no longer concerned with small corrections in the behaviour of these particles which require measurements of high precision, but the relativistic features of their motion have large effects which it is quite impossible to overlook. To put the point in its crudest form: in developing machines for physical research in which particles are accelerated to very high speeds, engineers have to incorporate devices, costing many thousands of pounds, which are required only because of the relativistic features of the particle motion. They are scarcely inclined to regard such features as a result of idle or mistaken speculation. To-day no physicist

---

R. E. Peierls, "Relativity," in *The Laws of Nature.* New York: Charles Scribner's Sons, 1956, chapter 6. Reprinted with permission of Charles Scribner's Sons.

who has practical knowledge of work with fast particles would question the principles of relativity.

The opposition from philosophers arose because the theory of relativity called in question statements that had been regarded as the concern of philosophers. It was doubted whether the physicist had the right to query ideas which the philosophers had regarded as evident and unquestionable truth.

By now it appears to be widely recognized that our ideas about space and time are derived ultimately from our experience of the outside world and that many statements which we regard as evident are true only within the limitations of our practical experience. They may turn out to be unjustified prejudices when extended to situations not normally familiar to us.

In the history of physics such situations are common. The discovery that the earth was revolving about its own axis and about the sun ran counter to mechanical intuition built on simple experience. When we learn at school about the opposition first encountered by this idea we tend to regard the scholars of the Middle Ages, who failed to accept it, as incredibly narrow-minded in their outlook, but the only reason why we ourselves find the thought easier is that we have become familiar with it through our early education and have accepted it before our critical power was developed far enough to question it seriously.

The idea of Galileo's law that a moving body tends to continue in its motion runs counter to everyday experience and was at first quite legitimately questioned by the scientists of the time. It was accepted only with difficulty by thinkers who generalized abstract ideas about motion.

The fact that light travels with finite speed is another fact which runs counter to our intuition. It is hard even to get used to the fact that the speed of sound is finite, and even the simple experience of watching a man chop wood at a distance, when the sound of the blow reaches us later than the sight of the falling axe, requires a conscious adjustment of our senses to an unfamiliar situation. We have learned even more strongly to rely on our eyes as telling us what goes on in the outer world and to regard the information they give us as true information about what goes on at the instant we see it. There is nothing in everyday experience on the surface of the earth which will disprove this illusion, but three hundred years of experience in astronomy and more recent methods for measuring times to enormous precision have got us to accept the finite speed of light. . . .

The acceptance of relativity was probably delayed by its name, which suggested a superficial connection with the philosophical concept of relativity, according to which all truth was regarded as relative. As we shall see, nothing is further from the contents of the new development. In relativity, the laws of physics have a precise and absolute form, only certain specific statements that our intuition leads us to regard as absolute, turn out to be prejudiced.

MOTION AND REST

Long before the developments which led to Einstein's formulation of the theory of relativity it was known that many of the laws of nature, in particular the laws of mechanics, did not alter their appearance if the observer, instead of standing still, was moving with uniform velocity and direction.

It is well known that the passenger in a railway train is not aware of the motion of the train unless the vibration of the coach, due to the roughness of the track, or the acceleration caused by starting or slowing down, or by the curvature of the track, show him that the train is moving. Frequently, when we look out of the train window at another train in the station we are confused as to whether it is our train or the adjacent one which has started moving. The question is settled when we catch a glimpse of the station buildings or other objects, which, we know from experience, usually stand still. If a fellow passenger asserted that in fact the station, with the track and all the landscape, was moving and the train was standing still, and if he was not impressed by the argument that stations and landscapes don't usually behave like that, we would not find it easy to prove him wrong. Any mechanical experiment which we carry out inside the compartment would proceed precisely in the same way as if the train was standing still (apart from the vibrations and curves to which reference has already been made). And this fact is usually expressed by saying that the laws of mechanics are the same for two observers who move relatively to each other with a uniform velocity. . . .

While the laws of mechanics therefore do not allow us to distinguish uniform motion from rest, this would appear to be different when our description of nature includes electricity and light. In discussing the laws of the electromagnetic field, . . . we found that electromagnetic waves, which include light waves, always travel with a fixed velocity, which we called c. We should therefore expect to find some difference in the apparent speed of propagation of light, owing to the fact that we are moving ourselves. For instance, if the only motion was due to the earth's rotation, light from a lamp placed to the east of us should reach our eye more quickly because our eye is in fact moving towards the lamp and therefore coming to meet the light waves. Similarly light from a lamp placed in a westerly direction should appear to travel more slowly because it has to overtake our eye, which is moving away. If u is the velocity with which the surface of the earth is moving, the apparent velocity from a light source in the east should in fact be $c + u$ and that from a light source in the west $c - u$. The difference is of course very small, because u is about a million times smaller than c, but careful experiment ought to reveal the difference.

This expected result was sometimes called the ether wind, because one thought at that time that all space was filled with a hypothetical substance called ether which acted as a carrier for electromagnetic waves. The effect of our motion through the ether on the propagation of light was similar to the effect of a strong wind on the propagation of sound, when it is well known that sound will appear to travel more rapidly down-wind than up-wind.

In their famous experiment Michelson and Morley set out to detect this "ether wind." To avoid the difficulty of measuring distances and times to the required high precision they used in fact a light beam which was split in two. One part was travelling, say, in an east-west direction and back after reflection from a mirror, and another north-south to another mirror and back. On being combined again these two light rays would form interference fringes. . . . The position of these fringes depends on how many oscillations each light wave has undergone during its travel. It is not difficult to work out the times taken for the return journey by either light beam. This calculation, which we shall omit here, shows that the effect of the earth's motion is the same as if the east-west arm of the apparatus was lengthened in the ratio of 1 to $\sqrt{1 - \frac{u^2}{c^2}}$. It is difficult to measure the length of the light paths to the required accuracy, but Michelson's reasoning was that on turning his apparatus through a right angle, so that now the other arm pointed in the east-west direction, this was bound to lengthen the time of travel of the one light ray and shorten the other, with a resultant shift of the interference fringes. Since the earth not merely rotates, but also revolves in its orbit about the sun, the effective speed with which we are moving should also alter its direction between day and night. Therefore even if the apparatus was left still, the fringes should alter their position in the course of a day.

The result of this experiment was completely negative. Whichever way the apparatus was turned and however long one looked at it no shift in the fringes was observed. All other attempts to observe the "ether wind" failed similarly. And hence the one kind of experiment which appeared to make it possible to distinguish motion from rest had failed.

THE LORENTZ CONTRACTION

It took a long time to understand completely the implications of this result. Various explanations were tried. It was for example suggested that the velocity of light might be dependent on the motion of the light source. Since the Michelson experiment employed a lamp which was fixed on the earth's surface one would then expect the light to travel with a given speed relatively to this source, i.e. relatively to the surface of the earth. But this explanation had to be given up on reflection, since we know that the light from the sun, for example, really comes from atoms in the atmosphere of the sun, which . . . are moving about irregularly and with great speed. The light from different atoms should therefore, if this explanation were right, have different speeds. If we looked not at the sun but at a distant star, which is similar to the sun, but further away, we should see this star not as a point, but as a streak, since light from different atoms on the star should have come to us with different speeds and should therefore have started when the star was in quite different positions in relation to our telescope.

It was also suggested that in the neighbourhood of the earth the ether should take some part in the motion of the earth, so that there was an "atmosphere" of ether around us, which moved in the same way as the earth itself. This hypothesis, too, was untenable, because if it were true there had to be somewhere outside the earth a transition region where the ether's speed changed from that of the earth to that of free space. This would lead to a refraction of light rays and an apparent displacement in the position of stars, which is not observed.

More such alternatives were considered and found equally unsuccessful. Nearer to the truth came a suggestion by H. A. Lorentz, which at first sight seems very far fetched. Lorentz suggested that every moving object contracts in the direction of motion in the ratio of $\sqrt{1 - \frac{u^2}{c^2}}$ to 1, if u is its speed. If the arm of Michelson's apparatus which points in the direction of motion had in fact contracted by that amount this would just make up for the effect of the "ether wind." When one first hears of this "Lorentz contraction" it sounds most artificial and unreasonable. It sounds, in fact, as if nature had conspired to adjust its laws in such a way as to make it hard for us to observe the ether wind. There seemed no reason in mechanics why bodies should undergo this peculiar contraction.

However, there was a much better basis for the suggestion than appeared at first sight, because Lorentz knew already that all matter consists of atoms . . . and that atoms contain positively and negatively charged particles, the forces holding them together being electric forces. Now in the laws of electricity and magnetism the velocity of light occurs as a constant. If charged particles move with a speed not very small compared with that of light, then their electric fields become more complicated. A point charge, for example, such as that of an atomic nucleus, has an electric field in which the potential depends only on the distance from the charge, so that all points on a sphere centered on the charge are at the same potential. If the charge is moving it produces not merely an electric, but also a magnetic field, because the transport of electric charges means the same thing as an electric current. In addition the electric and magnetic fields at any point in space near which the charge passes vary with time. We saw [earlier] that complications arise in the laws of the electric and magnetic field if the fields vary in time. This more complicated mathematical problem can be solved, and the result is that the surfaces of constant potential are no longer spheres, but are flattened in the direction of motion, by just the same amount that would be required by the Lorentz contraction.

This change in the electric forces no doubt would cause a change in the size and shape of an atom and in the distances of different atoms from each other. Not enough was known at the time about atomic structure to be sure what the overall effect would be. At least the possibility emerged that this Lorentz contraction could just come out as a consequence of the effect of the motion on the electric forces which hold atoms together.

This idea of the Lorentz contraction would, if confirmed, account for the negative result of the Michelson experiment, but at first sight it would open up a new possibility of finding out whether a body was at rest or in motion. Consider again for simplicity only the motion due to the rotation of the earth, and take a very accurately measured cube, which has one of its edges in the north-south direction. If we turn it through a right angle, the edge which was in a north-south direction before, and now is in an east-west direction, will have undergone the Lorentz contraction, whereas the other horizontal edge which previously was east-west and therefore was contracted, will have expanded. In the new position measurement should show us therefore that the body was no longer a cube. Now first of all, the amount of the Lorentz contraction is so small that it would be extremely hard to detect the change, but we can imagine that we had measuring instruments of sufficient accuracy to do so.

We should, however, then find a much more fundamental difficulty. The only way to measure a length is by comparing it directly or indirectly with a measuring rod or tape. This measuring rod itself, like any other rigid body, would be subject to the Lorentz contraction and therefore, as we turn it to place it first against the one edge and then against the other edge it would follow the Lorentz contraction. The two edges would still appear to be of the same length in terms of the marks on the measuring rod.

If we therefore apply consistently the postulate of Lorentz to all solid bodies without exception then it would be impossible in principle to detect the contraction by any direct measurement.

HOW TO COMPARE LENGTHS AND TIMES

But one might think that the Lorentz contraction would still be a suitable way to find out which of the two passing trains with which we started this discussion, is standing still, and which is moving. Suppose that the coaches of both trains are of identical construction, then that of the moving train should have contracted and therefore be shorter than that of the stationary train. To make matters easier imagine for the moment that the speed of the moving train was not one-millionth, but one-third of the velocity of light; then the Lorentz contraction would shorten it by 5 per cent, which should be easy to see. Thus all we would have to do is to watch carefully as two coaches pass each other and to see which is the longer.

However, this also is much more involved than it seems, if we remember that we are dealing with very high velocities and very short times. To make this point clear consider Figure 1. *A* is meant to indicate the stationary coach and *B* the moving one.

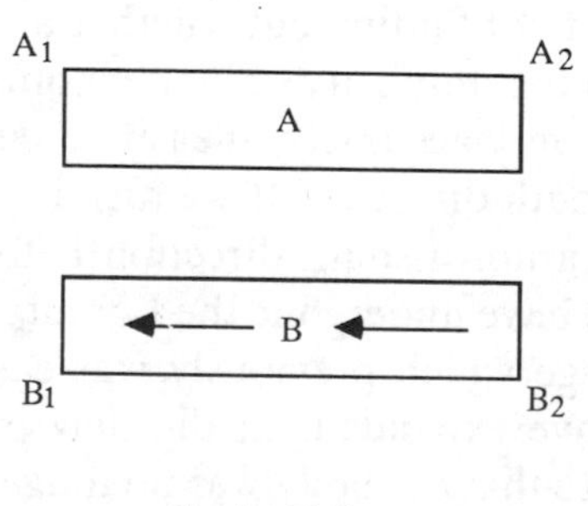

Passing Trains

Fig. 1.

If B is really shorter, then at the instant when its front end B_1 is in line with the end A_1 of the stationary coach, its rear end B_2 will already have passed the end A_2 of the stationary coach. But to be sure of this we must be very sure of our time. If we look at B_2 a very little too early it will not yet have passed A_2 and we would then think the coach too long. In other words, comparing the lengths of the two coaches boils down to being able to tell with great accuracy the time of passage of the ends past each other. The moving coach is longer, or shorter, than the other if its end B_2 passes A_2 after, or before, the end B_1 passes A_1.

We may, of course, look at the two ends simultaneously, but we then get involved in complications due to the finite speed of propagation of light. Supposing we stand at A_1. At the time when we see B_1 passing, we shall also see the end B_2, not as it is at that instant, but as it was a very short time earlier, namely earlier by the amount of time which the light has taken to travel from B_2 to A_1. It is true that we can correct for this, since we know the speed of propagation of light, but in working out this correction we must firstly know the length of our own coach, i.e. we must assume that our own coach is standing still, and thus no Lorentz contraction has to be applied, and also that the light travels with its normal speed and no ether wind has to be allowed for.

We might try to get some time signal across from A_2 to A_1 in some other way. We could imagine that an assistant stationed at A_2 presses a button when B_2 passes and that this gives some kind of signal at A_1, but this transmission must either be by mechanical means or by an electric current or some other device, and all these means of communication are in fact slower than the speed of light. To work out their speed of propagation with the required accuracy gets us into the same difficulty.

Let us then limit ourselves to light signals and consider how the whole process would appear from the point of view of an observer on the other train who is trying to apply the same test. He would also watch the passage of A_2 past B_2 and A_1 past B_1 and he would also transmit a light signal from one end of the coach to the other and apply a correction for the time this has taken. He would naturally assume that his train was at rest and would apply no correction for ether wind or for the Lorentz contraction. The interesting result of working this out in detail is that on the same basis he would come to the conclusion that our coach has been shortened by the Lorentz contraction, i.e. that A_1 and B_1 are in line *before* A_2 is in line with B_2.

The important point to realize is that the observations of both observers are completely interchangeable. We are coming to the conclusion that the coach B is moving and therefore shortened by the Lorentz contraction, whereas the other observer, starting from the assumption that his coach is at rest, would conclude that our coach *A* is moving and therefore shortened. No doubt we will criticize his measurement because he has made assumptions about the length of his coach and about the absence of an ether drift which to us appear wrong, but he will answer by pointing out that from his point of view the same criticism applies to our method.

The essence of the argument, which is a simplified version of that used by Einstein, is that the length of an object has no absolute meaning, but depends on the state of motion of the observer by whom it is determined. Similarly the question of time, i.e. which of two events occurring at two distant points is the earlier, has no absolute meaning, but depends on the point of view of the observer.

These ideas certainly are in conflict with our intuition, which regards the length of an object or the time at which an event takes place as absolute concepts, not necessarily related to any actual process of measurement. One has to get accustomed to the realization that the certainty with which we take these things for granted comes only from the experience of everyday life. In practice any doubt about the meaning of the exact length of a moving object, such as a railway coach, is completely negligible within the limits of error of any practical determination. In practice also we never get into trouble by assuming that any event which we can see takes place when we see it, since for practical purposes the speed of light may be regarded as infinitely large.

It is one of the merits of Einstein's theory to have brought out clearly the fact that our ideas of space and time are abstractions from experience, and that in applying them to situations involving greater speeds or shorter times than those to which we are normally accustomed we cannot take their meaning for granted without specifying them precisely. This is done only by formulating questions which can, in principle, be answered by an actual observation.

So far we find, then, that the postulate of the Lorentz contraction is consistent with the principle that the laws of physics are precisely the same for two observers moving uniformly with respect to each other and that there is no observation by which the point of view of one observer can be preferred over that of another. . . .

COMPOSITION OF VELOCITIES

With this new relationship between the records of time and distance kept by different observers there must also go a new rule for the composition of velocities. If a man on a train moving with velocity u fires a gun in the forward direction, which gives the bullet a speed v with respect to the train, then the total velocity of the bullet as seen from the track would be $u + v$. But in relativity we must translate the path of the bullet, which to the observer on the train is just motion with velocity v, by correcting for the difference in viewpoint about times and distances and if this is carried out we find the result $\frac{u + v}{1 + \frac{uv}{c^2}}$. If u and v are both small compared to c, the velocity of light, then the second part of the denominator is very small compared to 1, and the result is again that the velocities just add. But if u and v are not so small then the answer is different. For example if u and v are each one-half of the velocity of light the combined effect of the two velocities is $\frac{4}{5}c$. One sees easily that if u and v are both less than the velocity of light, however close they may be to it, the resultant velocity will still be less than c.

This rule is correct only when the two velocities are in the same direction. For the more general case the rule is more complicated and I shall not give it here.

One particular consequence of this law of composition of velocities is evident. If $v = c$, i.e. if instead of firing a gun the man on the train sends out a light signal, then the combined velocity is again equal to c, and this brings us back to the starting point, namely that light appears to be travelling to both observers with the same velocity.

This principle of the constant velocity of light is the most basic law of relativity. To avoid misunderstanding it should be stressed that in all this when we have spoken of the velocity of light we ought really to have said "the velocity of light in empty space." In any material medium, even in air, and to a greater extent in such dense substances as water or glass, we know from the refraction of light . . . that the velocity of light is different. This is caused by the effect of the light wave on the electric charges contained in the atoms. A light ray passing through glass will therefore not appear to different observers as propagating with the same speed. This is not surprising since the situations seen by the two observers are then different. For the one, glass or water is at rest, whereas for the other one it is moving. It was in fact in a study of the propagation of electromagnetic waves through moving bodies that Einstein was first led to a clear understanding of the arguments I have sketched above.

MECHANICS OF FAST-MOVING OBJECTS

This new application of the properties of distances and times also requires us to change the laws of mechanics as applied to fast-moving bodies; to see what modifications are required one can reason in the following way: look at the second law of Newton in the form that the force equals the rate of change of momentum. Before we can generalize this law we must be clear that both the ideas of momentum and force may need revision; the form which the mechanical law will take depends to some extent on which of the two, if either, we want to keep unchanged.

It turns out that the clearest statement of the laws is obtained if we retain the law of action and reaction, which says that the forces which two objects exert on each other must be opposite and equal. In consequence, the interaction between two objects does not change the total momentum. We also want to retain the law of conservation of energy, and in particular of mechanical energy, if we are dealing with an elastic collision. Consider then the collision between two elastic identical objects, for example two billiard balls. If these collide with equal and opposite speeds their combined momentum must be zero, and if momentum is to be conserved it must be zero even after the collision. This means after the collision the two billiard balls must still travel with opposite and equal velocity. . . .

Now suppose that they have collided in such a way that they separate again in a direction at right angles to the original. . . . The speeds must again be equal. Moreover, energy conservation requires the speed with which the balls separate to be equal to that with which they collided: If the velocity of approach is u the total kinetic energy before the collision is twice the kinetic energy of a ball moving with velocity u, and if they separate again with velocity u this will also be the final energy. Note that this argument is valid regardless of how the kinetic energy depends on velocity.

We conclude that the situation . . . is mechanically possible, i.e. preserves momentum and energy. If the mechanical laws are to be the same for all observers, the laws of conservation of energy and momentum must also appear satisfied if this same process is seen by a moving observer. Supposing that the moving observer is travelling with speed v from right to left, so that to him the ball A will appear to travel faster and B more slowly than before; after the collision both will apparently not be travelling at right angles to the original line, but more towards the right. The important thing is that from the law of composition of velocity we can work out the directions and magnitudes of all the velocities as seen by this observer. . . . Now the momentum and energy must still be conserved, and this tells us something about the momentum and energy of objects moving with various velocities. I shall not give the details of this calculation here, which requires a little more mathematics than we are using, but from it the dependence of energy and momentum on the speed of a particle can be determined quite unambiguously.

The result is that the momentum p of an object of mass m moving with velocity u is $p = \frac{mu}{\sqrt{1 - \frac{u^2}{c^2}}}$. . . . It is seen that the relativistic momentum for small u is practically the same as before, but for velocities near that of light, very much larger than the momentum of Newtonian mechanics. This same result can be expressed differently; we may still retain the definition of momentum as mass times velocity, provided we are prepared to regard the mass of the object as variable. If we denote by M the quantity $M = \frac{m}{\sqrt{1 - \frac{u^2}{c^2}}}$ then indeed the momentum is $p = Mu$. To distinguish the two definitions of mass one sometimes calls m the "rest mass" of the object, i.e. the mass which we determine by dynamical measurements when the object is at rest or moving very slowly. Even if we work with this idea of a variable mass we must, however, remember that we should not express the dynamical law in the form that force equals mass times acceleration. The force must be taken as the rate of change of the momentum, i.e. the rate of change of mass times velocity. These two are not the same when the mass also changes in the course of the motion.

This result says that as the speed of an object increases towards that of light, it becomes harder and harder to accelerate it. In fact, to make it reach the velocity of light in any finite time an infinitely strong force would be required. It is therefore evident that it is impossible for any material object ever to reach the velocity of light, let alone exceed it. This result is satisfactory for the consistency of our views. If it were possible to apply our considerations about the relations between different observers to a man on a train travelling with more than the velocity of light we would immediately become involved in paradoxes.

As regards the kinetic energy, the argument which I have sketched indicates that $E_{kin} = \frac{mc^2}{\sqrt{1 - \frac{u^2}{c^2}}} - mc^2$ . . . . As before, the Newtonian law is adequate at low speeds, but at high speed the energy of the moving object is much greater than it would be according to the old mechanics, and a body moving actually with the velocity of light would have to have an infinite amount of energy.

Using the definition of the variable mass, M, we may express our result for the kinetic energy as $E_{kin} = (M - m)c^2$ or, as the mass of the object increases the kinetic energy increases. The amount is exactly the increase of the mass times the square of the light velocity. This suggests a close relationship between mass and energy, though up to now only the kinetic energy entered into this. One can easily prove that also energy of any other form must lead to an increase in mass by precisely the same amount. To see this it is only necessary to consider the collision, not of two elastic objects, but of two completely inelastic objects, say two lumps of clay, of the same mass. If these collide with equal and opposite velocities they will just stick together. As the total momentum is zero the fused lump has no momentum and therefore is standing still. We may look at this process from the point of view of a moving observer and again require that momentum be conserved. If we work out the apparent velocities for the new observer we find that momentum would not be conserved if the fused lump had simply the combined rest mass of the separate lumps, but only if its rest mass had increased precisely by the amount of energy which had been converted into heat, divided by c^2. Hence the relationship between mass and energy includes also heat energy, and since we may arrange for the heat to cause some chemical reaction or to generate electricity by some internal mechanism, the same must apply to energy in any other form. . . .

APPLICATION AND CONFIRMATION

From reading so far the reader will have got the impression that we have built an elaborate structure of reasoning on one experimental fact, namely the negative result of the Michelson experiment. However reliable the experiment, and however attractive the general principle of the independence of all laws of physics of the state of motion of the observer, one would not have accepted such far-reaching conclusions without a great deal of further support. In fact other physicists have claimed from time to time that on repeating the Michelson experiment they did find a positive answer, but in the meantime other evidence for the theory of relativity had become so strong that we would have no cause to change our views if some fundamental flaw was discovered in Michelson's reasoning.

Such support comes from practically any observation of the behaviour of particles moving at high velocity. Historically the earliest observations made were on electrons, since, being the lightest particles, these can most easily be accelerated to velocities near those of light. For example, an electron reaches a speed of $0.8c$ after passing through an electric field with a potential difference of about 300,000 volts, which is now commonplace in any modern laboratory. At this speed its kinetic energy is about twice as large as it would be in nonrelativistic mechanics and the difference in its behaviour very marked.

Earlier experiments carried out with much lower voltages, and in which the deflection of electrons in electric and magnetic fields was measured accurately, showed up in the variable mass and therefore confirmed relativistic mechanics. Perhaps the most striking demonstration of relativistic behaviour results from a collision of a fast particle with one at rest. Supposing an electron has been accelerated somehow to a velocity near that of light and in passing through matter collides with a stationary electron. Since the two particles have the same mass and since usually in such collisions no mechanical energy is converted into any other form . . . we should conclude from nonrelativistic mechanics that after the collisions the two electrons would move in directions at right angles to each other. On the other hand, relativistic mechanics says that both electrons should still travel forward, their lines of motion making only a small angle with each other, this angle becoming less and less as the velocity of the incident electron approaches that of light. . . . After the collision the tracks form a very small angle with each other. . . .

Another drastic example is provided by modern particle accelerators, whose principle and purpose we shall discuss later. In these we have particles moving either along a straight line or (under the influence of a magnetic field) in a circle, and electric impulses are applied to them at suitable times. The performance of such machines depends therefore on knowing the speed of the particles so that the electric impulse can be arranged to be applied just as a particle passes a suitable point. Now there exist, for example, machines in which electrons can be given an energy of 300 MeV. By an MeV or million electron volts, we mean the energy which an electron would acquire on passing through an electric potential difference of one million volts. The rest energy of an electron is about ½ MeV and therefore we are dealing with electrons whose energy is about 600 times its rest energy. A little arithmetic shows that this makes their velocity less than that of light only by a little over one part in a million.

Without relativity, electrons of such energy should move with a speed of 35 times that of light, and it is evidently impossible to mistake one for the other.

These are only a few examples of the many tests to which relativistic mechanics has been subjected in recent years and the combined weight of these tests is sufficient to give us as much confidence in the relativistic laws, for fast-moving particles, as we have in the validity of Newton's laws in the domain of small velocities. . . .

L.7

Einstein's Principle of Relativity

HERMANN WEYL

HERMANN WEYL was born in 1885 in the town of Elmshorn (now in West Germany) and died in Switzerland at the age of seventy. While attending the University of Göttingen, he was a student of the great mathematician David Hilbert, whom he later succeeded as Professor of Mathematics in 1930. His tenure was short-lived because he joined the exodus of German intellectuals after the rise of Adolf Hitler. He accepted a position at the Institute for Advanced Study at Princeton University, where he worked until his retirement in 1955. As Professor Emeritus, he divided his time between Princeton and Zurich.

Weyl, an extraordinarily talented man, had broad scientific and philosophical interests and forged a critical link between the abstractions of pure mathematics and the realities of the new physics of relativity and quantum mechanics. His contributions in both areas were notable. His book *Raum, Zeit, Materie* (*Space, Time, Matter*) was published in 1918, the outgrowth of lectures on relativity, in which his deep interest in philosophy is a crucial element. His interests ranged from the foundation of mathematics to mathematical physics. Weyl was more interested in the coherent analysis of mathematical concepts than in repetitive computations. He was a pioneer in the development of group theory and developed a key theorem on the applications of representations in the fertile field of Lie algebras. Weyl was particularly interested in the phenomenon of symmetry and its relationship to group theory. He also made important contributions to function theory, topology and differential equations; his work was used by Schrödinger and Heisenberg in their attempts to systematize the developing theories of quantum mechanics. That Weyl was regarded as one of the most brilliant and versatile mathematicians of the twentieth century was demonstrated by his former colleague, Freeman Dyson, in his memorial tribute to Weyl in the 10 March 1956 issue of *Nature:*

> He alone could stand comparison with the last great universal mathematicians of the nineteenth century, Hilbert and Poincaré. So long as he was alive, he embodied a living contact between the main lines of advance in pure mathematics and in theoretical physics. Now he is dead, the contact is broken, and our hopes of comprehending the physical universe by a direct use of creative mathematical imagination are for the time being ended.

Weyl's brief chapter, "Einstein's Principle of Relativity," from *Space, Time, Matter* is a useful sequel to the selection by Peierls. It examines some of the same material from a different perspective and includes slightly more technical mathematics. Weyl's chapter also provides a description of the celebrated "light cones," a four-dimensional space-time representation of the "active future" and the "passive past" in a relativistic universe. He was also the first to introduce the concept of "gauge invariance" as a key to the unification of gravity and electromagnetism.

In space-time, everything which for each of us constitutes the past, the present, and the future is given en bloc. . . . *Each observer, as his time passes, discovers, so to speak, new slices of space-time which appear to him as successive aspects of the material world, though in reality the ensemble of events constituting space-time exist prior to his knowledge of them.*

—LOUIS DE BROGLIE

Everything we know about Nature is in accord with the idea that the fundamental processes of Nature lie outside space-time . . . but generate events that can be located in space-time.

—HENRY P. STAPP

LET US FOR the present retain our conception of the æther. It should be possible to determine the motion of a body, for example, the earth, relative to the fixed or motionless æther. We are not helped by aberration, for this only shows that this relative motion changes in the course of a year. Let A_1, O, A_2 be three fixed points on the earth that share in its motion. Suppose them to lie in a straight line along the direction of the earth's motion and to be equidistant, so that $A_1O = OA_2 = l$ and let v be the velocity of translation of the earth through the æther; let $\frac{v}{c} = q$, which we shall assume to be a very small quantity. A light-signal emitted at O will reach A_2 after a time $\frac{l}{c - v}$ has elapsed, and A_1 after a time $\frac{l}{c + v}$. Unfortunately, this difference cannot be demonstrated, as we have no signal that is more rapid than light and that we could use to communicate the time to another place.

It might occur to us to transmit time from one world-point to another by carrying a clock that is marking time from one place to the other. In practice, this process is not sufficiently accurate for our purpose. Theoretically . . . this transmission is dependent on the traversed path . . .

We have recourse to Fizeau's idea, and set up little mirrors at A_1 and A_2

Hermann Weyl, "Einstein's Principle of Relativity," in *Space, Time, Matter*. Trans. Henry L. Brose. New York: Dover Publications, Inc. 1922, pp. 169–177.

which reflect the light-ray back to O. If the light-signal is emitted at the moment O, then the ray reflected from A_2 will reach A after a time

$$\frac{l}{c-v}+\frac{l}{c+v}=\frac{2lc}{c^2-v^2}$$

whereas that reflected from A_1 reaches O after a time

$$\frac{l}{c+v}+\frac{l}{c-v}=\frac{2lc}{c^2-v^2}.$$

There is now no longer a difference in the times. Let us, however, now assume a third point A which participates in the translational motion through the æther, such that $OA = l$, but that OA makes a 90° angle with the direction of OA. Let O, O', O'' be the successive positions of the point O at the time 0 at which the signal is emitted, at the time t' at which it is reflected from the mirror A placed at A', and finally at the time $t' + t''$ at which it again reaches O, respectively. From $OA' = ct'$, $O''A' = ct''$, $OO' = vt'$, and $O'O'' = vt''$, we get the proportion

$$OA' : O''A' = OO' : O''O',$$

Consequently the two angles at A' are equal to one another. The reflecting mirror must be placed, just as when the system is at rest, perpendicularly to the rigid connecting line OA, in order that the light-ray may return to O. An elementary trigonometrical calculation gives for the *apparent speed in the 90° direction*

$$\frac{2l}{t'+t''}=\frac{c^2-v^2}{\sqrt{c^2-v^2}} \qquad (*)$$

Observations should enable us to determine the direction and magnitude of v.

These observations were attempted in the celebrated *Michelson-Morley experiment.* In this, two mirrors A, A' are rigidly fixed to O at distances l, l', the one along the line of motion, the other perpendicular to it. The whole apparatus may be rotated about O. By means of a transparent glass plate, one-half of which is silvered and which bisects the right angle at O, a light-ray is split up into two halves, one of which travels to A, the other to A'. They are reflected at these two points; and at O, owing to the partly silvered mirror, they are again combined to a single composite ray. We take l and l' approximately equal; then, owing to the difference in path given by (*), namely,

$$\frac{2l}{1-q^2} - \frac{2l'}{\sqrt{1-q^2}},$$

interference occurs. If the whole apparatus is now turned slowly through 90° about O until A' comes into the direction of motion, this difference of path becomes

$$\frac{2l}{\sqrt{1-q^2}} - \frac{2l'}{1-q^2}.$$

Consequently, there is a shortening of the path by an amount

$$2(l+l')\left(\frac{1}{1-q^2} - \frac{1}{\sqrt{1-q^2}}\right) \backsim (l+l')q^2.$$

This should express itself in a shift of the initial interference fringes. *Although conditions were such that, numerically, even only 1 percent of the displacement of the fringes expected by Michelson could not have escaped detection, no trace of it was to be found when the experiment was performed. . . .*

Not only the Michelson-Morley experiment but a whole series of further experiments designed to demonstrate that the earth's motion has an influence on combined mechanical and electromagnetic phenomena, have led to a null result. Æther mechanics has thus to account not only for Maxwell's laws but also for this remarkable interaction between matter and æther. It seems that the æther has betaken itself to the land of the shades in a final effort to elude the inquisitive search of the physicist!

The only reasonable answer that has been given to the question as to why a translation in the æther cannot be distinguished from rest is that of Einstein, namely, that *there is no æther!* (The æther has since the very beginning remained a vague hypothesis and one, moreover, that has acted very poorly in the face of facts.) The position is then this: for mechanics we get Galilei's Theorem of Relativity, for electrodynamics, Lorentz's Theorem. If this is really the case, they neutralise one another and thereby define an absolute space of reference in which mechanical laws have the Newtonian form, electrodynamical laws that given by Maxwell.

The difficulty of explaining the null result of the experiments whose purpose was to distinguish translation from rest, is overcome only by regarding *one or other* of these two principles of relativity as being valid for *all* physical phenomena. That of Galilei does not come into question for electrodynamics as this would mean that, in Maxwell's theory, those terms by which we distinguish moving fields from stationary ones would not occur: there would be no induction, no light, and no wireless telegraphy. On the other hand, even the contraction theory of Lorentz-Fitzgerald suggests that Newton's mechanics may be modified so that it satisfies the Lorentz-Einstein Theorem of Relativity, the deviations that occur being only of the order $(\frac{v}{c})^2$; they are then easily within reach of observation for all velocities v of planets or on the earth. The solution of Einstein, which at one stroke overcomes all difficulties, is then this: *the world is a four-dimensional affine space whose metrical structure is determined by a non-definite quadratic form* $Q(\mathbf{x}) = (\mathbf{xx})$ *which has one negative and three positive dimensions.* All physical quantities are scalars and tensors of this four-dimensional world, and all physical laws express invariant relations between them. The simple concrete meaning of the form $Q(\mathbf{x})$ is that a light-signal which has been emitted at the world-point O arrives at all those and only those world-points A for which $\mathrm{x} = \overrightarrow{OA}$ belongs to the one of the two conical sheets defined by the equation $Q(\mathbf{x}) = 0$. Hence that sheet (of the two cones) which "opens into the future" namely, $Q(\mathbf{x}) \leqq 0$ is distinguished objectively from that which opens into the past. By introducing an appropriate "normal" co-ordinate system consisting of the zero point O and the fundamental vectors e_i, we may bring $Q(\mathbf{x})$ into the normal form $(\mathbf{xx}) = -x_0^2 + x_1^2 + x_2^2 + x_3^2$, in which the x_i's are the co-ordinates of A; in addition, the fundamental vector e_0 is to belong to the cone opening into the future. *It is impossible to narrow down the selection from these normal co-ordinate systems any farther:* that is, none are specially favoured; they are all equivalent. If we make use of a particular one, then x_0 must be regarded as the time; x_1, x_2, x_3 as the Cartesian space co-ordinates; and all the ordinary expressions referring to space and time are to be used in this system of reference as usual. The adequate mathematical formulation of Einstein's discovery was first given by Minkowski: to him we are indebted for the idea of four-dimensional world-geometry, on which we based our argument from the outset.

How the null result of the Michelson-Morley experiment comes about is now clear. For if the interactions of the cohesive forces of matter as well as the transmission of light takes place according to Einstein's Principle of Relativity, measuring rods must behave so that no difference between rest and translation can be discovered by means of objective determinations. Seeing that Maxwell's equations satisfy Einstein's Principle of Relativity, as was recognised even by Lorentz, we must indeed regard *the Michelson-Morley experiment as a proof that the mechanics of rigid bodies must, strictly speaking, be in accordance not with that of Galilei's Principle of Relativity, but with that of Einstein.*

It is clear that the latter is mathematically much simpler and more intelligible than the former: world-geometry has been brought into closer touch with Euclidean space-geometry through Einstein and Minkowski. Moreover, as may easily be shown, Galilei's principle is found to be a limiting case of Einstein's world-geometry by making c converge to ∞. The physical purport of this is that *we are to discard our belief in the objective meaning of simultaneity; it was the great achievement of Einstein in the field of the theory of knowledge that he banished this dogma from our minds,* and this is what leads us to rank his name with that of Copernicus. The graphical picture given at the end of the preceding paragraph discloses immediately that the planes x'_0 = const. no longer coincide with the planes x_0 = const. In consequence of the metrical structure of the world, which is based on $Q(\mathbf{x})$, each plane x'_0 = const. has a measure-determination such that the ellipse in which it intersects the "light-cone," is a circle, and that Euclidean geometry holds for it. The point at which it is punctured by the $\mathbf{x}'_0$-axis is the mid-point of the elliptical section. So the propagation of light takes place in the "accented" system of reference, too, in concentric circles.

We shall next endeavour to eradicate the difficulties that seem to our intuition, our inner knowledge of space and time, to be involved in the revolution caused by Einstein in the conception of time. According to the ordinary view the following is true. If I shoot bullets out with all possible velocities in all directions from a point O, they will all reach world-points that are later than O; I cannot shoot back into the past. Similarly, an event which happens at O has an influence only on what happens at later world-points, whereas "one can no longer undo" the past: the extreme limit is reached by gravitation, acting according to Newton's law of attraction, as a result of which, for example, by extending my arm, I at the identical moment produce an effect on the planets, modifying their orbits ever so slightly. If we again suppress a space co-ordinate and use our graphical mode of representation, then the absolute meaning of the plane t = 0 which passes through O consists in the fact that it separates the "future" world-points, which can be influenced by actions at O, from the "past" world-points from which an effect may be conveyed to or conferred on O.

According to Einstein's Principle of Relativity, we get in place of the plane of separation $t = 0$ the light cone

$$x_1^2 + x_2^2 - c^2t^2 = 0$$

(which degenerates to the above double plane when $c = \infty$). This makes the position clear in this way. The direction of all bodies projected from *O* must point into the forward-cone, opening into the future (so also the direction of the world-line of my own body, my "life-curve" if I happen to be at *O*). Events at *O* can influence only happenings that occur at world-points that lie within this forward-cone: the limits are marked out by the resulting propagation of light into empty space. If I happen to be at *O,* then *O* divides my life-curve into past and future; no change is thereby caused. As far as my relationship to the world is concerned, however, the forward-cone comprises all the world-points which are affected by my active or passive doings at *O,* whereas all events that are complete in the past, that can no longer be altered, lie externally to this cone. *The sheet of the forward-cone separates my active future from my active past.* On the other hand, the interior of the backward-cone includes all events in which I have participated (either actively or as an observer) or of which I have received knowledge of some kind or other, for only such events may have had an influence on me; outside this cone are all occurrences that I may yet experience or would yet experience if my life were everlasting and nothing were shrouded from my gaze. *The sheet of the backward-cone separates my passive past from my passive future.* The sheet itself contains everything on its surface that I see at this moment, or can see; it is thus properly the picture of my external surroundings. In the fact that we must

Active future.

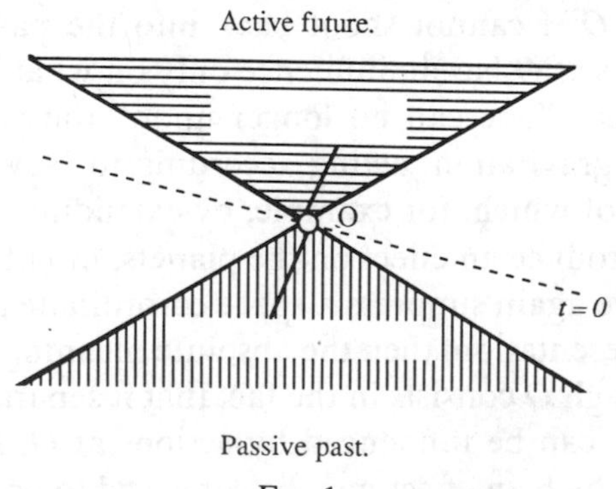

Passive past.

FIG. 1

in this way distinguish between *active* and *passive,* past and future, there lies the fundamental importance of Römer's discovery of the finite velocity of light to which Einstein's Principle of Relativity first gave full expression. The plane $t = 0$ passing through O in an allowable co-ordinate system may be placed so that it cuts the light-cone $Q(x) = 0$ only at O and thereby separates the cone of the active future from the cone of the passive past.

For a body moving with uniform translation it is always possible to choose an allowable co-ordinate system (= normal co-ordinate system) such that the body is at rest in it. The individual parts of the body are then separated by definite distances from one another, the straight lines connecting them make definite angles with one another, and so forth, all of which may be calculated by means of the formula of ordinary analytical geometry from the space co-ordinates x_1, x_2, x_3 of the points under consideration in the allowable co-ordinate system chosen. I shall term them the *static measures* of the body (this defines, in particular, the *static length* of a measuring rod). If this body is a clock, in which a periodical event occurs, there will be associated with this period in the system of reference, in which the clock is at rest, a definite time, determined by the increase of the co-ordinate x_0 during a period; we shall call this the "proper time" of the clock. If we push the body at one and the same moment at different points, these points will begin to move, but as the effect can at most be propagated with the velocity of light, the motion will only gradually be communicated to the whole body. As long as the expanding spheres encircling each point of attack and travelling with the velocity of light do not overlap, the parts surrounding these points that are dragged along move independently of one another. It is evident from this that, according to the theory of relativity, there cannot be rigid bodies in the old sense; that is, no body exists which remains objectively always the same no matter to what influences it has been subjected. How is it that in spite of this we can use our measuring rods for carrying out measurements in space? We shall use an analogy. If a gas that is in equilibrium in a closed vessel is heated at various points by small flames and is then removed adiabatically, it will at first pass through a series of complicated stages, which will not satisfy the equilibrium laws of thermo-dynamics. Finally, however, it will attain a new state of equilibrium corresponding to the new quantity of energy it contains, which is now greater owing to the heating. We require of a rigid body that is to be used for purposes of measurement (in particular, a linear *measuring rod*) that, *after coming to rest in an allowable system of reference,* it shall always remain exactly the same as before, that is, that it shall have *the same static measures* (or *static length*); and we require of *a clock* that goes correctly *that it shall always have the same proper-time when it has come to rest* (as a whole) *in an allowable system of reference.* We may assume that the measuring rods and clocks which we shall use satisfy this condition to a sufficient degree of approximation. It is only when, in our analogy, the gas is warmed sufficiently slowly (strictly speaking, infinitely slowly) that it will pass through a series of thermo-dynamic states of equilibrium; only when we move the measuring rods and clocks steadily, without jerks, will they preserve their static lengths and proper times. The limits of acceleration within which this assumption may be made without appreciable errors arising are certainly very wide. Definite and exact statements about this point can be made only when we have built up a *dynamics* based on physical and mechanical laws. . . .

M

General Relativity

INTRODUCTION

THE SPECIAL THEORY of relativity, that bombshell that Albert Einstein threw into the world of science when he quietly submitted his three papers to the editor of the *Annalen der Physik* in 1905, was based, as we saw in the previous section, on two hypotheses that are extremely simple in language, no matter how influential they were in changing our perceptions of the Universe: (1) All motion must be considered in relation to some object or system arbitrarily taken as being at rest, with the understanding that any such frame of reference is as acceptable as any other; and (2) the velocity of light in a vacuum, as measured by any observer, is independent of the velocity of the observer relative to the light source; that is, the speed of light in a vacuum is an absolute constant.

While the scientific world mulled over the special theory of relativity, Einstein devoted a decade of his young adulthood to working out his general theory of relativity, which was published in 1916, after the European powers had plunged into the horror of World War I. Almost as soon as the conflict was over, an international expedition was organized to test one of Einstein's predictions—the curving by gravity of the path of a light ray as it passes close to a massive heavenly body. Einstein's general theory had been applied earlier to the motion of Mercury and it had solved the age-old problem of the precession of the perihelion of the orbit of Mercury, which varies from that predicted by Newtonian theory by only 43.03 seconds of arc per century (a second of arc is the 1.296-millionth part of a complete rotation or 360°). Among other predictions made by Einstein in his general theory was that light beams (composed of photons) are affected by gravity—specifically, that the path of light from a star that passes very close to the edge of the sun in its journey to earth, is bent by 1.75 seconds of arc. Normally such a prediction cannot be tested because a beam of light is not visible against the far more intense light of the sun, but during a total eclipse the phenomenon is observable.

In 1919, a total eclipse of the sun was expected to be visible from an island in the Gulf of Guinea off the west coast of Africa. The British organized a full-scale expedition to test Einstein's prediction, captained by the astrophysicist Sir Arthur Eddington. The astronomers photographed the positions of certain stars in the direction of the sun and then repeated the observation precisely one-half year later when the sun was in a diametrically opposed spot in the firmamemt. The results confirmed Einstein's prediction, and the spectacular event brought Einstein immediate fame.

Repeated experiments have reconfirmed the general theory in so many ways that it has been essentially accepted, together with quantum mechanics, as the core of modern physical doctrine. This does not mean that Einstein's theory has been accepted in toto without question; some physicists, most notably R. H. Dicke, continue to challenge some of its predictions, although they remain a distinct minority within the professional community.

As soon as Einstein's name became generally known in Europe and America, an endless series of "popular" explanations of his theory were published. Among the earliest popularizations were very worthwhile versions written by Sir Arthur Eddington and Bertrand Russell, as well as mathematical and nonmathematical treatments by Einstein himself. More recently, Martin Gardner and Lincoln Barrett have published books that explain Einstein's unprecedented achievements clearly and concisely.

M.1

Foundations of Geometry

Georg Riemann

Georg Friedrich Bernhard Riemann, one of the most original thinkers in the history of mathematics, was born in 1826 in Breselenz (Hanover), Germany, and died at the age of forty in Italy. His father was a Lutheran pastor, and Riemann originally intended to become a theologian. He learned Hebrew and made a valiant effort to demonstrate the validity of Genesis by using mathematics. He did not succeed in this effort, but the experience led him to abandon theology for mathematics. His university studies in Göttingen were interrupted by the revolutionary upheavals of 1848, but with the victory of Kaiser Frederick Wilhelm IV over the radicals, Riemann returned to the university to finish his schooling. His thesis in mathematics met with the approval of the venerable mathematician Karl Gauss.

Among the many important contributions made to mathematics by Riemann in his brief life, the most important is a version of non-Euclidean geometry presented to the public in 1854. The general public, insofar as it understood what Riemann was talking about, was shocked, for the brash newcomer proposed a geometry in which the Euclidean axioms on parallels—that through a given point adjacent to a line, one and only one parallel line can be drawn—is abolished in favor of a new axiom stating that through such a point no parallels can be drawn. Euclid had also taken as one of his axioms the principle that two points determine one and only one straight line; Riemann argued that any number of straight lines can be determined by such conditions. And finally, in Riemannian geometry, the sum of the angles of a triangle always exceeds 180°.

Riemann's geometry was shocking to those familiar with the ideal world of Euclid even though it is more appropriate to our own world experiences. As we live on the surface of a spherical planet, the shortest distance between two points is a segment of a great circle. Two perpendiculars or lines of longitude erected on the equator meet at the North or South Pole, thus creating a triangle whose base angles, where the longitudinal line and the equator meet, are each of 90°; the third angle at the Pole, gives the polygon more than 180°. At best, Riemann's new geometry was considered a brilliant mathematical curiosity, but fifty years after Riemann's death, Einstein demonstrated that his version of geometry fits reality more closely than did Euclid's.

More specifically, Einstein in his general theory of relativity argued that space is positively curved due to the gravitational forces of the bodies distributed throughout the Universe. In Einstein's Universe, a line on an imaginary sphere surrounding a mass is not straight in the Euclidean sense but is, indeed, the shortest distance between two space-time points. Although the predictions of the general theory differ only slightly from those of Newton's theory, the latter's theory of gravitation suggested that the fact that two bodies travel around curved paths in each other's presence indicates an interacting gravitational field. Einstein, however, argued that what we perceive to be the gravitational attraction between two bodies is actually due to the geometric properties of space itself.

Our selection from Riemann is taken from a paper that examines some of the hypotheses that lie at the foundations of geometry. It was first published in English in *Nature*, and is representative of the clarity of Riemann's original approach to mathematics.

There are two worlds; the world that we can measure with line and rule, and the world that we feel with our hearts and imagination.

—LEIGH HUNT, *Men, Women and Books*

. . . Form is emptiness, emptiness is form.

—The Heart Sutra in *Prajnaparamita Sutras*

. . . MEASURE-DETERMINATIONS REQUIRE that quantity should be independent of position, which may happen in various ways. The hypothesis which first presents itself, and which I shall here develop, is that according to which the length of lines is independent of their position, and consequently every line is measurable by means of every other. Position-fixing being reduced to quantity-fixings, and the position of a point in an n-dimensional manifold being consequently expressed by means of n variables $x_1, x_2, x_3, \ldots x_n$, the determination of a line comes to the giving of these quantities as functions of one variable. The problem consists then in establishing a mathematical expression for the length of a line, and to this end we must consider the quantities x as expressible in terms of certain units. I shall treat this problem only under certain restrictions, and I shall confine myself in the first place to lines in which the ratios of the increments dx of the respective variables vary continuously. We may then conceive these lines broken up into elements, within which the ratios of the quantities dx may be regarded as constant; and the problem is then reduced to establishing for each point a general expression for the linear element ds starting from that point, an expression which will thus contain the quantities x and the quantities dx. I shall suppose, secondly, that the length of the linear element, to the first order, is unaltered when all the points of this element undergo the same infinitesimal displacement, which implies at the same time that if all the quantities dx are increased in the same ratio, the linear element will vary also in the same ratio. On these suppositions, the linear element may be any homogeneous function of the first degree of the quantities dx, which is unchanged

---

G. F. B. Riemann, "On the Hypotheses Which Lie at the Bases of Geometry," *Nature*, Vol. VIII, pp. 14–17, 36, 37.

when we change the signs of all the dx, and in which the arbitrary constants are continuous functions of the quantities x. To find the simplest cases, I shall seek first an expression for manifoldnesses of $n - 1$ dimensions which are everywhere equidistant from the origin of the linear element; that is, I shall seek a continuous function of position whose values distinguish them from one another. In going outwards from the origin, this must either increase in all directions or decrease in all directions; I assume that it increases in all directions, and therefore has a minimum at that point. If, then, the first and second differential coefficients of this function are finite, its first differential must vanish, and the second differential cannot become negative; I assume that it is always positive. This differential expression, then, of the second order remains constant when ds remains constant, and increases in the duplicate ratio when the dx, and therefore also ds, increase in the same ratio; it must therefore be ds^2 multiplied by a constant, and consequently ds is the square root of an always positive integral homogeneous function of the second order of the quantities dx, in which the coefficients are continuous functions of the quantities x. For Space, when the position of points is expressed by rectilinear coordinates, $ds = \sqrt{\Sigma(dx)^2}$; Space is therefore included in this simplest case. The next case in simplicity includes those manifoldnesses in which the line-element may be expressed as the fourth root of a quartic differential expression. The investigation of this more general kind would require no really different principles, but would take considerable time and throw little new light on the theory of space, especially as the results cannot be geometrically expressed; I restrict myself, therefore, to those manifoldnesses in which the line-element is expressed as the square root of a quadric differential expression. Such an expression we can transform into another similar one if we substitute for the n independent variables functions of n new independent variables. In this way, however, we cannot transform any expression into any other; since the expression contains $\frac{1}{2} n (n + 1)$ coefficients which are arbitrary functions of the independent variables; now by the introduction of new variables we can only satisfy n conditions, and therefore make no more than n of the coefficients equal to given quantities. The remaining $\frac{1}{2} n (n - 1)$ are then entirely determined by the nature of the continuum to be represented, and consequently $\frac{1}{2} n (n - 1)$ functions of positions are required for the determination of its measure-relations. Manifoldnesses in which, as in the Plane and in Space, the line-element may be reduced to the form $\sqrt{\Sigma dx^2}$, are therefore only a particular case of the manifoldnesses to be here investigated; they require a special name, and therefore these manifoldnesses in which the square of the line-element may be expressed as the sum of the squares of complete differentials I will call *flat*. In order now to review the true varieties of all the continua which may be represented in the assumed form, it is necessary to get rid of difficulties arising from the mode of representation, which is accomplished by choosing the variables in accordance with a certain principle.

For this purpose let us imagine that from any given point the system of shortest lines going out from it is constructed; the position of an arbitrary point may then be determined by the initial direction of the geodesic in which it lies, and by its distance measured along that line from the origin. It can therefore be expressed in terms of the ratios dx_0 of the quantities dx in this geodesic, and of the length s of this line. Let us introduce now instead of the dx_0 linear functions dx of them, such that the initial value of the square of the line-element shall equal the sum of the squares of these expressions, so that the independent variables are now the length s and the ratios of the quantities dx. Lastly, take instead of the dx quantities x_1, x_2, x_3, . . . x_n proportional to them, but such that the sum of their squares $=s^2$. When we introduce these quantities, the square of the line-element is Σdx^2 for infinitesimal values of the x, but the term of next order in it is equal to a homogeneous function of the second order of the $\frac{1}{2}n(n-1)$ quantities $(x_1dx_2-x_2dx_1)$, $(x_1dx_3-x_3dx_1)$ . . . , an infinitesimal, therefore, of the fourth order; so that we obtain a finite quantity on dividing this by the square of the infinitesimal triangle, whose vertices are (0,0,0, . . .), $(x_1\ x_2\ x_3, \ldots)$, $(dx_1, dx_2, dx_3, \ldots)$. This quantity retains the same value so long as the x and the dx are included in the same binary linear form, or so long as the two geodesics from 0 to x and from 0 to dx remain in the same surface-element; it depends therefore only on place and direction. It is obviously zero when the manifold represented is flat, *i.e.*, when the squared line-element is reducible to Σdx^2, and may therefore be regarded as the measure of the deviation of the manifoldness from flatness at the given point in the given surface-direction. Multiplied by $-\frac{3}{4}$ it becomes equal to the quantity which Privy Councillor Gauss has called the total curvature of a surface. For the determination of the measure-relations of a manifoldness capable of representation in the assumed form we found that $\frac{1}{2}n(n-1)$ place-functions were necessary; if, therefore, the curvature at each point in $\frac{1}{2}n(n-1)$ surface-directions is given, the measure-relations of the continuum may be determined from them—provided there be no identical relations among these values, which in fact, to speak generally, is not the case. In this way the measure-relations of a manifoldness in which the line-element is the square root of a quadric differential may be expressed in a manner wholly independent of the choice of independent variables. A method entirely similar may for this purpose he applied also to the manifoldness in which the line-element has a less simple expression, *e.g.*, the fourth root of a quartic differential. In this case the line-element, generally speaking, is no longer reducible to the form of the square root of a sum of squares, and therefore the deviation from flatness in the squared line-element is an infinitesimal of the second order, while in those manifoldnesses it was of the fourth order. This property of the last-named continua may thus be called flatness of the smallest parts. The most important property of these continua for our present purpose, for whose sake alone they are here investigated, is that the relations of the twofold ones may be geometrically represented by surfaces, and of the morefold ones may be reduced to those of the surfaces included in them; which now requires a short further discussion.

In the idea of surfaces, together with the intrinsic measure-relations in which only the length of lines on the surfaces is considered, there is always mixed up the position of points lying out of the surface. We may, however, abstract from external relations if we consider such deformations as leave unaltered the length of lines—*i.e.,* if we regard the surface as bent in any way without stretching, and treat all surfaces so related to each other as equivalent. Thus, for example, any cylindrical or conical surface counts as equivalent to a plane, since it may be made out of one by mere bending, in which the intrinsic measure-relations remain, and all theorems about a plane—therefore the whole of planimetry—retain their validity. On the other hand they count as essentially different from the sphere, which cannot be changed into a plane without stretching. According to our previous investigation the intrinsic measure-relations of a twofold extent in which the line-element may be expressed as the square root of a quadric differential, which is the case with surfaces, are characterized by the total curvature. Now this quantity in the case of surfaces is capable of a visible interpretation, viz., it is the product of the two curvatures of the surface, or multiplied by the area of a small geodesic triangle, it is equal to the spherical excess of the same. The first definition assumes the proposition that the product of the two radii of curvature is unaltered by mere bending; the second, that in the same place the area of a small triangle is proportional to its spherical excess. To give an intelligible meaning to the curvature of an n-fold extent at a given point and in a given surface-direction through it, we must start from the fact that a geodesic proceeding from a point is entirely determined when its initial direction is given. According to this we obtain a determinate surface if we prolong all the geodesics proceeding from the given point and lying initially in the given surface-direction; this surface has at the given point a definite curvature, which is also the curvature of the n-fold continuum at the given point in the given surface-direction.

Before we make the application to space, some considerations about flat manifoldnesses in general are necessary; *i.e.*, about those in which the square of the line-element is expressible as a sum of squares of complete differentials.

In a flat n-fold extent the total curvature is zero at all points in every direction; it is sufficient, however (according to the preceding investigation), for the determination of measure-relations, to know that at each point the curvature is zero in $\frac{1}{2} n (n - 1)$ independent surface-directions. Manifoldnesses whose curvature is constantly zero may be treated as a special case of those whose curvature is constant. The common character of these continua whose curvature is constant may be also expressed thus, that figures may be moved in them without stretching. For clearly figures could not be arbitrarily shifted and turned round in them if the curvature at each point were not the same in all directions. On the other hand, however, the measure-relations of the manifoldness are entirely determined by the curvature; they are therefore exactly the same in all directions at one point as at another, and consequently the same constructions can be made from it: whence it follows that in aggregates with constant curvature figures may have any arbitrary position given them. The measure-relations of these manifoldnesses depend only on the value of the curvature, and in relation to the analytic expression it may be remarked that if this value is denoted by α, the expression for the line-element may be written

$$\frac{1}{1 + \frac{1}{4}\alpha\Sigma x^2}\sqrt{\Sigma dx^2}.$$

The theory of *surfaces* of constant curvature will serve for a geometric illustration. It is easy to see that surfaces whose curvature is positive may always be rolled on a sphere whose radius is unity divided by the square root of the curvature; but to review the entire manifoldness of these surfaces, let one of them have the form of a sphere and the rest the form of surfaces of revolution touching it at the equator. The surfaces with greater curvature than this sphere will then touch the sphere internally, and take a form like the outer portion (from the axis) of the surface of a ring; they may be rolled upon zones of spheres having less radii, but will go round more than once. The surfaces with less positive curvature are obtained from spheres of larger radii, by cutting out the lune bounded by two great half-circles and bringing the section-lines together. The surface with curvature zero will be a cylinder standing on the equator; the surfaces with negative curvature will touch the cylinder externally and be formed like the inner portion (towards the axis) of the surface of a ring. If we regard these surfaces as *locus in quo* for surface-regions moving in them, as Space is *locus in quo* for bodies, the surface-regions can be moved in all these surfaces without stretching. The surfaces with positive curvature can always be so formed that surface-regions may also be moved arbitrarily about upon them without *bending*, namely (they may be formed) into sphere-surfaces; but not those with negative curvature. Besides this independence of surface-regions from position there is in surfaces of zero curvature also an independence of *direction* from position, which in the former surfaces does not exist.

APPLICATION TO SPACE

By means of these inquiries into the determination of the measure-relations of an *n*-fold extent the conditions may be declared which are necessary and sufficient to determine the metric properties of space, if we assume the independence of line-length from position and expressibility of the line-element as the square root of a quadric differential, that is to say, flatness in the smallest parts.

First, they may be expressed thus: that the curvature at each point is zero in three surface-directions; and thence the metric properties of space are determined if the sum of the angles of a triangle is always equal to two right angles.

Secondly, if we assume with Euclid not merely an existence of lines independent of position, but of bodies also, it follows that the curvature is everywhere constant; and then the sum of the angles is determined in all triangles when it is known in one.

Thirdly, one might, instead of taking the length of lines to be independent of position and direction, assume also an independence of their length and direction from position. According to this conception changes or differences of position are complex magnitudes expressible in three independent units.

In the course of our previous inquiries, we first distinguished between the relations of extension or partition and the relations of measure, and found that with the same extensive properties, different measure-relations were conceivable; we then investigated the system of simple size-fixings by which the measure-relations of space are completely determined, and of which all propositions about them are a necessary consequence; it remains to discuss the question how, in what degree, and to what extent these assumptions are borne out by experience. In this respect there is a real distinction between mere extensive relations, and measure-relations; in so far as in the former, where the possible cases form a discrete manifoldness, the declarations of experience are indeed not quite certain, but still not inaccurate; while in the latter, where the possible cases form a continuous manifoldness, every determination from experience remains always inaccurate: be the probability ever so great that it is nearly exact. This consideration becomes important in the extensions of these empirical determinations beyond the limits of observation to the infinitely great and infinitely small; since the latter may clearly become more inaccurate beyond the limits of observation, but not the former.

In the extension of space-construction to the infinitely great, we must distinguish between *unboundedness* and *infinite extent,* the former belongs to the extent relations, the latter to the measure-relations. That space is an unbounded threefold manifoldness, is an assumption which is developed by every conception of the outer world; according to which every instant the region of real perception is completed and the possible positions of a sought object are constructed, and which by these applications is forever confirming itself. The unboundedness of space possesses in this way a greater empirical certainty than any external experience. But its infinite extent by no means follows from this; on the other hand if we assume independence of bodies from position, and therefore ascribe to space constant curvature, it must necessarily be finite provided this curvature has ever so small a positive value. If we prolong all the geodesics starting in a given surface-element, we should obtain an unbounded surface of constant curvature, *i.e.,* a surface which in a *flat* manifoldness of three dimensions would take the form of a sphere, and consequently be finite.

The questions about the infinitely great are for the interpretation of nature useless questions. But this is not the case with the questions about the infinitely small. It is upon the exactness with which we follow phenomena into the infinitely small that our knowledge of their causal relations essentially depends. The progress of recent centuries in the knowledge of mechanics depends almost entirely on the exactness of the construction which has become possible through the invention of the infinitesimal calculus, and through the simple principles discovered by Archimedes, Galileo, and Newton, and used by modern physic. But in the natural sciences which are still in want of simple principles for such constructions, we seek to discover the causal relations by following the phenomena into great minuteness, so far as the microscope permits. Questions about the measure-relations of space in the infinitely small are not therefore superfluous questions.

If we suppose that bodies exist independently of position, the curvature is everywhere constant, and it then results from astronomical measurements that it cannot be different from zero; or at any rate its reciprocal must be an area in comparison with which the range of our telescopes may be neglected. But if this independence of bodies from position does not exist, we cannot draw conclusions from metric relations of the great, to those of the infinitely small; in that case the curvature at each point may have an arbitrary value in three directions, provided that the total curvature of every measurable portion of space does not differ sensibly from zero. Still more complicated relations may exist if we no longer suppose the linear element expressible as the square root of a quadric differential. Now it seems that the empirical notions on which the metrical determinations of space are founded, the notion of a solid body and of a ray of light, cease to be valid for the infinitely small. We are therefore quite at liberty to suppose that the metric relations of space in the infinitely small do not conform to the hypotheses of geometry; and we ought in fact to suppose it, if we can thereby obtain a simpler explanation of phenomena.

The question of the validity of the hypotheses of geometry in the infinitely small is bound up with the question of the basis of the metric relations of space. In this last question, which we may still regard as belonging to the doctrine of space, is found the application of the remark made above; that in a discrete manifoldness, the basis of its metric relations is given in the notion of it, while in a continuous manifoldness, this basis must come from outside. Either therefore the reality which underlies space must form a discrete manifoldness, or we must seek the basis of its metric relations outside it, in binding forces which act upon it.

The answer to these questions can only be got by starting from the conception of phenomena which has hitherto been justified by experience, and which Newton assumed as a foundation, and by making in this conception the successive changes required by facts which it cannot explain. Researches starting from general notions, like the investigation we have just made, can only be useful in preventing this work from being hampered by too narrow views, and progress in knowledge of the interdependence of things from being checked by traditional prejudices.

This leads us into the domain of another science, of physics, into which the object of this work does not allow us to go for the moment.

M.2

The Principles of Dynamics

Ernst Mach

Ernst Mach was an Austrian physicist and philosopher who lived from 1836 to 1916. He studied mathematics at the University of Vienna and accepted in succession academic appointments to Graz, Prague, and Vienna. According to his philosophy of science, all concepts are byproducts of sense experiences. He viewed scientific laws as mere descriptions of earlier experiences that provide us with the means to predict future experiences. Mach also held that all branches of science are essentially studies of sensations and that common patterns can be found in each branch.

Mach's name is known to the general public chiefly through the term *Mach number*, which is the ratio of the velocity of a moving object to the velocity of sound. A Mach number does not represent a fixed speed, but one which varies with the temperature and density of the medium through which a given object is moving. In a normal atmosphere at 20 °C (68 °F), Mach 1 is 758 miles per hour (1,220 kilometers per hour). The expression became popular when, on 14 October 1947, the pilot Charles Yeager first flew a plane at a speed in excess of Mach 1, thus "breaking the sound barrier." Mach is also remembered among scientists as one of the last die-hard disbelievers in atoms.

Mach was a brilliant theoretician whose researches had a strong influence on Albert Einstein's development of the special theory of relativity and, later, his general theory. Mach believed that the stodgy image of a static Universe consisting of countless independent systems contradicted the logical conclusion of Newton's law of gravitation—that is, that an infinite-range gravitational field exists. Mach was convinced that the Universe is interconnected by inertial properties of matter in such a way that the local motion of any object would be influenced, however imperceptibly, by the rest of the objects in the Universe. He argued that the crucial element in this cosmic field is acceleration, not velocity. Consequently, the field generated by the gravitational interactions of all objects would accelerate the motion of any affected object and, as a result, increase its inertial momentum. For Mach, relative motion was the only motion that could exist in a Universe of constantly shifting frames of reference. The pertinence of these principles to Einstein's own theory of relativity is evident.

Although a coherent theory of relativity did not appear until 1905, the notion of relative space and motion had appeared with irregular frequency from the time that Bishop Berkeley, a contemporary of Newton, had suggested that every motion or place is relative to some other motion or place. But the absence of a theoretical framework to support this theory caused it to languish for nearly two hundred years until the 1883 publication of Mach's *Science of Mechanics*. This work offered Mach's unique inertial view of matter in a relativistic Universe. Einstein finished Mach's work in 1916 with the publication of his general theory of relativity.

[T]he axioms of geometry therefore are neither a priori *judgments nor experimental facts. They are conventions; our choice among all possible conventions is guided by experimental facts; but it remains free and is limited only by the necessity of avoiding all contradictions.*

—HENRI POINCARÉ, *Foundations of Science*

The world doesn't yield to us directly, the description of the world stands in between.

—CARLOS CASTANEDA, *Tales of Power*

. . . IF, IN A material spatial system, there are masses with different velocities, which can enter into mutual relations with one another, these masses present to us forces. We can only decide how great these forces are when we know the velocities to which those masses are to be brought. *Resting* masses too are forces if *all* the masses do not rest. Think, for example, of Newton's rotating bucket in which the water is not yet rotating. If the mass m has the velocity v_1 and it is to be brought to the velocity v_2, the force which is to be spent on it is $p = m(v_1 - v_2)/t$, or the work which is to be expended is $ps = ½\, m(v_1^2 - v_2^2)$. *All* masses and *all* velocities, and consequently *all* forces, are relative. There is no decision about relative and absolute which we can possibly meet, to which we are forced, or from which we can obtain any intellectual or other advantage. When quite modern authors let themselves be led astray by the Newtonian arguments which are derived from the bucket of water, to distinguish between relative and absolute motion, they do not reflect that the system of the world is only given *once* to us, and the Ptolemaic or Copernican view is *our* interpretation, but both are equally actual. Try to fix Newton's bucket and rotate the heaven of fixed stars and then prove the absence of centrifugal forces.

It is scarcely necessary to remark that in the reflections here presented Newton has again acted contrary to his expressed intention only to investigate *actual facts*. No one is competent to predicate things about absolute space and absolute motion; they are pure things of thought, pure mental constructs, that cannot be produced in experience. All our principles of

---

Reprinted from *The Science of Mechanics* by Ernst Mach by permission of The Open Court Publishing Company, La Salle, Illinois, 1960, pp. 279–284.

mechanics are, as we have shown in detail, experimental knowledge concerning the relative positions and motions of bodies. Even in the provinces in which they are now recognized as valid, they could not, and were not, admitted without previously being subjected to experimental tests. No one is warranted in extending these principles beyond the boundaries of experience. In fact, such an extension is meaningless, as no one possesses the requisite knowledge to make use of it.

We must suppose that the change in the point of view from which the system of the world is regarded which was initiated by Copernicus, left deep traces in the thought of Galileo and Newton. But while Galileo, in his theory of the tides, quite naïvely chose the sphere of the fixed stars as the basis of a new system of co-ördinates, we see doubts expressed by Newton as to whether a given fixed star is at rest only apparently or really. . . . This appeared to him to cause the difficulty of distinguishing between true (absolute) and apparent (relative) motion. By this he was also impelled to set up the conception of *absolute space*. By further investigations in this direction—the discussion of the experiment of the rotating spheres which are connected together by a cord and that of the rotating water-bucket . . . —he believed that he could prove an absolute rotation, though he could not prove any absolute translation. By absolute rotation he understood a rotation relative to the fixed stars, and here centrifugal forces can always be found. "But how we are to collect," says Newton in the Scholium at the end of the Definitions, "the true motions from their causes, effects, and apparent differences, and *vice versa;* how from the motions, either true or apparent, we may come to the knowledge of their causes and effects, shall be explained more at large in the following Tract." The resting sphere of fixed stars seems to have made a certain impression on Newton as well. The natural system of reference is for him that which has any uniform motion or translation without rotation (relatively to the sphere of fixed stars). But do not the words quoted in inverted commas give the impression that Newton was glad to be able now to pass over to less precarious questions that could be tested by experience?

Let us look at the matter in detail. When we say that a body K alters its direction and velocity solely through the influence of another body K', we have asserted a conception that it is impossible to come at unless other bodies A, B, C . . . . are present with reference to which the motion of the body K has been estimated. In reality, therefore, we are simply cognizant of a relation of the body K to A, B, C . . . . If now we suddenly neglect A, B, C . . . . and attempt to speak of the deportment of the body K in absolute space, we implicate ourselves in a twofold error. In the first place, we cannot know how *K* would act in the absence of *A, B, C* . . . . ; and in the second place, every means would be wanting of forming a judgment of the behavior of *K* and of putting to the test what we had predicated—which latter therefore would be bereft of all scientific significance.

Two bodies *K* and *K'*, which gravitate toward each other, impart to each other in the direction of their line of junction accelerations inversely proportional to their masses *m, m'*. In this proposition is contained, not only a relation of the bodies *K* and *K'* to one another, but also a relation of them to other bodies. For the proposition asserts, not only that *K* and *K'* suffer with respect to one another the acceleration designated by $x(\overline{m + m}/r^2)$, but also that *K* experiences the acceleration $-x\, m'/r^2$ and K' the acceleration $+\, x\, m/r^2$ in the direction of the line of junction; facts which can be ascertained only by the presence of other bodies.

The motion of a body *K* can only be estimated by reference to other bodies *A, B, C* . . . . But since we always have at our disposal a sufficient number of bodies, that are as respects each other relatively fixed, or only slowly change their positions, we are, in such reference, restricted to no one *definite* body and can alternately leave out of account now this one and now that one. In this way the conviction arose that these bodies are indifferent generally.

It might be, indeed, that the isolated bodies *A, B, C* . . . . play merely a collateral role in the determination of the motion of the body *K,* and that this motion is determined by a *medium* in which *K* exists. In such a case we should have to substitute this medium for Newton's absolute space. Newton certainly did not entertain this idea. Moreover, it is easily demonstrable that the atmosphere is not this motion-determinative medium. We should, therefore, have to picture to ourselves some other medium, filling, say, all space, with respect to the constitution of which and its kinetic relations to the bodies placed in it we have at present no adequate knowledge. In itself such a state of things would not belong to the impossibilities. It is known, from recent hydrodynamical investigations, that a rigid body experiences resistance in a frictionless fluid only when its velocity *changes.* True, this result is derived theoretically from the notion of inertia; but it might, conversely, also be regarded as the primitive fact from which we have to start. Although, practically, and at present, nothing is to be accomplished with this conception, we might still hope to learn more in the future concerning this hypothetical medium; and from the point of view of science it would be in every respect a more valuable acquisition than the forlorn idea of absolute space. When we reflect that we cannot abolish the isolated bodies *A, B, C* . . . ., that is, cannot determine by experiment whether the part they play is fundamental or collateral, that hitherto they have been the sole and only competent means of the orientation of motions and of the description of mechanical facts, it will be found expedient provisionally to regard all motions as determined by these bodies.

Let us now examine the point on which Newton, apparently with sound reasons, rests his distinction of absolute and relative motion. If the earth is affected with an *absolute* rotation about its axis, centrifugal forces are set up in the earth: it assumes an oblate form, the acceleration of gravity is diminished at the equator, the plane of Foucault's pendulum rotates, and so on. All these phenomena disappear if the earth is at rest and the other heavenly bodies are affected with absolute motion round it, such that the same *relative* rotation is produced. This is, indeed, the case, if we start *ab initio* from the idea of absolute space. But if we take our stand on the basis of facts, we shall find we have knowledge only of *relative* spaces and motions. *Relatively,* not considering the unknown and neglected medium of space, the motions of the universe are the same whether we adopt the Ptolemaic or the Copernican mode of view. Both views are, indeed, equally *correct;* only the latter is more simple and more *practical.* The universe is not *twice* given, with an earth at rest and an earth in motion; but only *once,* with its *relative* motions, alone determinable. It is, accordingly, not permitted us to say how things would be if the earth did not rotate. We may interpret the one case that is given us, in different ways. If, however, we so interpret it that we come into conflict with experience, our interpretation is simply wrong. The principles of mechanics can, indeed, be so conceived, that even for relative rotations centrifugal forces arise.

Newton's experiment with the rotating vessel of water simply informs us, that the relative rotation of the water with respect to the sides of the vessel produces *no* noticeable centrifugal forces, but that such forces *are* produced by its relative rotation with respect to the mass of the earth and the other celestial bodies. No one is competent to say how the experiment would turn out if the sides of the vessel increased in thickness and mass till they were ultimately several leagues thick. The one experiment only lies before us, and our business is, to bring it into accord with the other facts known to us, and not with the arbitrary fictions of our imagination.

M.3, M.4

The Fundamental Ideas and Problems of the Theory of Relativity

ALBERT EINSTEIN

Geometry and Experience

ALBERT EINSTEIN

WHEN ALBERT EINSTEIN finally received a Nobel Prize in 1921, it was not for his special or general theory of relativity; instead it was for his early work on the photoelectric effect. Einstein was not present at the awards ceremony in Sweden, and he delivered his official Nobel lecture on 11 July 1923, to the Nordic Assembly of Naturalists at a meeting held in Gothenburg, a small town in southwestern Sweden. Einstein took advantage of the fact that his lecture was not being delivered at the time of the award and discussed the subject closest to his heart at the time: "The Fundamental Ideas and Problems of the Theory of Relativity."

Einstein often argued that mathematics is the linchpin of the physical sciences and in a famous 1921 lecture entitled "Geometry and Experience," delivered before the Prussian Academy of Sciences, he pointed to the consistency of mathematics as a feature elevating it above all sciences. He argued that mathematics provides science with a greater internal stability than it would otherwise have, since mathematical axioms such as Euclid's geometry are considerably more permanent than most scientific theories. Einstein said that mathematics gives precision and certainty to the sciences. However, he recognized that blind reliance on axioms could purge mathematics of all extraneous real-world elements to the point that when mathematics is internally propagated and intellectually pure, its applicability to the external world might no longer be possible. Certainly Einstein was aware of discrepancies between the perfectly straight lines of Euclid and the not-so-perfect lines found in the real world. He distinguished between practical geometry, which is largely dependent on induction from experience, and axiomatic geometry, which is based on logical inferences. Einstein appeared to favor the reliability of practical geometry over the exactness of axiomatic mathematics in his theories. Relativity depends on an inductive approach, which reduces the need for the absolutes of axiomatic geometry.

An earlier selection (L.3) provides a good idea of how Einstein viewed his work on relativity when speaking formally. The present selections describe his theories in well-organized, densely packed reviews of the ideas and problems of relativity some eighteen years after the appearance of his famous article in 1905.

The Fundamental Ideas and Problems of the Theory of Relativity

ALBERT EINSTEIN

From a systematic theoretical point of view, we may imagine the process of evolution of an empirical science to be a continuous process of induction. Theories are evolved, and are expressed in short compass as statements of a large number of individual observations in the form of empirical laws, from which the general laws can be ascertained by comparison. Regarded in this way, the development of a science bears some resemblance to the compilation of a classified catalogue. It is, as it were, a purely empirical enterprise.

—ALBERT EINSTEIN, *Relativity*, Appendix III

A theory is the more impressive the greater the simplicity of its premises is, the more different kinds of things it relates, and the more extended is its area of applicability.

—*Autobiographical Notes*

IF WE CONSIDER that part of the theory of relativity which may nowadays in a sense be regarded as bona fide scientific knowledge, we note two aspects which have a major bearing on this theory. The whole development

Albert Einstein, Nobel Prize in Physics Award Address, 1921. Reprinted with permission of Elsevier Publishing Company, and the Nobel Foundation.

of the theory turns on the question of whether there are physically preferred states of motion in Nature (physical relativity problem). Also, concepts and distinctions are only admissible to the extent that observable facts can be assigned to them without ambiguity (principle that concepts and distinctions should have meaning). This postulate, pertaining to epistemology, proves to be of fundamental importance.

These two aspects become clear when applied to a special case, e.g. to classical mechanics. Firstly we see that at any point filled with matter there exists a preferred state of motion, namely that of the substance at the point considered. Our problem starts however with the question whether physically preferred states of motion exist in reference to *extensive* regions. From the viewpoint of classical mechanics the answer is in the affirmative; the physically preferred states of motion from the viewpoint of mechanics are those of the inertial frames.

This assertion, in common with the basis of the whole of mechanics as it generally used to be described before the relativity theory, far from meets the above "principle of meaning." Motion can only be conceived as the relative motion of bodies. In mechanics, motion relative to the system of coordinates is implied when merely motion is referred to. Nevertheless this interpretation does not comply with the "principle of meaning" if the coordinate system is considered as something purely imaginary. If we turn our attention to experimental physics we see that there the coordinate system is invariably represented by a "practically rigid" body. Furthermore it is assumed that such rigid bodies can be positioned in rest relative to one another in common with the bodies of Euclidian geometry. Insofar as we think of the rigid measuring body as existing as an object which can be experienced, the "system of coordinates" concept as well as the concept of the motion of matter relative thereto can be accepted in the sense of the "principle of meaning." At the same time Euclidian geometry, by this conception, has been adapted to the requirements of the physics of the "principle of meaning." The question whether Euclidian geometry is valid becomes physically significant; its validity is assumed in classical physics and also later in the special theory of relativity.

In classical mechanics the inertial frame and time are best defined together by a suitable formulation of the law of inertia: It is possible to fix the time and assign a state of motion to the system of coordinates (inertial frame) such that, with reference to the latter, force-free material points undergo no acceleration; furthermore it is assumed that this time can be measured without disagreement by identical clocks (systems which run down periodically) in any arbitrary state of motion. There are then an infinite number of inertial frames which are in uniform translational motion relative to each other, and hence there is also an infinite number of mutually equivalent, physically preferred states of motion. Time is absolute, i.e. independent of the choice of the particular inertial frame; it is defined by more characteristics than logically necessary, although—as implied by mechanics—this should not lead to contradictions with experience. Note in passing that the logical weakness of this exposition from the point of view of the principle of meaning is the lack of an experimental criterion for whether a material point is force-free or not; therefore the concept of the inertial frame remains rather problematical. This deficiency leads to the general theory of relativity. We shall not consider it for the moment.

The concept of the rigid body (and that of the clock) has a key bearing on the foregoing consideration of the fundamentals of mechanics, a bearing which there is some justification for challenging. The rigid body is only approximately achieved in Nature, not even with desired approximation; this concept does not therefore strictly satisfy the "principle of meaning." It is also logically unjustifiable to base all physical consideration on the rigid or solid body and then finally reconstruct that body anatomically by means of elementary physical laws which in turn have been determined by means of the rigid measuring body. I am mentioning these deficiencies of method because in the same sense they are also a feature of the relativity theory in the schematic exposition which I am advocating here. Certainly it would be logically more correct to begin with the whole of the laws and to apply the "principle of meaning" to this whole first, i.e. to put the unambiguous relation to the world of experience last instead of already fulfilling it in an imperfect form for an artificially isolated part, namely the space-time metric. We are not, however, sufficiently advanced in our knowledge of Nature's elementary laws to adopt this more perfect method without going out of our depth. At the close of our considerations we shall see that in the most recent studies there is an attempt, based on ideas by Levi-Civita, Weyl, and Eddington, to implement that logically purer method.

It more clearly follows from the above what is implied by "preferred states of motion." They are preferred as regards the laws of Nature. States of motion are preferred when, relative to the formulation of the laws of Nature, coordinate systems within them are distinguished in that with respect to them those laws assume a form preferred by simplicity. According to classical mechanics the states of motion of the inertial frames in this sense are physically preferred. Classical mechanics permits a distinction to be made between (absolutely) unaccelerated and accelerated motions; it also claims that velocities have only a relative existence (dependent on the selection of the inertial frame), while accelerations and rotations have an absolute existence (independent of the selection of the inertial frame). This state of affairs can be expressed thus: According to classical mechanics "velocity relativity" exists, but not "acceleration relativity." After these preliminary considerations we can pass to the actual topic of our contemplations, the relativity theory, by characterizing its development so far in terms of principles.

The special theory of relativity is an adaptation of physical principles to Maxwell-Lorentz electrodynamics. From earlier physics it takes the assumption that Euclidian geometry is valid for the laws governing the position of rigid bodies, the inertial frame and the law of inertia. The postulate of equivalence of inertial frames for the formulation of the laws of Nature is assumed to be valid for the whole of physics (special relativity principle). From Maxwell-Lorentz electrodynamics it takes the postulate of invariance of the velocity of light in a vacuum (light principle).

To harmonize the relativity principle with the light principle, the assumption that an absolute time (agreeing for all inertial frames) exists, had to be abandoned. Thus the hypothesis is abandoned that arbitrarily moved and suitably set identical clocks function in such a way that the times shown by two of them, which meet, agree. A specific time is assigned to each inertial frame; the state of motion and the time of the inertial frame are defined, in accordance with the principle of meaning, by the requirement that the light principle should apply to it. The existence of the inertial frame thus defined and the validity of the law of inertia with respect to it are assumed. The time for each inertial frame is measured by identical clocks that are stationary relative to the frame.

The laws of transformation for space coordinates and time for the transition from one inertial frame to another, the Lorentz transformations as they are termed, are unequivocally established by these definitions and the hypotheses concealed in the assumption that they are free from contradiction. Their immediate physical significance lies in the effect of the motion relative to the used inertial frame on the form of rigid bodies (Lorentz contraction) and on the rate of the clocks. According to the special relativity principle the laws of Nature must be covariant relative to Lorentz transformations; the theory thus provides a criterion for general laws of nature. It leads in particular to a modification of the Newtonian point motion law in which the velocity of light in a vacuum is considered the limiting velocity, and it also leads to the realization that energy and inertial mass are of like nature.

The special relativity theory resulted in appreciable advances. It reconciled mechanics and electrodynamics. It reduced the number of logically independent hypotheses regarding the latter. It enforced the need for a clarification of the fundamental concepts in epistemological terms. It united the momentum and energy principle, and demonstrated the like nature of mass and energy. Yet it was not entirely satisfactory—quite apart from the quantum problems, which all theory so far has been incapable of really solving. In common with classical mechanics the special relativity theory favours certain states of motion—namely those of the inertial frames—to all other states of motion. This was actually more difficult to tolerate than the preference for a single state of motion as in the case of the theory of light with a stationary ether, for this imagined a real reason for the preference, i.e. the light ether. A theory which from the outset prefers no state of motion should appear more satisfactory. Moreover the previously mentioned vagueness in the definition of the inertial frame or in the formulation of the law of inertia raises doubts which obtain their decisive importance, owing to the empirical principle for the equality of the inertial and heavy mass, in the light of the following consideration.

Let K be an inertial frame without a gravitational field, K′ a system of coordinates accelerated uniformly relative to K. The behaviour of material points relative to K′ is the same as if K′ were an inertial frame in respect of which a homogeneous gravitational field exists. On the basis of the empirically known properties of the gravitational field, the definition of the inertial frame thus proves to be weak. The conclusion is obvious that any arbitrarily moved frame of reference is equivalent to any other for the formulation of the laws of Nature, that there are thus no physically preferred states of motion at all in respect of regions of finite extension (general relativity principle).

The implementation of this concept necessitates an even more profound modification of the geometric-kinematical principles than the special relativity theory. The Lorentz contraction, which is derived from the latter, leads to the conclusion that with regard to a system K′ arbitrarily moved relative to a (gravity field free) inertial frame K, the laws of Euclidian geometry governing the position of rigid (at rest relative to K′) bodies do not apply. Consequently the Cartesian system of coordinates also loses its significance in terms of the principle of meaning. Analogous reasoning applies to time; with reference to K′ the time can no longer meaningfully be defined by the indication on identical clocks at rest relative to K′, nor by the law governing the propagation of light. Generalizing, we arrive at the conclusion that gravitational field and metric are only different manifestations of the same physical field.

We arrive at the formal description of this field by the following consideration. For each infinitesimal point-environment in an arbitrary gravitational field a local frame of coordinates can be given for such a state of motion that relative to this local frame no gravitational field exists (local inertial frame). In terms of this interial frame we may regard the results of the special relativity theory as correct to a first approximation for this infinitesimally small region. There are an infinite number of such local inertial frames at any space-time point; they are associated by Lorentz transformations. These latter are characterised in that they leave invariant the "distance" ds of two infinitely adjacent point events—defined by the equation:

$$ds^2 = c^2dt^2 - dx^2 - dy^2 - dz^2$$

which distance can be measured by means of scales and clocks. For, x, y, z, t represent coordinates and time measured with reference to a local inertial frame.

To describe space-time regions of finite extent arbitrary point coordinates in four dimensions are required which serve no other purpose than to provide an unambiguous designation of the space-time points by four numbers each, x_1, x_2, x_3 and x_4, which takes account of the continuity of this four-dimensional manifold (Gaussian coordinates). The mathematical expression of the general relativity principle is then, that the systems of equations expressing the general laws of Nature are equal for all such systems of coordinates.

Since the coordinate differentials of the local inertial frame are expressed linearly by the differentials dx_ν of a Gaussian system of coordinates, when the latter is used, for the distance ds of two events an expression of the form

$$ds^2 = \Sigma g_{\mu\nu}\, dx_\mu dx_\nu \; (g_{\mu\nu} = g_{\nu\mu})$$

is obtained. The $g_{\mu\nu}$, which are continuous functions of x_ν, determine the metric in the four-dimensional manifold where ds is defined as an (absolute) parameter measurable by means of rigid scales and clocks. These same parameters $g_{\mu\nu}$ however also describe with reference to the Gaussian system of coordinates the gravitational field which we have previously found to be identical with the physical cause of the metric. The case in which the special relativity theory is valid for finite regions is characterized by the fact that when the system of coordinates is suitably chosen, the values of $g_{\mu\nu}$ for finite regions are independent of x_ν.

In accordance with the general theory of relativity the law of point motion in the pure gravitational field is expressed by the equation for the geodesic curve. Actually the geodesic curve is the simplest mathematically that in the special case of constant $g_{\mu\nu}$ becomes rectilinear. Here therefore we are confronted with the transfer of Galileo's law of inertia to the general theory of relativity.

In mathematical terms the search for the field equations amounts to ascertaining the simplest generally covariant differential equations to which the gravitational potentials $g_{\mu\nu}$ can be subjected. By definition these equations should not contain higher derivatives of $g_{\mu\nu}$ with respect to x_ν than the second, and these only linearly, which condition reveals these equations to be a logical transfer of the Poisson field equation of the Newtonian theory of gravity to the general theory of relativity.

The considerations mentioned led to the theory of gravity which yields the Newtonian theory as a first approximation and furthermore it yields the motion of the perihelion of Mercury, the deflection of light by the sun, and the red shift of spectral lines in agreement with experience.*

To complete the basis of the general theory of relativity, the electromagnetic field must still be introduced into it which, according to our present conviction, is also the material from which we must build up the elementary structures of matter. The Maxwellian field equations can readily be adopted into the general theory of relativity. This is a completely unambiguous adoption provided it is assumed that the equations contain no differential quotients of $g_{\mu\nu}$ higher than the first, and that in the customary Maxwellian form they apply in the local inertial frame. It is also easily possible to supplement the gravitational field equations by electromagnetic terms in a manner specified by the Maxwellian equations so that they contain the gravitational effect of the electromagnetic field.

These field equations have not provided a theory of matter. To incorporate the field generating effect of ponderable masses in the theory, matter had therefore (as in classical physics) to be introduced into the theory in an approximate, phenomenological representation.

*As regards the red shift, the agreement with experience is not yet completely assured, however.

And that exhausts the direct consequences of the relativity principle. I shall turn to those problems which are related to the development which I have traced. Already Newton recognized that the law of inertia is unsatisfactory in a context so far unmentioned in this exposition, namely that it gives no real cause for the special physical position of the states of motion of the inertial frames relative to all other states of motion. It makes the observable material bodies responsible for the gravitational behaviour of a material point, yet indicates no material cause for the inertial behaviour of the material point but devises the cause for it (absolute space or inertial ether). This is not logically inadmissible although it is unsatisfactory. For this reason E. Mach demanded a modification of the law of inertia in the sense that the inertia should be interpreted as an acceleration resistance of the bodies against *one another* and not against "space." This interpretation leads to the expectation that accelerated bodies have concordant accelerating action in the same sense on other bodies (acceleration induction).

This interpretation is even more plausible according to general relativity which eliminates the distinction between inertial and gravitational effects. It amounts to stipulating that, apart from the arbitrariness governed by the free choice of coordinates, the $g_{\mu\nu}$-field shall be completely determined by the matter. Mach's principle is favoured in general relativity by the circumstance that acceleration induction in accordance with the gravitational field equations really exists, although of such slight intensity that direct detection by mechanical experiments is out of the question.

Mach's principle can be accounted for in the general theory of relativity by regarding the world in spatial terms as finite and self-contained. This hypothesis also makes it possible to assume the mean density of matter in the world as *finite,* whereas in a spatially infinite (quasi-Euclidian) world it should disappear. It cannot, however, be concealed that to satisfy Mach's postulate in the manner referred to a term with no experimental basis whatsoever must be introduced into the field equations, which term logically is in no way determined by the other terms in the equations. For this reason this solution of the "cosmological problem" will not be completely satisfactory for the time being.

A second problem which at present is the subject of lively interest is the identity between the gravitational field and the electromagnetic field. The mind striving after unification of the theory cannot be satisfied that two fields should exist which, by their nature, are quite independent. A mathematically unified field theory is sought in which the gravitational field and the electromagnetic field are interpreted only as different components or manifestations of the same uniform field, the field equations where possible no longer consisting of logically mutually independent summands.

The gravitational theory, considered in terms of mathematical formalism, i.e. Riemannian geometry, should be generalized so that it includes the laws of the electromagnetic field. Unfortunately we are unable here to base ourselves on empirical facts as when deriving the gravitational theory (equality of the inertial and heavy mass), but we are restricted to the criterion of mathematical simplicity which is not free from arbitrariness. The attempt which at present appears the most successful is that, based on the ideas of Levi-Civita, Weyl and Eddington, to replace Riemannian metric geometry by the more general theory of affine connection.

The characteristic assumption of Riemannian geometry is the attribution to two infinitely adjacent points of a "distance" ds, the square of which is a homogeneous second order function of the coordinate differentials. It follows from this that (apart from certain conditions of reality) Euclidian geometry is valid in any infinitely small region. Hence to every line element (or vector) at a point P is assigned a parallel and equal line element (or vector) through any given infinitesimally adjacent point P′ (affine connection). A Riemannian metric determines an affine connection. Conversely, however, when an affine connection (law of infinitesimal parallel displacement) is mathematically given, generally no Riemannian metric exists from which it can be derived.

The most important concept of Riemannian geometry, "space curvature" on which the gravitational equations are also based, is based exclusively on the "affine correction." If one is given in a continuum, without first proceeding from a metric, it constitutes a generalization of Riemannian geometry but one which still retains the most important derived parameters. By seeking the simplest differential equations which can be obeyed by an affine connection there is reason to hope that a generalization of the gravitation equations will be found which includes the laws of the electromagnetic field. This hope has in fact been fulfilled although I do not know whether the formal connection so derived can really be regarded as an enrichment of physics as long as it does not yield any new physical connections. In particular a field theory can, to my mind, only be satisfactory when it permits the elementary electrical bodies to be represented as solutions free from singularities.

Moreover it should not be forgotten that a theory relating to the elementary electrical structures is inseparable from the quantum theory problems. So far also relativity theory has proved ineffectual in relation to this most profound physical problem of the present time. Should the form of the general equations some day, by the solution of the quantum problem, undergo a change however profound, even if there is a complete change in the parameters by means of which we represent the elementary process, the relativity principle will not be relinquished and the laws previously derived therefrom will at least retain their significance as limiting laws.

Geometry and Experience

ALBERT EINSTEIN

All the objects of human reason or enquiry may naturally be divided into two kinds, to wit, Relations of Ideas, *and* Matters of Fact. *Of the first kind are the sciences of Geometry, Algebra, and Arithmetic; and in short, every affirmation which is either intuitively or demonstratively certain.* That the square of the hypothenuse is equal to the square of the two sides, *is a proposition which expresses a relation between these figures.* That three times five is equal to the half of thirty, *expresses a relation between these numbers. Propositions of this kind are discoverable by the mere operation of thought, without dependence on what is anywhere existent in the universe. Though there never were a circle or triangle in nature, the truths demonstrated by Euclid would for ever retain their certainty and evidence.*

—DAVID HUME, *An Inquiry Concerning Human Understanding*

That all our knowledge begins with experience, there is indeed no doubt . . . but although our knowledge originates with experience, it does not all arise out of experience.

—IMMANUEL KANT, *Critique of Pure Reason*

ONE REASON WHY mathematics enjoys special esteem, above all other sciences, is that its laws are absolutely certain and indisputable, while those of all other sciences are to some extent debatable and in constant

Albert Einstein, "Geometry and Experience," in *Sidelights on Relativity.* New York: Dover Publications, Inc., 1983, pp. 27–56. Reprinted with permission of the Hebrew University of Jerusalem, Israel.

danger of being overthrown by newly discovered facts. In spite of this, the investigator in another department of science would not need to envy the mathematician if the laws of mathematics referred to objects of our mere imagination, and not to objects of reality. For it cannot occasion surprise that different persons should arrive at the same logical conclusions when they have already agreed upon the fundamental laws (axioms), as well as the methods by which other laws are to be deduced therefrom. But there is another reason for the high repute of mathematics, in that it is mathematics which affords the exact natural sciences a certain measure of security, to which without mathematics they could not attain.

At this point an enigma presents itself which in all ages has agitated inquiring minds. How can it be that mathematics, being after all a product of human thought which is independent of experience, is so admirably appropriate to the objects of reality? Is human reason, then, without experience, merely by taking thought, able to fathom the properties of real things.

In my opinion the answer to this question is, briefly, this:—As far as the laws of mathematics refer to reality, they are not certain; and as far as they are certain, they do not refer to reality. It seems to me that complete clearness as to this state of things first became common property through that new departure in mathematics which is known by the name of mathematical logic or "Axiomatics." The progress achieved by axiomatics consists in its having neatly separated the logical-formal from its objective or intuitive content; according to axiomatics the logical-formal alone forms the subject-matter of mathematics, which is not concerned with the intuitive or other content associated with the logical-formal.

Let us for a moment consider from this point of view any axiom of geometry, for instance, the following:—Through two points in space there always passes one and only one straight line. How is this axiom to be interpreted in the older sense and in the more modern sense?

The older interpretation:—Every one knows what a straight line is, and what a point is. Whether this knowledge springs from an ability of the human mind or from experience, from some collaboration of the two or from some other source, is not for the mathematician to decide. He leaves the question to the philosopher. Being based upon this knowledge, which precedes all mathematics, the axiom stated above is, like all other axioms, self-evident, that is, it is the expression of a part of this *a priori* knowledge.

The more modern interpretation:—Geometry treats of entities which are denoted by the words straight line, point, etc. These entities do not take for granted any knowledge or intuition whatever, but they presuppose only the validity of the axioms, such as the one stated above, which are to be taken in a purely formal sense, i.e. as void of all content of intuition or experience. These axioms are free creations of the human mind. All other propositions of geometry are logical inferences from the axioms (which are to be taken in the nominalistic sense only). The matter of which geometry treats is first defined by the axioms. Schlick in his book on epistemology has therefore characterised axioms very aptly as "implicit definitions."

This view of axioms, advocated by modern axiomatics, purges mathematics of all extraneous elements, and thus dispels the mystic obscurity which formerly surrounded the principles of mathematics. But a presentation of its principles thus clarified makes it also evident that mathematics as such cannot predicate anything about perceptual objects or real objects. In axiomatic geometry the words "point," "straight line," etc., stand only for empty conceptual schemata. That which gives them substance is not relevant to mathematics.

Yet on the other hand it is certain that mathematics generally, and particularly geometry, owes its existence to the need which was felt of learning something about the relations of real things to one another. The very word geometry, which, of course, means earth-measuring, proves this. For earth-measuring has to do with the possibilities of the disposition of certain natural objects with respect to one another, namely, with parts of the earth, measuring-lines, measuring-wands, etc. It is clear that the system of concepts of axiomatic geometry alone cannot make any assertions as to the relations of real objects of this kind, which we will call practically-rigid bodies. To be able to make such assertions, geometry must be stripped of its merely logical-formal character by the co-ordination of real objects of experience with the empty conceptual frame-work of axiomatic geometry. To accomplish this, we need only add the proposition:—Solid bodies are related, with respect to their possible dispositions, as are bodies in Euclidean geometry of three dimensions. Then the propositions of Euclid contain affirmations as to the relations of practically-rigid bodies.

Geometry thus completed is evidently a natural science; we may in fact regard it as the most ancient branch of physics. Its affirmations rest essentially on induction from experience, but not on logical inferences only. We will call this completed geometry "practical geometry," and shall distinguish it in what follows from "purely axiomatic geometry." The question whether the practical geometry of the universe is Euclidean or not has a clear meaning, and its answer can only be furnished by experience. All linear measurement in physics is practical geometry in this sense, so too is geodetic and astronomical linear measurement, if we call to our help the law of experience that light is propagated in a straight line, and indeed in a straight line in the sense of practical geometry.

I attach special importance to the view of geometry which I have just set forth, because without it I should have been unable to formulate the theory of relativity. Without it the following reflection would have been impossible:—In a system of reference rotating relatively to an inert system, the laws of disposition of rigid bodies do not correspond to the rules of Euclidean geometry on account of the Lorentz contraction; thus if we admit non-inert systems we must abandon Euclidean geometry. The decisive step in the transition to general co-variant equations would certainly not have been taken if the above interpretation had not served as a stepping-stone. If we deny the relation between the body of axiomatic Euclidean geometry and the practically-rigid body of reality, we readily arrive at the following view, which was entertained by that acute and profound thinker, H. Poincaré:—Euclidean geometry is distinguished above all other imaginable axiomatic geometries by its simplicity. Now since axiomatic geometry by itself contains no assertions as to the reality which can be experienced, but can do so only in combination with physical laws, it should be possible and reasonable—whatever may be the nature of reality—to retain Euclidean geometry. For if contradictions between theory and experience manifest themselves, we should rather decide to change physical laws than to change axiomatic Euclidean geometry. If we deny the relation between the practically-rigid body and geometry, we shall indeed not easily free ourselves from the convention that Euclidean geometry is to be retained as the simplest. Why is the equivalence of the practically-rigid body and the body of geometry—which suggests itself so readily—denied by Poincaré and other investigators? Simply because under closer inspection the real solid bodies in nature are not rigid, because their geometrical behaviour, that is, their possibilities of relative disposition, depend upon temperature, external forces, etc. Thus the original, immediate relation between geometry and physical reality appears destroyed, and we feel impelled toward the following more general view, which characterizes Poincaré's standpoint. Geometry (G) predicates nothing about the relations of real things, but only geometry together with the purport (P) of physical laws can do so. Using symbols, we may say that only the sum of (G) + (P) is subject to the control of experience. Thus (G) may be chosen arbitrarily, and also parts of (P); all these laws are conventions. All that is necessary to avoid contradictions is to choose the remainder of (P) so that (G) and the whole of (P) are together in accord with experience. Envisaged in this way, axiomatic geometry and the part of natural law which has been given a conventional status appear as epistemologically equivalent.

Sub specie aeterni Poincaré, in my opinion, is right. The idea of the measuring-rod and the idea of the clock co-ordinated with it in the theory of relativity do not find their exact correspondence in the real world. It is also clear that the solid body and the clock do not in the conceptual edifice of physics play the part of irreducible elements, but that of composite structures, which may not play any independent part in theoretical physics. But it is my conviction that in the present stage of development of theoretical physics these ideas must still be employed as independent ideas; for we are still far from possessing such certain knowledge of theoretical principles as to be able to give exact theoretical constructions of solid bodies and clocks.

Further, as to the objection that there are no really rigid bodies in nature, and that therefore the properties predicated of rigid bodies do not apply to physical reality,—this objection is by no means so radical as might appear from a hasty examination. For it is not a difficult task to determine the physical state of a measuring-rod so accurately that its behaviour relatively to other measuring-bodies shall be sufficiently free from ambiguity to allow it to be substituted for the "rigid" body. It is to measuring-bodies of this kind that statements as to rigid bodies must be referred.

All practical geometry is based upon a principle which is accessible to experience, and which we will now try to realise. We will call that which is enclosed between two boundaries, marked upon a practically-rigid body, a tract. We imagine two practically-rigid bodies, each with a tract marked out on it. These two tracts are said to be "equal to one another" if the boundaries of the one tract can be brought to coincide permanently with the boundaries of the other. We now assume that:

If two tracts are found to be equal once and anywhere, they are equal always and everywhere.

Not only the practical geometry of Euclid, but also its nearest generalisation, the practical geometry of Riemann, and therewith the general theory of relativity, rests upon this assumption. Of the experimental reasons which warrant this assumption I will mention only one. The phenomenon of the propagation of light in empty space assigns a tract, namely, the appropriate path of light, to each interval of local time, and conversely.

Thence it follows that the above assumption for tracts must also hold good for intervals of clock-time in the theory of relativity. Consequently it may be formulated as follows:—If two ideal clocks are going at the same rate at any time and at any place (being then in immediate proximity to each other), they will always go at the same rate, no matter where and when they are again compared with each other at one place.—If this law were not valid for real clocks, the proper frequencies for the separate atoms of the same chemical element would not be in such exact agreement as experience demonstrates. The existence of sharp spectral lines is a convincing experimental proof of the above-mentioned principle of practical geometry. This is the ultimate foundation in fact which enables us to speak with meaning of the mensuration, in Reimann's sense of the word, of the four-dimensional continuum of space-time.

The question whether the structure of this continuum is Euclidean, or in accordance with Riemann's general scheme, or otherwise, is, according to the view which is here being advocated, properly speaking a physical question which must be answered by experience, and not a question of a mere convention to be selected on practical grounds. Riemann's geometry will be the right thing if the laws of disposition of practically-rigid bodies are transformable into those of the bodies of Euclid's geometry with an exactitude which increases in proportion as the dimensions of the part of space-time under consideration are diminished.

It is true that this proposed physical interpretation of geometry breaks down when applied immediately to spaces of sub-molecular order of magnitude. But nevertheless, even in questions as to the constitution of elementary particles, it retains part of its importance. For even when it is a question of describing the electrical elementary particles constituting matter, the attempt may still be made to ascribe physical importance to those ideas of fields which have been physically defined for the purpose of describing the geometrical behaviour of bodies which are large as compared with the molecule. Success alone can decide as to the justification of such an attempt, which postulates physical reality for the fundamental principles of Riemann's geometry outside of the domain of their physical definitions. It might possibly turn out that this extrapolation has no better warrant than the extrapolation of the idea of temperature to parts of a body of molecular order of magnitude.

It appears less problematical to extend the ideas of practical geometry to spaces of cosmic order of magnitude. It might, of course, be objected that a construction composed of solid rods departs more and more from ideal rigidity in proportion as its spatial extent becomes greater. But it will hardly be possible, I think, to assign fundamental significance to this objection. Therefore the question whether the universe is spatially finite or not seems to me decidedly a pregnant question in the sense of practical geometry. I do not even consider it impossible that this question will be answered before long by astronomy. Let us call to mind what the general theory of relativity teaches in this respect. It offers two possibilities:

1. The universe is spatially infinite. This can be so only if the average spatial density of the matter in universal space, concentrated in the stars, vanishes, i.e. if the ratio of the total mass of the stars to the magnitude of the space through which they are scattered approximates indefinitely to the value zero when the spaces taken into consideration are constantly greater and greater.

2. The universe is spatially finite. This must be so, if there is a mean density of the ponderable matter in universal space differing from zero. The smaller that mean density, the greater is the volume of universal space.

I must not fail to mention that a theoretical argument can be adduced in favour of the hypothesis of a finite universe. The general theory of relativity teaches that the inertia of a given body is greater as there are more ponderable masses in proximity to it; thus it seems very natural to reduce the total effect of inertia of a body to action and reaction between it and the other bodies in the universe, as indeed, ever since Newton's time, gravity has been completely reduced to action and reaction between bodies. From the equations of the general theory of relativity it can be deduced that this total reduction of inertia to reciprocal action between masses—as required by E. Mach, for example—is possible only if the universe is spatially finite.

On many physicists and astronomers this argument makes no impression. Experience alone can finally decide which of the two possibilities is realised in nature. How can experience furnish an answer? At first it might seem possible to determine the mean density of matter by observation of that party of the universe which is accessible to our perception. This hope is illusory. The distribution of the visible stars is extremely irregular, so that we on no account may venture to set down the mean density of star-matter in the universe as equal, let us say, to the mean density in the Milky Way. In any case, however great the space examined may be, we could not feel convinced that there were no more stars beyond that space. So it seems impossible to estimte the mean density.

But there is another road, which seems to me more practicable, although it also presents great difficulties. For if we inquire into the deviations shown by the consequences of the general theory of relativity which are accessible to experience, when these are compared with the consequences of the Newtonian theory, we first of all find a deviation which shows itself in close proximity to gravitating mass, and has been confirmed in the case of the planet Mercury. But if the universe is spatially finite there is a second deviation from the Newtonian theory, which, in the language of the Newtonian theory, may be expressed thus:—The gravitational field is in its nature such as if it were produced, not only by the ponderable masses, but also by a mass-density of negative sign, distributed uniformly throughout space. Since this factitious mass-density would have to be enormously small, it could make its presence felt only in gravitating systems of very great extent.

Assuming that we know, let us say, the statistical distribution of the stars in the Milky Way, as well as their masses, then by Newton's law we can calculate the gravitational field and the mean velocities which the stars must have, so that the Milky Way should not collapse under the mutual attraction of its stars, but should maintain its actual extent. Now if the actual velocities of the stars, which can, of course, be measured, were smaller than the calculated velocities, we should have a proof that the actual attractions at great distances are smaller than by Newton's law. From such a deviation it could be proved indirectly that the universe is finite. It would even be possible to estimate its spatial magnitude.

Can we picture to ourselves a three-dimensional universe which is finite, yet unbounded?

The usual answer to this question is "No," but that is not the right answer. The purpose of the following remarks is to show that the answer should be "Yes." I want to show that without any extraordinary difficulty we can illustrate the theory of a finite universe by means of a mental image to which, with some practice, we shall soon grow accustomed.

First of all, an observation of epistemological nature. A geometrical-physical theory as such is incapable of being directly pictured, being merely a system of concepts. But these concepts serve the purpose of bringing a multiplicity of real or imaginary sensory experiences into connection in the end. To "visualise" a theory, or bring it home to one's mind, therefore means to give a representation to that abundance of experiences for which the theory supplies the schematic arrangement. In the present case we have to ask ourselves how we can represent that relation of solid bodies with respect to their reciprocal disposition (contact) which corresponds to the theory of a finite universe. There is really nothing new in what I have to say about this; but innumerable questions addressed to me prove that the requirements of those who thirst for knowledge of these matters have not yet been completely satisfied. So, will the initiated please pardon me, if part of what I shall bring forward has long been known?

What do we wish to express when we say that our space is infinite? Nothing more than that we might lay any number whatever of bodies of equal sizes side by side without ever filling space. Suppose that we are provided with a great many wooden cubes all of the same size. In accordance with Euclidean geometry we can place them above, beside, and behind one another so as to fill a part of space of any dimensions; but this construction would never be finished; we could go on adding more and more cubes without ever finding that there was no more room. That is what we wish to express when we say that space is infinite. It would be better to say that space is infinite in relation to practically-rigid bodies, assuming that the laws of disposition for these bodies are given by Euclidean geometry.

Another example of an infinite continuum is the plane. On a plane surface we may lay squares of cardboard so that each side of any square has the side of another square adjacent to it. The construction is never finished; we can always go on laying squares—if their laws of disposition correspond to those of plane figures of Euclidean geometry. The plane is therefore infinite in relation to the cardboard squares. Accordingly we say that the plane is an infinite continuum of two dimensions, and space an infinite continuum of three dimensions. What is here meant by the number of dimensions, I think I may assume to be known.

Now we take an example of a two-dimensional continuum which is finite, but unbounded. We imagine the surface of a large globe and a quantity of small paper discs, all of the same size. We place one of the discs anywhere on the surface of the globe. If we move the disc about, anywhere we like, on the surface of the globe, we do not come upon a limit or boundary anywhere on the journey. Therefore we say that the spherical surface of the globe is an unbounded continuum. Moreover, the spherical surface is a finite continuum. For if we stick the paper discs on the globe, so that no disc overlaps another, the surface of the globe will finally become so full that there is no room for another disc. This simply means that the spherical surface of the globe is finite in relation to the paper discs. Further, the spherical surface is a non-Euclidean continuum of two dimensions, that is to say, the laws of disposition for the rigid figures lying in it do not agree with those of the Euclidean plane. This can be shown in the following way. Place a paper disc on the spherical sur-

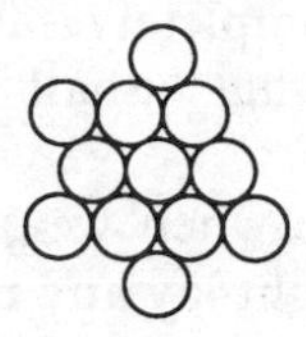

FIG. 1

face, and around it in a circle place six more discs, each of which is to be surrounded in turn by six discs, and so on. If this construction is made on a plane surface, we have an uninterrupted disposition in which there are six discs touching every disc except those which lie on the outside. On the spherical surface the construction also seems to promise success at the outset and the smaller the radius of the disc in proportion to that of the sphere, the more promising it seems. But as the construction progresses it becomes more and more patent that the disposition of the discs in the manner indicated, without interruption, is not possible, as it should be possible by Euclidean geometry of the plane surface. In this way creatures which cannot leave the spherical surface, and cannot even peep out from the spherical surface into three-dimensional space, might discover, merely by experimenting with discs, that their two-dimensional "space" is not Euclidean, but spherical space.

From the latest results of the theory of relativity it is probable that our three-dimensional space is also approximately spherical, that is, that the laws of disposition of rigid bodies in it are not given by Euclidean geometry, but approximately by spherical geometry, if only we consider parts of space which are sufficiently great. Now this is the place where the reader's imagination boggles. "Nobody can imagine this thing," he cries indignantly. "It can be said, but cannot be thought. I can represent to myself a spherical surface well enough, but nothing analogous to it in three dimensions."

We must try to surmount this barrier in the mind, and the patient reader will see that it is by no means a particularly difficult task. For this purpose we will first give our attention once more to the geometry of two-dimensional spherical surfaces. In the adjoining figure let *K* be the spherical surface, touched at *S* by a plane, *E*, which, for facility of presentation, is shown in the drawing as a bounded surface. Let *L* be a disc on the spherical surface. Now let us imagine that at the point *N* of the spherical surface:

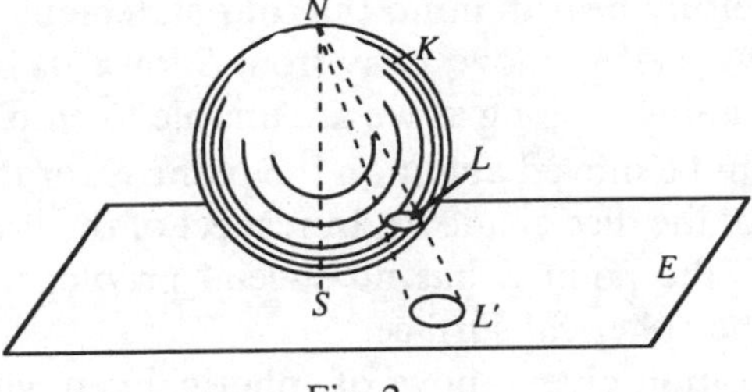

Fig. 2.

diametrically opposite to *S*, there is a luminous point, throwing a shadow *L′* of the disc *L* upon the plane *E*. Every point on the sphere has its shadow on the plane. If the disc on the sphere *K* is moved, its shadow *L′* on the plane *E* also moves. When the disc *L* is at *S*, it almost exactly coincides with its shadow. If it moves on the spherical surface away from *S* upwards, the disc shadow *L′* on the plane also moves away from *S* on the plane outwards, growing bigger and bigger. As the disc *L* approaches the luminous point *N*, the shadow moves off to infinity, and becomes infinitely great.

Now we put the question, What are the laws of disposition of the disc-shadows *L′* on the plane *E*? Evidently they are exactly the same as the laws of disposition of the discs *L* on the spherical surface. For to each original figure on *K* there is a corresponding shadow figure on *E*. If two discs on *K* are touching, their shadows on *E* also touch. The shadow-geometry on the plane agrees with the disc-geometry on the sphere. If we call the disc-shadows rigid figures, then spherical geometry holds good on the plane *E* with respect to these rigid figures. Moreover, the plane is finite with respect to the disc-shadows, since only a finite number of the shadows can find room on the plane.

At this point somebody will say, "That is nonsense. The disc-shadows are *not* rigid figures. We have only to move a two-foot rule about on the plane *E* to convince ourselves that the shadows constantly increase in size as they move away from *S* on the plane towards infinity." But what if the two-foot rule were to behave on the plane *E* in the same way as the disc-shadows *L′*? It would then be impossible to show that the shadows increase in size as they move away from *S*; such an assertion would then no longer have any meaning whatever. In fact the only objective assertion that can be made about the disc-shadows is just this, that they are related in exactly the same way as are the rigid discs on the spherical surface in the sense of Euclidean geometry.

We must carefully bear in mind that our statement as to the growth of the disc-shadows, as they move away from *S* towards infinity, has in itself no objective meaning, as long as we are unable to employ Euclidean rigid bodies which can be moved about on the plane *E* for the purpose of comparing the size of the disc-shadows. In respect of the laws of disposition of the shadows *L′*, the point *S* has no special privileges on the plane any more than on the spherical surface.

The representation given above of spherical geometry on the plane is important for us, because it readily allows itself to bc transferred to the three-dimensional case.

Let us imagine a point *S* of our space, and a great number of small spheres, *L′*, which can all be brought to coincide with one another. But these spheres are not to be rigid in the sense of Euclidean geometry; their radius is to increase (in the sense of Euclidean geometry) when they are moved away from *S* towards infinity, and this increase is to take place in exact accordance with the same law as applies to the increase of the radii of the disc-shadows *L′* on the plane.

After having gained a vivid mental image of the geometrical behaviour of our *L′* spheres, let us assume that in our space there are no rigid bodies at all in the sense of Euclidean geometry, but only bodies having the behaviour of our *L′* spheres. Then we shall have a vivid representation of three-dimensional spherical space, or, rather of three-dimensional spherical geometry. Here our spheres must be called "rigid" spheres. Their increase in size as they depart from *S* is not to be detected by measuring with measuring-rods, any more than in the case of the disc-shadows on *E*, because the standards of measurement behave in the same way as the spheres. Space is homogeneous, that is to say, the same spherical configurations are possible in the environment of all points.* Our space is finite, because, in consequence of the "growth" of the spheres, only a finite number of them can find room in space.

In this way, by using as stepping-stones the practice in thinking and visualisation which Euclidean geometry gives us, we have acquired a mental picture of spherical geometry. We may without difficulty impart more depth and vigour to these ideas by carrying out special imaginary constructions. Nor would it be difficult to represent the case of what is called elliptical geometry in an analogous manner. My only aim to-day has been to show that the human faculty of visualisation is by no means bound to capitulate to non-Euclidean geometry.

---

* This is intelligible without calculation—but only for the two-dimensional case—if we revert once more to the case of the disc on the surface of the sphere.

M.5

The New Law of Gravitation and the Old Law

ARTHUR S. EDDINGTON

ARTHUR STANLEY EDDINGTON was born in 1882 in Kendal, England. His family moved to Somerset in 1884 after the death of his father, a headmaster at a local school. Although the Eddington family had little money, Arthur's mother managed to send him to private schools. Eddington graduated from what is now the University of Manchester in 1902 and received his doctorate from Cambridge University in 1905. He worked as chief assistant to the Royal Astronomer until 1913 before returning to Cambridge to accept an appointment as Professor of Astronomy, a post that he held until his death in 1944.

Eddington's study of the general theory of relativity began when Willem de Sitter, the Dutch astronomer, forwarded a copy of Einstein's theory to Eddington, who was then secretary of the Royal Astronomical Society. For several years, it remained the only copy of the theory in England. Eddington immediately recognized its importance and began to teach himself the intricacies of its mathematical details. At the request of the Physical Society of London, Eddington prepared his *Report on the Relativity Theory of Gravitation,* which was published in 1918 and "is a masterpiece of concise and elegant exposition." Eddington's report was the first complete account of general relativity written in English. He revised the report in 1920 to include the results of his own 1919 eclipse expedition to the Isle of Principe in the Gulf of Guinea, which had confirmed a central prediction of Einstein's general theory: the bending of light by the gravitational field of matter. (Eddington himself was so busy changing photographic plates during the eclipse that he did not actually see it.)

In 1920, Eddington published a less technical account of relativity, *Space, Time and Gravitation,* from which the present selection has been excerpted. This book provided one of the earliest nontechnical treatments of relativity theory. Three years later Eddington published his most famous book on Einstein's theories, *Mathematical Theory of Relativity.* Einstein himself considered this book to be the finest presentation of relativity theory in any language, although its mathematics limited its accessibility. Eddington highlighted the work of Einstein, de Sitter and Weyl in his book. His own original contributions to the subject included the interpretation and derivation of relevant equations and modest modification of Weyl's geometry of world structure based on the notion of parallel displacement.

Having got what we want out of it . . . space curvature no longer interests us. We turn to flat space. . . . The scale of uncertainty, instead of being disguised as curvature, will be taken into account openly.

—ARTHUR S. EDDINGTON, *Fundamental Theory*

I DON'T KNOW what I may seem to the world, but, as to myself, I seem to have been only as a boy playing on the sea-shore, and diverting myself in now and then finding a smoother pebble or a prettier shell than ordinary, whilst the great ocean of truth lay all undiscovered before me.

SIR ISAAC NEWTON

Was there any reason to feel dissatisfied with Newton's law of gravitation?

Observationally it had been subjected to the most stringent tests, and had come to be regarded as the perfect model of an exact law of nature. The cases, where a possible failure could be alleged, were almost insignificant. There are certain unexplained irregularities in the moon's motion; but astronomers generally looked—and must still look—in other directions for the cause of these discrepancies. One failure only had led to a serious questioning of the law; this was the discordance of motion of the perihelion of Mercury. How small was this discrepancy may be judged from the fact that, to meet it, it was proposed to amend square of the distance to the 2.00000016 power of the distance. Further it seemed possible, though unlikely, that the matter causing the zodiacal light might be of sufficient mass to be responsible for this effect.

The most serious objection against the Newtonian law as an exact law was that it had become ambiguous. The law refers to the product of the masses of the two bodies, but the mass depends on the velocity—a fact unknown in Newton's day. Are we to take the variable mass or the mass reduced to rest? Perhaps a learned judge, interpreting Newton's statement like a last will and testament, could give a decision, but that is scarcely the way to settle an important point in scientific theory.

Arthur S. Eddington, "The New Law of Gravitation and the Old Law," in *Space, Time and Gravitation.* London: Cambridge University Press, 1920, pp. 93–109.

Further, *distance,* also referred to in the law, is something relative to an observer. Are we to take the observer travelling with the sun or with the other body concerned, or at rest in the æther or in some gravitational medium? . . .

It is often urged that Newton's law of gravitation is much simpler than Einstein's new law. That depends on the point of view; and from the point of view of the four-dimensional world Newton's law is far more complicated. Moreover, it will be seen that if the ambiguities are to be cleared up, the statement of Newton's law must be greatly expanded.

Some attempts have been made to expand Newton's law on the basis of the restricted principle of relativity alone. This was insufficient to determine a definite amendment. Using the principle of equivalence, or relativity of force, we have arrived at a definite law proposed in the last chapter. Probably the question has risen in the reader's mind, why should it be called the law of gravitation? It may be plausible as a law of nature; but what has the degree of curvature of space-time to do with attractive forces, whether real or apparent?

A race of flat-fish once lived in an ocean in which there were only two dimensions. It was noticed that in general fishes swam in straight lines, unless there was something obviously interfering with their free courses. This seemed a very natural behaviour. But there was a certain region where all the fish seemed to be bewitched; some passed through the region but changed the direction of their swim, others swam round and round indefinitely. One fish invented a theory of vortices, and said that there were whirlpools in that region which carried everything round in curves. By-and-by a far better theory was proposed; it was said that the fishes were all attracted towards a particular large fish—a sun-fish—which was lying asleep in the middle of the region; and that was what caused the deviation of their paths. The theory might not have sounded particularly plausible at first; but it was confirmed with marvellous exactitude by all kinds of experimental tests. All fish were found to possess this attractive power in proportion to their sizes; the law of attraction was extremely simple, and yet it was found to explain all the motions with an accuracy never approached before in any scientific investigations. Some fish grumbled that they did not see how there could be such an influence at a distance; but it was generally agreed that the influence was communicated through the ocean and might be better understood when more was known about the nature of water. Accordingly, nearly every fish who wanted to explain the attraction started by proposing some kind of mechanism for transmitting it through the water.

But there was one fish who thought of quite another plan. He was impressed by the fact that whether the fish were big or little they always took the same course, although it would naturally take a bigger force to deflect the bigger fish. He therefore concentrated attention on the courses rather than on the forces. And then he arrived at a striking explanation of the whole thing. There was a mound in the world round about where the sun-fish lay. Flat-fish could not appreciate it directly because they were two-dimensional; but whenever a fish went swimming over the slopes of the mound, although he did his best to swim straight on, he got turned round a bit. (If a traveller goes over the left slope of a mountain he must consciously keep bearing away to the left if he wishes to keep to his original direction relative to the point of the compass.) This was the secret of the mysterious attraction, or bending of the paths, which was experienced in the region.

The parable is not perfect, because it refers to a hummock in space alone, whereas we have to deal with hummocks in space-time. But it illustrates how a curvature of the world we live in may give an illusion of attractive force, and indeed can only be discovered through some such effect. How this works out in detail must now be considered.

In the form $G_{\mu\nu} = 0$, Einstein's law expresses conditions to be satisfied in a gravitational field produced by any arbitrary distribution of attracting matter. An analogous form of Newton's law was given by Laplace in his celebrated expression $\nabla^2 V = 0$. A more illuminating form of the law is obtained if, instead of putting the question what kinds of space-time can exist under the most general conditions in an empty region, we ask what kind of space-time exists in the region round a single attracting particle? We separate out the effect of a single particle, just as Newton did. . . .

We need only consider space of two dimensions—sufficient for the so-called plane orbit of a planet—time being added as the third dimension. The remaining dimension of space can always be added, if desired, by conditions of symmetry. The result of long algebraic calculations is that round a particle

$$ds^2 = -\frac{1}{\gamma}\,dr^2 - r^2 d\theta^2 + \gamma dt^2 \qquad (1)$$

where

$$\gamma = 1 - \frac{2m}{r}$$

The quantity m is the gravitational mass of the particle—but we are not supposed to know that at present, r and θ are polar coordinates, or rather they are the nearest thing to polar coordinates that can be found in space which is not truly flat.

The fact is that this expression for ds^2 is found in the first place simply as a particular solution of Einstein's equations of the gravitational field; it is a variety of hummock (apparently the simplest variety) which is not curved beyond the first degree. There *could* be such a state of the world under suitable circumstances. To find out what those circumstances are, we have to trace some of the consequences, find out how any particle moves when ds^2 is of this form, and then examine whcther we know of any case in which these consequences are found observationally. It is only after having ascertained that this form of ds^2 does correspond to the leading observed effects attributable to a particle of mass m at the origin that we have the right to identify this particular solution with the one we hoped to find.

It will be a sufficient illustration of this procedure, if we indicate how the position of the matter causing this particular solution is located. Wherever the formula (1) holds good there can be no matter, because the law which applies to empty space is satisfied. But if we try to approach the origin ($r = 0$), a curious thing happens. Suppose we take a measuring-rod, and, laying it radially, start marking off equal lengths with it along a radius, gradually approaching the origin. keeping the time t constant, and $d\theta$ being zero for radial measurements, the formula (1) reduces to

$$ds^2 = -\frac{1}{\gamma}dr^2$$

or

$$dr^2 = -\gamma ds^2.$$

We start with r large. By-and-by we approach the point where $r = 2m$. But here, from its definition, γ is equal to 0. So that, however large the measured interval ds may be, $dr = 0$. We can go on shifting the measuring-rod through its own length time after time, but dr is zero; that is to say, we do not reduce r. There is a magic circle which no measurement can bring us inside. It is not unnatural that we should picture something obstructing our closer approach, and say that a particle of matter is filling up the interior.

The fact is that so long as we keep to space-time curved only in the first degree, we can never round off the summit of the hummock. It must end in an infinite chimney. In place of the chimney, however, we round it off with a small region of greater curvature. This region cannot be empty because the law applying to empty space does not hold. We describe it therefore as containing matter—a procedure which practically amounts to a definition of matter. Those familiar with hydrodynamics may be reminded of the problem of the irrotational rotation of a fluid; the conditions cannot be satisfied at the origin, and it is necessary to cut out a region which is filled by a vortex-filament.

A word must also be said as to the co-ordinates r and t used in (1). They correspond to our ordinary notion of radial distance and time—as well as any variables in a non-Euclidean world can correspond to words which, as ordinarily used, presuppose a Euclidean world. We shall thus call r and t, distance and time. But to give names to coordinates does not give more information—and in this case gives considerably less information—than is already contained in the formula for ds^2. If any question arises as to the exact significance of r and t it must always be settled by reference to equation (1).

The want of flatness in the gravitational field is indicated by deviation of the coefficient γ from unity. If the mass $m = 0$, $\gamma = 1$, and space-time is perfectly flat. Even in the most intense gravitational fields known, the deviation is extremely small. For the sun, the quantity m, called the gravitational mass, is only 1.47 kilometres, for earth it is 5 millimetres. In any practical problem the ratio $2m/r$ must be exceedingly small. Yet it is on the small corresponding difference in γ that the whole of the phenomena of gravitation depend. . . .

The mathematical reader should find no difficulty in proving that for a particle with small velocity the acceleration towards the sun is approximately m/r^2, agreeing with the Newtonian law.

The result that the expression found for the geometry of the gravitational field of a particle leads to Newton's law of attraction is of great importance. It shows that the law $G_{\mu\nu} = 0$, proposed on theoretical grounds, agrees with observation at least approximately. It is no drawback that the Newtonian law applies only when the speed is small; all planetary speeds are small compared with the velocity of light, and the considerations mentioned at the beginning of this chapter suggest that some modification may be needed for speeds comparable with that of light.

Another important point to notice is that the attraction of gravitation is simply a geometrical deformation of the straight tracks. It makes no difference what body or influence is pursuing the track, the deformation is a general discrepancy between the "mental picture" and the "true map" of the portion of space-time considered. Hence light is subject to the same disturbance of path as matter. This is involved in the Principle of Equivalence; otherwise we could distinguish between the acceleration of a lift and a true increase of gravitation by optical experiments; in that case the observer for whom light-rays appear to take straight tracks might be described as absolutely unaccelerated and there could be no relativity theory. Physicists in general have been prepared to admit the likelihood of an influence of gravitation on light similar to that exerted on matter: and the problem whether or not light has "weight" has often been considered.

The appearance of γ as the coefficient of dt^2 is responsible for the main features of Newtonian gravitation; the appearance of $1/\gamma$ as the coefficient of dr^2 is responsible for the principal deviations of the new law from the old. This classification seems to be correct; but the Newtonian law is ambiguous and it is difficult to say exactly what are to be regarded as discrepancies from it. Leaving aside now the time-term as sufficiently discussed, we consider the space-terms alone*

$$ds^2 = \frac{1}{\gamma}\, dr^2 + r^2 d\theta^2.$$

The expression shows that space considered alone is non-Euclidean in the neighbourhood of an attracting particle. This is something entirely outside the scope of the old law of gravitation. Time can only be explored by something moving, whether a free particle or the parts of a clock, so that the non-Euclidean character of space-time can be covered up by introducing a field of force, suitably modifying the motion, as a convenient fiction. But space can be explored by static methods; and theoretically its non-Euclidean character could be ascertained by sufficient precise measures with rigid scales.

If we lay our measuring scale transversely and proceed to measure the circumference of a circle of nominal radius *r*, we see from the formula that the measured length *ds* is equal to $r/d\theta$, so that, when we have gone right round the circle, θ has increased by 2π and the measured circumference is $2\pi r$. But when we lay the scale radially the measured length *ds* is equal to $dr\sqrt{\gamma}$, which is always greater than *dr*. Thus, in measuring a diameter, we obtain a result greater than $2r$, each portion being greater than the corresponding change of *r*.

Thus if we draw a circle, placing a massive particle near the centre, so as to produce a gravitational field, and measure with a rigid scale the

* We change the sign of $ds,^2$ so that *ds*, when real, means measured space instead of measured time.

circumference and the diameter, the ratio of the measured circumference to the measured diameter will not be the famous number

$$\pi = 3.14159265358979323846264338327 9\ldots$$

but a little smaller. Or if we inscribe a regular hexagon in this circle its sides will not be exactly equal to the radius of the circle. Placing the particle near, instead of at, the centre, avoids measuring the diameter *through* the particle, and so makes the experiment a practical one. But though practical, it is not practicable to determine the non-Euclidean character of space in this way. Sufficient refinement of measures is not attainable. If the mass of a ton were placed inside a circle of five yards radius, the defect in the value of π would only appear in the twenty-fourth or twenty-fifth place of decimals.

It is of value to put the result in this way, because it shows that the relativist is not talking metaphysics when he says that space in the gravitational field is non-Euclidean. His statement has a plain physical meaning, which we may some day learn how to test experimentally. Meanwhile we can test it by indirect methods. . . .

[A body passing near a massive particle has its path bent owing to the non-Euclidean character of space.]

This bending of the path is additional to that due to the Newtonian force of gravitation which depends on the second appearance of γ in the formula. As already explained it is in general a far smaller effect and will appear only as a minute correction to Newton's law. The only case where the two rise to equal importance is when the track is that of a light wave, or of a particle moving with a speed approaching that of light; for then dr^2 rises to the same order of magnitude as dt^2.

To sum up, a ray of light passing near a heavy particle will be bent, firstly, owing to the non-Euclidean character of the combination of time with space. This bending is equivalent to that due to Newtonian gravitation, and may be calculated in the ordinary way on the assumption that light has weight like a material body. Secondly, it will be bent owing to the non-Euclidean character of space alone, and this curvature is additional to that predicted by Newton's law. If then we can observe the amount of curvature of a ray of light, we can make a crucial test of whether Einstein's or Newton's theory is obeyed. . . .

It is not difficult to show that the total deflection of a ray of light passing at a distance r from the centre of the sun is (in circular measure) $\dfrac{4m}{r}$, whereas the deflection of the same ray calculated on the Newtonian theory would be $\frac{2m}{r}$. For a ray grazing the surface of the sun the numerical value of this deflection is

$$1''.75 \text{ (Einstein's theory)},$$
$$0''.87 \text{ (Newtonian theory)}. \ldots$$

The bending affects stars seen near the sun, and accordingly the only chance of making the observation is during a total eclipse when the moon cuts off the dazzling light. Even then there is a great deal of light from the sun's corona which stretches far above the disc. It is thus necessary to have rather bright stars near the sun, which will not be lost in the glare of the corona. Further the displacements of these stars can only be measured relatively to other stars, preferably more distant from the sun and less displaced; we need therefore a reasonable number of outer bright stars to serve as reference points.

In a superstitious age a natural philosopher wishing to perform an important experiment would consult an astrologer to ascertain an auspicious moment for the trial. With better reason, an astronomer to-day consulting the stars would announce that the most favourable day of the year for weighing light is May 29. The reason is that the sun in its annual journey round the ecliptic goes through fields of stars of varying richness, but on May 29 it is in the midst of a quite exceptional patch of bright stars—part of the Hyades—by far the best star-field encountered. Now if this problem had been put forward at some other period of history, it might have been necessary to wait some thousands of years for a total eclipse of the sun to happen on the lucky date. But by strange good fortune an eclipse did happen on May 29, 1919. Owing to the curious sequence of eclipses a similar opportunity will recur in 1938; we are in the midst of the most favourable cycle. It is not suggested that it is impossible to make the test at other eclipses, but the work will necessarily be more difficult.

Attention was called to this remarkable opportunity by the Astronomer Royal in March, 1917; and preparations were begun by a Committee of the Royal Society and Royal Astronomical Society for making the observations. Two expeditions were sent to different places on the line of totality to minimise the risk of failure by bad weather. Dr. A. C. D. Crommelin and Mr. C. Davidson went to Sobral in North Brazil; Mr. E. T. Cottingham and the writer went to the Isle of Principe in the Gulf of Guinea, West Africa. . . .

It will be remembered that Einstein's theory predicts a deflection of 1″. 74 at the edge of the sun,* the amount falling off inversely as the distance from the sun's centre. The simple Newtonian deflection is half this, 0″. 87. The final results (reduced to the edge of the sun) obtained at Sobral and Principe with their "probable accidental errors" were

Sobral 1″. 98 ± 0″. 12,
Principe 1″. 61 ± 0″. 30.

It is usual to allow a margin of safety of about twice the probable error on either side of the mean. The evidence of the Principe plates is thus just about sufficient to rule out the possibility of the "half-deflection," and the Sobral plates exclude it with practical certainty. The value of the material found at Principe cannot be put higher than about one-sixth of that at Sobral; but it certainly makes it less easy to bring criticism against this confirmation of Einstein's theory seeing that it was obtained independently with two different instruments at different places and with different kinds of checks.

The best check on the results obtained with the 4-inch lens at Sobral is the striking internal accordance of the measures for different stars. The theoretical deflection should vary inversely as the distance from the sun's centre; hence, if we plot the mean radial displacement found for each star separately against the inverse distance, the points should lie on a straight line. This is shown in Figure 1 where the broken line shows the theoretical prediction of Einstein, the deviations being within the accidental errors of the determinations. A line of half the slope representing the half-deflection would clearly be inadmissible. . . .

We have seen that the swift-moving light-waves possess great advantages as a means of exploring the non-Euclidean property of space. But there is an old fable about the hare and the tortoise. The slow-moving planets have qualities which must not be overlooked. The light-wave traverses the region in a few minutes and makes its report; the planet plods on and on for centuries, going over the same ground again and again. Each time it goes round it reveals a little about the space, and the knowledge slowly accumulates.

According to Newton's law a planet moves round the sun in an ellipse, and if there are no other planets disturbing it, the ellipse remains the same for ever. According to Einstein's law the path is very nearly an el-

* The predicted deflection of light from infinity to infinity is just over 1″. 745, from infinity to the earth it is just under.

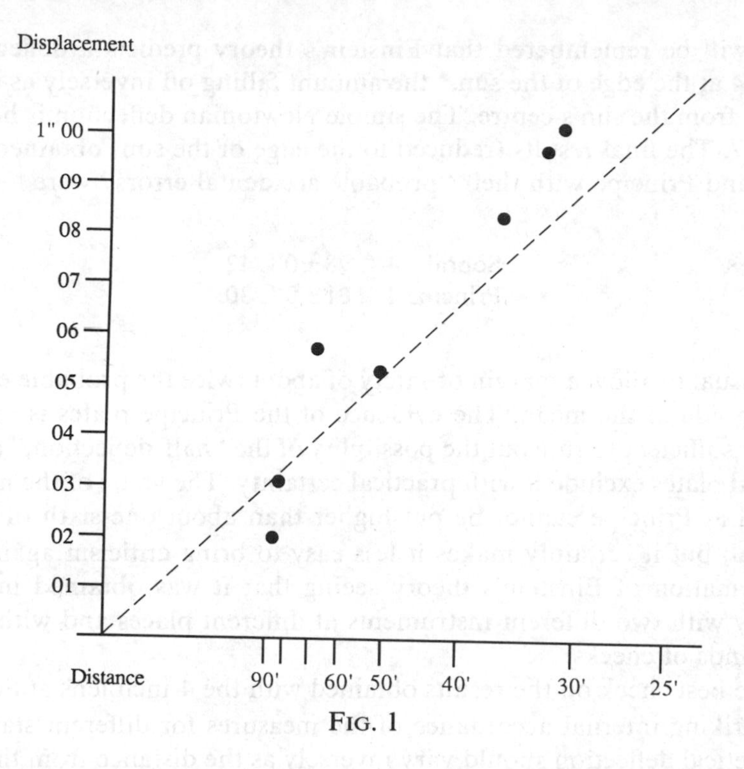

FIG. 1

lipse, but it does not quite close up; and in the next revolution the path has advanced slightly in the same direction as that in which the planet was moving. The orbit is thus an ellipse which very slowly revolves.

The exact prediction of Einstein's law is that in one revolution of the planet the orbit will advance through a fraction of a revolution equal to $3v^2/C^2$, where v is the speed of the planet and C the speed of light. The earth has 1/10,000 of the speed of light; thus in one revolution (one year) the point where the earth is at greatest distance from the sun will move on 3/100,000,000 of a revolution, or 0″ . 038. We could not detect this difference in a year, but we can let it add up for a century at least. It would then be observable but for one thing—the earth's orbit is very blunt, very nearly circular, and so we cannot tell accurately enough which way it is pointing and how its sharpest apses move. We can choose a planet with higher speed so that the effect is increased, not only because v^2 is increased, but because the revolutions take less time; but, what is perhaps more important, we need a planet with a sharp elliptical orbit, so that it is easy to observe how its apses move round. Both these conditions are fulfilled in the case of Mercury. It is the fastest of the planets, and the predicted advance of the orbit amounts to 43″ per century; further the eccentricity of its orbit is far greater than that of any of the other seven planets.

Now an unexplained advance of the orbit of Mercury had long been known. It had occupied the attention of Le Verrier, who, having successfully predicted the planet Neptune from the disturbances of Uranus, thought that the anomalous motion of Mercury might be due to an interior planet, which was called Vulcan in anticipation. But, though thoroughly sought for, Vulcan has never turned up. Shortly before Einstein arrived at his law of gravitation, the accepted figures were as follows. The actual observed advance of the orbit was 574″ per century; the calculated perturbations produced by all the known planets amounted to 532″ per century. The excess of 42″ per century remained to be explained. Although the amount could scarcely be relied on to a second of arc, it was at least thirty times as great as the probable accidental error.

The big discrepancy from the Newtonian gravitational theory is thus in agreement with Einstein's prediction of an advance of 43″ per century. . . .

The theory of relativity has passed in review the whole subject-matter of physics. It has unified the great laws, which by the precision of their formulation and the exactness of their application have won the proud place in human knowledge which physical science holds to-day. And yet, in regard to the nature of things, this knowledge is only an empty shell—a form of symbols. It is knowledge of structural form, and not knowledge of content. All through the physical world runs that unknown content, which must surely be the stuff of our consciousness. Here is a hint of aspects deep within the world of physics, and yet unattainable by the methods of physics. And, moreover, we have found that where science has progressed the farthest, the mind has but regained from nature that which the mind has put into nature.

We have found a strange foot-print on the shores of the unknown. We have devised profound theories, one after another, to account for its origin. At last, we have succeeded in reconstructing the creature that made the foot-print. And Lo! it is our own.

M.6

Relativity and Modern Theories of the Universe

WILLEM DE SITTER

WILLEM DE SITTER, a leading Dutch astronomer in the early decades of the twentieth century, was born in the town of Sneek (Friesland) in 1872 and died in Leiden in 1934. He was the son of a judge. In 1901, he took his degree in mathematics and physics at the University of Gröningen, where his teacher was the eminent astronomer Jacobus Cornelis Kapteyn, famed as one of the first mappers of the Milky Way. De Sitter spent two years at the Cape Observatory in South Africa and later joined the faculty of the University of Leiden in 1908. He became director of the Leiden Observatory in 1919 where he remained until his death. He made major contributions to celestial mechanics—especially his painstaking research on the dynamics of Jupiter's satellites (published in 1925 as *New Mathematical Theory of Jupiter's Satellites*)—and to the study of the rotation of the earth. De Sitter also determined several fundamental astronomical constants and applied the theory of relativity to cosmology.

One of the earliest astronomers to appreciate the significance of Einstein's special theory, de Sitter was instrumental in arousing sufficient interest in Einstein's work in England so that Arthur Eddington and others were prepared to test Einstein's general theory during the eclipse of the sun in 1919. In 1917 de Sitter introduced an alternative to Einstein's static Universe, offering what became known as the "de Sitter Universe." By 1928 it was possible to transform the de Sitter Universe mathematically into an expanding Universe. Attempts by astronomers to reconcile the Einstein and de Sitter models with astronomical observations gave impetus to the fledgling science of modern cosmology.

In November 1931, de Sitter was invited to give a course of six lectures at the Lowell Institute in Boston on developing an understanding of the structure of the Universe. These lectures were published by the Harvard University Press under the title *Kosmos,* from which the concluding lecture has been selected.

De Sitter's lectures were delivered to a scientifically sophisticated audience whose knowledge of tensor calculus was assumed. His lectures are lucid, but far from elementary expositions of his concept of the nonstatic, expanding Universe in which he presented three possible variants: his two expanding versions and one oscillating model.

Present-day developments in cosmology are coming to suggest rather insistently that everyday conditions could not persist but for the distant parts of the Universe, that all our ideas of space and geometry would become entirely invalid if the distant parts of the Universe were taken away. Our everyday experience even down to the smallest details seems to be so closely integrated to the grand-scale features of the Universe that it is well-nigh impossible to contemplate the two being separated.

—FRED HOYLE, *Frontiers of Astronomy*

THE LAST FIFTEEN years have seen the most remarkable development of physical and astronomical theories, and of our ideas about the universe. It is outside the scope of the present series of lectures to consider the new theories of atomic physics and the corresponding new development of astrophysics, but we must say some words about the theory of relativity, which is now about sixteen years old. . . .

The sequence of different positions of the same particle at different times forms a one-dimensional continuum in the four-dimensional space-time, which is called the *world-line* of the particle. All that physical experiments or observations can teach us refers to intersections of world-lines of different material particles, light-pulsations, etc., and how the course of the world-line in between these points of intersection is entirely irrelevant and outside the domain of physics. The system of intersecting world-lines can thus be twisted about at will, so long as no points of intersection are destroyed or created, and their order is not changed. It follows that the equations expressing the physical laws must be invariant for arbitrary transformations.

This is the mathematical formulation of the theory of relativity. The metric properties of the four-dimensional continuum are described, as is shown in treatises on differential geometry, by a certain number (ten, in fact) of quantities denoted by $g_{\alpha\beta}$, and commonly called "potentials." The physical status of matter and energy, on the other hand, is described by ten other quantities, denoted by $T_{\alpha\beta}$, the set of which is called the "mate-

---

Reprinted by permission of the publishers from *Kosmos* by Willem de Sitter, Cambridge, Mass.: Harvard University Press, Copyright 1932 by the President and Fellows of Harvard College, © renewed 1960 by Elonore de Sitter.

rial tensor." This special tensor has been selected because it has the property which is mathematically expressed by saying that its divergence vanishes, which means that it represents something permanent. The fundamental fact of mechanics is the law of inertia, which can be expressed in its most simple form by saying that it requires the fundamental laws of nature to be differential equations of the second order. Thus the problem was to find a differential equation of the second order giving a relation between the metric tensor $g_{\alpha\beta}$ and the material tensor $T_{\alpha\beta}$. This is a purely mathematical problem, which can be solved without any reference to the physical meaning of the symbols. The simplest possible equation (or rather set of ten equations, because there are ten g's) of that kind that can be found was adopted by Einstein as the fundamental equation of his theory. It defines the space-time continuum, or the "field." The world-lines of material particles and light quanta are the geodesics in the four-dimensional continuum defined by the solutions $g_{\alpha\beta}$ of these field-equations. The equations of the geodesic thus are equivalent to the equations of motion of mechanics. When we come to solve the field-equations and substitute the solutions in the equations of motion, we find that in the first approximation, i.e. for small material velocities (small as compared with the velocity of light), these equations of motion are the same as those resulting from Newton's theory of gravitation. The distinction between gravitation and inertia has disappeared; the gravitational action between two bodies follows from the same equations, and *is* the same thing, as the inertia of one body. A body, when not subjected to an extraneous force (i.e. a force other than gravitation), describes a geodesic in the continuum, just as it described a geodesic, or straight line, in the absolute space of Newton under the influence of inertia alone.

The field-equations and the equations of the geodesic together contain the whole science of mechanics, including gravitation.

In the first approximation, as has been said just now, the new theory gives the same results as Newton's theory of gravitation. The enormous wealth of experimental verification of Newton's law, which has been accumulated during about two and a half centuries, is therefore at the same time an equally strong verification of the new theory. In the second approximation there are small differences, which have been confirmed by observations, so far as they are large enough for such a confirmation to be possible. Thus especially the anomalous motion of the perihelion of Mercury, which had baffled all attempts at explanation for over half a century, is now entirely accounted for. Further the theory of relativity has predicted some new phenomena, such as the deflection of rays of light that pass near the sun, which has actually been observed on several occasions during eclipses; and the redshift of spectral lines originating in a strong gravitational field, which is also confirmed by observations, e.g. in the spectrum of the sun, and also in the spectrum of the companion of Sirius, which, being a so-called white dwarf, i.e. a small star with very high density and consequently a strong gravitational field, gives a considerable red-shift. We cannot stop to explain these phenomena in detail. It must suffice just to mention them.

Two points should be specially emphasised in connection with the general theory of relativity.

First that it is a purely *physical* theory, invented to explain empirical physical facts, especially the identity of gravitational and inertial mass, and to coördinate and harmonise different chapters of physical theory, and simplify the enunciation of the fundamental laws. There is nothing metaphysical about its origin. It has, of course, largely attracted the attention of philosophers, and has, on the whole, had a very wholesome influence on metaphysical theories. But that is not what it set out to do, that is only a by-product.

Second that it is a pure generalisation, or abstraction, like Newton's system of mechanics and law of gravitation. It contains *no hypothesis,* as contrasted with other modern physical theories, electron theory, quantum theory, etc., which are full of hypotheses. It is, as has already been said, to be considered as the logical sequence and completion of Newton's Principia.

A special feature of the development of physics in the nineteenth century has been the arising of general principles beside the special laws, such as the principles of conservation of mass and of energy, the principle of least action, and the like. These differ from the special laws, not only by being more general, but they aspire, so to say, to a higher status than the laws. Their claim is that they express fundamental facts of nature, general rules, to which all special laws have to conform. And they accordingly exclude a priori all attempts at "explanation" by hypotheses or mechanical models. It is characteristic of the theory of relativity that it enables us to include all these principles of conservation in one single equation.

We have a direct knowledge only of that part of the universe of which we can make observations. I have already called this "our neighbourhood." Even within the confines of this province our knowledge decreases very rapidly as we get away from our own particular position in space and time. It is only within the solar system that our empirical knowledge of the quantities determining the state of the universe, the potentials $g_{\alpha\beta}$, extends to the second order of smallness (and that only for g_{44}, and not for the others), the first order corresponding to about one unit in the eighth decimal place. . . .

During the last years the limits of our "neighbourhood" have been enormously extended by the observations of extragalactic nebulae, made chiefly at the Mount Wilson Observatory. These wonderful observations have enabled us to make fairly reliable estimates of the distances of these objects and to say something about their distribution in space. It appears that they are distributed approximately evenly over "our neighbourhood." They also are all of roughly the same size, so that we can make an estimate of the density of matter in space. Further the observations have disclosed the remarkable fact that in their spectra there is a displacement of the lines towards the red corresponding to a receding velocity increasing with the distance, and, so far as the determinations of the distances are reliable, proportional with it. If the velocity is proportional to the distance, then not only the distance of any nebula from us is increasing, but *all* mutual distances between any two of them are increasing at the same rate. Our own galactic system is only one of a great many, and observations made from any of the others would show exactly the same thing: all systems are receding, not from any particular centre, but *from each other:* the whole system of galactic systems is *expanding.*

It is perhaps somewhat difficult to imagine the expansion of three-dimensional space. A two-dimensional analogy may help to make it clear. Let the universe have only two dimensions, and let it be the surface of an india rubber ball. It is only the *surface* that is the universe, not the ball itself. Observations can only be made, distances can only be measured, along the surface, and evidently no point of the surface is different from any other point. Let there be specks of dust fixed to the surface to represent the different galactic systems. If the ball is inflated, the universe expands, and these specks of dust will recede from each other, their mutual distances, measured along the surface, will increase in the same rate as the radius of the ball. An observer in any one of the specks will see all the others receding from himself, but it does not follow that he is the centre of the universe. The universe (which is the surface of the ball, not the ball itself) has no centre. . . .

These then are the two observational facts about our neighbourhood, which have to be accounted for by the theory: there is a finite density of matter, and there is expansion, i.e. the mutual distances are increasing, and therefore the density is decreasing. Of course we can only be certain of these facts so far as our observations reach, i.e. for our "neighbourhood," but, in agreement with our principle of extrapolation, we extend these statements to the whole of the universe.

We have thus to find a universe—i.e. a set of potentials $g_{\alpha\beta}$ satisfying the field-equations of the general theory of relativity—that has both a finite density of matter and an expansion. And, since we only consider the universe on a very large scale, and make abstraction of all details and local irregularities, our universe must be homogeneous and isotropic. It follows at once from this condition of homogeneity and isotropy that the three-dimensional space of it must be what mathematicians call a space of constant curvature. Even so mathematics offers us a free choice between different kinds of space. The curvature may be positive, negative, or zero. It is not possible to picture, or imagine, the different kinds of three-dimensional space. We think that we have a mental picture of Euclidian, or flat, space, i.e. space of which the curvature is everywhere zero, but I am not sure that this is not a self-deception, caused by the fact that the geometry of this special space has been taught in the schools for the last two thousand or more years. It is certain that for physical phenomena on the scale which our sense organs are able to perceive, i.e. neither too small nor too large, the Euclidian space is a very close approximation to the true physical space, but for the electron, and for the universe, the approximation breaks down. To help us to understand three-dimensional spaces, two-dimensional analogies may be very useful (though also sometimes misleading). We can imagine different kinds of two-dimensional space, since we are able to place ourselves outside them. A two-dimensional space of zero curvature is a plane, say a sheet of paper. The two-dimensional space of positive curvature is a convex surface, such as the shell of an egg. It is bent away from the plane towards the same side in all directions. The curvature of the egg, however, is not constant: it is strongest at the small end. The surface of constant positive curvature is the sphere, say our india rubber ball of a moment ago. The two-dimensional space of negative curvature is a surface that is convex in some directions and concave in others, such as the surface of a saddle or the middle part of an hour glass. Of these two-dimensional surfaces we can form a mental picture because we can view them from outside, living, as we do, in three-dimensional space. But for a being, who would be unable to leave the surface on which he was living, that would be impossible. He could only decide of which kind his surface was by studying the properties of geometrical figures drawn on it. For the geometrical figures have different properties on the different surfaces. On the sheet of paper the sum of the three angles of a triangle is equal to two right angles, on the egg, or the sphere, it is larger, on the saddle it is smaller. On the flat paper—and on the saddle-shaped surface—we can proceed indefinitely in the same direction; on the egg or the sphere, if we continue to move in the same direction we ultimately come back to our starting point. The spaces of zero and negative curvature are infinite, that of positive curvature is finite. Thus the inhabitant of the two-dimensional surface could determine its curvature if he were able to study very large triangles or very long straight lines. If the curvature were so minute that the sum of the angles of the largest triangle that he could measure would still differ from two right angles by an amount too small to be appreciable with the means at his disposal then he would be unable to determine the curvature, unless he had some means of communicating with somebody living in the third dimension. Now our case with reference to three-dimensional space is exactly similar. We have no intuitive knowledge of the kind of space we live in. So we must find out which kind it is by studying the triangles and other geometrical figures in it. As we are concerned with *physical* space, the triangles that we must investigate are those formed by the tracks of material particles and rays of light, and naturally, in order to be able to distinguish different kinds of space, we must study very large triangles and rays of light coming from very great distances. Thus the decision must necessarily depend on astronomical observations. . . .

Let us begin by considering the finite density of matter in the universe. The average density is very small. Matter is actually distributed very unevenly, it is conglomerated into stars and galactic systems. The average density is the density that we should get if all these great systems could be evaporated into atoms of hydrogen, or protons, and these distributed evenly over the whole of space. There would then probably not be more than three or four protons in every cubic foot. That is a very small density indeed: it is about a million million times less than that of the most perfect vacuum that we can produce in our physical laboratories. The universe thus consists mostly of emptiness, and it appears natural to consider a universe without any matter at all, an empty universe, as a good approximation to begin with for our grand scale model. The galactic systems are details which can be put in afterwards. But we may also take as our first approximation a universe containing the same amount of matter as the actual one, but equally distributed, i.e. having a finite average density of three or four protons per cubic foot. The local deviations from the average, caused by the conglomeration of matter into stars and stellar systems, are then disregarded in the grand scale model, and are only taken into account when we come to study details.

Now fifteen years ago, in the beginning of 1917, two solutions of the field-equations for a homogeneous isotropic universe had been found, which I shall provisionally call the solutions "A" and "B." It should be mentioned that at that time only *static* solutions were looked for. It was thought that the universe must be a stable structure, which would retain its large scale properties unchanged for all time, or at least change them so slowly that the change could be disregarded. In one of these solutions (B) the average density was zero, it was empty; the other one (A) had a finite density. Both, of course, were, as was well appreciated, only approximations to the actual universe. In B, to get the real universe, we should have to put in a few galactic systems, in A we should have to condense the evenly distributed matter into galactic systems. The universe A is really and essentially static, there can be no systematic motions in it. It has an average density, but no expansion. It is therefore called the *static universe.* B, on the other hand, is not really static, it expands, and it could only parade in the garb of a static universe because there is nothing in it to show the expansion. B is therefore called the *empty universe.* Thus we had two approximations: the static universe with matter and without expansion, and the empty one without matter and with expansion. The actual universe, as we have seen, has both matter and expansion, and can, therefore, be neither A nor B. In 1917 this dilemma had not yet become urgent, and was hardly realised. The actual value of the density was still entirely unknown, and the expansion had not yet been discovered.

Now in both the solutions A and B the curvature is positive, in both three-dimensional space is finite: the universe has a definite size, we can speak of its radius, and, in the case A, of its total mass. In the case A, the static universe, there is a definite relation between the curvature and the density, in fact the density is proportional to the curvature, the factor of proportionality being a pure number ($1/4\pi$, if appropriate units are used). Thus, if we wish to have a finite density in a static universe, we must have a finite positive curvature.

At this point we must say a few words about the famous *lambda.* The field-equations, in their most general form, contain a term multiplied by a constant, which is denoted by the Greek letter λ (lambda), and which is sometimes called the "cosmical constant."... At first, in Einstein's paper of November 1915, in which the theory reached its final form, the term with λ was simply omitted; in other words λ was supposed to have the special value *zero.* ... But fifteen months later, in February 1917, it was found that a static solution with a positive curvature—the solution A—was not possible without the λ. In fact the curvature is proportional to λ (in solution A, λ is equal to the curvature; in B, treated as a static solution, it is three times the curvature). . . .

We must now take up the thread of the narrative where we left it a little while ago. We were in the position of having two possible solutions: the static universe with matter but without expansion, and the empty universe with expansion but without matter.

Now the observed rate of expansion is large: the universe doubles its size in about fifteen hundred million years, which is a short time, astronomically speaking. In the "static" universe expansion is impossible, the "empty" universe does expand. Therefore we may be tempted to consider the empty universe as the most likely approximation; and we can proceed to compute the radius of curvature of the universe, supposing it to be of the empty type, from the observed rate of expansion. It comes out as about two thousand million lightyears.

The universe, however, is not empty, but contains matter. The point is how much matter. Is the density anywhere near that corresponding to the static universe, or is it so small that we can consider the empty universe as a good approximation? We have seen that the universe is some million million times as empty as our most perfect vacuum. But this is not the correct way to measure the emptiness of the universe. We must use as a standard of comparison, not our terrestrial experience, but the theoretical density of the static universe. It is easy to compute the density of a static universe of a radius of two thousand million lightyears, and it comes out only very little larger than the observed density. The actual universe is thus very far from empty, it is, on the contrary, nearly full.

We thus come to the conclusion, which was already foreshadowed above, that the actual universe is neither the static nor the empty one. It differs so much from both of these that neither can be used as an appropriate grand scale model. We must thus look for other solutions of the general field-equations. On account of the expansion our solution must necessarily be a non-static one, and it must have a finite density. There is only one possible static solution possessing a finite density, viz. our old friend A, but of non-static solutions with finite density there exists a great variety. I will now depart from the strictly historical narrative and enumerate these different possible solutions, not in the order in which they have been discovered, but in the sequence of a natural classification.

In the solutions A and B the curvature of three-dimensional space was necessarily positive, and the mysterious "cosmical constant" λ was also positive. In the non-static solutions this is not so. At first this was not realised. We had become so accustomed to think of λ as an essentially positive quantity, and of a finite world with positive curvature, that the idea of investigating the possibility of solutions with negative or zero values of λ and of the curvature simply did not arise. But when this oversight was corrected, it appeared at once that in the non-static case both λ and the curvature need not be positive, but can be negative or zero quite as well. I will therefore use the value of λ and the sign of the curvature as the principles of classification. The instantaneous state of the universe is characterised by a certain quantity occurring in the equations, which is denoted by the letter R, and which, if there is a curvature, can be interpreted as the radius of curvature, or the "radius" for short. The way in which the universe expands is determined by the variation of this R with the time. There are three types, or families, of non-static universes, which I will call the oscillating universes, and the expanding universes of the first and of the second kind. . . .

In the oscillating universes the "radius" R increases from zero to a certain maximum size, which is different for each member of the family, and then decreases again to zero. The period of oscillation has a certain finite (and rather short) value, different for each member of the family. In the expanding family of the first kind the radius is continually increasing from a certain initial time, when it was zero, to become infinitely large after an infinite time. In the expanding series of the second type the radius has at the initial time a certain minimum value, different for the different members of the family, and increases to become infinite after an infinite time. . . .

TABLE 1

	Curvature		
λ	**Negative**	**Zero**	**Positive**
Negative	Oscillating	Oscillating	Oscillating
Zero	Expanding I	Expanding I	Oscillating
Positive	Expanding I	Expanding I	Oscillating Expanding I Expanding II

We do not know to which of the three possible families our own universe belongs, and there is nothing in our observational data to guide us in making the choice. And even if we have decided on the family, we have still the freedom to select any particular member of it. This is not because the data are not accurate enough, but because they are deficient in number. The observations give us *two* data, viz. the rate of expansion and the average density, and there are *three* unknowns: the value of λ, the sign of the curvature, and the scale of the figure, i. e. the units of *R* and of the time. The problem is indeterminate. . . . We might make the hypothesis that the true value of λ is zero. In that case the data of observation will allow us, if they are sufficiently accurate, to determine the curvature.

As a matter of fact neither the average density nor the rate of expansion are at the present time known with sufficient accuracy to make an actual *determination* possible, even if an hypothesis of this kind is adopted. All we can say is that, if the curvature is small (as we know it must be, because it is imperceptible by ordinary geometric methods in our neighbourhood), then λ must be small, and *if* the curvature is *very* small, then λ must be very small. On the other hand, *if* λ is very small, or zero, then the curvature must be very small, and may even be zero, for aught we can say at present. . . .

There is one very serious difficulty presented by the theory of the expanding universe, which we shall have to face with careful deliberation.

In all solutions there is a certain minimum value of the "radius" *R*, either zero, or in the expanding family of the second kind a finite value, which the universe had at a definite time in the past. There appears to be a definite "beginning of time," a few thousand million years back in history, as there is a definite "absolute zero" of temperature, corresponding to *minus* 273 degrees on the ordinary scale. What is the meaning of this?

The temptation is strong to identify the epoch of the beginning of the expansion with the "beginning of the world," whatever that may mean. Now astronomically speaking this beginning of the expansion took place only yesterday, not much longer ago than the formation of the oldest rocks on the earth. According to all our modern views the evolution of a star, of a double star, or a star cluster, requires intervals of time which are enormously longer. The stars and the stellar systems must be some thousands of times older than the universe!

What must be our attitude with regard to this paradox? It would appear that, if two theories are in contradiction, we must give up either the one or the other. The conflict apparently is between the modern theories of stellar evolution and the dynamical theories of the evolution of double stars and star clusters on the one hand, and the general theory of relativity on the other hand. If this were the real contest, there could be no doubt about the issue: the theory of relativity would come out of the trial victorious, and the theories of evolution would have to be revised. This seems to be Sir Arthur Eddington's standpoint, as he writes: "we must accept this alarmingly rapid dispersal of the nebulae with its important consequences in limiting the time available for evolution." I am afraid, however, that very few astronomers, not to speak of geophysicists, will be prepared to accept this drastic reduction of the time scale.

M.7

Geometry as a Branch of Physics

H. P. Robertson

H. P. Robertson, an American cosmologist, presented the solution of Einstein's field equations in a form that is particularly applicable to cosmology. This solution leads to an expression for the space-time metric known as the Robertson-Walker solution. The beauty and simplicity of this solution is that it leads to two differential equations for the radius of the Universe as a function of time and the density of the Universe. The solution contains a curvature constant which can have one of three values: −1, 0, or 1. If the curvature constant is −1, the curvature of space is negative (like a saddle) and the Universe is hyperbolic. A curvature constant of 0 means that space is flat and that the Universe is open and infinite. If the curvature constant is 1, the curvature of space is positive and the Universe is closed. The curvature constant depends on the density of matter in the cosmos. If the density is sufficiently high, space is positively curved and the present expansionary surge of the Universe will eventually be reversed. If the density is insufficient to brake the expansion, the Universe will continue to spread outward forever. As astronomers are presently unsure about the actual amount of mass in the cosmos, they are not yet able to predict its ultimate fate.

In 1951, Robertson, then a physicist at the California Institute of Technology, contributed a chapter on the fundamental importance of geometry to physics in an anthology edited by P. A. Schilpp and published under the title *Albert Einstein: Philosopher-Scientist.* In this paper, Robertson discusses various efforts that had been made to describe space in geometrical terms and explores the alternative geometries that have been extensively employed in cosmology. Robertson argues that the physical geometry of general relativity has replaced the abstract and artificial axioms of Euclid.

According to Robertson, mathematics no longer sets universal standards that are independent of conditions in the physical Universe; geometry is now a consequence of the actual state of the cosmos. As a result, determining the future of the Universe may depend on how precisely we can measure its density and hence calculate the positive or negative curvature of space. Robertson also discusses the interplay of object and extension in the formulation of a geometrized space and the basis for the universality of general relativity: Gravity acts equally on all matter, since the gravitational and inertial masses of any body are proportional to each other and may be taken as equal.

The principles highlighted in Robertson's article are still valid, but the 500-million-light-year radius he gives for the "radius of curvature" of the Universe is outdated. In the several decades since his article was published, increasingly powerful devices have been built that can now sight objects as far away as five billion light years; this new vantage point represents a tenfold increase in the radius of curvature and a thousandfold increase in the total volume of the observable Universe. Only time will tell how much more of our increasingly accessible Universe will be revealed.

In Einstein's conception space is no longer the stage on which the drama of physics is performed: it is itself one of the performers; for gravitation, which is a physical property, is entirely controlled by curvature, which is a geometrical property of space.

—EDMUND WHITTAKER, *Space and Spirit*

There is nothing in the world except empty curved space. Matter, charge, electromagnetism, and other fields are only manifestations of the bending of space. Physics is geometry.

—JOHN ARCHIBALD WHEELER, "Superspace and the Nature of Quantum Geometrodynamics," in *Battelle Rencontres* 1967

IS SPACE REALLY curved? That is a question which, in one form or another, is raised again and again by philosophers, scientists, T. C. Mits and readers of the weekly comic supplements. A question which has been brought into the limelight above all by the genial work of Albert Einstein, and kept there by the unceasing efforts of astronomers to wrest the answer from a curiously reluctant Nature.

But what is the meaning of the question? What, indeed, is the meaning of each word in it? Properly to formulate and adequately to answer the question would require a critical excursus through philosophy and mathematics into physics and astronomy, which is beyond the scope of the present modest attempt. Here we shall be content to examine the roles of deduction and observation in the problem of physical space, to exhibit certain high points in the history of the problem and in the end to illustrate the viewpoint adopted by presenting a relatively simple caricature of Einstein's general theory of relativity. It is hoped that this certainly incomplete and possibly naïve description will present the essentials of the problem from a neutral mathematico-physical viewpoint in a form suitable for incorporation into any otherwise tenable philosophical position. Here, for example, we shall not touch directly upon the important problem of form versus substance—but if one wishes to interpret the geometrical substratum here considered as a formal backdrop against which the

---

H. P. Robertson, "Geometry as a Branch of Physics." Reprinted from P.A. Schilpp's *Albert Einstein: Philosopher-Scientist* by permission of The Open Court Publishing Company, La Salle, Illinois, 1951, pp. 313–332.

contingent relations of nature are exhibited, one should be able to do so without distorting the scientific content.

First, then, we consider geometry as a deductive science, a branch of mathematics in which a body of theories is built up by logical processes from a postulated set of axioms (not "self-evident truths"). In logical position geometry differs not in kind from any other mathematical discipline—say the theory of numbers or the calculus of variations. As mathematics, it is not the science of measurement, despite the implications of its name—even though it did, in keeping with the name, originate in the codification of rules for land surveying. The principal criterion of its validity as a mathematical discipline is whether the axioms as written down are self-consistent, and the sole criterion of the truth of a theorem involving its concepts is whether the theorem can be deduced from the axioms. This truth is clearly relative to the axioms; the theorem that the sum of the three interior angles of a triangle is equal to two right angles, true in Euclidean geometry, is false in any of the geometries obtained on replacing the parallel postulate by one of its contraries. In the present sense it suffices for us that geometry is a body of theorems, involving among others the concepts of point, angle and a unique numerical relation called distance between pairs of points, deduced from a set of self-consistent axioms.

What, then, distinguishes Euclidean geometry as a mathematical system from those logically consistent systems, involving the same category of concepts, which result from the denial of one or more of its traditional axioms? This distinction cannot consist in its "truth" in the sense of observed fact in physical science; its truth, or applicability, or still better appropriateness, in this latter sense is dependent upon observation, and not upon deduction alone. The characteristics of Euclidean geometry, as mathematics, are therefore to be sought in its internal properties, and not in its relation to the empirical.

First, Euclidean geometry is a *congruence geometry,* or equivalently the space comprising its elements is *homogeneous and isotropic;* the intrinsic relations between points and other elements of a configuration are unaffected by the position or orientation of the configuration. As an example, in Euclidean geometry all intrinsic properties of a triangle—its angles, area, etc.,—are uniquely determined by the lengths of its three sides; two triangles whose three sides are respectively equal are "congruent"; either can by a "motion" of the space into itself be brought into complete coincidence with the other, whatever its original position and orientation may be. These motions of Euclidean space are the familiar translations and rotations, use of which is made in proving many of the theorems of Euclid. That the existence of these motions (the axiom of "free mobility") is a desideratum, if not indeed a necessity, for a geometry applicable to physical space, has been forcibly argued on *a priori* grounds by Von Helmholtz, Whitehead, Russell and others; for only in a homogeneous and isotropic space can the traditional concept of a rigid body be maintained.

But the Euclidean geometry is only one of several congruence geometries; there are in addition the "hyperbolic" geometry of Bolyai and Lobatchewsky, and the "spherical" and "elliptic" geometries of Riemann and Klein. Each of these geometries is characterized by a real number K, which for the Euclidean geometry is zero, for the hyperbolic negative, and for the spherical and elliptic geometries positive. In the case of 2-dimensional congruence spaces, which *may* (but need not) be conceived as surfaces embedded in a 3-dimensional Euclidean space, the constant K may be interpreted as the *curvature* of the surface into the third dimension—whence it derives its name. This name and this representation are for our purposes at least psychologically unfortunate, for we propose ultimately to deal exclusively with properties intrinsic to the space under consideration—properties which in the later physical applications can be measured within the space itself—and are not dependent upon some extrinsic construction, such as its relation to an hypothesized higher dimensional embedding space. We must accordingly seek some determination of K—which we nevertheless continue to call curvature—in terms of such inner properties.

In order to break into such an intrinsic characterization of curvature, we first relapse into a rather naïve consideration of measurements which may be made on the surface of the earth, conceived as a sphere of radius R. This surface is an example of a 2-dimensional congruence space of positive curvature $K = 1/R^2$ on agreeing that the abstract geometrical concept "distance" r between any two of its points (not the extremities of a diameter) shall correspond to the lesser of the two distances *measured on the surface* between them along the unique great circle which joins the two points. Consider now a "small circle" of radius r (measured on the surface!) about a point P on the surface; its perimeter L and area A (again measured on the surface!) are clearly less than the corresponding measures $2\pi r$ and πr^2 of the perimeter and area of a circle of radius r in the Euclidean plane. An elementary calculation shows that for sufficiently small r (i.e., small compared with R) these quantities on the sphere are given approximately by:

$$L = 2\pi r\,(1 - Kr^2/6 + \ldots),$$
(1)
$$A = \pi r^2\,(1 - Kr^2/12 + \ldots).$$

Thus, the ratio of the area of a small circle of radius 400 miles on the surface of the earth to that of a circle of radius 40 miles is found to be only 99.92, instead of 100.00 as in the plane.

Another consequence of possible interest for astronomical applications is that in spherical geometry the sum σ of the three angles of a triangle (whose sides are arcs of great circles) is *greater* than 2 right angles; it can in fact be shown that this "spherical excess" is given by

(2)
$$\sigma - \pi = K\delta,$$

where δ is the area of the spherical triangle and the angles are measured in radians (in which $180° = \pi$). Further, each full line (great circle) is of finite length $2\pi R$, and any two full lines meet in two points—there are no parallels!

In the above paragraph we have, with forewarning, slipped into a non-intrinsic quasi-physical standpoint in order to present the formulae (1) and (2) in a more or less intuitive way. But the essential point is that these formulae are in fact independent of this mode of presentation; they are relations between the mathematical concepts distance, angle, perimeter and area which follow as logical consequences from the axioms of this particular kind of non-Euclidean geometry. And since they involve the space-constant K, this "curvature" may in principle at least be determined *by measurements made on the surface,* without recourse to its embedment in a higher dimensional space.

Further, these formulae may be shown to be valid for a circle or triangle in the hyperbolic plane, a 2-dimensional congruence space for which $K < 0$. Accordingly here the perimeter and area of a circle are *greater,* and the sum of the three angles of a triangle *less,* than the corresponding quantities in the Euclidean plane. It may also be shown that each full line is of infinite length, that through a given point outside a given line an infinity of full lines may be drawn which do not meet the given line (the two lines bounding the family are said to be "parallel" to the given line), and that two full lines which meet do so in but one point.

The value of the intrinsic approach is especially apparent in considering 3-dimensional congruence spaces, where our physical intuition is of little use in conceiving them as "curved" in some higher-dimensional space. The intrinsic geometry of such a space of curvature K provides formulae for the surface area S and the volume V of a "small sphere" of radius r, whose leading terms are

$$(3) \quad \begin{aligned} S &= 4\pi r^2 (1 - Kr^2/3 + \ldots), \\ V &= 4/3\pi r^3 (1 - Kr^2/5 + \ldots). \end{aligned}$$

It is to be noted that in all these congruence geometries, except the Euclidean, there is at hand a natural unit of length $R = 1/K^{1/2}$; this length we shall, without prejudice, call the "radius of curvature" of the space.

So much for the congruence geometries. If we give up the axiom of free mobility we may still deal with the geometry of spaces which have only limited or no motions into themselves. Every smooth surface in 3-dimensional Euclidean space has such a 2-dimensional geometry; a surface of revolution has a 1-parameter family of motions into itself (rotations about its axis of symmetry), but not enough to satisfy the axiom of free mobility. Each such surface has at a point $P(x, y)$ of it an intrinsic "total curvature" $K(x, y)$, which will in general vary from point to point; knowledge of the curvature at all points essentially determines all intrinsic properties of the surface. The determination of $K(x, y)$ by measurements on the surface is again made possible by the fact that the perimeter L and area A of a closed curve, every point of which is at a given (sufficiently small) distance r from $P(x, y)$, are given by the formulae (1), where K is no longer necessarily constant from point to point. Any such variety for which $K = 0$ throughout is a ("developable") surface which may, on ignoring its macroscopic properties, be rolled out without tearing or stretching onto the Euclidean plane.

From this we may go on to the contemplation of 3- or higher-dimensional ("Riemannian") spaces, whose intrinsic properties vary from point to point. But these properties are no longer describable in terms of a single quantity, for the "curvature" now acquires at each point a directional character which requires in 3-space 6 components (and in 4-space 20) for its specification. We content ourselves here to call attention to a single combination of the 6, which we call the "mean curvature" of the space at the point $P(x, y, z)$, and which we again denote by K—or more fully by $K(x, y, z)$; it is in a sense the mean of the curvatures of various surfaces passing through P, and reduces to the previously contemplated space-constant K when the space in question is a congruence space. This concept is useful in physical applications, for the surface area S and the volume V of a sphere of radius r about the point $P(x, y, z)$ as center are again given by formulae (3), where now K is to be interpreted as the mean curvature $K(x, y, z)$ of the space at the point P. In four and higher dimensions similar concepts may be introduced and similar formulae developed, but for them we have no need here.

We have now to turn our attention to the world of physical objects about us, and to indicate how an ordered description of it is to be obtained in accordance with accepted, preferably philosophically neutral, scientific method. These objects, which exist for us in virtue of some prescientific concretion of our sense-data, are positioned in an extended manifold which we call physical space. The mind of the individual, retracing at an immensely accelerated pace the path taken by the race, bestirs itself to an analysis of the interplay between object and extension. There develops a notion of the permanence of the object and of the ordering and the change in time—another form of extension, through which object and subject appear to be racing together—of its extensive relationships. The study of the ordering of actual and potential relationships, the physical problem of space and time, leads to the consideration of geometry and kinematics as a branch of physical science. To certain aspects of this problem we now turn our attention.

We consider first that proposed solution of the problems of space which is based upon the postulate that space is an *a priori* form of the understanding. Its geometry must then be a congruence geometry, independent of the physical content of space; and since for Kant, the propounder of this view, there existed but one geometry, space must be Euclidean—and the problem of physical space is solved on the epistemological, prephysical, level.

But the discovery of other congruence geometries, characterized by a numerical parameter K, perforce modifies this view, and restores at least in some measure the objective aspect of physical space; the *a posteriori* ground for this space-constant K is then to be sought in the contingent. The means for its intrinsic determination is implicit in the formulae presented above; we have merely (!) to measure the volume V of a sphere of radius r or the sum σ of the angles of a triangle of measured area δ, and from the results to compute the value of K. On this modified Kantian view, which has been expounded at length by Russell, it is inconceivable that K might vary from point to point—for according to this view the very possibility of measurement depends on the constancy of space-structure, as guaranteed by the axiom of free mobility. It is of interest to mention in passing, in view of recent cosmological findings, the possibility raised by

A. Calinon (in 1889!) that the space-constant K might vary with time. But this possibility is rightly ignored by Russell, for the same arguments which would on this *a priori* theory require the constancy of K in space would equally require its constancy in time.

In the foregoing sketch we have dodged the real hook in the problem of measurement. As physicists we should state clearly those aspects of the physical world which are to correspond to elements of the mathematical system which we propose to employ in the description ("realization" of the abstract system). Ideally this program should prescribe fully the operations by which numerical values are to be assigned to the physical counterparts of the abstract elements. How is one to achieve this in the case in hand of determining the numerical value of the space-constant K?

Although K. F. Gauss, one of the spiritual fathers of non-Euclidean geometry, at one time proposed a possible test of the flatness of space by measuring the interior angles of a terrestrial triangle, it remained for his Göttingen successor K. Schwarzschild to formulate the procedure and to attempt to evaluate K on the basis of astronomical data available at the turn of the century. Schwarzschild's pioneer attempt is so inspiring in its conception and so beautiful in its expression that I cannot refrain from giving here a few short extracts from his work. After presenting the possibility that physical space may, in accordance with the neo-Kantian position outlined above, be non-Euclidean, Schwarzschild states (in free translation):

> One finds oneself here, if one but will, in a geometrical fairyland, but the beauty of this fairy tale is that one does not know but what it may be true. We accordingly bespeak the question here of how far we must push back the frontiers of this fairyland; of how small we must choose the curvature of space, how great its radius of curvature.

In furtherance of this program Schwarzschild proposes:

> A triangle determined by three points will be defined as the paths of light-rays from one point to another, the lengths of its sides a, b, c, by the times it takes light to traverse these paths, and the angles α β, γ will be measured with the usual astronomical instruments.

Applying Schwarzschild's prescription to observations on a given star, we consider the triangle ABC defined by the position A of the star and by two positions B, C of the earth—say six months apart—at which the angular positions of the star are measured. The base $BC = a$ is known, by measurements within the solar sytem consistent with the prescription, and the interior angles β, γ which the light-rays from the star make with the base-line are also known by measurement. From these the *parallax* $p = \pi - (\beta + \gamma)$ may be computed; in Euclidean space this parallax is simply the inferred angle α subtended at the star by the diameter of the earth's orbit. In the other congruence geometries the parallax is seen, with the aid of formula (2) above, to be equal to

$$(2') \qquad p = \pi - (\beta + \gamma) = \alpha - K\delta,$$

where α is the (unknown) angle at the star A, and δ is the (unknown) area of the triangle ABC. Now in spite of our incomplete knowledge of the elements on the far right, certain valid conclusions may be drawn from this result. First, if space is hyperbolic ($K < 0$), then for distant stars (for which $\alpha \sim 0$), the parallax p will remain positive; hence if stars are observed whose parallax is zero to within the errors of observation, this estimated error will give an upper limit to the absolute value $-K$, of the curvature. Second, if space is spherical ($K > 0$), for a sufficiently distant star (more distant than one-quarter the circumference of a Euclidean sphere of radius $R = 1/K^{1/2}$, as may immediately be seen by examining a globe) the sum $\beta + \gamma$ will exceed two right angles; hence the parallax p of such a star should be negative, and if no stars are in fact observed with negative parallax, the estimated error of observation will give an upper limit to the curvature K. Also, in this latter case the light sent out by the star must return to it after traversing the full line of length $2\pi R$ (πR in elliptic space), and hence we should, but for absorption and scattering, be able to observe the returning light as an anti-star in a direction opposite to that of the star itself!

On the basis of the evidence then available, Schwarzschild concluded that if space is hyperbolic its radius of curvature $R = 1/(-K)^{1/2}$ cannot be less than 64 light-years (i.e., the distance light travels in 64 years, and that if the space is elliptic its radius of curvature $R = 1/K^{1/2}$ is at least 1600 light-years. Hardly imposing figures for us today, who believe on other astronomical grounds that objects as distant as 500 million light-years have been sighted in the Mt. Wilson telescope, and who are expecting to find objects at twice that distance with the new Mt. Palomar mirror! But the value for us of the work of Schwarzschild lies in its sound operational approach to the problem of physical geometry—in refreshing contrast to the pontifical pronouncement of H. Poincaré, who after reviewing the subject stated:

> If therefore negative parallaxes were found, or if it were demonstrated that all parallaxes are superior to a certain limit, two courses would be open to us; we might either renounce Euclidean geometry, or else modify laws of optics and suppose that light does not travel rigorously in a straight line.
>
> It is needless to add that all the world would regard the latter solution as the more advantageous.
>
> The Euclidean geometry has, therefore, nothing to fear from fresh experiments.

So far we have tied ourselves into the neo-Kantian doctrine that space must be homogeneous and isotropic, in which case our proposed operational approach is limited in application to the determination of the numerical value of the space-constant K. But the possible scope of the operational method is surely broader than this; what if we do apply it to triangles and circles and spheres in various positions and at various times and find that the K so determined is in fact dependent on position in space and time? Are we, following Poincaré, to attribute these findings to the influence of an external force postulated for the purpose? Or are we to take our findings at face value, and accept the geometry to which we are led as a natural geometry for physical science?

The answer to this methodological question will depend largely on the *universality* of the geometry thus found—whether the geometry found in one situation or field of physical discourse may consistently be extended to others—and in the end partly on the predilection of the individual or of his colleagues or of his times. Thus Einstein's special theory of relativity, which offers a physical kinematics embracing measurements in space and time, has gone through several stages of acceptance and use, until at present it is a universal and indispensable tool of modern physics. Thus Einstein's general theory of relativity, which offers an extended kinematics which includes in its geometrical structure the universal force of gravitation, was long considered by some contemporaries to be a *tour de force,* at best amusing but in practice useless. And now, in extending this theory to the outer bounds of the observed universe, the kind of geometry suggested by the present marginal data seems to many so repugnant that they would follow Poincaré in postulating some *ad hoc* force, be it a double standard of time or a secular change in the velocity of light or Planck's constant, rather than accept it.

But enough of this general and historical approach to the problem of physical geometry! While we should like to complete this discussion with a detailed operational analysis of the solution given by the general theory of relativity, such an undertaking would require far more than the modest mathematical background which we have here presupposed. Further, the field of operations of the general theory is so unearthly and its *experimenta crucis* so delicate that an adequate discussion would take us far out from the familiar objects and concepts of the workaday world, and obscure the salient points we wish to make in a welter of unfamiliar and esoteric astronomical and mathematical concepts. What is needed is a homely experiment which could be carried out in the basement with parts from an old sewing machine and an Ingersoll watch, with an old file of *Popular Mechanics* standing by for reference! This I am, alas, afraid we have not achieved, but I do believe that the following example of a simple theory of measurement in a heat-conducting medium is adequate to expose the principles involved with a modicum of mathematical background. The very fact that it will lead to a rather bad and unacceptable physical theory will in itself be instructive, for its very failure will emphasize the requirement of universality of application—a requirement most satisfactorily met by the general theory of relativity.

The background of our illustration is an ordinary laboratory, equipped with Bunsen burners, clamps, rulers, micrometers and the usual miscellaneous impedimenta there met—at the turn of the century, no electronics required! In it the practical Euclidean geometry reigns (hitherto!) unquestioned, for even though measurements are there to be carried out with quite reasonable standards of accuracy, there is no need for sophisticated qualms concerning the effect of gravitational or magnetic or other general extended force-fields on its metrical structure. Now that we feel at home in these familiar, and disarming, surroundings, consider the following experiment:

Let a thin, flat metal plate be heated in any way—just so that the temperature T is not uniform over the plate. During the process clamp or otherwise constrain the plate to keep it from buckling, so that it can reasonably be said to remain flat by ordinary standards. Now proceed to make simple geometrical measurements on the plate with a short metal rule, which has a certain coefficient of expansion c, taking care that the rule is allowed to come into thermal equilibrium with the plate at each setting before making the measurement. The question now is, what is the geometry of the plate *as revealed by the results of these measurements?*

It is evident that, unless the coefficient of expansion c of the rule is zero, the geometry will not turn out to be Euclidean, for the rule will expand more in the hotter regions of the plate than in the cooler, distorting the

(Euclidean) measurements which would be obtained by a rule whose length did not change according to the usual laboratory standards. Thus the perimeter L of a circle centered at a point at which a burner is applied will surely turn out to be greater than π times its measured diameter $2r$, for the rule will expand in measuring through the hotter interior of the circle and hence give a smaller reading than if the temperature were uniform. On referring to the first of formulae (1) above it is seen that the plate would seem to have a negative curvature K at the center of the circle—the kind of structure exhibited by an ordinary twisted surface in the neighborhood of a "saddle-point." In general the curvature will vary from point to point in a systematic way; a more detailed mathematical analysis of the situation shows that, on removing heat sources and neglecting radiation losses from the faces of the plate, K is everywhere negative and that the "radius of curvature" $R = 1/(-K)^{1/2}$ at any point P is inversely proportional to the rate s at which heat flows past P. (R is in fact equal to k/cs, where k is the coefficient of heat conduction *of the plate* and c is as before the coefficient of expansion *of the rule.*) The hyperbolic geometry is accordingly realized when the heat flow is constant throughout the plate, as when the long sides of an elongated rectangle are kept at different fixed temperatures.

And now comes the question, what is the true geometry of the plate? The flat Euclidean geometry we had uncritically agreed upon at the beginning of the experiment, or the un-Euclidean geometry revealed by measurement? It is obvious that the question is improperly worded; the geometry is determinate only when we prescribe the method of measurement, i.e., when we set up a correspondence between the physical aspects (here readings on a definite rule obtained in a prescribed way) and the elements (here distances, in the abstract sense) of the mathematical system. Thus our original common-sense requirement that the plate not buckle, or that it be measured with an invar rule (for which $c \sim 0$), leads to Euclidean geometry, while the use of a rule with a sensible coefficient of expansion leads to a locally hyperbolic type of Riemannian geometry, which is in general not a congruence geometry.

There is no doubt that anyone examining this situation will prefer Poincaré's common-sense solution of the problem of the physical geometry of the plate—i.e., to attribute to it Euclidean geometry and to consider the measured deviations from this geometry, as due to the action of a force (thermal stresses in the rule). Most compulsive to this solution is the fact that this disturbing force lacks the requirement of universality; on employing a brass rule in place of one of steel we would find that the local curvature is trebled—and an ideal rule (c = 0) would, as we have noted, lead to the Euclidean geometry.

In what respect, then, does the general theory of relativity differ in principle from this geometrical theory of the hot plate? The answer is: *in its universality;* the force of gravitation which it comprehends in the geometrical structure acts equally on all matter. There is here a close analogy between the gravitational mass M of the field-producing body (Sun) and the inertial mass m of the test-particle (Earth) on the one hand, and the heat conduction K of the field (plate) and the coefficient of expansion c of the test-body (rule) on the other. *The success of the general relativity theory of gravitation as a physical geometry of space-time is attributable to the fact that the gravitational and inertial masses of any body are observed to be rigorously proportional for all matter.* Whereas in our geometrical theory of the thermal field the ratio of heat conductivity to coefficient of expansion varies from substance to substance, resulting in a change of the geometry of the field on changing the test-body.

From our present point of view the great triumph of the theory of relativity lies in its absorbing the universal force of gravitation into the geometrical structure; its success in accounting for minute discrepancies in the Newtonian description of the motions of test-bodies in the solar field, although gratifying, is nevertheless of far less moment to the philosophy of physical science. Einstein's achievements would be substantially as great even though it were not for these minute observational tests.

Our final illustration of physical geometry consists in a brief reference to the cosmological problem of the geometry of the observed universe as a whole—a problem considered in greater detail elsewhere in this volume. *If* matter in the universe can, taken on a sufficiently large scale (spatial gobs millions of light-years across), be considered as uniformly distributed, and if (as implied by the general theory of relativity) its geometrical structure is conditioned by matter, then to this approximation our 3-dimensional astronomical space must be homogeneous and isotropic, with a spatially-constant K which may however depend upon time. Granting this hypothesis, how do we go about measuring K, using of course only procedures which can be operationally specified, and to which congruence geometry are we thereby led? The way to the answer is suggested by the second of the formulae (3), for if the nebulae are by-and-large uniformly distributed, then the number N within a sphere of radius r must be proportional to the volume V of this sphere. We have then only to examine the dependence of this number N, as observed in a sufficiently powerful telescope, on the distance r to determine the deviation from the Euclidean value. But how is r operationally to be defined?

If all the nebulae were of the same intrinsic brightness, then their apparent brightness as observed from the Earth should be an indication of their distance from us; we must therefore examine the exact relation to be expected between apparent brightness and the abstract distance r. Now it is the practice of astronomers to assume that brightness falls off inversely with the square of the "distance" of the object—as it would do in Euclidean space, if there were no absorption, scattering, and the like. We must therefore examine the relation between this astronomer's "distance" d, as inferred from apparent brightness, and the distance r which appears as an element of the geometry. It is clear that *all* the light which is radiated at a given moment from the nebula will, after it has traveled a distance r, lie on the surface of a sphere whose area S is given by the first of the formulae (3). And since the practical procedure involved in determining d is equivalent to assuming that all this light lies on the surface of a Euclidean sphere of radius d, it follows immediately that the relationship between the "distance" d used in practice and the distance r dealt with in the geometry is given by the equation

$$4\pi d^2 = S = 4\pi r^2 \, (1 - Kr^2/3 + \ldots);$$

whence, to our approximation

$$(4) \qquad \begin{aligned} d &= r \, (1 - Kr^2/6 + \ldots), \text{ or} \\ r &= d \, (1 + Kd^2/6 + \ldots). \end{aligned}$$

But the astronomical data give the number N of nebulae counted out to a given inferred "distance" d, and in order to determine the curvature from them we must express N, or equivalently V, to which it is assumed proportional, in terms of d. One easily finds from the second of the formulae (3) and the formula (4) just derived that, again to the approximation here adopted,

$$(5) \qquad V = 4/3 \; \pi \; d^3 \; (1 + 3/10 \; Kd^2 + \ldots).$$

And now on plotting N against inferred "distance" d and comparing this empirical plot with the formula (5), it should be possible operationally to determine the "curvature" K.

The search for the curvature K indicates that, after making all known corrections, the number N seems to increase faster with d than the third power, which would be expected in Euclidean space, hence K is positive. The space implied thereby is therefore bounded, of finite total volume, and of a present "radius of curvature" $R = 1/K^{1/2}$ which is found to be of the order of 500 million light-years. Other observations, on the "red-shift" of light from these distant objects, enable us to conclude with perhaps more assurance that this radius is increasing in time at a rate which, if kept up, would double the present radius in something less than 2000 million years. . . .

N

Quanta

INTRODUCTION

CLASSICAL PHYSICS, which consists of Newtonian mechanics and the electromagnetic theory of Faraday and Maxwell, deals with the world around us—the earth, the solar system—and the two basic forces which affect it: electromagnetism and gravity.

Newtonian mechanics, consisting of Newton's laws of motion and gravity in a Euclidean framework of absolute space and absolute time, quite adequately explained nature for well over two centuries. In 1900, however, the discovery of the constancy of the speed of light, of subatomic particles and of Planck's quantum of action which means that energy is not a continuous flow of waves but an ensemble of discrete packets that Planck named *quanta,* led to a sharp division between classical and modern physics.

Quantum physics is mostly concerned with the world of the almost inconceivably small particles within the atom, which are affected by the electromagnetic force (familiar in Newtonian physics) but rarely by gravity. Two other forces—the strong force, which holds the nucleus of the atom together, and the weak force which is involved in other subatomic phenomena, particularly neutrinos and electrons—also influence subatomic particles. The relative strengths of the four forces, arbitrarily taking unity as the index for the strong force, are:

strong	1
electromagnetic	10^{-2}
weak	10^{-13}
gravitational	10^{-36}

The ranges of the electromagnetic and gravitational forces are infinite; the range of the strong and weak are subatomic—that is, practically infinitesimal, extending not much further beyond nuclear dimensions.

From 1900 to the postwar period, quantum physics was developed by the devoted search of thousands of physicists, among whom the principal figures are Max Planck, Albert Einstein, Niels Bohr, Werner Heisenberg, Louis de Broglie, Erwin Schrödinger, Max Born and Paul Dirac. Other names later joined this list and included Eugene Wigner, Enrico Fermi, I.I. Rabi, and, more recently, Julian Schwinger, Richard Feynman, Murray Gell-Mann, Sheldon Glashow, Steven Weinberg, and Abdus Salam, as well as a host of others whose work is described in the section on particle physics.

Einstein's researches at the beginning of the century led to the discovery that light consists of both waves and particles, the latter of which Einstein named *photons.* This remarkable discovery, which began as a daring intuition, was later confirmed by many experiments, and in 1923 the French physicist de Broglie theorized that this duality might also be characteristic of phenomena other than light waves and photons. Starting from Einstein's special theory of relativity, and using the relationship between the frequency of a photon and its energy, he developed a similar relationship for electrons, attributing waves to them (de Broglie waves). This theory was proved in 1927 in experiments on electron diffraction of electrons passing through and reflected by crystals (similar to Von Laue's x-ray diffraction). In subatomic physics, de Broglie's "matter waves" have been used for decades to explain the behavior of electrons.

The development of quantum mechanics in the 1920s and 1930s was a unique scientific revolution. The strict determination of classical physics, which is capable of being expressed in "ordinary language," gave way to statistical probability concepts, which unfortunately for the general public, can only be expressed in the most difficult mathematical terms. These concepts are usually known as the Copenhagen Interpretation of quantum mechanics, and were for the most part the product of Heisenberg and Bohr, who was head of the Institute of Theoretical Physics in Copenhagen. In its totality, it was unacceptable to Einstein, who argued for decades against the indeterministic concepts that stemmed from the Copenhagen school. The Danish school held these concepts to be basic, while its opponents considered them to be defects due to the fragmentary nature of present knowledge and, for that reason, considered the quantum mechanics as an incomplete physical theory.

Bohr's great contribution to quantum theory was his theory of atomic spectra based on his atomic model, the "Bohr atom," in which an atom consists of a very small, positively charged nucleus surrounded by electrons equal in number to the charge on the nucleus. This number Z is called the atomic number of the atom. The crucial feature of the Bohr atom, which incorporates the quantum theory, is that the electrons can only move in a discrete set of orbits. When all the electrons are in their lowest permitted orbits, the atom is said to be in its ground state (or unexcited). If an electron absorbs energy (either from a photon or by a collision), the atom is said to be excited; when the electron returns to its ground state again, it emits exactly one photon. If the number of electrons in the atom is less than Z, the atom is said to be ionized.

Later extensions of Bohr's theory to many elements (his original research had been based mainly on the hydrogen atom) led to an analogy between the behavior of electrons in an atom and acoustic vibrations (tonality, harmonies, etc). This in turn led Erwin Schrödinger to postulate a standing wave for each electron that varies with its position in the atom. Thus he replaced the electrons by de Broglie waves. His paper, published in 1926, marked the beginning of *wave mechanics,* a wholly new way of introducing the quantum theory into the atom. Schrödinger asserted that the electron is not a particle, but a wave, a matter wave just as a water wave is a wave. His theory was opposed by the Göttingen physicists headed by Max Born, who argued that since individual electrons can be counted by a Geiger counter or tracked in a bubble chamber, they must perforce be particles. Born's response pleased some but mystified a good many more of his contemporaries. The wave amplitude, he said, determines the probability of finding an electron (admittedly a particle) in a given region of space. His paper came out in June 1926 and created the image of a dice-rolling God, much to the anguish of Einstein and the world of empirical physicists in general. The American physicist Heinz Pagels, who has a particular talent for expressing difficult subjects in simple language, has written a caption for a drawing explaining wave amplitude:

> According to Max Born's statistical interpretation of the de Broglie-Schrödinger wave, the height or amplitude of the wave when squared gives the probability for finding the particle at that position. All that quantum theory could do was predict the wave shape and hence the probability that a quantum particle would have certain properties; it could not predict with certainty the outcome of single measurements of those properties, as did the old classical physics.

Curiously, Schrödinger, a philosopher as well as a physicist, was not greatly pleased by Born's statistical interpretation of his waves and remarked that had he known the consequences, he might not have published his paper. Einstein wrote to Born, "Quantum mechanics is certainly imposing, but an inner voice tells me it is not yet the real thing. The theory says much but does not really bring us closer to the secret of the Old One."

Born later answered, "If God has made the universe a perfect mechanism, He has at least conceded so much to our imperfect intellect that in order to predict little parts of it, we need not solve innumerable differential equations, but can use dice with fair success."

Classical physicists were convinced that sufficiently accurate measurements of phenomena would necessarily produce highly accurate predictions of future events. In 1927, however, Werner Heisenberg, in studying the electron's motion, found that, owing to the existence of the quantum of action h, any attempt to measure accurately the electron's momentum will automatically destroy the precise knowledge of its position. This discovery is the Heisenberg uncertainty principle.

Another important discovery in the development of quantum theory was the exclusion principle offered by Wolfgang Pauli, an Austrian physicist trained by the distinguished professor Arthur Sommerfeld. His principle asserts that no two identical electron wave patterns can exist in the same atom; it was this finding that led to the discovery that only two electrons can occupy the lowest energy level in Bohr's quantized atom, eight in the next lowest energy level and so on. These statistics are the basis for Linus Pauling's theory of the nature of the chemical bond and the acceptance of chemistry as a branch of physics.

The application of quantum mechanics to the understanding of "gross" molecular structure and to subatomic structures that involve the so-called ultimate particles of matter and where durations are counted in the trillionths of trillionths of seconds has resulted in vast changes. From Newton's world in its human scale, we have gone to a world of computers, satellites, and the incredible particle accelerators or "colliders," which whirl particles around enormous circular tracks, accelerating them to the speed of light and smashing them into each other to reveal their constituents.

N.1

The Genesis and Present State of Development of the Quantum Theory

MAX PLANCK

THE QUANTUM THEORY was introduced to the world by the German physicist Max Planck in his celebrated lecture of 14 December 1900. When he was awarded the Nobel Prize in physics for his pioneer work in this strikingly new area of science, he delivered a lecture in 1918, in which he reviewed the development of quantum studies up to that point.

Planck tells of his various false starts in his study of black-body radiation—the radiation emitted by a body having no reflecting power—and how the experimental observations on the wavelength distribution of black-body energy emission as a function of temperature and wavelength failed to agree with the results that were expected from the principles of classical physics. Elsewhere, Planck had told how, almost in desperation at not being able to solve the black-body problem, he hypothesized the existence of discrete packets of energy to which he gave the name quanta.

The study of quantum theory originated with the discovery of Planck's constant *h*—a term which is measured in units of action (energy × time). By using this constant one could determine an energy value for a subatomic system with a given quantity of action. Planck's constant is characteristic of quantum mechanics as a whole because the essence of any quantum mechanical system lies in its changes in action. However, these changes are not continuous over a broad range of values but occur in discrete intervals. Unlike the classical atomic model which presumed that the radiation or absorption of energy by an electron would allow it to drop or rise continuously to a different energy level, quantum mechanics asserts that electrons can emit or absorb only discrete quantities of energy because an electron's action must change in integral multiples of *h*. There is therefore a lowest energy level. Of course the energy emissions or absorptions may be small and the subsequent displacement of the electrons puny but, contrary to classical atomic theory, the amount of energy and the subsequent displacement is not infinitely small.

Quantum theory developed from the incompatibility in classical physics of the simultaneous particle theory and wave theory of light, the wave-particle duality. The inability of either theory to explain satisfactorily all characteristics of electromagnetic phenomena encouraged the formulation of a theory of wave-particle duality in which light possesses both undulatory and corpuscular properties.

One of the most striking characteristics of Planck's paper is the extraordinary generosity with which he acknowledges the contributions of literally dozens of colleagues and predecessors. He also includes a rare bit of nontrivial trivia—that the thermodynamic constant known as Boltzmann's constant was not actually introduced by Ludwig Boltzmann, who, to the best of Planck's knowledge, "never gave thought to the possibility of carrying out an exact measurement of the constant."

In his closing remarks, Planck expressed confidence that the many extremely difficult problems facing scientists as they explored the potential of his original theory would be solved, and "that which appears today so unsatisfactory," he wrote, "will in fact eventually, seen from a higher vantage point, be distinguished by its special harmony and simplicity." His words are a concise expression of the basic principle underlying all experimental and theoretical physics, a principle that inspires in our own age those physicists seeking the elusive grand unified theory that will explain in a single framework the strong, weak, electromagnetic, and gravitational forces.

Physical concepts are free creations of the human mind, and are not, however it may seem, uniquely determined by the external world. In our endeavor to understand reality we are somewhat like a man trying to understand the mechanism of a closed watch. He sees the face and the moving hands, even hears its ticking but he has no way of opening the case. If he is ingenious he may form some picture of a mechanism which could be responsible for all the things he observes but he may never be quite sure his picture is the only one which could explain his observations. He will never be able to compare his picture with the real mechanism and he cannot even imagine the possibility of the meaning of such a comparison.

—ALBERT EINSTEIN, *The Evolution of Physics*

. . . WHEN I LOOK back to the time, already twenty years ago, when the concept and magnitude of the physical quantum of action began, for the first time, to unfold from the mass of experimental facts, and again, to the long and ever tortuous path which led, finally, to its disclosure, the whole development seems to me to provide a fresh illustration of the long-since proved saying of Goethe's that man errs as long as he strives. And the whole strenuous intellectual work of an industrious research worker would appear, after all, in vain and hopeless, if he were not occasionally through some striking facts to find that he had, at the end of all his criss-cross journeys, at last accomplished at least one step which was conclusively nearer the truth. An indispensable hypothesis, even though still far from being a guarantee of success, is however the pursuit of a specific aim, whose lighted beacon, even by initial failures, is not betrayed.

For many years, such an aim for me was to find the solution to the problem of the distribution of energy in the normal spectrum of radiating heat. Since Gustav Kirchhoff has shown that the state of the heat radiation which takes place in a cavity bounded by any emitting and absorbing substances of uniform temperature is entirely independent upon the nature of the substances, a universal function was demonstrated which was dependent only upon temperature and wavelength, but in no way upon

Max Planck, Nobel Prize in Physics Award Address, 1918. Reprinted with permission of Elsevier Publishing Co., and the Nobel Foundation.

the properties of any substance. And the discovery of this remarkable function promised deeper insight into the connection between energy and temperature which is, in fact, the major problem in thermodynamics and thus in the whole of molecular physics. To attain this there was no other way but to seek out from all the different substances existing in Nature one of known emissive and absorptive power, and to calculate the properties of the heat radiation in stationary energy exchange with it. According to Kirchhoff's Law, this would have to prove independent of the nature of the body.

Heinrich Hertz's linear oscillator, whose laws of emission, for a given frequency, Hertz had just previously completely developed, seemed to me to be a particularly suitable device for this purpose. If a number of such Hertzian oscillators are set up within a cavity surrounded by a sphere of reflecting walls, then by analogy with audio oscillators and resonators, energy will be exchanged between them by the output and absorption of electromagnetic waves, and finally stationary radiation corresponding to Kirchhoff's Law, the so-called black-body radiation, should be set up within the cavity. I was filled at that time with what would be thought today naively charming and agreeable expectations, that the laws of classical electrodynamics would, if approached in a sufficiently general manner with the avoidance of special hypotheses, be sufficient to enable us to grasp the most significant part of the process to be expected, and thus to achieve the desired aim. I, therefore, developed first the laws of emission and absorption of a linear resonator on the most general basis, in fact I proceeded on such a detour which could well have been avoided had I made use of the existing electron theory of H. A. Lorentz, already basically complete. But since I did not quite trust the electron hypothesis, I preferred to observe that energy which flowed in and out through an enclosing spherical surface around the resonator at a suitable distance from it. By this method, only processes in a pure vacuum came into account, but a knowledge of these was sufficient to draw the necessary conclusions, however, about the energy changes in the resonator.

The fruit of this long series of investigations, of which some, by comparison with existing observations, mainly the vapour measurements by V. Bjerknes, were susceptible to checking, and were thereby confirmed, was the establishment of the general connection between the energy of a resonator of specific natural period of vibration and the energy radiation of the corresponding spectral region in the surrounding field under conditions of stationary energy exchange. The noteworthy result was found that this connection was in no way dependent upon the nature of the resonator, particularly its attenuation constants—a circumstance which I welcomed happily since the whole problem thus became simpler, for instead of the energy of radiation, the energy of the resonator could be taken and, thereby, a complex system, composed of many degrees of freedom, could be replaced by a simple system of one degree of freedom.

Nevertheless, the result meant no more than a preparatory step towards the initial onslaught on the particular problem which now towered with all its fearsome height even steeper before me. The first attempt upon it went wrong, for my original secret hope that the radiation emitted from the resonator can be in some characteristic way or other distinguished from the absorbed radiation and thereby allow a differential equation to be set up, from the integration of which one could gain some special condition for the properties of stationary radiation, proved false. The resonator reacted only to those rays which it also emitted, and was not in the slightest bit sensitive to the adjacent spectral region.

Furthermore, my hypothesis that the resonator could exercise a unilateral, i.e. irreversible, effect upon the energy in the surrounding radiation field, was strongly contested by Ludwig Boltzmann, who, with his riper experience in these problems, proved that according to the laws of classical dynamics each of the processes observed by me can proceed in exactly the opposite direction, in such a way, that a spherical wave emitted from the resonator, returns and contracts in steadily diminishing concentric spherical surfaces inwards to the resonator and is again absorbed by it, thereby allowing the formerly absorbed energy to be re-transmitted into space in the direction from which it came. And when I excluded this kind of singular process, such as an inwardly directed wave, by means of the introduction of a limiting definition, the hypothesis of natural radiation, all these analyses still showed ever more clearly that an important connecting element or term, essential for the complete grasp of the core of the problem, must be missing.

So there was nothing left for me but to tackle the problem from the opposite side, that of thermodynamics, in which field I felt, moreover, more confident. In fact my earlier studies of the Second Law of Heat Theory stood me in good stead, so that from the start I tried to get a connection, not between the temperature but rather the entropy of the resonator and its energy, and in fact, not its entropy exactly but the second derivative with respect to the energy since this has a direct physical meaning for the irreversibility of the energy exchange between resonator and radiation. Since I was, however, at that time still too far oriented towards the phenomenological aspect to come to closer quarters with the connection between entropy and probability, I saw myself, at first, relying solely upon the existing results of experience. In the foreground of interest at that time, in 1899, was the energy distribution law established by W. Wien shortly before, whose experimental proof was taken up, on the one hand, by F. Paschen at the Technische Hochschule in Hannover, and, on the other hand, by O. Lummer and E. Pringsheim at the State Institution in Charlottenburg. This law brought out the dependence of the radiation intensity on the temperature, representing it by an exponential function. If one calculates the connection between the entropy and the energy of a resonator, determined by the above law, the remarkable result is obtained that the reciprocal value of the above-mentioned differential coefficient, which I will call *R*, is proportional to the energy. This extremely simple relationship can be considered as the completely adequate expression of Wien's energy distribution law; for with the dependence upon the energy, the dependence upon the wavelength is always directly given through the general, well-established displacement law by Wien.

Since the whole problem concerned a universal law of Nature, and since at that time, as still today, I held the unshakeable opinion that the simpler the presentation of a particular law of Nature, the more general it is—though at the same time, which formula to take as the simpler, is a problem which cannot always be confidently and finally decided—I believed for a long time that the law that the quantity *R* is proportional to the energy, should be looked upon as the basis for the whole energy distribution law. This concept could not be maintained for long in the face of fresh measurements. Whilst for small values of the energy and for short waves, Wien's law was satisfactorily confirmed, noteworthy deviations for larger wavelengths were found, first by O. Lummer and E. Pringsheim, and finally by H. Rubens and F. Kurlbaum, whose measurements on the infrared residual rays of fluorite and rock salt revealed a totally different, though still extremely simple relationship, characterized by the fact that the quantity *R* is not proportional to the energy, but to the square of the energy, and in fact this holds with increasing accuracy for greater energies and wavelengths.

So, through direct experiment, two simple limits were determined for the function *R:* for small energies, proportionality with the energy; for greater energies, proportionality with the square of the energy. There was no better alternative but to make, for the general case, the quantity *R* equal to the sum of two terms, one of the first power, and one of the second power of the energy, so that for small energies the first is predominant, whilst for the greater energies the second is dominant. Thus the new radiation formula was found, which, in the face of its experimental proof, has stood firm to a reasonable extent until now. Even today, admittedly, we cannot talk of final exact confirmation. In fact, a fresh attempt at proof is urgently required.

However, even if the radiation formula should prove itself to be absolutely accurate, it would still only have, within the significance of a happily chosen interpolation formula, a strictly limited value. For this reason, I busied myself, from then on, that is, from the day of its establishment, with the task of elucidating a true physical character for the formula, and this problem led me automatically to a consideration of the connection between entropy and probability, that is, Boltzmann's trend of ideas; until after some weeks of the most strenuous work of my life, light came into the darkness, and a new undreamed-of perspective opened up before me.

I must make a small intercalation at this point. According to Boltzmann, entropy is a measure for physical probability, and the nature and essence of the Second Law of Heat Theory is that in Nature a state occurs more frequently, the more probable it is. Now one always measures in Nature the difference in entropies, never the entropy itself, and to this extent one cannot speak of the absolute entropy of a state, without a certain arbitrariness. Nevertheless, it is useful to introduce the suitably defined absolute value of entropy, namely for the reason that with its help certain general laws can be particularly easily formulated. The case seems to be parallel, as I see it, with that of energy. Energy itself cannot be measured, only its difference. For that reason one used to deal, not with energy, but with work, and even Ernst Mach, who had so much to do with the Law of Conservation of Energy, and who in principle kept away from all speculations beyond the field of observation, has always avoided speaking of energy itself. Likewise, in thermochemistry, one has always stuck to the thermal effect, that is, to energy differences, until Wilhelm Ostwald in particular emphatically showed that many detailed considerations could be significantly abbreviated if one dealt with energy itself instead of with calorimetric numbers. The additive constant, which was at first still undetermined in the expression of energy, has later been finally determined through the relativistic law of the proportionality between energy and inertia.

In a similar way to that for energy, an absolute value can be defined also for entropy and, as a result thereof, for the physical probability too, e.g. by so fixing the additive constant that energy and entropy disappear together. On the basis of a consideration of this kind a specific, relatively simple combinational method was obtained for the calculation of the physical probability of a specified energy distribution in a system of resonators, which led exactly to that entropy expression determined by the radiation law, and it brought me much-valued satisfaction for the many disappointments when Ludwig Boltzmann, in the letter returning my essay, expressed his interest and basic agreement with the train of thoughts expounded in it.

For the numerical treatment of the indicated consideration of probability, knowledge of two universal constants is required, both of which have an independent physical meaning, and whose subsequent evaluation from the law of radiation must provide proof as to whether the whole method is to be looked upon as a mere artifice for calculation, or whether it has an inherent real physical sense and interpretation. The first constant is of a more formal nature and is connected with the definition of temperature. If temperature were to be defined as the average kinetic energy of a molecule in an ideal gas, that is, as a tiny, little quantity, then the constant would have the value ⅔. In conventional temperature measure, on the contrary, the constant has an extremely small value which stands, naturally, in close association with the energy of a single molecule, and an exact knowledge of which leads, therefore, to the calculation of the mass of a molecule and those parameters related to it. This constant is often referred to as Boltzmann's constant, although, to my knowledge, Boltzmann himself never introduced it—a peculiar state of affairs, which can be explained by the fact that Boltzmann, as appears from his occasional utterances, never gave thought to the possibility of carrying out an exact measurement of the constant. Nothing can better illustrate the positive and hectic pace of progress which the art of experimenters has made over the past twenty years, than the fact that since that time, not only one, but a great number of methods have been discovered for measuring the mass of a molecule with practically the same accuracy as that attained for a planet.

At the time when I carried out the corresponding calculation from the radiation law, an exact proof of the number obtained was quite impossible, and not much more could be done than to determine the order of magnitude which was admissible. It was shortly afterward that E. Rutherford and H. Geiger succeeded in determining, by direct counting of the alpha particles, the value of the electrical elementary charge, which they found to be 4.65×10^{-10} electrostatic units; and the agreement of this figure with the number calculated by me, 4.69×10^{-10}, could be taken as decisive confirmation of the usefulness of my theory. Since then, more sophisticated methods have led to a slightly higher value, these measurements being carried out by E. Regener, R. A. Millikan, and others.

The explanation of the second universal constant of the radiation law was not so easy. Because it represents the product of energy and time (according to the first calculation it was 6.55×10^{-27} erg sec), I described it as the elementary quantum of action. Whilst it was completely indispensable for obtaining the correct expression for entropy—since only with its help could the magnitude of the "elementary regions" or "free rooms for action" of the probability, decisive for the assigned probability consideration, be determined—it proved elusive and resistant to all efforts to fit it into the framework of classical theory. As long as it was looked upon as infinitely small, that is, for large energies or long periods of time, everything went well; but in the general case, however, a gap yawned open in some place or other, which was the more striking, the weaker and faster the vibrations that were considered. The foundering of all efforts to bridge the chasm soon left little doubt. Either the quantum of action was a fictional quantity, then the whole deduction of the radiation law was in the main illusory and represented nothing more than an empty nonsignificant play on formulae, or the derivation of the radiation law was based on a sound physical conception. In this case the quantum of action must play a fundamental role in physics, and here was something entirely new, never before heard of, which seemed called upon to basically revise all our physical thinking, built as this was, since the establishment of the infinitesimal calculus by Leibniz and Newton, upon the acceptance of the continuity of all causative connections.

Experiment has decided for the second alternative. That the decision could be made so soon and so definitely was due not to the proving of the energy distribution law of heat radiation, still less to the special derivation of that law devised by me, but rather should it be attributed to the restless forward-thrusting work of those research workers who used the quantum of action to help them in their own investigations and experiments. The first impact in this field was made by A. Einstein who, on the one hand, pointed out that the introduction of the energy quanta, determined by the quantum of action, appeared suitable for obtaining a simple explanation for a series of noteworthy observations during the action of light, such as Stokes' Law, electron emission, and gas ionization, and, on the other hand, derived a formula for the specific heat of a solid body through the identification of the expression for the energy of a system of resonators with that of the energy of a solid body, and this formula expresses, more or less correctly, the changes in specific heat, particularly its reduction with falling temperature. The result was the emergence, in all directions, of a number of problems whose more accurate and extensive elaboration in the course of time brought to light a mass of valuable material. . . .

To be sure, the introduction of the quantum of action has not yet produced a genuine quantum theory. In fact, the path the research worker must yet tread to it is not less than that from the discovery of the velocity of light by Olaf Römer to the establishment of Maxwell's theory of light. The difficulties which the introduction of the quantum of action into the well-tried classical theory has posed right from the start have already been mentioned by me. During the course of the years they have increased rather than diminished, and if, in the meantime, the impetuous forward-driving research has passed to the order of the day for some of these, temporarily, the gaps left behind, awaiting subsequent filling, react even harder upon the conscientious systematologist. What serves in Bohr's theory as a basis to build up the laws of action, is assembled out of specific hypotheses which, up to a generation ago, would undoubtedly have been flatly rejected altogether by every physicist. The fact that in the atom, certain quite definite quantum-selected orbits play a special role, might be taken still as acceptable, less easily however, that the electrons, circulating in these orbits with definite acceleration, radiate no energy at all. The fact that the quite sharply defined frequency of an emitted photon should be different from the frequency of the emitting electron must seem to a theoretical physicist, brought up in the classical school, at first sight to be a monstrous and, for the purpose of a mental picture, a practically intolerable demand.

But numbers decide, and the result is that the roles, compared with earlier times, have gradually changed. What initially was a problem of fitting a new and strange element, with more or less gentle pressure, into what was generally regarded as a fixed frame has become a question of coping with an intruder who, after appropriating an assured place, has gone over to the offensive; and today it has become obvious that the old framework must somehow or other be burst asunder. It is merely a question of where and to what degree. If one may make a conjecture about the expected escape from this tight corner, then one could remark that all the signs suggest that the main principles of thermodynamics from the classical theory will not only rule unchallenged but will more probably become correspondingly extended. What the armchair experiments meant for the foundation of classical thermodynamics, the adiabatic hypothesis of

P. Ehrenfest means, provisionally, to the quantum theory; and in the same way as R. Clausius, as a starting point for the measurement of entropy, introduced the principle that, when treated appropriately, any two states of a material system can, by a reversible process, undergo a transition from one to the other, now the new ideas of Bohr's open up a very similar path into the interior of a wonderland hitherto hidden from him.

There is in particular one problem whose exhaustive solution could provide considerable elucidation. What becomes of the energy of a photon after complete emission? Does it spread out in all directions with further propagation in the sense of Huygens' wave theory, so constantly taking up more space, in boundless progressive attenuation? Or does it fly out like a projectile in one direction in the sense of Newton's emanation theory? In the first case, the quantum would no longer be in the position to concentrate energy upon a single point in space in such a way as to release an electron from its atomic bond, and in the second case, the main triumph of the Maxwell theory—the continuity between the static and dynamic fields and, with it, the complete understanding we have enjoyed, until now, of the fully investigated interference phenomena—would have to be sacrificed, both being very unhappy consequences for today's theoreticians.

Be that as it may, in any case no doubt can arise that science will master the dilemma, serious as it is, and that which appears today so unsatisfactory will in fact eventually, seen from a higher vantage point, be distinguished by its special harmony and simplicity. Until this aim is achieved, the problem of the quantum of action will not cease to inspire research and fructify it, and the greater the difficulties which oppose its solution, the more significant it finally will show itself to be for the broadening and deepening of our whole knowledge in physics.

N.2

Development of Our Conception of the Nature and Constitution of Radiation

ALBERT EINSTEIN

ONE OF THE outstanding characteristics of the mind of Albert Einstein was its extraordinary versatility. There were few fields of avant-garde physics in the first half of the twentieth century in which his voice was not heard. His relationship to quantum mechanics is a complex one. Although he was one of the first to apply Planck's quantum concept (in his 1905 paper on the photoelectric effect), in the 1920s and 1930s he often found himself at odds with many of his most brilliant contemporaries—Bohr and Heisenberg, for example—because Einstein—unlike most other physicists—was philosophically opposed to the statistical nature of the quantum mechanics. He simply could not accept the idea that the behavior of atoms—and hence the entire Universe—is governed by purely random physical processes. Einstein's deterministic outlook caused him to argue that there is an underlying order in the Universe though he was never able effectively to reconcile his belief with the probabilistic nature of the quantum mechanics.

As early as 1905, Einstein showed that radiant energy is itself quantized—a step that Planck himself had not taken. He demonstrated that, at a given temperature, radiation in thermodynamic equilibrium with the walls of its container behaves exactly like a perfect gas and consists of a discrete number of particles, each of energy $h\nu$ where h is Planck's constant and v is the frequency of the radiation. He called these particles of radiation *photons.* Although he was the formulator of the concept of quanta of energy, Planck had not considered the nature of radiant energy. A few years later, Einstein delivered a paper—our next selection—which was later published in the *Physikalische Zeitschrift* 22 (1909), in which he reviewed the history of the nature and constitution of radiation, concluding with the statement "I . . . wished to illustrate briefly that the two structural phenomena [wave structure and quantized (i.e. particle) structure of radiation] do not have to be regarded as incompatible with each other." Einstein's lecture was followed by a discussion by Einstein and several other physicists, including Planck. It is also included.

Beware when the great God lets loose a thinker on this planet.

—Ralph Waldo Emerson, "Circles," *Essays*

Quantum mechanics is very impressive but I am convinced that God does not play dice.

—Albert Einstein

When it was discovered that light exhibits the phenomena of interference and refraction, it seemed hardly to be doubted that light had to be considered as a wave propagation. Because light is also able to propagate through vacuum one had to imagine also that there is a special kind of matter present which mediates the propagation of light waves. It was necessary for the laws of light propagation in ponderable bodies to assume that this matter, which was called luminiferous ether, was also present in those, and that it was mainly the luminiferous ether which mediates the propagation of light. The existence of that luminiferous ether seemed indubitable. In the introduction about the ether in the first volume of the excellent textbook in physics by Chwolson, published in 1902, this sentence can be found: "The probability of the hypothesis of the existence of this element is extremely close to certainty."

But today we have to regard the ether hypothesis as an outdated point of view. It is undeniable that there exist an extensive group of facts concerning radiation which show that light has some fundamental properties which can be understood more easily from the viewpoint of Newton's corpuscular theory of light rather than the viewpoint of the undulation theory. My opinion is therefore that the next step of the development of theoretical physics will lead us to a theory of light which can be interpreted as a kind of fusion of undulation and corpuscular theory. The purpose of the following exposition is to justify this opinion and to show that a deep change in our conception about the nature and the constitution of light is indispensable.

The biggest step which theoretical optics has achieved since the introduction of the undulation theory likely consists of Maxwell's brilliant dis-

---

Albert Einstein, "Development of Our Conception of the Nature and Constitution of Radiation," *Physikalische Zeitschrift* 22, 1909. Translated by Christian Holm. Reprinted with permission of the Hebrew University of Jerusalem, Israel.

covery of the possibility of conceiving of light as an electromagnetic process. Instead of a mechanical quantity, namely deformation and velocity of the parts of the ether, this theory introduces the electromagnetic state of ether and matter into consideration and thereby reduces the optical problems to electromagnetic ones. The more the electromagnetic theory was developed, the more the question of whether electromagnetic processes could be traced back to mechanical ones was moved into the background; one got used to treating the conception of electric and magnetic field-strength, electric charge density, etc. as elementary conceptions, which need not be interpreted mechanically.

The foundations of theoretical optics were simplified by the introduction of the electromagnetic theory; the number of arbitrary hypotheses was diminished. The old question of the direction of the oscillation of polarised light became irrelevant. The difficulties concerning the boundary conditions at the boundary of two media had arisen from the foundations of the theory. There was no longer a need for an arbitrary hypothesis to incorporate longitudinal light waves. Light pressure, which has only been established lately and which plays such an important role in the theory of radiation, came out as a consequence of the theory. I don't want to try to present here a complete list of the well-known achievements, but stick to the main point with regard to which the electromagnetic theory agrees with the kinematical theory, or better expressed, seems to agree.

According to both theories light waves seem to be mainly an inherent property of states of a hypothetical medium, the ether, which is also to be present in the absence of radiation. It was therefore reasonable that the motion of this medium had an influence on the optical and electromagnetic phenomena. The search for the laws that underlie this influence started a change of our basic conception concerning the nature of radiation, a development we want to examine briefly.

The basic question one was forced to was the following: Does the luminiferous ether follow the motion of matter, or does it move differently in the interior of moving bodies, or, finally, does it perhaps not participate at all in the movement of matter, but stay always at rest? To decide this question Fizeau did an important experiment with interference which is based on the following consideration. If the body is at rest, light propagates in it with velocity V. If the body takes its ether completely with it when it is moved, light propagates relative to the body in the same way in this case as if the body is in rest. In this case the velocity of propagation relative to the body is V also. Taken absolutely, that is, relative to an observer who does not move with the body, the velocity of propagation of a light-ray is equal to the geometrical sum of V and the moving velocity v of the body. If velocity of propagation and velocity of motion have the same direction and sense, then V_{abs} is simply the sum of both velocities, that is

$$V_{abs} = V + v.$$

To test if this consequence of the hypothesis of the completely comoving luminiferous ether holds true, Fizeau made two coherent monochromatic lightwaves each pass axially through a tube filled with water and afterwards let them interfere. If the water now moved axially through the tubes, through one tube in the direction of the light, through the other in the opposite direction, a displacement of the interference fringes took place, by which he could draw a conclusion about the influence of the velocity of the body on the absolute velocity.

As is well known and expected, he found that there is an influence of the velocity of the body, but always smaller than that expected according to the hypothesis of the completely comoving ether. He finds that

$$V_{\text{abs}} = V + av,$$

where a is always smaller than one. By neglecting dispersion it follows that

$$a = 1 - \frac{1}{n^2}.$$

From this experiment it follows that a complete comoving of the ether does not take place and therefore in general there is relative motion of the ether with respect to the matter. But the earth is a body which has velocities with different directions during one year, and it was not supposed that the ether in our laboratories participates completely in the movement of our earth, any more than it seemed to participate completely in the movement of the water in Fizeau's experiment. One was therefore to conclude that there exists a relative motion of the ether with respect to our apparatus which changes with the time of day and year and one had to expect that this relative velocity leads to an apparent anisotropy of space in optical experiments, which means that the optical phenomena are dependent on the orientation of the apparatus. The most diverse experiments were carried out to establish this anisotropy without being able to establish the expected dependence of the phenomena upon the direction of the apparatus.

This contradiction was for the most part removed by H. A. Lorentz's pioneering work in the year 1895. Lorentz showed one gets a theory which satisfies almost all phenomena without setting up other hypotheses. In particular the results of the above indicated experiment of Fizeau and the negative result of the mentioned experiments to establish the earth's movement against the ether were explained. Lorentz's theory seemed inconsistent with only one experiment, namely with Michelson's and Morley's interference experiment.

Lorentz has shown that no influence of a common translational movement of the apparatus is present in his theory that affects the ray-paths of optical experiments, except for terms which contained the quotient

$$\frac{\text{velocity of the body}}{\text{velocity of light}}$$

to the second or higher powers as a factor. But at that time the Michelson-Morley's interference experiment was already known, which showed that in one special case even terms of second order with respect to the quotient

$$\frac{\text{velocity of the body}}{\text{velocity of light}}$$

are not noticeable as was expected from the viewpoint of the theory of the stationary luminiferous ether. To cover this experiment with the theory, Lorentz and FitzGerald introduced, as is known, the assumption that all bodies, and therefore also those which connected the parts of Michelson and Morley's experimental arrangement, change their shape in a certain way if they are moved relative to the ether.

This situation was now highly unsatisfactory. Lorentz's theory was the only theory which was useful and clear in its foundations. It rested on the assumption of an absolutely motionless ether. The earth had to be regarded as moving relative to this ether. But all experiments to detect this relative motion were without result, and therefore one was forced to set up this quite peculiar hypothesis to be able to comprehend that the relative motion is not noticeable.

Michelson's experiment leads to the assumption that all phenomena relative to a comoving earth coordinate system, or more generally to every unaccelerated system, obey the same laws. In the following we want to call this assumption the "principle of relativity," for short. Before we touch upon the question of whether it is possible to adhere to the principle of relativity, we want to consider briefly what becomes of the ether hypothesis if we adhere to this principle.

With the ether hypothesis as a basis, the experiment led to the assumption of a motionless ether. The principle of relativity then states that, according to a coordinate system K′, moving uniformly relative to the ether, all laws of nature are equivalent to the corresponding laws in a coordinate system K, at rest relative to the ether. If this holds true, then it is as reasonable to think of the ether as being at rest relative to K′ as to K. It is then generally unnatural to distinguish one of both coordinate systems K and K′ in a way that one introduces an ether which is at rest relative to it. From that it follows that a satisfactory theory is only achieved by renouncing the ether hypothesis. The electromagnetic fields that constitute light appear not as states of a hypothetical medium but as independent phenomena which are emitted by light sources, the same as in Newton's corpuscular theory of light. According to this theory a space which is not filled by radiation and which is free of ponderable matter appears to be really empty.

In a superficial consideration it seems to be impossible to bring the essence of Lorentz's theory into harmony with the principle of relativity. If a light ray propagates in vacuum then, according to Lorentz's theory, with respect to the coordinate system K at rest in the ether, it always does so with the fixed velocity c, independent of the state of motion of the emitting body. We wish to call this statement the principle of constant light speed. According to the addition theorem for velocities the same light ray is not propagated with the same velocity c in a coordinate system K′, uniformly translated relative to the ether. The laws of light propagation seem to be different in both coordinate systems and therefore it seems to follow that the principle of relativity is incompatible with the laws of light propagation.

The addition theorem for velocities, however, is based on the arbitrary presupposition that the statements about time and about the shape of moving bodies have a meaning which is independent of the state of motion of the coordinate system used. But one convinces oneself that the introduction of clocks that are at rest relative to the coordinate system used is needed to define the time and the shape of moving bodies. These conceptions have to be therefore specially fixed for each coordinate system and it is not self-evident that this definition leads to the same time t and t' for single events for two coordinate systems, K and K′, moving relative to each other; likewise one cannot say *a priori* that every statement about the shape of bodies which is valid in a coordinate system K is also valid in the frame K′ moving relative to K.

From this it follows that the transformation equations used till now for the transition from one coordinate system to another coordinate system moving uniformly relative to it are based on arbitrary assumptions. Giving these up, it can be shown that the basis of Lorentz's theory, or more general, the principle of constant light speed can be brought into harmony with the principle of relativity. In this way new equations of the coordinate transformation are obtained which are definitely determined by the two principles and which are, with a suitable choice of the origin of coordinates and time, characterized by making an identity of the equation

$$x^2 + y^2 + z^2 - c^2t^2 = x'^2 + y'^2 + z'^2 - c^2t'^2$$

Here c is the velocity of light in vacuum, x, y, z, t are the space-time coordinates with respect to K, x', y', z', t' with respect to K′.

This path leads to the so-called theory of relativity, from which I only want to cite one consequence, because it leads to a certain modification in the field of physics. It can be shown that the initial mass of a body decreases by L/c^2 if the body emits the energy of radiation L. One can obtain this in the following manner.

We consider a motionless free suspending body which emits the same amount of energy by radiation in two opposite directions. Yet the body remains at rest. If we characterize the energy of the body before the emission with E_0, the energy after the emission with E_1, the amount of emitted radiation with L, according to the principle of energy conservation it follows that

$$E_0 = E_1 + L.$$

Let us consider now the body and the radiation emitted by it with respect to a coordinate system to which the body is moved with the velocity v. The theory of relativity gives us then the tools to calculate the energy of the emitted radiation with respect to the new coordinate system. One obtains for it the value

$$L' = L \cdot \frac{1}{\sqrt{1 - \frac{v^2}{c^2}}}.$$

Because the principle of conservation of energy also has to be true in the new coordinate system, by an analogous consideration one gets

$$E'_0 = E'_1 + L \frac{1}{\sqrt{1 - \frac{v^2}{c^2}}}.$$

By subtracting and neglecting terms in v/c of fourth and higher order, one gets

$$(E'_0 - E_0) = (E'_1 - E_1) + \frac{1}{2} \frac{L}{c^2} v^2.$$

But $E'_0 - E_0$ is nothing but the kinetic energy of the body before the emission of light, and $E'_1 - E_1$ is nothing but the kinetic energy after the emission of light. If we call M_0 the body's mass before the emission, M_1 the body's mass after the emission, then by neglecting terms of higher order than the second we can set

$$\frac{1}{2} M_0 v^2 = \frac{1}{2} M_1 v^2 + \frac{1}{2} \frac{L}{c^2} v^2$$

or

$$M_0 = M_1 + \frac{L}{c^2}.$$

Hence the inertial mass of a body is diminished by light emission. The emitted energy figures as part of the body's mass. We can draw the further conclusion that every absorption and emission of energy goes with an increase or decrease, respectively, of the mass of the body concerned. Energy and mass appear as equivalent quantities just as do heat and mechanical energy.

The theory of relativity has therefore changed our conception of the nature of light insofar as it no longer conceives light as a consequence of the states of a hypothetical medium but as an independent entity like matter. Furthermore, this theory has in common with the corpuscular theory of light the feature that it transfers inertial mass from the emitting body to the absorbing body. The theory of relativity does not change anything in our conception about the structure of radiation, in particular about the distribution of energy in the volume being radiated. However it is my opinion that we are standing at the beginning of a not yet transparent, but nevertheless highly important, development with respect to this side of the question. What I will put forward in the following is mostly bare personal opinion and also the results of reflections which have not yet passed a sufficient check by others. I still put them forward here, not because I have an excessive trust in my own view but because I hope that one or another of you will be motivated to deal with the questions concerned.

Without going deeper into any theoretical considerations, one notices that our theory of light is not able to explain certain fundamental properties of the phenomena of light. Why is the occurrence of a certain photochemical reaction only dependent on the color, but not on the intensity of light? Why are short-wave rays in general more effective than long-wave rays? Why is the velocity of the photo-emitted cathode-rays independent of the intensity of light? Why are high temperatures needed, or high molecular energies, to get short-wave components in the radiation emitted by bodies?

The undulation theory in today's version gives no answer to any of these questions. In particular it is not yet comprehensible why cathode-rays, created by x-rays or photo-emitted, attain such a remarkable velocity independent of beam intensity. Competent physicists have been forced to take refuge in a quite remote hypothesis by the appearance of such huge amounts of energy at a molecular structure under the influence of a source in which the energy is distributed so much less densely, as we must presuppose for light and x-rays according to the undulation theory. They have assumed that light plays only an initiatory role during this process, that the molecular energy that appears is of radioactive nature. But since this hypothesis is already as good as abandoned I do not wish to bring up any arguments against it.

The basic properties of the undulation theory which cause these difficulties seem to lie in the following. While in the kinetic theory of molecules for every process in which only a few elementary particles participate, for instance, in every molecular collision, there exists the inverse process, this is not the case for the elementary processes of radiation according to the undulation theory. An oscillating ion produces a spherical wave propagating outwardly according to the current theory. The inverse process does not exist as an *elementary process.* The spherical wave propagating inwardly is mathematically realizable but one needs for its approximate realization an incredible number of elementary emitting structures. Hence the elementary process of light emission does not have the character of invertibility. I think in this case our undulation theory does not describe the truth. It seems that Newton's corpuscular theory of light contains more truth than the undulation theory, because according to the former the energy that a corpuscle is given during emission is not dispersed over infinite space but remains available for an elementary absorption process. Think of the laws for the generation of secondary cathode-rays by x-rays.

If primary cathode-rays hit a metal plate P_1, they produce x-rays. If these hit another metal plate P_2, they again produce cathode-rays whose velocities are of the same order as the velocities of the primary cathode-rays. As far as we know today the velocity of the secondary cathode-rays is neither dependent on the distance of the plates P_1 and P_2 nor of the intensity of the primary cathode-rays, but exclusively on the velocity of the primary cathode-rays. Suppose this is strictly correct. What will happen if we let the intensity of the primary cathode-rays, or the size of the plate P_1 upon which they impinge, decrease in such a way that one can interpret the collision of an electron of the primary cathode-rays as an isolated process? If the preceding is really true, we will have to suppose, because the velocity of the secondary rays is independent of the intensity of the primary cathode-rays, that either nothing is created at P_2 (as a consequence of the collision of this electron on P_1) or there occurs a secondary emission of an electron on P_2 with a velocity which is of the same order as the one of the electron which hits P_1. In other words, the elementary process of radiation seems to happen in such a way that the energy of the primary electron is not distributed and dispersed by a spherical wave propagating in all directions, as the undulation theory requires. On the contrary, at least a large part of this energy seems to be available at any point of P_2 or elsewhere. *The elementary process of radiation emission seems to be directed.* Furthermore I have the impression that the processes of creation of the x-ray at P_1 and creation of the secondary cathode-ray at P_2 are essentially inverse processes.

The constitution of radiation seems to be different than what follows from the undulation theory. The theory of heat radiation has provided us with important hints about this, and it was primarily that theory on which

Mr. Planck has based his formula for radiation. Because I cannot assume this theory to be generally known I will briefly present the main points.

In the interior of a cavity of temperature T there is radiation of definite composition that is independent of the nature of the body. Per unit volume the amount of radiation present in the cavity whose frequency varies between ν and $\nu + d\nu$ is $\rho d\nu$. The problem consists in seeking ρ as a function of ν and T. If the cavity contains an electric resonator of eigenfrequency r_0 and negligible damping, the electromagnetic theory of radiation allows us to calculate the temporal mean of the energy ($\bar{E}$) of the resonator as a function of $\rho(r_0)$. The problem is therefore reduced to that of calculating $\bar{E}$ as a function of temperature. But the latter problem can be again reduced to the following. Suppose very many (N) resonators of the frequency ν_0 are present in the cavity. How does the entropy of this resonator system depend on its energy?

To solve this question, Mr. Planck applies the general relation between entropy and probability which was obtained by Boltzmann from his research on gas theory. In general

$$\text{entropy} = \text{k} \cdot \log \text{W},$$

where k is a universal constant and W is the probability of the chosen state. This probability is measured by the "number of complexions," a number which indicates in how many different ways the chosen state is realizable. In the case of the above problem, the state of the resonator system is defined by its total energy, so that the question to be solved is: In how many possible ways can the given total energy be distributed among the N resonators? To solve this, Mr. Planck divided the total energy into equal quanta of given size ε. A complexion is determined by the number of particles ε which belong to each resonator. The number of such complexions which give the total energy has to be determined and will be set equal to W. By use of Wien's displacement law, which can be derived from thermodynamics, Mr. Planck then further concludes that we must set $\varepsilon = h\nu$, where h is a number independent of ν. He finds in this way his radiation formula

$$\rho = \frac{8\pi h\nu^3}{c^3} \cdot \frac{1}{\text{c}^{\frac{h\nu}{kr}} - 1}.$$

which coincides so far with all previous experience.

According to this derivation it might appear that Planck's formula has to be regarded as a consequence of today's electromagnetic theory of radiation. However, this is not the case, in particular because of the following reason. One can regard the number of complexions under discussion as an expression for the multiplicity of possible distributions of the total energy among N resonators only if every possible distribution of the energy appears, at least with a certain approximation, among the complexions used to calculate W. A necessary condition is that for all ν which correspond to a reasonable energy density ρ, the energy quantum ε is small compared with the mean energy $\bar{E}$ of the resonator. By a simple calculation, one finds, however, that $\varepsilon|\bar{E}$, for the wavelength 0.5 micrometer and an absolute temperature T = 1700, is not only not small compared to 1 but even very large compared to 1. It has a value of about $6.5 \cdot 10^7$. In the given numerical example of the counting of the complexions one proceeds as if the energy of the resonator can take only the values zero, $6.5 \cdot 10^7$ times this mean energy, or a multiple of that. It is clear that with this procedure only an infinitely small part of the theoretically possible distributions of the energy is used for the calculation of the entropy. The number of these complexions is therefore, according to the basic theory, not an expression for the probability of the states in the sense of Boltzmann. To take over Planck's theory means, in my opinion, to give up the foundations of our theory of radiation.

I have already tried to show you that we have to abandon our present foundations of the theory of radiation. In any case, one cannot think of denying Planck's theory because it does not fit in these foundations. This theory has led to a determination of the elementary quanta which is brilliantly confirmed by the latest measurement of this value on the basis of counting alpha-particles. For the elementary quantum of electricity Rutherford and Geiger got the mean value $4.65 \cdot 10^{-10}$, Regener $4.79 \cdot 10^{-10}$, while Mr. Planck calculated the intermediate value $4.69 \cdot 10^{-10}$ out of the constants of his formula by means of his theory of radiation.

Planck's theory leads to the following assumption. If it is really true that a radiating resonator can take up only such values of energy that are multiples of $h\nu$, then we are led to the assumption that emission and absorption of radiation take place only in quanta of this energy size. On the basis of this hypothesis, the hypothesis of light quanta, the above mentioned questions about absorption and emission of radiation can be answered. As far as our knowledge extends, the quantitative predictions of this hypothesis are verified. But the following question comes up. Is it possible, that Planck's formula is correct, but that there exists a derivation which is not based on such incredible-seeming assumptions as Planck's theory? Is it possible to replace the hypothesis of light quanta by another assumption which can explain the known phenomena just as well? If it is necessary to modify the elements of the theory, is it still possible to keep the equation for the propagation of radiation and to change our conception about the elementary processes of emission and absorption?

To clarify this we want to try to proceed in the direction inverse to what Mr. Planck did with his theory of radiation. We regard Planck's formula as correct and ask if something can be concluded from it with respect to the constitution of the radiation. I want to sketch here only one of two considerations which I have carried out in this sense because I found it especially convincing due to its intuitive nature.

In a cavity let there be an ideal gas and also a plate made of a rigid substance that can be moved freely only perpendicular to its plane. Because of the irregularity of the collisions between gas molecules and plate, the latter will start moving in such a way that its mean kinetic energy is equal to one third of the mean kinetic energy of a monatomic gas molecule. This is a prediction of statistical mechanics. Let's suppose now that besides the gas, which we can think of as consisting of only a few molecules, radiation is present in the cavity and that this so called heat radiation has the same temperature as the gas. This will be the case if the walls of the cavity have the definite temperature T, are not permeable to the radiation, and are not everywhere completely reflecting toward the cavity. Furthermore we suppose for the moment that our plate is completely reflecting on both sides. With these preliminaries not only the gas but also the radiation will effect the plate. The radiation will exhibit a pressure on both sides of the plate. The pressure forces of both sides are equal to one another if the plate is in rest. But if it is moved, the area which precedes the movement (the front surface) will reflect more radiation than the back surface. The pressure force acting on the front surface in a backward direction will be larger than the pressure force acting on the back surface. As a resultant of the two, there remains a force which opposes the movement of the plate and which increases with the velocity of the plate. Let us call this resultant "radiation friction" for short.

If we suppose for a moment that we have taken into account all of the mechanical influences of the radiation on the plate, then we come to the following conclusion. Through collisions with the gas molecules the plate is imparted momenta at irregular intervals into irregular directions. Between two such collisions the velocity of the plate decreases because of the radiation friction, whereby kinetic energy is converted into radiation energy. We would have the consequence that the energy of the gas molecules is constantly converted into energy of radiation by the plate, up to the moment when all the energy present is converted into energy of radiation. There would not be an equilibrium of temperature between gas and radiation.

This consideration is erroneous because one cannot regard the pressure forces exerted by the radiation on the plate as temporally constant and free of irregular fluctuations any more then the pressure force exerted by the gas on the plate. To allow thermal equilibrium these fluctuations of the pressure forces of the radiation have to be constituted in such a way that they compensate in the mean for the losses of velocity of the plate by radiation friction, where the mean kinetic energy of the plate is equal to one third of the mean kinetic energy of a monatomic gas molecule. If the law of radiation is known, the radiational friction can be calculated, and hence the mean value of the momentum that the plate has to receive because of the fluctuations of the radiation pressure in order that statistical equilibrium can exist.

The consideration gets still more interesting if we so choose the plate that it reflects only radiation in the range of frequencies $d\nu$, but allows radiation of different frequencies to pass without any absorption; one obtains then the fluctuations of the pressure of the radiation in the range of frequencies $d\nu$. For this case I want to give the result of the calculation. If I denote by Δ the value of the quantity of motion [momentum] which gets transferred to the plate during the time τ because of the irregular fluctuations of the pressure of radiation, I get the expression [f = surface area]

$$\overline{\Delta^2} = \frac{1}{c}\left[h\rho\nu + \frac{c^3}{8\pi}\frac{\rho^2}{\nu^2}\right] d\nu f t$$

for the mean value of the square of Δ.

First of all notice the simplicity of this expression. There cannot be any radiation formula that coincides with the experiments within observational error and gives such a simple expression for the statistical properties of the radiation pressure as Planck's formula.

As for the interpretation, notice first that the expression for the mean square of the fluctuations is a sum of two terms. It is as if there are two independent different causes present that produce the fluctuations of the radiation pressure. Out of the proportionality of Δ^2 to f one concludes that the fluctuations of the pressure for parts of the plate that are side by side and whose linear size is large compared to the wavelength of the reflected radiation, give results which are independent of each other.

The undulation theory gives an explanation only for the second term of the expression for Δ^2. According to the undulation theory bunches of rays that differ only infinitesimally in direction, frequency and state of polarization, have to interfere, and the sum of these interferences, which arise in the most disordered fashion, has to correspond to a fluctuation of the radiation pressure. By a simple consideration of dimensions one can see that these fluctuations must have a form like the second term of our formula. It is seen that the undulating structure of the radiation indeed gives rise to the expected fluctuations of the radiation pressure.

But how can the first term of the formula be explained? This is not at all negligible but on the contrary in the range of validity of the so-called Wien radiation law it is solely dominant. For a wavelength of 0.5 micrometer and $T = 1700$ this term is about $6.5 \cdot 10^7$ times bigger than the second term. If the radiation consists only of very few extended complexes of energy $h\nu$, that propagate independently through space and are reflected independently of each other—an assumption which is the crudest picture of the hypothesis of light quanta—then our plate would be effected due to fluctuations of the radiation pressure by an amount given by the first term of our formula.

In my opinion, the following must be concluded from the above formula, which itself is a consequence of Planck's formula. *Besides the spatial irregularities in the distribution of the quantity of motion of radiation which arise out of the undulation theory, there are also other irregularities in the spatial distribution of the quantity of motion present, which greatly exceed the first mentioned irregularities at low energy densities of the radiation.* I would like to add that a different treatment considering the spatial distribution of the energy gives the same result as above with respect to the spatial distribution of the quantity of motion.

As far as I know nobody has succeeded in setting up a mathematical theory of radiation which yields both the undulational structure and the structure that follows from the first term of the above formula (quantized structure). The difficulties are mainly that the properties of the fluctuations of radiation, as they are expressed in the above formula, give few formal hints for setting up a theory. Suppose the phenomena of refraction and interference were still unknown, but one knew that the mean value of the irregular fluctuations of the radiation pressure were determined by the second term of the above equation, where ν is a parameter of unknown meaning which determines the color. Who would have enough imagination to build up the undulation theory of light on this basis?

To me, the most natural concept is that the occurrence of the electromagnetic fields of light is bound to singular points in the same way as the occurrence of electrostatic fields according to electron theory. It cannot be excluded that in such a theory the total energy of the electromagnetic field has to be regarded as being localized in these singularities, in the same way as in the old theory of action at a distance. I think of every singular point as being surrounded by a force field, which has in essence the character of a plane wave and whose amplitude decreases with the distance to the singular point. If many such singularities are present at distances which are small compared to the size of the force field of a singular point, then the force fields will start to superimpose and the sum will give rise to an undulational force field, which probably differs only slightly from an undulational field in the sense of the present electromagnetic theory of light. I do not have to emphasize that such a picture has no worth as long as it does not lead to an exact theory. I only wished to illustrate briefly that the two structural phenomena (undulational structure and quantized structure) that the radiation possesses according to Planck's formula, do not have to be regarded as incompatible with each other.

DISCUSSION

Planck: I would like to add some comments about the lecture, but before doing so I can only join in the gratitude of all participants who listened with the greatest interest to that which Mr. Einstein has presented, and who were inspired to further investigation whenever a contradiction perhaps emerged. I will restrict myself naturally to the points where I have a different opinion than that of the speaker. Most of what the speaker has pointed out will not be opposed. I too stress the introduction of some kind of quanta. We do not get on with the whole theory of radiation without splitting energy in some sense into quanta, which ought to be thought as "atoms of action." There is now the question of where to look for these quanta. According to the latest explanations of Mr. Einstein it would be necessary to think of free radiation in vacuum, that is the light waves themselves, as atomistically constituted, and therefore to give up Maxwell's equations. This seems to me to be a step that is in my opinion not yet necessary. I do not want to go into details but just state the following: in his last consideration Mr. Einstein infers fluctuations of free radiation in pure vacuum from the movement of matter. This reasoning seems to me totally unobjectionable only if one knows completely the interaction between radiation in the vacuum and the motion of matter. If this is not the case then there is lacking the bridge that is necessary in order to go from the motion of the mirror to the intensity of the incident radiation. But to me this interaction between free electric energy in vacuum and the motion of the atoms of matter seems to be known very poorly. Essentially it is based on emission and absorption of light. The radiation pressure too consists essentially of this, at least according to the generally accepted theory of dispersion, which reduces also reflection to absorption and emission. But emission and absorption is exactly the dark spot we know very little about. We might know perhaps something about absorption, but what about emission? It is thought to be caused by the acceleration of electrons. But this is the weakest point of the whole theory of electrons. The electron is considered to possess some definite volume and a specific finite charge density, be it spatial density or surface density; without this one does not get along. But this contradicts the atomistic conception of electricity in some sense. These are not impossibilities but problems and I am almost astonished that there did not arise more opposition to this.

I think that the quantum theory can set to work usefully at that point. We can express the laws only for large times. But for short times and high accelerations one still stands in front of a gap, whose filling requires new hypotheses. Perhaps one may assume that an oscillating resonator does not possess a constant variable energy range but that its energy is a multiple of an elementary quantum. I think that one can achieve a satisfactory theory of radiation if one uses this statement. There is now always the question: how to picture such a thing? That means, one requires a mechanical or electrodynamical model of such a resonator. But we have no discrete "action elements" in mechanics and present electrodynamics and therefore cannot produce such a mechanical or electrodynamical model. This appears now to be mechanically impossible and we have to get accustomed to this fact. Our trials to describe the luminiferous ether mechanically have also failed completely. One wished to picture the electric current mechanically too and one thought about the analogy of a water current, but this had to be abandoned too, and as we got used to this we have to get used to such a resonator also. Of course, the new theory has to be worked out in much greater detail than has been done until now. Perhaps someone else will be more lucky than I. In any case I think that we should first try to transfer the problem of quantum theory into the area of *interaction* between matter and radiating energy. The processes in the pure vacuum could then still be explained by Maxwell's equations.

H. Ziegler: If one thinks about the basic particles of matter as invisible little spheres which possess an invariable speed of light, then all interactions of matter like states and electrodynamic phenomena can be described and thus we would have erected the bridge between the material and immaterial world that Mr. Planck wanted.

Stark: Mr. Planck has pointed out that so far we have no reason to go over to Einstein's conclusion that we should regard radiation in space, where it occurs separated from matter, as concentrated. Originally I too was of the opinion that one can restrict oneself for the present to tracing the elementary law back to a certain behavior of the resonators. I think, however, that there is one phenomenon which forces us to think of electromagnetic radiation as separated from matter and concentrated in space. This is the phenomenon that electromagnetic radiation that emerges from a Röntgen-tube into the surrounding space can still come concentratedly to act on a single atom, even at large distances of up to ten meters. I think that this phenomenon is reason enough to investigate the question of whether electromagnetic radiation energy has to be conceived as concentrated even where it occurs separated from matter. The conception as advocated by Mr. Einstein would result in a practical consequence, which can be verified experimentally. As is well known not only alpha-rays but also beta-rays give rise to scintillations on a fluorescent screen. According to the developed concept the same has also to hold true for gamma-rays and Röntgen-rays.

Planck: There is something unique to Röntgen-rays, and I do not wish to claim too much in this case. Stark has mentioned something in favor of the quantum theory, I wish to add a remark against it, that is, the interferences at the enormous path differences of hundreds of thousands wavelengths. If a quantum interferes with itself it must have a spatial extension of hundreds of thousands wavelengths. This is also a definite difficulty.

Stark: Interference phenomena can easily be contraposed to the quantum theory. But I would like to express the hope that, if one will treat them with greater care for the quantum hypothesis, one will also attain an explanation. Concerning the experimental part, it has been emphasized that the experiments that Mr. Planck referred to were performed with very dense radiation so that very many quanta of the same frequency were concentrated in the light bundle. That has to be taken into consideration in dealing with those interference phenomena. The interference phenomena would probably be different with very low radiation density.

Einstein: The interference phenomena would probably not be so difficult to arrange as one imagines, and that for the following reason: One should not think that radiation consists of quanta that do not interact with each other; this would be impossible for an explanation of the interference phenomena. I think of a quantum as a singularity, surrounded by a large vector field. With a large number of quanta a vector field can be composed that differs little from the one we presume for radiation. I can imagine that when radiation hits a boundary there occurs a separation of the quanta by processes at the boundary, say according to the phase of the resulting field at which the quanta reach the separating surface. The equations for the resulting field would differ little from those of the previous theory. It is not asserted that we have to change very many of our conceptions about the phenomena of interference. I wish to compare this with the process of molecularisation of the carriers of the electrostatic field. The field as produced by atomistic electric particles is not very essentially distinguished from the previous conception, and it is not to be excluded that something similar will take place in the radiation theory. I do not see a difficulty of principle with the phenomena of interference.

N.3

The Structure of the Atom

NIELS BOHR

NIELS BOHR, one of the prime figures in the development of twentieth-century physics, was the son of a professor of physiology. He was born in Copenhagen in 1885 and died there at the age of seventy-seven in 1962. After receiving his doctorate in physics at Copenhagen in 1911, he moved to Cambridge, where he worked briefly with Joseph John Thomson. When he found himself in disagreement with Thomson on matters particularly close to the heart of the British physicist—the structure of the atom—Bohr left Cambridge for the University of Manchester. There he found a far more compatible master in Ernest Rutherford, whose theories on atomic structure—specifically his discovery of the atomic nucleus—Bohr combined with aspects of quantum theory to construct a more useful atomic model. In 1916, Bohr was appointed Professor of Theoretical Physics at Copenhagen, becoming the head of the Institute for Theoretical Physics there in 1920. After 1930, he turned his attention to the transmutations and disintegrations of atomic nuclei, offering the liquid drop nuclear model, which helped to illustrate the process of nuclear fission after Hahn and Strassmann discovered the splitting of the uranium nucleus. He also developed his principle of complementarity, which stated that a system such as an electron can be described either in terms of particles or wave motion. After the German occupation of Denmark in World War II, Bohr fled first to Sweden and then to the United States, where he worked in the field of atomic energy. He later advocated its peaceful use and wrote about the political problems engendered by atomic weapons. During the last few years of his life, he became interested in the field of molecular biology; his last article dealt with the problems of life and was published after his death.

From the time of Democritus until the advent of Thomson, scientists had regarded the atom—if indeed it existed—as the solid, irreducible component of matter. After the discovery of the electron by Thomson in 1897, the British physicist proposed a theory of atomic structure in which electrons were embedded in a diffuse sphere of positive charge, rather like raisins in a pudding. By 1911, Rutherford had discovered the nucleus of the atom, and speculated that the atom was composed of an extremely minute, positively charged nucleus containing most of the atom's mass, around which a number of electrons (the number varying from element to element) "orbited," rather like the planets of the solar system around the sun. The apparent flaw in Rutherford's theory was that James Clerk Maxwell had shown that an electrically charged particle, if accelerated, produces radiation and, in doing so, loses energy. This meant that Rutherford's electrons would spiral in toward the positively charged nucleus in a fraction of a second. Bohr introduced a hypothesis according to which an orbiting electron does not radiate energy if it is in one of the discrete orbits determined by Planck's constant. He also proposed that when an orbiting electron jumps to another orbit, it either absorbs a photon (jumps to an orbit farther from the nucleus) or emits a photon (drops to a lower orbit). In presenting his theory of atomic structure, Bohr also explained in detail certain phenomena long associated with the spectrum of hydrogen. On hearing how Bohr's theory fit so precisely with the observed facts in the case of the hydrogen spectrum, Einstein is reported to have commented, "Then this is one of the greatest discoveries ever made."

Physicists argued over the accuracy of Bohr's theory for many years, but after an impressive amount of experimental evidence accumulated by spectroscopy essentially confirmed his conclusions. He was awarded the Nobel Prize in physics in 1922.

Our selection is excerpted from his Nobel lecture, which he delivered on 11 December 1922, and deals with his model of the atom.

The great extension of our experience in recent years has brought to light the insufficiency of our simple mechanical conceptions and, as a consequence, has shaken the foundation on which the customary interpretation of observation was based.

—NIELS BOHR, *Atomic Theory and the Description of Nature*

. . . In quantum mechanics, we are not dealing with an arbitrary renunciation of a more detailed analysis of atomic phenomena, but with a recognition that such an analysis is in principle excluded.

—NIELS BOHR, *Atomic Physics and Human Knowledge*

. . .THE PRESENT STATE of atomic theory is characterized by the fact that we not only believe the existence of atoms to be proved beyond a doubt, but also we even believe that we have an intimate knowledge of the constituents of the individual atoms. . . . The discovery of the electron towards the close of the last century furnished the direct verification and led to a conclusive formulation of the conception of the atomic nature of electricity which had evolved since the discovery by Faraday of the fundamental laws of electrolysis and Berzelius's electrochemical theory, and had its greatest triumph in the electrolytic dissociation theory of Arrhenius. This discovery of the electron and elucidation of its properties was the result of the work of a large number of investigators, among whom Lenard and John Joseph Thomson may be particularly mentioned. The latter especially has made very important contributions to our subject by his ingenious attempts to develop ideas about atomic constitution on the basis of the electron theory. The present state of our knowledge of the elements of atomic structure was reached, however, by the discovery of the atomic nucleus, which we owe to Rutherford, whose work on the radioactive substances discovered towards the close of the last century has much enriched physical and chemical science.

According to our present conceptions, an atom of an element is built up of a nucleus that has a positive electrical charge and is the seat of by far the greatest part of the atomic mass, together with a number of elec-

---

Niels Bohr, Nobel Prize in Physics Award Address, 1922. Reprinted with permission of Elsevier Publishing Co., and the Nobel Foundation.

trons, all having the same negative charge and mass, which move at distances from the nucleus that are very great compared to the dimensions of the nucleus or of the electrons themselves. In this picture we at once see a striking resemblance to a planetary system, such as we have in our own solar system. Just as the simplicity of the laws that govern the motions of the solar system is intimately connected with the circumstance that the dimensions of the moving bodies are small in relation to the orbits, so the corresponding relations in atomic structure provide us with an explanation of an essential feature of natural phenomena in so far as these depend on the properties of the elements. It makes clear at once that these properties can be divided into two sharply distinguished classes.

To the first class belong most of the ordinary physical and chemical properties of substances, such as their state of aggregation, colour, and chemical reactivity. These properties depend on the motion of the electron system and the way in which this motion changes under the influence of different external actions. On account of the large mass of the nucleus relative to that of the electrons and its smallness in comparison to the electron orbits, the electronic motion will depend only to a very small extent on the nuclear mass, and will be determined to a close approximation solely by the total electrical charge of the nucleus. Especially the inner structure of the nucleus and the way in which the charges and masses are distributed among its separate particles will have a vanishingly small influence on the motion of the electron system surrounding the nucleus. On the other hand, the structure of the nucleus will be responsible for the second class of properties that are shown in the radioactivity of substances. In the radioactive processes we meet with an explosion of the nucleus, whereby positive or negative particles, the so-called α- and β-particles, are expelled with very great velocities.

Our conceptions of atomic structure afford us, therefore, an immediate explanation of the complete lack of interdependence between the two classes of properties, which is most strikingly shown in the existence of substances which have to an extraordinarily close approximation the same ordinary physical and chemical properties, even though the atomic weights are not the same, and the radioactive properties are completely different. Such substances, of the existence of which the first evidence was found in the work of Soddy and other investigators on the chemical properties of the radioactive elements, are called isotopes, with reference to the classification of the elements according to ordinary physical and chemical properties. It is not necessary for me to state here how it has been shown in recent years that isotopes are found not only among the radioactive elements, but also among ordinary stable elements; in fact, a large number of the latter that were previously supposed simple have been shown by Aston's well-known investigations to consist of a mixture of isotopes with different atomic weights.

The question of the inner structure of the nucleus is still but little understood, although a method of attack is afforded by Rutherford's experiments on the disintegration of atomic nuclei by bombardment with α-particles. Indeed, these experiments may be said to open up a new epoch in natural philosophy in that for the first time the artificial transformation of one element into another has been accomplished. In what follows, however, we shall confine ourselves to a consideration of the ordinary physical and chemical properties of the elements and the attempts which have been made to explain them on the basis of the concepts just outlined.

It is well known that the elements can be arranged as regards their ordinary physical and chemical properties in a *natural system* which displays most suggestively the peculiar relationships between the different elements. It was recognized for the first time by Mendeleev and Lothar Meyer that when the elements are arranged in an order which is practically that of their atomic weights, their chemical and physical properties show a pronounced periodicity. A diagrammatic representation of this so-called Periodic Table is given in Fig. 1, where, however, the elements are not arranged in the ordinary way but in a somewhat modified form of a table first given by Julius Thomsen, who has also made important contributions to science in this domain. In the figure the elements are denoted by their usual chemical symbols, and the different vertical columns indicate the so-called periods. The elements in successive columns which possess homologous chemical and physical properties are connected with lines. The meaning of the squared brackets around certain series of elements in the later periods, the properties of which exhibit typical deviations from the simple periodicity in the first periods, will be discussed later.

In the development of the theory of atomic structure the characteristic features of the natural system have found a surprisingly simple interpretation. Thus we are led to assume that the ordinal number of an element in the Periodic Table, the so-called atomic number, is just equal to the number of electrons which move about the nucleus in the neutral atom. In an imperfect form, this law was first stated by Van den Broek; it was, however, foreshadowed by J.J. Thomson's investigations of the number of electrons in the atom, as well as by Rutherford's measurements of the

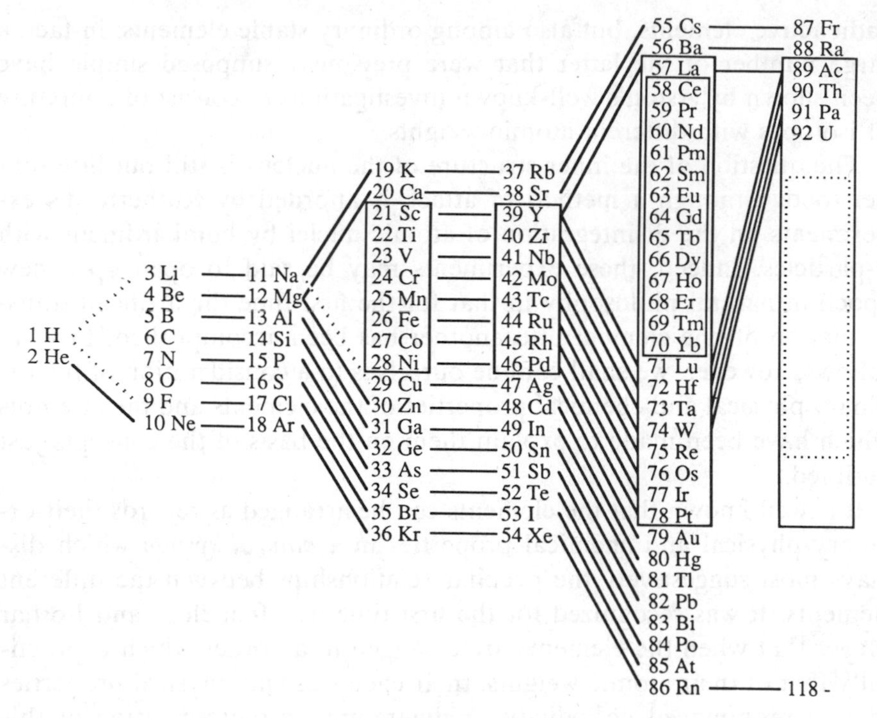

Fig. 1

charge on the atomic nucleus. As we shall see, convincing support for this law has since been obtained in various ways, especially by Moseley's famous investigations of the X-ray spectra of the elements. We may perhaps also point out, how the simple connexion between atomic number and nuclear charge offers an explanation of the laws governing the changes in chemical properties of the elements after expulsion of α- or β-particles, which found a simple formulation in the so-called radioactive displacement law.

ATOMIC STABILITY AND ELECTRODYNAMIC THEORY

As soon as we try to trace a more intimate connexion between the properties of the elements and atomic structure, we encounter profound difficulties, in that essential differences between an atom and a planetary system show themselves here in spite of the analogy we have mentioned.

The motions of the bodies in a planetary system, even though they obey the general law of gravitation, will not be completely determined by this law alone, but will depend largely on the previous history of the system. Thus the length of the year is not determined by the masses of the sun and the earth alone, but depends also on the conditions that existed during the formation of the solar system, of which we have very little knowledge. Should a sufficiently large foreign body some day traverse our solar system, we might among other effects expect that from that day the length of the year would be different from its present value.

It is quite otherwise in the case of atoms. The definite and unchangeable properties of the elements demand that the state of an atom cannot undergo permanent changes due to external actions. As soon as the atom is left to itself again, its constituent particles must arrange their motions in a manner which is completely determined by the electric charges and masses of the particles. We have the most convincing evidence of this in spectra, that is, in the properties of the radiation emitted from substances in certain circumstances, which can be studied with such great precision. It is well known that the wavelengths of the spectral lines of a substance, which can in many cases be measured with an accuracy of more than one part in a million, are, in the same external circumstances, always exactly the same within the limit of error of the measurements, and quite independent of the previous treatment of this substance. It is just to this circumstance that we owe the great importance of spectral analysis, which has been such an invaluable aid to the chemist in the search for new elements, and has also shown us that even on the most distant bodies of the universe there occur elements with exactly the same properties as on the earth.

On the basis of our picture of the constitution of the atom it is thus impossible, so long as we restrict ourselves to the ordinary mechanical laws, to account for the characteristic atomic stability which is required for an explanation of the properties of the elements.

The situation is by no means improved if we also take into consideration the well-known electrodynamic laws which Maxwell succeeded in formulating on the basis of the great discoveries of Oersted and Faraday in the first half of the last century. Maxwell's theory has not only shown itself able to account for the already known electric and magnetic phenomena in all their details, but has also celebrated its greatest triumph in the prediction of the electromagnetic waves which were discovered by Hertz, and are now so extensively used in wireless telegraphy.

For a time it seemed as though this theory would also be able to furnish a basis for an explanation of the details of the properties of the elements, after it had been developed, chiefly by Lorentz and Larmor, into a form consistent with the atomistic conception of electricity. I need only remind you of the great interest that was aroused when Lorentz, shortly after the discovery by Zeeman of the characteristic changes that spectral lines undergo when the emitting substance is brought into a magnetic field, could give a natural and simple explanation of the main features of the phenomenon. Lorentz assumed that the radiation which we observe in a spectral line is sent out from an electron executing simple harmonic vibrations about a position of equilibrium, in precisely the same manner as the electromagnetic waves in radiotelegraphy are sent out by the electric oscillations in the antennae. He also pointed out how the alteration observed by Zeeman in the spectral lines corresponded exactly to the alteration in the motion of the vibrating electron which one would expect to be produced by the magnetic field.

It was, however, impossible on this basis to give a closer explanation of the spectra of the elements, or even of the general type of the laws holding with great exactness for the wavelengths of lines in the spectra, which had been established by Balmer, Rydberg, and Ritz. After we obtained details as to the constitution of the atom, this difficulty became still more manifest; in fact, so long as we confine ourselves to the classical electrodynamic theory we cannot even understand why we obtain spectra consisting of sharp lines at all. This theory can even be said to be incompatible with the assumption of the existence of atoms possessing the structure we have described, in that the motions of the electrons would claim a continuous radiation of energy from the atom, which would cease only when the electrons had fallen into the nucleus.

THE ORIGIN OF THE QUANTUM THEORY

It has, however, been possible to avoid the various difficulties of the electrodynamic theory by introducing concepts borrowed from the so-called quantum theory, which marks a complete departure from the ideas that have hitherto been used for the explanation of natural phenomena. This theory was originated by Planck, in the year 1900, in his investigations on the law of heat radiation, which, because of its independence of the individual properties of substances, lent itself peculiarly well to a test of the applicability of the laws of classical physics to atomic processes.

Planck considered the equilibrium of radiation between a number of systems with the same properties as those on which Lorentz had based his theory of the Zeeman effect, but he could now show not only that classical physics could not account for the phenomena of heat radiation, but also that a complete agreement with the experimental law could be obtained if—in pronounced contradiction to classical theory—it were assumed that the energy of the vibrating electrons could not change continuously, but only in such a way that the energy of the system always remained equal to a whole number of so-called energy-quanta. The magnitude of this quantum was found to be proportional to the frequency of oscillation of the particle, which, in accordance with classical concepts, was supposed to be also the frequency of the emitted radiation. The proportionality factor had to be regarded as a new universal constant, since termed Planck's constant, similar to the velocity of light, and the charge and mass of the electron.

Planck's surprising result stood at first completely isolated in natural science, but with Einstein's significant contributions to this subject a few years after, a great variety of applications was found. In the first place, Einstein pointed out that the condition limiting the amount of vibrational energy of the particles could be tested by investigation of the specific heat of crystalline bodies, since in the case of these we have to do with similar vibrations, not of a single electron, but of whole atoms about positions of equilibrium in the crystal lattice. Eintstein was able to show that the experiment confirmed Planck's theory, and through the work of later investigators this agreement has proved quite complete. Furthermore, Einstein emphasized another consequence of Planck's results, namely, that radiant energy could only be emitted or absorbed by the oscillating particle in so-called "quanta of radiation," the magnitude of each of which was equal to Planck's constant multiplied by the frequency.

In his attempts to give an interpretation of this result, Einstein was led to the formulation of the so-called "hypothesis of light-quanta," according to which the radiant energy, in contradiction to Maxwell's electromagnetic theory of light, would not be propagated as electromagnetic waves, but rather as concrete light atoms, each with an energy equal to that of a quantum of radiation. This concept led Einstein to his well-known theory of the photoelectric effect. This phenomenon, which had been entirely unexplainable on the classical theory, was thereby placed in a quite different light, and the predictions of Einstein's theory have received such exact experimental confirmation in recent years, that perhaps the most exact determination of Planck's constant is afforded by measurements on the photoelectric effect. In spite of its heuristic value, however, the hypothesis of light-quanta, which is quite irreconcilable with so-called interference phenomena, is not able to throw light on the nature of radiation. I need only recall that these interference phenomena constitute our only means of investigating the properties of radiation and therefore of assigning any closer meaning to the frequency which in Einstein's theory fixes the magnitude of the light-quantum. . . .

THE QUANTUM THEORY OF ATOMIC CONSTITUTION

The question of further development of the quantum theory was in the meantime placed in a new light by Rutherford's discovery of the atomic nucleus (1911). As we have already seen, this discovery made it quite clear that by classical conceptions alone it was quite impossible to understand the most essential properties of atoms. One was therefore led to seek for a formulation of the principles of the quantum theory that could immediately account for the stability in atomic structure and the properties of the radiation sent out from atoms, of which the observed properties of substances bear witness. Such a formulation I proposed (1913) in the form of two postulates, which may be stated as follows:

(1). Among the conceivably possible states of motion in an atomic system there exist a number of so-called *stationary states* which, in spite of the fact that the motion of the particles in these states obeys the laws of classical mechanics to a considerable extent, possess a peculiar, mechanically unexplainable stability, of such a sort that every permanent change in the motion of the system must consist in a complete transition from one stationary state to another.

(2). While in contradiction to the classical electromagnetic theory no radiation takes place from the atom in the stationary states themselves, a process of transition between two stationary states can be accompanied by the emission of electromagnetic radiation, which will have the same properties as that which would be sent out according to the classical theory from an electrified particle executing an harmonic vibration with constant frequency. This frequency ν has, however, no simple relation to the motion of the particles of the atom, but is given by the relation

$$h\nu = E' - E'',$$

where h is Planck's constant, and $E' - E''$ are the values of the energy of the atom in the two stationary states that form the initial and final state of the radiation process. Conversely, irradiation of the atom with electromagnetic waves of this frequency can lead to an absorption process, whereby the atom is transformed back from the latter stationary state to the former.

While the first postulate has in view the general stability of the atom, the second postulate has chiefly in view the existence of spectra with sharp lines. Furthermore, the quantum-theory condition entering in the last postulate affords a starting-point for the interpretation of the laws of series spectra. The most general of these laws, the combination principle enunciated by Ritz, states that the frequency ν for each of the lines in the spectrum of an element can be represented by the formula

$$\nu = T'' - T',$$

where T'' and T' are two so-called "spectral terms" belonging to a manifold of such terms characteristic of the substance in question.

According to our postulates, this law finds an immediate interpretation in the assumption that the spectrum is emitted by transitions between a number of stationary states in which the numerical value of the energy of the atom is equal to the value of the spectral term multiplied by Planck's constant. This explanation of the combination principle is seen to differ fundamentally from the usual ideas of electrodynamics, as soon as we consider that there is no simple relation between the motion of the atom and the radiation sent out. The departure of our considerations from the ordinary ideas of natural philosophy becomes particularly evident, however, when we observe that the occurrence of two spectral lines, corresponding to combinations of the same spectral term with two other different terms, implies that the nature of the radiation sent out from the atom is not determined only by the motion of the atom at the beginning of the radiation process, but also depends on the state to which the atom is transferred by the process.

At first glance one might, therefore, think that it would scarcely be possible to bring our formal explanation of the combination principle into direct relation with our views regarding the constitution of the atom, which, indeed, are based on experimental evidence interpreted on classical mechanics and electrodynamics. A closer investigation, however, should make it clear that a definite relation may be obtained between the spectra of the elements and the structure of their atoms on the basis of the postulates.

THE HYDROGEN SPECTRUM

The simplest spectrum we know is that of hydrogen. The frequencies of its lines may be represented with great accuracy by means of Balmer's formula:

$$\nu = K\left(\frac{1}{n''^2} - \frac{1}{n'^2}\right),$$

where K is a constant and n' and n'' are two integers. In the spectrum we accordingly meet a single series of spectral terms of the form K/n^2, which decrease regularly with increasing term number n. In accordance with the postulates, we shall therefore assume that each of the hydrogen lines is emitted by a transition between two states belonging to a series of stationary states of the hydrogen atom in which the numerical value of the atom's energy is equal to hK/n^2.

Following our picture of atomic structure, a hydrogen atom consists of a positive nucleus and an electron which—so far as ordinary mechanical conceptions are applicable—will with great approximation describe a periodic elliptical orbit with the nucleus at one focus. The major axis of the orbit is inversely proportional to the work necessary completely to remove the electron from the nucleus, and, in accordance with the above, this work in the stationary states is just equal to $hK{\neq}n^2$. We thus arrive at a manifold of stationary states for which the major axis of the electron orbit takes on a series of discrete values proportional to the squares of the whole numbers. The accompanying Fig. 2 shows these relations diagrammatically. For the sake of simplicity the electron orbits in the stationary states are represented by circles, although in reality the theory places no restriction on the eccentricity of the orbit, but only determines the length of the major axis. The arrows represent the transition processes that correspond to the red and green hydrogen lines, H_α and H_β, the frequency of which is given by means of the Balmer formula when we put $n'' = 2$ and $n' = 3$ and 4 respectively. The transition processes are also represented which correspond to the first three lines of the series of ultraviolet lines found by Lyman in 1914, of which the frequencies are given by the formula when n is put equal to 1, as well as to the first line of the infrared series discovered some years previously by Paschen, which are given by the formula if n'' is put equal to 3.

This explanation of the origin of the hydrogen spectrum leads us quite

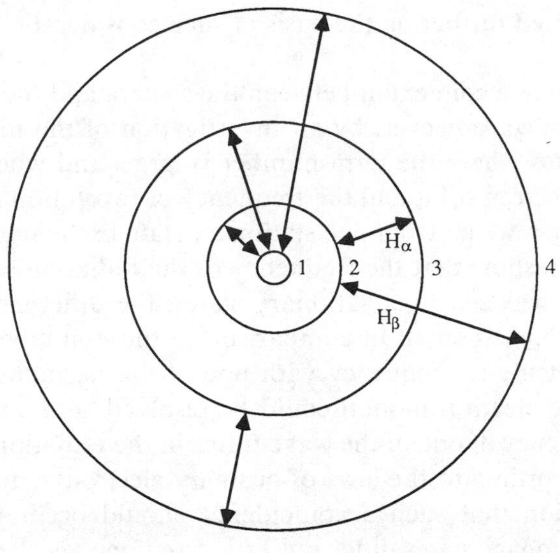

Fig. 2

naturally to interpret this spectrum as the manifestation of a process whereby the electron is bound to the nucleus. While the largest spectral term with term number 1 corresponds to the final stage in the binding process, the small spectral terms that have larger values of the term number correspond to stationary states which represent the initial states of the binding process, where the electron orbits still have large dimensions, and where the work required to remove an electron from the nucleus is still small. The final stage in the binding process we may designate as the normal state of the atom, and it is distinguished from the other stationary states by the property that, in accordance with the postulates, the state of the atom can only be changed by the addition of energy whereby the electron is transferred to an orbit of larger dimensions corresponding to an earlier stage of the binding process.

The size of the electron orbit in the normal state calculated on the basis of the above interpretation of the spectrum agrees roughly with the value for the dimensions of the atoms of the elements that have been calculated by the kinetic theory of matter from the properties of gases. Since, however, as an immediate consequence of the stability of the stationary states that is claimed by the postulates, we must suppose that the interaction between two atoms during a collision cannot be completely described with the aid of the laws of classical mechanics, such a comparison as this cannot be carried further on the basis of such considerations as those just outlined.

A more intimate connexion between the spectra and the atomic model has been revealed, however, by an investigation of the motion in those stationary states where the term number is large, and where the dimensions of the electron orbit and the frequency of revolution in it vary relatively little when we go from one stationary state to the next following. It was possible to show that the frequency of the radiation sent out during the transition between two stationary states, the difference of the term numbers of which is small in comparison to these numbers themselves, tended to coincide in frequency with one of the harmonic components into which the electron motion could be resolved, and accordingly also with the frequency of one of the wave trains in the radiation which would be emitted according to the laws of ordinary electrodynamics.

The condition that such a coincidence should occur in this region where the stationary states differ but little from one another proves to be that the constant in the Balmer formula can be expressed by means of the relation

$$K = \frac{2\pi^2 e^4 m}{h^3},$$

where e and m are respectively the charge and mass of the electron, while h is Planck's constant. This relation has been shown to hold to within the considerable accuracy with which, especially through the beautiful investigations of Millikan, the quantities e, m, and h are known.

This result shows that there exists a connexion between the hydrogen spectrum and the model for the hydrogen atom which, on the whole, is as close as we might hope considering the departure of the postulates from the classical mechanical and electrodynamic laws. At the same time, it affords some indication of how we may perceive in the quantum theory, in spite of the fundamental character of this departure, a natural generalization of the fundamental concepts of the classical electrodynamic theory. To this most important question we shall return later, but first we will discuss how the interpretation of the hydrogen spectrum on the basis of the postulates has proved suitable in several ways, for elucidating the relation between the properties of the different elements.

RELATIONSHIPS BETWEEN THE ELEMENTS

The discussion above can be applied immediately to the process whereby an electron is bound to a nucleus with any given charge. The calculations show that, in the stationary state corresponding to a given value of the number *n,* the size of the orbit will be inversely proportional to the nuclear charge, while the work necessary to remove an electron will be directly proportional to the square of the nuclear charge. The spectrum that is emitted during the binding of an electron by a nucleus with charge *N* times that of the hydrogen nucleus can therefore be represented by the formula:

$$\nu = N^2 K\left(\frac{1}{n''^2} - \frac{1}{n'^2}\right).$$

If in this formula we put $N = 2$, we get a spectrum which contains a set of lines in the visible region which was observed many years ago in the spectrum of certain stars. Rydberg assigned these lines to hydrogen because of the close analogy with the series of lines represented by the Balmer formula. It was never possible to produce these lines in pure hydrogen, but just before the theory for the hydrogen spectrum was put forward, Fowler succeeded in observing the series in question by sending a strong discharge through a mixture of hydrogen and helium. This investigator also assumed that the lines were hydrogen lines, because there existed no experimental evidence from which it might be inferred that two different substances could show properties resembling each other so much as the spectrum in question and that of hydrogen. After the theory was put forward, it became clear, however, that the observed lines must belong to a spectrum of helium, but that they were not like the ordinary helium spectrum emitted from the neutral atom. They came from an ionized helium atom which consists of a single electron moving about a nucleus with double charge. In this way there was brought to light a new feature of the relationship between the elements, which corresponds exactly with our present ideas of atomic structure, according to which the physical and chemical properties of an element depend in the first instance only on the electric charge of the atomic nucleus.

Soon after this question was settled the existence of a similar general relationship between the properties of the elements was brought to light by Moseley's well-known investigations on the characteristic X-ray spectra of the elements, which was made possible by Laue's discovery of the interference of X-rays in crystals and the investigations of W. H. and W. L. Bragg on this subject. It appeared, in fact, that the X-ray spectra of the different elements possessed a much simpler structure and a much greater mutual resemblance than their optical spectra. In particular, it appeared that the spectra changed from element to element in a manner that corresponded closely to the formula given above for the spectrum emitted during the binding of an electron to a nucleus, provided N was put equal to the atomic number of the element concerned. This formula was even capable of expressing, with an approximation that could not be without significance, the frequencies of the strongest X-ray lines, if small whole numbers were substituted for n' and n''.

This discovery was of great importance in several respects. In the first place, the relationship between the X-ray spectra of different elements proved so simple that it became possible to fix without ambiguity the atomic number for all known substances, and in this way to predict with certainty the atomic number of all such hitherto unknown elements for which there is a place in the natural system. Indeed, we find that the square root of the frequency for certain characteristic X-ray lines depends linearly on the atomic number. These lines belong to the group of so-called K-lines, which are the most penetrating of the characteristic rays. With very close approximation the points lie on straight lines, and the fact that they do so is conditioned not only by our taking account of known elements, but also by our leaving an open place between molybdenum (42) and ruthenium (44), just as in Mendeleev's original scheme of the natural system of the elements.

Further, the laws of X-ray spectra provide a confirmation of the general theoretical conceptions, both with regard to the constitution of the atom and the ideas that have served as a basis for the interpretation of spectra. Thus the similarity between X-ray spectra and the spectra emitted during the binding of a single electron to a nucleus may be simply interpreted from the fact that the transitions between stationary states with which we are concerned in X-ray spectra are accompanied by changes in the motion of an electron in the inner part of the atom, where the influence of the attraction of the nucleus is very great compared with the repulsive forces of the other electrons.

The relations between other properties of the elements are of a much more complicated character, which originates in the fact that we have to do with processes concerning the motion of the electrons in the outer part of the atom, where the forces that the electrons exert on one another are of the same order of magnitude as the attraction towards the nucleus, and where, therefore, the details of the interaction of the electrons play an important part. A characteristic example of such a case is afforded by the spatial extension of the atoms of the elements. Lothar Meyer himself directed attention to the characteristic periodic change exhibited by the ratio of the atomic weight to the density, the so-called atomic volume, of the elements in the natural system. . . . While the X-ray spectra vary uniformly with the atomic number, the atomic volumes show a characteristic periodic change which corresponds exactly to the change in the chemical properties of the elements.

Ordinary optical spectra behave in an analogous way. In spite of the dissimilarity between these spectra, Rydberg succeeded in tracing a certain general relationship between the hydrogen spectrum and other spectra. Even though the spectral lines of the elements with higher atomic number appear as combinations of a more complicated manifold of spectral terms which is not so simply co-ordinated with a series of whole numbers, still the spectral terms can be arranged in series each of which shows a strong similarity to the series of terms in the hydrogen spectrum. This similarity appears in the fact that the terms in each series can, as Rydberg pointed out, be very accurately represented by the formula $K/(n+\alpha)^2$, where K is the same constant that occurs in the hydrogen spectrum, often called the Rydberg constant, while n is the term number, and α a constant which is different for the different series.

This relationship with the hydrogen spectrum leads us immediately to regard these spectra as the *last step of a process whereby the neutral atom is built up by the capture and binding of electrons to the nucleus,* one by one. In fact, it is clear that the last electron captured, so long as it is in that stage of the binding process in which its orbit is still large compared to the orbits of the previously bound electrons, will be subjected to a force from the nucleus and these electrons, that differs but little from the force with which the electron in the hydrogen atom is attracted towards the nucleus while it is moving in an orbit of corresponding dimensions.

The spectra so far considered, for which Rydberg's laws hold, are excited by means of electric discharge under ordinary conditions and are often called arc spectra. The elements emit also another type of spectrum, the so-called spark spectra, when they are subjected to an extremely powerful discharge. Hitherto it was impossible to disentangle the spark spectra in the same way as the arc spectra. Shortly after the above view on the origin of arc spectra was brought forward, however, Fowler found (1914) that an empirical expression for the spark spectrum lines could be established which corresponds exactly to Rydberg's laws with the single difference that the constant *K* is replaced by a constant four times as large. Since, as we have seen, the constant that appears in the spectrum sent out during the binding of an electron to a helium nucleus is exactly equal to 4 *K,* it becomes evident that spark spectra are due to the ionized atom, and that their emission corresponds to *the last step but one in the formation of the neutral atom* by the successive capture and binding of electrons. . . .

THE QUANTUM THEORY OF MULTIPLY-PERIODIC SYSTEMS

While it was thus possible by means of the fundamental postulates of the quantum theory to account directly for certain general features of the properties of the elements, a closer development of the ideas of the quantum theory was necessary in order to account for these properties in further detail. In the course of the last few years a more general theoretical basis has been attained through the development of formal methods that permit the fixation of the stationary states for electron motions of a more general type than those we have hitherto considered. For a simply periodic motion such as we meet in the pure harmonic oscillator, and at least to a first approximation, in the motion of an electron about a positive nucleus, the manifold of stationary states can be simply co-ordinated to a series of whole numbers. For motions of the more general class mentioned above, the so-called *multiply-periodic* motions, however, the stationary states compose a more complex manifold, in which, according to these formal methods, each state is characterized by several whole numbers, the so-called "quantum numbers."

In the development of the theory a large number of physicists have taken part, and the introduction of several quantum numbers can be traced back to the work of Planck himself. But the definite step which gave the impetus to further work was made by Sommerfeld (1915) in his explanation of the fine structure shown by the hydrogen lines when the spectrum is observed with a spectroscope of high resolving power. The occurrence of this fine structure must be ascribed to the circumstance that we have to deal, even in hydrogen, with a motion which is not exactly simply periodic. In fact, as a consequence of the change in the electron's mass with velocity that is claimed by the theory of relativity, the electron orbit will undergo a very slow precession in the orbital plane. The motion will therefore be doubly periodic, and besides a number characterizing the term in the Balmer formula, which we shall call the *principal quantum number* because it determines in the main the energy of the atom, the fixation of the stationary states demands another quantum number which we shall call the *subordinate quantum number.*

A survey of the motion in the stationary states thus fixed is given in the diagram (Fig. 3), which reproduces the relative size and form of the electron orbits. Each orbit is designated by a symbol n_k, where n is the principal quantum number and k the subordinate quantum number. All orbits with the same principal quantum number have, to a first approximation, the same major axis, while orbits with the same value of k have the same parameter, i.e. the same value for the shortest chord through the focus. Since the energy values for different states with the same value of n but different values of k differ a little from each other, we get for each hydrogen line corresponding to definite values of n' and n'' in the Balmer formula a number of different transition processes, for which the frequencies of the emitted radiation as calculated by the second postulate are not exactly the same. As Sommerfeld was able to show, the components this gives for each hydrogen line agree with the observations on the fine structure of hydrogen lines to within the limits of experimental error. In the figure the arrows designate the processes that give rise to the components of the red and green lines in the hydrogen spectrum, the frequencies of which are obtained by putting $n'' = 2$ and $n' = 3$ or 4 respectively in the Balmer formula.

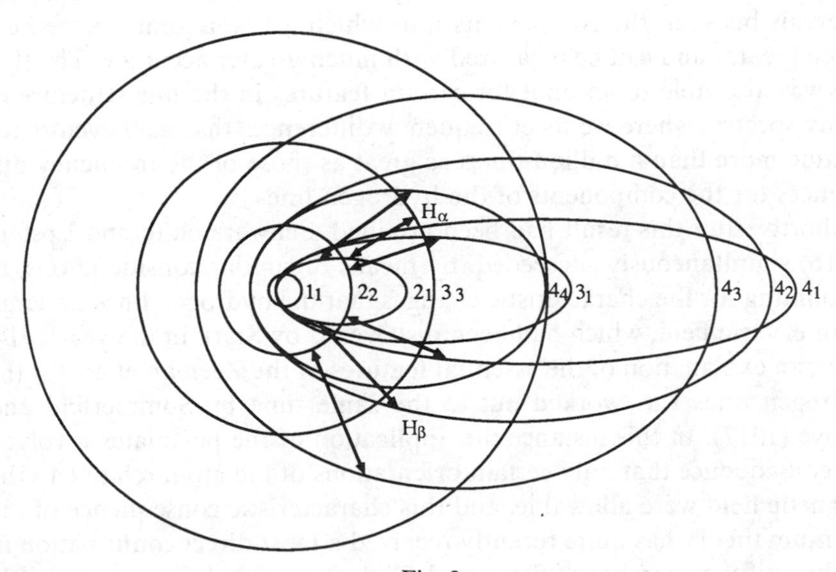

Fig. 3

In considering the figure it must not be forgotten that the description of the orbit is there incomplete, in so much as with the scale used the slow precession does not show at all. In fact, this precession is so slow that even for the orbits that rotate most rapidly the electron performs about 40,000 revolutions before the perihelion has gone round once. Nevertheless, it is this precession alone that is responsible for the multiplicity of the stationary states characterized by the subordinate quantum number. If, for example, the hydrogen atom is subjected to a small disturbing force which perturbs the regular precession, the electron orbit in the stationary states will have a form altogether different from that given in the figure. This implies that the fine structure will change its character completely, but the hydrogen spectrum will continue to consist of lines that are given to a close approximation by the Balmer formula, due to the fact that the approximately periodic character of the motion will be retained. Only when the disturbing forces become so large that even during a single revolution of the electron the orbit is appreciably disturbed, will the spectrum undergo essential changes. The statement often advanced that the introduction of two quantum numbers should be a necessary condition for the explanation of the Balmer formula must therefore be considered as a misconception of the theory.

Sommerfeld's theory has proved itself able to account not only for the fine structure of the hydrogen lines, but also for that of the lines in the helium spark spectrum. Owing to the greater velocity of the electron, the intervals between the components into which a line is split up are here much greater and can be measured with much greater accuracy. The theory was also able to account for certain features in the fine structure of X-ray spectra, where we meet frequency differences that may even reach a value more than a million times as great as those of the frequency differences for the components of the hydrogen lines.

Shortly after this result had been attained, Schwarzschild and Epstein (1916) simultaneously succeeded, by means of similar considerations, in accounting for the characteristic changes that the hydrogen lines undergo in an electric field, which had been discovered by Stark in the year 1914. Next, an explanation of the essential features of the Zeeman effect for the hydrogen lines was worked out at the same time by Sommerfeld and Debye (1917). In this instance the application of the postulates involved the consequence that only certain orientations of the atom relative to the magnetic field were allowable, and this characteristic consequence of the quantum theory has quite recently received a most direct confirmation in the beautiful researches of Stern and Gerlach on the deflexion of swiftly moving silver atoms in a nonhomogenous magnetic field.

THE CORRESPONDENCE PRINCIPLE

While the development of the theory of spectra was based on the working out of formal methods for the fixation of stationary states, I succeeded shortly afterwards in throwing light on the theory from a new viewpoint, by pursuing further the characteristic connexion between the quantum theory and classical electrodynamics already traced out in the hydrogen spectrum. In connexion with the important work of Ehrenfest and Einstein these efforts led to the formulation of the so-called *correspondence principle,* according to which the occurrence of transitions between the stationary states accompanied by emission of radiation is traced back to the harmonic components into which the motion of the atom may be resolved and which, according to the classical theory, determine the properties of the radiation to which the motion of the particles gives rise.

According to the correspondence principle, it is assumed that every transition process between two stationary states can be co-ordinated with a corresponding harmonic vibration component in such a way that the probability of the occurrence of the transition is dependent on the amplitude of the vibration. The state of polarization of the radiation emitted during the transition depends on the further characteristics of the vibration, in a manner analogous to that in which on the classical theory the intensity and state of polarization in the wave system emitted by the atom as a consequence of the presence of this vibration component would be determined respectively by the amplitude and further characteristics of the vibration.

With the aid of the correspondence principle it has been possible to confirm and to extend the above-mentioned results. Thus it was possible to develop a complete quantum theory explanation of the Zeeman effect for the hydrogen lines, which, in spite of the essentially different character of the assumptions that underlie the two theories, is very similar throughout to Lorentz's original explanation based on the classical theory. In the case of the Stark effect, where, on the other hand, the classical theory was completely at a loss, the quantum theory explanation could be so extended with the help of the correspondence principle as to account for the polarization of the different components into which the lines are split, and also for the characteristic intensity distribution exhibited by the components. . . .

Besides the principal quantum integer *n,* the stationary states are further characterized by a subordinate quantum integer *s,* which can be negative as well as positive and has a meaning quite different from that of the quantum number *k* occurring in the relativity theory of the fine structure of the hydrogen lines, which fixed the form of the electron orbit in the undisturbed atom. Under the influence of the electric field both the form of the orbit and its position undergo large changes, but certain properties of the orbit remain unchanged, and the subordinate quantum number *s* is connected with these. . . . The theory reproduces completely the main feature of the experimental results, and in the light of the correspondence principle we can say that the Stark effect reflects down to the smallest details the action of the electric field on the orbit of the electron in the hydrogen atom, even though in this case the reflection is so distorted that, in contrast with the case of the Zeeman effect, it would scarcely be possible directly to recognize the motion on the basis of the classical ideas of the origin of electromagnetic radiation.

Results of interest were also obtained for the spectra of elements of higher atomic number, the explanation of which in the meantime had made important progress through the work of Sommerfeld, who introduced several quantum numbers for the description of the electron orbits. Indeed, it was possible, with the aid of the correspondence principle, to account completely for the characteristic rules which govern the seemingly capricious occurrence of combination lines, and it is not too much to say that the quantum theory has not only provided a simple interpretation of the combination principle, but has further contributed materially to the clearing up of the mystery that has long rested over the application of this principle. . . .

THE NATURAL SYSTEM OF THE ELEMENTS

The ideas of the origin of spectra outlined in the preceding have furnished the basis for a theory of the structure of the atoms of the elements which has shown itself suitable for a general interpretation of the main features of the properties of the elements, as exhibited in the natural system. This theory is based primarily on considerations of the manner in which the atom can be imagined to be built up by the capture and binding of electrons to the nucleus, one by one. As we have seen, the optical spectra of elements provide us with evidence on the progress of the last steps in this building-up process.

An insight into the kind of information that the closer investigation of the spectra has provided in this respect may be obtained from Fig. 4, which gives a diagrammatic representation of the orbital motion in the stationary states corresponding to the emission of the arc-spectrum of

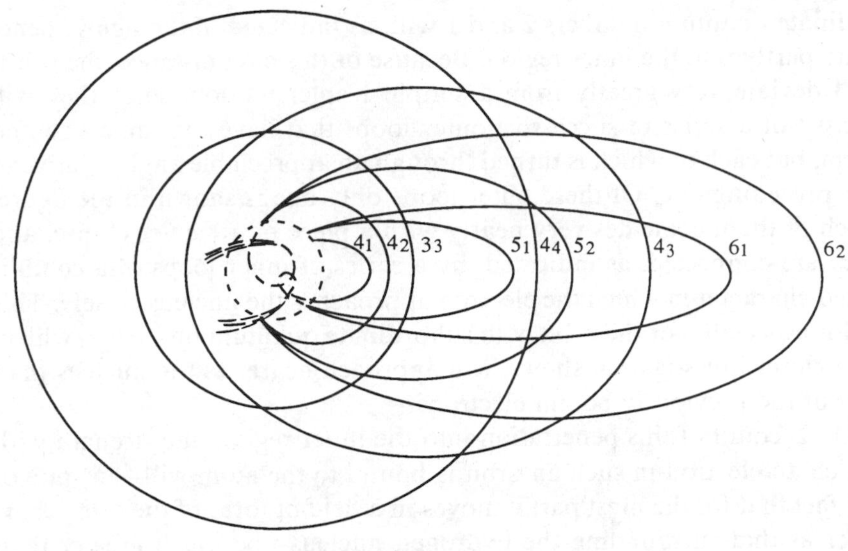

Fig. 4

potassium. The curves show the form of the orbits described in the stationary states by the last electron captured in the potassium atom, and they can be considered as stages in the process whereby the 19th electron is bound after the 18 previous electrons have already been bound in their normal orbits. In order not to complicate the figure, no attempt has been made to draw any of the orbits of these inner electrons, but the region in which they move is enclosed by a dotted circle. In an atom with several electrons the orbits will, in general, have a complicated character. Because of the symmetrical nature of the field of force about the nucleus, however, the motion of each single electron can be approximately described as a plane periodic motion on which is superimposed a uniform rotation in the plane of the orbit. The orbit of each electron will therefore be to a first approximation doubly periodic, and will be fixed by two quantum numbers, as are the stationary states in a hydrogen atom when the relativity precession is taken into account.

In Fig. 4 . . . the electron orbits are marked with the symbol n_k, where n is the principal quantum number and k the subordinate quantum number. While for the initial states of the binding process, where the quantum numbers are large, the orbit of the last electron captured lies completely outside of those of the previously bound electrons, this is not the case for the last stages. Thus, in the potassium atom, the electron orbits with subordinate quantum numbers 2 and 1 will, as indicated in the figure, penetrate partly into the inner region. Because of this circumstance, the orbits will deviate very greatly from a simple Kepler motion, since they will consist of a series of successive outer loops that have the same size and form, but each of which is turned through an appreciable angle relative to the preceding one. Of these outer loops only one is shown in the figure. Each of them coincides very nearly with a piece of a Kepler ellipse, and they are connected, as indicated, by a series of inner loops of a complicated character in which the electron approaches the nucleus closely. This holds especially for the orbit with subordinate quantum number 1, which, as a closer investigation shows, will approach nearer to the nucleus than any of the previously bound electrons.

On account of this penetration into the inner region, the strength with which an electron in such an orbit is bound to the atom will—in spite of the fact that for the most part it moves in a field of force of the same character as that surrounding the hydrogen nucleus—be much greater than for an electron in a hydrogen atom that moves in an orbit with the same principal quantum number, the maximum distance of the electron from the nucleus at the same time being considerably less than in such a hydrogen orbit. As we shall see, this feature of the binding process in atoms with many electrons is of essential importance in order to understand the characteristic periodic way in which many properties of the elements as displayed in the natural system vary with the atomic number.

In the accompanying table (Fig. 5) is given a summary of the results concerning the structure of the atoms of the elements to which the author has been led by a consideration of successive capture and binding of electrons to the atomic nucleus. The figures before the different elements are the atomic numbers, which give the total number of electrons in the neutral atom. The figures in the different columns give the number of electrons in orbits corresponding to the values of the principal and subordinate quantum numbers standing at the top. . . .

I hope that I have succeeded in giving a summary of some of the most important results that have been attained in recent years in the field of atomic theory, and I should like, in concluding, to add a few general remarks concerning the viewpoint from which these results may be judged, and particularly concerning the question of how far, with these results, it is possible to speak of an explanation, in the ordinary sense of the word. By a theoretical explanation of natural phenomena we understand in general a classification of the observations of a certain domain with the help of analogies pertaining to other domains of observation, where one presumably has to do with simpler phenomena. The most that one can

		1_1	2_1	2_2	3_1	3_2	3_3	4_1	4_2	4_3	4_4	5_1	5_2	5_3	5_4	5_5	6_1	6_2	6_3	6_4	6_5	6_6	7_1	7_2
1	H	1																						
2	He	2																						
3	Li	2	1																					
4	Be	2	2																					
5	B	2	2	(1)																				
–	–	–	–	–																				
10	Ne	2	4	4																				
11	Na	2	4	4	1																			
12	Mg	2	4	4	2																			
13	Al	2	4	4	2	1																		
–	–	–	–	–	–	–																		
18	A	2	4	4	4	4																		
19	K	2	4	4	4	4		1																
20	Ca	2	4	4	4	4		2																
21	Sc	2	4	4	4	4	1	(2)																
22	Ti	2	4	4	4	4	2	(2)																
–	–	–	–	–	–	–	–	–																
29	Cu	2	4	4	6	6	6	1																
30	Zn	2	4	4	6	6	6	2																
31	Ga	2	4	4	6	6	6	2	1															
–	–	–	–	–	–	–	–	–	–															
36	Kr	2	4	4	6	6	6	4	4															
37	Rb	2	4	4	6	6	6	4	4			1												
38	Sr	2	4	4	6	6	6	4	4			2												
39	Y	2	4	4	6	6	6	4	4	1		(2)												
40	Zr	2	4	4	6	6	6	4	4	2		(2)												
–	–	–	–	–	–	–	–	–	–	–		–												
47	Ag	2	4	4	6	6	6	6	6	6		1												
48	Cd	2	4	4	6	6	6	6	6	6		2												
49	In	2	4	4	6	6	6	6	6	6		2	1											
–	–	–	–	–	–	–	–	–	–	–		–	–											
54	X	2	4	4	6	6	6	6	6	6		4	4											
55	Cs	2	4	4	6	6	6	6	6	6		4	4				1							
56	Ba	2	4	4	6	6	6	6	6	6		4	4				2							
57	La	2	4	4	6	6	6	6	6	6		4	4	1			(2)							
58	Ce	2	4	4	6	6	6	6	6	6	1	4	4	1			(2)							
59	Pr	2	4	4	6	6	6	6	6	6	1	4	4	1			(2)							
–	–	–	–	–	–	–	–	–	–	–	–	–	–	–			–							
71	Cp	2	4	4	6	6	6	8	8	8	8	4	4	1			(2)							
72	–	2	4	4	6	6	6	8	8	8	8	4	4	2			(2)							
–	–	–	–	–	–	–	–	–	–	–	–	–	–	–			–							
79	Au	2	4	4	6	6	6	8	8	8	8	6	6	6			1							
80	Hg	2	4	4	6	6	6	8	8	8	8	6	6	6			2							
81	Tl	2	4	4	6	6	6	8	8	8	8	6	6	6			2	1						
–	–	–	–	–	–	–	–	–	–	–	–	–	–	–			–	–						
86	Em	2	4	4	6	6	6	8	8	8	8	6	6	6			4	4						
87	–	2	4	4	6	6	6	8	8	8	8	6	6	6			4	4					1	
88	Ra	2	4	4	6	6	6	8	8	8	8	6	6	6			4	4					2	
89	Ac	2	4	4	6	6	6	8	8	8	8	6	6	6			4	4	1				(2)	
90	Th	2	4	4	6	6	6	8	8	8	8	6	6	6			4	4	2				(2)	
–	–	–	–	–	–	–	–	–	–	–	–	–	–	–			–	–	–				–	
118	?	2	4	4	6	6	6	8	8	8	8	8	8	8	8		6	6	6				4	4

Fig. 5

demand of a theory is that this classification can be pushed so far that it can contribute to the development of the field of observation by the prediction of new phenomena.

When we consider the atomic theory, we are, however, in the peculiar position that there can be no question of an explanation in this last sense, since here we have to do with phenomena which from the very nature of the case are simpler than in any other field of observation, where the phenomena are always conditioned by the combined action of a large number of atoms. We are therefore obliged to be modest in our demands and content ourselves with concepts which are formal in the sense that they do not provide a visual picture of the sort one is accustomed to require of the explanations with which natural philosophy deals. Bearing this in mind I have sought to convey the impression that the results, on the other hand, fulfil, at least in some degree, the expectations that are entertained of any theory; in fact, I have attempted to show how the development of atomic theory has contributed to the classification of extensive fields of observation, and by its predictions has pointed out the way to the completion of this classification. It is scarcely necessary, however, to emphasize that the theory is yet in a very preliminary stage, and many fundamental questions still await solution.

N.4

The Fundamental Idea of Wave Mechanics

ERWIN SCHRÖDINGER

ERWIN SCHRÖDINGER was born in Vienna in 1887, the only child in a wealthy family. The son of a gifted chemist and botanist, Schrödinger was a brilliant child whose interests during his early school years ranged from the sciences to ancient grammar and German poetry. He went to the University of Vienna, where he earned his doctorate in physics in 1910 at the age of twenty-three. When World War I broke out, he was inducted into military service; he served with distinction as an artillery officer and emerged from the war unscathed. He had planned to give up physics for philosophy, but when the university town in which he had found a job was removed from Austria as part of the peace treaty, Schrödinger decided to continue his work as a physicist. He served as assistant to Max Wien in 1920; then he worked successively at Stuttgart, Breslau, and the University of Zurich where he replaced Max von Laue. Schrödinger spent his most productive years at Zurich where he developed warm friendships with Hermann Weyl and Peter Debye. He worked on problems dealing with the specific heats of solids, thermodynamics, and atomic spectra, as well as on the physiology of color.

In the early 1920s the French physicist Louis de Broglie had hypothesized what he called *matter waves,* that is, waves, not of electromagnetism, but of particles of matter, and suggested that the same relationship that existed between the wavelength of light and its momentum might apply to matter waves. Schrödinger had read about de Broglie's matter waves in a footnote to a paper by Einstein. It occurred to him that de Broglie's work might be applied to the Bohr version of the atom, with its levels of electron orbits. Schrödinger buttressed his basic ideas with a strict mathematical support and developed what is known as the *Schrödinger wave equation,* a major discovery in the development of quantum mechanics.

In the Schrödinger atom, an electron can have any orbit provided this orbit can accommodate an integral number of its matter wavelengths—what is called a *standing wave;* such a wave represents a stationary state of the electron, which thus does not radiate light nor violate Maxwell's equations. The difficulty encountered by the Rutherford atom was thus eliminated. The wave equation became the cornerstone of quantum mechanics, especially after its consistency with the matrix mechanics announced a year earlier in 1925 by Werner Heisenberg was confirmed.

Schrödinger shared the Nobel Prize in physics with the British physicist Paul Dirac. Our selection is his Nobel lecture, which was delivered on 12 December 1933, and which examined his own work in formulating his theory of wave mechanics.

Although Schrödinger had become Planck's successor at the University of Berlin in 1927, the rise of Hitler led him temporarily to take up residence at Oxford University in 1933 before accepting a position at Graz in 1936. He escaped to Italy following the annexation of Austria in 1938 and then spent some time at Princeton University before becoming director of the School for Theoretical Physics at the Institute for Advanced Studies in Dublin, where he remained until retiring in 1955. He worked on a theory to unite electromagnetism and gravitation and wrote the well-known book *What Is Life?* in 1944. Schrödinger returned to Vienna in 1955 where he died in 1961 after a long illness.

[Quantum mechanics requires] a final renunciation of the classical idea of causality and a radical revision of our attitude toward the problem of physical reality.

—NIELS BOHR, *Atomic Theory and Human Knowledge*

The world thus appears as a complicated tissue of events, in which connections of different kinds alternate or overlap or combine and thereby determine the texture of the whole.

—WERNER HEISENBERG, *Physics and Philosophy*

ON PASSING THROUGH an optical instrument, such as a telescope or a camera lens, a ray of light is subjected to a change in direction at each refracting or reflecting surface. The path of the rays can be constructed if we know the two simple laws which govern the changes in direction: the law of refraction which was discovered by Snellius a few hundred years ago, and the law of reflection with which Archimedes was familiar more than 2,000 years ago. As a simple example, Fig. 1 shows a ray A–B which is subjected to refraction at each of the four boundary surfaces of two lenses in accordance with the law of Snellius.

Fermat defined the total path of a ray of light from a much more general point of view. In different media, light propagates with different velocities, and the radiation path gives the appearance as if the light must arrive at its destination *as quickly as possible.* (Incidentally, it is permissible here to consider *any two* points along the ray as the starting- and endpoints.) The least deviation from the path actually taken would mean a delay. This is the famous Fermat *principle of the shortest light time,* which in a marvellous manner determines the entire fate of a ray of light by a single statement and also includes the more general case, when the nature of the medium varies not suddenly at individual surfaces, but gradually from place to place. The atmosphere of the earth provides an example. The more deeply a ray of light penetrates into it from outside, the more slowly it progresses in an increasingly denser air. Although the differences in the speed of propagation are infinitesimal, Fermat's principle in these circumstances demands that the light ray should curve earthward (see Fig. 2), so that it remains a little longer in the higher "faster" layers

---

Erwin Schrödinger, Nobel Prize in Physics Award Address, 1933. Reprinted with permission of Elsevier Publishing Co., and the Nobel Foundation.

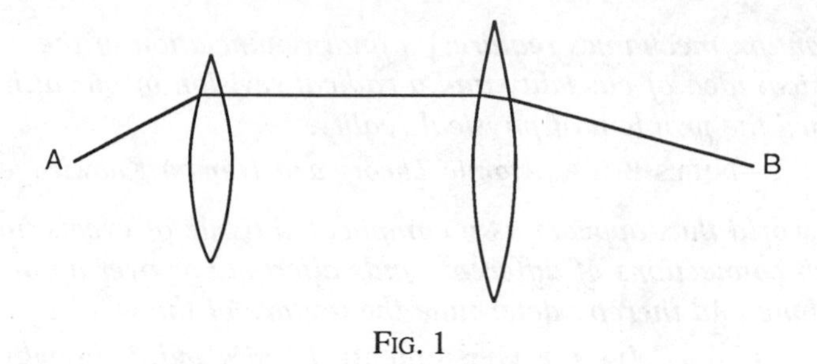

FIG. 1

and reaches its destination more quickly than by the shorter straight path (broken line in the figure; disregard the square, ẂWW¹W¹ for the time being). I think, hardly any of you will have failed to observe that the sun when it is deep on the horizon appears to be not circular but flattened: its vertical diameter looks to be shortened. This is a result of the curvature of the rays.

According to the wave theory of light, the light rays, strictly speaking, have only fictitious significance. They are not the physical paths of some particles of light, but are a mathematical device, the so-called orthogonal trajectories of wave surfaces, imaginary guide lines as it were, which point in the direction normal to the wave surface in which the latter advances (cf. Fig. 3 which shows the simplest case of concentric spherical wave surfaces and accordingly rectilinear rays, whereas Fig. 4 illustrates the case of curved rays). It is surprising that a general principle as important as Fermat's relates directly to these mathematical guide lines, and not to the wave surfaces, and one might be inclined for this reason to consider it a mere mathematical curiosity. Far from it. It becomes properly

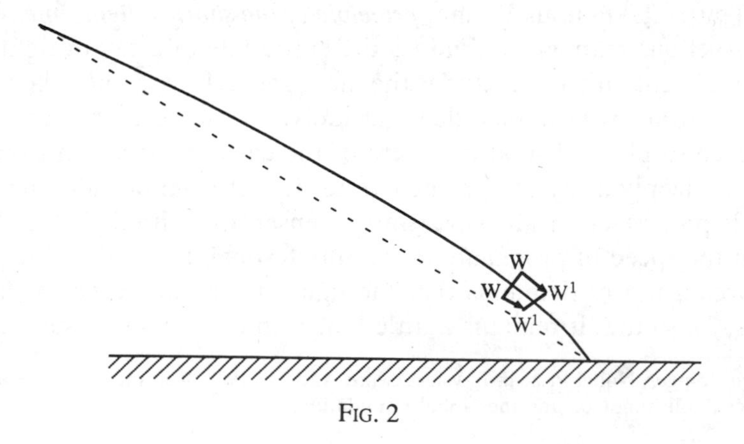

FIG. 2

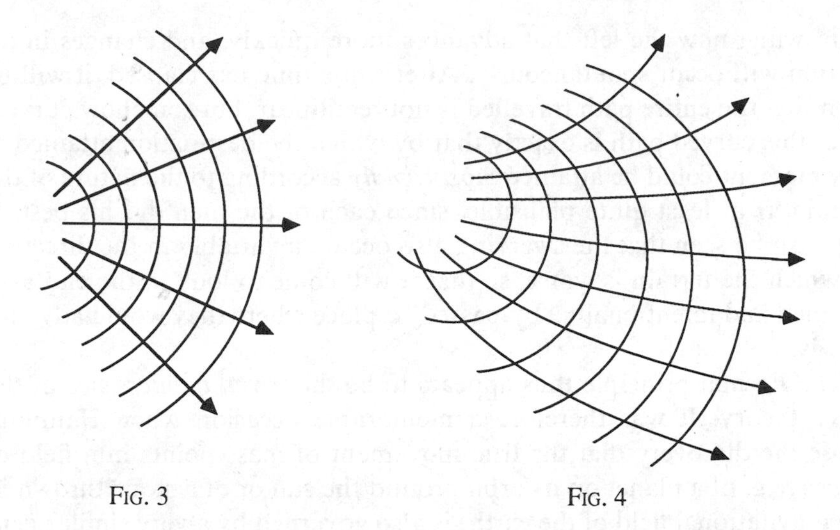

FIG. 3 FIG. 4

understandable only from the point of view of wave theory and ceases to be a divine miracle. From the wave point of view, the so-called *curvature* of the light ray is far more readily understandable as a *swerving* of the wave surface, which must obviously occur when neighbouring parts of a wave surface advance at different speeds; in exactly the same manner as a company of soldiers marching forward will carry out the order "right incline" by the men taking steps of varying lengths, the right-wing man the smallest, and the left-wing man the longest. In atmospheric refraction of radiation for example (Fig. 2) the section of wave surface WW must necessarily swerve to the right towards W^1W^1 because its left half is located in slightly higher, thinner air and thus advances more rapidly than the right part at lower point. (In passing, I wish to refer to one point at which the *Snellius'* view fails. A horizontally emitted light ray should remain horizontal because the refraction index does not vary in the horizontal direction. In truth, a horizontal ray curves more strongly than any other, which is an obvious consequence of the theory of a swerving wave front.) On detailed examination the Fermat principle is found to be completely *tantamount* to the trivial and obvious statement that—given local distribution of light velocities—the wave front must swerve in the manner indicated. I cannot prove this here, but shall attempt to make it plausible. I would again ask you to visualize a rank of soldiers marching forward. To ensure that the line remains dressed, let the men be connected by a long rod which each holds firmly in his hand. No orders as to direction are given; the only order is: let each man march or run as fast as he can. If the nature of the ground varies slowly from place to place, it will be now the right wing, now the left that advances more quickly, and changes in direction will occur spontaneously. After some time has elapsed, it will be seen that the entire path travelled is not rectilinear, but somehow curved. That this curved path is exactly that by which the destination attained at any moment could be attained *most rapidly* according to the nature of the terrain, is at least quite plausible, since each of the men did his best. It will also be seen that the swerving also occurs invariably in the direction in which the terrain is worse, so that it will come to look in the end as if the men had intentionally "by-passed" a place where they would advance slowly.

The Fermat principle thus appears to be the *trivial quintessence* of the wave theory. It was therefore a memorable occasion when Hamilton made the discovery that the true movement of mass points in a field of forces (e.g. of a planet on its orbit around the sun or of a stone thrown in the gravitational field of the earth) is also governed by a very similar general principle, which carries and has made famous the name of its discoverer since then. Admittedly, the Hamilton principle does not say exactly that the mass point chooses the quickest way, but it does say something *so* similar—the analogy with the principle of the shortest travelling time of light is *so* close, that one was faced with a puzzle. It seemed as if Nature had realized one and the same law twice by entirely different means: first in the case of light, by means of a fairly obvious play of rays; and again in the case of the mass points, which was anything but obvious, unless somehow wave nature were to be attributed to them also. And this, it seemed impossible to do. Because the "mass points" on which the laws of mechanics had really been confirmed experimentally at that time were only the large, visible, sometimes *very* large bodies, the planets, for which a thing like "wave nature" appeared to be out of the question.

The smallest, elementary components of matter which we today, much more specifically, call "mass points," were purely hypothetical at the time. It was only after the discovery of radioactivity that constant refinements of methods of measurement permitted the properties of these particles to be studied in detail, and now permit the paths of such particles to be photographed and to be measured very exactly (stereophotogrammetrically) by the brilliant method of C.T.R. Wilson. As far as the measurements extend they confirm that the same mechanical laws are valid for particles as for large bodies, planets, etc. However, it was found that neither the molecule nor the individual atom can be considered as the "ultimate component": but even the atom is a system of highly complex structure. Images are formed in our minds of the structure of atoms *consisting of* particles, images which seem to have a certain similarity with the planetary system. It was only natural that the attempt should at first be made to consider as valid the same laws of motion that had proved themselves so amazingly satisfactory on a large scale. In other words, Hamilton's mechanics, which, as I said above, culminates in the Hamilton principle, were applied also to the "inner life" of the atom. That there is a very close analogy between Hamilton's principle and Fermat's optical principle had meanwhile become all but forgotten. If it was remembered, it was considered to be nothing more than a curious trait of the mathematical theory.

Now, it is very difficult, without further going into details, to convey a proper conception of the success or failure of these classical-mechanical images of the atom. On the one hand, Hamilton's principle in particular proved to be the most faithful and reliable guide, which was simply indispensable; on the other hand one had to suffer, to do justice to the facts, the rough interference of entirely new incomprehensible postulates, of the so-called quantum conditions and quantum postulates. Strident disharmony in the symphony of classical mechanics—yet strangely familiar—played as it were on the same instrument. In mathematical terms we can formulate this as follows: whereas the Hamilton principle merely postulates that a given integral must be a minimum, without the numerical value of the minimum being established by this postulate, it is now demanded that the numerical value of the minimum should be restricted to integral multiples of a universal natural constant, Planck's quantum of action. This incidentally. The situation was fairly desperate. Had the old mechanics failed completely, it would not have been so bad. The way would then have been free to the development of a new system of mechanics. As it was, one was faced with the difficult task of saving the *soul* of the old system, whose inspiration clearly held sway in this microcosm, while at the same time flattering it as it were into accepting the quantum conditions not as gross interference but as issuing from its own innermost essence.

The way out lay just in the possibility, already indicated above, of attributing to the Hamilton principle, also, the operation of a wave mechanism on which the point-mechanical processes are essentially based, just as one had long become accustomed to doing in the case of phenomena relating to light and of the Fermat principle which governs them. Admittedly, the individual path of a mass point loses its proper physical significance and becomes as fictitious as the individual isolated ray of light. The essence of the theory, the minimum principle, however, remains not only intact, but reveals its true and simple meaning only under the wave-like aspect, as already explained. Strictly speaking, the new theory is in fact not *new,* it is a completely organic development, one might almost be tempted to say a more elaborate exposition, of the old theory.

How was it then that this new more "elaborate" exposition led to notably different results; what enabled it, when applied to the atom, to obviate difficulties which the old theory could not solve? What enabled it to render gross interference acceptable or even to make it its own?

Again, these matters can best be illustrated by analogy with optics. Quite properly, indeed, I previously called the Fermat principle the quintessence of the wave theory of light: nevertheless, it cannot render dispensable a more exact study of the wave process itself. The so-called refraction and interference phenomena of light can only be understood if we trace the wave process in detail because what matters is not only the eventual destination of the wave, but also whether at a given moment it arrives there with a wave peak or a wave trough. In the older, coarser experimental arrangements, these phenomena occurred as small details only and escaped observation. Once they were noticed and were interpreted correctly, by means of waves, it was easy to devise experiments in which the wave nature of light finds expression not only in small details, but on a very large scale in the entire character of the phenomenon.

Allow me to illustrate this by two examples, first, the example of an optical instrument, such as telescope, microscope, etc. The object is to obtain a sharp image, i.e. it is desired that all rays issuing from a point should be reunited in a point, the so-called focus (cf. Fig. 5a). It was at first believed that it was only geometrical-optical difficulties which prevented this: they are indeed considerable. Later it was found that even in the best designed instruments focussing of the rays was considerably inferior than would be expected if each ray exactly obeyed the Fermat principle independently of the neighbouring rays. The light which issues from a point and is received by the instrument is reunited behind the instrument not in a single point any more, but is distributed over a small circular area, a so-called diffraction disc, which, otherwise, is in most cases a circle only because the apertures and lens contours are generally circular. For, the cause of the phenomenon which we call *diffraction* is that not all the spherical waves issuing from the object point can be accommodated by the instrument. The lens edges and any apertures merely cut out a part of the wave surfaces (cf. Fig. 5b) and—if you will permit me to use a more suggestive expression—the injured margins resist rigid unification in a point and produce the somewhat blurred or vague image. The degree of blurring is closely associated with the *wavelength* of the light and is completely inevitable because of this deep-seated theoretical relationship.

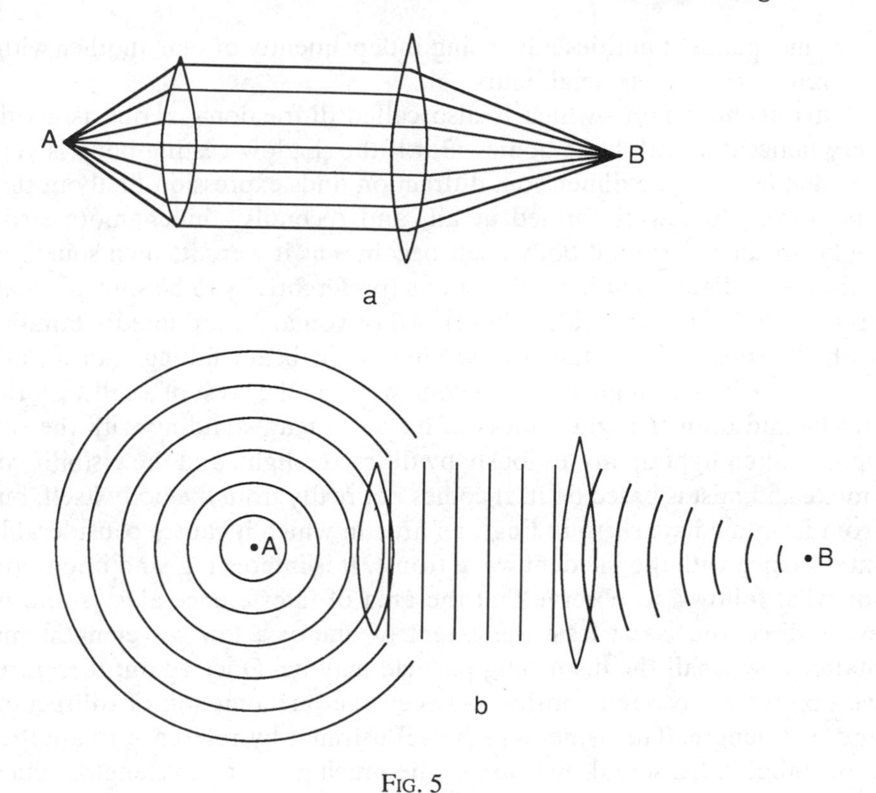

FIG. 5

Hardly noticed at first, it governs and restricts the performance of the modern microscope which has mastered all other errors of reproduction. The images obtained of structures not much coarser or even still finer than the wavelengths of light are only remotely or not at all similar to the original.

A second, even simpler example is the shadow of an opaque object cast on a screen by a small point light source. In order to construct the shape of the shadow, each light ray must be traced and it must be established whether or not the opaque object prevents it from reaching the screen. The *margin* of the shadow is formed by those light rays which only just brush past the edge of the body. Experience has shown that the shadow margin is not absolutely sharp even with a point-shaped light source and a sharply defined shadow-casting object. The reason for this is the same as in the first example. The wave front is as it were bisected by the body (cf. Fig. 6) and the traces of this injury result in blurring of the margin of the shadow which would be incomprehensible if the individual light rays were independent entities advancing independently of one another without reference to their neighbours.

This phenomenon—which is also called diffraction—is not as a rule very noticeable with large bodies. But if the shadow-casting body is very small at least in one dimension, diffraction finds expression firstly in that no proper shadow is formed at all, and secondly—much more strikingly—in that the small body itself becomes as it were its own source of light and radiates light in all directions (preferentially to be sure, at small angles relative to the incident light). All of you are undoubtedly familiar with the so-called "motes of dust" in a light beam falling into a dark room. Fine blades of grass and spiders' webs on the crest of a hill with the sun behind it, or the errant locks of hair of a man standing with the sun behind often light up mysteriously by diffracted light, and the visibility of smoke and mist is based on it. It comes not really from the body itself, but from its immediate surroundings, an area in which it causes considerable interference with the incident wave fronts. It is interesting, and important for what follows, to observe that the area of interference always and in every direction has at least the extent of one or a few wavelengths, no matter how small the disturbing particle may be. Once again, therefore, we observe a close relationship between the phenomenon of diffraction and wavelength. This is perhaps best illustrated by reference to another wave process, i.e. sound. Because of the much greater wavelength, which

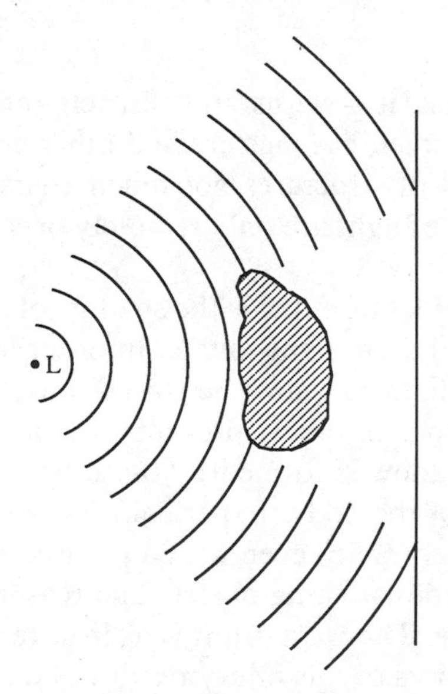

FIG. 6

is of the order of centimetres and metres, shadow formation recedes in the case of sound, and diffraction plays a major, and practically important, part: we can easily *hear* a man calling from behind a high wall or around the corner of a solid house, even if we cannot *see* him.

Let us return from optics to mechanics and explore the analogy to its fullest extent. In optics the *old* system of mechanics corresponds to intellectually operating with isolated mutually independent light rays. The new undulatory mechanics corresponds to the wave theory of light. What is gained by changing from the old view to the new is that the diffraction phenomena can be accommodated, or better expressed, what is gained is something that is strictly analogous to the diffraction phenomena of light and which on the whole must be very unimportant, otherwise the old view of mechanics would not have given full satisfaction so long. It is, however, easy to surmise that the neglected phenomenon may in some circumstances make itself very much felt, will entirely dominate the mechanical process, and will face the old system with insoluble riddles, if *the entire mechanical system is comparable in extent with the wavelengths of the "waves of matter"* which play the same part in mechanical processes as that played by the light waves in optical processes.

This is the reason why in these minute systems, the atoms, the old view was bound to fail, which though remaining intact as a close approximation for gross mechanical processes, but is no longer adequate for the delicate interplay in areas of the order of magnitude of one or a few wavelengths. It was astounding to observe the manner in which all those strange additional requirements developed spontaneously from the new undulatory view, whereas they had to be forced upon the old view to adapt them to the inner life of the atom and to provide some explanation of the observed facts.

Thus, the salient point of the whole matter is that the diameters of the atoms and the wavelength of the hypothetical material waves are of approximately the same order of magnitude. And now you are bound to ask whether it must be considered mere chance that in our continued analysis of the structure of matter we should come upon the order of magnitude of the wavelength at this of all points, or whether this is to some extent comprehensible. Further, you may ask, how we know that this is so, since the material waves are an entirely new requirement of this theory, unknown anywhere else. Or is it simply that this is an *assumption* which had to be made?

The agreement between the orders of magnitude is no mere chance, nor is any special assumption about it necessary; it follows automatically from the theory in the following remarkable manner. That the heavy *nu-*

cleus of the atom is very much smaller than the atom and may therefore be considered as a point centre of attraction in the argument which follows may be considered as experimentally established by the experiments on the scattering of alpha rays done by Rutherford and Chadwick. Instead of the *electrons* we introduce hypothetical waves, whose wavelengths are left entirely open, because we know nothing about them yet. This leaves a letter, say *a,* indicating a still unknown figure, in our calculation. We are, however, used to this in such calculations and it does not prevent us from calculating that the nucleus of the atom must produce a kind of diffraction phenomenon in these waves, similarly as a minute dust particle does in light waves. Analogously, it follows that there is a close relationship between the extent of the area of interference with which the nucleus surrounds itself and the wavelength, and that the two are of the same order of magnitude. What this is, we have had to leave open; but the most important step now follows: *we identify the area of interference, the diffraction halo, with the atom; we assert that the atom in reality is merely the diffraction phenomenon of an electron wave captured as it were by the nucleus of the atom.* It is no longer a matter of chance that the size of the atom and the wavelength are of the same order of magnitude: it is a matter of course. We know the numerical value of neither, because we still have in our calculation the *one* unknown constant, which we called *a.* There are two possible ways of determining it, which provide a mutual check on one another. First, we can so select it that the manifestations of life of the atom, above all the spectrum lines emitted, come out correctly quantitatively; these can after all be measured very accurately. Secondly, we can select *a* in a manner such that the diffraction halo acquires the size required for the atom. These two determinations of *a* (of which the second is admittedly far more imprecise because "size of the atom" is no clearly defined term) *are in complete agreement with one another.* Thirdly, and lastly, we can remark that the constant remaining unknown, physically speaking, does not in fact have the dimension of a length, but of an action, i.e. energy × time. It is then an obvious step to substitute for it the numerical value of Planck's universal quantum of action, which is accurately known from the laws of heat radiation. It will be seen that *we return,* with the full, now considerable accuracy, *to the first* (most accurate) *determination.*

Quantitatively speaking, the theory therefore manages with a minimum of new assumptions. It contains a single available constant, to which a numerical value familiar from the older quantum theory must be given, first to attribute to the diffraction halos the right size so that they can be reasonably identified with the atoms, and secondly, to evaluate

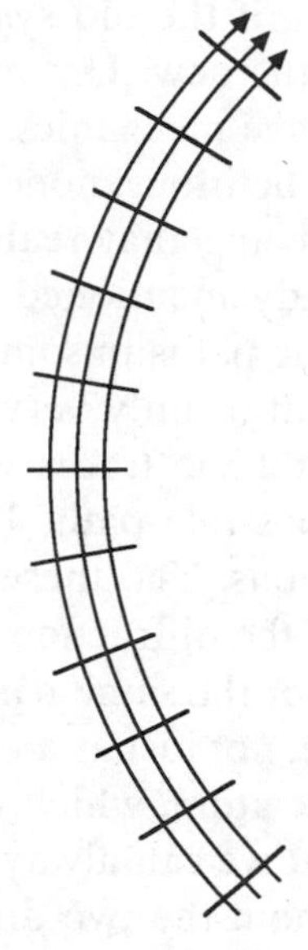

FIG. 7

quantitatively and correctly all the manifestations of life of the atom, the light radiated by it, the ionization energy, etc.

I have tried to place before you the fundamental idea of the wave theory of matter in the simplest possible form. I must admit now that in my desire not to tangle the ideas from the very beginning, I have painted the lily. Not as regards the high degree to which all sufficiently, carefully drawn conclusions are confirmed by experience, but with regard to the conceptual ease and simplicity with which the conclusions are reached. I am not speaking here of the mathematical difficulties, which always turn out to be trivial in the end, but of the conceptual difficulties. It is, of course, easy to say that we turn from the concept of a *curved path* to a system of wave surfaces normal to it. The wave surfaces, however, even if we consider only small parts of them (see Fig. 7) include at least a narrow *bundle* of possible curved paths, to all of which they stand in the same relationship. According to the old view, but not according to the new, one of them in each concrete individual case is distinguished from all the others which are "only possible," as that "really travelled." We are faced here with the full force of the logical opposition between an

either–or (point mechanics)

and a

both–and (wave mechanics)

This would not matter much, if the old system were to be dropped entirely and to be *replaced* by the new. Unfortunately, this is not the case. From the point of view of wave mechanics, the infinite array of possible point paths would be merely fictitious, none of them would have the prerogative over the others of being that really travelled in an individual case. I have, however, already mentioned that we have yet really observed such individual particle paths in some cases. The wave theory can represent this, either not at all or only very imperfectly. We find it confoundedly difficult to interpret the traces we *see* as nothing more than narrow bundles of equally possible paths between which the wave surfaces establish cross-connections. Yet, these cross-connections are necessary for an understanding of the diffraction and interference phenomena which can be demonstrated for the same particle with the same plausibility—and that on a large scale, not just as a consequence of the theoretical ideas about the interior of the atom, which we mentioned earlier. Conditions are admittedly such that we can always manage to make do in each concrete individual case without the two different aspects leading to different expectations as to the result of certain experiments. We cannot, however, manage to make do with such old, familiar, and seemingly indispensable terms as "real" or "only possible"; we are never in a position to say what really *is* or what really *happens,* but we can only say what will be *observed* in any concrete individual case. Will we have to be permanently satisfied with this . . .? On principle, yes. On principle, there is nothing new in the postulate that in the end exact science should aim at nothing more than the description of what can really be observed. The question is only whether from now on we shall have to refrain from tying description to a clear hypothesis about the real nature of the world. There are many who wish to pronounce such abdication even today. But I believe that this means making things a little too easy for oneself.

I would define the present state of our knowledge as follows. The ray or the particle path corresponds to a *longitudinal* relationship of the propagation process (i.e. *in the direction* of propagation), the wave surface on the other hand to a *transversal* relationship (i.e. *normal* to it). *Both* relationships are without doubt real; one is proved by photographed particle paths, the other by interference experiments. To combine both in a uniform system has proved impossible so far. Only in extreme cases does either the transversal, shell-shaped or the radial, longitudinal relationship predominate to such an extent that we *think* we can make do with the wave theory alone or with the particle theory alone.

N.5

The Development of Quantum Mechanics

WERNER HEISENBERG

WERNER HEISENBERG, discoverer of the famous uncertainty principle, was born in Würzburg, Germany, in 1901. His father was a history professor at the University of Munich. Young Heisenberg studied with Arnold Sommerfeld at the University of Munich, taking his degree in 1923. He served as an assistant to Max Born at Göttingen and spent time with Niels Bohr in Copenhagen; thus he was thrust into the forefront of the nascent discipline of quantum physics. The 1920s was the decade during which the concept of the atom as developed by Rutherford and Bohr was being refined. Physicists were attacking the problem by treating electrons as electromagnetically interacting particles whose dynamics were governed by three integral quantum numbers for which some empirical rules had been established.

Heisenberg devised a new strategy. His predecessors had worked on the atom in an attempt to interpret the spectral lines, or rather their positions in the spectra. Why not, he reasoned, use the lines as the point of departure and devise a mathematical model to explain their relationships? To do this task, Heisenberg, at Max Born's suggestion, revived a forgotten mathematical device known as matrix mathematics. Heisenberg's efforts led him to the same conclusion reached independently by Schrödinger even though both physicists had used completely different means. (A distinctive feature of matrix mathematics, incidentally, is that although the commutative law holds for addition: $a + b = b + a$, the same does not hold true for multiplication: $a \times b \neq b \times a$.) In applying matrix mathematics to atomic structure, Heisenberg represented each dynamical variable of an electron—e.g., its energy, its momentum, its position and its angular momentum—by a matrix. Each term representing a dynamical variable in the classical equations of motion of the electron was then replaced by its appropriate matrix and this set of equations was solved. The matrix for the electron's energy was then constructed and its diagonal matrix elements divided by Planck's constant gave the frequencies of the spectral lines. Born and Heisenberg brought in a third collaborator, a brilliant young physicist named Pascual Jordan, and the trio produced, in less than a month, one of the major innovations in the study of quantum mechanics.

The uncertainty principle, the doctrine that inspired Einstein's remark that he could not believe that God played dice with the Universe, was announced in 1927. The simplest statement of the principle can be expressed as follows: It is not possible to determine (measure or know) simultaneously the position and momentum of a particle: the more precise the measurement of position, for example, the less precise the knowledge of its momentum and vice versa. That is, measuring its position alters its momentum unpredictably. It is something like a situation in the normal world, where to comment "How silent it is here!" is to destroy the silence.

Heisenberg showed that since, according to classical optics, the precise measurement of the position of an electron with a microscope required the use of light of infinitely short wavelength (infinitely high frequency), the electron's momentum is altered by an infinite amount (actually by an unknown amount) because the photon that is reflected from the electron into the microscope imparts an infinite momentum to the electron. If we do not measure the electron's position, we can determine its momentum, but then we no longer know its position since it must be observed over a finite distance. This example illustrates the usefulness of the wave-particle duality in explaining the behavior of subatomic phenomena. However, the example also underscores the central theme of the uncertainty principle: As we venture further into the subatomic world, our vision becomes less clear because there are limits beyond which we cannot measure accurately. Any attempt to clarify one aspect of the observed phenomenon, the position of an electron, only blurs another aspect of the system, the energy level of the electron. Consequently, we can no longer pursue the absolute truths of classical physics but must accept the tradeoffs that are part of the subatomic world. The contrast between quantum theory and the strict causal laws of classical physics, which presumed the complete separation of object and observer, is evident.

When Heisenberg was awarded the Nobel Prize in 1932, he chose as his Nobel lecture a general history of the development of quantum mechanics. The following selection has been excerpted from his speech. He was later appointed professor at the University of Berlin in 1941 and director of the Kaiser Wilhelm Institute for Physics. At the end of World War II Heisenberg was taken prisoner by American troops and sent to England for a brief period. He then returned to Germany to reorganize with his colleagues what was to be named the Max Planck Institute for Physics at Göttingen. In 1954–55, he gave the Gifford Lectures at St. Andrews University in Scotland, which eventually appeared in book form. From 1957 until his death in 1976, Heisenberg worked on problems of plasma physics and thermonuclear processes and worked with the International Institute of Atomic Physics at Geneva. His theoretical work during the latter years of his life centered on devising a unified field theory of elementary particles. He also found time to write a series of lively nontechnical books that explored many areas of modern physics.

The ingenious but nevertheless somewhat artificial assumptions of [Bohr's model of the atom] . . . are replaced by a much more natural assumption in de Broglie's wave phenomena. The wave phenomena forms the real "body" of the atom. It replaces the individual punctiform electrons, which in Bohr's model swarm around the nucleus.

—ERWIN SCHRÖDINGER, "Images of Matter," *On Modern Physics*

If we ask, for instance, whether the position of the electron remains the same, we must say "no"; if we ask whether the electron's position changes with time, we must say "no"; if we ask whether it is in motion, we must say "no."

—J. ROBERT OPPENHEIMER, *Science and the Common Understanding*

QUANTUM MECHANICS AROSE, in its formal content, from the endeavour to expand Bohr's principle of correspondence to a complete mathematical scheme by refining his assertions. The physically new viewpoints that distinguish quantum mechanics from classical physics were prepared by the researches of various investigators engaged in analysing the difficulties posed in Bohr's theory of atomic structure and in the radiation theory of light.

In 1900, through studying the law of black-body radiation which he had discovered, Planck had detected in optical phenomena a discontinuous phenomenon totally unknown to classical physics which, a few years later, was most precisely expressed in Einstein's hypothesis of light quanta. The impossibility of harmonizing the Maxwellian theory with the pronouncedly visual concepts expressed in the hypothesis of light quanta subsequently compelled research workers to the conclusion that radiation phenomena can only be understood by largely renouncing their immediate visualization. The fact, already found by Planck and used by Einstein, Debye, and others, that the element of discontinuity detected in radiation phenomena also plays an important part in material processes, was expressed systematically in Bohr's basic postulates of the quantum theory

---

Werner Heisenberg, Nobel Prize in Physics Award Address, 1932. Reprinted with permission of Elsevier Publishing Co., and the Nobel Foundation.

which, together with the Bohr-Sommerfeld quantum conditions of atomic structure, led to a qualitative interpretation of the chemical and optical properties of atoms. The acceptance of these basic postulates of the quantum theory contrasted uncompromisingly with the application of classical mechanics to atomic systems, which, however, at least in its qualitative affirmations, appeared indispensable for understanding the properties of atoms. This circumstance was a fresh argument in support of the assumption that the natural phenomena in which Planck's constant plays an important part can be understood only by largely forgoing a visual description of them. Classical physics seemed the limiting case of visualization of a fundamentally unvisualizable microphysics, the more accurately realizable the more Planck's constant vanishes relative to the parameters of the system. This view of classical mechanics as a limiting case of quantum mechanics also gave rise to Bohr's principle of correspondence which, at least in qualitative terms, transferred a number of conclusions formulated in classical mechanics to quantum mechanics. In connection with the principle of correspondence there was also discussion whether the quantum-mechanical laws could in principle be of a statistical nature; the possibility became particularly apparent in Einstein's derivation of Planck's law of radiation. Finally, the analysis of the relation between radiation theory and atomic theory by Bohr, Kramers, and Slater resulted in the following scientific situation:

According to the basic postulates of the quantum theory, an atomic system is capable of assuming discrete, stationary states, and therefore discrete energy values; in terms of the energy of the atom the emission and absorption of light by such a system occurs abruptly, in the form of impulses. On the other hand, the visualizable properties of the emitted radiation are described by a wave field, the frequency of which is associated with the difference in energy between the initial and final states of the atom by the relation

$$E_1 - E_2 = h\nu$$

To each stationary state of an atom corresponds a whole complex of parameters which specify the probability of transition from this state to another. There is no direct relation between the radiation classically emitted by an orbiting electron and those parameters defining the probability of emission; nevertheless Bohr's principle of correspondence enables a specific term of the Fourier expansion of the classical path to be assigned to each transition of the atom, and the probability for the particular transition follows qualitatively similar laws as the intensity of those Fourier components. Although therefore in the researches carried out by Rutherford, Bohr, Sommerfeld, and others, the comparison of the atom with a planetary system of electrons leads to a qualitative interpretation of the optical and chemical properties of atoms, nevertheless the fundamental dissimilarity between the atomic spectrum and the classical spectrum of an electron system imposes the need to relinquish the concept of an electron path and to forgo a visual description of the atom.

The experiments necessary to define the electron-path concept also furnish an important aid in revising it. The most obvious answer to the question how the orbit of an electron in its path within the atom could be observed namely, will perhaps be to use a microscope of extreme resolving power. But since the specimen in this microscope would have to be illuminated with light having an extremely short wavelength, the first light quantum from the light source to reach the electron and pass into the observer's eye would eject the electron completely from its path in accordance with the laws of the Compton effect. Consequently only one point of the path would be observable experimentally at any one time.

In this situation, therefore, the obvious policy was to relinquish at first the concept of electron paths altogether, despite its substantiation by Wilson's experiments, and, as it were, to attempt subsequently how much of the electron-path concept can be carried over into quantum mechanics.

In the classical theory the specification of frequency, amplitude, and phase of all the light waves emitted by the atom would be fully equivalent to specifying its electron path. Since from the amplitude and phase of an emitted wave the coefficients of the approprate term in the Fourier expansion of the electron path can be derived without ambiguity, the complete electron path therefore can be derived from a knowledge of all amplitudes and phases. Similarly, in quantum mechanics, too, the whole complex of amplitudes and phases of the radiation emitted by the atom can be regarded as a complete description of the atomic system, although its interpretation in the sense of an electron path inducing the radiation is impossible. In quantum mechanics, therefore, the place of the electron coordinates is taken by a complex of parameters corresponding to the Fourier coefficients of classical motion along a path. These, however, are no longer classified by the energy of state and the number of the corresponding harmonic vibration, but are in each case associated with two stationary states of the atom, and are a measure for the transition probability of the atom from one stationary state to another. A complex of coefficients of this type is comparable with a matrix such as occurs in linear algebra. In exactly the same way each parameter of classical mechanics, e.g. the momentum or the energy of the electrons, can then be assigned a corresponding matrix in quantum mechanics. To proceed from here beyond a mere description of the empirical state of affairs it was necessary to associate systematically the matrices assigned to the various parameters in the same way as the corresponding parameters in classical mechanics are associated by equations of motions. When, in the interest of achieving the closest possible correspondence between classical and quantum mechanics, the addition and multiplication of Fourier series were tentatively taken as the example for the addition and multiplication of the quantum-theory complexes, the product of two parameters represented by matrices appeared to be most naturally represented by the product matrix in the sense of linear algebra—an assumption already suggested by the formalism of the Kramers-Ladenburg dispersion theory.

It thus seemed consistent simply to adopt in quantum mechanics the equations of motion of classical physics, regarding them as a relation between the matrices representing the classical variables. The Bohr-Sommerfeld quantum conditions could also be re-interpreted in a relation between the matrices, and together with the equations of motion they were sufficient to define all matrices and hence the experimentally observable properties of the atom.

Born, Jordan, and Dirac deserve the credit for expanding the mathematical scheme outlined above into a consistent and practically usable theory. These investigators observed in the first place that the quantum conditions can be written as commutation relations between the matrices representing the momenta and the coordinates of the electrons, to yield the equations (p_r, momentum matrices; q_r, coordinate matrices):

$$p_r q_s - q_s p_r = \frac{h}{2\pi i}\delta_{rs} \qquad q_r q_s - q_s q_r = 0 \qquad p_r p_s - p_s p_r = 0$$

$$\partial_{rs} = \begin{cases} 1 \text{ for } r = s \\ 0 \text{ for } r \neq s \end{cases}$$

By means of these commutation relations they were able to detect in quantum mechanics as well the laws which were fundamental to classical mechanics: the invariability of time of energy, momentum, and angular momentum.

The mathematical scheme so derived thus ultimately bears an extensive formal similarity to that of the classical theory, from which it differs outwardly by the commutation relations which, moreover, enabled the equations of motion to be derived from the Hamiltonian function.

In the physical consequences, however, there are very profound differences between quantum mechanics and classical mechanics which impose the need for a thorough discussion of the physical interpretation of quantum mechanics. As hitherto defined, quantum mechanics enables the radiation emitted by the atom, the energy values of the stationary states, and other parameters characteristic for the stationary states to be treated. The theory hence complies with the experimental data contained in atomic spectra. In all those cases, however, where a visual description is required of a transient event, e.g. when interpreting Wilson photographs, the formalism of the theory does not seem to allow an adequate representation of the experimental state of affairs. At this point Schrödinger's wave mechanics, meanwhile developed on the basis of de Broglie's theses, came to the assistance of quantum mechanics.

In the course of the studies which Mr. Schrödinger will report here himself he converted the determination of the energy values of an atom into an eigenvalue problem defined by a boundary-value problem in the coordinate space of the particular atomic system. After Schrödinger had shown the mathematical equivalence of wave mechanics, which he had discovered, with quantum mechanics, the fruitful combination of these two different areas of physical ideas resulted in an extraordinary broadening and enrichment of the formalism of the quantum theory. Firstly it was only wave mechanics which made possible the mathematical treatment of complex atomic systems, secondly analysis of the connection between the two theories led to what is known as the transformation theory developed by Dirac and Jordan. As it is impossible within the limits of the present lecture to give a detailed discussion of the mathematical structure of this theory, I should just like to point out its fundamental physical significance. Through the adoption of the physical principles of quantum mechanics into its expanded formalism, the transformation theory made it possible in completely general terms to calculate for atomic systems the probability for the occurrence of a particular, experimentally ascertainable, phenomenon under given experimental conditions. The hypothesis conjectured in the studies on the radiation theory and enunciated in precise terms in Born's collision theory, namely that the wave function governs the probability for the presence of a corpuscle, appeared to be a special case of a more general pattern of laws and to be a natural consequence of the fundamental assumptions of quantum mechanics. Schrödinger, and in later studies Jordan, Klein, and Wigner as well, had succeeded in developing as far as permitted by the principles of the quantum theory de Broglie's original concept of visualizable matter waves occurring in space and time, a concept formulated even before the development of quantum mechanics. But for that the connection between Schrödinger's concepts and de Broglie's original thesis would certainly have seemed a looser one by this statistical interpretation of wave mechanics and by the greater emphasis on the fact that Schrödinger's theory is concerned with waves in multidimensional space. Before proceeding to discuss the explicit significance of quantum mechanics it is perhaps right for me to deal briefly with this question as to the existence of matter waves in three-dimensional space, since the solution to this problem was only achieved by combining wave and quantum mechanics.

A long time before quantum mechanics was developed Pauli had inferred from the laws in the Periodic System of the elements the well-known principle that a particular quantum state can at all times be occupied by only a single electron. It proved possible to transfer this principle to quantum mechanics on the basis of what at first sight seemed a surprising result: the entire complex of stationary states which an atomic system is capable of adopting breaks down into definite classes such that an atom in a state belonging to one class can never change into a state belonging to another class under the action of whatever perturbations. As finally clarified beyond question by the studies of Wigner and Hund, such a class of states is characterized by a definite symmetry characteristic of the Schrödinger eigenfunction with respect to the transposition of the coordinates of two electrons. Owing to the fundamental identity of electrons, any external perturbation of the atom remains unchanged when two electrons are exchanged and hence causes no transitions between states of various classes. The Pauli principle and the Fermi-Dirac statistics derived from it are equivalent with the assumption that only that class of stationary states is achieved in nature in which the eigenfunction changes its sign when two electrons are exchanged. According to Dirac, selecting the symmetrical system of terms would lead not to the Pauli principle, but to Bose-Einstein electron statistics.

Between the classes of stationary states belonging to the Pauli principle or to Bose-Einstein statistics, and de Broglie's concept of matter waves there is a peculiar relation. A spatial wave phenomenon can be treated according to the principles of the quantum theory by analysing it using the Fourier theorem and then applying to the individual Fourier component of the wave motion, as a system having one degree of freedom, the normal laws of quantum mechanics. Applying this procedure for treating wave phenomena by the quantum theory, a procedure that has also proved fruitful in Dirac's studies of the theory of radiation, to de Broglie's matter waves, exactly the same results are obtained as in treating a whole complex of material particles according to quantum mechanics and selecting the symmetrical system of terms. Jordan and Klein hold that the two methods are mathematically equivalent even if allowance is also made for the interaction of the electrons, i.e. if the field energy originating from the continuous space charge is included in the calculation in de Broglie's wave theory. Schrödinger's considerations of the energy-momentum tensor assigned to the matter waves can then also be adopted in this theory as consistent components of the formalism. The studies of Jordan and Wigner show that modifying the commutation relations underlying this quantum theory of waves results in a formalism equivalent to that of quantum mechanics based on the assumption of Pauli's exclusion principle.

These studies have established that the comparison of an atom with a planetary system composed of nucleus and electrons is not the only visual picture of how we can imagine the atom. On the contrary, it is apparently no less correct to compare the atom with a charge cloud and use the correspondence to the formalism of the quantum theory borne by this concept to derive qualitative conclusions about the behaviour of the atom. However, it is the concern of wave mechanics to follow these consequences.

Reverting therefore to the formalism of quantum mechanics; its application to physical problems is justified partly by the original basic assumptions of the theory, partly by its expansion in the transformation theory on the basis of wave mechanics, and the question is now to expose the explicit significance of the theory by comparing it with classical physics.

In classical physics the aim of research was to investigate objective processes occurring in space and time, and to discover the laws governing their progress from the initial conditions. In classical physics a problem was considered solved when a particular phenomenon had been proved to occur objectively in space and time, and it had been shown to obey the general rules of classical physics as formulated by differential equations. The manner in which the knowledge of each process had been acquired, what observations may possibly have led to its experimental determination, was completely immaterial, and it was also immaterial for the consequences of the classical theory, which possible observations were to verify the predictions of the theory. In the quantum theory, however, the situation is completely different. The very fact that the formalism of quantum mechanics cannot be interpreted as visual description of a phenomenon occurring in space and time shows that quantum mechanics is in no way concerned with the objective determination of space-time phenomena. On the contrary, the formalism of quantum mechanics should be used in such a way that the probability for the outcome of a further experiment may be concluded from the determination of an experimental situation in an atomic system, providing that the system is subject to no perturbations other than those necessitated by performing the two experiments. The fact that the only definite known result to be ascertained after the fullest possible experimental investigation of the system is the probability for a certain outcome of a second experiment shows, however, that each observation must entail a discontinuous change in the formalism describing the atomic process and therefore also a discontinuous change in the physical phenomenon itself. Whereas in the classical theory the kind of observation has no bearing on the event, in the quantum theory the disturbance associated with each observation of the atomic phenomenon has a decisive role. Since, furthermore, the result of an observation as a rule leads only to assertions about the probability of certain results of subsequent observations, the fundamentally unverifiable part of each perturbation must, as shown by Bohr, be decisive for the non-contradictory operation of quantum mechanics. This difference between classical and atomic physics is understandable, of course, since for heavy bodies such as the planets moving around the sun the pressure of the sunlight which is reflected at their surface and which is necessary for them to be observed is negligible; for the smallest building units of matter, however, owing to their low mass, every observation has a decisive effect on their physical behaviour.

The perturbation of the system to be observed caused by the observation is also an important factor in determining the limits within which a visual description of atomic phenomena is possible. If there were experiments which permitted accurate measurement of all the characteristics of an atomic system necessary to calculate classical motion, and which, for example, supplied accurate values for the location and velocity of each electron in the system at a particular time, the result of these experiments could not be utilized at all in the formalism, but rather it would directly contradict the formalism. Again, therefore, it is clearly that fundamentally unverifiable part of the perturbation of the system caused by the measurement itself which hampers accurate ascertainment of the classical characteristics and thus permits quantum mechanics to be applied. Closer examination of the formalism shows that between the accuracy with which the location of a particle can be ascertained and the accuracy with which its momentum can simultaneously be known, there is a relation according to which the product of the probable errors in the measurement of the location and momentum is invariably at least as large as Planck's constant divided by 4π. In a very general form, therefore, we should have

$$\Delta p \, \Delta q \geqslant \frac{h}{4\pi}$$

where p and q are canonically conjugated variables. These uncertainty relations for the results of the measurement of classical variables form the necessary conditions for enabling the result of a measurement to be expressed in the formalism of the quantum theory. Bohr has shown in a series of examples how the perturbation necessarily associated with each observation indeed ensures that one cannot go below the limit set by the uncertainty relations. He contends that in the final analysis an uncertainty introduced by the concept of measurement itself is responsible for part of that perturbation remaining fundamentally unknown. The experimental determination of whatever space-time events invariably necessitates a fixed frame—say the system of coordinates in which the observer is at rest—to which all measurements are referred. The assumption that this frame is "fixed" implies neglecting its momentum from the outset, since "fixed" implies nothing other, of course, than that any transfer of momentum to it will evoke no perceptible effect. The fundamentally necessary uncertainty at this point is then transmitted via the measuring apparatus into the atomic event.

Since in connection with this situation it is tempting to consider the possibility of eliminating all uncertainties by amalgamating the object, the measuring apparatuses, and the observer into one quantum-mechanical system, it is important to emphasize that the act of measurement is necessarily visualizable, since, of course, physics is ultimately only concerned with the systematic description of space-time processes. The behaviour of the observer as well as his measuring apparatus must therefore be discussed according to the laws of classical physics, as otherwise there is no further physical problem whatsoever. Within the measuring apparatus, as emphasized by Bohr, all events in the sense of the classical theory will therefore be regarded as determined, this also being a necessary condition before one can, from a result of measurements, unequivocally conclude what has happened. In quantum theory, too, the scheme of classical physics which objectifies the results of observation by assuming in space and time processes obeying laws is thus carried through up to the point where the fundamental limits are imposed by the unvisualizable character of the atomic events symbolized by Planck's constant. A visual description for the atomic events is possible only within certain limits of accuracy—but within these limits the laws of classical physics also still apply. Owing to these limits of accuracy as defined by the uncertainty relations, moreover, a visual picture of the atom free from ambiguity has not been determined. On the contrary the corpuscular and the wave concepts are equally serviceable as a basis for visual interpretation.

The laws of quantum mechanics are basically statistical. Although the parameters of an atomic system are determined in their entirety by an experiment, the result of a future observation of the system is not generally accurately predictable. But at any later point of time there are observations which yield accurately predictable results. For the other observations only the probability for a particular outcome of the experiment can be given. The degree of certainty which still attaches to the laws of quantum mechanics is, for example, responsible for the fact that the principles of conservation for energy and momentum still hold as strictly as ever. They can be checked with any desired accuracy and will then be valid according to the accuracy with which they are checked. The statistical character of the laws of quantum mechanics, however, becomes apparent in that an accurate study of the energetic conditions renders it impossible to pursue at the same time a particular event in space and time.

For the clearest analysis of the conceptual principles of quantum mechanics we are indebted to Bohr who, in particular, applied the concept of complementarity to interpret the validity of the quantum-mechanical laws. The uncertainty relations alone afford an instance of how in quantum mechanics the exact knowledge of one variable can exclude the exact knowledge of another. This complementary relationship between different aspects of one and the same physical process is indeed characteristic for the whole structure of quantum mechanics. I had just mentioned that, for example, the determination of energetic relations excludes the detailed description of space-time processes. Similarly, the study of the chemical properties of a molecule is complementary to the study of the motions of the individual electrons in the molecule, or the observation of interference phenomena is complementary to the observation of individual light quanta. Finally, the areas of validity of classical and quantum mechanics can be marked off one from the other as follows: Classical physics represents that striving to learn about Nature in which essentially we seek to draw conclusions about objective processes from observations and so ignore the consideration of the influences which every observation has on the object to be observed; classical physics, therefore, has its limits at the point from which the influence of the observation on the event can no longer be ignored. Conversely, quantum mechanics makes possible the treatment of atomic processes by partially forgoing their space-time description and objectification.

So as not to dwell on assertions in excessively abstract terms about the interpretation of quantum mechanics, I would like briefly to explain with a well-known example how far it is possible through the atomic theory to achieve an understanding of the visual processes with which we are concerned in daily life. The interest of research workers has frequently been focused on the phenomenon of regularly shaped crystals suddenly forming from a liquid, e.g. a supersaturated salt solution. According to the atomic theory the forming force in this process is to a certain extent the symmetry characteristic of the solution to Schrödinger's wave equation, and to that extent crystallization is explained by the atomic theory. Nevertheless this process retains a statistical and—one might almost say—historical element which cannot be further reduced: even when the state of the liquid is completely known before crystallization, the shape of the crystal is not determined by the laws of quantum mechanics. The formation of regular shapes is just far more probable than that of a shapeless lump. But the ultimate shape owes its genesis partly to an element of chance which in principle cannot be analysed further.

Before closing this report on quantum mechanics, I may perhaps be allowed to discuss very briefly the hopes that may be attached to the further development of this branch of research. It would be superfluous to mention that the development must be continued, based equally on the studies of de Broglie, Schrödinger, Born, Jordan, and Dirac. Here the attention of the research workers is primarily directed to the problem of reconciling the claims of the special relativity theory with those of the quantum theory. The extraordinary advances made in this field by Dirac about which Mr. Dirac will speak here, meanwhile leave open the question whether it will be possible to satisfy the claims of the two theories without at the same time determining the Sommerfeld fine-structure constant. The attempts made hitherto to achieve a relativistic formulation of the quantum theory are all based on visual concepts so close to those of classical physics that it seems impossible to determine the fine-structure constant within this system of concepts. The expansion of the conceptual system under discussion here should, furthermore, be closely associated with the further development of the quantum theory of wave fields, and it appears to me as if this formalism, notwithstanding its thorough study by a number of workers (Dirac, Pauli, Jordan, Klein, Wigner, Fermi) has still not been completely exhausted. Important pointers for the further development of quantum mechanics also emerge from the experiments involving the structure of the atomic nuclei. From their analysis by means of the Gamow theory, it would appear that between the elementary particles of the atomic nucleus forces are at work which differ somewhat in type from the forces determining the structure of the atomic shell; Stern's experiments seem, furthermore, to indicate that the behaviour of the heavy elementary particles cannot be represented by the formalism of Dirac's theory of the electron. Future research will thus have to be prepared for surprises which may otherwise come both from the field of experience of nuclear physics as well as from that of cosmic radiation. But however the development proceeds in detail, the path so far traced by the quantum theory indicates that an understanding of those still unclarified features of atomic physics can only be acquired by forgoing visualization and objectification to an extent greater than that customary hitherto. We have probably no reason to regret this, because the thought of the great epistemological difficulties with which the visual atom concept of earlier physics had to contend gives us the hope that the abstracter atomic physics developing at present will one day fit more harmoniously into the great edifice of Science.

N.6

The Statistical Interpretation of Quantum Mechanics

Max Born

Max Born, the son of a professor of anatomy, was born in 1882 in Breslau, a town now part of Poland and known as Wroclaw. He earned his doctorate at Göttingen in 1907 and did further graduate work with J. J. Thomson at Cambridge. His subsequent teaching career took him to Chicago in 1912, at the invitation of Albert Michelson, and then to the Universities of Berlin and Göttingen. The rise of Nazism in the 1930s forced Born, like his colleague Schrödinger, to leave Germany for Great Britain. Heisenberg, by contrast, a staunch nationalist, chose to remain in Germany and continued his work at the University of Berlin for the duration of World War II.

Born, along with Dirac, Schrödinger, and Heisenberg, helped to fashion the mathematical basis for quantum mechanics. Although Schrödinger's wave equation had given a mathematical description of the motion of electrons in atoms, it struck Born as being somewhat arbitrary. Schrödinger had suggested that matter is made up of waves; he theorized that electrons are vibrating clouds—not discrete particles—and that the radiation of electrons results from vibrations associated with the electron waves. Born was the first to point out that Schrödinger's cloud model was inherently unstable because the undulatory properties of the wave packets, as hypothesized, would cause the clouds to dissipate.

In considering the question of de Broglie's matter waves, Born turned to optics for the analogy that the intensity of light at a given point can be considered an index to the probability of a photon's presence. Born hypothesized that the square of the absolute value of the matter-wave amplitude of a particle at a point is a measure of the probability of the particle's presence. The new wave interpretation of subatomic phenomena offered by Schrödinger was thought by Born to be a misnomer as the waves were not actually real but were instead probability waves. Born wrote that "the whole course of events is determined by the laws of probability; to a state in space there corresponds a definite probability, which is given by the de Broglie wave associated with the state." Physics was, for Born, "in the nature of the case indeterminate, and therefore the affair of statistics."

Born was belatedly awarded the Nobel Prize in 1954 for his contributions to quantum mechanics. His Nobel lecture, entitled "The Statistical Interpretation of Quantum Mechanics," provides a brief history of the origins of quantum theory, and then outlines the contributions made by Born and his Göttingen colleagues, Heisenberg and Jordan.

[T]he whole course of events is determined by the laws of probability; to a state in space there corresponds a definite probability, which is given by the de Broglie wave associated with the state.

—MAX BORN, *Atomic Physics*

In a probabilistic world we no longer deal with quantities and statements which concern a specific, real universe as a whole but ask instead questions which may find their answers in a large number of similar universes. Thus chance has been admitted, not merely as a mathematical tool for physics, but as part of its warp and cleft.

—NORBERT WIENER, *The Human Use of Human Beings*

[*The Many Worlds Interpretation of Quantum Mechanics reveals a universe that*] *is constantly splitting into a stupendous number of branches, all resulting from the measurement like interactions between its myriad of components. Moreover, every quantum transition taking place on every star, in every galaxy, in every remote corner of the universe is splitting our local world into myriads of copies of itself.*

—BRYCE S. DEWITT, "Quantum Mechanics and Reality," *Physics Today*

THE WORK I shall discuss here contains no discovery of a fresh natural phenomenon, but rather the basis for a new mode of thought in regard to natural phenomena. This way of thinking has permeated both experimental and theoretical physics to such a degree that it hardly seems possible to say anything more about it that has not been already so often said. However, there are some particular aspects which I should like to discuss. . . . The first point is this: the work at the Göttingen school, which I directed at that time (1926–1927), contributed to the solution of an intellectual crisis into which our science had fallen as a result of Planck's discovery of the quantum of action in 1900. Today, physics finds itself in a similar crisis—I do not mean here its entanglement in politics and economics as a result of the mastery of a new and frightful force of Nature,

---

Max Born, Nobel Prize in Physics Award Address, 1954. Reprinted with permission of Elsevier Publishing Co., and the Nobel Foundation.

but I am considering more the logical and epistemological problems posed by nuclear physics. Perhaps it is well at such a time to recall what took place earlier in a similar situation, especially as these events are not without a definite dramatic flavour.

The second point I wish to make is that when I say that the physicists had accepted the concepts and mode of thought developed by us at the time, I am not quite correct. There are some very noteworthy exceptions, particularly among the very workers who have contributed most to building up the quantum theory. Planck, himself, belonged to the sceptics until he died. Einstein, de Broglie, and Schrödinger have unceasingly stressed the unsatisfactory features of quantum mechanics and called for a return to the concepts of classical, Newtonian physics while proposing ways in which this could be done without contradicting experimental facts. Such weighty views cannot be ignored. Niels Bohr has gone to a great deal of trouble to refute the objections. I, too, have ruminated upon them and believe I can make some contribution to the clarification of the position. The matter concerns the borderland between physics and philosophy, and so my physics lecture will partake of both history and philosophy, for which I must crave your indulgence.

First of all, I will explain how quantum mechanics and its statistical interpretation arose. At the beginning of the twenties, every physicist, I think, was convinced that Planck's quantum hypothesis was correct. According to this theory *energy* appears in finite quanta of magnitude $h\nu$ in oscillatory processes having a specific frequency ν (e.g. in light waves). Countless experiments could be explained in this way and always gave the same value of Planck's constant h. Again, Einstein's assertion that light quanta have *momentum* $h\nu/c$ (where c is the speed of light) was well supported by experiment (e.g. through the Compton effect). This implied a revival of the corpuscular theory of light for a certain complex of phenomena. The wave theory still held good for other processes. Physicists grew accustomed to this *duality* and learned how to cope with it to a certain extent.

In 1913 Niels Bohr had solved the riddle of *line spectra* by means of the quantum theory and had thereby explained broadly the amazing stability of the atoms, the structure of their electronic shells, and the Periodic System of the elements. For what was to come later, the most important assumption of his teaching was this: an atomic system cannot exist in all mechanically possible states, forming a continuum, but in a series of discrete "stationary" states. In a transition from one to another, the difference in energy $E_m - E_n$ is emitted or absorbed as a light quantum $h\nu_{mn}$ (according to whether E_m is greater or less than E_n). This is an interpretation in terms of energy of the fundamental law of spectroscopy discovered some years before by W. Ritz. The situation can be taken in at a glance by writing the energy levels of the stationary states twice over, horizontally and vertically. This produces a square array

$$\begin{array}{ccccc} & E_1 & E_2 & E_3 & \cdots \\ E_1 & 11 & 12 & 13 & - \\ E_2 & 21 & 22 & 23 & - \\ E_3 & 31 & 32 & 33 & - \\ \vdots & - & - & - & - \end{array}$$

in which positions on a diagonal correspond to states, and non-diagonal positions correspond to transitions.

It was completely clear to Bohr that the law thus formulated is in conflict with mechanics, and that therefore the use of the energy concept in this connection is problematical. He based this daring fusion of old and new on his *principle of correspondence.* This consists in the obvious requirement that ordinary classical mechanics must hold to a high degree of approximation in the limiting case where the numbers of the stationary states, the so-called quantum numbers, are very large (that is to say, far to the right and to the lower part in the above array) and the energy changes relatively little from place to place, in fact practically continuously.

Theoretical physics maintained itself on this concept for the next ten years. The problem was this: an harmonic oscillation not only has a frequency, but also an intensity. For each transition in the array there must be a corresponding intensity. The question is how to find this through the considerations of correspondence? It meant guessing the unknown from the available information on a known limiting case. Considerable success was attained by Bohr himself, by Kramers, Sommerfeld, Epstein, and many others. But the decisive step was again taken by Einstein who, by a fresh derivation of Planck's radiation formula, made it transparently clear that the classical concept of intensity of radiation must be replaced by the statistical concept of *transition probability.* To each place in our pattern or array there belongs (together with the frequency $\nu_{mn} = (E_n - E_m)/h$) a definite probability for the transition coupled with emission or absorption.

In Göttingen we also took part in efforts to distil the unknown mechanics of the atom from the experimental results. The logical difficulty became ever sharper. Investigations into the scattering and dispersion of light showed that Einstein's conception of transition probability as a measure of the strength of an oscillation did not meet the case, and the idea of an *amplitude* of oscillation associated with each transition was indispensable. In this connection, work by Ladenburg, Kramers, Heisenberg, Jordan and me should be mentioned. The art of guessing correct formulae, which deviate from the classical formulae, yet contain them as a limiting case according to the correspondence principle, was brought to a high degree of perfection. A paper of mine, which introduced, for the first time I think, the expression *quantum mechanics* in its title, contains a rather involved formula (still valid today) for the reciprocal disturbance of atomic systems.

Heisenberg, who at that time was my assistant, brought this period to a sudden end. He cut the Gordian knot by means of a philosophical principle and replaced guess-work by a mathematical rule. The principle states that concepts and representations that do not correspond to physically observable facts are not to be used in theoretical description. Einstein used the same principle when, in setting up his theory of relativity, he eliminated the concepts of absolute velocity of a body and of absolute simultaneity of two events at different places. Heisenberg banished the picture of electron orbits with definite radii and periods of rotation because these quantities are not observable, and insisted that the theory be built up by means of the square arrays mentioned above. Instead of describing the motion by giving a co-ordinate as a function of time, $x(t)$, an array of transition amplitudes x_{mn} should be determined. To me the decisive part of his work is the demand to determine a rule by which from a given

$$\text{array} \begin{bmatrix} x_{11}\ x_{12} & \cdots\cdots \\ x_{21}\ x_{22} & \cdots\cdots \\ \cdots & \cdots \end{bmatrix} \text{ the array for the square } \begin{bmatrix} (x^2)_{11}\ (x^2)_{12} & \cdots\cdots \\ (x^2)_{21}\ (x^2)_{22} & \cdots\cdots \\ \cdots & \cdots \end{bmatrix}$$

can be found (or, more general, the *multiplication rule* for such arrays).

By observation of known examples solved by guess-work he found this rule and applied it successfully to simple examples such as the harmonic and anharmonic oscillator.

This was in the summer of 1925. Heisenberg, plagued by hay fever, took leave for a course of treatment by the sea and gave me his paper for publication if I thought I could do something with it.

The significance of the idea was at once clear to me and I sent the manuscript to the *Zeitschrift für Physik.* I could not take my mind off Heisenberg's multiplication rule, and after a week of intensive thought and trial I suddenly remembered an algebraic theory which I had learned from my teacher, Professor Rosanes, in Breslau. Such square arrays are well known to mathematicians and, in conjunction with a specific rule for multiplication, are called matrices. I applied this rule to Heisenberg's quantum condition and found that this agreed in the diagonal terms. It was easy to guess what the remaining quantities must be, namely, zero; and at once there stood before me the peculiar formula

$$pq - qp = h/2\pi i$$

This meant that coordinates q and momenta p cannot be represented by figure values but by symbols, the product of which depends upon the order of multiplication—they are said to be "non-commuting."

I was as excited by this result as a sailor would be who, after a long voyage, sees from afar, the longed-for land, and I felt regret that Heisenberg was not there. I was convinced from the start that we had stumbled on the right path. Even so, a great part was only guess-work, in particular, the disappearance of the non-diagonal elements in the above-mentioned expression. For help in this problem I obtained the assistance and collaboration of my pupil Pascual Jordan, and in a few days we were able to demonstrate that I had guessed correctly. The joint paper by Jordan and myself contains the most important principles of quantum mechanics including its extension to electrodynamics. There followed a hectic period of collaboration among the three of us, complicated by Heisenberg's absence. There was a lively exchange of letters; my contributions to these, unfortunately, have been lost in the political disorders. The result was a three-author paper which brought the formal side of the investigation to a definite conclusion. Before this paper appeared, came the *first dramatic surprise:* Paul Dirac's paper on the same subject. The inspiration afforded by a lecture of Heisenberg's in Cambridge had led him to similar results as we had obtained in Göttingen except that he did not resort to the known matrix theory of the mathematicians, but discovered the tool for himself and worked out the theory of such non-commutating symbols.

The first non-trivial and physically important application of quantum mechanics was made shortly afterwards by W. Pauli who calculated the stationary energy values of the *hydrogen atom* by means of the matrix method and found complete agreement with Bohr's formulae. From this moment onwards there could no longer be any doubt about the correctness of the theory.

What this formalism really signified was, however, by no means clear.

Mathematics, as often happens, was cleverer than interpretative thought. While we were still discussing this point there came the *second dramatic surprise,* the appearance of Schrödinger's famous papers. He took up quite a different line of thought which had originated from Louis de Broglie.

A few years previously, the latter had made the bold assertion, supported by brilliant theoretical considerations, that wave-corpuscle duality, familiar to physicists in the case of light, must also be valid for electrons. To each electron moving free of force belongs a plane wave of a definite wavelength which is determined by Planck's constant and the mass. This exciting dissertation by de Broglie was well known to us in Göttingen. One day in 1925 I received a letter from C. J. Davisson giving some peculiar results on the reflection of electrons from metallic surfaces. I, and my colleague on the experimental side, James Franck, at once suspected that these curves of Davisson's were crystal-lattice spectra of de Broglie's electron waves, and we made one of our pupils, Elsasser, investigate the matter. His result provided the first preliminary confirmation of the idea of de Broglie's, and this was later proved independently by Davisson and Germer and G. P. Thomson by systematic experiments.

But this acquaintance with de Broglie's way of thinking did not lead us to an attempt to apply it to the electronic structure in atoms. This was left to Schrödinger. He extended de Broglie's wave equation which referred to force-free motion, to the case where the effect of force is taken into account, and gave an exact formulation of the *subsidiary conditions,* already suggested by de Broglie, to which the wave function ψ must be subjected, namely that it should be single-valued and finite in space and time. And he was successful in deriving the stationary states of the hydrogen atom in the form of those monochromatic solutions of his wave equation which do not extend to infinity.

For a brief period at the beginning of 1926, it looked as though there were, suddenly, two self-contained but quite distinct systems of explanation extant: matrix mechanics and wave mechanics. But Schrödinger himself soon demonstrated their complete equivalence.

Wave mechanics enjoyed a very great deal more popularity than the Göttingen or Cambridge version of quantum mechanics. It operates with a wave function ψ, which in the case of *one* particle at least, can be pictured in space, and it uses the mathematical methods of partial differential equations which are in current use by physicists. Schrödinger thought that his wave theory made it possible to return to deterministic classical physics. He proposed (and he has recently emphasized his proposal anew), to dispense with the particle representation entirely, and instead of speaking of electrons as particles, to consider them as a continuous density distribution $|\psi^2|$ (or electric density $e|\psi|^2$).

To us in Göttingen this interpretation seemed unacceptable in face of well established experimental facts. At that time it was already possible to count particles by means of scintillations or with a Geiger counter, and to photograph their tracks with the aid of a Wilson cloud chamber. . . . Almost all experiments lead to statements about relative frequencies of events, even when they occur concealed under such names as effective cross section or the like.

How does it come about then, that great scientists such as Einstein, Schrödinger, and de Broglie are nevertheless dissatisfied with the situation? Of course, all these objections are levelled not against the correctness of the formulae, but against their interpretation. Two closely knitted points of view are to be distinguished: the question of determinism and the question of reality.

Newtonian mechanics is deterministic in the following sense:

If the initial state (positions and velocities of all particles) of a system is accurately given, then the state at any other time (earlier or later) can be calculated from the laws of mechanics. All the other branches of classical physics have been built up according to this model. Mechanical determinism gradually became a kind of article of faith: the world as a machine, an automaton. As far as I can see, this idea has no forerunners in ancient and medieval philosophy. The idea is a product of the immense success of Newtonian mechanics, particularly in astronomy. In the 19th century it became a basic philosophical principle for the whole of exact science. I asked myself whether this was really justified. Can absolute predictions really be made for all time on the basis of the classical equations of motion? It can easily be seen, by simple examples, that this is only the case when the possibility of absolutely exact measurement (of position, velocity, or other quantities) is assumed. Let us think of a particle moving without friction on a straight line between two end-points (walls), at which it experiences completely elastic recoil. It moves with constant speed equal to its initial speed v_o backwards and forwards, and it can be stated exactly where it will be at a given time provided that v_o is accurately known. But if a small inaccuracy Δv_o is allowed, then the inaccuracy of prediction of the position at time t is $t\Delta v_o$ which increases with t. If one waits long enough until time $t_c = l \neq \Delta v_o$ where l is the distance between the elastic walls, the inaccuracy Δx will have become equal to the whole space l. Thus it is impossible to forecast anything about the position at a time which is later than t_c. Thus determinism lapses completely into indeterminism as soon as the slightest inaccuracy in the data on velocity is permitted. Is there any sense—and I mean any physical sense, not metaphysical sense—in which one can speak of absolute data? Is one justified in saying that the coordinate $x = \pi$ cm where $\pi = 3.1415...$ is the familiar transcendental number that determines the ratio of the circumference of a circle to its diameter? As a mathematical tool the concept of a real number represented by a nonterminating decimal fraction is exceptionally important and fruitful. As the measure of a physical quantity it is nonsense. If π is taken to the 20th or the 25th place of decimals, two numbers are obtained which are indistinguishable from each other and the true value of π by any measurement. According to the heuristic principle used by Einstein in the theory of relativity, and by Heisenberg in the quantum theory, concepts which correspond to no conceivable observation should be eliminated from physics. This is possible without difficulty in the present case also. It is only necessary to replace statements like $x = \pi$ cm by: the probability of distribution of values of x has a sharp maximum at $x = \pi$ cm; and (if it is desired to be more accurate) to add: of such and such a breadth. In short, ordinary mechanics must also be statistically formulated. I have occupied myself with this problem a little recently, and have realized that it is possible without difficulty. This is not the place to go into the matter more deeply. I should like only to say this: the determinism of classical physics turns out to be an illusion, created by overrating mathematico-logical concepts. It is an idol, not an ideal in scientific research and cannot, therefore, be used as an objection to the essentially indeterministic statistical interpretation of quantum mechanics.

Much more difficult is the objection based on reality. The concept of a particle, e.g. a grain of sand, implicitly contains the idea that it is in a definite position and has definite motion. But according to quantum mechanics it is impossible to determine simultaneously with any desired accuracy both position and velocity (more precisely: momentum, i.e. mass times velocity). Thus two questions arise: what prevents us, in spite of the theoretical assertion, to measure both quantities to any desired degree of accuracy by refined experiments? Secondly, if it really transpires that this is not feasible, are we still justified in applying to the electron the concept of particle and therefore the ideas associated with it?

Referring to the first question, it is clear that if the theory is correct—and we have ample grounds for believing this—the obstacle to simultaneous measurement of position and motion (and of other such pairs of so-called conjugate quantities) must lie in the laws of quantum mechanics themselves. In fact, this is so. But it is not a simple matter to clarify the situation. Niels Bohr himself has gone to great trouble and ingenuity to develop a theory of measurements to clear the matter up and to meet the most refined and ingenious attacks of Einstein, who repeatedly tried to think out methods of measurement by means of which position and motion could be measured simultaneously and accurately. The following emerges: to measure space coordinates and instants of time, rigid measuring rods and clocks are required. On the other hand, to measure momenta and energies, devices are necessary with movable parts to absorb the impact of the test object and to indicate the size of its momentum. Paying regard to the fact that quantum mechanics is competent for dealing with the interaction of object and apparatus, it is seen that no arrangement is possible that will fulfil both requirements simultaneously. There exist, therefore, mutually exclusive though complementary experiments which only as a whole embrace everything which can be experienced with regard to an object.

This idea of *complementarity* is now regarded by most physicists as the key to the clear understanding of quantum processes. Bohr has generalized the idea to quite different fields of knowledge, e.g. the connection between consciousness and the brain, to the problem of free will, and other basic problems of philosophy. To come now to the last point: can we call something with which the concepts of position and motion cannot be associated in the usual way, a thing, or a particle? And if not, what is the reality which our theory has been invented to describe?

The answer to this is no longer physics, but philosophy, and to deal with it thoroughly would mean going far beyond the bounds of this lecture. I have given my views on it elsewhere. Here I will only say that I am emphatically in favour of the retention of the particle idea. Naturally, it is necessary to redefine what is meant. For this, well-developed concepts are available which appear in mathematics under the name of invariants in transformations. Every object that we perceive appears in innumerable aspects. The concept of the object is the invariant of all these aspects. From this point of view, the present universally used system of concepts in which particles and waves appear simultaneously, can be completely justified.

The latest research on nuclei and elementary particles has led us, however, to limits beyond which this system of concepts itself does not appear to suffice. The lesson to be learned from what I have told of the origin of quantum mechanics is that probable refinements of mathematical methods will not suffice to produce a satisfactory theory, but that somewhere in our doctrine is hidden a concept, unjustified by experience, which we must eliminate to open up the road. . . .

N.7

Exclusion Principle and Quantum Mechanics

Wolfgang Pauli

George Gamow, the Russian-American physicist who wrote brilliant popularizations of many aspects of science, described Wolfgang Pauli as one of the most colorful of all the pioneers of modern quantum theory. Pauli was born in Vienna in 1900. By the time he went to the University of Munich in 1918, he already had earned a reputation as a mathematical prodigy. He wrote a paper on the general theory of relativity that attracted Einstein's interest; it was published when Pauli was only nineteen years old. In 1920, he was given the assignment of writing an article on relativity for an encyclopedia of mathematics. The article was well received and enhanced the reputation of the youthful scientist in the physics community. In a letter to Einstein, Born commented on the paper just before it was to be typeset: "Pauli's article for the Encyclopedia is apparently finished and is said to weigh two and one-half kilos. This should give some idea of its intellectual weight. The little chap is not only clever but industrious as well."

After obtaining his doctorate in 1921, Pauli spent a year at the University of Göttingen working as assistant to Max Born and then a year with Niels Bohr in Copenhagen. After five years as lecturer at the University of Hamburg, he was appointed professor at the Federal Institute of Technology in Zurich in 1928. He then worked as a visiting professor at various U.S. universities including Princeton, Michigan, and Purdue. Although elected to the Chair of Theoretical Physics at Princeton, Pauli returned to Zurich at the end of World War II. He hypothesized the existence of the neutrino, an uncharged massless particle which carries off energy during radioactive beta decay. He also helped to lay the foundation for quantum field theory and showed the relationship between spin and statistics of elementary particles. After a very productive career, he died in Zurich in 1958.

Pauli's most enduring claim to fame is his identification of a general property of electron waves, which is called the Pauli exclusion principle. It states that no two identical electron wave patterns can exist in the same atom. There are only two possible variants of the wave pattern for electrons at the lowest energy level; thus, there can be only two electrons there. For the next highest level of energy, there are eight possible variations, and thus there can be eight electrons in that level. In atoms of high atomic weight, with many electrons, the wave patterns become less distinct, but Pauli's principle endures. Since it is at the highest energy level where the chemical bonding of one atom to another occurs to form a molecule, the value of Pauli's work to quantizing chemistry (which thus becomes a branch of physics) can scarcely be overstated.

Pauli was awarded the Nobel Prize in 1945, and his Nobel lecture, our next selection, was delivered on 13 December 1946. As it deals rigorously with the exclusion principle, it is not a simple essay for the reader. All during his life, Pauli was known as a person of the highest scientific and intellectual standards, who was capable of treating with scorn those who did not meet his own high standards.

Those who are not shocked when they first come across quantum theory cannot possibly have understood it.

—NIELS BOHR

The ultimate origin of the difficulty [of explaining quantum theory] lies in the fact (or philosophical principle) that we are compelled to use words of common language when we wish to describe a phenomena, not by logical or mathematical analysis, but by a picture appealing to the imagination. Common language has grown by everyday experience and can never surpass these limits. Classical physics has restricted itself to the use of concepts of this kind; by analyzing visible motions it has developed two ways of representing them by elementary processes: moving particles and waves. There is no other way of giving a pictorial description of motions—we have to apply it even in the region of atomic process, where classical physics breaks down.

—MAX BORN, *Atomic Physics*

THE HISTORY OF the discovery of the "exclusion principle" goes back to my students days in Munich. While, in school in Vienna, I had already obtained some knowledge of classical physics and the then new Einstein relativity theory, it was at the University of Munich that I was introduced by Sommerfeld to the structure of the atom—somewhat strange from the point of view of classical physics. I was not spared the shock which every physicist, accustomed to the classical way of thinking, experienced when he came to know of Bohr's "basic postulate of quantum theory" for the first time. At that time there were two approaches to the difficult problems connected with the quantum of action. One was an effort to bring abstract order to the new ideas by looking for a key to translate classical mechanics and electrodynamics into quantum language which would form a logical generalization of these. This was the direction which was taken by Bohr's "correspondence principle." Sommerfeld, however, preferred, in view of the difficulties which blocked the use of the concepts of kinematical models, a direct interpretation, as independent of models as possible, of the laws of spectra in terms of integral numbers, following, as

Wolfgang Pauli, Nobel Prize in Physics Award Address, 1946. Reprinted with permission of Elsevier Publishing Co., and the Nobel Foundation.

Kepler once did in his investigation of the planetary system, an inner feeling for harmony. Both methods, which did not appear to me irreconcilable, influenced me. The series of whole numbers 2, 8, 18, 32 . . . giving the lengths of the periods in the natural system of chemical elements, was zealously discussed in Munich, including the remark of the Swedish physicist, Rydberg, that these numbers are of the simple form $2n^2$, if n takes on all integer values. Sommerfeld tried especially to connect the number 8 and the number of corners of a cube.

A new phase of my scientific life began when I met Niels Bohr personally for the first time. This was in 1922, when he gave a series of guest lectures at Göttingen, in which he reported on his theoretical investigations on the Periodic System of Elements. I shall recall only briefly that the essential progress made by Bohr's considerations at that time was in explaining, by means of the spherically symmetric atomic model, the formation of the intermediate shells of the atom and the general properties of the rare earths. The question, as to why all electrons for an atom in its ground state were not bound in the innermost shell, had already been emphasized by Bohr as a fundamental problem in his earlier works. In his Göttingen lectures he treated particularly the closing of this innermost K-shell in the helium atom and its essential connection with the two non-combining spectra of helium, the ortho- and para-helium spectra. However, no convincing explanation for this phenomenon could be given on the basis of classical mechanics. It made a strong impression on me that Bohr at that time and in later discussions was looking for a *general* explanation which should hold for the closing of *every* electron shell and in which the number 2 was considered to be as essential as 8 in contrast to Sommerfeld's approach.

Following Bohr's invitation, I went to Copenhagen in the autumn of 1922, where I made a serious effort to explain the so-called "anomalous Zeeman effect," as the spectroscopists called a type of splitting of the spectral lines in a magnetic field which is different from the normal triplet. On the one hand, the anomalous type of splitting exhibited beautiful and simple laws and Landé had already succeeded to find the simpler splitting of the spectroscopic terms from the observed splitting of the lines. The most fundamental of his results thereby was the use of half-integers as magnetic quantum numbers for the doublet-spectra of the alkali metals. On the other hand, the anomalous splitting was hardly understandable from the standpoint of the mechanical model of the atom, since very general assumptions concerning the electron, using classical theory as well as quantum theory, always led to the same triplet. A closer investigation of this problem left me with the feeling that it was even more unapproachable. We know now that at that time one was confronted with two logically different difficulties simultaneously. One was the absence of a general key to translate a given mechanical model into quantum theory which one tried in vain by using classical mechanics to describe the stationary quantum states themselves. The second difficulty was our ignorance concerning the proper classical model itself which could be suited to derive at all an anomalous splitting of spectral lines emitted by an atom in an external magnetic field. It is therefore not surprising that I could not find a satisfactory solution of the problem at that time. I succeeded, however, in generalizing Landé's term analysis for very strong magnetic fields, a case which, as a result of the magneto-optic transformation (Paschen-Back effect), is in many respects simpler. This early work was of decisive importance for the finding of the exclusion principle.

Very soon after my return to the University of Hamburg, in 1923, I gave there my inaugural lecture as *Privatdozent* on the Periodic System of Elements. The contents of this lecture appeared very unsatisfactory to me, since the problem of the closing of the electronic shells had been clarified no further. The only thing that was clear was that a closer relation of this problem to the theory of multiplet structure must exist. I therefore tried to examine again critically the simplest case, the doublet structure of the alkali spectra. According to the point of view then orthodox, which was also taken over by Bohr in his already mentioned lectures in Göttingen, a non-vanishing angular momentum of the atomic core was supposed to be the cause of this doublet structure.

In the autumn of 1924 I published some arguments against this point of view, which I definitely rejected as incorrect and proposed instead of it the assumption of a new quantum theoretic property of the electron, which I called a "two-valuedness not describable classically." At this time a paper of the English physicist, Stoner, appeared which contained, besides improvements in the classification of electrons in subgroups, the following essential remark: For a given value of the principal quantum number is the number of energy levels of a single electron in the alkali metal spectra in an external magnetic field the same as the number of electrons in the closed shell of the rare gases which corresponds to this principal quantum number.

On the basis of my earlier results on the classification of spectral terms in a strong magnetic field the general formulation of the exclusion principle became clear to me. The fundamental idea can be stated in the following way: The complicated numbers of electrons in closed subgroups are reduced to the simple number *one* if the division of the groups by giving the values of the four quantum numbers of an electron is carried so far that every degeneracy is removed. An entirely non-degenerate energy level is already "closed," if it is occupied by a single electron; states in contradiction with this postulate have to be excluded. The exposition of this general formulation of the exclusion principle was made in Hamburg in the spring of 1925, after I was able to verify some additional conclusions concerning the anomalous Zeeman effect of more complicated atoms during a visit to Tübingen with the help of the spectroscopic material assembled there.

With the exception of experts on the classification of spectral terms, the physicists found it difficult to understand the exclusion principle, since no meaning in terms of a model was given to the fourth degree of freedom of the electron. The gap was filled by Uhlenbeck and Goudsmit's idea of electron spin, which made it possible to understand the anomalous Zeeman effect simply by assuming that the spin quantum number of one electron is equal to ½ and that the quotient of the magnetic moment to the mechanical angular moment has for the spin a value twice as large as for the ordinary orbit of the electron. Since that time, the exclusion principle has been closely connected with the idea of spin. Although at first I strongly doubted the correctness of this idea because of its classical-mechanical character, I was finally converted to it by Thomas' calculations on the magnitude of doublet splitting. On the other hand, my earlier doubts as well as the cautious expression "classically non-describable two-valuedness" experienced a certain verification during later developments, since Bohr was able to show on the basis of wave mechanics that the electron spin cannot be measured by classically describable experiments (as, for instance, deflection of molecular beams in external electromagnetic fields) and must therefore be considered as an essentially quantum-mechanical property of the electron.

The subsequent developments were determined by the occurrence of the new quantum mechanics. In 1925, the same year in which I published my paper on the exclusion principle, de Broglie formulated his idea of matter waves and Heisenberg the new matrix-mechanics, after which in the next year Schrödinger's wave mechanics quickly followed. It is at present unnecessary to stress the importance and the fundamental character of these discoveries, all the more as these physicists have themselves explained, here in Stockholm, the meaning of their leading ideas. Nor does time permit me to illustrate in detail the general epistemological significance of the new discipline of quantum mechanics, which has been done, among others, in a number of articles by Bohr, using hereby the idea of "complementarity" as a new central concept. I shall only recall that the statements of quantum mechanics are dealing only with possibilities, not with actualities. They have the form "This is not possible" or "Either this or that is possible," but they can never say "That will actually happen then and there." The actual observation appears as an event outside the range of a description by physical laws and brings forth in general a discontinuous selection out of the several possibilities foreseen by the statistical laws of the new theory. Only this renouncement concerning the old claims for an objective description of the physical phenomena, independent of the way in which they are observed, made it possible to reach again the self-consistency of quantum theory, which actually had been lost since Planck's discovery of the quantum of action. Without discussing further the change of the attitude of modern physics to such concepts as "causality" and "physical reality" in comparison with the older classical physics I shall discuss more particularly in the following the position of the exclusion principle on the new quantum mechanics.

As it was first shown by Heisenberg, wave mechanics leads to qualitatively different conclusions for particles of the same kind (for instance for electrons) than for particles of different kinds. As a consequence of the impossibility to distinguish one of several like particles from the other, the wave functions describing an ensemble of a given number of like particles in the configuration space are sharply separated into different classes of symmetry which can never be transformed into each other by external perturbations. In the term "configuration space" we are including here the spin degree of freedom, which is described in the wave function of a single particle by an index with only a finite number of possible values. For electrons this number is equal to two; the configuration space of N electrons has therefore 3 N space dimensions and N indices of "two-valuedness." Among the different classes of symmetry, the most important ones (which moreover for two particles are the only ones) are the symmetrical class, in which the wave function does not change its value when the space and spin coordinates of two particles are permuted, and the antisymmetrical class, in which for such a permutation the wave function changes its sign. At this stage of the theory three different hypotheses turned out to be logically possible concerning the actual ensemble of several like particles in Nature.

I. This ensemble is a mixture of all symmetry classes.
II. Only the symmetrical class occurs.
III. Only the antisymmetrical class occurs.

As we shall see, the first assumption is never realized in Nature. Moreover, it is only the third assumption that is in accordance with the exclusion principle, since an antisymmetrical function containing two particles in the same state is identically zero. The assumption III can therefore be considered as the correct and general wave mechanical formulation of the exclusion principle. It is this possibility which actually holds for electrons.

This situation appeared to me as disappointing in an important respect. Already in my original paper I stressed the circumstance that I was unable to give a logical reason for the exclusion principle or to deduce it from more general assumptions. I had always the feeling and I still have it today, that this is a deficiency. Of course in the beginning I hoped that the new quantum mechanics, with the help of which it was possible to deduce so many half-empirical formal rules in use at that time, will also rigorously deduce the exclusion principle. Instead of it there was for electrons still an exclusion: not of particular states any longer, but of whole classes of states, namely the exclusion of all classes different from the antisymmetrical one. The impression that the shadow of some incompleteness fell here on the bright light of success of the new quantum mechanics seems to me unavoidable. We shall resume this problem when we discuss relativistic quantum mechanics but wish to give first an account of further results of the application of wave mechanics to systems of several like particles.

In the paper of Heisenberg, which we are discussing, he was also able to give a simple explanation of the existence of the two non-combining spectra of helium which I mentioned in the beginning of this lecture. Indeed, besides the rigorous separation of the wave functions into symmetry classes with respect to space-coordinates and spin indices together, there exists an approximate separation into symmetry classes with respect to space coordinates alone. The latter holds only so long as an interaction between the spin and the orbital motion of the electron can be neglected. In this way the para- and ortho-helium spectra could be interpreted as belonging to the class of symmetrical and antisymmetrical wave functions respectively in the space coordinates alone. It became clear that the energy difference between corresponding levels of the two classes has nothing to do with magnetic interactions but is of a new type of much larger order of magnitude, which one called exchange energy.

Of more fundamental significance is the connection of the symmetry classes with general problems of the statistical theory of heat. As is well known, this theory leads to the result that the entropy of a system is (apart from a constant factor) given by the logarithm of the number of quantum states of the whole system on a so-called energy shell. One might first expect that this number should be equal to the corresponding volume of the multidimensional phase space divided by h^f, where h is Planck's constant and f the number of degrees of freedom of the whole system. However, it turned out that for a system of N like particles, one had still to divide this quotient by $N!$ in order to get a value for the entropy in accordance with the usual postulate of homogeneity that the entropy has to be proportional to the mass for a given inner state of the substance. In this way a qualitative distinction between like and unlike particles was already preconceived in the general statistical mechanics, a distinction which Gibbs tried to express with his concepts of a generic and a specific phase. In the light of the result of wave mechanics concerning the symmetry classes, this division by $N!$, which had caused already much discussion, can easily be interpreted by accepting one of our assumptions II and III, according to both of which only *one* class of symmetry occurs in Nature. The density of quantum states of the whole system then really becomes smaller by a factor $N!$ in comparison with the density which had to be expected according to an assumption of the type I admitting all symmetry classes.

Even for an ideal gas, in which the interaction energy between molecules can be neglected, deviations from the ordinary equation of state have to be expected for the reason that only *one* class of symmetry is possible as soon as the mean de Broglie wavelength of a gas molecule becomes of an order of magnitude comparable with the average distance between two molecules, that is, for small temperatures and large densities. For the antisymmetrical class the statistical consequences have been derived by Fermi and Dirac, for the symmetrical class the same had been done already before the discovery of the new quantum mechanics by Einstein and Bose. The former case could be applied to the electrons in a metal and could be used for the interpretation of magnetic and other properties of metals.

As soon as the symmetry classes for electrons were cleared, the question arose which are the symmetry classes for other particles. One example for particles with symmetrical wave functions only (assumption II) was already known long ago, namely the photons. This is not only an immediate consequence of Planck's derivation of the spectral distribution of the radiation energy in the thermodynamical equilibrium, but it is also necessary for the applicability of the classical field concepts to light waves in the limit where a large and not accurately fixed number of photons is present in a single quantum state. We note that the symmetrical class for photons occurs together with the integer value 1 for their spin, while the antisymmetrical class for the electron occurs together with the half-integer value ½ for the spin.

The important question of the symmetry classes for nuclei, however, had still to be investigated. Of course the symmetry class refers here also to the permutation of both the space coordinates and the spin indices of two like nuclei. The spin index can assume $2I + 1$ values if I is the spin-quantum number of the nucleus which can be either an integer or a half-integer. I may include the historical remark that already in 1924, before the electron spin was discovered, I proposed to use the assumption of a nuclear spin to interpret the hyperfine-structure of spectral lines. This proposal met on the one hand strong opposition from many sides but influenced on the other hand Goudsmit and Uhlenbeck in their claim of an electron spin. It was only some years later that my attempt to interpret the hyperfine-structure could be definitely confirmed experimentally by investigations in which also Zeeman himself participated and which showed the existence of a magneto-optic transformation of the hyperfine-structure as I had predicted it. Since that time the hyperfine-structure of spectral lines became a general method of determining the nuclear spin.

In order to determine experimentally also the symmetry class of the nuclei, other methods were necessary. The most convenient, although not the only one, consists in the investigation of band spectra due to a molecule with two like atoms. It could easily be derived that in the ground state of the electron configuration of such a molecule the states with even and odd values of the rotational quantum number are symmetric and antisymmetric respectively for a permutation of the space coordinates of the two nuclei. Further there exist among the $(2I + 1)^2$ spin states of the pair of nuclei, $(2I + 1)(I + 1)$ states symmetrical and $(2I + 1)I$ states antisymmetrical in the spins, since the $(2I + 1)$ states with two spins in the same direction are necessarily symmetrical. Therefore the conclusion was reached: If the total wave function of space coordinates and spin indices of the nuclei is symmetrical, the ratio of the weight of states with an even rotational quantum number to the weight of states with an odd rotational quantum number is given by $(I + 1) : I$. In the reverse case of an antisymmetrical total wave function of the nuclei, the same ratio is $I : (I + 1)$. Transitions between one state with an even and another state with an odd rotational quantum number will be extremely rare as they can only be caused by an interaction between the orbital motions and the spins of the nuclei. Therefore the ratio of the weights of the rotational states with different parity will give rise to two different systems of band spectra with different intensities, the lines of which are alternating.

The first application of this method was the result that the protons have the spin ½ and fulfill the exclusion principle just as the electrons. The initial difficulties to understand quantitatively the specific heat of hydrogen molecules at low temperatures were removed by Dennison's hypothesis, that at this low temperature the thermal equilibrium between the two modifications of the hydrogen molecule (ortho-H_2: odd rotational quantum numbers, parallel proton spins; para-H_2: even rotational quantum numbers, antiparallel spins) was not yet reached. As you know, this hypothesis was later confirmed by the experiments of Bonhoeffer and Harteck and of Eucken, which showed the theoretically predicted slow transformation of one modification into the other.

Among the symmetry classes for other nuclei those with a different party of their mass number M and their charge number Z are of a particular interest. If we consider a compound system consisting of numbers A_1, A_2, . . . of different constituents, each of which is fulfilling the exclusion principle, and a number S of constituents with symmetrical states, one has to expect symmetrical or antisymmetrical states if the sum $A_1 + A_2 +$ . . . is even or odd. This holds regardless of the parity of S. Earlier one tried the assumption that nuclei consist of protons and electrons, so that M is the number of protons, $M - Z$ the number of electrons in the nucleus. It had to be expected then that the parity of Z determines the symmetry class of the whole nucleus. Already for some time the counter-example of nitrogen has been known to have the spin 1 and symmetrical states. After the discovery of the neutron, the nuclei have been considered, however, as composed of protons and neutrons in such a way that a nucleus with mass number M and charge number Z should consist of Z protons and $M - Z$ neutrons. In case the neutrons would have symmetrical states, one should again expect that the parity of the charge number Z determines the symmetry class of the nuclei. If, however, the neutrons fulfill the exclusion principle, it has to be expected that the parity of M determines the symmetry class: For an even M, one should always have symmetrical states, for an odd M, antisymmetrical ones. It was the latter rule that was confirmed by experiment without exception, thus proving that the neutrons fulfill the exclusion principle.

The most important and most simple crucial example for a nucleus with a different parity of M and Z is the heavy hydrogen or deuteron with $M = 2$ and $Z = 1$ which has symmetrical states and the spin $I = 1$, as could be proved by the investigation of the band spectra of a molecule with two deuterons. From the spin value 1 of the deuteron can be concluded that the neutron must have a half-integer spin. The simplest possible assumption that this spin of the neutron is equal to ½, just as the spin of the proton and of the electron, turned out to be correct.

There is hope, that further experiments with light nuclei, especially with protons, neutrons, and deuterons will give us further information about the nature of the forces between the constituents of the nuclei, which, at present, is not yet sufficiently clear. Already now we can say, however, that these interactions are fundamentally different from electromagnetic interactions. The comparison between neutron-proton scattering and proton-proton scattering even showed that the forces between these particles are in good approximation the same, that means independent of their electric charge. If one had only to take into account the magnitude of the interaction energy, one should therefore expect a stable di-proton or ${}^{2}_{2}\mathrm{He}(M = 2, Z = 2)$ with nearly the same binding energy as the deuteron. Such a state is, however, forbidden by the exclusion principle in accordance with experience, because this state would acquire a wave function symmetric with respect to the two protons. This is only the simplest example of the application of the exclusion principle to the structure of compound nuclei, for the understanding of which this principle is indispensable, because the constituents of these heavier nuclei, the protons and the neutrons, fulfill it.

In order to prepare for the discussion of more fundamental questions, we want to stress here a law of Nature which is generally valid, namely, the connection between spin and symmetry class. *A half-integer value of the spin quantum number is always connected with antisymmetrical states (exclusion principle), an integer spin with symmetrical states.* This law holds not only for protons and neutrons but also for protons and electrons. Moreover, it can easily be seen that it holds for compound systems, if it holds for all of its constituents. If we search for a theoretical explanation of this law, we must pass to the discussion of relativistic wave mechanics, since we saw that it can certainly not be explained by non-relativistic wave mechanics.

We first consider classical fields, which, like scalars, vectors, and tensors transform with respect to rotations in the ordinary space according to a one-valued representation of the rotation group. We may, in the following, call such fields briefly "one-valued" fields. So long as interactions of different kinds of field are not taken into account, we can assume that all field components will satisfy a second-order wave equation, permitting a superposition of plane waves as a general solution. Frequency and wave number of these plane waves are connected by a law which, in accordance with de Broglie's fundamental assumption, can be obtained from the relation between energy and momentum of a particle claimed in relativistic mechanics by division with the constant factor h equal to Planck's constant divided by 2π. Therefore, there will appear in the classical field equations, in general, a new constant μ with the dimension of a reciprocal length, with which the rest-mass m in the particle picture is connected by $m = h\,\mu/c$, where c is the vacuum-velocity of light. From the assumed property of one-valuedness of the field it can be concluded, that the number of possible plane waves for a given frequency, wave number and direction of propagation, is for a non-vanishing μ always odd. Without going into details of the general definition of spin, we can consider this property of the polarization of plane waves as characteristic for fields which, as a result of their quantization, give rise to integer spin values.

The simplest cases of one-valued fields are the scalar field and a field consisting of a four-vector and an antisymmetric tensor like the potentials and field strengths in Maxwell's theory. While the scalar field is simply fulfilling the usual wave equation of the second order in which the term proportional to μ^2 has to be included, the other field has to fulfill equations due to Proca which are a generalization of Maxwell's equations which become in the particular case $\mu = 0$. It is satisfactory that for these simplest cases of one-valued fields the energy density is a positive definite quadratic form of the field-quantities and their first derivatives at a certain point. For the general case of one-valued fields it can at least be achieved that the total energy after integration over space is always positive.

The field components can be assumed to be either real or complex. For a complex field, in addition to energy and momentum of the field, a four-vector can be defined which satisfies the continuity equation and can be interpreted as the four-vector of the electric current. Its fourth component determines the electric charge density and can assume both positive and negative values. It is possible that the charged mesons observed in cosmic rays have integral spins and thus can be described by such a complex field. In the particular case of real fields this four-vector of current vanishes identically.

Especially in view of the properties of the radiation in the thermodynamical equilibrium in which specific properties of the field sources do not play any role, it seemed to be justified first to disregard in the formal process of field quantization the interaction of the field with the sources. Dealing with this problem, one tried indeed to apply the same mathematical method of passing from a classical system to a corresponding system governed by the laws of quantum mechanics which has been so successful in passing from classical point mechanics to wave mechanics. It should not be forgotten, however, that a field can only be observed with help of its interaction with test bodies which are themselves again sources of the field.

The result of the formal process of field quantization were partly very encouraging. The quantized wave fields can be characterized by a wave function which depends on an infinite sequence of (non-negative) integers as variables. As the total energy and the total momentum of the field and, in case of complex fields, also its total electric charge turn out to be linear functions of these numbers, they can be interpreted as the number of particles present in a specified state of a single particle. By using a sequence of configuration spaces with a different number of dimensions corresponding to the different possible values of the total number of particles present, it could easily be shown that this description of our system by a wave function depending on integers is equivalent to an ensemble of particles with wave functions symmetrical in their configuration spaces.

Moreover Bohr and Rosenfeld proved in the case of the electromagnetic field that the uncertainty relations which result for the average values of the field strengths over finite space-time regions from the formal commutation rules of this theory have a direct physical meaning so long as the sources can be treated classically and their atomistic structure can be disregarded. We emphasize the following property of these commutation rules: All physical quantities in two world points, for which the four-vector of their joining straight line is spacelike commute with each other. This is indeed necessary for physical reasons because any disturbance by measurements in a world point P_1 can only reach such points P_2 for which the vector P_1P_2 is timelike, that is, for which $c(t_1 - t_2) > r_{12}$. The points P_2 with a spacelike vector P_1P_2 for which $c(t_1 - t_2) < r_{12}$ cannot be reached by this disturbance and measurements in P_1 and P_2 can then never influence each other.

This consequence made it possible to investigate the logical possibility of particles with integer spin which would obey the exclusion principle. Such particles could be described by a sequence of configuration spaces with different dimensions and wave functions antisymmetrical in the coordinates of these spaces or also by a wave function depending on integers again to be interpreted as the number of particles present in specified states which now can only assume the values 0 or 1. Wigner and Jordan proved that also in this case operators can be defined which are functions of the ordinary space-time coordinates and which can be applied to such a wave function. These operators do not fulfill any longer commutation rules: instead of the difference, the *sum* of the two possible products of two operators, which are distinguished by the different order of its factors, is now fixed by the mathematical conditions the operators have to satisfy. The simple change of the sign in these conditions changes entirely the physical meaning of the formalism. In the case of the exclusion principle there can never exist a limiting case where such operators can be replaced by a classical field. Using this formalism of Wigner and Jordan I could prove under very general assumptions that a relativistic invariant theory describing systems of like particles with integer spin obeying the exclusion principle would always lead to the non-commutability of physical quantities joined by a spacelike vector. This would violate a reasonable physical principle which holds good for particles with symmetrical states. In this way, by combination of the claims of relativistic invariance and the properties of field quantization, one step in the direction of an understanding of the connection of spin and symmetry class could be made.

The quantization of one-valued complex fields with a non-vanishing four-vector of the electric current gives the further result that particles both with positive and negative electric charge should exist and that they can be annihilated and generated in external electromagnetic fields. This pair-generation and annihilation claimed by the theory makes it necessary to distinguish clearly the concept of charge density and of particle density. The latter concept does not occur in a relativistic wave theory either for fields carrying an electric charge or for neutral fields. This is satisfactory since the use of the particle picture and the uncertainty relations (for instance by analyzing imaginative experiments of the type of the γ-ray microscope) gives also the result that a localization of the particle is only possible with limited accuracy. This holds both for the particles with integer and with half-integer spins. In a state with a mean value E of its energy, described by a wave packet with a mean frequency $\nu = E/h$, a particle can only be localized with an error $\Delta x > hc/E$ or $\Delta x > c/\nu$. For photons, it follows that the limit for the localization is the wavelength; for a particle with a finite rest-mass m and a characteristic length $\mu^{-1} = h/mc$, this limit is in the rest system of the center of the wave packet that describes the state of the particles given by $\Delta x > h/mc$ or $\Delta x > \mu^{-1}$.

Until now I have mentioned only those results of the application of quantum mechanics to classical fields which are satisfactory. We say that the statements of this theory about averages of field strength over finite space-time regions have a direct meaning while this is not so for the values of the field strength at a certain point. Unfortunately in the classical expression of the energy of the field there enter averages of the squares of the field strengths over such regions which cannot be expressed by the averages of the field strengths themselves. This has the consequence that the zero-point energy of vacuum derived from the quantized field becomes infinite, a result which is directly connected with the fact that the system considered has an infinite number of degrees of freedom. It is clear that this zero-point energy has no physical reality, for instance it is not the source of a gravitational field. Formally it is easy to subtract constant infinite terms which are independent of the state considered and never change; nevertheless it seems to me that already this result is an indication that a fundamental change in the concepts underlying the present theory of quantized fields will be necessary.

In order to clarify certain aspects of relativistic quantum theory, I have discussed here, different from the historical order of events, the one-valued fields first. Already earlier Dirac had formulated his relativistic wave equations corresponding to material particles with spin ½ using a pair of so-called spinors with two components each. He applied these equations to the problem of one electron in an electromagnetic field. In spite of the great success of this theory in the quantitative explanation of the fine structure of the energy levels of the hydrogen atom and in the computation of the scattering cross section of one photon by a free electron, there was one consequence of this theory which was obviously in contradiction with experience. The energy of the electron can have, according to the theory, both positive and negative values, and, in external electromagnetic fields, transitions should occur from states with one sign of energy to states with the other sign. On the other hand there exists in this theory a four-vector satisfying the continuity equation with a fourth component corresponding to a density which is definitely positive.

It can be shown that there is a similar situation for all fields, which, like the spinors, transform for rotations in ordinary space according to two-valued representations, thus changing their sign for a full rotation. We shall call briefly such quantities "two-valued." From the relativistic wave equations of such quantities one can always derive a four-vector bilinear in the field components which satisfies the continuity equation and for which the fourth component, at least after integration over the space, gives an essentially positive quantity. On the other hand, the expression for the total energy can have both the positive and the negative sign.

Is there any means to shift the minus sign from the energy back to the density of the four-vector? Then the latter could again be interpreted as charge density in contrast to particle density and the energy would become positive as it ought to be. You know that Dirac's answer was that this could actually be achieved by application of the exclusion principle. In his lecture delivered here in Stockholm he himself explained his proposal of a new interpretation of his theory, according to which in the actual vacuum all the states of negative energy should be occupied and only deviations of this state of smallest energy, namely holes in the sea of these occupied states are assumed to be observable. It is the exclusion principle which guarantees the stability of the vacuum, in which all states of negative energy are occupied. Furthermore the holes have all properties of particles with positive energy and positive electric charge, which in external electromagnetic fields can be produced and annihilated in pairs. These predicted positrons, the exact mirror images of the electrons, have been actually discovered experimentally.

The new interpretation of the theory obviously abandons in principle the standpoint of the one-body problem and considers a many-body problem from the beginning. It cannot any longer be claimed that Dirac's relativistic wave equations are the only possible ones but if one wants to have relativistic field equations corresponding to particles, for which the value ½ of their spin is known, one has certainly to assume the Dirac equations. Although it is logically possible to quantize these equations like classical fields, which would give symmetrical states of a system consisting of many such particles, this would be in contradiction with the postulate that the energy of the system has actually to be positive. This postulate is fulfilled on the other hand if we apply the exclusion principle and Dirac's interpretation of the vacuum and the holes, which at the same time substitutes the physical concept of charge density with values of both signs for the mathematical fiction of a positive particle density. A similar conclusion holds for all relativistic wave equations with two-valued quantities as field components. This is the other step (historically the earlier one) in the direction of an understanding of the connection between spin and symmetry class.

I can only shortly note that Dirac's new interpretaton of empty and occupied states of negative energy can be formulated very elegantly with the help of the formalism of Jordan and Wigner mentioned before. The transition from the old to the new interpretation of the theory can indeed be carried through simply by interchanging the meaning of one of the operators with that of its hermitian conjugate if they are applied to states originally of negative energy. The infinite "zero charge" of the occupied states of negative energy is then formally analogous to the infinite zero-point energy of the quantized one-valued fields. The former has no physical reality either and is not the source of an electromagnetic field.

In spite of the formal analogy between the quantization of the one-valued fields leading to ensembles of like particles with symmetrical states and to particles fulfilling the exclusion principle described by two-valued operator quantities, depending on space and time coordinates, there is of course the fundamental difference that for the latter there is no limiting case, where the mathematical operators can be treated like classical fields. On the other hand we can expect that the possibilities and the limitations for the applications of the concepts of space and time, which find their expression in the different concepts of charge density and particle density, will be the same for charged particles with integer and with half-integer spins.

The difficulties of the present theory become much worse, if the interaction of the electromagnetic field with matter is taken into consideration, since the well-known infinities regarding the energy of an electron in its own field, the so-called self-energy, then occur as a result of the application of the usual perturbation formalism to this problem. The root of this difficulty seems to be the circumstance that the formalism of field quantization has only a direct meaning so long as the sources of the field can be treated as continuously distributed, obeying the laws of classical physics, and so long as only averages of field quantities over finite space-time regions are used. The electrons themselves, however, are essentially non-classical field sources.

At the end of this lecture I may express my critical opinion, that a correct theory should neither lead to infinite zero-point energies nor to infinite zero charges, that it should not use mathematical tricks to subtract infinities or singularities, nor should it invent a "hypothetical world" which is only a mathematical fiction before it is able to formulate the correct interpretation of the actual world of physics.

From the point of view of logic, my report on "Exclusion principle and quantum mechanics" has no conclusion. I believe that it will only be possible to write the conclusion if a theory will be established which will determine the value of the fine-structure constant and will thus explain the atomistic structure of electricity, which is such an essential quality of all atomic sources of electric fields actually occurring in Nature.

N.8

The Copenhagen Interpretation of Quantum Theory

WERNER HEISENBERG

IN 1948, sixteen years after being awarded the Nobel Prize, Werner Heisenberg published a nontechnical book entitled *Physics and Philosophy* that sketched not only many of the advances made by him and his colleagues in physics but also his philosophy of science. In the long history of the development of quantum theory, a major role was played by the scientists of the physics institute established for Niels Bohr in Copenhagen by the Danish government (with funding supplied by the Carlsberg Brewery, to which the world owes a debt of gratitude for its many contributions not only to science but to the arts as well). Gradually a complicated array of ideas, an "anthology" of concepts contributed piecemeal by the young physicists of the 1920s and 1930s, became known as the Copenhagen interpretation of quantum mechanics, although no such interpretation was ever actually presented as a formal doctrine.

Essentially, the Copenhagen interpretation represented an admission by the scientists gathered at the Fifth Solvay Congress in Brussels in 1927 that a complete understanding of reality was beyond the capacity of rational thought. In other words, physicists could forever mull over ideas about the nature of reality, but would never be able to consider the nature of reality itself. Alternatively, the validity of a concept is not how closely it corresponds with some absolute truth, but how well it agrees with our own experiences. It was this injection of subjective truth and the acknowledgment that any fundamental truth about the Universe would always be hidden from our inquiring minds that caused Einstein to protest that "the most incomprehensible thing about the Universe is that it is comprehensible."

The Copenhagen interpretation asserts that a probability function does not describe a certain event but describes a continuum of possible events until a measurement interferes with the isolation of the system and a single event is actualized. The Copenhagen interpretation eliminates the one-to-one correspondence between reality and theory by disregarding laws governing individual events. Henry Stapp, a physicist at the Lawrence Livermore Laboratory in Berkeley, California, has described the Copenhagen interpretation in the following terms:

> [It is] a rejection of the presumption that nature could be understood in terms of elementary space-time realities. According to the new view, the complete description of matter at the atomic level was given by the probability functions that referred, not to underlying microscopic space-time realities, but rather to the macroscopic objects of sense experience. The theoretical structure did not extend down and anchor itself on fundamental microscopic space-time realities. Instead it turned back and anchored itself in the concrete sense realities that form the basis of social life. . . . This pragmatic description is to be contrasted with descriptions that attempt to peer behind the scenes and tell us what is really happening.*

More recently, two other views have competed with the Copenhagen interpretation but have not seriously challenged its preeminent position. The Copenhagen interpretation, as we have seen, holds that all events are possible until a specific event is triggered that eliminates all other branching realities. The many worlds or Everett-Wheeler-Graham model offers a slightly different explanation, in which alternative realities are not suppressed once an interactive event such as a measurement occurs, but merely become inaccessible to the observer. The alternative realities of the many worlds model continue to branch indefinitely. A third view, offered by Eugene Wigner, asserts that consciousness is the force that organizes nature or at least what we perceive to be nature. Many physicists are especially reluctant to consider the idea of a pervasive consciousness

*Henry Stapp, "The Copenhagen Interpretation and the Nature of Space-Time," *American Journal of Physics,* 40, 1972, 1098ff.

structuring the Universe. Thus, it is not surprising that Wigner's model has been politely disregarded by most physicists.

Perhaps the closest thing we have to an authorative first-hand formulation of the Copenhagen interpretation is the third chapter of Heisenberg's book, *Physics and Philosophy.*

I remember discussions with Bohr which went through many hours till very late at night and ended almost in despair; and when at the end of the discussion I went alone for a walk in the neighboring park I repeated to myself again and again the question: Can nature possibly be so absurd as it seemed to us in these atomic experiments?

—WERNER HEISENBERG, *Physics and Philosophy*

THE COPENHAGEN INTERPRETATION of quantum theory starts from a paradox. Any experiment in physics, whether it refers to the phenomena of daily life or to atomic events, is to be described in the terms of classical physics. The concepts of classical physics form the language by which we describe the arrangement of our experiments and state the results. We cannot and should not replace these concepts by any others. Still the application of these concepts is limited by the relations of uncertainty. We must keep in mind this limited range of applicability of the classical concepts while using them, but we cannot and should not try to improve them.

For a better understanding of this paradox it is useful to compare the procedure for the theoretical interpretation of an experiment in classical physics and in quantum theory. In Newton's mechanics, for instance, we may start by measuring the position and the velocity of the planet whose motion we are going to study. The result of the observation is translated into mathematics by deriving numbers for the co-ordinates and the momenta of the planet from the observation. Then the equations of motion are used to derive from these values of the co-ordinates and momenta at a given time the values of these co-ordinates or any other properties of the system at a later time, and in this way the astronomer can predict the properties of the system at a later time. He can, for instance, predict the exact time for an eclipse of the moon.

In quantum theory the procedure is slightly different. We could for instance be interested in the motion of an electron through a cloud chamber and could determine by some kind of observation the initial position and

---

Werner Heisenberg, "The Copenhagen Interpretation of Quantum Theory, in *Physics and Philosophy,* Volume 19 of World Perspective Series, Planned and Edited by Ruth Nanda Anshen. New York: Harper & Row Publishers, Inc. 1958, pp. 44–58.

velocity of the electron. But this determination will not be accurate; it will at least contain the inaccuracies following from the uncertainty relations and will probably contain still larger errors due to the difficulty of the experiment. It is the first of these inaccuracies which allows us to translate the result of the observation into the mathematical scheme of quantum theory. A probability function is written down which represents the experimental situation at the time of the measurement, including even the possible errors of the measurement.

This probability function represents a mixture of two things, partly a fact and partly our knowledge of a fact. It represents a fact in so far as it assigns at the initial time the probability unity (i.e., complete certainty) to the initial situation: the electron moving with the observed velocity at the observed position; "observed" means observed within the accuracy of the experiment. It represents our knowledge in so far as another observer could perhaps know the position of the electron more accurately. The error in the experiment does—at least to some extent—not represent a property of the electron but a deficiency in our knowledge of the electron. Also this deficiency of knowledge is expressed in the probability function.

In classical physics one should in a careful investigation also consider the error of the observation. As a result one would get a probability distribution for the initial values of the co-ordinates and velocities and therefore something very similar to the probability function in quantum mechanics. Only the necessary uncertainty due to the uncertainty relations is lacking in classical physics.

When the probability function in quantum theory has been determined at the initial time from the observation, one can from the laws of quantum theory calculate the probability function at any later time and can thereby determine the probability for a measurement giving a specified value of the measured quantity. We can, for instance, predict the probability for finding the electron at a later time at a given point in the cloud chamber. It should be emphasized, however, that the probability function does not in itself represent a course of events in the course of time. It represents a tendency for events and our knowledge of events. The probability function can be connected with reality only if one essential condition is fulfilled: if a new measurement is made to determine a certain property of the system. Only then does the probability function allow us to calculate the probable result of the new measurement. The result of the measurement again will be stated in terms of classical physics.

Therefore, the theoretical interpretation of an experiment requires three distinct steps: (1) the translation of the initial experimental situation into a probability function; (2) the following up of this function in the course of time; (3) the statement of a new measurement to be made of the system, the result of which can then be calculated from the probability function. For the first step the fulfillment of the uncertainty relations is a necessary condition. The second step cannot be described in terms of the classical concepts; there is no description of what happens to the system between the initial observation and the next measurement. It is only in the third step that we change over again from the "possible" to the "actual."

Let us illustrate these three steps in a simple ideal experiment. It has been said that the atom consists of a nucleus and electrons moving around the nucleus; it has also been stated that the concept of an electronic orbit is doubtful. One could argue that it should at least in principle be possible to observe the electron in its orbit. One should simply look at the atom through a microscope of a very high resolving power, then one would see the electron moving in its orbit. Such a high resolving power could to be sure not be obtained by a microscope using ordinary light, since the inaccuracy of the measurement of the position can never be smaller than the wave length of the light. But a microscope using γ-rays with a wave length smaller than the size of the atom would do. Such a microscope has not yet been constructed but that should not prevent us from discussing the ideal experiment.

Is the first step, the translation of the result of the observation into a probability function, possible? It is possible only if the uncertainty relation is fulfilled after the observation. The position of the electron will be known with an accuracy given by the wave length of the γ-ray. The electron may have been practically at rest before the observation. But in the act of observation at least one light quantum of the γ-ray must have passed the microscope and must first have been deflected by the electron. Therefore, the electron has been pushed by the light quantum, it has changed its momentum and its velocity, and one can show that the uncertainty of this change is just big enough to guarantee the validity of the uncertainty relations. Therefore, there is no difficulty with the first step.

At the same time one can easily see that there is no way of observing the orbit of the electron around the nucleus. The second step shows a wave packet moving not around the nucleus but away from the atom, because the first light quantum will have knocked the electron out from the atom. The momentum of light quantum of the γ-ray is much bigger than the original momentum of the electron if the wave length of the γ-ray is much smaller than the size of the atom. Therefore, the first light quantum is sufficient to knock the electron out of the atom and one can never observe more than one point in the orbit of the electron; therefore, there is no orbit in the ordinary sense. The next observation—the third step—will show the electron on its path from the atom. Quite generally there is no way of describing what happens between two consecutive observations. It is of course tempting to say that the electron must have been somewhere between the two observations and that therefore the electron must have described some kind of path or orbit even if it may be impossible to know which path. This would be a reasonable argument in classical physics. But in quantum theory it would be a misuse of the language which, as we will see later, cannot be justified. We can leave it open for the moment, whether this warning is a statement about the way in which we should talk about atomic events or a statement about the events themselves, whether it refers to epistemology or to ontology. In any case we have to be very cautious about the wording of any statement concerning the behavior of atomic particles.

Actually we need not speak of particles at all. For many experiments it is more convenient to speak of matter waves; for instance, of stationary matter waves around the atomic nucleus. Such a description would directly contradict the other description if one does not pay attention to the limitations given by the uncertainty relations. Through the limitations the contradiction is avoided. The use of "matter waves" is convenient, for example, when dealing with the radiation emitted by the atom. By means of its frequencies and intensities the radiation gives information about the oscillating charge distribution in the atom, and there the wave picture comes much nearer to the truth than the particle picture. Therefore, Bohr advocated the use of both pictures, which he called "complementary" to each other. The two pictures are of course mutually exclusive, because a certain thing cannot at the same time be a particle (i.e., substance confined to a very small volume) and a wave (i.e., a field spread out over a large space), but the two complement each other. By playing with both pictures, by going from the one picture to the other and back again, we finally get the right impression of the strange kind of reality behind our atomic experiments. Bohr uses the concept of "complementarity" at several places in the interpretation of quantum theory. The knowledge of the position of a particle is complementary to the knowledge of its velocity or momentum. If we know the one with high accuracy we cannot know the other with high accuracy; still we must know both for determining the behavior of the system. The space-time description of the atomic events is complementary to their deterministic description. The probability function obeys an equation of motion as the co-ordinates did in Newtonian mechanics; its change in the course of time is completely determined by the quantum mechanical equation, but it does not allow a description in space and time. The observation, on the other hand, enforces the description in space and time but breaks the determined continuity of the probability function by changing our knowledge of the system.

Generally the dualism between two different descriptions of the same reality is no longer a difficulty since we know from the mathematical formulation of the theory that contradictions cannot arise. The dualism between the two complementary pictures—waves and particles—is also clearly brought out in the flexibility of the mathematical scheme. The formalism is normally written to resemble Newtonian mechanics, with equations of motion for the co-ordinates and the momenta of the particles. But by a simple transformation it can be rewritten to resemble a wave equation for an ordinary three-dimensional matter wave. Therefore, this possibility of playing with different complementary pictures has its analogy in the different transformations of the mathematical scheme; it does not lead to any difficulties in the Copenhagen interpretation of quantum theory.

A real difficulty in the understanding of this interpretation arises, however, when one asks the famous question: But what happens "really" in an atomic event? It has been said before that the mechanism and the results of an observation can always be stated in terms of the classical concepts. But what one deduces from an observation is a probability function, a mathematical expression that combines statements about possibilities or tendencies with statements about our knowledge of facts. So we cannot completely objectify the result of an observation, we cannot describe what "happens" between this observation and the next. This looks as if we had introduced an element of subjectivism into the theory, as if we meant to say: what happens depends on our way of observing it or on the fact that we observe it. Before discussing this problem of subjectivism it is necessary to explain quite clearly why one would get into hopeless difficulties if one tried to describe what happens between two consecutive observations.

For this purpose it is convenient to discuss the following ideal experiment: We assume that a small source of monochromatic light radiates toward a black screen with two small holes in it. The diameter of the holes may be not much bigger than the wave length of the light, but their distance will be very much bigger. At some distance behind the screen a photographic plate registers the incident light. If one describes this experiment in terms of the wave picture, one says that the primary wave penetrates through the two holes; there will be secondary spherical waves starting from the holes that interfere with one another, and the interference will produce a pattern of varying intensity on the photographic plate.

The blackening of the photographic plate is a quantum process, a chemical reaction produced by single light quanta. Therefore, it must also be possible to describe the experiment in terms of light quanta. If it would be permissible to say what happens to the single light quantum between its emission from the light source and its absorption in the photographic plate, one could argue as follows: The single light quantum can come through the first hole or through the second one. If it goes through the first hole and is scattered there, its probability for being absorbed at a certain point of the photographic plate cannot depend upon whether the second hole is closed or open. The probability distribution on the plate will be the same as if only the first hole was open. If the experiment is repeated many times and one takes together all cases in which the light quantum has gone through the first hole, the blackening of the plate due to these cases will correspond to this probability distribution. If one considers only those light quanta that go through the second hole, the blackening should correspond to a probability distribution derived from the assumption that only the second hole is open. The total blackening, therefore, should just be the sum of the blackenings in the two cases; in other words, there should be no interference pattern. But we know this is not correct, and the experiment will show the interference pattern. Therefore, the statement that any light quantum must have gone *either* through the first *or* through the second hole is problematic and leads to contradictions. This example shows clearly that the concept of the probability function does not allow a description of what happens between two observations. Any attempt to find such a description would lead to contradictions; this must mean that the term "happen" is restricted to the observation.

Now, this is a very strange result, since it seems to indicate that the observation plays a decisive role in the event and that the reality varies, depending upon whether we observe it or not. To make this point clearer we have to analyze the process of observation more closely.

To begin with, it is important to remember that in natural science we are not interested in the universe as a whole, including ourselves, but we direct our attention to some part of the universe and make that the object of our studies. In atomic physics this part is usually a very small object, an atomic particle or a group of such particles, sometimes much larger—the size does not matter; but it is important that a large part of the universe, including ourselves, does *not* belong to the object.

Now, the theoretical interpretation of an experiment starts with the two steps that have been discussed. In the first step we have to describe the arrangement of the experiment, eventually combined with the first observation, in terms of classical physics and translate this description into a probability function. This probability function follows the laws of quantum theory, and its change in the course of time, which is continuous, can be calculated from the initial conditions; this is the second step. The probability function combines objective and subjective elements. It contains statements about possibilities or better tendencies ("potentia" in Aristotelian philosophy), and these statements are completely objective, they do not depend on any observer; and it contains statements about our knowledge of the system, which of course are subjective in so far as they may be different for different observers. In ideal cases the subjective element in the probability function may be practically negligible as compared with the objective one. The physicists then speak of a "pure case."

When we now come to the next observation, the result of which should be predicted from the theory, it is very important to realize that our object has to be in contact with the other part of the world, namely, the experimental arrangement, the measuring rod, etc., before or at least at the moment of observation. This means that the equation of motion for the probability function does now contain the influence of the interaction with the measuring device. This influence introduces a new element of uncertainty, since the measuring device is necessarily described in the terms of classical physics; such a description contains all the uncertainties concerning the microscopic structure of the device which we know from thermodynamics, and since the device is connected with the rest of the world, it contains in fact the uncertainties of the microscopic structure of the whole world. These uncertainties may be called objective in so far as they are simply a consequence of the description in the terms of classical physics and do not depend on any observer. They may be called subjective in so far as they refer to our incomplete knowledge of the world.

After this interaction has taken place, the probability function contains the objective element of tendency and the subjective element of incomplete knowledge, even if it has been a "pure case" before. It is for this reason that the result of the observation cannot generally be predicted with certainty; what can be predicted is the probability of a certain result of the observation, and this statement about the probability can be checked by repeating the experiment many times. The probability function does—unlike the common procedure in Newtonian mechanics—not describe a certain event but, at least during the process of observation, a whole ensemble of possible events.

The observation itself changes the probability function discontinuously; it selects of all possible events the actual one that has taken place. Since through the observation our knowledge of the system has changed discontinuously, its mathematical representation also has undergone the discontinuous change and we speak of a "quantum jump." When the old adage "Natura non facit saltus" is used as a basis for criticism of quantum theory, we can reply that certainly our knowledge can change suddenly and that this fact justifies the use of the term "quantum jump."

Therefore, the transition from the "possible" to the "actual" takes place during the act of observation. If we want to describe what happens in an atomic event, we have to realize that the word "happens" can apply only to the observation, not to the state of affairs between two observations. It applies to the physical, not the psychical act of observation, and we may say that the transition from the "possible" to the "actual" takes place as soon as the interaction of the object with the measuring device, and thereby with the rest of the world, has come into play; it is not connected with the act of registration of the result by the mind of the observer. The discontinuous change in the probability function, however, takes place with the act of registration, because it is the discontinuous change of our knowledge in the instant of registration that has its image in the discontinuous change of the probability function.

To what extent, then, have we finally come to an objective description of the world, especially of the atomic world? In classical physics science started from the belief—or should one say from the illusion?—that we could describe the world or at least parts of the world without any reference to ourselves. This is actually possible to a large extent. We know that the city of London exists whether we see it or not. It may be said that classical physics is just that idealization in which we can speak about parts of the world without any reference to ourselves. Its success has led to the general ideal of an objective description of the world. Objectivity has become the first criterion for the value of any scientific result. Does the Copenhagen interpretation of quantum theory still comply with this ideal? One may perhaps say that quantum theory corresponds to this ideal as far as possible. Certainly quantum theory does not contain genuine subjective features, it does not introduce the mind of the physicist as a part of the atomic event. But it starts from the division of the world into the "object" and the rest of the world, and from the fact that at least for the rest of the world we use the classical concepts in our description. This division is arbitrary and historically a direct consequence of our scientific method; the use of the classical concepts is finally a consequence of the general human way of thinking. But this is already a reference to ourselves and in so far our description is not completely objective.

It has been stated in the beginning that the Copenhagen interpretation of quantum theory starts with a paradox. It starts from the fact that we describe our experiments in the terms of classical physics and at the same time from the knowledge that these concepts do not fit nature accurately. The tension between these two starting points is the root of the statistical character of quantum theory. Therefore, it has sometimes been suggested that one should depart from the classical concepts altogether and that a radical change in the concepts used for describing the experiments might possibly lead back to a nonstatical, completely objective description of nature.

This suggestion, however, rests upon a misunderstanding. The concepts of classical physics are just a refinement of the concepts of daily life and are an essential part of the language which forms the basis of all natural science. Our actual situation in science is such that we *do* use the classical concepts for the description of the experiments, and it was the problem of quantum theory to find theoretical interpretation of the experiments on this basis. There is no use in discussing what could be done if we were other beings than we are. At this point we have to realize, as von Weizsäcker has put it, that "Nature is earlier than man, but man is earlier than natural science." The first part of the sentence justifies classical physics, with its ideal of complete objectivity. The second part tells us why we cannot escape the paradox of quantum theory, namely, the necessity of using the classical concepts.

We have to add some comments on the actual procedure in the quantum-theoretical interpretation of atomic events. It has been said that we always start with a division of the world into an object, which we are going to study, and the rest of the world, and that this division is to some extent arbitrary. It should indeed not make any difference in the final result if we, e.g., add some part of the measuring device or the whole device to the object and apply the laws of quantum theory to this more complicated object. It can be shown that such an alteration of the theoretical treatment would not alter the predictions concerning a given experiment. This follows mathematically from the fact that the laws of quantum theory are for the phenomena in which Planck's constant can be considered as a very small quantity, approximately identical with the classical laws. But it would be a mistake to believe that this application of the quantum-theoretical laws to the measuring device could help to avoid the fundamental paradox of quantum theory.

The measuring device deserves this name only if it is in close contact with the rest of the world, if there is an interaction between the device and the observer. Therefore, the uncertainty with respect to the microscopic behavior of the world will enter into the quantum-theoretical system here just as well as in the first interpretation. If the measuring device would be isolated from the rest of the world, it would be neither a measuring device nor could it be described in the terms of classical physics at all.

With regard to this situation Bohr has emphasized that it is more realistic to state that the division into the object and the rest of the world is not arbitrary. Our actual situation in research work in atomic physics is usually this: we wish to understand a certain phenomenon, we wish to recognize how this phenomenon follows from the general laws of nature. Therefore, that part of matter or radiation which takes part in the phenomenon is the natural "object" in the theoretical treatment and should be separated in this respect from the tools used to study the phenomenon. This again emphasizes a subjective element in the description of atomic events, since the measuring device has been constructed by the observer, and we have to remember that what we observe is not nature in itself but nature exposed to our method of questioning. Our scientific work in physics consists in asking questions about nature in the language that we possess and trying to get an answer from experiment by the means that are at our disposal. In this way quantum theory reminds us, as Bohr has put it, of the old wisdom that when searching for harmony in life one must never forget that in the drama of existence we are ourselves both players and spectators. It is understandable that in our scientific relation to nature our own activity becomes very important when we have to deal with parts of nature into which we can penetrate only by using the most elaborate tools.

N.9

The Principle of Superposition

Paul A. M. Dirac

Paul Adrien Maurice Dirac was the son of Swiss parents who had emigrated to England. He was born in Bristol in 1902 and died in 1983 in Tallahassee, Florida. He originally trained as an engineer, but when no satisfactory employment was forthcoming, he took a graduate degree in mathematics in 1926. His grasp of the subject as well as his creativity and insight were such that after only six years he was honored by being named to the position once held by Sir Isaac Newton—the Lucasian Chair of Mathematics at Cambridge University.

In 1930, Dirac deduced the existence of a positive counterpart to the electron: a particle with the positive charge of a proton but with the precise mass of the electron. Such theorizing was hard to accept at the time Dirac's article was published in 1928, but in 1932 Carl Anderson announced the discovery of an antielectron (later named the positron), and thus the existence of antimatter became known. Dirac's prediction was the first time any physicist had foreseen the discovery of a new particle; within two years Pauli had repeated the feat by predicting the existence of what Fermi named the *neutrino*—a stable elementary particle now found in all interactions involving the weak force.

Dirac's greatest achievement was to combine in a new equation the ideas of de Broglie, Schrödinger, Heisenberg, and Born with Einstein's special theory of relativity. This new equation yields the spin of the electron since it is really a combination of four distinct equations, two of which represent the electron in a state of positive energy and two of which represent the electron in a state of negative energy. The positive energy electron has two states of spin; the same is true for the negative energy electron. His famous wave equation (the Dirac equation of the electron) introduces special relativity into Schrödinger's equation; Dirac's work helped to reconcile to some extent the opposing philosophies of relativity theory and quantum theory.

C. P. Snow, who has written charmingly about most of the physicists involved in the development of quantum theory, pointed out that Dirac was brought up by his Swiss parents to be bilingual in English and French, except that he was singularly silent in both languages and truly eloquent only in his manipulation of mathematical symbols. This latter skill has been invaluable in the development of modern physics, because not all who are blessed by extraordinary intuition in physics have the mathematical skills to complement that instinct—the outstanding example being Einstein, who, although certainly no slouch in dealing with mathematics, did not have mathematical skills comparable to those displayed by some other physicists.

Dirac eventually published a book entitled *The Principles of Quantum Mechanics,* which was first issued in 1930 by the Clarendon Press at Oxford and subsequently republished in many editions. Our selection from Dirac's volume includes his discussions of the need for a quantum theory, the polarization of photons, the interference of photons, and superposition and indeterminacy, all facets of what he called "the principle of superposition."

Despite its intimidating designation, Dirac's principle is quite simple: It holds that two or more vibrations or waves may be superimposed upon each other—that is, their respective amplitudes summed in order to give a single wave in which the component waves interfere constructively or destructively depending on their phases. We observe this principle in everyday life when we observe any object which is illuminated by combinations or superpositions of many different waves.

[The probability wave] meant a tendency for something. It was a quantitative version of the old concept of "Potentia" in Aristotelian philosophy. It introduced something standing in the middle between the idea of an event and the actual event, a strange kind of physical reality just in the middle between possibility and reality.

—WERNER HEISENBERG, *Physics and Philosophy*

. . . An independent reality in the ordinary sense can be ascribed neither to the phenomena nor to the agencies of observation.

—NIELS BOHR, *Atomic Theory and the Description of Nature*

THE NEED FOR A QUANTUM THEORY

CLASSICAL MECHANICS HAS been developed continuously from the time of Newton and applied to an ever-widening range of dynamical systems, including the electromagnetic field of interaction with matter. The underlying ideas and the laws governing their application form a simple and elegant scheme, which one would be inclined to think could not be seriously modified without having all its attractive features spoilt. Nevertheless it has been found possible to set up a new scheme, called quantum mechanics, which is more suitable for the description of phenomena on the atomic scale and which is in some respects more elegant and satisfying than the classical scheme. This possibility is due to the changes which the new scheme involves being of a very profound character and not clashing with the features of the classical theory that make it so attractive, as a result of which all these features can be incorporated in the new scheme.

The necessity for a departure from classical mechanics is clearly shown by experimental results. In the first place the forces known in classical electrodynamics are inadequate for the explanation of the remarkable stability of atoms and molecules, which is necessary in order that materials may have any definite physical and chemical properties at all. The introduction of new hypothetical forces will not save the situation, since there exist general principles of classical mechanics, holding for all kinds of forces, leading to results in direct disagreement with observation. For example, if an atomic system has its equilibrium disturbed in any way and is then left alone, it will be set in oscillation and the oscillations will get impressed on the surrounding electromagnetic field, so that their frequencies may be observed with a spectroscope. Now whatever the laws of force governing the equilibrium, one would expect to be able to include the various frequencies in a scheme comprising certain fundamental frequencies and their harmonies. This is not observed to be the case. Instead, there is observed a new and unexpected connexion between the frequencies, called Ritz's Combination Law of Spectroscopy, according to which all the frequencies can be expressed as differences between certain terms, the number of terms being much less than the number of frequencies. This law is quite unintelligible from the classical standpoint.

One might try to get over the difficulty without departing from classical mechanics by assuming each of the spectroscopically observed frequencies to be a fundamental frequency with its own degree of freedom, the laws of force being such that the harmonic vibrations do not occur. Such a theory will not do, however, even apart from the fact that it would give no explanation of the Combination Law, since it would immediately bring one into conflict with the experimental evidence on specific heats. Classical statistical mechanics enables one to establish a general connexion between the total number of degrees of freedom of an assembly of vibrating systems and its specific heat. If one assumes all the spectroscopic frequencies of an atom to correspond to different degrees of freedom, one would get a specific heat for any kind of matter very much greater than the observed value. In fact the observed specific heats at ordinary temperatures are given fairly well by a theory that takes into account merely the motion of each atom as a whole and assigns no internal motion to it at all.

This leads us to a new clash between classical mechanics and the results of experiment. There must certainly be some internal motion in an atom to account for its spectrum, but the internal degrees of freedom, for some classically inexplicable reason, do not contribute to the specific heat. A similar clash is found in connexion with the energy of oscillation of the electromagnetic field in a vacuum. Classical mechanics requires the specific heat corresponding to this energy to be infinite, but it is observed to be quite finite. A general conclusion from experimental results is that oscillations of high frequency do not contribute their classical quota to the specific heat.

As another illustration of the failure of classical mechanics we may consider the behavior of light. We have, on the one hand, the phenomena of interference and diffraction, which can be explained only on the basis of a wave theory; on the other, phenomena such as photo-electric emission and scattering by free electrons, which show that light is composed of small particles. These particles, which are called photons, have each a definite energy and momentum, depending on the frequency of the light, and appear to have just as real an existence as electrons, or any other particles known in physics. A fraction of a photon is never observed.

Experiments have shown that this anomalous behaviour is not peculiar to light, but is quite general. All material particles have wave properties, which can be exhibited under suitable conditions. We have here a very striking and general example of the breakdown of classical mechanics—not merely an inaccuracy in its laws of motion, but *an inadequacy of its concepts to supply us with a description of atomic events.*

The necessity to depart from classical ideas when one wishes to account for the ultimate structure of matter may be seen, not only from experimentally established facts, but also from general philosophical grounds. In a classical explanation of the constitution of matter, one would assume it to be made up of a large number of small constituent parts and one would postulate laws for the behaviour of these parts, from which the laws of the matter in bulk could be deduced. This would not complete the explanation, however, since the question of the structure and stability of the constituent parts is left untouched. To go into this question, it becomes necessary to postulate that each constituent part is itself made up of smaller parts, in terms of which its behaviour is to be explained. There is clearly no end to this procedure, so that one can never arrive at the ultimate structure of matter on these lines. So long as *big* and *small* are merely relative concepts, it is no help to explain the big in terms of the small. It is therefore necessary to modify classical ideas in such a way as to give an absolute meaning to size.

At this stage it becomes important to remember that science is concerned only with observable things and that we can observe an object only by letting it interact with some outside influence. An act of observation is thus necessarily accompanied by some disturbance of the object observed. We may define an object to be big when the disturbance accompanying our observation of it may be neglected, and small when the disturbance cannot be neglected. This definition is in close agreement with the common meanings of big and small.

It is usually assumed that, by being careful, we may cut down the disturbance accompanying our observation to any desired extent. The concepts of big and small are then purely relative and refer to the gentleness of our means of observation as well as to the object being described. In order to give an absolute meaning to size, such as is required for any theory of the ultimate structure of matter, we have to assume that *there is a limit to the fineness of our powers of observation and the smallness of the accompanying disturbance—a limit which is inherent in the nature of things and can never be surpassed by improved technique or increased skill on the part of the observer.* If the object under observation is such that the unavoidable limiting disturbance is negligible, then the object is big in the absolute sense and we may apply classical mechanics to it. If, on the other hand, the limiting disturbance is not negligible, then the object is small in the absolute sense and we require a new theory for dealing with it.

A consequence of the preceding discussion is that we must revise our ideas of causality. Causality applies only to a system which is left undisturbed. If a system is small, we cannot observe it without producing a serious disturbance and hence we cannot expect to find any causal connexion between the results of our observations. Causality will still be assumed to apply to undisturbed systems and the equations which will be set up to describe an undisturbed system will be differential equations expressing a causal connexion between conditions at one time and conditions at a later time. These equations will be in close correspondence with the equations of classical mechanics, but they will be connected only indirectly with the results of observations. There is an unavoidable indeterminacy in the calculation of observational results, the theory enabling us to calculate in general only the probability of our obtaining a particular result when we make an observation.

THE POLARIZATION OF PHOTONS

The discussion in the preceding section about the limit to the gentleness with which observations can be made and the consequent indeterminacy in the results of those observations does not provide any quantitative basis for the building up of quantum mechanics. For this purpose a new set of accurate laws of nature is required. One of the most fundamental and most drastic of these is the *Principle of Superposition of States.* We shall lead up to a general formulation of this principle through a consideration of some special cases, taking first the example provided by the polarization of light.

It is known experimentally that when plane-polarized light is used for ejecting photo-electrons, there is a preferential direction for the electron emission. Thus the polarization properties of light are closely connected with its corpuscular properties and one must ascribe a polarization to the photons. One must consider, for instance, a beam of light plane-polarized in a certain direction as consisting of photons each of which is plane-polarized in that direction and a beam of circularly polarized light as consisting of photons each circularly polarized. Every photon is in a certain *state of polarization,* as we shall say. The problem we must now consider is how to fit in these ideas with the known facts about the resolution of light into polarized components and the recombination of these components.

Let us take a definite case. Suppose we have a beam of light passing through a crystal of tourmaline, which has the property of letting through only light plane-polarized perpendicular to its optic axis. Classical electrodynamics tells us what will happen for any given polarization of the incident beam. If this beam is polarized perpendicular to the optic axis, it will all go through the crystal; if parallel to the axis, none of it will go through; while if polarized at an angle α to the axis, a fraction $\sin^2\alpha$ will go through. How are we to understand these results on a photon basis?

A beam that is plane-polarized in a certain direction is to be pictured as made up of photons each plane-polarized in that direction. This picture leads to no difficulty in the cases when our incident beam is polarized perpendicular or parallel to the optic axis. We merely have to suppose that each photon polarized perpendicular to the axis passes unhindered and unchanged through the crystal, while each photon polarized parallel to the axis is stopped and absorbed. A difficulty arises, however, in the case of the obliquely polarized incident beam. Each of the incident photons is then obliquely polarized and it is not clear what will happen to such a photon when it reaches the tourmaline.

A question about what will happen to a particular photon under certain conditions is not really very precise. To make it precise one must imagine some experiment performed having a bearing on the question and inquire what will be the result of the experiment. Only questions about the results of experiments have a real significance and it is only such questions that theoretical physics has to consider.

In our present example the obvious experiment is to use an incident beam consisting of only a single photon and to observe what appears on the back side of the crystal. According to quantum mechanics the result of this experiment will be that sometimes one will find a whole photon, of energy equal to the energy of the incident photon, on the back side and other times one will find nothing. When one finds a whole photon, it will be polarized perpendicular to the optic axis. One will never find only a part of a photon on the back side. If one repeats the experiment a large number of times, one will find the photon on the back side in a fraction $\sin^2\alpha$ of the total number of times. Thus we may say that the photon has a probability $\sin^2\alpha$ of passing through the tourmaline and appearing on the back side polarized perpendicular to the axis and a probability $\cos^2\alpha$ of being absorbed. These values for the probabilities lead to the correct classical results for an incident beam containing a large number of photons.

In this way we preserve the individuality of the photon in all cases. We are able to do this, however, only because we abandon the determinacy of the classical theory. The result of an experiment is not determined, as it would be according to classical ideas, by the conditions under the control of the experimenter. The most that can be predicted is a set of possible results, with a probability of occurrence for each.

The foregoing discussion about the result of an experiment with a single obliquely polarized photon incident on a crystal of tourmaline answers all that can legitimately be asked about what happens to an obliquely polarized photon when it reaches the tourmaline. Questions about what decides whether the photon is to go through or not and how it changes its direction of polarization when it does go through cannot be investigated by experiment and should be regarded as outside the domain of science. Nevertheless some further description is necessary in order to correlate the results of this experiment with the results of other experiments that might be performed with photons and to fit them all into a general scheme. Such further description should be regarded, not as an attempt to answer questions outside the domain of science, but as an aid to the formulation of rules for expressing concisely the results of large numbers of experiments.

The further description provided by quantum mechanics runs as follows. It is supposed that a photon polarized obliquely to the optic axis may be regarded as being partly in the state of polarization parallel to the axis and partly in the state of polarization perpendicular to the axis. The state of oblique polarization may be considered as the result of some kind of superposition process applied to the two states of parallel and perpendicular polarization. This implies a certain special kind of relationship between the various states of polarization, a relationship similar to that between polarized beams in classical optics, but which is now to be applied, not to beams, but to the states of polarization of one particular photon. This relationship allows any state of polarization to be resolved into, or expressed as a superposition of, any two mutually perpendicular states of polarization.

When we make the photon meet a tourmaline crystal, we are subjecting it to an observation. We are observing whether it is polarized parallel or perpendicular to the optic axis. The effect of making this observation is to force the photon entirely into the state of parallel or entirely into the state of perpendicular polarization. It has to make a sudden jump from being partly in each of these two states to being entirely in one or other of them. Which of the two states it will jump into cannot be predicted, but is governed only by probability laws. If it jumps into the parallel state it gets absorbed and if it jumps into the perpendicular state it passes through the crystal and appears on the other side preserving this state of polarization.

INTERFERENCE OF PHOTONS

In this section we shall deal with another example of superposition. We shall again take photons, but shall be concerned with their position in space and their momentum instead of their polarization. If we are given a beam of roughly monochromatic light, then we know something about the location and momentum of the associated photons. We know that each of them is located somewhere in the region of space through which the beam is passing and has a momentum in the direction of the beam of magnitude given in terms of the frequency of the beam by Einstein's photo-electric law—momentum equals frequency multiplied by a universal constant. When we have such information about the location and momentum of a photon we shall say that it is in a definite *translational state.*

We shall discuss the description which quantum mechanics provides of the interference of photons. Let us take a definite experiment demonstrating interference. Suppose we have a beam of light which is passed through some kind of interferometer, so that it gets split up into two components and the two components are subsequently made to interfere. We may, as in the preceding section, take an incident beam consisting of only a single photon and inquire what will happen to it as it goes through the apparatus. This will present to us the difficulty of the conflict between the wave and corpuscular theories of light in an acute form.

Corresponding to the description that we had in the case of the polarization, we must now describe the photon as going partly into each of the two components into which the incident beam is split. The photon is then, as we may say, in a translational state given by the superposition of the two translational states associated with the two components. We are thus led to a generalization of the term "translational state" applied to a photon. For a photon to be in a definite translational state it need not be associated with one single beam of light, but may be associated with two or more beams of light which are the components into which one original beam has been split. In the accurate mathematical theory each translational state is associated with one of the wave functions of ordinary wave optics, which wave function may describe either a single beam or two or more beams into which one original beam has been split. Translational states are thus superposable in a similar way to wave functions.

Let us consider now what happens when we determine the energy in one of the components. The result of such a determination must be either the whole photon or nothing at all. Thus the photon must change suddenly from being partly in one beam and partly in the other to being entirely in one of the beams. This sudden change is due to the disturbance in the translational state of the photon which the observation necessarily makes. It is impossible to predict in which of the two beams the photon will be found. Only the probability of either result can be calculated from the previous distribution of the photon over the two beams.

One could carry out the energy measurement without destroying the component beam by, for example, reflecting the beam from a movable mirror and observing the recoil. Our description of the photon allows us to infer that, *after* such an energy measurement, it would not be possible to bring about any interference effects between the two components. So long as the photon is partly in one beam and partly in the other, interference can occur when the two beams are superposed, but this possibility disappears when the photon is forced entirely into one of the beams by an observation. The other beam then no longer enters into the description of the photon, so that it counts as being entirely in the one beam in the ordinary way for any experiment that may subsequently be performed on it.

On these lines quantum mechanics is able to effect a reconciliation of the wave and corpuscular properties of light. The essential point is the association of each of the translational states of a photon with one of the wave functions of ordinary wave optics. The nature of this association cannot be pictured on a basis of classical mechanics, but is something entirely new. It would be quite wrong to picture the photon and its associated wave as interacting in the way in which particles and waves can interact in classical mechanics. The association can be interpreted only statistically, the wave function giving us information about the probability of our finding the photon in any particular place when we make an observation of where it is.

Some time before the discovery of quantum mechanics people realized that the connexion between light waves and photons must be of a statistical character. What they did not clearly realize, however, was that the wave function gives information about the probability of *one* photon being in a particular place and not the probable number of photons in that place. The importance of the distinction can be made clear in the following way. Suppose we have a beam of light consisting of a large number of photons split up into two components of equal intensity. On the assumption that the intensity of a beam is connected with the probable number of photons in it, we should have half the total number of photons going into each component. If the two components are now made to interfere, we should require a photon in one component to be able to interfere with one in the other. Sometimes these two photons would have to annihilate one another and other times they would have to produce four photons. This would contradict the conservation of energy. The new theory, which connects the wave function with probabilities for one photon, gets over the difficulty by making each photon go partly into each of the two components. Each photon then interferes only with itself. Interference between two different photons never occurs.

The association of particles with waves discussed above is not restricted to the case of light, but is, according to modern theory, of universal applicability. All kinds of particles are associated with waves in this way and conversely all wave motion is associated with particles. Thus all particles can be made to exhibit interference effects and all wave motion has its energy in the form of quanta. The reason why these general phenomena are not more obvious is on acount of a law of proportionality between the mass or energy of the particles and the frequency of the waves, the coefficient being such that for waves of familiar frequencies the associated quanta are extremely small, while for particles even as light as electrons the associated wave frequency is so high that it is not easy to demonstrate interference.

SUPERPOSITION AND INDETERMINACY

The reader may possibly feel dissatisfied with the attempt in the two preceding sections to fit in the existence of photons with the classical theory of light. He may argue that a very strange idea has been introduced—the possibility of a photon being partly in each of two states of polarization, or partly in each of two separate beams—but even with the help of this strange idea no satisfying picture of the fundamental single-photon processes has been given. He may say further that this strange idea did not provide any information about experimental results for the experiments discussed, beyond what could have been obtained from an elementary consideration of photons being guided in some vague way by waves. What, then, is the use of the strange idea?

In answer to the first criticism it may be remarked that the main object of physical science is not the provision of pictures, but is the formulation of laws governing phenomena and the application of these laws to the discovery of new phenomena. If a picture exists, so much the better; but whether a picture exists or not is a matter of only secondary importance. In the case of atomic phenomena no picture can be expected to exist in the usual sense of the word "picture," by which is meant a model functioning essentially on classical lines. One may, however, extend the meaning of the word "picture" to include any *way of looking at the fundamental laws which makes their self-consistency obvious.* With this extension, one may gradually acquire a picture of atomic phenomena by becoming familiar with the laws of the quantum theory.

With regard to the second criticism, it may be remarked that for many simple experiments with light, an elementary theory of waves and photons connected in a vague statistical way would be adequate to account for the results. In the case of such experiments quantum mechanics has no further information to give. In the great majority of experiments, however, the conditions are too complex for an elementary theory of this kind to be applicable and some more elaborate scheme, such as is provided by quantum mechanics, is then needed. The method of description that quantum mechanics gives in the more complex cases is applicable also to the simple cases and although it is then not really necessary for accounting for the experimental results, its study in these simple cases is perhaps a suitable introduction to its study in the general case.

There remains an overall criticism that one may make to the whole scheme, namely, that in departing from the determinacy of the classical theory a great complication is introduced into the description of Nature, which is a highly undesirable feature. This complication is undeniable, but it is offset by a great simplification, provided by the general *principle of superposition of states,* which we shall now go on to consider. But first it is necessary to make precise the important concept of a "state" of a general atomic system.

Let us take any atomic system, composed of particles or bodies with specified properties (mass, moment of inertia, etc.) interacting according to specified laws of force. There will be various possible motions of the particles or bodies consistent with the laws of force. Each such motion is called a *state* of the system. According to classical ideas one could specify a state by giving numerical values to all the coordinates and velocities of the various component parts of the system at some instant of time, the whole motion being then completely determined. Now the argument shows that we cannot observe a *small* system with that amount of detail which classical theory supposes. The limitation in the power of observation puts a limitation on the number of data that can be assigned to a state. Thus a state of an atomic system must be specified by fewer or more indefinite data than a complete set of numerical values for all the coordinates and velocities at some instant of time. In the case when the system is just a single photon, a state would be completely specified by a given translational state . . . together with a given state of polarization. . . .

A state of a system may be defined as an undisturbed motion that is restricted by as many conditions or data as are theoretically possible without mutual interference or contradiction. In practice the conditions could be imposed by a suitable preparation of the system, consisting perhaps in passing it through various kinds of sorting apparatus, such as slits and polarimeters, the system being left undisturbed after the preparation. The word "state" may be used to mean either the state at one particular time (after the preparation), or the state throughout the whole of time after the preparation. To distinguish these two meanings, the latter will be called a "state of motion" when there is liable to be ambiguity.

The general principle of superposition of quantum mechanics applies to the states, with either of the above meanings, of any one dynamical system. It requires us to assume that between these states there exist peculiar relationships such that whenever the system is definitely in one state we can consider it as being partly in each of two or more other states. The original state must be regarded as the result of a kind of *superposition* of the two or more new states, in a way that cannot be conceived on classical ideas. Any state may be considered as the result of a superposition of two or more other states, and indeed in an infinite number of ways. Conversely any two or more states may be superposed to give a new state. The procedure of expressing a state as the result of superposition of a number of other states is a mathematical procedure that is always permissible, independent of any reference to physical conditions, like the procedure of resolving a wave into Fourier components. Whether it is useful in any particular case, though, depends on the special physical conditions of the problem under consideration.

In the two preceding sections examples were given of the superposition principle applied to a system consisting of a single photon. The first dealt with states differing only with regard to the polarization and the second with states differing only with regard to the motion of the photon as a whole.

The nature of the relationships which the superposition principle requires to exist between the states of any system is of a kind that cannot be explained in terms of familiar physical concepts. One cannot in the classical sense picture a system being partly in each of two states and see the equivalence of this to the system being completely in some other state. There is an entirely new idea involved, to which one must get accustomed and in terms of which one must proceed to build up an exact mathematical theory, without having any detailed classical picture.

When a state is formed by the superposition of two other states, it will have properties that are in some vague way intermediate between those of the two original states and that approach more or less closely to those of either of them according to the greater or less "weight" attached to this state in the superposition process. The new state is completely defined by the two original states when their relative weights in the superposition process are known, together with a certain phase difference, the exact meaning of weights and phases being provided in the general case by the mathematical theory. In the case of the polarization of a photon their meaning is that provided by classical optics, so that, for example, when two perpendicularly plane-polarized states are superposed with equal weights, the new state may be circularly polarized in either direction, or linearly polarized at an angle $\frac{1}{4}\pi$, or else elliptically polarized, according to the phase difference.

The non-classical nature of the superposition process is brought out clearly if we consider the superposition of two states, A and B, such that there exists an observation which, when made on the system in state A, is certain to lead to one particular result, a say, and when made on the system in state B is certain to lead to some different result, b say. What will be the result of the observation when made on the system in the superposed state? The answer is that the result will be sometimes a and sometimes b, according to a probability law depending on the relative weights of A and B in the superposition process. It will never be different from both a and b. *The intermediate character of the state formed by superposition thus expresses itself through the probability of a particular result for an observation being intermediate between the corresponding probabilities for the original states, not through the result itself being intermediate between the corresponding results for the original states.*

In this way we see that such a drastic departure from ordinary ideas as the assumption of superposition relationships between the states is possible only on account of the recognition of the importance of the disturbance accompanying an observation and of the consequent indeterminacy in the result of the observation. When an observation is made on any atomic system that is in a given state, in general the result will not be determinate, i.e., if the experiment is repeated several times under identical conditions several different results may be obtained. It is a law of nature, though, that if the experiment is repeated a large number of times, each particular result will be obtained in a definite fraction of the total number of times, so that there is a definite *probability* of its being obtained. This probability is what the theory sets out to calculate. Only in special cases when the probability for some result is unity is the result of the experiment determinate.

The assumption of superposition relationships between the states leads to a mathematical theory in which the equations that define a state are linear in the unknowns. In consequence of this, people have tried to establish analogies with systems in classical mechanics, such as vibrating strings or membranes, which are governed by linear equations and for which, therefore, a superposition principle holds. Such analogies have led to the name "Wave Mechanics" being sometimes given to quantum mechanics. It is important to remember, however, that *the superposition that occurs in quantum mechanics is of an essentially different nature from any occurring in the classical theory,* as is shown by the fact that the quantum superposition principle demands indeterminacy in the results of observations in order to be capable of a sensible physical interpretation. The analogies are thus liable to be misleading. . . .

N.10

Tomonaga, Schwinger, and Feynman

FREEMAN DYSON

FREEMAN DYSON is a Cambridge University-trained physicist who emigrated to the United States in 1951. He was born in Crowthorne, England, in 1923 and became an American citizen in 1957. He was Professor of Physics at Cornell University for several years before moving to the Institute for Advanced Study at Princeton University. Dyson's scholarly specialty has been the theory of quantum electrodynamics, but he has also gained fame among a wider audience for his well-written science books, including one on nuclear weapons and public policy, and, particularly, for his speculations on the possibility of life on other planets. Dyson has bravely formulated the theory that an extremely advanced civilization might find a way of building absorbing structures within a spherical globe around a star to serve as a trap for the star's radiation. The star might then be used as a virtually infinite energy source, its waste heat vented into interstellar space. The presence of such a structure might be indicated by the sudden dimming of an ordinary star, a symptom of the intensive exploitation of the star's radiative energy.

The 1965 Nobel Prize in physics was awarded to Shin'ichirō Tomonaga of Tokyo, Julian Schwinger of Harvard, and Richard Feynman of the California Institute of Technology for their contributions (developed independently for the most part) to the modern theory of quantum electrodynamics or QED. In the paper which follows, Dyson admits the difficulty of describing QED to the layman. Other writers have called QED "the most exact scientific theory ever produced . . . one of the major intellectual triumphs of this century."

Dyson's paper explains how Tomonaga, isolated in Japan during World War II, maintained a flourishing school of young theoretical physicists. Tomonaga published his own fundamental paper on QED in 1943 in Japanese but an English version did not become available in America until the winter of 1947–1948. It was then that Schwinger and Feynman were beginning to combine their respective labors into a basic theory that would prove to be unrivaled in dealing accurately with the motions and interactions of electrons, muons, and photons, that is, the entire range of electromagnetic interactions.

Nothing is more important about quantum physics than this: it has destroyed the concept of the world as "sitting out there." The universe afterwards will never be the same.

—JOHN ARCHIBALD WHEELER

THE 1965 NOBEL Prize for physics has been awarded to three theorists, Shin'ichirō Tomonaga of Tokyo, Julian Schwinger of Harvard, and Richard Feynman of the California Institute of Technology. The prize was given for their creation of the modern theory of quantum electrodynamics. This is the theory which brought order and harmony into the vast middle ground of physics, excluding gravitation on the one side and nuclear forces on the other, but including the laws of atomic structure, radiation, creation and annihilation of particles, solid-state physics, plasma physics, maser and laser technology, optical and microwave spectroscopy, electronics, and chemistry. Quantum electrodynamics unifies all these diverse phenomena into a small number of principles of great generality and elegance, weaving together special relativity with quantum mechanics in a seamless fabric. It is in a certain sense the most perfect and the most highly developed part of physics.

Since its completion in 1948, the theory has been tested by means of a succession of experiments of steadily increasing accuracy. For example, the magnetic moment of the electron was recently measured by Crane at the University of Michigan with an error of less than 1 part in 10 million. This was a beautiful and formidably difficult experiment, but unfortunately the result attracted little attention; it only proved that quantum electrodynamics was right to two more places of decimals.

Just this year there have been experimental indications of a possible deviation from the theory in the behavior of electron-positron pairs produced at energies of billions of volts. If confirmed, this deviation will by no means invalidate the theory, but will only show for the first time where the boundary lies between quantum electrodynamics and the world of high-energy particles. It is still one of the major mysteries of physics how quantum electrodynamics, a theory which deliberately excludes from

---

Freeman Dyson, "Tomonaga, Schwinger, and Feynman," *Science* Vol. 150 (October 29, 1965), pp. 588–591. Reprinted with permission of the American Association for the Advancement of Science.

consideration all particles except the well-known electron, positron, and photon, can give so amazingly accurate a representation of reality over so wide a range of conditions.

The three creators of the theory did their work independently and not simultaneously. Tomonaga kept alive in Japan during World War II a school of theoretical physics which was in some ways ahead of the rest of the world. In these conditions of total isolation he published his fundamental paper in Japanese in 1943. Schwinger and Feynman were meanwhile fully occupied with the development of radar and nuclear energy, respectively. When they returned to academic life after the war, their interest was aroused by a series of new experiments on the fine details of the hydrogen atom. The experiments had become possible as a result of the wartime development of microwave techniques, and were about a thousand times more accurate than the best prewar measurements. The new experiments made glaringly obvious the lack of a satisfactory theory of radiative processes, and so Schwinger and Feynman were led along different paths to invent such a theory. Each of them completed his work during the winter of 1947–48, just at the time the first English-language translations of the papers of Tomonaga and his students began to arrive from Japan. It was interesting to find that, although the new experiments had played a decisive role in the thinking of Schwinger and Feynman, Tomonaga had been able to reach an essentially identical insight on the basis of theoretical considerations alone.

The fact that the theory had three discoverers rather than one proved very fruitful for its further development. Each of the three brought a different viewpoint and a different style, and so the theory gained in breadth and richness. Tomonaga was most concerned with basic physical principles; his papers were simple, clear, and free from elaboration of detail. Schwinger was most concerned with the construction of a complete and massive mathematical formulation; his papers were monuments of formal ingenuity. An unkind critic once said: "Other people publish to show how to do it, but Julian Schwinger publishes to show you that only he can do it." It was in fact Schwinger who was the first to hack his way through the mathematical jungle and arrive at a definite numerical value for the magnetic moment of the electron.

Feynman's approach was the most original of the three: he was willing to take nothing for granted, and so he was forced to reconstruct almost the whole of quantum mechanics and electrodynamics from his own point of view. He was concerned with deriving simple rules for the direct calculation of physically observable quantities. His invention of "Feynman graphs" and "Feynman integrals" made it easy to apply the theory to concrete problems. In the end, Feynman's rules of calculation have become standard tools of theoretical analysis, not only in quantum electrodynamics but in high-energy physics as a whole. And Feynman's insistence on discussing directly observable quantities led to the growth of the "S-matrix point of view," which now dominates current thinking about the fundamental particles and their interactions.

The theory which came to triumph in 1948 is not an easy one to describe in nontechnical language. It must be placed in the context of some earlier history. The pioneers of quantum mechanics—Dirac, Heisenberg, Pauli, and Fermi—had worked out the physical basis for quantum electrodynamics during the late 1920's. The basis consisted in a direct application of the methods of quantum mechanics to the Maxwell equations describing the electromagnetic field. The resulting theory seemed to give a qualitatively correct account of radiation processes, but it failed to give exact predictions. When pushed beyond the first approximation, it always gave infinite or meaningless answers. In the face of this situation, the physicists of the 1930's mostly looked for radical changes in the theory. It was generally believed that the "divergence difficulties" were symptoms of fundamental errors, and were only to be escaped by altering the theory drastically. So from 1935 to 1945 there was a succession of fruitless attempts to cure quantum electrodynamics of the divergence disease by methods of radical surgery.

Tomonaga, Schwinger, and Feynman rescued the theory without making any radical innovations. Their victory was a victory of conservatism. They kept the physical basis of the theory precisely as it had been laid down by Dirac, and only changed the mathematical superstructure. By polishing and refining with great skill the mathematical formalism, they were able to show that the theory does in fact give meaningful predictions for all observable quantities. The predictions are in all cases finite, unambiguous, and in agreement with experiment. The divergent and meaningless quantities are indeed present in the theory, but they appear in such a way that they automatically eliminate themselves from any quantity which is in principle observable. The exact correspondence between quantities which are unambiguously calculable and quantities which are observable becomes, in the end, the theory's most singular virtue.

The theory, as Tomonaga, Schwinger, and Feynman left it, has stood the test of time for seventeen years. It describes only a part of physical reality, and it makes no claim to finality. But its success within its area of applicability has been so complete that it seems sure to survive, at least as a special limiting case, within any more comprehensive theory that may come later to supersede it.

N.11

The Development of Space-Time View of Quantum Electrodynamics

RICHARD FEYNMAN

THROUGH HIS MASTERFUL text *Feynman's Lectures on Physics* Richard Feynman, one of the most distinguished American physicists of the twentieth century, has taught, directly or indirectly, an entire generation of younger physicists. In a well-known Public Broadcasting television interview, he has been presented to the world as something of a celebrity. And in a recent best-selling book of autobiographical reminiscences and philosophical musings, he bids well to become an observer of what C. P. Snow once referred to as "the culture of science."

Feynman was born in New York City in 1918. In 1939, he earned his undergraduate degree at the Massachusetts Institute of Technology; then in 1942 his doctorate was awarded to him at Princeton. After serving as Professor of Theoretical Physics at Cornell University, he accepted a similar position at the California Institute of Technology.

Feynman's approach to physics is original in that he builds his own intellectual conception from the ground up. In the following lecture we see his mind at work, struggling against the vast, but necessarily precise theory of quantum electrodynamics. Feynman also voices his wonder that scientific truth (or at least our understanding of that truth) can so often be written in different ways—"the differential equations of Maxwell . . . the Schrödinger equation and the Heisenberg formulation of quantum mechanics. . . . There is always another way to say the same thing that doesn't look at all like the way you said it before. I don't know what the reason for this is [but] I think it is somehow a representation of the simplicity of nature."

Quantum electrodynamics is a relativistic theory of quantum mechanics concerned with electromagnetic interactions. What is known as the "Feynman propagator approach" describes the scattering of electrons and photons in terms of an integral that sums up contributions to the interactions from all possible ways in which the particles can interact by the exchange of what are called virtual photons and electron-positron pairs. The existence of virtual particles is made possible by the Heisenberg uncertainty principle's allowance for brief violations of the law of conservation of mass and energy during which, for short periods of time, particles may be created that would otherwise be forbidden. Quantum electrodynamics combines the electromagnetic field with the particle manifestations of electromagnetic waves:

> Since photons are also electromagnetic waves, and since these waves are vibrating fields, the photons must be manifestations of electromagnetic fields. Hence the concept of a quantum field: that is, of a field which takes the form of quanta, or particles. This is indeed an entirely new concept which has been extended to describe all subatomic particles and their interactions, each type of particle corresponding to a different field. In these quantum field theories, the classical contrast between the solid particles and the space surrounding them is completely overcome. The quantum field is seen as the fundamental physical entity: a continuous medium which is present everywhere in space. Particles are merely local condensations of the field; concentrations of energy which come and go, thereby losing their individual character and dissolving into the underlying field.*

In this quantum field theory, there is a representation, or Feynman diagram, for every particle interaction. Each diagram corresponds to a mathematical formula and predicts with great precision the probability that a particular interaction will occur. Even though quantum electrodynamics has been criticized as being too much a product of blackboard physics in that virtual particles are not necessarily found in nature but are

---

* F. Capra, *The Tao of Physics.* New York: Bantam, 1977, pp. 196–197.

instead a convenient abstraction, it does predict fairly accurately what nature will do as verified by experiments.

Indeed, physicists can only be grateful that Feynman chose, over twenty years ago, to say the same thing in a way that does not look at all like the ways it had been said before.

[W]hen new groups of phenomena compel changes in the pattern of thought . . . even the most eminent of physicists find immense difficulties. For the demand for change in the thought pattern may engender the feeling that the ground is to be pulled from under one's feet. . . . Once one has experienced the desperation with which clever and conciliatory men of science react to the demand for a change in the thought pattern, one can only be amazed that such revolutions in science have actually been possible at all.

—WERNER HEISENBERG, *Across the Frontiers*

WE HAVE A habit in writing articles published in scientific journals to make the work as finished as possible, to cover all the tracks, to not worry about the blind alleys or to describe how you had the wrong idea first, and so on. So there isn't any place to publish, in a dignified manner, what you actually did in order to get to do the work, although there has been, in these days, some interest in this kind of thing. Since winning the prize is a personal thing, I thought I could be excused in this particular situation, if I were to talk personally about my relationship to quantum electrodynamics, rather than to discuss the subject itself in a refined and finished fashion. Furthermore, since there are three people who have won the prize in physics, if they are all going to be talking about quantum electrodynamics itself, one might become bored with the subject. So, what I would like to tell you about today are the sequence of events, really the sequence of ideas, which occurred, and by which I finally came out the other end with an unsolved problem for which I ultimately received a prize.

I realize that a truly scientific paper would be of greater value, but such a paper I could publish in regular journals. So, I shall use this Nobel Lecture as an opportunity to do something of less value, but which I cannot do elsewhere. I ask your indulgence in another manner. I shall include details of anecdotes which are of no value either scientifically, nor for understanding the development of ideas. They are included only to make the lecture more entertaining.

Richard P. Feynman, Nobel Prize in Physics Award Address, 1965. Reprinted with permission of Elsevier Publishing Co., and the Nobel Foundation.

I worked on this problem about eight years until the final publication in 1947. The beginning of the thing was at the Massachusetts Institute of Technology, when I was an undergraduate student reading about the known physics, learning slowly about all these things that people were worrying about, and realizing ultimately that the fundamental problem of the day was that the quantum theory of electricity and magnetism was not completely satisfactory. This I gathered from books like those of Heitler and Dirac. I was inspired by the remarks in these books; not by the parts in which everything was proved and demonstrated carefully and calculated, because I couldn't understand those very well. At the young age what I could understand were the remarks about the fact that this doesn't make any sense, and the last sentence of the book of Dirac I can still remember, "It seems that some essentially new physical ideas are here needed." So, I had this as a challenge and an inspiration. I also had a personal feeling, that since they didn't get a satisfactory answer to the problem I wanted to solve, I don't have to pay a lot of attention to what they did do.

I did gather from my readings, however, that two things were the source of the difficulties with the quantum electrodynamical theories. The first was an infinite energy of interaction of the electron with itself. And this difficulty existed even in the classical theory. The other difficulty came from some infinities which had to do with the infinite numbers of degrees of freedom in the field. As I understood it at the time (as nearly as I can remember) this was simply the difficulty that if you quantized the harmonic oscillators of the field (say in a box) each oscillator has a ground state energy of $(1/2)\hbar\omega$ and there is an infinite number of modes in a box of every increasing frequency ω, and therefore there is an infinite energy in the box. I now realize that that wasn't a completely correct statement of the central problem; it can be removed simply by changing the zero from which energy is measured. At any rate, I believed that the difficulty arose somehow from a combination of the electron acting on itself and the infinite number of degrees of freedom of the field.

Well, it seemed to me quite evident that the idea that a particle acts on itself, that the electrical force acts on the same particle that generates it, is not a necessary one—it is a sort of a silly one, as a matter of fact. And, so I suggested to myself, that electrons cannot act on themselves, they can only act on other electrons. That means there is no field at all. You see, if all changes contribute to making a single common field, and if that common field acts back on all the charges, then each charge must act back on itself. Well, that was where the mistake was, there was no field. It was just that when you shook one charge, another would shake later. There was a direct interaction between charges, albeit with a delay. The law of force connecting the motion of one charge with another would just involve a delay. Shake this one, that one shakes later. The sun atom shakes; my eye electron shakes eight minutes later, because of a direct interaction across.

Now, this has the attractive feature that it solves both problems at once. First, I can say immediately, I don't let the electron act on itself, I just let this act on that, hence, no self-energy! Secondly, there is not an infinite number of degrees of freedom in the field. There is no field at all; or if you insist on thinking in terms of ideas like that of a field, this field is always completely determined by the action of the particles which produce it. You shake this particle, it shakes that one, but if you want to think in a field way, the field, if it's there, would be entirely determined by the matter which generates it, and therefore, the field does not have any *independent* degrees of freedom and the infinities from the degrees of freedom would then be removed. As a matter of fact, when we look out anywhere and see light, we can always "see" some matter as the source of the light. We don't just see light (except recently some radio reception has been found with no apparent material source).

You see then that my general plan was to first solve the classical problem, to get rid of the infinite self-energies in the classical theory, and to hope that when I made a quantum theory of it, everything would just be fine.

That was the beginning, and the idea seemed so obvious to me and so elegant that I fell deeply in love with it. And, like falling in love with a woman, it is only possible if you do not know much about her, so you cannot see her faults. The faults will become apparent later, but after the love is strong enough to hold you to her. So, I was held to this theory, in spite of all difficulties, by my youthful enthusiasm.

Then I went to graduate school and somewhere along the line I learned what was wrong with the idea that an electron does not act on itself. When you accelerate an electron it radiates energy and you have to do extra work to account for that energy. The extra force against which this work is done is called the force of radiation resistance. The origin of this extra force was identified in those days, following Lorentz, as the action of the electron itself. The first term of this action, of the electron on itself, gave a kind of inertia (not quite relativistically satisfactory). But that inertia-like term was infinite for a point-charge. Yet the next term in the sequence gave an energy loss rate, which for a point-charge agrees exactly with the rate you get by calculating how much energy is radiated. So, the force of radiation resistance, which is absolutely necessary for the conservation of energy would disappear if I said that a charge could not act on itself.

So, I learned in the interim when I went to graduate school the glaringly obvious fault of my own theory. But, I was still in love with the original theory, and was still thinking that with it lay the solution to the difficulties of quantum electrodynamics. So, I continued to try on and off to save it somehow. I must have some action develop on a given electron when I accelerate it to account for radiation resistance. But, if I let electrons only act on other electrons the only possible source for this action is another electron in the world. So, one day, when I was working for Professor Wheeler and could no longer solve the problem that he had given me, I thought about this again and I calculated the following. Suppose I have two charges—I shake the first charge, which I think of as a source and this makes the second one shake, but the second one shaking produces an effect back on the source. And so, I calculated how much that effect back on the first charge was, hoping it might add up the force of radiation resistance. It didn't come out right, of course, but I went to Professor Wheeler and told him my ideas. He said,—yes, but the answer you get for the problem with the two charges that you just mentioned will, unfortunately, depend upon the charge and the mass of the second charge and will vary inversely as the square of the distance R, between the charges, while the force of radiation resistance depends on none of these things. I thought, surely, he had computed it himself, but now having become a professor, I know that one can be wise enough to see immediately what some graduate student takes several weeks to develop. He also pointed out something that also bothered me, that if we had a situation with many charges all around the original source at roughly uniform density and if we added the effect of all the surrounding charges the inverse R square would be compensated by the R^2 in the volume element and we would get a result proportional to the thickness of the layer, which would go to infinity. That is, one would have an infinite total effect back at the source. And, finally he said to me, and you forgot something else, when you accelerate the first charge, the second acts later, and then the reaction back here at the source would be still later. In other words, the action occurs at the wrong time. I suddenly realized what a stupid fellow I am, for what I had described and calculated was just ordinary reflected light, not radiation reaction.

But, as I was stupid, so was Professor Wheeler that much more clever. For he then went on to give a lecture as though he had worked this all out before and was completely prepared, but he had not, he worked it out as he went along. First, he said, let us suppose that the return action by the charges in the absorber reaches the source by advanced waves as well as by the ordinary retarded waves of reflected light; so that the law of interaction acts backward in time, as well as forward in time. I was enough of a physicist at that time not to say, "Oh, no, how could that be?" For today all physicists know from studying Einstein and Bohr, that sometimes an idea which looks completely paradoxical at first, if analyzed to completion in all detail and in experimental situations, may, in fact, not be paradoxical. So, it did not bother me any more than it bothered Professor Wheeler to use advance waves for the back reaction—a solution of Maxwell's equations, which previously had not been physically used.

Professor Wheeler used advanced waves to get the reaction back at the right time and then he suggested this: If there were lots of electrons in the absorber, there would be an index of refraction *n*, so, the retarded waves coming from the source would have their wavelengths slightly modified in going through the absorber. Now, if we shall assume that the advanced waves come back from the absorber without an index—why? I don't know, let's assume they come back without an index—then, there will be a gradual shifting in phase between the return and the original signal so that we would only have to figure that the contributions act as if they come from only a finite thickness, that of the first wave zone. (More specifically, up to that depth where the phase in the medium is shifted appreciably from what it would be in vacuum, a thickness proportional to $\lambda(n-1)$.) Now, the less the number of electrons in here, the less each contributes, but the thicker will be the layer that effectively contributes because with less electrons, the index differs less from 1. The higher the charges of these electrons, the more each contributes, but the thinner the effective layer, because the index would be higher. And when we estimated it (calculated without being careful to keep the correct numerical factor), sure enough, it came out that the action back at the source was completely independent of the properties of the charges that were in the surrounding absorber. Further, it was of just the right character to represent radiation resistance, but we were unable to see if it was just exactly the right size. He sent me home with orders to figure out exactly how much advanced and how much retarded wave we need to get the thing to come out numerically right, and after that, figure out what happened to the advanced effects that you would expect if you put a test charge here close to the source. For if all charges generate advanced, as well as retarded effects, why would that test not be effected by the advanced waves from the source?

I found that you get the right answer if you use half-advanced and half-retarded as the field generated by each charge. That is, one is to use the solution of Maxwell's equation which is symmetrical in time and that the reason we got no advanced effects at a point close to the source in spite of the fact that the source was producing an advanced field is this: Suppose the source *s* surrounded by a spherical absorbing wall ten light seconds away, and that the test charge is one second to the right of the source. Then the source is as much as eleven seconds away from some parts of the wall and only nine seconds away from other parts. The source acting at time $t = 0$ induces motions in the wall at time $+10$. Advanced effects from this can act on the test charge as early as eleven seconds earlier, or at $t = -1$. This is just at the time that the direct advanced waves from the source should reach the test charge, and it turns out the two effects are exactly equal and opposite and cancel out! At the later time $+1$ effects on the test charge from the source and from the walls are again equal, but this time are of the same sign and add to convert the half-retarded wave of the source to full retarded strength.

Thus, it became clear that there was the possibility that if we assume all actions are *via* half-advanced and half-retarded solutions of Maxwell's equations and assume that all sources are surrounded by material absorbing all the light which is emitted, then we could account for radiation resistance as a direct action of the charges of the absorber acting back by advanced waves on the source.

Many months were devoted to checking all these points. I worked to show that everything is independent of the shape of the container, and so on, that the laws are exactly right, and that the advanced effects really cancel in every case. We always tried to increase the efficiency of our demonstrations, and to see with more and more clarity why it works. I won't bore you by going through the details of this. Because of our using advanced waves, we also had many apparent paradoxes, which we gradually reduced one by one, and saw that there was in fact no logical difficulty with the theory. It was perfectly satisfactory.

We also found that we could reformulate this thing in another way, and that is by a principle of least action. Since my original plan was to describe everything directly in terms of particle motions, it was my desire to represent this new theory without saying anything about fields. It turned out that we found a form for an action directly involving the motions of the charges only, which upon variation would give the equations of motion of these charges. The expression for this action A is

$$A = \Sigma_i m_i \int \left(\dot{X}^i\mu \dot{X}^i\mu \right)^{1/2} d\alpha_i + \tfrac{1}{2} \sum_{\substack{ij \\ i \neq j}} e_i e_j \iint \delta(I_{ij}^2) \dot{X}^i\mu(\alpha_i) \dot{X}^j\mu(\alpha_j) d\alpha_i d\alpha_j \quad (1)$$

where

$$I_{ij}^2 = [X^i\mu(\alpha_i) - X^j\mu(\alpha_j)]\ [X^i\mu(\alpha_i) - X^j\mu(\alpha_j)]$$

where $X^i\mu\ (\alpha_i)$ is the four-vector position of the ith particle as a function of some parameter α_i, $\dot{X}^i\mu\ (\alpha_i)$ is $dX^i\mu(\alpha)/d\alpha_i$. The first term is the integral of proper time, the ordinary action of relativistic mechanics of free particles of mass m_i. (We sum in the usual way on the repeated index μ.) The second term represents the electrical interaction of the charges. It is summed over each pair of charges (the factor ½ is to count each pair once, the term $i = j$ is omitted to avoid self-action). The interaction is a double integral over a delta function of the square of space-time interval I^2 between two points on the paths. Thus, interaction occurs only when this interval vanishes, that is, along light cones.

The fact that the interaction is exactly one-half advanced and half-retarded meant that we could write such a principle of least action, whereas interaction *via* retarded waves alone cannot be written in such a way.

So, all of classical electrodynamics was contained in this very simple form. It looked good, and therefore, it was undoubtedly true, at least to the beginner. It automatically gave half-advanced and half-retarded effects and it was without fields. By omitting the term in the sum when $i = j$, *I* omit self-interaction and no longer have any infinite self-energy. This then was the hoped-for solution to the problem of ridding classical electrodynamics of the infinities.

It turns out, of course, that you can reinstate fields if you wish to, but you have to keep track of the field produced by each particle separately. This is because to find the right field to act on a given particle, you must exclude the field that it creates itself. A single universal field to which all contribute will not do. This idea had been suggested earlier by Frenkel and so we called these Frenkel fields. This theory which allowed only particles to act on each other was equivalent to Frenkel's fields using half-advanced and half-retarded solutions.

There were several suggestions for interesting modifications of electrodynamics. We discussed lots of them, but I shall report on only one. It was to replace this delta function in the interaction by another function, say, $f(I^2_{ij})$, which is not infinitely sharp. Instead of having the action occur only when the interval between the two charges is exactly zero, we would replace the delta function of I^2 by a narrow peaked thing. Let's say that $f(Z)$ is large only near $Z = 0$ width of order a^2. Interactions will now occur when $T^2 - R^2$ is of order a^2 roughly where T is the time difference and R is the separation of the charges. This might look like it disagrees with experience, but if a is some small distance, like 10^{-13} cm, it says that the time delay T in action is roughly $\sqrt{R^2 \pm a^2}$ or approximately,—if R is much larger than a, $T = R \pm a^2/2R$. This means that the deviation of time T from the ideal theoretical time R of Maxwell, gets smaller and smaller, the further the pieces are apart. Therefore, all theories involved in analyzing generators, motors, etc., in fact, all of the tests of electrodynamics that were available in Maxwell's time, would be adequately satisfied if a were 10^{-13} cm. If R is of the order of a centimeter this deviation in T is only 10^{-26} parts. So, it was possible, also, to change the theory in a simple manner and to still agree with all observations of classical electrodynamics. You have no clue of precisely what function to put in for f, but it was an interesting possibility to keep in mind when developing quantum electrodynamics.

It also occurred to us that if we did that (replace δ by f) we could not reinstate the term $i = j$ in the sum because this would now represent in a relativistically invariant fashion a finite action of a charge on itself. In fact, it was possible to prove that if we did do such a thing, the main effect of the self-action (for not too rapid accelerations) would be to produce a modification of the mass. In fact, there need be no mass m_i term, all the mechanical mass could be electromagnetic self-action. So, if you would like, we could also have another theory with a still simpler expression for the action A. In expression (1) only the second term is kept, the sum extended over all i and j, and some function f replaces δ. Such a simple form could represent all of classical electrodynamics which aside from gravitation is essentially all of classical physics.

Although it may sound confusing, I am describing several different alternative theories at once. The important thing to note is that at this time we had all these in mind as different possibilities. There were several possible solutions of the difficulty of classical electrodynamics, any one of which might serve as a good starting point to the solution of the difficulties of quantum electrodynamics.

I would also like to emphasize that by this time I was becoming used to a physical point of view different from the more customary point of view. In the customary view, things are discussed as a function of time in very great detail. For example, you have the field at this moment, a differential equation gives you the field at the next moment and so on; a method, which I shall call the Hamilton method, the time differential method. We have, instead (in (1) say) a thing that describes the character of the path throughout all of space and time. The behavior of nature is determined by saying her whole space-time path has a certain character. For an action like (1) the equations obtained by variation (of $X^i\mu(\alpha_i)$) are no longer at all easy to get back into Hamiltonian form. If you wish to use as variables only the coordinates of particles, then you can talk about the property of the paths—but the path of one particle at a given time is affected by the path of another at a different time. If you try to describe, therefore, things differentially, telling what the present conditions of the particles are, and how these present conditions will affect the future—you see, it is impossible with particles alone, because something the particle did in the past is going to affect the future.

Therefore, you need a lot of bookkeeping variables to keep track of what the particle did in the past. These are called field variables. You will, also, have to tell what the field is at this present moment, if you are to be able to see later what is going to happen. From the overall space-time view of the least action principle, the field disappears as nothing but bookkeeping variables insisted on by the Hamiltonian method.

As a by-product of this same view, I received a telephone call one day at the graduate college at Princeton from Professor Wheeler, in which he said, "Feynman, I know why all electrons have the same charge and the same mass" "Why?" "Because, they are all the same electron!" And, then he explained on the telephone, "suppose that the world lines which we were ordinarily considering before in time and space—instead of only going up in time were a tremendous knot, and then, when we cut through the knot, by the plane corresponding to a fixed time, we would see many, many world lines and that would represent many electrons, except for one thing. If in one section this is an ordinary electron world line, in the section in which it reversed itself and is coming back from the future we have the wrong sign to the proper time—to the proper four velocities—and that's equivalent to changing the sign of the charge, and, therefore, that part of a path would act like a positron." "But, Professor," I said, "there aren't as many positrons as electrons." "Well, maybe they are hidden in the protons or something," he said. I did not take the idea that all the electrons were the same from him as seriously as I took the observation that positrons could simply be represented as electrons going from the future to the past in a back section of their world lines. That, I stole!

To summarize, when I was done with this, as a physicist I had gained two things. One, I knew many different ways of formulating classical electrodynamics, with many different mathematical forms. I got to know how to express the subject every which way. Second, I had a point of view—the overall space-time point of view—and a disrespect for the Hamiltonian method of describing physics.

I would like to interrupt here to make a remark. The fact that electrodynamics can be written in so many ways—the differential equations of Maxwell, various minimum principles with fields, minimum principles without fields, all different kinds of ways, was something I knew, but I have never understood. It always seems odd to me that the fundamental laws of physics, when discovered, can appear in so many different forms that are not apparently identical at first, but, with a little mathematical fiddling you can show the relationship. An example of that is the Schrödinger equation and the Heisenberg formulation of quantum mechanics. I don't know why this is—it remains a mystery, but it was something I learned from experience. There is always another way to say the same thing that doesn't look at all like the way you said it before. I don't know what the reason for this is. I think it is somehow a representation of the simplicity of nature. A thing like the inverse square law is just right to be represented by the solution of Poisson's equation, which, therefore, is a very different way to say the same thing that doesn't look at all like the way you said it before. I don't know what it means, that nature chooses these curious forms, but maybe that is a way of defining simplicity. Perhaps a thing is simple if you can describe it fully in several different ways without immediately knowing that you are describing the same thing.

I was now convinced that since we had solved the problem of classical electrodynamics (and completely in accordance with my program from M. I. T., only direct interaction between particles, in a way that made fields unnecessary) that everything was definitely going to be all right. I was convinced that all I had to do was make a quantum theory analogous to the classical one and everything would be solved.

So, the problem is only to make a quantum theory, which has as its classical analog, this expression (1). Now, there is no unique way to make a quantum theory from classical mechanics, although all the textbooks make believe there is. What they would tell you to do, was find the momentum variables and replace them by $(h/i)(\partial/\partial x)$, but I couldn't find a momentum variable, as there wasn't any.

The character of quantum mechanics of the day was to write things in the famous Hamiltonian way—in the form of a differential equation, which described how the wave function changes from instant to instant, and in terms of an operator, H. If the classical physics could be reduced to a Hamiltonian form, everything was all right. Now, least action does not imply a Hamiltonian form if the action is a function of anything more than positions and velocities at the same moment. If the action is of the form of the integral of a function (usually called the Lagrangian) of the velocities and positions at the same time

$$S = \int L(\dot{x}, x)\,dt \tag{2}$$

then you can start with the Lagrangian and then create a Hamiltonian and work out the quantum mechanics, more or less uniquely. But this thing (1) involves the key variables, positions, at two different times and therefore, it was not obvious what to do to make the quantum-mechanical analogue.

I tried—I would struggle in various ways. One of them was this: if I had harmonic oscillators interacting with a delay in time, I could work out what the normal modes were and guess that the quantum theory of the normal modes was the same as for simple oscillators and kind of work my way back in terms of the original variables. I succeeded in doing that, but I hoped then to generalize to other than a harmonic oscillator, but I learned to my regret something, which many people have learned. The harmonic oscillator is too simple; very often you can work out what it should do in quantum theory without getting much of a clue as to how to generalize your results to other systems.

So that didn't help me very much, but when I was struggling with this problem, I went to a beer party in the Nassau Tavern in Princeton. There was a gentleman, newly arrived from Europe (Herbert Jehle) who came and sat next to me. Europeans are much more serious than we are in America because they think that a good place to discuss intellectual matters is a beer party. So, he sat by me and asked, "What are you doing" and so on, and I said, "I'm drinking beer." Then I realized that he wanted to know what work I was doing and I told him I was struggling with this problem, and I simply turned to him and said, "Listen, do you know any way of doing quantum mechanics, starting with action—where the action integral comes into the quantum mechanics?" "No," he said, "but Dirac has a paper in which the Lagrangian, at least, comes into quantum mechanics. I will show it to you tomorrow."

Next day we went to the Princeton Library, they have little rooms on the side to discuss things, and he showed me this paper. What Dirac said was the following: There is in quantum mechanics a very important quantity which carries the wave function from one time to another, besides the differential equation but equivalent to it, a kind of a kernel, which we might call $K(x',x)$, which carries the wave function $\psi(x)$ known at time t, to the wave function $\psi(x')$ at time, $t + \varepsilon$. Dirac points out that this function K was *analogous* to the quantity in classical mechanics that you would calculate if you took the exponential of $i\epsilon$, multiplied by the Lagrangian $L(\dot{x},x)$ imagining that these two positions x,x' corresponded to t and $t + \varepsilon$. In other words,

$$K\,(x',x) \text{ is analogous to } e^{i\epsilon L\left(\frac{x'-x}{\epsilon},\,x\right)/\hbar}$$

Professor Jehle showed me this, I read it, he explained it to me, and I said, "What does he mean, they are analogous; what does that mean, *analogous?* What is the use of that?" He said, "You Americans! You always want to find a use for everything!" I said that I thought that Dirac must mean that they were equal. "No," he explained, "he doesn't mean they are equal." "Well," I said, "let's see what happens if we make them equal."

So I simply put them equal, taking the simplest example where the Lagrangian is ½ $Mx^2 - V(x)$ but soon found I had to put a constant of proportionality A in, suitably adjusted. When I substituted $Ae^{i\epsilon L/\hbar}$ for K to get

$$\psi\,(x', t + \varepsilon) = \int A \exp\left[\frac{i\varepsilon}{\hbar} L\left(\frac{x' - x}{\epsilon}, x\right)\right] \psi\,(x, t)\, dx \qquad (3)$$

and just calculated things out by Taylor series expansion, out came the Schrödinger equation. So, I turned to Professor Jehle, not really understanding, and said, "Well, you see Professor Dirac meant that they were proportional." Professor Jehle's eyes were bugging out—he had taken out a little notebook and was rapidly copying it down from the blackboard, and said, "No, no, this is an important discovery. You Americans are always trying to find out how something can be used. That's a good way to discover things!" So, I thought I was finding out what Dirac meant, but, as a matter of fact, had made the discovery that what Dirac thought was analogous, was, in fact, equal. I had then, at least, the connection between the Lagrangian and quantum mechanics, but still with wave functions and infinitesimal times.

It must have been a day or so later, when I was lying in bed thinking about these things, that I imagined what would happen if I wanted to calculate the wave function at a finite interval later.

I would put one of these factors $e^{i\varepsilon L/\hbar}$ in here, and that would give me the wave functions the next moment, $t + \varepsilon$ and then I could substitute that back into (3) to get another factor of $e^{i\varepsilon L/\hbar}$ and give me the wave function the next moment, $t + 2\varepsilon$, and so on and so on. In that way I found myself thinking of a large number of integrals, one after the other in sequence. In the integrand was the product of the exponentials, which, of course, was the exponential of the sum of terms like εL. Now, L is the Lagrangian and ε is like the time interval dt, so that if you took a sum of such terms, that's exactly like an integral. That's like Riemann's formula for the integral $\int L dt$, you just take the value at each point and add them together. We are to take the limit as $\varepsilon \rightarrow 0$, of course. Therefore, the connection between the wave function of one instant and the wave function of another instant a finite time later could be obtained by an infinite number of integrals (because ε goes to zero, of course), of exponential $(iS/\hbar)$ where S is the action expression (2). At last, I had succeeded in representing quantum mechanics directly in terms of the action S.

This led later on to the idea of the amplitude for a path; that for each possible way that the particle can go from one point to another in space-time, there's an amplitude. That amplitude is e to the $i/\hbar$ times the action for the path. Amplitudes from various paths superpose by addition. This then is another, a third way, of describing quantum mechanics, which looks quite different than that of Schrödinger or Heisenberg, but which is equivalent to them.

Now immediately after making a few checks on this thing, what I wanted to do, of course, was to substitute the action (1) for the other (2). The first trouble was that I could not get the thing to work with the relativistic case of spin one-half. However, although I could deal with the matter only non-relativistically, I could deal with the light or the photon interactions perfectly well by just putting the interaction terms of (1) into any action, replacing the mass terms by the non-relativistic $(Mx^2/2)dt$. When the action has a delay, as it now had, and involved more than one time, I had to lose the idea of a wave function. That is, I could no longer describe the program as: given the amplitude for all positions at a certain time to compute the amplitude at another time. However, that didn't cause very much trouble. It just meant developing a new idea. Instead of wave functions we could talk about this; that if a source of a certain kind emits a particle, and a detector is there to receive it, we can give the amplitude that the source will emit and the detector receive. We do this without specifying the exact instant that the source emits or the exact instant that any detector receives, without trying to specify the state of anything at any particular time in between, but by just finding the amplitude for the complete experiment. And, then we could discuss how the amplitude would change if you had a scattering sample in between, as you rotated and changed angles, and so on, without really having any wave functions.

It was also possible to discover what the old concepts of energy and momentum would mean with this generalized action. And, so I believed that I had a quantum theory of classical electrodynamics—or rather of this new classical electrodynamics described by action (1). I made a number of checks. If I took the Frenkel field point of view, which you remember was more differential, I could convert it directly to quantum mechanics in a more conventional way. The only problem was how to specify in quantum mechanics the classical boundary conditions to use only half-advanced and half-retarded solutions. By some ingenuity in defining what that meant, I found that the quantum mechanics with Frenkel fields, plus a special boundary condition, gave me back this action, (1) in the new form of quantum mechanics with a delay. So, various things indicated that there wasn't any doubt I had everything straightened out.

It was also easy to guess how to modify the electrodynamics, if anybody ever wanted to modify it. I just changed the delta to an *f*, just as I would for the classical case. So, it was very easy, a simple thing. To describe the old retarded theory without explicit mention of fields I would have to write probabilities, not just amplitudes. I would have to square my amplitudes and that would involve double path integrals in which there are two *S*'s and so forth. Yet, as I worked out many of these things and studied different forms and different boundary conditions, I got a kind of funny feeling that things weren't exactly right. I could not clearly identify the difficulty and in one of the short periods during which I imagined I had laid it to rest, I published a thesis and received my Ph.D.

During the war, I didn't have time to work on these things very extensively, but wandered about on buses and so forth, with little pieces of paper, and struggled to work on it and discovered indeed that there was something wrong, something terribly wrong. I found that if one generalized the action from the nice Lagrangian forms (2) to the forms (1) then the quantities which I defined as energy, and so on, would be complex. The energy values of stationary states wouldn't be real and probabilities of events wouldn't add up to 100%. That is, if you took the probability that this would happen and that would happen—everything you could think of would happen, it would not add up to one.

Another problem on which I struggled very hard, was to represent relativistic electrons with this new quantum mechanics. I wanted to do a unique and different way—and not just by copying the operators of Dirac into some kind of an expression and using some kind of Dirac algebra instead of ordinary complex numbers. I was very much encouraged by the fact that in one space dimension, I did find a way of giving an amplitude to every path by limiting myself to paths, which only went back and forth at the speed of light. The amplitude was simple ($i\varepsilon$) to a power equal to the number of velocity reversals where I have divided the time into steps ε and I am allowed to reverse velocity only at such a time. This gives (as ε approaches zero) Dirac's equation in two dimensions—one dimension of space and one of time ($\hbar = M = c = 1$).

Dirac's wave function has four components in four dimensions, but in this case, it has only two components and this rule for the amplitude of a path automatically generates the need for two components. Because if this is the formula for the amplitude of a path, it will not do you any good to know the total amplitude of all paths, which come into a given point to find the amplitude to reach the next point. This is because for the next time, if it came in from the right, there is no new factor $i\varepsilon$ if it goes out to the right, whereas, if it came in from the left there was a new factor $i\varepsilon$. So, to continue this same information forward to the next moment, it was not sufficient information to know the total amplitude to arrive, but you had to know the amplitude to arrive from the right and the amplitude to arrive to the left, independently. If you did, however, you could then compute both of those again independently and thus you had to carry two amplitudes to form a differential equation (first order in time).

And, so I dreamed that if I were clever, I would find a formula for an amplitude of a path that was beautiful and simple for three dimensions of space and one of time, which would be equivalent to the Dirac equation, and for which the four components, matrices, and all those other mathematical funny things would come out as a simple consequence—I have never succeeded in that either. But, I did want to mention some of the unsuccessful things on which I spent almost as much effort, as on the things that did work.

To summarize the situation a few years after the way, I would say, I had much experience with quantum electrodynamics, at least in the knowledge of many different ways of formulating it, in terms of path integrals of actions and in other forms. One of the important by-products, for example, of much experience in these simple forms, was that it was easy to see how to combine together what was in those days called the longitudinal and transverse fields, and in general, to see clearly the relativistic invariance of the theory. Because of the need to do things differentially there had been, in the standard quantum electrodynamics, a complete split of the field into two parts, one of which is called the longitudinal part and the other mediated by the photons, or transverse waves. The longitudinal part was described by a Coulomb potential acting instantaneously in the Schrödinger equation, while the transverse part had entirely different description in terms of quantization of the transverse waves. This separation depended upon the relativistic tilt of your axes in space-time. People moving at different velocities would separate the same field into longitudinal and transverse fields in a different way. Furthermore, the entire formulation of quantum mechanics insisting, as it did, on the wave function at a given time, was hard to analyze relativistically. Somebody else in a different coordinate system would calculate the succession of events in terms of wave functions on differently cut slices of space-time, and with a different separation of longitudinal and transverse parts. The Hamiltonian theory did not look relativistically invariant, although, of course, it was. One of the great advantages of the overall point of view, was that you could see the relativistic invariance right away—or as Schwinger would say—the covariance was manifest. I had the advantage, therefore, of having a manifestedly covariant form for quantum electrodynamics with suggestions for modifications and so on. I had the disadvantage that if I took it too seriously—I mean, if I took it seriously at all in this form,—I got into trouble with these complex energies and the failure of adding probabilities to one and so on. I was unsuccessfully struggling with that.

Then Lamb did his experiment, measuring the separation of the $2S½$ and $2P½$ levels of hydrogen, finding it to be about 1000 megacycles of frequency difference. Professor Bethe, with whom I was then associated at Cornell, is a man who has this characteristic: If there's a good experimental number you've got to figure it out from theory. So, he forced the quantum electrodynamics of the day to give him an answer to the separation of these two levels. He pointed out that the self-energy of an electron itself is infinite, so that the calculated energy of a bound electron should also come out infinite. But, when you calculated the separation of the two energy levels in terms of the corrected mass instead of the old mass, it would turn out, he thought, that the theory would give convergent finite answers. He made an estimate of the splitting that way and found out that it was still divergent, but he guessed that was probably due to the fact that he used an unrelativistic theory of the matter. Assuming it would be convergent if relativistically treated, he estimated he would get about a thousand megacycles for the Lamb-shift, and thus, made the most important discovery in the history of the theory of quantum electrodynamics. He worked this out on the train from Ithaca, New York, to Schenectady and telephoned me excitedly from Schenectady to tell me the result, which I don't remember fully appreciating at the time.

Returning to Cornell, he gave a lecture on the subject, which I attended. He explained that it gets very confusing to figure out exactly which infinite term corresponds to what in trying to make the correction for the infinite change in mass. If there were any modifications whatever, he said, even though not physically correct (that is not necessarily the way nature actually works), but any modification whatever at high frequencies, which would make this correction finite, then there would be no problem at all to figuring out how to keep track of everything. You just calculate the finite mass correction Δm to the electron mass m_o, substitute the numerical values of $m_o + \Delta m$ for m in the results for any other problem and all these ambiguities would be resolved. If, in addition, this method were relativistically invariant, then we would be absolutely sure how to do it without destroying relativistically invariant.

After the lecture, I went up to him and told him, "I can do that for you, I'll bring it in for you tomorrow." I guess I knew every way to modify quantum electrodynamics known to man, at the time. So, I went in next day, and explained what would correspond to the modification of the delta-function to f and asked him to explain to me how you calculate the self-energy of an electron, for instance, so we can figure out if it's finite.

I want you to see an interesting point. I did not take the advice of Professor Jehle to find out how it was useful. I never used all that machinery which I had cooked up to solve a single relativistic problem. I hadn't even calculated the self-energy of an electron up to that moment, and was studying the difficulties with the conservation of probability, and so on, without actually doing anything, except discussing the general properties of the theory.

But now I went to Professor Bethe, who explained to me on the blackboard, as we worked together, how to calculate the self-energy of an electron. Up to that time when you did the integrals they had been logarithmically divergent. I told him how to make the relativistically invariant modifications that I thought would make everything all right. We set up the integral which then diverged at the sixth power of the frequency instead of logarithmically!

So, I went back to my room and worried about this thing and went around in circles trying to figure out what was wrong because I was sure physically everything had to come out finite, I couldn't understand how it came out infinite. I became more and more interested and finally realized

I had to learn how to make a calculation. So, ultimately, I taught myself how to calculate the self-energy of an electron working my patient way through the terrible confusion of those days of negative energy states and holes and longitudinal contributions and so on. When I finally found out how to do it and did it with the modifications I wanted to suggest, it turned out that it was nicely convergent and finite, just as I had expected. Professor Bethe and I have never been able to discover what we did wrong on the blackboard two months before, but apparently we just went off somewhere and we have never been able to figure out where. It turned out, that what I had proposed, if we had carried it out without making a mistake would have been all right and would have given a finite correction. Anyway, it forced me to go back over all this and to convince myself physically that nothing can go wrong. At any rate, the correction to mass was now finite, proportional to $\ln(ma/\hbar)$ where a is the width of the function f which was substituted for δ. If you wanted an unmodified electrodynamics, you would have to take a equal to zero, getting an infinite mass correction. But, that wasn't the point. Keeping a finite, I simply followed the program outlined by Professor Bethe and showed how to calculate all the various things, the scatterings of electrons from atoms without radiation, the shifts of levels and so forth, calculating everything in terms of the experimental mass, and noting that the results as Bethe suggested, were not sensitive to a in this form and even had a definite limit as $a \to 0$.

The rest of my work was simply to improve the techniques then available for calculations, making diagrams to help analyze perturbation theory quicker. Most of this was first worked out by guessing—you see, I didn't have the relativistic theory of matter. For example, it seemed to me obvious that the velocities in non-relativistic formulas have to be replaced by Dirac's matrix α or in the more relativistic forms by the operators γ_μ. I just took my guesses from the forms that I had worked out using path integrals for non-relativistic matter, but relativistic light. It was easy to develop rules of what to substitute to get the relativistic case. I was very surprised to discover that it was not known at that time, that every one of the formulas that had been worked out so patiently by separating longitudinal and transverse waves could be obtained from the formula for the transverse waves alone, if instead of summing over only the two perpendicular polarization directions you would sum over all four possible directions of polarization. It was so obvious from the action (1) that I thought it was general knowledge and would do it all the time. I would get into arguments with people, because I didn't realize they didn't know that; but, it turned out that all their patient work with longitudinal waves was always equivalent to just extending the sum on the two transverse directions of polarization over all four directions. This was one of the amusing advantages of the method. In addition, I included diagrams for the various terms of the perturbation series, improved notations to be used, worked out easy ways to evaluate integrals, which occurred in these problems, and so on, and made a kind of handbook on how to do quantum electrodynamics.

But one step of importance that was physically new was involved with the negative energy sea of Dirac, which caused me so much logical difficulty. I got so confused that I remembered Wheeler's old idea about the positron being, maybe, the electron going backward in time. Therefore, in the time dependent perturbation theory that was usual for getting self-energy, I simply supposed that for a while we could go backward in the time, and looked at what terms I got by running the time variables backward. They were the same as the terms that other people got when they did the problem a more complicated way, using holes in the sea, except, possibly, for some signs. These, I, at first, determined empirically by inventing and trying some rules.

I have tried to explain that all the improvements of relativistic theory were at first more or less straightforward, semi-empirical shenanigans. Each time I would discover something, however, I would go back and I would check it so many ways, compare it to every problem that had been done previously in electrodynamics (and later, in weak coupling meson theory) to see if it would always agree, and so on, until I was absolutely convinced of the truth of the various rules and regulations which I concocted to simplify all the work.

During this time, people had been developing meson theory, a subject I had not studied in any detail. I became interested in the possible application of my methods to perturbation calculations in meson theory. But, what was meson theory? All I knew was that meson theory was something analogous to electrodynamics, except that particles corresponding to the photon had a mass. It was easy to guess the δ-function in (1), which was a solution of d'Alembertian equals zero, was to be changed to the corresponding solution of d'Alembertian equals m^2. Next, there were different kinds of mesons—the ones in closest analogy to photons, coupled *via* $\gamma\mu\gamma\mu$, are called vector mesons—there were also scalar mesons. Well, maybe that corresponds to putting unity in place of the $\gamma\mu$, I would here then speak of "pseudo vector coupling" and I would guess what that probably was. I didn't have the knowledge to understand the way these were defined in the conventional papers because they were expressed at that time in terms of creation and annihilation operators, and so on, which I had not successfully learned. I remember that when someone had started to teach me about creation and annihilation operators, that this operator creates an electron, I said, "How do you create an electron? It disagrees with the conservation of charge," and in that way, I blocked my mind from learning a very practical scheme of calculation. Therefore, I had to find as many opportunities as possible to test whether I guessed right as to what the various theories were.

One day a dispute arose at a Physical Society meeting as to the correctness of a calculation by Slotnick of the interaction of an electron with a neutron using pseudo scalar theory with pseudo vector coupling and also, pseudo scalar theory with pseudo scalar coupling. He had found that the answers were not the same, in fact, by one theory, the result was divergent, although convergent with the other. Some people believed that the two theories must give the same answer for the problem. This was a welcome opportunity to test my guesses as to whether I really did understand what these two couplings were. So, I went home, and during the evening I worked out the electron neutron scattering for the pseudo scalar and pseudo vector coupling, saw they were not equal and subtracted them, and worked out the difference in detail. The next day at the meeting, I saw Slotnick and said, "Slotnick, I worked it out last night, I wanted to see if I got the same answers you do. I got a different answer for each coupling—but, I would like to check in detail with you because I want to make sure of my methods." And, he said, "What do you mean you worked it out last night, it took me six months!" And, when we compared the answers he looked at mine and he asked, "What is that Q in there, that variable Q?" (I had expressions like $(\tan^{-1}Q)/Q$ etc.). I said, "That's the momentum transferred by the electron, the electron deflected by different angles." "Oh," he said, "no, I only have the limiting value as Q approaches zero; the forward scattering." Well, it was easy enough to just substitute Q equals zero in my form and I then got the same answers as he did. But, it took him six months to do the case of zero momentum transfer, whereas, during one evening I had done the finite and arbitrary momentum transfer. That was a thrilling moment for me, like receiving the Nobel Prize, because that convinced me, at last, I did have some kind of method and technique and understood how to do something that other people did not know how to do. That was my moment of triumph in which I realized I really had succeeded in working out something worthwhile.

At this stage, I was urged to publish this because everybody said it looks like an easy way to make calculations, and wanted to know how to do it. I had to publish it, missing two things; one was proof of every statement in a mathematically conventional sense. Often, even in a physicist's sense, I did not have a demonstration of how to get all of these rules and equations from conventional electrodynamics. But, I did know from experience, from fooling around, that everything was, in fact, equivalent to the regular electrodynamics and had partial proofs of many pieces, although, I never really sat down, like Euclid did for the geometers of Greece, and made sure that you could get it all from a single simple set of axioms. As a result, the work was criticized, I don't know whether favorably or unfavorably, and the "method" was called the "intuitive method." For those who do not realize it, however, I should like to emphasize that there is a lot of work involved in using this "intuitive method" successfully. Because no simple clear proof of the formula or idea presents itself, it is necessary to do an unusually great amount of checking and rechecking for consistency and correctness in terms of what is known, by comparing to other analogous examples, limiting cases, etc. In the face of the lack of direct mathematical demonstration, one must be careful and thorough to make sure of the point, and one should make a perpetual attempt to demonstrate as much of the formula as possible. Nevertheless, a very great deal more truth can become known than can be proven.

It must be clearly understood that in all this work, I was representing the conventional electrodynamics with retarded interaction, and not my half-advanced and half-retarded theory corresponding to (1). I merely use (1) to guess at forms. And, one of the forms I guessed at corresponded to changing δ to a function f of width a^2, so that I could calculate finite results for all of the problems. This brings me to the second thing that was missing when I published the paper, an unresolved difficulty. With δ replaced by f the calculations would give results which were not "unitary," that is, for which the sum of the probabilities of all alternatives was not unity. The deviation from unity was very small, in practice, if a was very small. In the limit that I took a very tiny, it might not make any difference. And, so the process of the renormalization could be made, you could calculate everything in terms of the experimental mass and then take the limit and the apparent difficulty that the unitary is violated temporarily seems to disappear. I was unable to demonstrate that, as a matter of fact, it does.

It is lucky that I did not wait to straighten out that point, for as far as I know, nobody has yet been able to resolve this question. Experience with meson theories with stronger couplings and with strongly coupled vector photons, although not proving anything, convinces me that if the coupling were stronger, or if you went to a higher order (137th order of perturbation theory for electrodynamics), this difficulty would remain in the limit and there would be real trouble. That is, I believe there is really no satisfactory quantum electrodynamics, but I'm not sure. And, I believe, that one of the reasons for the slowness of present-day progress in understanding the strong interactions is that there isn't any relativistic theoretical model, from which you can really calculate everything. Although, it is usually said, that the difficulty lies in the fact that *strong* interactions are too hard to calculate, I believe, it is really because strong interactions in field theory have no solution, have no sense—they're either infinite, or, if you try to modify them, the modification destroys the unitarity. I don't think we have a completely satisfactory relativistic quantum-mechanical model, even one that doesn't agree with nature, but, at least, agrees with the logic that the sum of probability of all alternatives has to be 100%. Therefore, I think that the renormalization theory is simply a way to sweep the difficulties of the divergences of electrodynamics under the rug. I am, of course, not sure of that.

This completes the story of the development of the space-time view of quantum electrodynamics. I wonder if anything can be learned from it. I doubt it. It is most striking that most of the ideas developed in the course of this research were not ultimately used in the final result. For example, the half-advanced and half-retarded potential was not finally used, the action expression (1) was not used, the idea that charges do not act on themselves was abandoned. The path-integral formulation of quantum mechanics was useful for guessing at final expressions and at formulating the general theory of electrodynamics in new ways—although, strictly it was not absolutely necessary. The same goes for the idea of the positron being a backward moving electron, it was very convenient, but not strictly necessary for the theory because it is exactly equivalent to the negative energy sea point of view.

We are struck by the very large number of different physical viewpoints and widely different mathematical formulations that are all equivalent to one another. The method used here, of reasoning in physical terms, therefore, appears to be extremely inefficient. On looking back over the work, I can only feel a kind of regret for the enormous amount of physical reasoning and mathematically re-expression which ends by merely re-expressing what was previously known, although in a form which is much more efficient for the calculation of specific problems. Would it not have been much easier to simply work entirely in the mathematical framework to elaborate a more efficient expression? This would certainly seem to be the case, but it must be remarked that although the problem actually solved was only such a reformulation, the problem originally tackled was the (possibly still unsolved) problem of avoidance of the infinities of the usual theory. Therefore, a new theory was sought, not just a modification of the old. Although the quest was unsuccessful, we should look at the question of the value of physical ideas in developing a *new* theory.

Many different physical ideas can describe the same physical reality. Thus, classical electrodynamics can be described by a field view, or an action at a distance view, etc. Originally, Maxwell filled space with idler wheels, and Faraday with fields lines, but somehow the Maxwell equations themselves are pristine and independent of the elaboration of words attempting a physical description. The only true physical description is that describing the experimental meaning of the quantities in the equation—or better, the way the equations are to be used in describing experimental observations. This being the case perhaps the best way to proceed is to try to guess equations, and disregard physical models or descriptions. For example, McCullough guessed the correct equations for light propagation in a crystal long before his colleagues using elastic models could make head or tail of the phenomena, or again, Dirac obtained his equation for the description of the electron by an almost purely mathematical proposition. A simple physical view of which all the contents of this equation can be seen is still lacking.

Therefore, I think equation guessing might be the best method to proceed to obtain the laws for the part of physics which is presently unknown. Yet, when I was much younger, I tried this equation guessing and I have seen many students try this, but it is very easy to go off in wildly incorrect and impossible directions. I think the problem is not to find the *best* or most efficient method to proceed to a discovery, but to find any method at all. Physical reasoning does help some people to generate suggestions as to how the unknown may be related to the known. Theories of the known, which are described by different physical ideas, may be equivalent in all their predictions and are hence scientifically indistinguishable. However, they are not psychologically identical when trying to move from that base into the unknown. For different views suggest different kinds of modifications which might be made and hence are not equivalent in the hypotheses one generates from them in one's attempt to understand what is not yet understood. I, therefore, think that a good theoretical physicist today might find it useful to have a wide range of physical viewpoints and mathematical expressions of the same theory (for example, of quantum electrodynamics) available to him. This may be asking too much of one man. Then new students should as a class have this. If every individual student follows the same current fashion in expressing and thinking about electrodynamics or field theory, then the variety of hypotheses being generated to understand strong interactions, say, is limited. Perhaps rightly so, for possibly the chance is high that the truth lies in the fashionable direction. But, on the off-chance that it is in another direction—a direction obvious from an unfashionable view of field theory—who will find it? Only someone who has sacrificed himself by teaching himself quantum electrodynamics from a peculiar and unusual point of view; one that he may have to invent for himself. I say sacrificed himself because he most likely will get nothing from it, because the truth may lie in another direction, perhaps even the fashionable one.

But, if my own experience is any guide, the sacrifice is really not great because if the peculiar viewpoint taken is truly experimentally equivalent to the usual in the realm of the known there is always a range of applications and problems in this realm for which the special viewpoint gives one a special power and clarity of thought, which is valuable in itself. Furthermore, in the search for new laws, you always have the psychological excitement of feeling that possible nobody has yet thought of the crazy possibility you are looking at right now.

So what happened to the old theory that I fell in love with as a youth? Well, I would say it's become an old lady, that has very little attractive left in her and the young today will not have their hearts pound when they look at her anymore. But, we can say the best we can for any old woman, that she has been a very good mother and she has given birth to some very good children. . . .

N.12

Quantum Electrodynamics

JULIAN SCHWINGER

JULIAN SCHWINGER shares with Richard Feynman the distinction of being one of the finest theoretical physicists ever produced by America, as well as having a similar biography and professional interest in quantum electrodynamics. Schwinger was born in New York City in 1918, three months before Feynman. He was a child prodigy, self-taught in calculus, and able to solve differential equations by the time he entered high school. He attended Townsend Harris High School, and specialized in the classics, languages, and mathematics, although his main interest was in theoretical physics. As Townsend was affiliated with the City College of New York (CCNY), Schwinger's amazing abilities and almost photographic memory soon made him well known to the physics faculty there; he matriculated at CCNY at the age of fifteen. Although he attended the usual freshman courses, he was more often found in the physics library, hunched over advanced physics books or working on problems of relativity or quantum theory.

In 1935, Schwinger began attending a theoretical seminar at Columbia University and attracted the attention of the teacher, I. I. Rabi, a future Nobel laureate himself, by his explanation of a curious aspect of the Einstein-Podolsky-Rosen thought experiment. Rabi invited Schwinger to attend Columbia on scholarship and Schwinger accepted, entering as a junior. By the time he received his baccalaureate degree in the fall of 1936, Schwinger had already submitted his first original papers for publication and completed his Ph.D. thesis on the scattering of neutrons, although he did not receive his doctorate until completing the two-year graduate residence requirement. After graduating, Schwinger worked for brief periods at the Universities of Wisconsin and California. He worked with J. Robert Oppenheimer at California, studying nuclear forces. Schwinger worked as an assistant professor at Purdue for two years, then left in 1943 to join the staff of the Massachusetts Institute of Technology Radiation Laboratory. He remained there until 1945, developing his variational method for treating problems in electromagnetism. At the end of World War II, Schwinger became an associate professor of physics at Harvard. In 1947, he was promoted to full professor, becoming at the age of twenty-nine the youngest full professor in the history of Harvard University.

Schwinger then used his formidable mathematical talents to discover the mass and charge renormalization in quantum electrodynamics. His thorough grasp of quantum field theory, coupled with his mathematical background, led him to resolve some of the problems of the theory by recasting it in a completely covariant form that was in complete agreement with relativity theory. This revised and relativistically accurate version of quantum electrodynamics was used by Schwinger to derive the Lamb shift (which explains minor discrepancies in energy between the energy levels of different states of hydrogen as resulting from the interaction of quantized electromagnetic fields with matter), as well as to predict the calculated magnetic movement of the electron. Schwinger continued his work in quantum electrodynamics and published papers on all aspects of the theory. More recently, he has devoted himself to constructing a field theory of elementary particles somewhat similar to the electrodynamic field theory of photons.

For his work in quantum electrodynamics, Schwinger shared the 1965 Nobel Prize in physics with Feynman and Tomonaga.

It must be admitted that the meaning of quantum physics, in spite of all its achievements, is not yet clarified as thoroughly as, for instance, the ideas underlying relativity theory. The relation of reality and observation is the central problem. We seem to need a deeper epistemological analysis of what constitutes an experiment, a measurement, and what sort of language is used to communicate its result. Is it that of classical physics, as Niels Bohr seems to think, or is it the "natural language," in which everyone in the conduct of his daily life encounters the world, his fellow men, and himself? The analogy with Hilbert's mathematics, where the practical manipulation of concrete symbols rather than the data of some "pure consciousness" serves as the essential extralogical basis, seems to suggest the latter.

—HERMANN WEYL, *Philosophy, Mathematics and Natural Science*

THE DEVELOPMENT OF quantum mechanics in the years 1925 and 1926 had produced rules for the description of systems of microscopic particles, which involved promoting the fundamental dynamical variables of a corresponding classical system into operators with specified commutators. By this means, a system, described initially in classical particle language, acquires characteristics associated with the complementary classical wave picture. It was also known that electromagnetic radiation contained in an enclosure, when considered as a classical dynamical system, was equivalent energetically to a denumerably infinite number of harmonic oscillators. With the application of the quantization process to these fictitious oscillators the classical radiation field assumed characteristics describable in the complementary classical particle language. The ensuing theory of light quantum emission and absorption by atomic systems marked the beginning of quantum electrodynamics, as the theory of the quantum dynamical system formed by the electromagnetic field in interaction with charged particles (in a narrower sense, the lightest charged particles). The quantization procedure could be transferred from the

---

Julian Schwinger, "Quantum Electrodynamics," in *Selected Papers in Quantum Electrodynamics.* New York: Dover Publications, Inc. 1958, pp. vii–xvii.

variables of the fictitious oscillators to the components of the field in three-dimensional space, based upon the classical analogy between a field specified within small spatial cells, and equivalent particle systems. When it was attempted to quantize the complete electromagnetic field, rather than the radiation field that remains after the Coulomb interaction is separated, difficulties were encountered that stem from the gauge ambiguity of the potentials that appear in the Lagrangian formulation of the Maxwell equations. The only real dynamical degrees of freedom are those of the radiation part of the field. Yet one can employ additional degrees of freedom which are suppressed finally by imposing a consistent restriction on the admissible states of the system. To make more evident the relativistic invariance of the scheme, other equivalent forms were given to the theory by introducing different time coordinates for each of a fixed number of charged particles coupled to the electromagnetic field. This formal period of quantization of the electromagnetic field was terminated by a critical analysis of the limitations in the accuracy of simultaneous measurements of two field strengths, produced by the known quantum restrictions on the simultaneous measurability of properties of material test bodies. The complete agreement of these considerations with the formal implications of the operator commutation relations indicated the necessity and consistency of applying the quantum mechanical description to all dynamical systems. The synthesis of the complementary classical particle and field languages in the concept of the quantized field, as exemplified in the treatment of the electromagnetic field, was found to be of general applicability to systems formed by arbitrary numbers of identical particles, although the rules of field quantization derived by analogy from those of particle mechanics were too restrictive, yielding only systems obeying the Bose-Einstein statistics. The replacement of commutators by anti-commutators was necessary to describe particles, like the electron, that obey the Fermi-Dirac statistics. In the latter situation there is no realizable physical limit for which the system behaves as a classical field.

But, from the origin of quantum electrodynamics in the classical theory of point charges came a legacy of difficulties. The coupling of an electron with the electromagnetic field implied an infinite energy displacement, and, indeed, an infinite shift of all spectral lines emitted by an atomic system; in the reaction of the electromagnetic field stimulated by the presence of the electron, arbitrarily short wave lengths play a disproportionate and divergent role. The phenomenon of electron-positron pair creation, which finds a natural place in the relativistic electron field theory, contributes to this situation in virtue of the fluctuating densities of charge and current that occur even in the vacuum state as the matter-field counterpart of the fluctuations in electric and magnetic field strengths. In computing the energy of a single electron relative to that of the vacuum state, it is of significance that the presence of the electron tends to suppress the charge-current fluctuations induced by the fluctuating electromagnetic field. The resulting electron energy, while still divergent in its dependence upon the contributions of arbitrarily short wave lengths, exhibits only a logarithmic infinity; the combination of quantum and relativistic effects has destroyed all correspondence with the classical theory and its strongly structure-dependent electromagnetic mass. The existence of current fluctuations in the vacuum has other implications, since the introduction of an electromagnetic field induces currents that tend to modify the initial field; the "vacuum" acts as a polarizable medium. New non-linear electromagnetic phenomena appear, such as the scattering of one light beam by another, or by an electrostatic field. But, in the calculation of the current induced by weak fields, there occur terms that depended divergently upon the contributions of high-energy electron-positron pairs. These were generally considered to be completely without physical significance, although it was noticed that the contribution to the induced charge density that is proportional to the inducing density, with a logarithmically divergent coefficient, would result in an effective reduction of all densities by a constant factor which is not observable separately under ordinary circumstances. In contrast with the divergences at infinitely high energies, another kind of divergent situation was encountered in calculating the total probability that a photon be emitted in a collision of a charged particle. Here, however, the deficiency was evidently in the approximate method of calculation; in any deflection of a charged particle it is certain that "zero" frequency quanta shall be emitted, which fact must be taken into account if meaningful questions are to be asked. The concentration on photons of very low energy permitted a sufficiently accurate treatment to be developed, in which it was recognized that the correct quantum description of a freely moving charged particle includes an electromagnetic field that accompanies the particle, as in the classical picture. It also began to be appreciated that the quantum treatment of radiation processes was inconsistent in its identification of the mass of the electron, when decoupled from the electromagnetic field, with the experimentally observed mass. Part of the effect of the electromagnetic coupling is to generate the field that accompanies the charge, and which reacts on it to produce an electromagnetic mass. This is familiar classically, where the sum of the two mass contributions appears as the effective electron mass in an equation of motion which, under ordinary conditions, no longer refers to the detailed structure of the electron. Hence, it was concluded that a classical theory of the latter type should be the correspondence basis for a quantum electrodynamics.

Further progress came only with the spur of experimental discovery. Exploiting the wartime development of electronic and microwave techniques, delicate measurements disclosed that the electron possessed an intrinsic magnetic moment slightly greater than that predicted by the relativistic quantum theory of a single particle, while another prediction of the latter theory concerning the degeneracy of states in the excited levels of hydrogen was contradicted by observing a separation of the states. (Historically, the experimental stimulus came entirely from the latter measurement; the evidence on magnetic anomalies received its proper interpretation only in consequence of the theoretical prediction of an additional spin magnetic moment.) If these new electron properties were to be understood as electrodynamic effects, the theory had to be recast in a usable form. The parameters of mass and charge associated with the electron in the formalism of electrodynamics are not the quantities measured under ordinary conditions. A free electron is accompanied by an electromagnetic field which effectively alters the inertia of the system, and an electromagnetic field is accompanied by a current of electron-positron pairs which effectively alters the strength of the field and of all charges. Hence a process of renormalization must be carried out, in which the initial parameters are eliminated in favor of those with immediate physical significance. The simplest approximate method of accomplishing this is to compute the electrodynamic corrections to some property and then subtract the effect of the mass and charge redefinitions. While this is a possible non-relativistic procedure, it is not a satisfactory basis for relativistic calculations where the difference of two individually divergent terms is generally ambiguous. It was necessary to subject the conventional Hamiltonian electrodynamics to a transformation designed to introduce the proper description of single electron and photon states, so that the interactions among these particles would be characterized from the beginning by experimental parameters. As the result of this calculation, performed to the first significant order of approximation in the electromagnetic coupling, the electron acquired new electrodynamic properties, which were completely finite. These included an energy displacement in an external magnetic field corresponding to an additional spin magnetic moment, and a displacement of energy levels in a Coulomb field. Both predictions were in good accord with experiment, and later refinements in experiment and theory have only emphasized that agreement. However, the Coulomb calculation disclosed a serious flaw; the additional spin interaction that appeared in an electrostatic field was not that expected from the relativistic transformation properties of the supplementary spin magnetic moment, and had to be artificially corrected. Thus, a complete revision in the computational techniques of the relativistic theory could not be avoided. The electrodynamic formalism is invariant under Lorentz transformations and gauge transformations, and the concept of renormalization is in accord with these requirements. Yet, in virtue of the divergences inherent in the theory, the use of a particular coordinate system or gauge in the course of computation could result in a loss of covariance. A version of the theory was needed that manifested covariance at every stage of the calculation. The basis of such a formulation was found in the distinction between the elementary properties of the individual uncoupled fields, and the effects produced by the interaction between them. The application of these methods to the problems of vacuum polarization, electron mass, and the electromagnetic properties of single electrons now gave finite, covariant results which justified and extended the earlier calculations. Thus, to the first approximation at least, the use of a covariant renormalization technique had produced a theory that was devoid of divergences and in agreement with experience, all high energy difficulties being isolated in the renormalization constants. Yet, in one aspect of these calculations, the preservation of gauge invariance, the utmost caution was required, and the need was felt for less delicate methods of evaluation. Extreme care would not be necessary if, by some device, the various divergent integrals could be rendered convergent while maintaining their general covariant features. This can be accomplished by substituting, for the mass of the particle, a suitably weighted spectrum of masses, where all auxiliary masses eventually tend to infinity. Such a procedure has no meaning in terms of physically realizable particles. It is best understood, and replaced, by a description of the electron with the aid of an invariant proper-time parameter. Divergences appear only when one integrates over this parameter, and gauge invariant, Lorentz invariant results are automatically guaranteed merely by reserving this integration to the end of the calculation.

Throughout these developments the basic view of electromagnetism was that originated by Maxwell and Lorentz—the interaction between charges is propagated through the field by local action. In its quantum mechanical transcription it leads to formalisms in which charged particles and field appear on the same footing dynamically. But another approach is also familiar classically; the field produced by arbitrarily moving charges can be evaluated, and the dynamical problem reformulated as the purely mechanical one of particles interacting with each other, and themselves, through a propagated action at a distance. The transference of this line of thought into quantum language was accompanied by another shift in emphasis relative to the previously described work. In the latter, the effect on the particles of the coupling with the electromagnetic field was expressed by additional energy terms which could then be used to evaluate energy displacements in bound states, or to compute corrections to scattering cross-sections. Now the fundamental viewpoint was that of scattering, and in its approximate versions led to a detailed space-time description of the various interaction mechanisms. The two approaches are equivalent; the formal integration of the differential equations of one method supplying the starting point of the other. But if one excludes the consideration of bound states, it is possible to expand the elements of a scattering matrix in powers of the coupling constant, and examine the effect of charge and mass renormalization, term by term, to indefinitely high powers. It appeared that, for any process, the coefficient of each power in the renormalized coupling constant was completely finite. This highly satisfactory result did not mean, however, that the act of renormalization had, in itself, produced a more correct theory. The convergence of the power series is not established, and the series doubtless has the significance of an asymptotic expansion. Yet, for practical purposes, in which the smallness of the coupling parameter is relevant, this analysis gave assurance that calculations of arbitrary precision could be performed.

The evolutionary process by which relativistic field theory was escaping from the confines of its non-relativistic heritage culminated in a complete reconstruction of the foundations of quantum dynamics. The quantum mechanics of particles has been expressed as a set of operator prescriptions superimposed upon the structure of classical mechanics in Hamiltonian form. When extended to relativistic fields, this approach had the disadvantage of producing an unnecessarily great asymmetry between time and space, and of placing the existence of Fermi-Dirac fields on a purely empirical basis. But the Hamiltonian form is not the natural starting point of classical dynamics. Rather, this is supplied by Hamilton's action principle, and action is a relativistic invariant. Could quantum dynamics be developed independently from an action principle, which, being freed from the limitations of the correspondence principle, might automatically produce two distinct types of dynamical variables?

The correspondence relation between classical action, and the quantum mechanical description of time development by a transformation function, had long been known. It had also been observed that, for infinitesimal time intervals and sufficiently simple systems, this asymptotic connection becomes sharpened into an identity of the phase of the transformation function with the classically evaluated action. The general quantum dynamical principle was found in a differential characterization of transformation functions, involving the variation of an action operator. When the action operator is chosen to produce first order differential equations of motion, or field equations, it indeed predicts the existence of two types of dynamical variables, with operator properties described by commutators and anti-commutators, respectively. Furthermore, the connection between the statistics and the spin of the particles is inferred from invariance requirements, which strengthens the previous arguments based upon properties of non-interacting particles. The practical utility of this quantum dynamical principle stems from its very nature; it supplies differential equations for the construction of the transformation functions that contain all the dynamical properties of the system. It leads in particular to a concise expression of quantum electrodynamics in the form of coupled differential equations for electron and photon propagation functions. Such functions enjoy the advantages of space-time pictorializability, combined with general applicability to bound systems or scattering situations. Among these applications has been a treatment of that most electrodynamic of systems—positronium, the metastable atom formed by a positron and an electron. The agreement between theory and experiment on the finer details of this system is another quantitative triumph of quantum electrodynamics.

The post-war developments of quantum electrodynamics have been largely dominated by questions of formalism and technique, and do not contain any fundamental improvement in the physical foundations of the theory. Such a situation is not new in the history of physics; it took the labors of more than a century to develop the methods that express fully the mechanical principles laid down by Newton. But, we may ask, is there a fatal fault in the structure of field theory? Could it not be that the divergences—apparent symptoms of malignancy—are only spurious byproducts of an invalid expansion in powers of the coupling constant and that renormalization, which can change no physical implication of the theory, simply rectifies this mathematical error? This hope disappears on recognizing that the observational basis of quantum electrodynamics is self-contradictory. The fundamental dynamical variables of the electron-positron field, for example, have meaning only as symbols of the localized creation and annihilation of charged particles, to which are ascribed a definite mass without reference to the electromagnetic field. Accordingly it should be possible, in principle, to confirm these properties by measurements, which, if they are to be uninfluenced by the coupling of the particles to the electromagnetic field, must be performed instantaneously. But there appears to be nothing in the formalism to set a standard for arbitrarily short times and, indeed, the assumption that over sufficiently small intervals the two fields behave as though free from interaction is contradicted by evaluating the supposedly small effect of the coupling. Thus, although the starting point of the theory is the independent assignment of properties to the two fields, they can never be disengaged to give those properties immediate observational significance. It seems that we have reached the limits of the quantum theory of measurement, which asserts the possibility of instantaneous observations, without reference to specific agencies. The localization of charge with indefinite precision requires for its realization a coupling with the electromagnetic field that can attain arbitrarily large magnitudes. The resulting appearance of divergences, and contradictions, serves to deny the basic measurement hypothesis. We conclude that a convergent theory cannot be formulated consistently within the framework of present space-time concepts. To limit the magnitude of interactions while retaining the customary coördinate description is contradictory, since no mechanism is provided for precisely localized measurements.

In attempting to account for the properties of electron and positron, it has been natural to use the simplified form of quantum electrodynamics in which only these charged particles are considered. Despite the apparent validity of the basic assumption that the electron-positron field experiences no appreciable interaction with fields other than electromagnetic, this physically incomplete theory suffers from a fundamental limitation. It can never explain the observed value of the dimensionless coupling constant measuring the electron charge. Indeed, since charge renormalization is a property of the electromagnetic field, and the latter is influenced by the behavior of every kind of fundamental particle with direct or indirect electromagnetic coupling, a full understanding of the electron charge can exist only when the theory of elementary particles has come to a stage of perfection that is presently unimaginable. It is not likely that future developments will change drastically the practical results of the electron theory, which gives contemporary quantum electrodynamics a certain enduring value. Yet the real significance of the work of the past decade lies in the recognition of the ultimate problems facing electrodynamics, the problems of conceptual consistency and of physical completeness. No final solution can be anticipated until physical science has met the heroic challenge to comprehend the structure of the sub-microscopic world that nuclear exploration has revealed.

N.13

Bell's Inequality

HEINZ PAGELS

HEINZ R. PAGELS, Professor of Physics at Rockefeller University in New York, has also served as president of the New York Academy of Sciences. In 1982, he published a book entitled *The Cosmic Code: Quantum Physics as the Language of Nature,* in response, he tells us in his preface, to a remark by the outstanding physicist Isidor Rabi, who chided his fellow scientists for failing to bring the excitement of physics to a larger public. Physicists, according to Rabi, have done less than science fiction writers to communicate the spirit of science to a wider audience.

Pagels writes well and with infectious enthusiasm. His book begins with a chapter on the "last classical physicist"—Albert Einstein—and then goes on to cover Einstein's general theory of relativity, the origins of quantum theory in the work of Max Planck, and the development of quantum theory up to the present time. An outstanding chapter of the book deals with what is known as Bell's Inequality, which Pagels treats with all the excitement of a good detective story.

Over a long period of time, Einstein and Niels Bohr, chief spokesman for the Copenhagen interpretation of quantum mechanics, carried on a debate over the soundness of quantum theory. Many thought experiments were proposed by Einstein, who could never reconcile himself to the chance element—the probabilistic character—that quantum theory took on during the 1920s. In 1934–35, when Einstein was working at Princeton, he and two colleagues, Boris Podolsky and Nathan Rosen, published a paper on what is generally known as the "EPR paradox" (the initial letters of their names), which is a misnomer in that its theme is not actually a paradox. The paper was published in *Physical Review* under the title "Can quantum-mechanical descriptions of physical reality be considered complete?"

The EPR authors asked their readers to imagine two particles interacting and then flying apart with no further interaction with each other or any other particle. The rules of quantum theory allow exact measurement of the total momentum of the two particles. If the momentum of one particle is determined at a certain point, simple subtraction from the total momentum gives the momentum of the other particle. Determining the position of the one particle permits determining the position of the other particle. Thus, both position and momentum of the second particle have been determined, in violation of Heisenberg's uncertainty principle.

EPR set off a fairly tumultuous debate among physicists; one of them, John S. Bell, turned his mind to the problem, with fascinating results. The following selection is Heinz Pagels's account of Bell's work.

It is generally acknowledged that the quantum theory has many strikingly novel features, including discreteness of energy and momentum, discrete jumps in quantum processes, wave-particle duality, barrier penetration, etc. However, there has been too little emphasis on what is, in our view, the most fundamentally different new feature of all, i.e. the intimate interconnection of different systems that are not in spatial contact. This has been especially clearly revealed through the . . . well known experiment of Einstein, Podolsky and Rosen.

—D. BOHM AND B. HILEY, "On the Intuitive Understanding of Non-Locality as Implied by Quantum Theory"

PHYSICISTS RESPONDED TO the new quantum theory in two ways. The first and primary one was the application of the new theory to natural phenomena, which led to the development of the quantum theory of solids, quantum field theory, and nuclear physics. The second direction was more philosophically oriented and focused on the interpretative problems of the new theory.

It is fair to say that the majority of practicing physicists are not very interested in these interpretative problems. Pragmatic theoretical physicists take their motivation from new experiments and ideas related to experiments. They take the Copenhagen interpretation for granted until there is some experimental indication that they should not. The question of interpreting the quantum theory has had little impact on understanding nuclear physics, elementary particle physics, or the construction of transistors and other electronic devices.

In spite of the lack of impact on the practical problems of modern physics, the research done on interpretative questions persists. Physicists and philosophers cannot rid themselves of the question "What is quantum reality?" In asking this question a degree of clarification about the nature of quantum reality has emerged. Over the years a series of thought experiments such as the two-hole experiment and Schrödinger's cat, as well as actual experiments, have been devised that serve the purpose of

---

Heinz R. Pagels, "Bell's Inequality," in *The Cosmic Code: Quantum Physics as the Language of Nature.* New York: Simon and Schuster, 1982, pp. 160–176. Copyright © 1982 by Heinz R. Pagels. Reprinted by permission of Simon & Schuster, Inc.

exposing quantum weirdness—those features of quantum reality that differ from naive realism. Two of these, the EPR experiment and Bell's experiment, have been extensively discussed by physicists and philosophers. They are the basis for our discussion here about the nature of physical reality.

Following Bohr's initial presentation of the Copenhagen interpretation of the quantum theory in 1927, physicists began to recognize the radical nature of the interpretation of reality it proposed. The essence of the Copenhagen interpretation is that the world must be actually observed to be objective. Einstein was among the most prominent critics of this viewpoint. He eventually ceased to criticize the consistency of the interpretation. Instead he focused his attack on the issue of whether or not the quantum theory gave a complete description of reality.

In 1935, Einstein, Podolsky, and Rosen wrote a paper proposing a thought experiment leading to what is often called the EPR paradox. This is a misnomer, since no contradiction is involved; there is no paradox. The EPR article expressed Einstein's view that the standard Copenhagen interpretation of the quantum theory and objective reality are incompatible. He was right. But the main point of the EPR argument was to argue that the quantum theory as it stands is incomplete—there are objective elements of reality that it does not specify. As Einstein later summarized, "I am therefore inclined to believe that the description of [the] quantum mechanism . . . has to be regarded as an incomplete and indirect description of reality, to be replaced at some later date by a more complete and direct one."

Given that no logical flaw exists in the Copenhagen interpretation and no experiment exists that contradicts the predictions of the quantum theory, how did EPR come to this remarkable conclusion? In order to understand their conclusion, we will have to outline the assumptions the three authors made before describing their thought experiment.

We have discussed the assumption of objective reality—that the world exists in a definite state. Bohr in his Copenhagen interpretation and most physicists realized that quantum theory denies this assumption, but Einstein and his collaborators thought that they were too quick in rejecting the idea that at least some measurable properties of the microworld had an objective meaning. They felt that no reasonable idea of reality could completely reject objectivity, and thus objectivity was the first assumption of the EPR team.

Einstein, we know, was distressed by the indeterminism of the quantum theory. But that objection was not the main one that stood in the way of his accepting the theory's picture of reality. A principle of physics that he held even more dear than determinism was the principle of local causality—that distant events cannot instantaneously influence local objects without any mediation. What the EPR argument did without making any assumption about determinism or indeterminism was to show that quantum theory violated local causality. This finding startled most physicists, because they too held the principle of local causality sacred. Let us examine this concept of local causality more closely.

The basic idea of local causality is the following: Events far away cannot directly and instantaneously influence objects here. If a fire breaks out a hundred miles away there is no way it can directly influence you. A second after the fire breaks out a friend may telephone you and tell you about the fire—but that is ordinary causality. Information about the fire has been transmitted by an electromagnetic signal from your friend to you. We can precisely define causality if we imagine constructing an imaginary surface around any object. Then the principle of local causality asserts that whatever influences the object either is attributable to local changes in the state of the object itself or due to energy being transmitted through the surface. That this principle—accepted by all physicists—stands at the center of all our thinking about physics is expressed in Einstein's remarks:

> If one asks what, irrespective of quantum mechanics, is characteristic of the world of ideas of physics, one is first of all struck by the following: The concepts of physics relate to a real outside world. . . . It is further characteristic of these physical objects that they are thought of as arranged in a space-time continuum. An essential aspect of this arrangement of things in physics is that they lay claim, at a certain time, to an existence independent of one another, provided these objects "are situated in different parts of space."

What the EPR team did with their definition of objectivity was to point out that quantum theory had either to violate the principle of local causality or be incomplete. Since no one seriously wants to abandon causality, they concluded the quantum theory had to be incomplete. Here is their argument in a nutshell.

Two particles, call them 1 and 2, are sitting near each other with their positions from some common point given by q_1 and q_2 respectively. We assume the particles are moving and that their momenta are p_1 and p_2. Although the Heisenberg uncertainty relation implies that we cannot simultaneously measure p_1 and q_1 or p_2 and q_2 without uncertainty, it does allow us to simultaneously measure the *sum* of the momenta $p = p_1 + p_2$ and the *distance between* the two particles $q = q_1 - q_2$ without any uncertainty. The two particles interact, and then particle 2 flies off to London while 1 remains in New York. These two locations are so far apart that it seems reasonable to suppose that what we do to particle 1 in New York should in no way influence particle 2 in London—the principle of local causality. Since we know that the total momentum is conserved—it is the same before the interaction as after—if we measure the momentum p_1 of the particle in New York, then by subtracting this quantity from the known total momentum p, we deduce exactly the momentum $p_2 = p - p_1$ of particle 2 in London. Likewise by next exactly measuring the position q_1 of the particle in New York we can deduce the position of particle 2 in London by subtracting the known distance between the particles, $q_2 = q_1 - q$. Measuring the position q_1 of the New York particle will disturb our previous measurement of its momentum p_1, but it should not (if we believe in local causality) alter the momentum p_2 we just deduced for the particle far away in London. Hence we have deduced both the momentum p_2 and the position q_2 of the particle in London without any uncertainty. But the Heisenberg uncertainty principle says it is impossible to determine both the position and momentum of a single particle without uncertainty. By assuming local causality we have done what is impossible according to the quantum theory. It seems as if the quantum theory requires that by measuring particle 1 in New York we have instantaneously influenced particle 2 way off in London. On the basis of this argument EPR concluded that either we admit that quantum theory has such causality violating "spooky action-at-a-distance" or the quantum theory is incomplete and there is indeed a way of simultaneously measuring both position and momentum. Since few physicists like to admit the possibility of such " 'telepathic' means," we should all accept the conclusion that quantum theory is incomplete.

The EPR article caused a great stir among physicists and philosophers. The complacency of the standard interpretation of quantum reality was shaken. No one had previously emphasized these action-at-a-distance effects implied by the usual quantum theory. Were Einstein and his collaborators right in their conclusion that quantum theory could not be the last word on reality? Where is the flaw in their argument? There is no flaw. There is, however, an assumption in the EPR thought experiment that ought to be made explicit. The argument assumes that properties of particle 2 such as its position and momentum in London have objective existence without actually measuring them. The EPR team deduced these properties assuming they had objective significance purely by measurements on particle 1. They then concluded that if quantum theory was right there had to be action-at-a-distance. This conclusion of EPR is sound.

But there is an alternative interpretation of this experiment—the Copenhagen interpretation, which denies the objectivity of the world without actually measuring it. Bohr, who promoted this view, would maintain that the position and momentum of particle 2 have no objective meaning until they are directly measured. If such measurements are carried out they will obey the Heisenberg uncertainty relations in agreement with the quantum theory. Then one avoids the conclusion of action-at-a-distance—instantaneous nonlocal interactions. Einstein, in opposition to Bohr, could never accept the idea of an observer-created reality. Instead he showed that if reality was objective and quantum theory complete, then there had to be nonlocal effects. Since violating causality is so repugnant, Einstein concluded quantum theory was incomplete.

For over thirty years physicists debated the conclusions of the EPR article. Perhaps hiding behind quantum reality lay yet another reality? To attack this question, John Bell, a theoretical physicist at CERN near Geneva, took the next step on the road to quantum reality in 1965. In his paper he did not appeal to the formalism of quantum theory but directly to experiment—he proposed a real experiment, not just a thought experiment. What Bell showed was that the kind of incompleteness of quantum theory envisioned by EPR was not possible. There were only two physical interpretations of Bell's experiment—either the world was nonobjective and did not exist in a definite state, or it was nonlocal with instantaneous action-at-a-distance. Take your pick of weirdness.

Bell's paper addressed the question of hidden variables—the idea that somehow the usual quantum theory is incomplete and there exists a hypothetical *sub*quantum theory which specifies additional physical information about the state of the world in the form of these new hidden variables. If physicists knew these variables, they could predict the outcome of a particular measurement (not just the probabilities of various outcomes) and even determine the momentum and position of particles simultaneously. Such a subquantum theory would actually restore determinism and objectivity. If we imagine that reality is a deck of cards, all the quantum theory does is predict the probability of various hands dealt. If there were hidden variables it would be like looking into the deck and predicting the individual cards in each hand.

It would seem that if the usual quantum theory is experimentally correct, then it rules out any subquantum theory of hidden variables and a hidden reality. Von Neumann, the mathematician, had a proof that such variables, hiding behind the veil of quantum reality, could not exist, and because of his proof people stopped thinking about hidden variables. Von Neumann's proof was logically flawless, but as Bell first pointed out, one of the assumptions that went into Von Neumann's proof did not apply to quantum theory and therefore the proof was irrelevant. The question of whether quantum theory allowed for hidden variables and causal reality was still open. To this question Bell next turned.

Bell wanted to see what the quantum world would be like if local hidden variables really existed—and here the word "local" is important. Local hidden variables refer to physical quantities which locally determine the state of an object inside an imaginary surface. By contrast, nonlocal hidden variables could be instantaneously changed by events on the other side of the universe. Assuming that any hidden variables are "local" is the assumption of local causality. Using this assumption, Bell derived a mathematical formula, an inequality, which would be checked experimentally. The experiment has been done independently at least half a dozen times, and Bell's inequality—along with its central assumption of local causality—was violated. The world, it seemed, was *not* locally causal! We will subsequently scrutinize this amazing conclusion, but first we will describe Bell's experiment in considerable detail. As someone remarked, "God is in the details," and we find if we look at the details, the experiment reveals a remarkable act of sleight of hand by the God that plays dice.

Bell's inequality applies to a large class of quantum experiments. Before applying this inequality to the quantum world it is useful first to derive the inequality for a purely classical, visualizable experiment. There will be no quantum weirdness in this classical experiment—just as there wasn't any for the machine gun firing at the two holes. The reason for first deriving Bell's inequality for a classical physics experiment is that all the assumptions that go into its derivation can be seen explicitly. There are no "hidden" variables for a classical system—all the cards are laid out on the table.

Imagine that we have a special nail gun that shoots nails two at a time in exactly opposite directions along a fixed line. Unlike most nail guns, which shoot nails like arrows, this one shoots them sideways—a pair of nails flies away from the gun with their long axis perpendicular to the direction of motion. Although each nail in a pair has the same orientation, different pairs, shot off successively, have completely random orientations

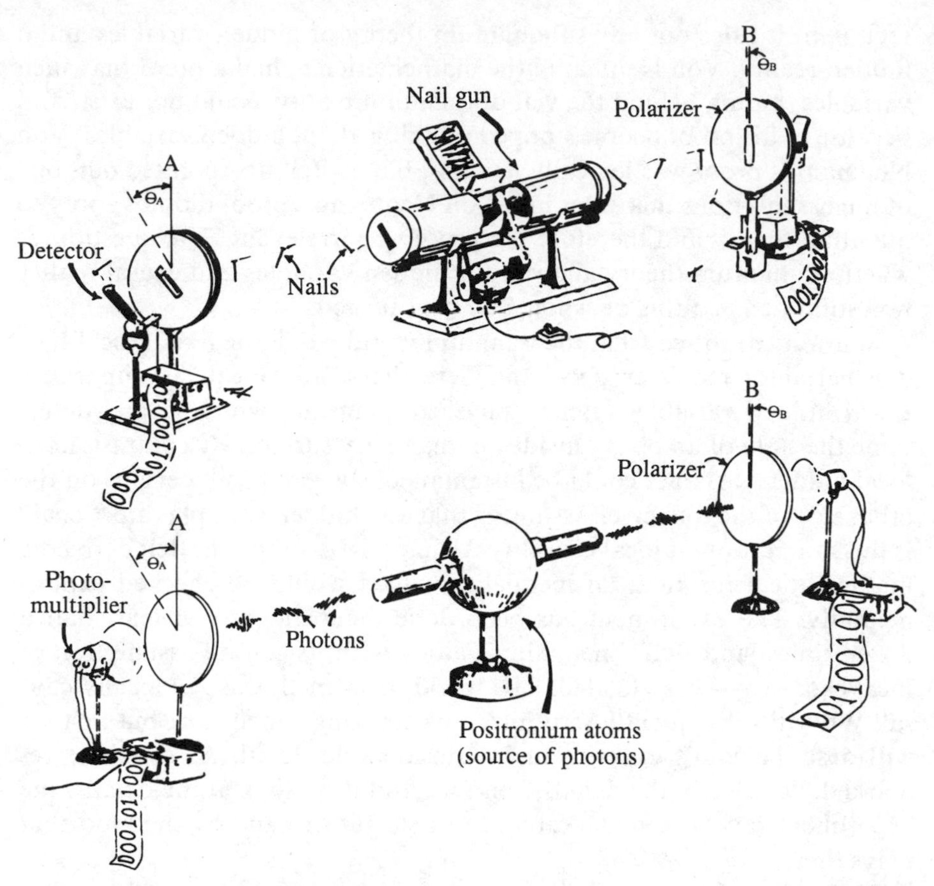

Bell's experiment: The flying nail gun and the positronium source of correlated pairs of photons. If the nails or photons are properly oriented then they pass through their respective polarizers at A and B and are detected. Hits are registered as a 1 and misses as a 0. The angle $\Theta = \Theta_A - \Theta_B$ is the relative angle between the polarizers at A and B.

relative to each other. The reason for all these peculiar requirements will become apparent when we consider a corresponding quantum system.

The flying nails are aimed at two metal sheets, A and B, each with a wide slot in it. These slots behave like real polarizers—devices which let objects with a specific orientation pass through them while blocking the passage of identical but improperly oriented objects. For example, polarized sunglasses let waves of light which are vibrating with a vertical orientation go through them while blocking light which vibrates horizontally. Since most reflected light, in contrast to direct light, is vibrating horizontally, the effect of the polarized sunglasses is to cut out glare. The slots we will call polarizers, because they only let flying nails which are aligned with the slot pass while blocking all others. We can adjust the orientation of these polarizers in the course of the experiment. At sheets A and B there are two observers who keep records of the nails that get through and those that don't. If a nail gets through the slot a "hit" is recorded as a 1 and if it fails a "miss" is recorded as a 0.

Initially the two polarizers are both oriented in the same direction as the gun fires its pairs of nails. Since each member of a pair has precisely the same orientation and the polarizers at A and B are aligned, each member of the pair either gets through the slot or it fails—hits and misses are exactly correlated at A and B. The record at A and B might look like

A: 0100011001000010110100110010110001000100 . . .

B: 0100011001000010110100110010110001000100 . . .

Each sequence of 0s and 1s is random, because the gun fires the pairs out at random orientations. But note that the two random sequences are precisely correlated.

The next step is to change the relative angle between the two polarizers by rotating the slot at A clockwise by a small angle Θ and holding B as a fixed standard. With this configuration a nail in a pair will sometimes get through at A but fail at B and vice versa. The hits and misses at A and B are no longer exactly correlated. The record might look like

A: 0001011000101011100011110010110010100100 . . .
↕ ↕ ↕ ↕
B: 0011001000101011100011010010010010100100 . . .

where the mismatches are indicated. These mismatches we can call "errors," for they may be thought of as errors in A's record relative to B's, which is the standard. In the above example there were 4 errors out of 40, so the error rate E(Θ) for the angle set at Θ is E(Θ) = 10%.

Suppose that we had left the polarizer at A untouched but rotated the one at B counterclockwise by the angle Θ. Now we might say the "errors" are in B's record and A's acts as the standard. The error rate will clearly be the same as before, E(Θ) = 10%, because the configuration is identical.

The final step is to rotate A's polarizer by an angle Θ clockwise so that the total relative angle between the two polarizers is now doubled to 2Θ. What is the error rate for this new configuration? This is easy to answer provided we assume that the errors at A are independent of the situation at B and vice versa. In making this assumption we are assuming local causality. After all, what does a nail getting through its polarizer at A have to do with the situation at B? Since the errors produced at B were previously E(Θ) we must add to this the errors produced by rotating the polarizer at A, which is also E(Θ). So it seems that the error rate with the new setting should be the sum of the two independent error rates, or E(Θ) + E(Θ) = 2E(Θ). But by shifting A by the small angle Θ we have lost the standard record for B's record, and likewise by shifting B we have lost A's standard. This means that from time to time an error will be produced at both A and B—a double error. But a double error is detected as no error at all. For example, suppose a pair of nails would have registered a 1 and 1 at A and B if the polarizers were perfectly aligned. But because A's polarizer is shifted the nail then misses and a 0 is registered. This shows up as an error. But since we have also shifted B's polarizer it is possible that the nail there also misses. This is a double error in which two hits, a 1 and 1, has been changed to two misses, a 0 and 0. The two misses are seen as no error. Because of the impossibility of detecting a double error, the error rate with an angle 2Θ between the two polarizers—E(2Θ)—will necessarily be less than the sum of the error rates for each of the separate shifts. This is expressed mathematically by the formula

$$E(2\Theta) \leq 2E(\Theta)$$

which is Bell's inequality.

No doubt if this odd experiment were performed, Bell's inequality would be satisfied. For example, with an angle of 2Θ the record might look like

```
A: 0010110011111000101010100111101011101000 . . .
         ↕    ↕              ↕      ↕  ↕
B: 0010100011011100101010100110101011001100 . . .
```

or 6 errors out of 40 so E(2Θ) = 15% ≤ 2 × 10% = 20%. Bell's inequality is satisfied for this classical physics experiment.

Let us examine closely the crucial assumptions that have gone into obtaining Bell's inequality. We have assumed that the nails are real objects flying through space and that the orientation of pairs of nails is the same. We aren't actually observing that the nails have a definite orientation because they fly by us so quickly. This seems like a safe assumption for nails, but we have indulged in the fantasy of objectivity. We are assuming that the nails exist like ordinary rocks, tables, and chairs. Suppose we are the observer at A. Then we are assuming that a nail flying toward B, even if B is on the moon, has a definite orientation. The notion that things exist in a definite state even if we do not observe them is the assumption of objectivity—and of classical physics.

The second crucial assumption in obtaining Bell's inequality was that the errors produced at A and at B were completely independent of each other. By shifting the polarizer at A we did not influence the physical situation at B and vice versa—the assumption of local causality.

These two assumptions—objectivity and local causality—are crucial in obtaining Bell's inequality. What happens if we now replace flying nails with photons—particles of light?

Instead of a nail gun we will use positronium atoms as our source of particles. Positronium is an atom consisting of a single electron bound to a positron (anti-electron), and this atom sometimes decays into two photons traveling in opposite directions. The important features of this positronium decay is that the two photons have their relative polarization precisely correlated—like the flying nails. The polarization of a photon is the orientation of its vibration in space. If one photon has its polarization in one direction, its companion flying off in the opposite direction has its polarization in the same direction. The absolute direction of the polarization of the two correlated photons changes from decay to decay in a random way, but the relative polarization between any pair of photons is fixed. That is the important feature of this source—it is like the nail gun.

The photons fly off in opposite directions and pass through separate polarizers at A and B, located far apart with observers stationed there. Behind the polarizers are photomultiplier tubes that can detect single photons. If a photomultiplier tube detects a photon, the event is recorded by a 1, and if it detects no photon the event is recorded by a 0. In the initial configuration the two polarizers at A and B are perfectly aligned relative to each other. Let the polarizer at B be fixed while the one at A is free to rotate and call the relative angle between the two polarizers Θ so that in this initial configuration $\Theta = 0$.

If a photon hits the polarizer it has a certain probability of getting through and being detected. If the photon's polarization happens to align parallel to that of the polarizer it gets through to the detector, and a 1 is registered. If the polarization of the photon is perpendicular to the polarizer, then it won't get through and a 0 is registered. With other orientations there is only a probability that it gets through.

The polarization of the photons relative to the polarizer is completely random, so that each detector, in the initial configuration with Θ = 0, will register a series of 0s and 1s. Suppose the series looks something like this at each detector:

A: 01101011000010110101110011000101110 . . .

B: 01101011000010110101110011000101110 . . .

This is just like the records with the nail gun. The series are identical because each pair of photons is polarized identically and the angle between the polarizers is zero. Further, each series has on the average an equal number of zeroes and ones, since it is as likely for a photon to get through the polarizer to the detector as not.

Next we rotate the polarizer at A, slightly, so the angle Θ is not zero. Set Θ = 25°. This slight shift means that the two photons in each pair have a slightly different probability of going through the polarizers and being detected. The series are not precisely identical but instead disagree from time to time. However, on the average, both the series at A or B have an equal number of zeroes and ones because the probability of getting through the polarizer is independent of its orientation. The new series looks like

A: 0010111011000111110110100111000101011100 . . .
↕ ↕ ↕ ↕
B: 0110011011000111010110100110000101011100 . . .

where we have indicated the mismatches. In the above example the rate of errors, since there are 4 errors out of 40 detections, is E(Θ) = 10%.

So far this experiment done with photons resembles that with the nails. Photons are behaving just like the perfectly visualizable experiment with the flying nails. If we assume the state of polarization the photons have at A and B is objective (objectivity assumption) and that what one measures at A does not influence what happens at B (local causality assumption), then Bell's inequality, E(2Θ) ≤ 2E(Θ), ought to hold for this experiment. If we double the angle to 2Θ = 50%, the following records are found:

A: 1000111001100110111001111110110101000100 . . .
↕↕ ↕ ↕ ↕↕↕ ↕ ↕ ↕↕↕
B: 1110111101000111001001110110110101101010 . . .

This is 12 errors out of 40, so that $E(2\Theta) = 30\%$. Now let us compare this result with the requirement of Bell's inequality. Since $E(\Theta) = 10\%$ we have $2E(\Theta) = 20\%$; but Bell's inequality requires that $E(2\Theta) \leq 2E(\Theta)$, so that 30% is supposed to be less than 20%—completely false—30% is greater than 20%! We conclude that Bell's inequality is violated by this experiment, as it is for real experiments with photons. Consequently, either the assumption of objectivity or locality or both are wrong. That is very remarkable!

We have described the experiment and Bell's inequality in some detail because it is rather elementary and illustrates the crux of quantum weirdness. Bell was motivated to find a way of testing if there were hidden variables that exist out there in the world of rocks, tables, and chairs. He showed that the violation of the inequality by quantum theory did not necessarily rule out an objective world described by hidden variables but the reality they represented had to be nonlocal. Behind quantum reality there could be another reality described by these hidden variables and in this reality there would be influences that move instantaneously an arbitrary distance without evident mediation. It is possible to believe the quantum world is objective—as Einstein wanted—but then you are forced into accepting nonlocal influences—something Einstein, and most physicists, would never accept.

To get an intuitive sense of how objectivity implies nonlocality, compare the records for the angle $\Theta = 25°$ and $\Theta = 50°$. There are just too many errors (12) for the 50° setting as compared to the number of errors (4) for the 25° setting. It seems that by moving A's polarizer we must have influenced the polarization of the photons about to be detected at B and that produces all those "extra" errors that violate Bell's inequality. Observer B could be on the earth and A light-years away, on another galaxy. A, by moving the polarizer, it seems, is sending a signal faster than the speed of light, thus instantaneously changing B's record. That certainly seems like action-at-a-distance and the end of locality.

Now that we see what we have been forced into we might want to look at this a bit further. Either alternative—a nonobjective or a nonlocal reality—is a bit hard to take. Some recent popularizers of Bell's work when confronted with this conclusion have gone on to claim that telepathy is verified or the mystical notion that all parts of the universe are instantaneously interconnected is vindicated. Others assert that this implies communication faster than the speed of light. That is rubbish; the quantum theory and Bell's inequality imply nothing of this kind. Individuals who make such claims have substituted a wish-fulfilling fantasy for understanding. If we closely examine Bell's experiment we will see a bit of sleight of hand by the God that plays dice which rules out actual nonlocal influences. Just as we think we have captured a really weird beast—like acausal influences—it slips out of our grasp. The slippery property of quantum reality is again manifested.

Bohr would be the first to point out an alternative interpretation of the experimental violation of Bell's inequality. In order to conclude that the photons were subject to nonlocal influences we have indulged in the fantasy that they exist in a definite state. Try and verify that, Bohr would insist. If we can verify that the photons actually exist in a definite state of polarization without altering that state, then indeed we must conclude from Bell's experiment that we have real nonlocal influences.

For the flying nails this verification is easy—we set up a high-speed camera and take pictures of them just as they arrive at the polarizers. This won't disturb their state. But then the experiment with the flying nails did not violate Bell's inequality as did the experiment with photons.

If we now try to verify the state of polarization of a photon we find that this is not possible without altering the requirement that both members of a pair of photons have identical polarization. In measuring the polarization of the photon we put it into a definite state, but this alters the initial conditions of the experiment. This is identical to the problem we faced in the two-hole experiment with the electron. By observing with light beams which hole the electron went through we changed the detected pattern. Likewise, the very act of establishing the objective state of the photon alters the conditions under which Bell's inequality was derived. The attempt to experimentally verify the objectivity assumption has the consequence that the conditions of the experiment are altered in just such a way that we can no longer use the violation of Bell's inequality to conclude that nonlocal influences exist.

Suppose then that we do not try to verify the state of the flying photons. After all, we have the records of hits and misses at A and B and these are part of the macroscopic world of tables, chairs, and cats and are certainly objective. Cannot the observer at B read his record, see that Bell's inequality is violated, and conclude that local causality has also been violated? The answer is no, because the God that plays dice has a trick to show us. Remember that the source of photons emits them in pairs with random polarization. This means that the records at A and B, no matter what the angle is, are completely random sequences of 0s and 1s. And that fact is what lets us slip out of the conclusion of real nonlocal influences.

At first you might think that by changing A's polarizer we have directly influenced the number of errors produced at B. Hence by altering A's polarizer to various settings in a sequence of moves, B could, by observing the alteration in the number of errors produced at B, get a message from A—a telegraph that would violate causality. But no information can possibly be transmitted from A to B using this device because holding a single record of events at either A or B would be like holding the message of a top-secret communication in a random code—you can't ever get the message. Because the sequences at A and B are always completely random there is no way to communicate between A and B. That is how real nonlocality is avoided by the God that plays dice; He is always shuffling the deck of nature.

Random stereograms . . . illustrate this trick. Each half of the stereogram is completely random, but two random sequences of dots if compared can yield nonrandom information. The information is in the cross-correlation gotten by comparing the two sequences. It is the same with the records at A and B—the information about the relative angle between the polarizers at A and B is in the cross-correlation of the two records; it is not in either record separately. All that happens when the polarizer angle is changed is for one random sequence to be changed into another random sequence, and there is no way to tell that happens by looking at only one record. Because such real random processes actually occur in nature—as they do in this experiment—we avoid the conclusion of real nonlocality.

What a marvelous trick nature has used to avoid real nonlocal influences! If we asked out of all things in this universe which one, if altered in a random way, would remain unchanged, the answer is: a random sequence. A random sequence changed in a random way remains random—a mess remains a mess. The random sequences at A and B are like that. But by comparing these sequences we can see that there has been a change due to moving the polarizers—the information is in the cross-correlation, not in the individual records. And that cross-correlation is completely predicted by the quantum theory.

We conclude that even if we accept the objectivity of the microworld then Bell's experiment does not imply actual nonlocal influences. It does imply that one can instantaneously change the cross-correlation of two random sequences of events on other sides of the galaxy. But the cross-correlation of two sets of widely separated events is not a local object and the information it may contain cannot be used to violate the principle of local causality. . . .

O

Symmetry

INTRODUCTION

SYMMETRY IS WITH us all our lives. The human form is symmetric, give or take a quirk here or there. Our first enduring impression, the awareness of our mother's face, is one of symmetry. In particle physics the term *symmetry* has a specialized meaning. We can deduce the basic laws of physics from symmetry principles: Spatial symmetry implies conservation of momentum; rotational symmetry implies conservation of angular momentum; temporal symmetry implies conservation of energy; and gauge symmetry implies conservation of charge. The McGraw-Hill scientific dictionary, which is a technological version of Webster's, defines symmetry in physics as follows:

> *Unitary symmetry:* An approximate symmetry law obeyed by the strong interaction of elementary particles; it may be described as the equivalence of [the interaction of the] three fundamental particles, termed quarks, out of which all hadrons could be assumed to be composed.

In the September 1961 issue of *Annals of Physics,* Murray Gell-Mann published a paper entitled "Gauge Theories of Vector Particles," a work about which he recently remarked, "The first half is a classic, the second a mess." In it he had recourse to a system of classification of the elements of quantum field theory that had been devised by Elie-Joseph Cartan, a pure mathematician of the French school, who was totally unaware at the time of the existence of quarks. (Cartan, incidentally, lived a modest life, and the great prestige his name now enjoys came only after his death.) He invented a simple and logical set of terms that describes mathematically what physicists call gauge fields, such as the electromagnetic field and the fields of the strong and weak interactions, and that was later used by Sheldon Glashow of Harvard to suggest the unified electroweak force. All of these weighty matters, of supreme importance in recent developments in subatomic physics, are treated fully in the selections by Steven Weinberg, Abdus Salam, and Glashow, and are included in the third volume of this anthology.

The concept of symmetry, as Heinz Pagels has pointed out, is remarkable in the sense that "as physicists came to understand the mathematical symmetries of field theory they soon discovered that these symmetries actually required the very interactions they had observed among the fields and their associated quanta."

Physicists today are concerned with three types of symmetry—spatial symmetry, symmetry in time, and gauge symmetry, which implies conservation of electric charge. Spatial symmetry means that any point in space can be conveniently defined by reference to any three mutually perpendicular coordinates. Often the axes are illustrated by using the thumb of the right hand, erect, to indicate the x coordinate; the forefinger, straightened out, as the y coordinate; and the middle finger, more or less relaxed, as the z coordinate (using the left hand changes the symmetry from right- to left-handed). These two sets of coordinates are mirror images of each other, and one passes from one set to the other simply by reversing the directions of all three of the coordinates or any one of them. In quantum physics this quality of the invariance of the laws to spatial inversion is refered to as *parity*. Although the conservation of parity has been and continues to be an important principle in quantum mechanics, in 1956 T. D. Lee and C. N. Yang demonstrated that in processes involving weak interactions (emission or absorption of neutrinos) parity is violated. Lee and Yang's results were later experimentally demonstrated by C. S. Wu and her collaborators.

Time symmetry implies that the laws of physics are unaltered if we go from the future to the past instead of from the past to the future. This symmetry, of course, is of little importance to the man who sets his alarm clock, but to the particle physicist it is vital. To date, some experimental tests that physicists have devised seem to indicate that time reversal invariance is violated.

The symmetry that implies conservation of electric charge is never broken because charge is always conserved.

O.1

Elementary Particles

Chen Ning Yang

All science since the time of the Babylonians has advanced by a series of progressions—that is, by steps that are themselves a type of quanta. Since the great watershed year of 1900, the year that separates classical physics from modern physics, the material involved in virtually every scientific discipline has become so complex, the mass of experimental data accumulated so enormous, that it becomes useful to stop and summarize its recent accomplishments. Chen Ning Yang, known to his fellow theoreticians as "Frank," is one of the great figures of twentieth-century physics. In 1961, he published a book entitled *Elementary Particles: A Short History of Some Discoveries in Atomic Physics,* which combined observations on both physical and artistic symmetry.

Yang, born in Hofei, China, in 1922, was reared in a scientific environment as his father was a mathematician. He emigrated to the United States in 1945 with the express intention of studying with Enrico Fermi. He received his Ph.D. in 1948 from the University of Chicago, and then moved on to the Institute of Advanced Study at Princeton, where he became a professor in 1955. He shared the 1957 Nobel Prize in physics with Tsung-Dao Lee for their demonstration that not all parity laws—fundamental physical principles dealing with the symmetry of nature with respect to right and left in their application to elementary particles and their reactions—are valid. Previously, most physicists had assumed that the principle of parity did not vary; that no fundamental distinction could be made between left and right; and that the laws of physics are the same in a right-handed system of coordinates as they are in a left-handed system of coordinates.

Yang and Lee were first inspired to question the assumption of parity invariance in particle reactions by unexpected observations of the K-meson particle. They wondered what experimental support existed for parity invariance and, after extensive laboratory work, found that a general symmetry assumption was not valid. They devised further experiments to test right-left symmetry principles in many other elementary particle interactions. For example, they investigated the distribution of the direction of electrons when the atomic nuclei of an isotope of cobalt were exposed to a magnetic field at very low temperatures; the downward distribution revealed it to be an asymmetrical reaction.

In a chapter on the Yang-Lee-Wu work on parity in his book *The Ambidexterous Universe,* Martin Gardner, one of the best writers on science for nonscientists, speculated on an idea originally suggested by John Campbell, Jr., the editor of a science fiction magazine. Campbell had written an editorial in which he wondered whether some crucial difference between occidental and oriental culture had predisposed two Chinese physicists, Yang and Lee, to question the symmetry of natural law that Western scientists had long taken for granted. "It is an interesting thought," Gardner wrote. "I myself pointed out . . . in March 1958 that the great religious symbol of the Orient . . . is the circle divided asymmetrically. The dark and light areas are known respectively as the Yin and the Yang . . . symbols of all the fundamental dualities of life: good and evil, beauty and ugliness, truth and falsehood, male and female, night and day, sun and moon, heaven and earth, pleasure and pain, odd and even, left and right, positive and negative . . . the list is endless."

Harmony would lose its attractiveness if it did not have a background of discord.

—TEHYI HSIEH, *Chinese Epigrams Inside Out and Proverbs*

INVESTIGATIONS CARRIED OUT in various laboratories have produced interesting information concerning the elementary particles. We shall try to describe some of these. . . . In all fields of research, when one is struggling with problems which are not within one's previous domain of experience, the very definition of the problem is often obscure. To be able to forge ahead, ingenuity and technical proficiency are of course required. But more important, there must be above all an independence of judgment which can come only from confidence and solidarity in the old knowledge, and persistence and boldness in the quest into the new. This is not easy to achieve—we should not expect it to be. In an article commemorating the seventieth birthday of Einstein, Philipp Frank said he had spoken to Einstein one day about a certain physicist who had had very little success in his research work. He consistently attacked problems which offered tremendous difficulties. He applied penetrating analysis and succeeded only in discovering more and more difficulties; he was not rated very highly by most of his colleagues. Einstein, according to Frank, said, "I admire this type of man. I have little patience with scientists who take a board of wood, look for its thinnest part and drill a great number of holes where drilling is easy."

Of course it is true that most of the time persistence in a new field of research leads only to more difficulties, or to blind alleys. Let us, however, look at one of these blind alleys, namely, Thomson's idea of an atom, which today we can view with a detachment that comes with the benefit of sixty years of hindsight. We remember that upon the discovery of the electron, Thomson had a picture of the atom in which the electrons reside at positions of equilibrium, around which they vibrate when disturbed. By measuring the amount of scattering of X-rays by various matter, he was led to a calculation of the number of electrons per atom in different chemical elements, and arrived at remarkably correct conclu-

---

sions. Thomson saw that the electronic structure of the atoms offered the enormously important possibility of explaining the chemical properties of the elements. He then asked, what should be the structure of an atom with one electron, with two electrons, three electrons, etc.? . . . Three electrons embedded in a uniformly and positively charged sphere form a triangle. For four electrons it is intuitively obvious that the positions of equilibrium for the electrons form the corners of a regular tetrahedron. However, when one considers atoms with more electrons, the mathematical problem of finding their equilibrium positions, though well defined, is difficult to solve. Thomson therefore had recourse to the following experimental arrangement to simulate mechanically the structure of his atom. By the use of corks, a number of long needle magnets were made to float in water. . . . The poles of the needle magnets were made parallel to each other so that forces between them were repulsive, and in fact followed the inverse square law in the same way that the electrons in an atom repel each other. To artificially produce the influence of the uniform positive charge in the atom which serves to hold the electrons in their equilibrium positions, he applied a magnetic field generated by an electromagnet high above the surface of the water. It is easy to demonstrate that the horizontal component of the magnetic force due to the electromagnet on any one of the needles, is approximately proportional to the distance from the needle to the point directly underneath the magnet. In Thomson's atom, the same is true of the force exerted by the uniform positive charge distribution on the electrons. The equilibrium configuration assumed by the floating magnetic needles is therefore an approximate solution of the electron configuration in Thomson's atom for the two-dimensional case. . . . In the configurations found by Thomson, one sees the interesting arrangement that for large numbers of needles, rings are formed. Here is a tabulation by Thomson of the number of needles in each ring. A comparison of this table with the periodic table then known, suggested itself. Thomson and his students were thus led to many investigations of such configurations and of the vibrational frequencies of the electrons embedded in the positions of equilibrium. We know today that these efforts were not along the right track. But we also know today that they were not in vain: as we have discussed previously, it was exactly out of such considerations that the eventually correct idea of Rutherford's atom grew.

Let us now try to describe in some detail four items from the study of the elementary particles in the last ten years. The first one concerns a puzzle presented by the experimental data around the years 1951–1952. It

TABLE 1

1.	2.	3.	4.	5.
1.5	2.6	3.7	4.8	5.9
1.6	2.7	3.8	4.9	—
1.7	—	—	—	—
1.5. 9	2.7.10	3.7.10	4.8.12	5.9.12
1.6. 9	2.8.10	3.7.11	4.8.13	5.9.13
1.6.10	2.7.11	3.8.10	4.9.12	—
1.6.11	—	3.8.11	4.9.13	—
—	—	3.8.12	—	—
—	—	3.8.13	—	—
1.5. 9.12	2.7.10.15	3.7.12.13	4.9.13.14	—
1.5. 9.13	2.7.12.14	3.7.12.14	4.9.13.15	—
1.6. 9.12	—	3.7.13.14	4.9.14.15	—
1.6.10.12	—	3.7.13.15	—	—
1.6.10.13	—	—	—	—
1.6.11.12	—	—	—	—
1.6.11.13	—	—	—	—
1.6.11.14	—	—	—	—
1.6.11.15	—	—	—	—

was found at that time that strange particles were produced quite copiously in collisions between high energy particles. Since the size, or extension of these particles is of the order of 10^{-13} centimeters, and since the velocity of approach between the particles is of the order of the velocity of light, namely 3×10^{10} centimeters per second, the collision time is obviously of the order of 10^{-23} seconds. This 10^{-23} seconds is therefore in many senses a unit of the time scale of these phenomena. It was also already known at that time that the strange particles are unstable, disintegrating into various kinds of particles. The average lifetime of each of these particles is measurable by many methods. For example . . . the Λ is a neutral particle and therefore does not leave any visible track in the chamber, only betraying its existence when it disintegrates into a proton and a π^- at B. The distance between the birth at A and the death at B of the Λ is easily measured. The velocity of the Λ can be inferred from the velocities of its disintegration products, which in turn are measurable by the curvatures of their tracks. From the distance AB and the velocity it is elementary to obtain the lifetime of this particular Λ. Taking the average of many such measurements, the mean lifetime of the Λ was determined to be a few times 10^{-10} seconds. This is a very short lifetime viewed from the human scale, but a tremendously long lifetime viewed from the time scale of 10^{-23} seconds mentioned above. . . .

The strange particles are thus produced in times of the order of 10^{-23} seconds, and disintegrate in about 10^{-10} seconds, a time interval longer than the nuclear time scale by a factor of 10^{13}. In other words, the production process clearly involves forces stronger than the disintegration process by a factor of 10^{13}. Since both processes seemed to involve similar particles it was very puzzling why they took place with such drastically different rates, or strengths.

The resolution of this puzzle came with the suggestion of A. Pais and of Y. Nambu that the strange particles are *produced in groups* of more than one, in which situation they interact *strongly* with themselves and with the other particles involved in the process. On the other hand, in cases where they participate singly, as in the *disintegration* processes, they interact only *weakly.* This idea of the "associated production" process later received detailed experimental confirmation. . . .

We shall now come to the second item to be discussed: the classification of the strengths of interactions. We have already seen that an associated production of strange particles is caused by a strong interaction, and a disintegration of a strange particle by a weak interaction. Such a classification into strong and weak interactions was first discussed in the years 1948–1949 and is now known to be applicable to all measured interactions. The resulting picture is illustrated in the following table.

TABLE 2

	Strength
1. STRONG INTERACTIONS (Nuclear Interactions)	1
2. ELECTROMAGNETIC INTERACTIONS	10^{-2}
3. WEAK INTERACTIONS (Decay Interactions)	10^{-13}
4. GRAVITATIONAL INTERACTIONS	10^{-38}

In this table the strong interactions include those responsible for the associated production of strange particles, those representing the forces that bind the particles in a nucleus together, and also those responsible for various π meson interactions, such as

$$p + p \rightarrow \pi^{+} + p + n.$$

The third category consists of the interactions responsible for the decays. The ratio of the strengths of the first and the third categories is the factor 10^{13} mentioned before, and is indicated in the column labelled "strength." Prior to the discovery of the $\pi \rightarrow \mu$ decay process, we recall, there was a large discrepancy of a factor 10^{13}. That was the same factor 10^{13} that separates the strong from the weak interactions: the π mesons interact with the atomic nuclei through the strong interactions, while the μ mesons in the experiment of Conversi, Pancini, and Piccioni interact with the atomic nuclei through weak interactions.

The other two entries in Table 2 are more familiar interactions. The electromagnetic interactions, through the theories of Faraday and Maxwell, are the best understood of all. The gravitational interactions, though important when very heavy bodies such as the sun or the earth are involved, are extremely weak between the elementary particles, as indicated in the column labelled "strength." Because of this, gravitational interactions have so far fallen outside the range of elementary particle studies, though few physicists doubt that ultimately they must be brought in to give a unified picture.

Let us emphasize that the classification of interactions depicted here gives us one of the very important orientations in the study of elementary particles. It allows for a clear cut separation of the complex manifestations of each of the four types of interactions. The deep and difficult question of the origin of such a clear cut separation, and in particular, the origin of the approximate equality of the strengths of the many unrelated weak interactions remains unsolved.

The third item concerns a phenomenon called isotopic spin symmetry. One observes that the particles occur very remarkably in "multiplets" or groups with different charges but almost the same mass. The oldest known of these multiplets is the one consisting of the neutron and the proton. That the neutron and the proton are similar to each other, except for their electric charge, already suggested itself in the phenomenon that nuclei tend to have equal numbers of neutrons and protons. This was observed and discussed in the early 1930's, and received increasingly detailed experimental support as time went on. Among these was the observation that the newly discovered particles also occurred in multiplets. To find a deeper reason behind this symmetry is another very striking and tantalizing problem that has so far defied all efforts of investigation.

The phenomena of associated production and isotopic spin symmetry are related to each other, as was pointed out independently by M. Gell-Mann and K. Nishijima around 1953. They observed that the center of charge of a multiplet of strange particles may be displaced from that of the non-strange particles; for example, the Λ is displaced by half a unit of charge toward the left from the neutron-proton multiplet. A new quantity called the strangeness was introduced, with a value equal to twice the amount of this displacement. It can be shown that isotopic spin symmetry implies that in a strong interaction the sum total of the strangenesses of the particles remains unchanged in the process. This rule is called the strangeness conservation rule. It has become one of the most fruitful concepts in the study of the strange particles. For example in the process

$$\pi^- + p \rightarrow \pi^0 + \Lambda$$

the sum total of the strangenesses of the particles before the process is 0. The sum total of the strangenesses of the particles after the process is −1. According to the rule this process is not strong, a conclusion in agreement with experiments. On the other hand, the process

$$\pi^- + p \rightarrow K^0 + \Lambda$$

. . . conserves strangeness. The rule predicts that this process is strong, again in agreement with experimental results.

The fourth item we shall discuss concerns the symmetry principles, of which the isotopic spin symmetry discussed above is a special kind. The word symmetry is a common word used in daily language. . . .

The concept of symmetry in physics stems directly from our everyday notion. By tracing symmetries in a dynamical problem important conclusions can be reached. For example, a circular orbit of an electron in a hydrogen atom is a consequence and an indication of the symmetry of the Coulomb force exerted by the nucleus on the electron. The symmetry in this case means that the force has the same magnitude in all directions. Symmetry principles like this have played a role in classical physics, but in quantum mechanics the role has greatly gained both in depth and breadth. For example, the elliptical orbits as well as the circular orbits now assume a role in symmetry considerations. Indeed it is scarcely possible to over-emphasize the importance of the symmetry principles in quantum mechanics. To give two examples: the general structure of the periodic table is essentially a direct and beautiful consequence of the symmetry referred to above, the isotropy of the Coulomb force; the existence of the antiparticles . . . were anticipated in Dirac's theory, which was built on the principle of relativistic symmetry. In both cases, as in other examples, nature seems to take advantage of the simple mathematical representation of the symmetry laws. The intrinsic elegance and beautiful perfection of the mathematical reasoning involved and the complexity and depth of the physical consequences are great sources of encouragement to physicists. One learns to hope that nature possesses an order that one may aspire to comprehend.

One of the symmetry principles, the symmetry between the left and the right, has been discussed since ancient times. The question whether nature exhibits such symmetry was debated at great length by philosophers of the past. In daily life, of course, right and left are quite different from each other. In biological phenomena, it was known since Pasteur's work in 1848 that organic compounds appear oftentimes in the form of only one of two kinds, both of which, however, occur in inorganic processes and are mirror images of each other. In fact, Pasteur had considered for a time the idea that the ability to produce only one of the two forms was the very prerogative of life.

The laws of physics, however, have always shown complete symmetry between the left and the right. This symmetry can also be formulated in quantum mechanics as a conservation law called the conservation of parity, which is completely identical to the principle of right-left symmetry. The first formulation of the concept of parity was due to E. P. Wigner. It rapidly became very useful in the analysis of atomic spectra. The concept was later extended to cover phenomena in nuclear physics and the physics of mesons and strange particles. One became accustomed to the idea of nuclear parities as well as atomic parities, and one discussed and measured the parities of mesons. Throughout these developments the concept of parity and the law of parity conservation proved to be extremely fruitful, and the success had in turn been taken as a support for the validity of right-left symmetry in physical laws.

In the years 1954–1956 a puzzle called the θ-τ puzzle developed. The θ and τ mesons are today known to be the same particle which is usually called *K*. In those years, however, one only knew that there were particles that disintegrate into two π mesons and particles that disintegrate into three π mesons. They were called respectively θ's and τ's, the τ being the name given to it by Powell in 1949. As time went on, measurements became more accurate and the increasing accuracy brought out more and more clearly a puzzlement. On the one hand it was clear that θ and τ had very accurately the same mass. They were also found to behave identically in other respects. So it looked as if θ and τ were really the same particle disintegrating in two different ways. On the other hand, increasingly accurate experiments also showed that θ and τ did not have the same parity and could not therefore be the same particle.

The resolution of the puzzle lay in a change in the concept of right-left symmetry. In the summer of 1956, T. D. Lee and I examined the then existing experimental foundation of this concept and came to the conclusion that, contrary to generally held belief, no experimental evidence of right-left symmetry actually existed for the weak interactions. If right-left symmetry does not hold for the weak interactions, the concept of parity is inapplicable to the decay mechanism of the θ and τ particles and they could therefore be one and the same particle, as we now know they are.

As a possible way out of the θ-τ puzzle, it was suggested that one should test experimentally whether right-left symmetry is violated for the weak interactions. The principle of the test is very simple: two sets of experimental arrangements which are mirror images of each other are set up. They must contain weak interactions and they must not be identical to each other. One then examines whether the two arrangements always give the same results. If they do not, one would have an unambiguous proof of the violation of right-left symmetry in this experiment. In Figure 1 the first such experiment, performed by C. S. Wu, E. Ambler, R. W. Hayward, D. D. Hoppes, and R. P. Hudson in 1956, is schematically illustrated. The cobalt nuclei disintegrate by weak interactions and the disintegration products are counted. Notice that the currents flowing in the loops are very essential elements of the experiment. Without these currents the two arrangements on the two sides of the imagined mirror would have been identical and would have always given the same results. To make the influence of the currents felt by the cobalt nuclei, however, it was necessary to eliminate the disturbance on the cobalt produced by thermal agitations. The experiments had to be done, therefore, at extremely low temperatures of the order of less than .01 degrees absolute. The technique of combining β-disintegration measurements with low temperature apparatus was a major difficulty. . . .

The result of the experiment was that there was a very large difference in the readings of the two meters shown in Figure 1. Since the behavior of the other parts of the apparatus observes right-left symmetry, the asymmetry must be attributed to the disintegration process of cobalt, which is due to a weak interaction.

Hermann Weyl said . . . that in art "seldom is asymmetry merely the absence of symmetry." This seems also to hold in physics, for with the discovery of the lack of right-left symmetry there arose two new aspects concerning the right-left symmetry-asymmetry of the elementary particles and their interactions. The first has to do with the *structure of the neutrino* and very interestingly is a revival of an idea originally formulated by Weyl in 1929. It was, however, rejected in the past because it did not pre-

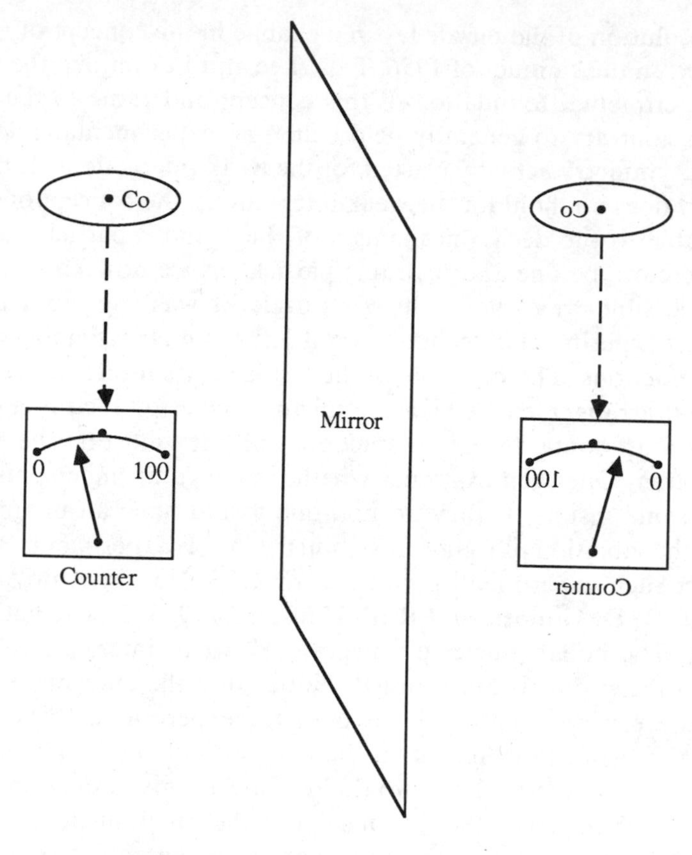

FIG. 1

serve right-left symmetry. Since the neutrino participates only in weak interactions, the overthrow of right-left symmetry in weak interactions invalidated this reason of rejection and revived Weyl's idea. Many experiments were done in 1957 on the neutrino and confirmed its predictions. Let us emphasize that Weyl's proposal was based on mathematical grounds of elegance and simplicity. It can hardly be an accident that nature again in this case, as in other cases, betrays her partiality for the beauty of mathematical reasoning.

The second aspect concerns the question whether right-left symmetry is really lost in the light of the new development. The very interesting point here is that if one changes one's definition of a mirror reflection, then mirror reflection symmetry could be restored. To explain this point let us denote the readings of the meters on the left and on the right of Fig. 1

by L and R respectively. Let us further denote the readings of the same arrangements, but constructed with antimatter, by $\bar{L}$ and $\bar{R}$ respectively. Before the experiment of Wu, Ambler, Hayward, Hoppes, and Hudson, it was believed that

$$L = R, \text{ and } \bar{L} = \bar{R},$$

on grounds of right-left symmetry. Also it was believed that

$$L = \bar{L}, \text{ and } R = \bar{R},$$

on grounds of matter-antimatter symmetry. Consequently one believed that

$$L = R = \bar{L} = \bar{R}.$$

Their experiment demonstrated the fallacy of this belief by explicitly showing

$$L \neq R.$$

It can be proved from their quantitative results and from subsequent experiments in many laboratories that in fact

$$L = \bar{R} \neq \bar{L} = R.$$

There is thus clearly less symmetry than previously believed, but there is still *some* symmetry left in the relations

$$L = \bar{R}$$
$$\bar{L} = R.$$

Both of these can be summarized in the principle that if one performs a mirror reflection *and* converts all matter into antimatter, then physical laws remain unchanged. This combined transformation which leaves physical laws unchanged could thus be *defined* as the true mirror reflection process. According to this definition, mirror reflection symmetry is restored.

This new definition of mirror reflection is illustrated in Fig. 2 where antimatter is represented by white lines against a black background.

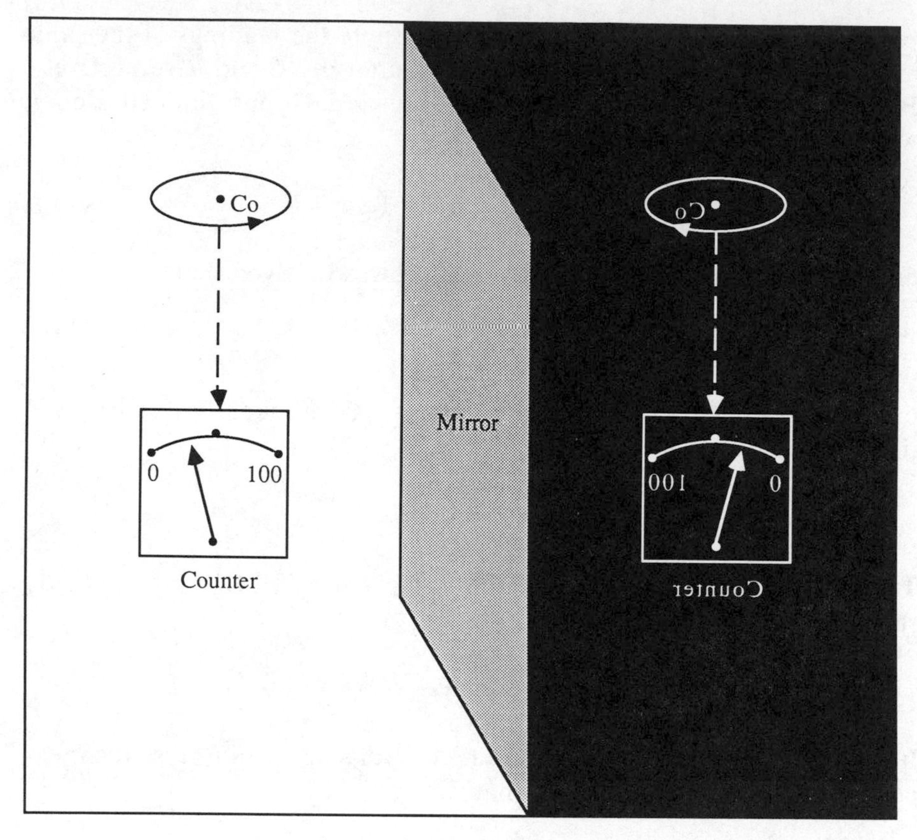

FIG. 2

In this figure the reading of the meter on the left is L, that of the meter on the right is $\bar{R}$. Experiments show that these two readings *are* the same.

Of course the question remains why it is necessary in order to have symmetry, to *combine* the operation of switching matter and antimatter with a mirror reflection. The answer to such a question can only be obtained through a deeper understanding of the relationship between matter and antimatter. No such understanding is in sight today. However, it is both interesting and useful to recall a somewhat similar question and its final resolution. We quote from Weyl's book *Symmetry:*

> Ernst Mach tells of the intellectual shock he received when he learned as a boy that a magnetic needle is deflected in a certain sense, to the left or to the right, if suspended parallel to a wire

> through which an electric current is sent in a definite direction. Since the whole geometric and physical configuration, including the electric current and the south and north poles of the magnetic needle, to all appearances, are symmetric with respect to the plane E laid through the wire and the needle, the needle should react like Buridan's ass between equal bundles of hay and refuse to decide between left and right . . .

Mach's difficulty lies in the apparent reflection symmetry of the arrangement . . . with respect to the plane E containing the wire and the magnetic needle. But a deeper *understanding of magnetism* showed that the symmetry is only apparent. A magnet is a magnet because it contains electrons making loop motions [in the plane normal to the magnetic needle]. Under a reflection . . . the polarity of the magnet changes. Thus the reflection symmetry with respect to the plane E is not real, and Mach's difficulty is removed. . . .

Merely for the sake of illustration the arguments and reasoning used have been qualitative in nature, but this should not be taken as a suggestion that new particles and new concepts are accepted in physics without detailed and quantitative formulation. Physics is a precise science and the physicists are deadly serious about their subject.

But what is the ultimate aim of the subject? Where does a physicist's vision lie? No one speaks with greater knowledge or greater authority on this topic than Einstein. He said that the aim of the subject is to construct concepts out of which a comprehensive workable system of theoretical physics can be formulated. The system must be as simple as possible, and yet must lead by deductive reasoning to conclusions that correspond with empirical experience. He wrote:

> A complete system of theoretical physics is made up of concepts, fundamental laws which are supposed to be valid for those concepts and conclusions to be reached by logical deduction. It is these conclusions which must correspond with our separate experiences. . . .
>
> These fundamental concepts and postulates, which cannot be further reduced logically, form the essential part of a theory, which reason cannot touch. It is the grand object of all theory to make these irreducible elements as simple and as few in number as possible, without having to renounce the adequate representation of any empirical content whatever.

On the subject of the realizability of this grand object, he concluded:

> . . . , can we ever hope to find the right way? Nay more, has this right way any existence outside our illusions? Can we hope to be guided in the right way by experience when there exist theories (such as classical mechanics) which to a large extent do justice to experience, without getting to the root of the matter? I answer without hesitation that there is, in my opinion, a right way, and that we are capable of finding it. Our experience hitherto justifies us in believing that nature is the realisation of the simplest conceivable mathematical ideas. I am convinced that we can discover by means of purely mathematical constructions the concepts and the laws connecting them with each other, which furnish the key to the understanding of natural phenomena.

The faith so movingly expressed by Einstein still sustains the physicists today.

O.2

Violations of Symmetry in Physics

EUGENE P. WIGNER

EUGENE PAUL WIGNER, born in Budapest in 1902, became an American citizen in 1937. He started his professional career as an industrial chemist. Wigner received his Ph.D. from the University of Berlin at the age of twenty-three, taught at his alma mater for a while, and then at Göttingen, where he and the eminent mathematician David Hilbert were coworkers for a brief period of time. In 1930, he came to the United States to join the faculty at Princeton University. Wigner was, incidentally, a brother-in-law of Paul Dirac.

Wigner devoted considerable attention to calculating the motion of nucleons by investigating the forces acting between them. Several experiments in 1933 led him to conclude that there is an attractive short-range force between two nucleons that is a million times stronger than the electric force between the electrons in the outer region of the atom. He showed that generally valid symmetries of the laws of motion could be used to uncover essential properties of the nuclei. Wigner's work assumed added importance in interpreting elementary particle experiments in the following decades.

During his stay at Princeton, Wigner extended the pioneering work on neutron absorption done by Fermi and his colleagues of the University of Rome. Wigner put it on a firm theoretical basis and developed a theory of the conservation of parity which, many years later, was undermined by the work of Tsung-Dao Lee and Chen Ning Yang. Wigner was one of the scientists who joined Leo Szilard and Albert Einstein in calling the attention of President Franklin Roosevelt to the possibility of developing the atomic bomb. He worked closely with both Szilard and Fermi on the Manhattan Project and was instrumental in developing the Hanford (Washington) atomic energy plant. Wigner served on the General Advisory Committee to the U.S. Atomic Energy Commission from 1952–64. He was awarded the Nobel Prize in physics in 1963, sharing it with Maria Goeppert-Mayer and Hans Jensen.

Nine years after Lee, Yang, Wu, and their colleagues demonstrated the violation of parity, Wigner wrote an article for *Scientific American.* In it he explained how three of the seven "mirrors" that physicists had devised to illustrate the symmetry of natural laws had been shattered by later discoveries. Wigner's article suggests that only one of the mirrors may remain intact—the so-called CPT (charge conjugation, parity, time) mirror.

. . . What immortal hand or eye,
Could frame thy fearful symmetry?

—WILLIAM BLAKE, *The Tiger*

IT WAS JUST nine years ago this month that physicists learned to their astonishment that left-handedness and right-handedness are built into the universe at the most fundamental level. Until December 1956, they had assumed that if an event is possible, its mirror image is also possible, and that if one looks at some real event in a mirror, what one sees could also actually happen. This was known as reflection symmetry, and it forms the basis of the parity principle. In the summer of 1956 certain puzzling phenomena in nuclear physics led T. D. Lee and C. N. Yang to question the principle's general validity. In a few months C. S. Wu, Ernest Ambler, Dale D. Hoppes and R. P. Hudson had demonstrated that the phenomena clearly violated the principle.

The parity principle was one of several symmetry principles that physicists had long accepted as axiomatic in developing their mathematical theories. With the fall of parity they became uneasy about the other principles and sought ways to test each of them in turn. As a result of this endeavor at least two more principles have fallen and a third has been called into serious question. This is time symmetry: the principle that nature is indifferent to the direction in which time flows. Physicists have believed deeply that nature is similar to an electric clock that will run forward or backward, depending on which way the starting knob is turned.

The various symmetries can be compared to mirrors that reflect natural events in carefully specified ways. The parity mirror, which we shall designate the *P* mirror, is simply the ordinary mirror of everyday life. It has one property, however, that may puzzle the layman; accordingly we shall also let "*P* mirror" stand for "physicist's mirror" in order to distinguish it from the layman's mirror, which we shall call the *L* mirror.

Everyone is familiar with the fact that when an electric current flows in a coil of wire, it induces a magnetic field. We learned in school that the direction of the field—the direction of its north magnetic pole—can be

---

Eugene P. Wigner, "Violations of Symmetry in Physics," *Scientific American* December 1965, pp. 28–36.

determined by the "right-hand rule." This rule states that if the forefinger of the right hand has the shape and direction of the current flow, the thumb of the right hand points to the north pole of the induced magnetic field.

The early students of electricity defined the direction of current flow as being from the positive terminal of a battery to the negative. Now that we understand that an electric current depends on the flow of electrons, it seems more reasonable to speak of the direction of the electron flow, which is from the negative terminal to the positive. Therefore if we are told the direction in which electrons flow in a coil of wire, we must use a *left-hand* rule to determine the direction of north in the magnetic field. (We realize, of course, that "north" and "south" are themselves conventions based on the fact that a compass needle is said to point toward the earth's North Pole.) In any case, in this article the direction of the magnetic field will be considered in relation to electron flow, which requires the left-hand rule. The reader is no doubt familiar with these elementary principles, but reviewing them may help to prevent confusion when we begin to look into mirrors.

Let us examine the parity mirror first. Imagine that we have before us on a table a coil of wire in which electrons are flowing clockwise as seen from above. The left-hand rule tells us that the induced magnetic field is pointing upward. . . . Now imagine that there is a mirror on the ceiling directly over the table; what will we see in it? If we have placed our left hand next to the coil and shaped it to form the left-hand rule, we shall see in the mirror what appears to be a right hand with the thumb pointing down. The hand in the mirror tells us (correctly) that the electron flow as seen in the mirror is counterclockwise, but it also tells us (incorrectly) that the magnetic field is pointing down. The hand in the mirror misinforms us about the direction of the magnetic field because it is (or appears to be) a right hand, and a real field is related to a real electron flow by a left hand. If we accept the right-hand (incorrect) view of the field direction, we can be said to be interpreting the mirror image as laymen; in this sense the mirror is an *L* mirror. If, however, we insist as physicists that the electron flow is the prime reality and that the magnetic field is secondary, we will insist on using the left-hand rule to determine the direction of the magnetic field in the mirror and conclude that it is actually pointing back into the mirror and not out of the mirror. Thus the magnetic field on the tabletop and the field in the mirror—the physicist's *P* mirror—are pointing in the same direction. . . .

* * *

We can now describe the parity experiment performed by Miss Wu and her collaborators. In the center of a ring of electric current they placed some radioactive cobalt, which emits electrons and neutrinos when it decays. The whole experimental arrangement has a plane of symmetry: the plane through the ring current. There is nothing to distinguish the upward from the downward direction. It nevertheless turned out that the electrons from the decaying cobalt atoms emerged asymmetrically: almost exclusively in the upward direction. If one were to place a mirror over the experiment, parallel to the plane of the ring current, the electrons would appear to be coming out of the mirror, or downward, which is not the direction they would travel if the current and radioactive material were real rather than mirror images. . . .

This unexpected result attracted a great deal of attention—and a Nobel prize. The fact that the electrons emerged with a preferential direction, a direction that can be represented by the thumb of the left hand if the forefinger has the shape and direction of the electron flow, meant that the radioactivity of cobalt is partial toward the left hand.

If I may recall days long past, nobody was very happy with this result. It is a fact, of course, that most of us are as partial toward our right hand as cobalt is toward its left. We feel, however, that radioactive cobalt is not entitled to be partial because it should have forgotten its past and because, at the time it emitted the decay particles, it was under no influence that would have favored one side of the plane of the ring of current over the other. This plane was a symmetry plane at the beginning of the experiment; it should have remained a symmetry plane as long as it was undisturbed by outside influences. This statement is equivalent to the postulate that a possible sequence of real events should remain a possible sequence of events if every event is replaced by its mirror image. Evidently this is not the case for the disintegration process of radioactive cobalt.

Before long a score of physicists had independently proposed a reinterpretation of the Wu experiment that salvaged the principle of reflection symmetry. In essence they proposed that nature does not see itself in the *P* mirror but in a "magic mirror" where the signs of all electric charges are reversed. In this mirror the mirror image of an electron is a positron (a positive electron) and the mirror image of a radioactive cobalt nucleus is a similar nucleus made of antimatter (antineutrons and antiprotons). If one could view the Wu experiment in this proper mirror, one would see positrons flowing in the direction that electrons had been assumed to flow. Since a flow of positrons is equivalent to a flow of positive current, one would have to use a *right-hand* rule to see how the magnetic field is pointing. One would then discover that the magnetic field is pointing out of the mirror, or downward, and thus directly opposite to the magnetic field in the real experiment. The decay particles emitted by the antimatter nuclei of radioactive cobalt would also tend to travel out of the mirror, or downward, thereby completing the mirror image of the Wu experiment. . . .

This reinterpretation of reflection symmetry was originally pure speculation, motivated solely by the desire to maintain the principle of reflection symmetry for the laws of nature. An experimental test was out of the question: even today we are far from being able to produce anticobalt, that is, a cobalt nucleus consisting of antiprotons and antineutrons. It was possible, however, to test the new hypothesis in other ways. The reinterpretation turned out to be relevant and was in agreement with all experimental findings until quite recently. These findings include the direction of flight of particles emitted in decay reactions other than that of radioactive cobalt. In particular there is a case in which a radioactive particle as well as its antiparticle can be produced and observed. These particles are the muon and the antimuon; the decay of the antimuon looks in all details as the image of the decay of the muon would look in the magic mirror just described.

We have not yet mentioned the role of the ring current in Miss Wu's experiment. Its purpose is to create a magnetic field perpendicular to the plane of the current. This field in turn orients the spins of the nuclei of the radioactive cobalt atoms. The direction of the decay particles is related directly to the spins of the radioactive nuclei emitting them, and only indirectly by means of the magnetic field to the direction of the flow of electrons in the ring.

The spins of the cobalt nuclei carry an angular momentum, a fundamental property associated with rotating motion. In all studies of rotating motion before Miss Wu's experiment, angular momentum was found to have a symmetry plane in the plane of rotation. If this plane were kept horizontal, one would expect the disintegration products (if any) of a rotating object to proceed with equal probability upward and downward. The fact that they do not when the rotating object happens to be a radioactive cobalt nucleus means that the total symmetry of the laws of nature is smaller than physicists had previously believed. The laws are not invariant if reflected in a *P* mirror. The magic mirror that gives a true reflection is called a *CP* mirror; it is a combination of the parity (*P*) mirror, which deflects the positions of particles, and a "charge conjugation" (*C*) mirror, which changes the sign of electric charges.

* * *

How many mirrors has the theoretician conceived all together? I hope that we will not be suspected of patent-preemption if we claim to have "invented" seven mirrors. They are essentially various composites of the *P* and *C* mirrors and a third mirror: the *T* mirror, which reflects the direction of time. The seven mirrors are *P, C, T, CP, CT, PT* and *CPT.*

We have already seen how electric currents and magnetic fields are reflected in the *P* mirror. Let us now consider how the *P* mirror reflects the path of a particle as it is scattered, or deflected, by another particle. . . . We can imagine that the scatterer is a heavy particle, such as an oxygen nucleus, and that the incident particle is a light particle, for example a positron. Thus each particle has a positive electric charge. . . . The positron is so light that it will hardly affect the position of the oxygen nucleus; we need be concerned only with the path of the positron as it approaches the oxygen nucleus and is scattered.

We must also take into account the fact that the incident particle has an angular momentum, or spin, and that the axis of spin is parallel to the particle's direction of motion. After the particle has been scattered its direction of motion has changed, and the new direction of motion will be found to correlate with what has happened to the particle's spin angular momentum. If the spin remains pointing in the original direction (remains parallel), the particle's direction of motion will be somewhat above the plane of the original direction. If the spin flips around to point in the opposite direction (becomes antiparallel), the particle's direction of motion will be below the original plane. The particle can traverse either path and will take each path in a certain fraction of all cases observed.

The scattering event in front of the mirror will always be the same, but the image will be different in different mirrors. We shall assume that in each case the mirror is vertical and to the right, that is, between the object and the image. If the mirror is the physicist's *P* mirror, the reflected paths are just what one would expect. The path seen curving to the right in the actual case is seen curving to the left in the mirror. Moreover, the particle's direction of spin is reversed, so that if the particle seems to be spinning clockwise, as seen from the rear in the actual experiment, it will appear to be spinning counterclockwise as seen from the rear in the mirror image.

If one carried out this experiment, one would unquestionably find that the *P* mirror is right; with the accuracy of measurement now available one could not detect the difference between a real path and the path as it appears in the mirror. Yet we know from Miss Wu's experiment with radioactive cobalt that the mirror is not really right: in her experiment the *P* mirror is not quite right in general and that actuality will deviate from what it shows, even though in our scattering experiment its error would be immeasurably small.

Let us consider now how the scattering experiment looks in the *C* mirror. . . . The *C* mirror is not, of course, a material mirror: it does not change the location of points, the direction of motion or the sense of spin direction. All it does is substitute negative electric charges for positive electric charges and vice versa; or, more generally, it substitutes antimatter for matter and vice versa. Thus when we "look" into the *C* mirror we see that the oxygen nucleus of our scattering experiment is replaced by an antinucleus consisting of antineutrons and antiprotons, and so has an overall negative charge, and that the positron is replaced by an electron, which is also negatively charged.

The situation with the *C* mirror is quite similar to that with the *P* mirror. Since no one knows how to make an antimatter nucleus as heavy as the nucleus of oxygen, however, our particular scattering experiment cannot be performed. But it is known from similar scattering experiments with antiparticles that there are no observable differences between actual scattering patterns and their reflections in the *C* mirror. Nevertheless, it has been established by other experiments that *C* reflection is no more an exact symmetry than *P* reflection is. Unfortunately the experimental demonstration of *C* violation is not as direct as Miss Wu's demonstration of *P* violation. The argument for the violation of *C* symmetry is a mathematical one based on the observed spin direction of electrons and positrons that are respectively emitted by negative and positive muons. The experiment was performed in 1957 by G. Culligan, S. G. F. Frank, J. R. Holt, J. C. Kluyver and T. Massam of the University of Liverpool.

If the *P* and *C* mirrors are known to be slightly defective when they are tested individually, is it possible that the magic *CP* mirror mentioned earlier still provides a faithful reflection of reality? The *CP* image can be obtained in either of two ways: by reflecting the *P* image in a *C* mirror or by reflecting the *C* image in a *P* mirror. The fact that the image obtained by two such reflections is an excellent picture of reality follows from the fact that the image produced by each mirror is extremely close to reality if reality is reflected in it. Indeed, until recently physicists believed the slight discrepancies in the individual mirrors canceled each other, so that the *CP* mirror was in exact accord with reality. This certainly seemed to be the case, at least for a number of phenomena that occur in radioactive decays and that violate *C* and *P* separately. Before discussing the experiment that has now cast doubt on the *CP* mirror, I should make a few comments on the *T* mirror.

Like the *C* mirror, the *T* mirror does not change the paths of particles. It merely reverses their direction, thus implying that the time axis is reversed. . . . In fact the designation *T* stands for time-reversal. . . . The concept is hard to accept intuitively because our everyday experience with events that are patently irreversible is so compelling; the pieces of a shattered teacup have never been known to reassemble themselves spontaneously. Irreversibility of this kind is not at issue. The physicist is concerned rather with the detailed reversibility of events at the atomic and subatomic scale. A model for this kind of irreversibility would be the behavior of a perfectly elastic ball. If such a ball were dropped on a perfectly elastic surface, it would bounce forever. If one were to make a moving picture of this ball as it bounced, there would be no way to tell whether the film were being run forward or backward; the time axis would be completely reversible.

Until recently the *T* mirror, like the *CP* mirror, was believed to be exact. And for reasons I shall describe later, physicists are forced to believe that the combination of the *T* mirror with the *CP* mirror—the *CPT* mirror—may still remain exact, even though the *C, P* and *T* mirrors appear to fail separately!

Let us turn to the experiment that has cast doubt on the validity of the *CP* mirror and, by implication, on the validity of the *T* mirror. The experiment was carried out a little over a year ago at the Brookhaven National Laboratory by James H. Christenson, James W. Cronin, Val L. Fitch and René Turlay of Princeton University. One of the original purposes of the experiment was the confirmation of *CP* invariance, not a demonstration of its failure. Experiments occasionally give surprising results, however; this one certainly did. Nonetheless, the evidence for the violation of *CP* invariance is not as direct as the evidence for the violation of *P* invariance furnished by Miss Wu's experiment or even the evidence for the violation of *C* invariance in the experiment of Culligan and his collaborators.

The evidence for the failure of the *CP* mirror stems from one mode of decay exhibited by the *K* meson, or *K* particle. *K* particles are readily produced in a high-energy particle accelerator when a proton beam is directed at a suitable target, such as beryllium. The interaction of a high-energy proton with a neutron (contained in the atomic nuclei of the target) invariably yields two heavy particles, one of which is usually a proton or a neutron. When the bombardment energy is around 30 billion electron volts, as it was in the Brookhaven experiment, the other heavy particle is likely to be one of the so-called strange particles, such as a lambda particle or a sigma particle. Simultaneously the interaction produces a *K* particle, which can be either positive (K^+), negative (K^-) or neutral (K^0).

The *K* meson is a very queer particle. It is the same particle whose puzzling decay behavior prompted Lee and Yang to question *P* invariance. Even before that it was ascertained that the K^0 is not a single particle but two particles that are antiparticles of each other. When a particle has an electric charge, it can easily be separated from its antiparticle because the latter must have an opposite charge. It is therefore impossible to create a charged particle that can be, with some probability, its own antiparticle.

The situation is different if a particle has no electric charge. In this case a state is quite conceivable in which a neutral particle such as a K^0 meson has a 50–50 chance of being either a particle or an antiparticle. An even more surprising result of present quantum-mechanical theory is that there is not one such state but a continuous manifold of "superpositions" of such states. For our purposes, however, we need be concerned only with the two states that are designated ($K^0 + \bar{K}^0$) and ($K^0 - \bar{K}^0$). The bar over the second *K* in each pair signifies antiparticle. It should be emphasized that each state stands for a single particle, but the properties of the two states are different and can be shown to be so by experiment.

The existence of such superposition states is a consequence of the wave nature of matter. Similar superpositions also play an important role in the low-energy region, in particular in the theory of optically active organic compounds, such as optically active amino acids and sugars. For example, one form of sugar can have the property of rotating the plane of polarization of polarized light to the right. Another sugar of identical chemical composition will have the property of rotating the plane of polarized light to the left. The difference in optical activity is accounted for solely by the fact that the two compounds have three-dimensional structures that are mirror images and thus bear to each other the relation of left and right hands [*see Fig. 1*].

The quantum-mechanical interpretation of the position of an atom that determines whether an organic compound is left-handed or right-handed is plotted in Fig. 2. The horizontal axis gives the position of the atom, in terms of left or right, in the optically active compound. The vertical axis is the "probability amplitude" for each position of the atom; the probability of finding the atom at any particular position is defined as the square of

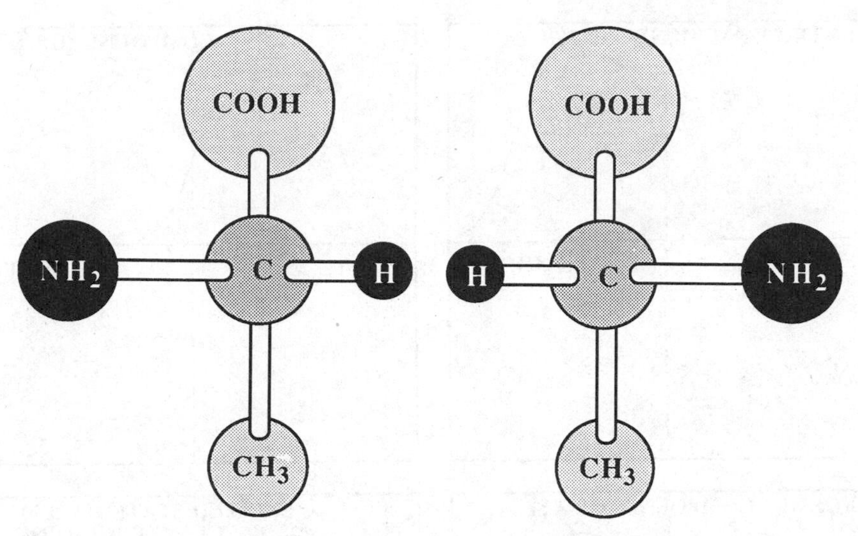

FIG. 1. Mirror-Image Molecules

the probability amplitude. When the curve of the probability amplitude lies entirely to the right of center, the atom is surely to the right, thereby creating an asymmetric situation. This corresponds to a right-handed, or dextro, compound. When the curve of probability amplitude lies entirely to the left, the atom is surely to the left, corresponding to the left-handed, or levo, compound.

The lower pair of curves in the illustration represents probability amplitudes for atoms in which left and right positions are equally probable, with the result that the rotational properties of the compound cancel each other and no optical activity is observed. In the curve at lower left the probability amplitude consists of two symmetrical humps, both positive. This state is optically inactive because it is a mirror image of itself. Such states are stable at low temperature and are described as racemic mixtures. In the curve at lower right the probability amplitude is asymmetric: the hump at the left is positive, whereas the hump at the right is negative. The probability, however, is the square of the vertical displacement of the asymmetric curve and therefore has a positive and equal value on both left and right. This state, although optically inactive also, has different properties from the first optically inactive state; in particular, its energy is very slightly higher.

These four states—two optically active and two inactive—have their

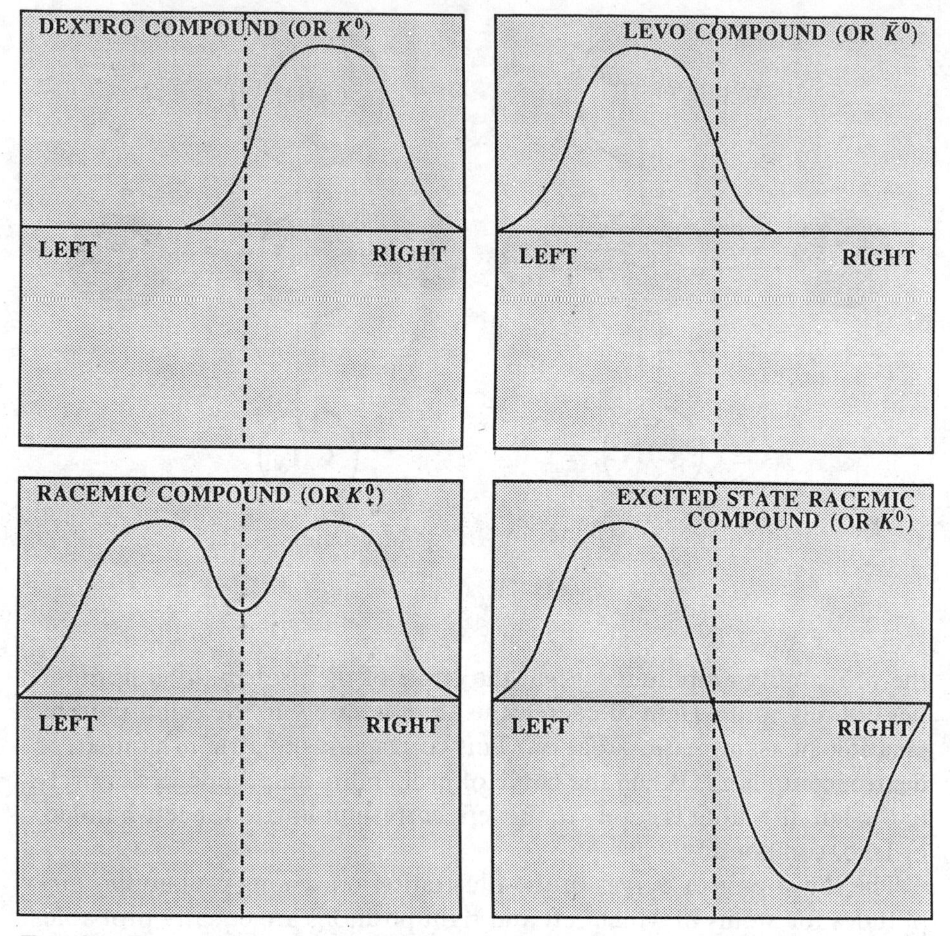

FIG. 2. Quantum-mechanical interpretation of dextro and levo organic compounds applies also to states of the neutral K meson.

counterpart in the neutral *K* mesons. When the neutral *K* meson is first created, it appears as the antiparticle $\bar{K}^0$ and corresponds to the optically active left-handed compound. This state can be considered as the sum of the symmetric and antisymmetric states representing the two optically inactive forms of the compound. In the case of the neutral *K* meson these two states can be designated $K_+{}^0$ and $K_-{}^0$ (also known as $K_1{}^0$ and $K_2{}^0$). The only essential difference between the two inactive states of the compound and the two neutral meson states $K_+{}^0$ and $K_-{}^0$ is that the reflections with respect to which the meson states are symmetric and asymmetric are not ordinary reflections but reflections in a *CP* mirror.

Now, the symmetric state K_+^0 can decay into two pi mesons and will do so in about 10^{-10} second. The asymmetric state K_-^0 can and should decay into three pi mesons, a decay process that takes about 600 times longer than the two-pi decay, or about 6×10^{-8} second. Therefore K_-^0 mesons will still be plentiful long after all K_+^0 mesons have disappeared. The K_+^0 can decay into two pi mesons because all the states of a system composed of two pi mesons are symmetric with respect to *CP* reflection, as is the K_+^0 state itself. The K_-^0 state, being antisymmetric with respect to *CP* reflection, should not be able to decay into a symmetric state, but it can decay into three pi mesons. A system of three pi mesons does have states that are antisymmetric with respect to *CP*. The three-pi decay is a slower process, hence the longer life of the K_-^0.

What Cronin, Fitch and their collaborators observed, however, is that a small fraction of the K_-^0 mesons do decay into two pi mesons, in defiance of *CP* symmetry. Since only about one in 500 of the K_-^0 mesons decays into two pi mesons, this mode of decay is more than 100,000 times slower for the K_-^0 than for the rapidly decaying K_+^0. Nevertheless, the "forbidden" decay does occur, and this is interpreted as a breakdown of CP symmetry.

One can see that the preceding argument is quite involved and is by no means so simple as that represented in the breakdown of *P* symmetry in Miss Wu's experiment. As a result some physicists have been reluctant to accept the Cronin-Fitch experiment as conclusive evidence for the failure of *CP* symmetry. There may be a way out that preserves *CP* symmetry—in fact, several ways have been suggested—but the weight of the argument is increasing that *CP* has failed.

What follows from the violation of *CP* symmetry? Physicists are left with the belief that the very last mirror, the *CPT* mirror, is a true mirror. This belief is not based on nature's innate preference for symmetry; it is based on the stubborn fact that we cannot formulate equations of motion in quantum field theory that lack this symmetry and still satisfy the postulates of Einstein's special theory of relativity. If the principle of *CPT* symmetry is valid, it is evidence for the correctness of the general framework of quantum electrodynamics and of the special theory of relativity, not for nature's preference for any additional symmetry.

It must be noted with some apprehension, however, that in order for the *CPT* mirror to remain valid the *T* mirror itself must be invalid. The reasoning, based on the Cronin-Fitch experiment, is this. The K_-^0 begins in an antisymmetric state and decays into a symmetric state when it is reflected in a *CP* mirror, thereby proving the mirror defective. If the image in the *CP* mirror is now reflected in the *T* mirror, the original asymmetry should be restored—provided that the *CPT* mirror (*C* plus *P* plus *T*) is valid. To turn a symmetric state into an antisymmetric one, however, the *T* mirror by itself must produce an antisymmetric image. This is equivalent to saying that time is not invariant under reflection and that time-reversal symmetry has failed.

Physicists have scarcely begun to examine the implications of this final breakdown. Leaving aside the apparent collapse of *T* symmetry, onc can conclude from the failure of *P, C* and *CP* symmetry that the laws of nature do show a preference for either the right or the left hand. We are surrounded by many phenomena that appear to show just such a preference, or, more precisely, such a distinction between right and left. Most of us are right-handed and our hearts are on the left side. On a large scale we observe that the earth rotates to the left (counterclockwise) as seen from above the North Pole and proceeds to the left around the sun. The sun, in turn, travels to the right around the galaxy as viewed from above the north galactic pole. Heretofore these asymmetries were attributed to asymmetries in the initial conditions. Now it is possible to attribute the same asymmetries to the laws of motion, that is, to assume that the universe was initially more symmetrical than it is now and that the present state evolved as a result of the asymmetry of the laws of motion. Few people are as yet ready to accept these speculations; I personally do not believe they are valid. Such speculations could nevertheless be tested if we had enough information about the sense of rotation of planets in other solar systems.

The fact that the laws of nature have no pure space-reflection symmetry has one consequence that is unpleasant to admit. It deprives us of the illusion that these laws are—in perhaps a subtle but nonetheless a real sense—the simplest laws that can be conceived and that are compatible with some obvious experience. If one law of nature is possible, an alternative law obtained by reflecting the first law on a plane would be equally possible and equally simple. We had previously thought the law obtained by reflection would be identical with the original law, just as the reflection of a sphere is also a sphere. Now we know that this is not so. The difficulty began when Miss Wu's experiment showed us that the preferential direction of the decay particles of radioactive cobalt was arbitrarily upward. We got out of this difficulty by postulating another substance: radioactive anticobalt that emits particles downward. This restores the symmetry because we can say that the laws of nature are symmetric but imply two kinds of substance, matter and antimatter. The apparent asymmetry in the laws of nature was thereby reduced to an asymmetry in the initial conditions that allowed matter to predominate over antimatter, at least in the only part of the universe we know at first hand.

The recent experiment of Cronin and Fitch indicates, however, that such an explanation is impossible. The indication, to be sure, is only indirect; we must explore all avenues that may yield another interpretation and preserve the spatial-reflection symmetry of the laws of nature. If these avenues do not lead out of the difficulty, we will have to admit that two absolutely equally simple laws of nature are conceivable, of which nature has chosen, in its grand arbitrariness, only one. The extent to which the laws of nature are the simplest conceivable laws has come to an end—no matter how subtly they may be formulated—as long as they are formulated in terms of concepts that are subject to the symmetry principles we are accustomed to associating with space-time.

The question naturally arises whether or not physics has experienced similar crises before. It has. In classical physics, matter was supposed to be infinitely subdivisible without change in its bulk properties such as specific gravity, viscosity or elasticity. The discovery of the atom put an end to this infinite subdivisibility. The atom therefore increased the complexity of the structure of matter, and this unavoidable new fact was for many people as obnoxious as the lack of reflection symmetry in the laws of nature is for us.

Most of the consequences of atomic structure that first appeared obnoxious were eliminated when physicists and chemists learned to use the atomic scale for their measurements and realized, for example, that atoms provide a natural unit of length. Without such a unit it would be difficult to understand why human beings have an average height somewhere between five and six feet; if all phenomena could be scaled up or down with impunity, men, mice and bacteria could be the same size. Atomic theory also provided explanations for the properties of matter, for its density, viscosity, elasticity and so on. Hence in its end result atomic theory enriched rather than complicated our picture of nature. There is hope, but as yet only a hope, that the present probing into the symmetry of space-time will have a similar result.

Particle Physics: Symmetry

TSUNG-DAO LEE

TSUNG-DAO LEE, born in 1926 in Shanghai, China, came to the United States as a recipient of a Chinese government scholarship, which took him to the University of Chicago in 1946. As an undergraduate at Chicago, he was permitted to work directly toward the Ph.D. He received his degree in 1950 under the direction of Edward Teller. While studying at Chicago, he found his future collaborator, Yang, who had been an acquaintance back in China and who was taking his degree under Fermi. Lee's interest in the hydrogen content of white dwarf states had culminated in his thesis topic, followed by a brief stint as a research associate at the Yerkes Astronomical Observatory. While working as a research associate and lecturer at Berkeley and the Institute of Advanced Study at Princeton, Lee became well-known for his work in statistical mechanics and nuclear physics. No less a figure than J. Robert Oppenheimer said that Lee's work was characterized by "remarkable freshness, versatility, and style."

Although their subsequent careers took them to different schools—Lee to Columbia University, Yang to Princeton—they kept in close touch and collaborated on research on *K* mesons, elementary particles that take part in strong interactions and have integral spin. This research eventually led to their discovery of the nonconservation of parity. At age thirty-one, Lee was the second youngest scientist ever to receive a Nobel Prize. He and Yang have also coauthored a number of prominent articles that have appeared in *Physical Review.*

Currently the holder of the Enrico Fermi Professorship of Physics at Columbia, Lee is the author of, among other works, *Particle Physics and*

Introduction to Field Theory, from which we have selected a chapter dealing with symmetries in particle physics.

Like Wigner, who had written his *Scientific American* article sixteen years before the publication of Lee's book, Lee recognizes that certain symmetries—permutation symmetry of the Bose-Einstein and Fermi-Dirac statistics, continuous space-time symmetries (translation, rotation, acceleration, etc.), some of the so-called U_1 symmetries related to conservation rules and possibly to that concerned with color symmetry—are still thought to be intact, while all the other symmetries seem to be broken.

The universe is built on a plan the profound symmetry of which is somehow present in the inner structure of our intellect.

—PAUL VALÉRY

SINCE THE BEGINNING of physics, symmetry considerations have provided us with an extremely powerful and useful tool in our effort to understand nature. Gradually they have become the backbone of our theoretical formulation of physical laws. In this chapter we shall review these symmetry operations and examine their foundation. Such an examination is useful, especially in view of the various asymmetries that have been discovered during the past quarter century.

There are four main groups of symmetries, or broken symmetries, that are found to be of importance in physics:

1. Permutation symmetry: Bose-Einstein and Fermi-Dirac statistics.
2. Continuous space-time symmetries, such as translation, rotation, acceleration, etc.
3. Discrete symmetries, such as space inversion, time reversal, particle-antiparticle conjugation, etc.
4. Unitary symmetries, which include U_1-symmetries such as those related to conservation of charge, baryon number, lepton number, etc., SU_2(isospin)-symmetry, SU_3(color)-symmetry, and SU_n(flavor)-symmetry.

Among these, the first two groups, together with some of the U_1-symmetries and perhaps the SU_3 (color)-symmetry in the last group, are believed to be exact. All the rest seem to be broken.

NON-OBSERVABLES, SYMMETRY TRANSFORMATIONS AND CONSERVATION LAWS

The root of all symmetry principles lies in the assumption that it is impossible to observe certain basic quantities; these will be called "non-

T. D. Lee, "Particle Physics: Symmetry," in *Particle Physics and Introduction to Field Theory.* London: Harwood Academic Publishers, 1981, pp. 177–188.

observables." Let us illustrate the relation between non-observables, symmetry transformations and conservation laws by a simple example. Consider the interaction energy V between two particles at positions $\vec{r}_1$ and $\vec{r}_2$. The assumption that the *absolute position is a non-observable* means that we can arbitrarily choose the origin O from which these position vectors are drawn; the interaction energy should be independent of O. In other words, V is invariant under an arbitrary space translation, changing O to O′;

$$\vec{r}_1 \rightarrow \vec{r}_1 + \vec{d} \qquad \text{and} \qquad \vec{r}_2 \rightarrow \vec{r}_2 + \vec{d}. \tag{1}$$

Consequently, V is a function only of the relative distance $\vec{r}_1 - \vec{r}_2$,

$$V = V(\vec{r}_1 - \vec{r}_2).$$

From this, we deduce that the total momentum of this system of two particles must be conserved, since its rate of change is equal to the force which, on account of (2), is zero.

This example illustrates the interdependence among three aspects of a symmetry principle: the physical assumption of a non-observable, the implied invariance under the connected mathematical transformation and the physical consequences of a conservation law or selection rule. In an entirely similar way, we may assume the absolute time to be a non-observable. The physical law must then be invariant under a time-translation

$$t \rightarrow t + \tau$$

which results in the conservation of energy. By assuming the absolute spatial direction to be a non-observable, we derive rotation invariance and obtain the conservation law of angular momentum. By assuming that absolute (uniform) velocity is not an observable, we derive the symmetry requirement of Lorentz invariance, and with it the conservation laws connected with the six generators of the Lorentz group. Similarly, the foundation of general relativity rests on the assumption that it is impossible to distinguish the difference between an acceleration and a suitably arranged gravitational field.

The following table summarizes these three fundamental aspects for some of the symmetry principles used in physics.

TABLE 1

Non-observables	Symmetry Transformations	Conservation Laws or Selection Rules
difference between identical particles	permutation	B.-E. or F.-D. statistics
absolute spatial position	space translation $\vec{r} \to \vec{r} + \vec{\Delta}$	momentum
absolute time	time translation $t \to t + \tau$	energy
absolute spatial direction	rotation $\hat{r} \to \hat{r}'$	angular momentum
absolute velocity	Lorentz transformation	generators of the Lorentz group
absolute right (or absolute left)	$\vec{r} \to -\vec{r}$	parity
absolute sign of electric charge	$e \to -e$ (or $\psi \to e^{i\phi} \psi^{\dagger}$)	charge conjugation (or particle anti-particle conjugation)
relative phase between states of different charge Q	$\psi \to e^{iQ\theta} \psi$	charge
relative phase between states of different baryon number N	$\psi \to e^{iN\theta} \psi$	baryon number
relative phase between states of different lepton number L	$\psi \to e^{iL\theta} \psi$	lepton number
difference between different coherent mixture of p and n states	$\begin{pmatrix} p \\ n \end{pmatrix} \to u \begin{pmatrix} p \\ n \end{pmatrix}$	isospin

ASYMMETRIES AND OBSERVABLES

Since the validity of all symmetry principles rests on the theoretical hypothesis of non-observables, the violation of symmetry arises whenever what was thought to be a non-observable turns out to be actually an observable. In a sense, the discovery of "violations" is not that surprising. It is not difficult to imagine that some of the "non-observables" may indeed be fundamental, but some may simply be due to the limitations of our present ability to measure things. As we improve our experimental techniques, our domain of observation also enlarges. It should not be completely unexpected that we may succeed in detecting one of those supposed "non-observables" at some time and therein lies the root of symmetry breaking.

The notable examples of such discoveries are the asymmetry of physical laws under the right-left mirror transformation, the particle-antiparticle conjugation and the change in the direction of time flow, past to future and future to past. It turns out that all these supposed non-observables can actually be observed. Let me illustrate the relation between "asymmetries and observables" by first considering the example of right-left asymmetry, commonly known as parity nonconservation.

Of course it is well known that even in daily life, right and left are distinct from each other. Our hearts, for example, are usually not on the right side. The word "right" also means correct, while the word "sinister" in its Latin root means left. In English, one says "right-left," but in Chinese 左右 : 左 (left) usually precedes 右 (right). However, such asymmetry in daily life is attributed to either the accidental asymmetry of our environment or initial conditions. Before the discovery of parity nonconservation in 1957, it was taken for granted that the laws of nature are symmetric under a right-left transformation.

One may wonder why, in spite of the clear difference between right and left in daily life, before 1956 practically all physicists could believe in right-left symmetry in physical laws.

Let us consider two cars which are made exactly alike, except that one is the mirror image of the other. Car a has a driver on the left front seat and the gas pedal to his right, while ɒ has the driver on the right front seat with the gas pedal on his left. Both cars are filled with the same amount of gasoline. (For the sake of discussion, we may ignore the fact that the gasoline molecule is not exactly mirror-symmetric.) Now, suppose the driver of Car a starts the car by turning the ignition key clockwise and stepping on the gas pedal with his right foot, causing the car to move forward at a certain speed, say 20 mph. The other does exactly the same thing, except that he interchanges right with left; i.e., he turns the ignition key counterclockwise and steps on the gas pedal with his left foot, but keeps the pedal at the same degree of inclination. What will the motion of Car ɒ be? The reader is encouraged to make a guess.

Probably your common sense will say that both cars should move forward at exactly the same speed. If so, you are just like the pre-1956 physicists. It would seem reasonable that two arrangements, identical except that one is the mirror image of the other, should behave in exactly the same way in all respects (except of course for the original right-left difference). This is precisely what was discovered to be untrue. The possibility of right-left asymmetry in natural laws was first suggested theoretically in 1956 in connection with the θ-τ puzzle. It was discovered experimentally within a few months in β-decay and in π- and μ-decays. In principle, in the above example of two cars one may install, say, Co^{60} β-decay as part of the ignition mechanism. It will then be possible to construct two cars that are exact mirror images of each other, but may nevertheless move in a completely different way; Car a may move forward at a certain speed while Car ɒ may move at a totally different speed, or may even move backwards. That is the essence of right-left asymmetry, or parity-nonconservation.

As we shall discuss, the discoveries made in 1957 established not only right-left asymmetry, but also the asymmetry between the positive and negative signs of electric charge. In the standard nomenclature, right-left asymmetry is referred to as P violation, or parity-nonconservation. The asymmetry between opposite signs of electric charge is called C violation, or charge conjugation violation, or sometimes particle-antiparticle asymmetry.

At the same time as the possibility of P and C violations was suggested, questions of possible asymmetries under time reversal and under CP were also raised. Actual experimental confirmation did not come until quite a few years later.

Since non-observables imply symmetry, these discoveries of asymmetry must imply observables. One may ask, what are the observables that have been discovered in connection with these symmetry-breaking phenomena? We recall that in our daily lives the sign of electric charge is merely a convention. The electron is considered to be negatively charged because we happened to assign the proton a positive charge, and the converse is also true. But now, with the discovery of asymmetry, is it possible to give an absolute definition? Can we find some absolute difference between the positive and negative signs of electricity, or between left and right?

As an illustration, let us consider the example of two imaginary, advanced civilizations A and B. These two civilizations are assumed to be completely separate from each other; nevertheless they manage to communicate, but only through neutral unpolarized messages, such as unpolarized light. After years of such communication these two civilizations may decide to increase their contact. Being very advanced, they realize that they must first agree on (i) the sign of electric charge, and (ii) the definition of a righthand screw.

The first is important in order to establish whether the proton in civilization A corresponds to the proton or the antiproton in civilization B. If the protons in A are the same as those in B, then a closer contact is possible. Otherwise it might lead to annihilation. The definition of a righthand screw is important if these two civilizations decide to have even closer contact, such as trading machinery. The academic problem that concerns us is whether it is possible to transmit both pieces of information by using only neutral and unpolarized messages. Without these discoveries of P, C and CP asymmetries, this would not be feasible. Now, assuming that these two civilizations are as advanced as ours, such an agreement can in principle be achieved.

First, both civilizations should establish high-energy physics laboratories which can produce the long-lived neutral kaon K^o_L. By analyzing the semileptonic three-body decay modes of K^o_L under a magnetic field, they can easily separate the decay $K^o_L \rightarrow e^- + \pi^+ + \bar{\nu}$ from $K^o_L \rightarrow e^+ + \pi^- + \nu$. They would discover that although the parent particle K^o_L is neutral and spherically symmetric, nevertheless these two decay rates are different

$$\frac{\text{rate } (K^o_L \rightarrow e^+ + \pi^- + \nu)}{\text{rate } (K^o_L \rightarrow e^- + \pi^+ + \bar{\nu})} = 1.00648 \pm 0.00035. \tag{3}$$

This is indeed remarkable since it means that by rate counting one can differentiate e^- from e^+. Thus, there is an absolute difference between the opposite signs of electric charge. Now, each civilization only needs to examine the faster decay mode in (3), and compare the charge of the final e with that of its "proton." If both civilizations have the same relative sign, then it means that they are made of the same matter.

Next, we come to the second task: the definition of a righthand screw. This can be done by measuring the spin and momentum direction of the neutrino or antineutrino in π-decay: $\pi^+ \rightarrow e^+ + \nu$ and $\pi^- \rightarrow e^- + \bar{\nu}$. Although $\pi^\pm$ is spherically symmetric, in its decay every ν defines a lefthand screw, while every $\bar{\nu}$ a righthand screw; i.e., the helicity of ν is always $-\frac{1}{2}$ and that of $\bar{\nu}$, $+\frac{1}{2}$. . . .

Consequently, we see that through neutral and unpolarized messages these two civilizations can indeed give an absolute meaning to + and − signs of charge as well as to right and left.

We note that in the decay $\pi^\pm \rightarrow e^\pm + \nu$ (or $\bar{\nu}$), both C and P symmetries are violated; but if we interchange + with −, and also right with left, then it might seem that symmetry could be regained (called CP symmetry). However, CP symmetry is also violated in the K^o_L-decay, because there is an absolute rate difference between the final states $e^+\pi^-\nu$ and $e^-\pi^+\bar{\nu}$. . . . From CP asymmetry and observed amplitudes of various K-decays we can deduce the asymmetry with respect to time-reversal T. At present, it appears that physical laws are not symmetrical with respect to C, P, T, CP, PT and TC. Nevertheless, all indications are that the joint action of CPT (i.e., particle $\leftrightarrows$ antiparticle, right $\leftrightarrows$ left and past $\leftrightarrows$ future) remains a good symmetry.

P

Molecules and Solids

INTRODUCTION

SOLID-STATE PHYSICS came into prominence as a separate discipline after World War II. The three states of matter—solid, liquid, and gaseous—had traditionally been recognized and taken for granted, but advances in electronics, such as the invention of new, solid devices to replace vacuum tubes, storage batteries, and the like, brought the term "solid state" to the fore. In a solid state the important properties are determined by whether there is a free movement of electrons. Metals unlike nonmetals are distinguished by the free movement of electrons. The discovery of transistors, superconductivity, Fermi conductor bonds, and atomic substructure expanded the discipline into one of the most important divisions of physics, from both a theoretical and practical point of view.

Specialists in solid-state physics, who are estimated to range in number from 25,000 to 30,000, concern themselves with such matters as density, structure, electrical and thermal conductivity, magnetic properties, compressibility, and such parameters as specific heat and coefficient of expansion. Since measurements of these properties inevitably lead to the classification of materials into groups of related data, solid-state physics, in practice, has developed a number of subspecialties covering such general classes as metals, conductors, insulators, and semiconductors. Largely because of the immediate practical uses to which advances in solid-state studies are put, research in important noncrystalline groups such as plastics, and natural organic material such as solid vegetable and animal matter (wood, bone, etc.) have, for the moment, been given short shrift.

The simplest of all metallic elements, from a structural point of view, is lithium, which has three electrons (the only simpler elements are hydrogen and helium). It is possible to study the trielectronic behavior of the individual lithium atoms with great precision; however, when the atoms are linked electronically to form lithium metal, the complexities begin to mount because of the repulsion existing among the electrons (which, of course, bear identical charges) and the electrostatic attraction of the nuclei. Thus, accurate calculation of all the properties of even so simple a metal becomes extremely difficult. With the increase in atomic number in more complex metals, the problems of the solid-state theoretician become increasingly difficult.

To the solid-state physicist, we owe the highly useful discovery that the electrical conductivity of a given metal decreases with a rise in temperature and conversely increases for such semiconductors as germanium and silicon. On the other hand, new techniques developed by atomic physicists have greatly enlarged the repertory of their solid-state colleagues; photon and neutron scattering and absorption are particularly productive analytical techniques for studying solids.

One of the major tasks of solid-state physics remains that of deepening our understanding of magnetism. Why is it that of all the natural metals in the periodic table only four (iron, cobalt, nickel, gadolinium) are naturally ferromagnetic? After twenty-five or thirty centuries, one of the first physical properties of metals to be detected by early natural philosophers in the lodestone, is still one of the least understood physical phenomena. Fermi and Dirac, both concerned primarily with the subatomic world, are among the few theoretical physicists to contribute anything fundamental to the understanding of ferromagnetism—that is, that only pairs of electrons, in each of which the two spins are opposite (anti-parallel spins), can occupy the same energy level within the atom. Another contribution from the theorists did little to clear up the mysteries of magnetism, but did a great deal to complicate the situation: In 1950, two researchers, the Americans Clifford Shull and James Smart, discovered the phenomenon of antiferromagnetism in manganese oxide. They found that of the twenty-five electrons in each atom of manganese, twenty pair off with opposite spins, while five continue stubbornly with their spins in the same orientation (parallel spins), and that the electrons of successive manganese atoms in the molecule are arranged in reverse alignment, thus rendering the sample magnetically neutral.

P.1

Modern Structural Chemistry

LINUS PAULING

IF A CHEMISTRY student trained in the 1930s picked up a modern chemistry textbook, he or she would feel quite lost. Instead of the familiar, simple, take-it-or-leave-it statements ("Carbon has a valence of four, oxygen of two; thus these two elements combine in the proportion of 2 to 1, and carbon dioxide has the formula CO_2, meaning that two atoms of oxygen are combined with one of carbon," etc.), the reader would find references to electron shells, levels of energy, and valence electrons in the outer shell of an atom—as far a cry from the valence theories of the 1930s as were those of John Dalton's atoms, which had hooks tangling with the hooks of other atoms.

The revolution in chemistry, which led to the reclassification of chemistry as a branch of physics, owes much to a classic of American scientific research and writing: Linus Pauling's *The Nature of the Chemical Bond.* Pauling was among the first (and certainly the most brilliant) theoreticians in chemistry to apply de Broglie's quantum mechanics to chemistry. Drawing upon the earlier work of F. London and W. Heitler, he proposed a theory whereby chemical bonding was achieved through the sharing of outer-shell electrons. For this work he was awarded the Nobel Prize for chemistry in 1954.

Pauling, the son of a pharmacist, was born in 1901 in Portland, Oregon. He graduated from Oregon State Agricultural College in 1922 and earned his Ph.D. at the California Institute of Technology in 1925. For two years, he worked in Europe for Arnold Sommerfeld, Niels Bohr, Erwin Schrödinger, and William Henry Bragg before returning to teach at Cal

Tech in 1927. Pauling became a full professor there in 1931. Aside from his work in applying quantum mechanics to bonding in chemical compounds, Pauling also introduced the idea of hybrid orbitals to account for the shapes of molecules. He also studied the properties of hemoglobin, which led to fruitful discoveries about the structures of protein. In 1940, he and Max Delbrück introduced a theory of antibody-antigen reactions. Pauling also devoted considerable efforts to studying the genetic disease sickle cell anemia.

Pauling did not rest on his laurels. He studied complex molecular structures, including those found in living tissues. He came very close to beating out Crick and Watson in the discovery of the famous double-helix structure of DNA. He was one of the first scientists to speak out strongly against the proliferation of nuclear weapons and fought against both Soviet and American tests of nuclear devices. For his political efforts he was awarded the Nobel Prize for peace—the second person (after Marie Curie) to win the prize twice and the only person to win it alone both times.

In recent years, Pauling has been both highly praised and furiously attacked for his research into vitamin C—he has endorsed heavy ingestion of vitamin C as a cure for the common cold, and he has maintained that vitamin C is valuable in treating cancer.

Electronic calculators can solve problems which the man who made them cannot solve; but no government-subsidized commission of engineers and physicists could create a worm.

—JOSEPH WOOD KRUTCH, "March," *The Twelve Seasons*

Technology made large populations possible; large populations now make technology indispensable.

—JOSEPH WOOD KRUTCH "The Nemesis of Power," *Human Nature and the HumanCondition*

The machine does not isolate man from the great problems of nature but plunges him more deeply into them.

—SAINT-EXUPÉRY, *Wind, Sand, and Stars*

A CENTURY AGO the structural theory of organic chemistry was developed. Frankland in 1852 suggested that an atom of an element has a definite capacity for combining with atoms of other elements—a definite valence. Six years later Kekulé and Couper, independently, introduced the idea of valence bonds between atoms, including bonds between two carbon atoms, and suggested that carbon is quadrivalent. In 1861 Butlerov, making use for the first time of the term "chemical structure," stated clearly that the properties of a compound are determined by its molecular structure, and reflect the way in which atoms are bonded to one another in the molecules of the compound. The development of the structure theory of organic chemistry then progressed rapidly, and this theory has been of inestimable value in aiding organic chemists to interpret their experimental results and to plan new experiments.

A most important early addition to organic structure theory was made by the first Nobel Laureate in Chemistry, Jacobus van't Hoff, who in 1874 recognized that the optical activity of carbon compounds can be explained by the postulate that the four-valence bonds of the carbon atom are directed in space toward the corners of a tetrahedron.

The structure theory of inorganic chemistry may be said to have been born only fifty years ago, when Alfred Werner, Nobel Laureate in Chem-

---

Linus Pauling, Nobel Prize in Chemistry Award Address, 1954. Reprinted with permission of Elsevier Publishing Co., and the Nobel Foundation.

istry in 1913, found that the chemical composition and properties of complex inorganic substances could be explained by assuming that metal atoms often coordinate about themselves a number of atoms different from their valence, usually four atoms at the corners either of a tetrahedron or of a square coplanar with the central atom, or six atoms at the corners of an octahedron. His ideas about the geometry of inorganic complexes were completely verified twenty years later, through the application of the technique of X-ray diffraction.

After the discovery of the electron many efforts were made to develop an electronic theory of the chemical bond. A great contribution was made in 1916 by Gilbert Newton Lewis, who proposed that the chemical bond, such as the single bond between two carbon atoms or a carbon atom and a hydrogen atom represented by a line in the customary structural formula for ethane, consists of a pair of electrons held jointly by the two atoms that are bonded together. Lewis also suggested that atoms tend to assume the electronic configuration of a noble gas, through the sharing of electrons with other atoms or through electron transfer, and that the eight outermost electrons in an atom with a noble-gas electronic structure are arranged tetrahedrally in pairs about the atom. Applications of the theory and additional contributions were made by many chemists, including Irving Langmuir and Nevil Vincent Sidgwick.

After the discovery of quantum mechanics in 1925 it became evident that the quantum mechanical equations constitute a reliable basis for the theory of molecular structure. It also soon became evident that these equations, such as the Schrödinger wave equation, cannot be solved rigorously for any but the simplest molecules. The development of the theory of molecular structure and the nature of the chemical bond during the past twenty-five years has been in considerable part empirical—based upon the facts of chemistry—but with the interpretation of these facts greatly influenced by quantum mechanical principles and concepts.

The solution of the wave equation for the hydrogen molecule-ion by Ø. Burrau . . . completely clarified the question of the nature of the one-electron bond in this molecule-ion. Two illuminating quantum mechanical discussions of the shared-electron-pair bond in the hydrogen molecule were then simultaneously published, one by Heitler and London . . . and the other by E. U. Condon. . . . In the approximate solution of the wave equation for the hydrogen molecule by Heitler and London a wave function is used that requires the two electrons to be separated, each being close to one of the two nuclei. The treatment by Condon permits the electrons to be distributed between the two nuclei independently of one another, each occupying a wave function similar to Burrau's function for the hydrogen molecule-ion. Condon's treatment is the prototype of the molecular-orbital treatment that has been extensively applied in the discussion of aromatic and conjugated molecules, and Heitler and London's treatment is the prototype of the valence-bond method. When the effort is made to refine the two treatments they tend to become identical.

Those early applications of quantum mechanics to the problem of the nature of the chemical bond made it evident that in general a covalent bond, involving the sharing of a pair of electrons between two atoms, can be formed if two electrons are available (their spins must be opposed, in order that the bond be formed), and if each atom has available a stable electronic orbital for occupancy by the electrons.

The equivalence of the four bonds formed by a carbon atom, which had become a part of chemical theory, was not at first easily reconciled with the quantum mechanical description of the carbon atom as having one $2s$ orbital and three $2p$ orbitals in its outer shell. The solution to this difficulty was obtained when it was recognized that as a result of the resonance phenomenon of quantum mechanics a tetrahedral arrangement of the four bonds of the carbon atom is achieved. The carbon atom can be described as having four equivalent tetrahedral bond orbitals, which are hybrids of the s and p orbitals. Further study of this problem led to the discovery of many sets of hybrid bond orbitals, which could be correlated with bond angles, magnetic moments, and other molecular properties. In particular it was found that sp^3, dsp^2, and d^2sp^3 hybrid orbitals correspond respectively to the tetrahedral, square planar, and octahedral configurations of inorganic complexes that had been discovered by Werner. Conclusions as to the utilization of atomic orbitals in bond formation can be drawn from experimental values of magnetic moments. For example, the theory of the dsp^2 square complexes of bipositive nickel, palladium, and platinum requires that these substances be diamagnetic. The square complexes of bipositive palladium and platinum had been recognized by Werner and their structure verified by Dickinson . . . ; but the assignment of the square configuration to the complexes of nickel which are diamagnetic had not been made until the development of the new theory.

Further detailed information about the chemical bond resulted from a consideration of the energy of single bonds in relation to the relative electronegativity of the bonded atoms. It was found that the elements can be assigned electronegativity values such as to permit the rough prediction of the heats of formation of compounds to which chemical structures involving only single bonds are conventionally assigned, and that many of the properties of substances can be discussed in a simple way with the use of the electronegativity values of the elements.

The idea that the properties of many organic compounds, especially the aromatic compounds, cannot be simply correlated with a single valence-bond structure, but require the assignment of a somewhat more complex electronic structure, was developed during the period 1923 to 1926 by a number of chemists, including Lowry, Lapworth, Robinson, and Ingold in England, Lucas in the United States, and Arndt and Eistert in Germany. It was recognized that the properties of aromatic and conjugated molecules can be described by the use of two or more valence-bond structures, as reflected in the names, the theory of mesomerism and the theory of intermediate states, proposed for the new chemical theory. In 1931 Slater, E. Hückel, and others recognized that these theories can be given a quantum mechanical interpretation: an approximate wave function for a molecule of this sort can be set up as the sum of wave functions representing the hypothetical structures corresponding to the individual valence-bond structures. The molecule can then be described as having a structure that is a hybrid of the individual valence-bond structures, or as resonating among these structures, and the theory itself is now usually called the resonance theory of chemical structure. Very many quantitative calculations, approximate solutions of the wave equation, for aromatic and conjugated molecules have been made, with results that are in general in good agreement with experiment. Perhaps more important than the quantitative calculations is the possibility of prediction by simple chemical arguments. For example, the amide group, an important structural feature of proteins, can be described as resonating between two structures, one with the double bond between the carbon atom and the oxygen atom, and the other with the double bond between the carbon atom and the nitrogen atom:

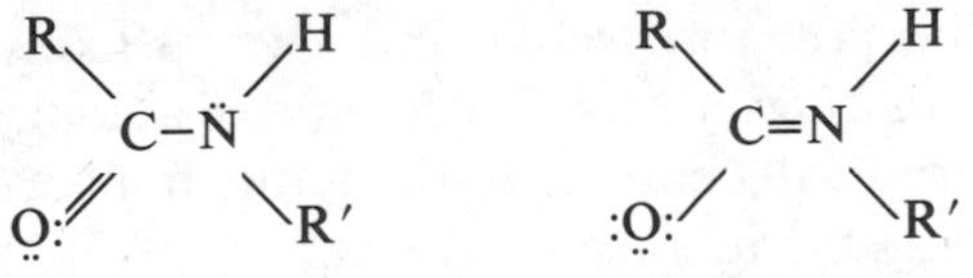

```
R       H          R       H
 \     /            \     /
  C—N̈               C=N
 //    \            /     \
Ö:      R'       :Ö:       R'
```

General arguments about the stability of alternative structures indicate that the structure with the double bond between carbon and oxygen should contribute somewhat more to the normal state of the amide group than the other structure; experience with other substances and acquaintance with the results of quantum mechanical calculations suggest the ratio 60%:40% for the respective contributions of these structures. A 40% contribution of the structure with the double bond between the carbon atom and the nitrogen atom would confer upon this bond the property of planarity of the group of six atoms; the resistance to deformation from the planar configuration would be expected to be 40% as great as for a molecule such as ethylene, containing a pure double bond, and it can be calculated that rotation of one end by 3° relative to the other end would introduce a strain energy of 100 cal/mole. The estimate of 40% double-bond character for the C-N bond is supported by the experimental value of the bond length, 1.32 A, interpreted with the aid of the empirical relation between double-bond character and interatomic distance. Knowledge of the structure of amides and also of the amino acids, provided by the theory of resonance and verified by extensive careful experimental studies made by R. B. Corey and his coworkers, has been of much value in the determination of the structure of proteins.

In the description of the theory of resonance in chemistry there has been a perhaps unnecessarily strong emphasis on its arbitrary character. It is true, of course, that a description of the benzene molecule can be given, in quantum mechanical language, without any reference to the two Kekulé structures, in which double bonds and single bonds alternate in the ring. An approximate wave function for the benzene molecule may be formulated by adding together two functions, representing the two Kekulé structures, and adding other terms, to make the wave function approximate the true wave function for the molecule more closely, or it may be constructed without explicit introduction of the wave functions representing the two Kekulé structures. It might be possible to develop an alternative simple way of discussing the structure of the amide group, for example, that would have permitted chemists to predict its properties, such as planarity; but in fact no simple way of discussing this group other than the way given above, involving resonance between two valence-bond structures, has been discovered, and it seems likely that the discussion of complex molecules in terms of resonance among two or more valence-bond structures will continue in the future to be useful to chemists, as it has been during the past twenty years.

The convenience and usefulness of the concept of resonance in the discussion of chemical problems are so great as to make the disadvantage of the element of arbitrariness of little significance. Also, it must not be forgotten that the element of arbitrariness occurs in essentially the same way in the simple structure theory of organic chemistry as in the theory of resonance—there is the same use of idealized, hypothetical structural elements. In the resonance discussion of the benzene molecule the two Kekulé structures have to be described as hypothetical: it is not possible to synthesize molecules with one or the other of the two Kekulé structures. In the same way, however, the concept of the carbon-carbon single bond is an idealization. The benzene molecule has its own structure, which cannot be exactly composed of structural elements from other molecules. The propane molecule also has its own structure, which cannot be composed of structural elements from other molecules—it is not possible to isolate a portion of the propane molecule, involving parts of two carbon atoms and perhaps two electrons in between them, and say that this portion of the propane molecule is the carbon-carbon single bond, identical with a portion of the ethane molecule. The description of the propane molecule as involving carbon-carbon single bonds and carbon-hydrogen single bonds is arbitrary: the concepts themselves are idealizations, in the same way as the concept of the Kekulé structures that are described as contributing to the normal state of the benzene molecule. Chemists have found that the simple structure theory of organic chemistry and also the resonance theory are valuable, despite their use of idealizations and their arbitrary character.

Other extensions of the theory of the chemical bond made in recent years involve the concept of fractional bonds. Twenty-five years ago it was discovered that a simple theory of complex crystals with largely ionic structures, such as the silicate minerals, can be developed on the basis of the assumption that each cation or metal atom divides its charge or valence equally among the anions that are coordinated about it. For example, in a crystal of topaz, $Al_2SiO_4F_2$, each silicon atom is surrounded by a tetrahedron of four oxygen atoms, and each aluminum atom is surrounded by an octahedron of four oxygen atoms and two fluorine atoms. The valence of silicon, 4, is assumed to be divided among four bonds, which then have the bond number 1—they are single bonds. The valence of aluminum, 3, is divided among six bonds, each of which is a half bond. A stable structure results when the atoms are arranged in such a way that each anion, oxygen or fluorine, forms bonds equal to its valence. In topaz each oxygen atom forms one single bond with silicon and two half bonds with aluminum, whereas each fluorine atom forms only two half bonds with aluminum. The distribution of the valences hence then corresponds to the bivalence of oxygen and the univalence of fluorine. It was pointed out by W. L. Bragg that if the metal atoms are idealized as cations (Si^{++++} and Al^{+++}) and the oxygen and fluorine atoms as anions (O^{--} and F^{-}), this distribution corresponds to having the shortest possible lines of force between the cations and the anions—the lines of force need to reach only from a cation to an immediately adjacent anion, which forms part of its coordination polyhedron. Occasionally ionic crystals are found in which there are small deviations from this requirement, but only rarely are the deviations larger than one quarter of a valence unit.

Another application of the concept of fractional valence bonds has been made in the field of metals and alloys. In the usual quantum mechanical discussion of metals, initiated by W. Pauli . . . and Sommerfeld . . . the assumption was made that only a small number of electrons contribute significantly to the binding together of the metal atoms. For example, it was customary to assume that only one electron, occupying a 4*s* orbital, is significantly involved in the copper-copper bonds in the metal copper. Sixteen years ago an analysis of the magnetic properties of the transition metals was made that indicated that the number of bonding electrons in the transition metals is much larger, of the order of magnitude of six. Iron, for example, can be described as having six valence electrons, which occupy hybrid d^3sp^2 orbitals. The six bonds, corresponding to these six valence electrons, resonate among the fourteen positions connecting an iron atom with its fourteen nearest neighbors. The bonds to the eight nearest neighbors have bond number approximately ⅝, and those to the six slightly more distant neighbors have bond number ⅙. In gamma iron, where each atom is surrounded by twelve equally distant neighbors, the bonds are half bonds. The concept that the structure of metals and intermetallic compounds can be described in terms of valence bonds that resonate among alternative positions, aided by an extra orbital on most or all of the atoms (the metallic orbital), has been found of value in the discussion of the properties of these substances. The resonating-bond theory of metals is supported especially strongly by the consideration of interatomic distances in metals and intermetallic compounds.

The iron atom has eight electrons outside of the argon shell of eighteen. Six of these electrons are assumed, in the resonating-valence-bond theory, to be valence electrons, and the remaining two are atomic electrons, occupying 3*d* orbitals, and contributing two Bohr magnetons to the magnetic moment of the atom. A theory of the ferromagnetism of iron has recently been developed, in which, as suggested by Zener . . . the interaction producing the Weiss field in the ferromagnetic metal is an interaction of the spin moments of the atomic electrons and uncoupled spins of some of the valence electrons. It has been found possible to use spectroscopic energy values to predict the number of uncoupled valence electrons, and hence the saturation magnetic moment for iron: the calculation leads to 0.26 uncoupled valence electrons per atom, and saturation magnetic moment 2.26 Bohr magnetons, which might be subject to correction by two or three percent because of the contribution of orbital moment. The experimental value is 2.22. A calculated value of the Curie temperature in rough agreement with experiment is also obtained.

The valence theory of metals and intermetallic compounds is still in a rather unsatisfactory state. It is not yet possible to make predictions about the composition and properties of intermetallic compounds with even a small fraction of the assurance with which they can be made about organic compounds and ordinary inorganic compounds. We may, however, hope that there will be significant progress in the attack on this problem during the next few years.

Let us now return to the subject of the structural chemistry of organic substances, especially the complex substances that occur in living organisms, such as protein. Recent work in this field has shown the value of the use of structural arguments that go beyond those of the classical structure theory of organic chemistry. The interatomic distances and bond angles in the polypeptide chains of proteins are precisely known, the bond distances to within about 0.02 Å and the bond angles to within about 2°. It is known that the amide groups must retain their planarity; the atoms are expected not to deviate from the planar configuration by more than perhaps 0.05 Å. There is rotational freedom about the single bonds connecting the alpha carbon atom with the adjacent amide carbon and nitrogen atoms, but there are restrictions on the configurations of the polypeptide chain that can be achieved by rotations about these bonds: atoms of different parts of the chain must not approach one another so closely as to introduce large steric repulsion, and in general the N-H and O atoms of different amide groups must be so located relative to one another as to permit the formation of hydrogen bonds, with N-H · · · O distance equal to 2.79 ± 0.10 Å and with the oxygen atom not far from the N-H axis. These requirements are stringent ones. Their application to a proposed hydrogen-bonded structure of a polypeptide chain cannot in general be made by the simple method of drawing a structural formula; instead, extensive numerical calculations must be carried out, or a model must be constructed. For the more complex structures, such as those that are now under consideration for the polypeptide chains of collagen and gelatin, the analytical treatment is so complex as to resist successful execution, and only the model method can be used. In order that the principles of modern structural chemistry may be applied with the power that their reliability justifies, molecular models must be constructed with great accuracy. For example, molecular models on the scale 2.5 cm = 1 Å have to be made with a precision better than 0.01 cm.

We may, I believe, anticipate that the chemist of the future who is interested in the structure of proteins, nucleic acids, polysaccharides, and other complex substances with high molecular weight will come to rely upon a new structural chemistry, involving precise geometrical relationships among the atoms in the molecules and the rigorous application of the new structural principles, and that great progress will be made, through this technique, in the attack, by chemical methods, on the problems of biology and medicine.

P.2

Semiconductor Research Leading to the Point Contact Transistor

John Bardeen

THE AMERICAN PHYSICIST John Bardeen was the first person ever to win two Nobel Prizes in Physics: his first in 1956, for the discovery of the transistor, he shared with William Shockley and Walter Brattain; the second in 1972, for his theoretical studies on superconductivity, he shared with Leon Cooper and John Schrieffer.

Bardeen was born in Madison, Wisconsin, in 1908 and completed his undergraduate studies at the University of Wisconsin before going on to take his Ph.D. under Eugene Wigner at Princeton in 1936. From 1938 until the outbreak of World War II, he taught at the University of Minnesota. He served in the U.S. Navy as a physicist during the war; after demobilization, he joined the Bell Telephone Laboratories in 1945. In 1951, he went back to teaching, serving as professor of physics at the University of Illinois, where his research was in the little-known field of superconductivity.

In accepting the Nobel award in 1956, Bardeen, Shockley, and Brattain each discussed a different aspect of their research in their lectures. After paying his respects to the many physicists who had worked on semiconductor theory over the years, Bardeen emphasized that the development of new and efficient techniques for the purification of germanium and silicon had had an enormous impact on the work done in this field. Following intensive government-sponsored research during the war, germanium and silicon had emerged as the most promising elements for use in research in solid-state devices. A curious irony is that crystals, similar to the old-fashioned ones used in the earliest "crystal sets" for radio reception, and rendered obsolete by the invention of the vacuum tube, returned to serve as the point of departure for a revolution in electronics that has radically transformed our world and our lives.

Bardeen's second Nobel Prize was awarded for his work in formulating a useful superconductivity theory. Although reduction in the electrical resistance of mercury cooled to a temperature of 4.2K had been observed as early as 1911 by H. Kamerlingh-Omnes, and superconductivity properties had subsequently been shown to be a property of many other elements, it fell to Bardeen and his colleagues to demonstrate that current in a superconductor is carried not by individual electrons but by bonded pairs of electrons which formed as a result of interactions between electrons and vibrations of the atoms in a crystal. Bardeen's work revived interest in superconductivity and spawned new efforts to devise practical applications for supercooled materials.

Only science can hope to keep technology in some sort of moral order.

—EDGAR Z. FRIEDENBERG, "The Impact of the School," *The Vanishing Adolescent*

The danger of the past was that men became slaves. The danger of the future is that men may become robots.

—ERICH FROMM, *The Sane Society*

The moment man cast off his age-long belief in magic, Science bestowed upon him the blessings of the Electric Current.

—JEAN GIRAUDOUX, *The Enchanted*

If there is technological advance without social advance, there is, almost automatically, an increase in human misery.

—MICHAEL HARRINGTON, appendix to *The Other America*

Is it a fact—or have I dreamt it—that, by means of electricity, the world of matter has become a great nerve, vibrating thousands of miles in a breathless point of time?

—NATHANIEL HAWTHORNE, *The House of the Seven Gables*

. . . THE DISCOVERY OF the transistor effect occurred in the course of a fundamental research program on semiconductors initiated at the Bell Telephone Laboratories in early 1946. . . .

The general aim of the program was to obtain as complete an understanding as possible of semiconductor phenomena, not in empirical terms, but on the basis of atomic theory. A sound theoretical foundation was available from work done during the thirties:

(1) Wilson's quantum mechanical theory, based on the energy band model, and describing conduction in terms of excess electrons and holes. It is fundamental to all subsequent developments. The theory shows how the concentration of carriers depends on the temperature and on impurities.

(2) Frenkel's theories of certain photoconductive phenomena (change of contact potential with illumination and the photomagneto electric effect) in which general equations were introduced which describe current flow

---

John Bardeen, Nobel Prize in Physics Award Address, 1956. Reprinted with permission of Elsevier Publishing Co., and the Nobel Foundation.

when non-equilibrium concentrations of both holes and conduction electrons are present. He recognized that flow may occur by diffusion in a concentration gradient as well as by an electric field.

(3) Independent and parallel developments of theories of contact rectification by Mott, Schottky and Davydov. The most complete mathematical theories were worked out by Schottky and his co-worker, Spenke.

Of great importance for our research program was the development during and since the war of methods of purification and control of the electrical properties of germanium and silicon. These materials were chosen for most of our work because they are well-suited to fundamental investigations with the desired close coordination of theory and experiment. Depending on the nature of the chemical impurities present, they can be made to conduct by either excess electrons or holes.

Largely because of commercial importance in rectifiers, most experimental work in the thirties was done on copper oxide (Cu_2O) and selenium. Both have complex structures and conductivities which are difficult to control. While the theories provided a good qualitative understanding of many semiconductor phenomena, they had not been subjected to really convincing quantitative checks. In some cases, particularly in rectification, discrepancies between experiment and theory were quite large. It was not certain whether the difficulties were caused by something missing in the theories or by the fact that the materials used to check the theories were far from ideal.

In the U.S.A., research on germanium and silicon was carried out during the war by a number of university, government and industrial laboratories in connection with the development of point-contact or "cat's whisker" detectors for radar. Particular mention should be made of the study of germanium by a group at Purdue University working under the direction of K. Lark-Horovitz and of silicon by a group at the Bell Telephone Laboratories. The latter study was initiated by R. S. Ohl before the war and carried out subsequently by him and by a group under J. H. Schaff. By 1946 it was possible to produce relatively pure polycrystalline materials and to control the electrical properties by introducing appropriate amounts of donor and acceptor impurities. Some of the earliest work (1915) on the electrical properties of germanium and silicon was done in Sweden by Prof. C. Benedicks.

Aside from intrinsic scientific interest, an important reason for choosing semiconductors as a promising field in which to work, was the many and increasing applications in electronic devices, which, in 1945, included diodes, varistors and thermistors. There had long been the hope of making a triode, or an amplifying device, with a semiconductor. Two possibilities had been suggested. One followed from the analogy between a metal semiconductor rectifying contact and a vacuum-tube diode. If one could somehow insert a grid in the space-charge layer at the contact, one should be able to control the flow of electrons across the contact. A major practical difficulty is that the width of the space-charge layer is typically only about 10^{-4} cm. That the principle is a sound one was demonstrated by Hilsch and Pohl, who built a triode in an alkali-halide crystal in which the width of the space-charge layer was of the order of one centimeter. Because amplification was limited to frequencies of less than one cycle per second, it was not practical for electronic applications.

The second suggestion was to control the conductance of a thin film or slab of semiconductor by application of a transverse electric field (called the *field effect*). In a simple form, the slab forms one plate of a parallel plate condenser, the control electrode being the other plate. When a voltage is applied across the condenser, charges are induced in the slab. If the induced charges are mobile carriers, the conductance should change with changes of voltage on the control electrode. This form was suggested by Shockley; his calculations indicated that, with suitable geometry and materials, the effect should be large enough to produce amplification of an a.c. signal.

Point-contact and junction transistors operate on a different principle than either of these two suggestions, one not anticipated at the start of the program. The transistor principle, in which both electrons and holes play a role, was discovered in the course of a basic research program on surface properties.

Shockley's field-effect proposal, although initially unsuccessful, had an important bearing on directing the research program toward a study of surface phenomena and surface states. Several tests which Shockley carried out at various times with J. R. Haynes, H. J. McSkimin, W. A. Yager and R. S. Ohl, using evaporated films of germanium and silicon, all gave negative results. In analyzing the reasons for this failure, it was suggested that there were states for electrons localized at the surface, and that a large fraction of the induced charge was immobilized in these states. Surface states also accounted for a number of hitherto puzzling features of germanium and silicon point-contact diodes.

In addition to the possibility of practical applications, research on surface properties appeared quite promising from the viewpoint of fundamental science. Although surface states had been predicted as a theoretical possibility, little was known about them from experiment. The decision was made, therefore, to stress research in this area. The study of surfaces initiated at that time (1946) has been continued at the Bell Laboratories and is now being carried out by many other groups as well.

It is interesting to note that the field effect, originally suggested for possible value for a device, has been an extremely fruitful tool for the fundamental investigation of surface states. Further, with improvements in semiconductor technology, it is now possible to make electronic amplifiers with high gain which operate on the field-effect principle.

Before discussing the research program, we shall give first some general background material on conduction in semiconductors and metal-semiconductor rectifying contacts.

NATURE OF CONDUCTION IN SEMICONDUCTORS

An electronic semiconductor is typically a valence crystal whose conductivity depends markedly on temperature and on the presence of minute amounts of foreign impurities. The ideal crystal at the absolute zero is an insulator. When the valence bonds are completely occupied and there are no extra electrons in the crystal, there is no possibility for current to flow. Charges can be transferred only when imperfections are present in the electronic structure, and these can be of two types: *excess electrons* which do not fit into the valence bonds and can move through the crystal, and *holes,* places from which electrons are missing in the bonds, which also behave as mobile carriers. While the excess electrons have the normal negative electronic charge, $-e$, holes have a positive charge, $+e$. It is a case of two negatives making a positive; a missing negative charge is a positive defect in the electron structure.

The bulk of a semiconductor is electrically neutral; there are as many positive charges as negative. In an *intrinsic* semiconductor, in which current carriers are created by thermal excitation, there are approximately equal numbers of excess electrons and holes. Conductivity in an *extrinsic* semiconductor results from impurity ions in the lattice. In n-type material, the negative charge of the excess electrons is balanced by a net positive space charge of impurity ions. In p-type, the *positive* charge of the holes is balanced by negatively charged impurities. Foreign atoms which can become positively charged on introduction to the lattice are called *donors;* atoms which become negatively ionized are called *acceptors.* Thus donors make a semiconductor n-type, acceptors p-type. When both donors and acceptors are present, the conductivity type depends on which is in excess. Mobile carriers then balance the *net* space charge of the impurity ions. Terminology used is listed in the table below:

TABLE 1

Designation of conductivity type		Majority carrier	Dominant impurity ion
n-type	excess	electron (n/cm^3)	donor
p-type	defect	hole (p/cm^3)	acceptor

These ideas can be illustrated quite simply for silicon and germanium, which, like carbon, have a valence of four and lie below carbon in the Periodic Table. Both crystallize in the diamond structure in which each atom is surrounded tetrahedrally by four others with which it forms bonds. Carbon in the form of a diamond is normally an insulator; the bond structure is complete and there are no excess electrons. If ultraviolet light falls on diamond, electrons can be ejected from the bond positions by the photoelectric effect. Excess electrons and holes so formed can conduct electricity; the crystal becomes photoconductive.

The energy required to free an electron from a bond position so that it and the hole left behind can move the crystal, is much less in silicon and germanium than for diamond. Appreciable numbers are released by thermal excitations at high temperatures; this gives intrinsic conductivity.

Impurity atoms in germanium and silicon with more than four valence electrons are usually donors, those with less than four acceptors. For example, Group V elements are donors, Group III elements acceptors. When an arsenic atom, a Group V element, substitutes for germanium in the crystal, only four of its valence electrons are required to form the bonds. The fifth is only weakly held by the forces of Coulomb attraction, greatly reduced by the high dielectric constant of the crystal. The energy required to free the extra electron is so small that the arsenic atoms are completely ionized at room temperature. Gallium, a typical Group III acceptor, has only three valence electrons. In order to fill the four bonds, Ga picks up another electron and enters the crystal in the form of a negative ion, Ga^-. The charge is balanced by a free hole.

While some of the general notions of excess and defect conductivity, donors and acceptors, go back earlier, Wilson was the first to formalize an adequate mathematical theory in terms of the band picture of solids. The band picture itself, first applied to metals, is a consequence of an application of quantum mechanics to the motion of electrons in the periodic

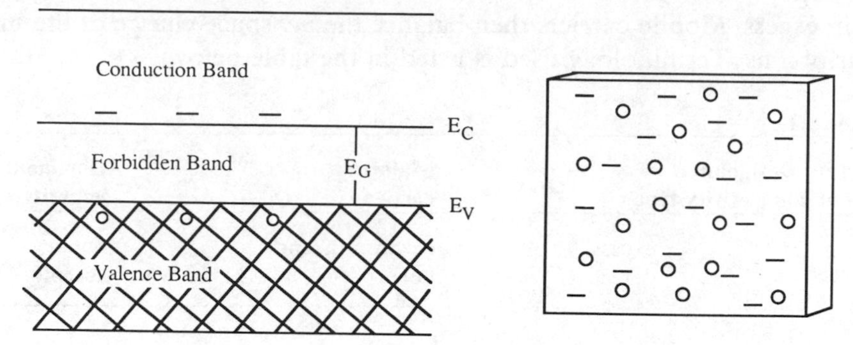

FIG. 1. Energy-level diagram of an intrinsic semiconductor. There is a random distribution of electrons and holes in equal numbers.

potential field of a crystal lattice. Energy levels of electrons in atoms are discrete. When the atoms are combined to form a crystal, the allowed levels form continuous bands. When a band is completely occupied, the net current of all of the electrons in the band is zero. Metals have incompletely filled bands. In insulators and semiconductors, there is an energy gap between the highest filled band and the next higher allowed band of levels, normally unoccupied.

The relations are most simply illustrated in terms of an energy-level diagram of the crystal. In Fig. 1 is shown a schematic energy-level diagram of an intrinsic semiconductor. Electrons taking part in the chemical bonds form a continuous band of levels called the valence band. Above these is an energy gap in which there are no allowed levels in the ideal crystal, and then another continuous band of levels called the conduction band. The energy gap, E_G, is the energy required to free an electron from the valence bonds. Excess, or conduction, electrons have energies in the lower part of the conduction band. The very lowest state in this band, E_C, corresponds to an electron at rest, the higher states to electrons moving through the crystal with additional energy of motion. Holes correspond to states near the top of the valence band, E_V, from which electrons are missing. In an intrinsic semiconductor, electrons and holes are created in equal numbers by thermal excitation of electrons from the valence to the conduction band, and they are distributed at random through the crystal.

In an n-type semiconductor, as illustrated in Fig. 2a, there is a large number of electrons in the conduction band and very few holes in the valence band. Energy levels corresponding to electrons localized around Group V donor impurity atoms are typically in the forbidden gap and a little below the conduction band. This means that only a small energy is required to ionize the donor and place the electron removed in the conduction band. The charge of the electrons in the conduction band is compensated by the positive space charge of the donor ions. Levels of Group III acceptors (Fig. 2b) are a little above the valence band. When occupied by thermal excitation of electrons from the valence band, they become negatively charged. The space charge of the holes so created is compensated by that of the negative acceptor ions.

Occupancy of the levels is given by the position of the Fermi level, E_F. The probability, f, that a level of energy E is occupied by an electron is given by the Fermi-Dirac function:

$$f = \frac{1}{1 + \exp\,(E - E_F)/kT}$$

The energy gap in a semiconductor is usually large compared with thermal energy, kT (~0.025 cV at room temperature), so that for levels well above E_F one can use the approximation

$$f \simeq \exp\,[-(E - E_F)/kT]$$

For levels below E_F, it is often more convenient to give the probability

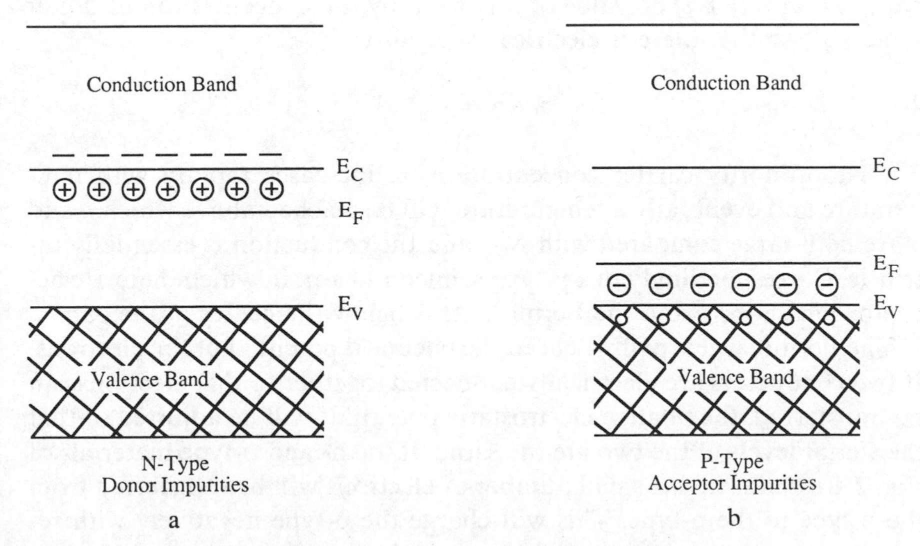

FIG. 2. Energy-level diagrams for n- and p-type semiconductors.

$$f_p = 1 - f = \frac{1}{1 + \exp[(E_F - E)/kT]}$$

that the level is unoccupied, or "occupied by a hole." Again, for levels well below E_F,

$$f_p \simeq \exp[-(E_F - E)/kT]$$

The expressions for the total electron and hole concentrations (number per unit volume), designated by the symbols n and p respectively, are of the form

$$n = N_C \exp[-(E_C - E_F)/kT]$$
$$p = N_V \exp[-(E_F - E_V)/kT]$$

where N_C and N_V vary slowly with temperature compared with the exponential factors. Note that the product np is independent of the position of the Fermi level and depends only on the temperature:

$$np = n_i^2 = N_C N_V \exp[-(E_C - E_V)/kT] = N_C N_V \exp[-E_C/kT]$$

Here n is the concentration in an intrinsic semiconductor for which $n = p$.

In an n-type semiconductor, the Fermi level is above the middle of the gap, so that $n \gg p$. The value of n is fixed by the concentration of donor ions, N_d^+, so that there is electrical neutrality:

$$n - p = N_d^+$$

. . . The minority carrier concentration, p, increases rapidly with temperature and eventually a temperature will be reached above which n and p are both large compared with N_d^+ and the conduction is essentially intrinsic. Correspondingly in a p-type semiconductor, in which there are acceptor ions, $p \gg n$, and the Fermi level is below the center of the gap.

The Fermi level is equivalent to the chemical potential of the electrons. If two conductors are electrically connected together so that electrons can be transferred, the relative electrostatic potentials will be adjusted so that the Fermi levels of the two are the same. If the n- and p-type materials of Fig. 2 are connected, a small number of electrons will be transferred from the n-type to the p-type. This will charge the p-type negatively with respect to the n-type and raise the electrostatic potential energy of the electrons accordingly. Electron transfer will take place until the energy levels of the p-type material are raised relative to those of the n-type by the amount required to make the Fermi levels coincide. . . .

Conduction electrons and holes are highly mobile, and may move through the crystal for distances of hundreds or thousands of the interatomic distance, before being scattered by thermal motion or by impurities or other imperfections. This is to be understood in terms of the wave property of the electron; a wave can travel through a perfect periodic structure without attenuation. In treating acceleration in electric or magnetic fields, the wave aspect can often be disregarded, and electrons and holes thought of as classical particles with an effective mass of the same order, but differing from the ordinary electron mass. The effective mass is often anisotropic, and different for different directions of motion in the crystal. This same effective mass picture can be used to estimate the thermal motion of the gas of electrons and holes. Average thermal velocities at room temperature are of the order of 10^7 cm/sec.

Scattering can be described in terms of a mean free path for the electrons and holes. In relatively pure crystals at ordinary temperatures, scattering occurs mainly by interaction with the thermal vibrations of the atoms of the crystal. In less pure crystals, or in relatively pure crystals at low temperatures, the mean free path may be determined by scattering by impurity atoms. Because of the scattering, the carriers are not uniformly accelerated by an electric field, but attain an average drift velocity proportional to the field. Ordinarily the drift velocity is much smaller than the average thermal velocity. Drift velocities may be expressed in terms of the mobilities, μ_n and μ_p of the electrons and holes respectively.

In an electric field E,

$$(V_d)_n = -\mu_n E$$
$$(V_d)_p = \mu_p E$$

Because of their negative charge, conduction electrons drift oppositely to the field. Values for pure germanium at room temperature are $\mu_n = 3{,}800$ cm^2/volt sec; $\mu_p = 1{,}800$ cm^2/volt sec. This means that holes attain a drift velocity of 1,800 cm/sec in a field of one volt/cm.

Expressions for the conductivity are:

$$\begin{aligned} &\text{n-type:} && \sigma_n = ne\mu_n \\ &\text{p-type:} && \sigma_p = pe\mu_p \\ &\text{intrinsic:} && \sigma = ne\mu_n + pe\mu_p \end{aligned}$$

. . . Carriers move not only under the influence of an electric field, but also by diffusion; the diffusion current is proportional to the concentration gradient. Expressions for the particle current densities of holes and electrons, respectively, are

$$j_p = p\mu_p E - D_p \operatorname{grad} p$$
$$j_n = n\mu_n E - D_n \operatorname{grad} n$$

Einstein has shown that mobilities and diffusion coefficients are related:

$$\mu = \frac{e}{kT} D$$

where k is Boltzmann's constant. Diffusion and conduction currents both play an important role in the transistor.

The diffusion term was first considered by Wagner in his theory of oxidation of metals. The equations were worked out more completely by Frenkel in an analysis of the diffusive flow which occurs when light is absorbed near one face of a slab. . . . The light quanta raise electrons from the valence to the conduction bands, creating conduction electrons and holes in equal numbers. These diffuse toward the interior of the slab. Because of recombination of conduction electron and hole pairs, the concentration drops as the diffusion occurs. Frenkel gave the general equations of flow when electrons and holes are present in attempting to account for the Dember effect (change in contact potential with light) and the photomagnetoelectric (PME) effect. The latter is a voltage analogous to a Hall voltage observed between the ends of a slab in a transverse magnetic field. . . . The Dember voltage was presumed to result from a difference of mobility, and thus of diffusion coefficient, between electrons and holes. Electrical neutrality requires that the concentrations and thus the concentration gradients be the same. Further, under steady-state conditions the flow of electrons to the interior must equal the flow of holes, so that there is no net electrical current. However, if D_n is greater than D_p, the diffusive flow of electrons would be greater than that of holes. What happens is that an electric field, E, is introduced which aids holes and retards the electrons so as to equalize the flows. The integral of E gives a voltage difference between the surface and the interior, and thus a change in contact potential. As we will mention later, much larger changes in contact potential with light may come from surface barrier effects.

CONTACT RECTIFIERS

In order to understand how a point-contact transistor operates, it is necessary to know some of the features of a rectifying contact between a metal and semiconductor. Common examples are copper-oxide and selenium rectifiers and germanium and silicon point-contact diodes which pass current much more readily for one direction of applied voltage than the opposite. We shall follow Schottky's picture, and use as an illustration a contact to an n-type semiconductor. Similar arguments apply to p-type rectifiers with appropriate changes of sign of the potentials and charges. It is most convenient to make use of an energy-level diagram in which the changes in energy bands resulting from changes in electrostatic potential are plotted along a line perpendicular to the contact, as in Fig. 3. Rectification results from the potential energy barrier at the interface which impedes the flow of electrons across the contact.

The Fermi level of the metal is close to the highest of the normally occupied levels of the conduction band. Because of the nature of the metal-semiconductor interface layers, a relatively large energy, χ, perhaps of

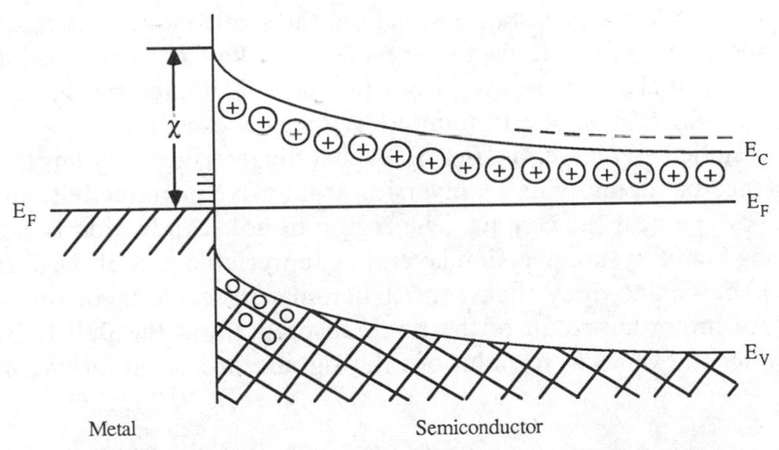

FIG. 3. Equilibrium energy-level diagram for a metal-semiconductor rectifying contact along a line perpendicular to the interface. Variations in the energy bands of the semiconductor result from changes in electrostatic potential due to the layer of uncompensated space-charge. The overall change in potential from the surface to the interior is such as to bring the Fermi level in the interior of the semiconductor into coincidence with that of the metal. In this example, there is an inversion from n-type conductance in the bulk to p-type at the surface.

the order of 0.5 cV, is required to take an electron from the Fermi level of the metal and place it in the conduction band in the semiconductor. In the interior of the semiconductor, which is electrically neutral, the position of the Fermi level relative to the energy bands is determined by the concentration of conduction electrons, and thus of donors. In equilibrium, with no voltage applied, the Fermi levels of the metal and semiconductor must be the same. This is accomplished by a region of space charge adjacent to the metal in which there is a variation of electrostatic potential, and thus of potential energy of the electron, as illustrated.

In the bulk of the semiconductor there is a balance between conduction electrons and positive donors. In the barrier region which is one of high potential energy for electrons, there are few electrons in the conduction band. The uncompensated space charge of the donors is balanced by a negative charge at the immediate interface. It is these charges, in turn, which produce the potential barrier. The width of the space-charge region is typically of the order of 10^{-5} to 10^{-4} cm.

When a voltage is applied, most of the drop occurs across the barrier layer. The direction of easy flow is that in which the semiconductor is negative relative to the metal. The bands are raised, the barrier becomes narrower, and electrons can flow more easily from the semiconductor to the metal. In the high resistance direction, the semiconductor is positive, the bands are lowered relative to the metal, and the barrier is broadened. The current of electrons flowing from the metal is limited by the energy barrier, χ, which must be surmounted by thermal excitation.

If χ is sufficiently large, the Fermi level at the interface may be close to the valence band, implying an inversion from n-type conductivity in the bulk to p-type near the contact. The region of hole conduction is called, following Schottky, an inversion layer. An appreciable part of the current flow to the contact may then consist of minority carriers, in this case holes. An important result of the research program at the Bell Laboratories after the war was to point out the significance of minority carrier flow.

EXPERIMENTS ON SURFACE STATES

We have mentioned in the introduction that the negative result of the field-effect experiment was an important factor in suggesting the existence of surface states on germanium and silicon, and directing the research program toward a study of surface properties. As is shown in Fig. 4, the experiment consists of making a thin film or slab one plate of a

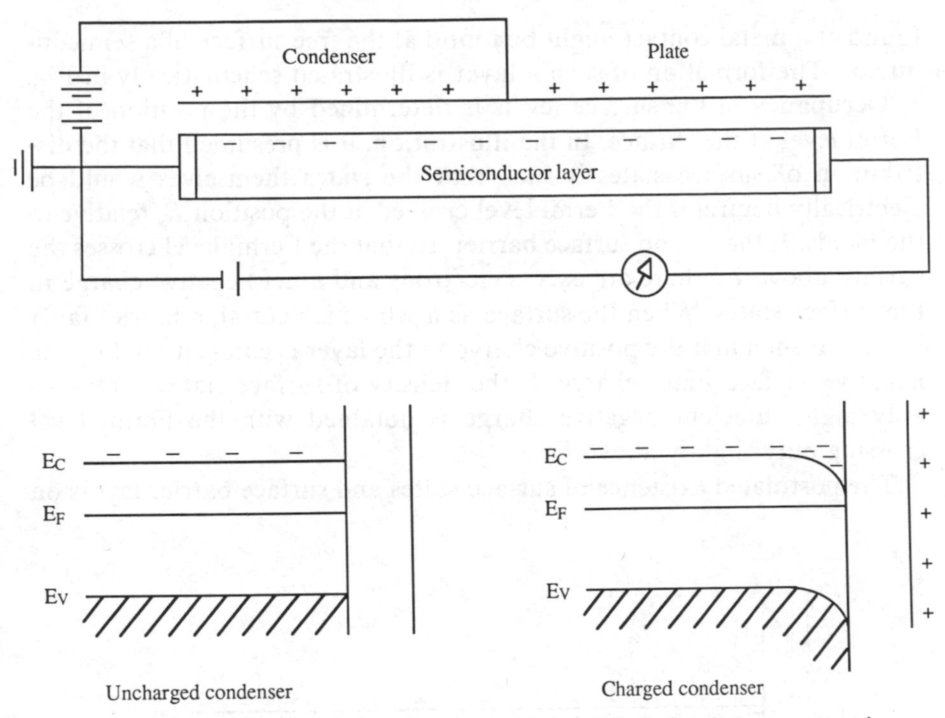

FIG. 4. Schematic diagram of a field-effect experiment for an n-type semiconductor with no surface states.

parallel plate condenser and then measuring the change in conductance of the slab with changes in voltage applied across the condenser. The hypothetical case illustrated is an n-type semiconductor with no surface states. When the field plate is positive, the negative charge induced on the semiconductor consists of added electrons in the conduction band. The amount of induced charge can be determined from the applied voltage and measured capacity of the system. If the mobility is known, the expected change in conductance can be calculated readily.

When experiments were performed on evaporated films of germanium and silicon, negative results were obtained; in some cases the predicted effect was more than one thousand times the experimental limit of detection. Analysis indicated that a large part of the discrepancy, perhaps a factor of 50 to 100, came from the very low mobility of electrons in the films as compared with bulk material. The remaining was attributed to shielding by surface states. . . .

It was predicted that if surface states exist, a barrier layer of the type found at a metal contact might be found at the free surface of a semiconductor. The formation of such a layer is illustrated schematically in Fig. 5. Occupancy of the surface levels is determined by the position of the Fermi level at the surface. In the illustration, it is presumed that the distribution of surface states is such that the states themselves would be electrically neutral if the Fermi level crossed at the position F_s relative to the bands. If there is no surface barrier, so that the Fermi level crosses the surface above F_s, there are excess electrons and a net negative charge in the surface states. When the surface as a whole is neutral, a barrier layer is formed such that the positive charge in the layer is compensated by the negative surface states charge. If the density of surface states is reasonably high, sufficient negative charge is obtained with the Fermi level crossing only slightly above F_s. . . .

The postulated existence of surface states and surface barrier layers on

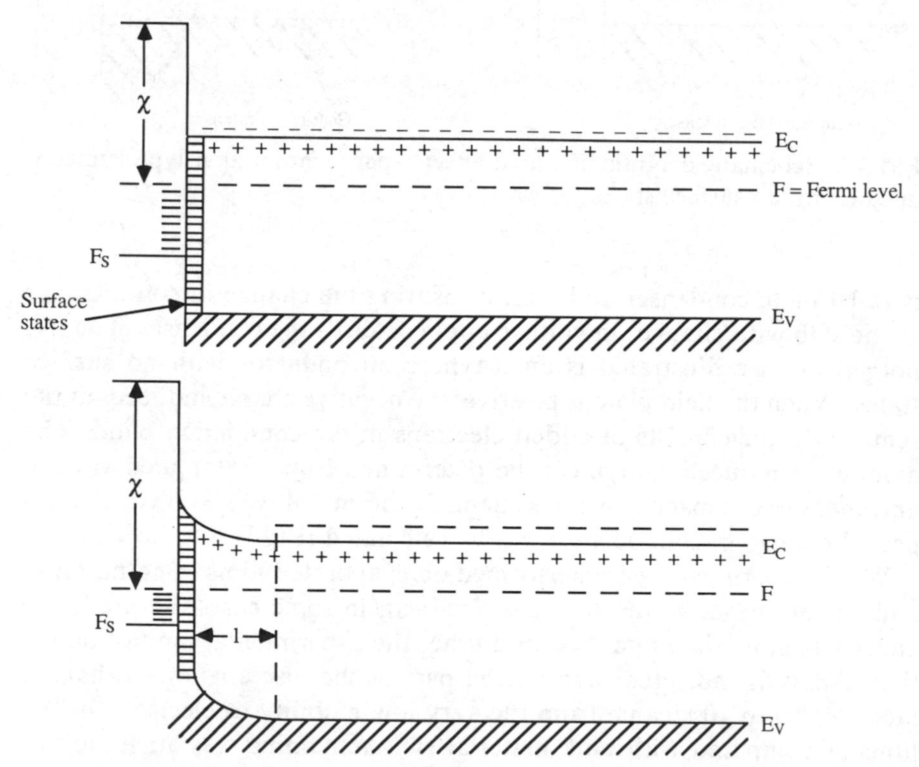

FIG. 5. Formation of a space-charge barrier layer at the free surface of a semiconductor.

the free surface of germanium and silicon accounted for several properties of germanium and silicon which had hitherto been puzzling. These included (1) lack of dependence of rectifier characteristics on the work function of the metal contact, (2) current voltage characteristics of a contact made with two pieces of germanium, and (3) the fact that there was found little or no contact potential difference between n- and p-type germanium and between n- and p-type silicon.

While available evidence for surface states was fairly convincing, it was all of an indirect nature. Further, none of the effects gave any evidence about the height of the surface barrier and of the distribution of surface states. A number of further experiments which might yield more concrete evidence about the surface barrier was suggested by Shockley, Brattain and myself. Shockley predicted that a difference in contact potential would be found between n- and p-type specimens with large impurity concentration. A systematic study of Brattain and Shockley using silicon specimens with varying amounts of donor and acceptor impurities showed that this was true, and an estimate was obtained for the density of surface states. Another experiment which indicated the presence of a surface barrier was a measurement of the change in contact potential with illumination of the surface. This is just the Dember effect, which Frenkel had attempted to account for by the difference in mobilities of the electrons and holes generated by the light and diffusing to the interior. It was found that the change is usually much larger and often of the opposite sign than predicted by Frenkel's theory, which did not take into account a surface barrier. . . .

The apparatus used by Brattain to measure contact potential and change in contact potential with illumination is shown in Fig. 6. The reference electrode, generally platinum, is in the form of a screen so that light can pass through it. By vibrating the electrode, the contact potential itself can be measured by the Kelvin method. If light chopped at an appropriate frequency falls on the surface and the electrode is held fixed, the change with illumination can be measured from the alternating voltage developed across the condenser. In the course of the study, Brattain tried several ambient atmospheres and different temperatures. He observed a large effect when a liquid dielectric filled the space between the electrode and semiconductor surface. He and Gibney then introduced electrolytes, and observed effects attributed to large changes in the surface barrier with voltage applied across the electrolyte. Evidently ions piling up at the surface created a very large field which penetrated through the surface states. . . .

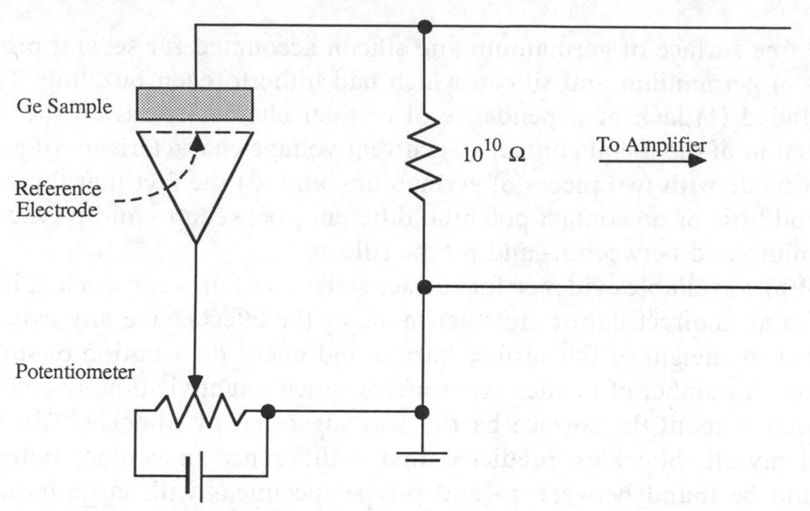

FIG. 6. Schematic diagram of apparatus used by Brattain to measure contact potential and change of contact potential with light.

EXPERIMENTS ON INVERSION LAYERS

Use of an electrolyte provided a method for changing the surface barrier, so that it should be possible to observe a field effect in a suitable arrangement. We did not want to use an evaporated film because of the poor structure and low mobility. With the techniques available at the time, it would have been difficult to prepare a slab from bulk material sufficiently thin to observe a sizable effect. It was suggested that one could get the effect of a thin film in bulk material by observing directly the flow in an inversion layer of opposite conductivity type near the surface. Earlier work of Ohl and Scaff indicated that one could get an inversion layer of n-type conductivity on p-type silicon by suitably oxidizing the surface. If a point contact is made which rectifies to the p-type base, it would be expected to make low resistance contact to the inversion layer.

In the arrangement which Brattain and I used in the initial tests, the point contact was surrounded by, but insulated from, a drop of electrolyte. An electrode in the electrolyte could be used to apply a strong field at the semiconductor surface in the vicinity of the contact. The reverse, or high resistance direction is that in which point is positive relative to the block. Part of the reverse current consists of electrons flowing through the n-type inversion layer to the contact. It was found that the magnitude of this current could be changed by applying a voltage on the electrolyte probe, and thus, by the field effect, changing the conductance of the inversion layer. Since under static conditions only a very small current flowed through the electrolyte, the set-up could be used as an amplifier. In the initial tests, current and power amplification, but not voltage amplification, was observed. As predicted from the expected decrease in number of electrons in the inversion layer, a negative voltage applied to the probe was found to decrease the current flowing in the reverse direction to the contact.

It was next decided to try a similar arrangement with a block of n-type germanium. Although we had no prior knowledge of a p-type inversion layer on the surface, the experiments showed definitely that a large part of the reverse current consisted of holes flowing in an inversion layer near the surface. A positive change in voltage on the probe decreased the reverse current. Considerable voltage as well as current and power amplification was observed.

Because of the long time constants of the electrolyte used, amplification was obtained only at very low frequencies. We next tried to replace the electrolyte by a metal control electrode insulated from the surface by either a thin oxide layer or by a rectifying contact. A surface was prepared by Gibney by anodizing the surface and then evaporating several gold spots on it. Although none made the desired high resistance contact to the block, we decided to see what effects would be obtained. A point contact was placed very close to one of the spots and biased in the reverse direction (see Fig. 7). A small effect on the reverse current was observed when the spot was biased positively, but of *opposite* direction to that observed with the electrolyte. An increase in positive bias *increased* rather than decreased the reverse current to the point contact. The effect was large enough to give some voltage, but no power amplification. This experiment suggested that holes were flowing into the germanium surface from the gold spot, and that the holes introduced in this way flowed into the point contact to enhance the reverse current. This was the first indication of the transistor effect.

It was estimated that power amplification could be obtained if the metal contacts were spaced at distances of the order of 0.005 cm. In the first attempt, which was successful, contacts were made by evaporating gold on a wedge, and then separating the gold at the point of the wedge with a razor blade to make two closely spaced contacts. After further experimentation, it appeared that the easiest way to achieve the desired close separation was to use two appropriately shaped point contacts placed very close together. Success was achieved in the first trials; the point-contact transistor was born.

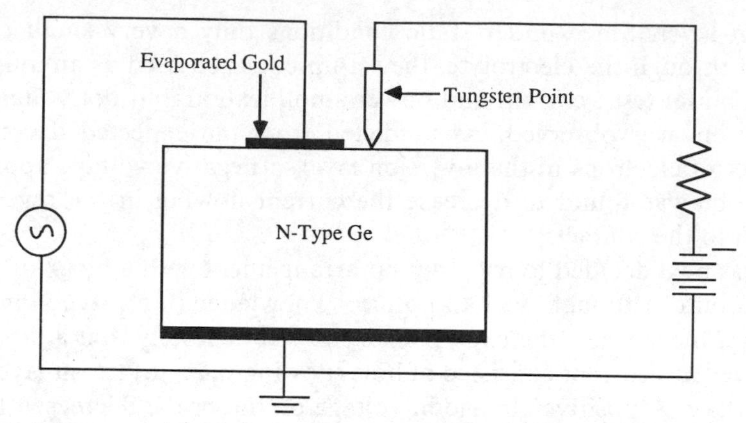

FIG. 7. Diagram of experiment in which the transistor effect was first observed. Positive voltage applied to the gold spot introduced holes into the n-type germanium block which flowed to the point contact biased in the reverse direction. It was found that an increase in positive voltage increased the reverse current. When connected across a high impedance, the change in voltage of the point contact was larger than the change at the gold spot, both measured relative to the base electrode.

It was evident from the experiments that a large part of both the forward and reverse currents from a germanium point contact is carried by minority carriers, in this case holes. If this fact had been recognized earlier, the transistor might have come sooner.

Operation of a point-contact transistor is illustrated in Fig. 8. When operated as an amplifier, one contact, the emitter, is biased with a d.c. voltage in the forward direction, the second, the collector, in the negative or high resistance direction. A third contact, the base electrode, makes a low resistance contact to the block. A large part of the forward current consists of holes flowing into the block. Current from the collector consists in part of electrons flowing from the contact and in part of holes flowing toward the contact. The collector current produces an electric field in the block which is in such a direction as to attract holes introduced at the emitter. A large part of the emitter current, introduced at low impedance, flows in the collector circuit. Biased in the reverse direction, the collector has high impedance and can be matched to a high impedance load. There is thus a large voltage amplification of an input signal. It is found that there is some current amplification as well, giving an overall power gain of 20 db. or more. An increase in hole current at the collector affects the barrier there in such a way as to enhance the current of electrons flowing from the contact.

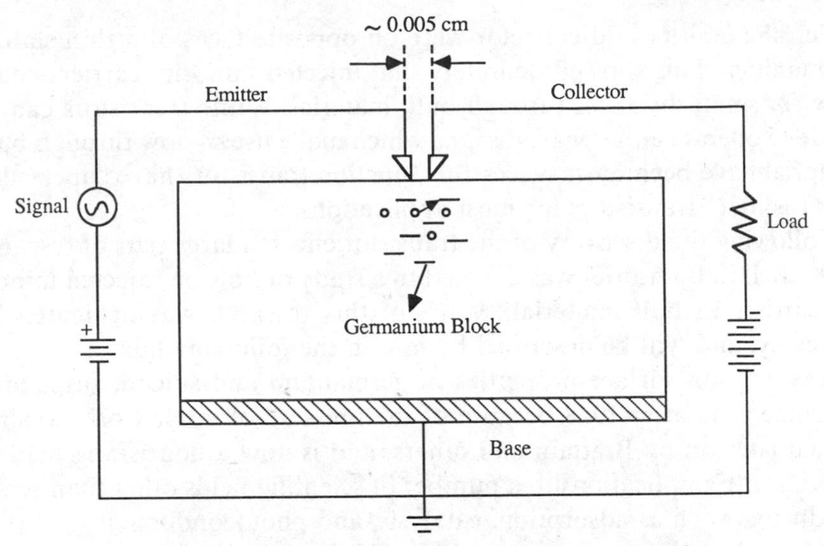

FIG. 8. Schematic diagram of point-contact transistor.

The collector current must be sufficiently large to provide an electric field to attract the holes from the emitter. The optimum impedance of the collector is considerably less than that of a good germanium diode in the reverse direction. In the first experiments, it was attempted to achieve this by treating the surface so as to produce a large inversion layer of p-type conductivity on the surface. In this case, a large fraction of the hole current may flow in the inversion layer. Later, it was found that better results could be obtained by electrically forming the collector by passing large current pulses through it. In this case the surface treatment is less critical, and most of the emitter current flows through the bulk.

Studies of the nature of the forward and reverse currents to a point contact to germanium were made by making probe measurements of the variation of potential in the vicinity of the contact. These measurements showed a large increase in conductivity when the contact was biased in the forward direction and in some cases evidence for a conducting inversion layer near the surface when biased in the reverse direction.

Before it was established whether the useful emitter current was confined to an inversion layer or could flow through the bulk, Shockley proposed a radically different design for a transistor based on the latter possibility. This is the junction transistor design in which added minority carriers from the emitter diffuse through a thin base layer to the collector. Independently of this suggestion, Shive made a point-contact transistor in which the emitter and collector were on opposite faces of a thin slab of germanium. This showed definitely that injected minority carriers could flow for small distances through bulk material. While transistors can be made to operate either way, designs which make use of flow through bulk material have been most successful. Junction transistors have superseded point-contact transistors for most applications.

Following the discovery of the transistor effect, a large part of research at the Bell Laboratories was devoted to a study of flow on injected minority carriers in bulk material. Much of this research was instigated by Shockley, and will be described by him in the following talk.

Research on surface properties of germanium and silicon, suspended for some time after 1948 because of the pressure of other work, was resumed later on by Brattain and others, and is now a flourishing field of activity with implications to a number of scientific fields other than semiconductors such as adsorption, catalysis, and photoconductivity. . . .

It is evident that many years of research by a great many people, both before and after the discovery of the transistor effect, has been required to bring our knowledge of semiconductors to its present development. We were fortunate enough to be involved at a particularly opportune time and to add another small step in the control of Nature for the benefit of mankind. . . .

P.3

Transistor Technology Evokes New Physics

WILLIAM SHOCKLEY

WILLIAM SHOCKLEY, cowinner of the Nobel Prize in physics in 1956, was born in London in 1910 of American parents. The family returned to the United States when Shockley was three years old. He received his undergraduate degree from the California Institute of Technology in 1932 and his Ph.D. from the Massachusetts Institute of Technology in 1936. Shockley joined the staff of the Bell Telephone Laboratories in 1936. It was there that he and his collaborators Bardeen and Brattain later discovered that certain elements in crystal form containing impurities were more efficient as rectifiers—devices permitting current to flow in only one direction, thus converting alternating into direct current—than the vacuum tubes of Fleming and De Forest. This work led to the invention of the transistor, the key component in the miniaturization of computers and virtually every other type of electronic gadget. The invention of the transistor so accelerated the development of successive generations of computers that by the 1980s, one could fit in a shirt pocket a computer equivalent in power to one that would have filled an entire room in the 1950s.

After a singularly successful career as a theoretical and experimental physicist and as a teacher, Shockley became a controversial figure in the 1970s. It was then that he argued that genetic factors in human intelligence implied qualitative differences in intelligence among the various races. In the 1980s, his name again made headlines when he donated a certain amount of his sperm to a frozen sperm bank set up by some California operators who had decided to ensure ample supplies of Nobel genes for the enrichment of future generations.

Long before he became controversial, however, Shockley gave his Nobel lecture on 11 December 1956. In it he stressed that research that takes as its point of departure a clearly defined practical end (in this case, the invention of a substitute for the bulky, fragile vacuum tube) is not necessarily, as some physicists feel, inferior science. In the case of the Bardeen-Shockley-Brattain invention of the transistor, experimental work aimed at developing a practical, commercially useful device, did indeed open an entire new branch of solid-state physics.

Applied Science is a conjuror, whose bottomless hat yields impartially the softest of Angora rabbits and the most petrifying of Medusas.

—ALDOUS HUXLEY, *Tomorrow and Tomorrow and Tomorrow*

A computer with as many vacuum tubes as a man has neurons in his head would require the Pentagon to house it, Niagara's power to run it, and Niagara's waters to cool it.

—WARREN S. MCCULLOCH, quoted in Robert Lindner's *Must You Conform?*

Man must be at once more humble and more confident—more humble in the face of the destructive potentials of what he can achieve, more confident of his own humanity as against computers and robots which are only engines to simulate him.

—MAX LERNER, "Manipulating Life," in the *New York Post,* Jan. 24, 1968

INTRODUCTION

THE OBJECTIVE OF producing useful devices has strongly influenced the choice of the research projects with which I have been associated. It is frequently said that having a more-or-less-specific practical goal in mind will degrade the quality of research. I do not believe that this is necessarily the case and to make my point in this lecture I have chosen my examples of the new physics of semiconductors from research projects which were very definitely motivated by practical considerations. . . .

. . . I would like to express some viewpoints about words often used to classify types of research in physics; for example, pure, applied, unrestricted, fundamental, basic, academic, industrial, practical, etc. It seems to me that all too frequently some of these words are used in a derogatory sense, on the one hand to belittle the practical objectives of producing something useful and, on the other hand, to brush off the possible long-range value of explorations into new areas where a useful outcome cannot be foreseen. Frequently, I have been asked if an experiment I have

William Shockley, Nobel Prize in Physics Award Address, 1956. Reprinted with permission of Elsevier Publishing Co., and the Nobel Foundation.

planned is pure or applied research; to me it is more important to know if the experiment will yield new and probably enduring knowledge about nature. If it is likely to yield such knowledge, it is, in my opinion, good fundamental research; and this is much more important than whether the motivation is purely esthetic satisfaction on the part of the experimenter on the one hand or the improvement of the stability of a high-power transistor on the other. It will take both types to "confer the greatest benefit on mankind" sought for in Nobel's will.

THE FIVE BASIC IMPERFECTIONS

Before discussing the selected group of experiments in transistor physics, I shall extend somewhat farther Dr. Bardeen's remarks about the characteristics and interactions of electrons, holes, donors and acceptors. For this purpose I shall use a reference to water solutions of acids, bases and salts as an aid to exposition.

The dissociation of pure water to positive hydrogen ions and negative hydroxyl ions satisfies the mass-action law

$$[H^+]\,[OH^-] = \text{function of } T$$

where the concentration of H_2O is so nearly constant for dilute solutions that it may be taken as constant and not shown explicitly. This equation has as its analogue

$$n \cdot p = f(T)$$

for a semiconductor where n and p are the electron and hole concentrations. The equation is accurate, provided neither n nor p is so large that its statistics become degenerate. The normal electron-pair bond here plays the role of an undissociated water molecule. In pure deionized water and in a pure semiconductor, electrical neutrality requires that the concentration of the positive charges equals that of the negative charges:

$$[H^+] = [OH^-] \text{ and } p = n$$

A semiconductor which is so pure that impurities do not affect the property being considered is called *intrinsic* in regard to such a property. Using the subscript i to designate this case, we have

$$p_i = n_i \quad \text{and} \quad n_i^2 = f(T)$$

The concentration n_i is referred to as the intrinsic concentration.

The chemical analogue to an n-type semiconductor is a base and the charge neutrality equations are

$$[H^+] + [Na^+] = [OH^-]$$
$$p + N_d = n$$

where N_d is the concentration of donors and it is assumed that all donors are fully ionized.

Similarly a p-type semiconductor is analogous to an acid:

$$[H^+] = [OH^-] + [Cl^-]$$
$$p = n + N_a$$

where N_a is the acceptor concentration.

A neutral salt also has its analogue which is called a *compensated* semiconductor. For this case the donor and acceptor concentrations are equal so that the equation for electrical neutrality

$$p + N_d = n + N_a$$

reduces to equality of n and p so that each is equal to n_i. The electrical conductivity of a perfectly compensated silicon or germanium crystal is almost equal to that of an intrinsic crystal; there may be a small difference due to the reduction of mobility by scattering by the charged ions. The difference between the low conductivity of the perfectly compensated semiconductor and the high conductivity of a neutral salt solution arises from the fact that the donors and acceptors are tied in place in a semiconductor while the cations and anions of a salt have mobilities comparable to that of an OH-ion.

Compensation, or rather overcompensation, plays a vital role in the manufacture of semiconductor devices. For example, it permits the conversion of n-type material to p-type by the addition of acceptors without the prior removal of donors. A crystal grown from an n-type melt may be changed to p-type by adding an excess of acceptors.

The words *majority* and *minority* are frequently useful in discussing semiconductors. In an n-type specimen, the majority carriers are electrons and the minority carriers are holes.

Holes, electrons, donors and acceptors represent four of the five classes of imperfections that must be considered in semiconductor crystals in order to understand semiconductor effects. The fifth imperfection has been given the name *deathnium.* The chemical analogue of deathnium is a catalyst. In the case of water as the analogue to a crystal, there is, so far as I know, no important corresponding catalyst. What deathnium does is to hasten the establishment of equilibrium between holes and electrons. If, due to the various possible disturbances important in transistor electronics, the concentration of minority carriers is, for example, substantially raised, then the minority carriers will combine with the majority carriers to produce normal electron-pair bonds, by this means restoring equilibrium. Deathnium catalyzes this recombination process. The symbols for the five imperfections are shown in Table 1.

TABLE 1

1.	—	(excess) electron
2.	+	hole
3.	□	deathnium
4.	⊕	donor
5.	⊖	acceptor

The role of deathnium can be illustrated in terms of the phenomenon of photoconductivity. If light shines on a germanium crystal, then the pairs of electrons and holes that are formed will impart a conductivity to the crystal. This conductivity is known as photoconductivity. If the source of light is removed, the photoconductivity will die away, due to the recombination of the holes and the electrons. Thus, if an electron falls into an incomplete bond, one hole-electron pair will be eliminated.

The photoconductivity dies away with a characteristic time known as the lifetime. Thus, after the light is turned off, the photoconductivity will drop to approximately one-half its value in one lifetime. This process continues with a reduction of approximately one-half in each subsequent period of one lifetime.

If the process of recombination of holes and electrons were a direct one, the lifetime would be the same in all germanium crystals. It is found experimentally, however, that two otherwise very similar germanium crystals will have lifetimes that differ by as much as 1,000 fold. In one crystal, the lifetime may be a millisecond, whereas in another it may be a microsecond. This variation in lifetime requires the presence of some sort of imperfection which catalyzes the recombination of the holes and the electrons.

Actually, there are several forms of deathnium. For example, if electrons having an energy of several million electron volts fall upon a germanium crystal, the lifetime is subsequently reduced . . . It is known that such bombardment produces disorder of the germanium atoms. A high-energy electron can eject a germanium atom bodily from its normal position in the crystal structure, thus leaving a vacancy behind, where there should be an atom, and causing the ejected atom to become either an extra atom or an interstitial atom fitting into a place in the structure which would normally be empty . . . These disordering effects function as deathnium. It has also been found that copper and nickel chemical impurities in the germanium produce marked reductions in lifetime.

The way in which deathnium catalyzes the recombination process is indicated in Fig. 1. In part (*b*) of this figure, an electron is captured by a deathnium center. The deathnium center thus becomes a baited trap which is ready to capture a hole. If a hole comes near to the deathnium center, the electron can drop into it, thus forming a normal covalent bond, and the deathnium center is then uncharged and ready to repeat the process.

It is characteristic of all microscopic processes that they may go backwards as well as forwards. Thus, the deathnium center may generate hole-electron pairs as well as eliminate them. The generation process is indicated in parts (*d*) and (*e*) of Fig. 1. In part (*d*) the deathnium center captures an electron from an adjoining normal electron-pair bond. This produces a hole which wanders off. Somewhat later, the deathnium center ejects the electron and thus reverts to its empty state in which it is ready either to recombine or to generate another hole-electron pair.

Under conditions of thermal equilibrium, both the recombination process and the generation process proceed continuously. The energy required to generate the hole-electron pair is furnished by the thermal energy of vibration of the atoms in the germanium crystal. The condition of thermal equilibrium is achieved when the two processes balance. For germanium at room temperature, this leads to a conductivity of about 0.02 $\text{ohm}^{-1}\ \text{cm}^{-1}$.

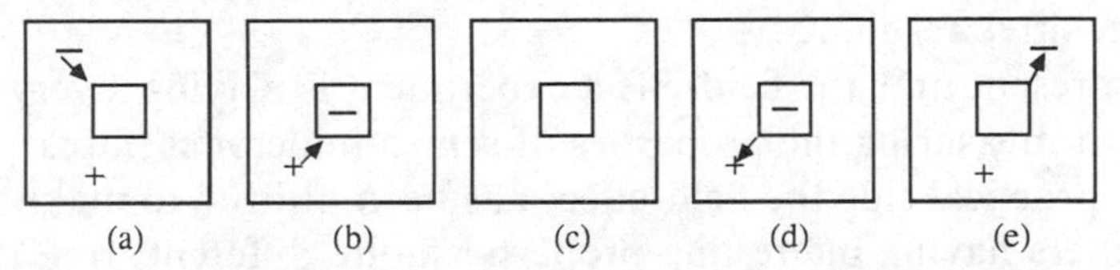

FIG. 1. A recombination center (deathnium) captures alternately an electron and a hole and thus catalyzes their recombination, as shown in parts (*a*), (*b*), and (*c*). The thermally activated generation process is shown in (*d*) and (*e*).

Since the concentration of holes and electrons under equilibrium conditions is governed by a mass-action law, the product *np* is independent of the concentration of deathnium. For example, if the concentration of deathnium is doubled, both the rate of generation and the rate of recombination are doubled, but the equilibrium concentrations of holes and electrons are unaltered.

Evidence that the deathnium mechanism shown in Fig. 1 is correct has been obtained by studying the dependence of the rate of recombination upon hole and electron densities. These studies are found to be in general agreement with the predictions based on the statistical model of Fig. 1.

THE FIELD EFFECT

The experiment which played the largest role in stimulating the transistor electronics program at Bell Telephone Laboratories was the so-called field-effect experiment. I devised this experiment as a direct result of trying to invent a semiconductor amplifier having separate input and output circuits. From the then known properties of semiconductors, I concluded that a thin film of silicon or germanium would have a substantial change in its conductivity if it were made into one of a pair of condenser plates and the condenser strongly charged. The surface charge, if represented as a change in mobile carriers, could appreciably increase or decrease the conductance of the film.

A number of experiments were initially tried using evaporated layers and other thin layers. These all gave negligible effects and progress was at a standstill until Bardeen proposed his theory of surface states to explain the lack of an observable effect.

Bardeen's model also explained a number of other previously mysterious phenomena and led to the suggestion of doing the field-effect experiment at liquid-air temperature to immobilize the surface states. This gave the first positive result. . . .

At the present time the field-effect experiment is playing a very important role in measuring the properties of semiconductor surfaces.

On the practical side the field-effect has been utilized to make transistor amplifiers having interesting properties quite different from those of junction transistors.

INJECTION AND DRIFT

At the time of the discovery of the point-contact transistor . . . there were a number of unresolved questions regarding its mode of operation. The original transistor showed evidence that the coupling between the input or emitter point and the output or collector point took place through conduction in a thin surface layer of conductivity type opposite to the underlying or base material. Somewhat later the idea that the emitter point might actually be injecting minority carriers into the body of the semiconductor developed. The development of this idea came as a result of two independent events: the invention of the junction-transistor structure . . . (as discussed below, injection plays an essential role in the junction transistor) and the observation . . . that transistor action could occur with the points on opposite sides of a thin slab of semiconductor.

In order to test the model of carrier injection, J. R. Haynes and the author collaborated in the drift-mobility experiment or "Haynes' experiment" on germanium specimens. In order to understand the significance of the experiment in elucidating transistor action the mechanism of current flow at the metal-semiconductor contact must be considered.

In Fig. 2, the metal is represented in a highly pictorial fashion. The valence electrons in a metal are thought of as forming an electron gas,

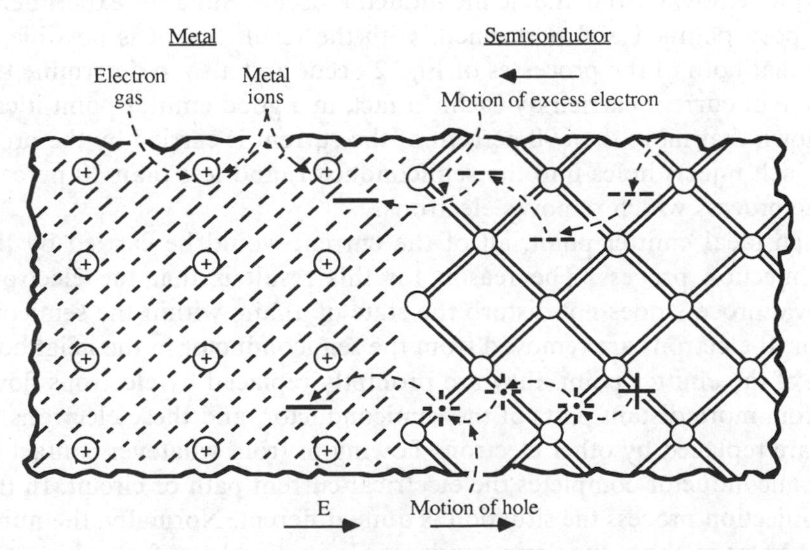

FIG. 2. Two possible mechanisms for current flow near an emitter point as described in text.

which permeates the entire structure. Thus, the electrons are not held in position in valence bonds as they are in an insulator. The electron gas can flow freely through the structure of the metal, and this fact accounts for the high conductivity of metals. In the upper part of Fig. 2 one of the processes for removing electrons from the semiconductor is represented. Since the semiconductor is n-type, it contains excess electrons; these excess electrons may be drawn to the metal by its positive charge and thus enter the metal to produce a current of electrons flowing out of the emitter point through the connecting lead.

Another possible mechanism for electron transfer from semiconductor to metal is shown in the lower part of Fig. 2. In this case, an electron is withdrawn from one of the valence bonds adjacent to the metal. This process also transfers an electron from the semiconductor to the metal, but when the transfer occurs, a hole is left behind. The hole is repelled by the positive charge on the emitter contact and moves deeper into the semiconductor.

Both of the processes discussed above have the same effect so far as the metal emitter point and conducting leads to the emitter point are concerned. Both produce a net flow of electrons from semiconductor to the emitter point and through the leads to the emitter terminal. It is thus evident that some more subtle experiment than simply measuring the current to the emitter point is necessary to show that both processes of electron removal from the semiconductor occur. Suitable experiments have been planned and performed, with the result that it is possible to show that both of the processes of Fig. 2 occur and also to determine the fraction of current carried by each. In fact, in a good emitter point it can be shown that more than 90 percent of the current is carried by the process which injects holes into the semiconductor, and less than 10 percent by the process which removes electrons.

In an ideal emitter point, all of the current would be carried by the hole-injection process. The reason for this result is that the electron-removal process does not disturb the state of affairs within the semiconductor. If electrons are removed from the semiconductor in the neighborhood of the emitter point, they are promptly replaced by electrons flowing from more distant parts of the semiconductor, and these electrons in turn are replaced by other electrons flowing in from whatever contact to the semiconductor completes the electrical-current path or circuit. In the hole-injection process the situation is quite different. Normally, the number of holes in the n-type semiconductor is negligible so far as a contribution to the conductivity is concerned. However, when electrons are removed from the valence bonds and holes are injected, relatively large numbers of holes will be introduced. The conductivity of the semiconductor will be increased in the neighborhood of the emitter point in much the same fashion that it would be if light were to shine on the semiconductor and produce hole-electron pairs. This disturbance in the electronic structure can be used to produce amplifying action in the transistor. . . .

A p-n junction is the simplest so-called *compositional* structure in semiconductor electronics. By a compositional structure is meant a single crystal of semiconductor in which the composition in terms of the donor and acceptor content varies in a systematic and controlled way. Before describing the theory of the functioning of a p-n junction, I would like to say something about the way in which p-n junctions whose behavior was satisfactory from a theoretical point of view were first achieved. This history also is an example of the interaction of practical needs on a research program.

To begin with, attempts were made to produce p-n junctions by allowing molten drops of germanium of one conductivity type to fall upon a heated block of germanium of the other conductivity type. Although p-n junctions were obtained by these means, their characteristics failed to live up to the predictions of theory. (The problems were almost certainly those of cleanliness—the importance of copper was not known at the time.)

. . . a program was set up to grow large single crystals of germanium. . . . An experimental apparatus [was made] for "pulling" single crystals of germanium from a melt of germanium heated in a graphite crucible into which was dipped a small single-crystal seed. The recent advances of transistor science and technology are founded on these crystals. The addition of acceptors to an n-type melt during pulling changed the solidifying material from n- to p-type and gave the first good p-n junctions.

Another extremely important development in the preparation of materials should be mentioned. This is the method of *zone refining*. . . . Noting that impurities tend to be more soluble in molten germanium than in solid germanium, Pfann devised a system of repetitive purification by crystallization. By making an elongated graphite boat and providing means to heat locally a small length or zone of it, he was able to pass molten zones successively from one end to the other of the germanium and by this means to sweep impurities clear of the major portion of the crystal and to concentrate them near one end. By this means germanium crystals having one impurity atom in about 10^{10} germanium atoms have been produced. The density of impurities in these crystals is thus comparable to the density of molecules in a gas at a pressure of 10^{-5} millimeters of mercury. It is appropriate to call zone refining the vacuum pump of transistor electronics.

Fig. 3 represents a p-n junction. In discussing its electrical properties, we will be concerned with the five kinds of imperfections shown in the lower part of the figure. From a mechanical point of view the crystal is practically homogeneous and perfect. A typical concentration for impurities in the crystal might be 10^{15} cm^{-3}. This density of imperfections is so low that if one were to traverse a row of atoms from end to end in the crystal one would, on the average, strike only about ten imperfections. Thus the crystal structure itself is only slightly altered by the presence of the imperfections. From the electrical point of view, on the other hand, the imperfections have profound effects.

As is shown in Fig. 3, the electrons are found chiefly in the n-region where they neutralize the chemical space charge of the donors, and the holes are similarly found in the p-region. In order for this situation to

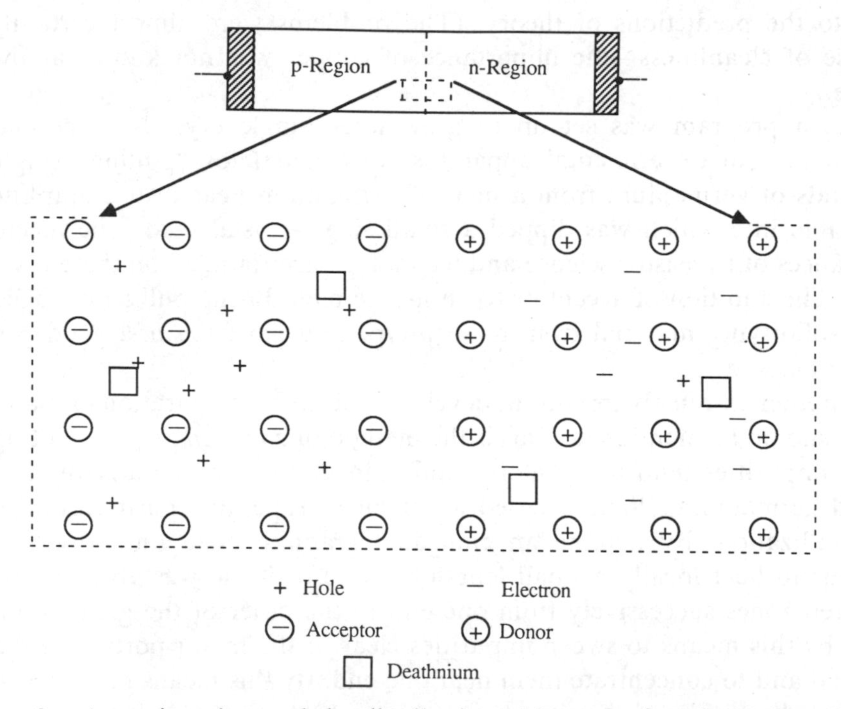

FIG. 3. A p-n junction and the distribution of imperfections in it. (For simplicity, compensating pairs of donors and acceptors are not shown.)

persist, as it does under conditions of thermal equilibrium, there must be an electric field present in the crystal. The idea that an electrical field is present in a conductor under conditions of thermal equilibrium is at first surprising. However, the necessity for such an electric field can readily be seen in terms of Fig. 3. Let us first of all suppose that no electric field is present across the junction; then as a result of diffusion, holes will cross the junction from left to right and electrons will similarly diffuse across the junction from right to left. As a result, there will be a net transfer to the right of positive charge across the junction. This will result in an electric field which will build up to just such a point that further current flow is prevented. . . .

Under conditions of thermal equilibrium no net current of either holes or electrons will flow across the junction. It is advantageous, however, to consider this equilibrium situation as arising from compensating currents. We shall illustrate this by considering the flow of holes back and forth across the junction. Although the density of holes is small in the n-region, it is still appreciable and plays a vital role in the behavior of the p-n junction. Let us consider the course of a hole that arrives in the n-region by climbing the potential hill. . . . Having climbed the hill and arrived at the plateau of constant electrostatic potential in the n-type region, it will then proceed to move by a random diffusive motion. The most probable outcome of this motion will be that it will diffuse to the steep part of the hill and slide back down into the p-type region. We shall not be concerned with holes which follow this particular course. On the other hand, it may, by chance, diffuse more deeply into the n-type region. In this event, it will on the average diffuse for lifetime τ, and subsequently it will be captured by a deathnium center in which it recombines with an electron.

The average depth to which holes diffuse in the n-type region depends upon the lifetime. The holes spread out in the region by diffusion. When the suitable differential equation describing this process is solved, it is found that the average depth to which they penetrate is given by the equation

$$L = \sqrt{D\tau}$$

where L is known as the diffusion length, D is the diffusion constant for holes, and τ is the lifetime for holes in the n-region. Thus under equilibrium conditions a current of holes flows from the p-region into the n-region and penetrates on the average one diffusion length L before recombining with electrons.

Under equilibrium conditions a *principle of detailed balance* holds. This principle of statistical mechanics says that each process and its opposite occur equally frequently. Hence we must conclude that the flow of holes from the p-region into the n-region, followed by recombination, must be exactly balanced by a reverse process. The reverse process is thermal generation of holes through deathnium centers, followed by diffusion to barrier where they slide down into p-type region.

The application of voltage to the terminals of the device shown in Fig. 3 destroys the exact balance of the two currents just discussed. In considering the application of voltage we shall neglect any voltage drops at the contacts between the metal electrodes of Fig. 3 and the semiconductors. At the end of this section we will return briefly to the reasons why such voltage drops may be neglected. The effect of the application of voltages upon the currents is represented in Fig. 4. In part (*a*) of this figure we show the thermal equilibrium condition. The two currents previously discussed are represented by I_f and I_g, these currents standing, respectively, for the current of holes entering the n-region and recombining and the current generated in the n-region and diffusing to the barrier. For the condition of thermal equilibrium these two currents are equal and opposite. In part (*b*) of the figure the situation for a large "reverse" bias is shown. For reverse bias, negative voltage is required to the p-region and positive to the n-region so that the electrostatic potential difference between the two regions is increased. If the electrostatic potential is sufficiently high, corresponding to the situation shown in part (*b*), then practically no holes can climb the potential hill and I_f drops nearly to zero. This situation is represented by showing I_f as a vector of negligible length, whereas I_g has practically the same value as it has for the case of thermal equilibrium. In general, the diffusion length L is large compared to the width of the dipole or space-charge region. Hence the region where I_g arises is practically unaffected by reverse bias and I_g is thus independent of reverse bias. This independence of current upon bias is referred to as *saturation.*

When forward bias is applied, the situation shown in Fig. 4(*c*) occurs and I_f increases. This increase is described in terms of the energy difference for a hole in the n-region and p-region. This energy difference is equal to the charge of the electron times the electrostatic potential differences between the two sides. We can apply a general theorem from statistical mechanics to a consideration of the number of holes, which, by chance, acquire sufficient energy to climb the potential hill. This theorem states that each time the potential hill is increased by one thermal unit of energy, kT, then the number of holes capable of climbing the higher hill

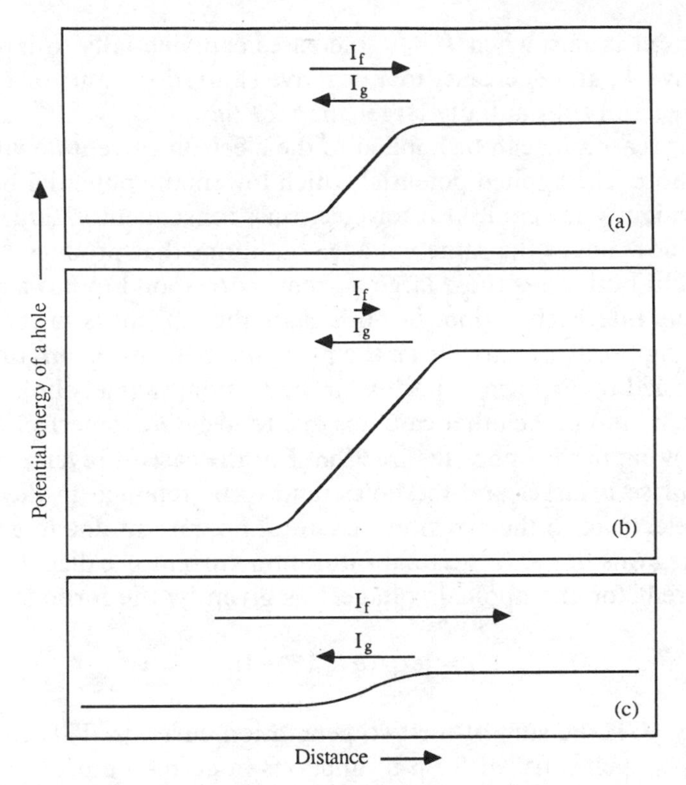

FIG. 4. Dependence of recombination and generation currents upon bias. (*a*) Thermal equilibrium; (*b*) reverse bias; (*c*) forward bias.

is reduced by a factor of 1/e. Since the potential barrier is already present under conditions of thermal equilibrium, it follows also that each lowering of the barrier by an amount kT will increase the current by a factor of e. The change in height of the barrier caused by the applied voltage V is $-qV$, where the polarity is so chosen that positive values correspond to plus potentials applied to the p-region and q is the absolute value of the charge of the electron. $V = 0$ is the equilibrium case, and for this case I_f is equal to I_g. Hence, in general, the recombination current is

$$I_f = I_g \exp qV/kT$$

This gives rise to a total current of holes from p-region to n-region, given by the difference

$$I_f - I_g = I_g [\exp (qV/kT) - 1]$$

This current is zero when $V = 0$, increases exponentially to large values for positive V, and decreases to a negative saturation value of I_g when V is negative and substantially larger than kT/q.

Similar reasoning can be applied to the electron current flowing across the junction. The applied potential which lowers the potential barrier for holes, evidently lowers it also for electrons; consequently, large electron currents flow under the same voltage conditions that produce large hole currents. In both cases these large currents correspond to flows of minority carriers into each region. In both cases the current is in the sense of transferring positive charge from the p-region to the n-region. In one case this is carried in the form of positive imperfections, namely holes, moving from p to n, and in the other case it is due to negative imperfections, electrons, moving in the opposite direction. For the case of reverse biases the potential rise is larger and the holes tend to be retained in the p-region and the electrons in the n-region. A saturation current due to generation in both regions flows. If the total saturation current is called I_s, then the total current for any applied voltage V is given by the formula

$$I = [\exp (qV/kT) - 1] I_s$$

Evidently, I_s is the sum of the two generation currents. This equation is found to be well-satisfied for p-n junctions in germanium. . . .

THE JUNCTION TRANSISTOR

The junction transistor is a compositional structure composed of two p-n junctions placed back to back. In its general behaviour as an amplifying device, a junction transistor shows great similarities to a vacuum-tube triode, or thermionic valve. Fig. 5 shows an n-p-n junction transistor in an amplifying circuit, the transistor being in the form of a sandwich with a layer of p-type germanium interposed between two layers of n-type germanium. Non-rectifying electrical contacts are made to the three layers. Under operating conditions the n-type region on the right, known as the collector, is biased positive so as to become attractive to electrons. As a result, a reverse bias appears between it and the middle region, known as the base. The current flowing across this reverse-biased junction can be controlled by an input signal applied between the base layer and the n-type region to the left, known as the emitter. As I shall describe below in more detail, the bias across the emitter junction controls the electron flow into the base region. In effect, the emitter junction acts like

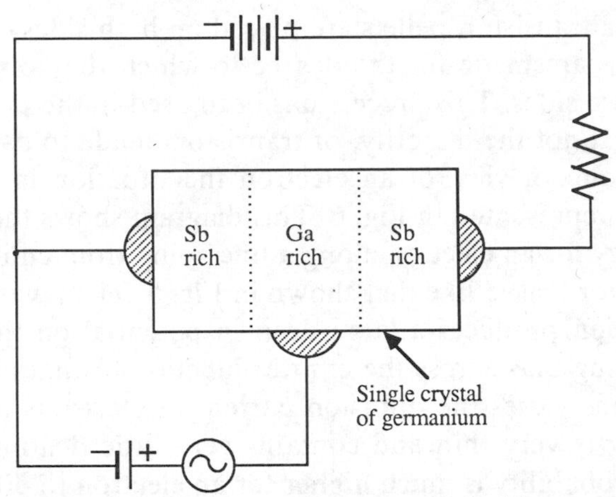

FIG. 5. The structure of a junction transistor and the bias supply for its operation in an amplifying circuit.

the region between the cathode and the grid in the vacuum tube. Electrons which enter the base region have a high probability of diffusing to the collector junction, and thus the flow of electrons from emitter to collector can be varied by varying the potential across the emitter base junction. The action is very similar to that controlling the flow of electrons from the cathode to the anode in the thermionic triode.

Junction transistors can be fabricated in a variety of ways. The compositional structure can be produced in a crystal-growing machine by techniques like those used for making simple p-n junctions. As the crystal is grown from a melt containing antimony, a pellet containing indium is dropped into the melt, and a second pellet containing antimony is dropped in a few seconds later. The portion of the crystal which grows between the dropping of the two pellets is rich in indium and is consequently p-type. The second pellet over-compensates the effect of the added indium and the subsequent material is again n-type. From such a single crystal small rods may be cut and contacts made. . . .

An alternative technique for producing the compositional structure starts with a thin plate of germanium which subsequently plays the role of the p-type region. A pellet of metal containing a donor is placed on this plate. The plate and pellet are then raised to such a temperature that the metal melts and dissolves a small amount of germanium. When the metal and germanium are subsequently cooled, the germanium precipitates from the metal and grows back onto the crystal structure of the base material. This regrown germanium carries with it some of the donors contained in the molten metal and thus grows an n-type region. In the fabrication of a transistor, pellets are placed on both sides of a thin plate and allowances are made for the degree to which they dissolve germanium on the two sides. This process has been used in the production of a large fraction, if not the majority, of transistors made to date.

From the point of view of an electron the situation in an operating transistor is as represented in Fig. 6. This diagram shows the variation in potential energy for an electron along a line going from emitter to collector in a transistor biased like that shown in Fig. 5. The reverse bias at the collector junction produces a large drop in potential on the right-hand side. The varying bias across the emitter junction changes the height of the hill and thus varies the diffusion current of electrons into the base. The base layer is very thin and contains very little deathnium. Consequently, the probability is much higher for an electron to diffuse through the base layer and arrive at the collector junction than for it to combine with a hole through a deathnium center in the base layer. In a well-built junction transistor, in fact, the electron flow through the base region proceeds so efficiently that the electron current flowing through the base layer to the collector may exceed the current combining in the base layer by a factor of 100 or more. This means that the input currents flowing to the base layer through the base lead may control the output currents 100 times larger flowing to the collector region. Although this situation is not nearly as ideal as in a vacuum triode, in which the grid current is still smaller compared with the anode current, it permits the junction transistor to operate as a highly efficient amplifier.

So far as voltage requirements are concerned, the junction transistor is

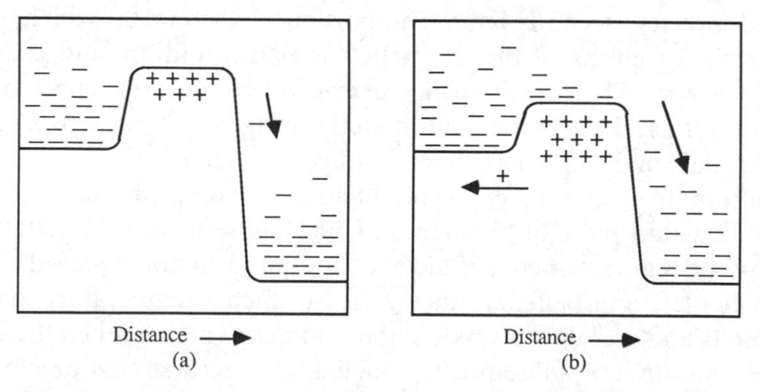

FIG. 6. The potential energy of an electron in an n-p-n junction transistor for two values of bias across the emitter junction. In (*b*) the forward bias is greater than in (*a*).

far superior to any vacuum tube, because a junction transistor can be brought fully into its operational range with voltages as small as 50–100 mV. Such voltages make the potential hill at the collector junction shown on Fig. 6 several times the thermal voltage ($kT/q = 25$ mV). As a consequence, any electron reaching the collector junction is sure to slide down the hill and has a negligible chance of returning.

The junction transistor has almost ideal pentode characteristics in terms of collector saturation. This also follows from the same considerations which make it operate well at low collector voltages. In fact, in some junction transistors there is a negligible variation in current over a voltage range of a hundredfold or more, say from 100 mV to 10 volts on the collector.

The junction transistor described in Figs. 5 and 6 is only one of a large family of transistors. The p-n-i-p and n-p-i-n transistors are variations of this form especially designed for low capacitance between the collector and the base. The so-called junction tetrode is a special form of junction transistor in which the current flow is controlled so as to occur only over a small region of the base. Junction transistors of these types have been used in oscillating circuits at frequencies as high as 1000 Mc/s.

In closing this section, I should like to point out that the junction transistor operates more flexibly in terms of power than do vacuum tubes. This is of great importance, since it means that junction transistors may be used efficiently in some cases where the power level to be amplified is very much smaller than the heater current required in the most efficient of vacuum tubes. The possibility of operating at very low power is due to the low voltages necessary to operate a junction transistor, and because of its small size and the high degree of purity possible, the current due to generation by deathnium centers can be made smaller than a microampere. It is thus possible to make a junction transistor which can be put into an amplifying condition with a power input substantially less than a microwatt. The same transistor, however, can operate at voltages as high as 10 volts and currents of the order of 10 mA. Thus, it can cover a power range effectively of 100,000. This fact indicates that there will probably be a much smaller diversity of transistor types than there are of vacuum tube types.

CONCLUSION

In a lecture of reasonable scope, it is no longer possible to do justice to the many branches of semiconductor physics that have developed since the discovery of the transistor effect. In the previous sections I have tried to illustrate how the stimulation of aiming for practical goals led to an experimental program which brought out many new and important aspects of electronic phenomena in semiconductors.

Not all of the important work has, of course, been motivated by highly practical ends which could be so immediately achieved once the fundamental science was understood. Much of the work on surface states should now be regarded as being in the pure-science class inasmuch as it is not possible to see at all directly how the scientific results will find utility. However, it seems highly probable that once the phenomena of surface states are thoroughly understood from a scientific point of view, many useful suggestions will arise as to how this knowledge may be employed to make better devices.

Also in the class of purely scientific work are the very beautiful determinations of the energy band structure in silicon and germanium that have resulted from *cyclotron resonance* in silicon and germanium. This work has led to a more detailed knowledge of energy band surfaces than has ever existed before, except possibly for some of the most simple and almost trivial cases.

One of the large new areas that has begun to grow as a result of the large amount of transistor technology may be called "The Theory of Solid Solutions in Silicon and Germanium." Recently, interactions between donors and acceptors of various types have been given active consideration. It is evident that interesting phenomena are associated with the incorporation of any foreign element from the Periodic Table in silicon and germanium. Important distinctions are being drawn between the same atoms in a substitutional position and an interstitial position. It is also being found that some atoms may exist in a semiconductor in not merely two states of charge as in the case of a simple donor and acceptor but in as many as four different conditions of charge. It seems highly probable that this area of work will become a field of active interest in physical chemistry in the years to come.

Among the phenomena motivated in considerable measure by practical aims we should mention the phenomena of avalanche breakdown which is useful in voltage limiting diodes and protection devices. A strong motivation for following up early leads which suggested that such effects might occur, came from the conversation I once had with Dr. O. E. Buckley, whom I have mentioned before in this lecture. Dr. Buckley pointed out the need for a device for protecting telephones from damage due to voltages induced by lightning near telephone lines. It was in large measure my knowledge of this need that gave emphasis to this work. The original interpretation which appeared to be consistent with the dielectric breakdown mechanism once proposed by Zener (direct rupture of electron-pair bonds by strong electric fields with the resultant production of hole-electron pairs) seems now to have been in error. The investigation of the same phenomena by McKay and his collaborators has led to a new branch of semiconductor electronics dealing with impact ionization of holes in electrons moving in intense electric fields in semiconductors. Both the effect and its theory are playing an active role in a certain class of switching transistors.

In closing . . . I would like to refer to a paragraph written in my book in 1950. I am pleased to see that the predictions of the paragraph appear to have been borne out to a considerable extent and I feel that it is now as applicable as it was then:

"It may be appropriate to speculate at this point about the future of transistor electronics. Those who have worked intensively in the field share the author's feeling of great optimism regarding the ultimate potentialities. It appears to most of the workers that an area has been opened up comparable to the entire area of vacuum and gas-discharge electronics. Already several transistor structures have been developed and many others have been explored to the extent of demonstrating their ultimate practicality, and still other ideas have been produced which have yet to be subjected to adequate experimental tests. It seems likely that many inventions unforeseen at present will be made based on the principles of carrier injection, the field effect, the Suhl effect, and the properties of rectifying junctions. It is quite probable that other new physical principles will also be utilized to practical ends as the art develops." . . .

More Is Different

PHILIP W. ANDERSON

PHILIP W. ANDERSON, a pioneer researcher in the field of superconductivity, is another of the brilliant physicists engaged in both theoretical and experimental science at the Bell Telephone Laboratories. He was born in 1923 in Indianapolis, Indiana. After receiving his undergraduate degree at Harvard in 1943, Anderson spent two years at the Naval Research Laboratory working on antenna engineering. Anderson took his doctorate in physics at Harvard University in 1949 and then joined Bell Laboratories, where he has done research in solid-state physics. He has also worked as a professor at Princeton University.

> In 1961 Anderson conceived a theoretical model to describe what happens when an impurity atom is present in a metal—now widely known and used as the *Anderson model.* Also named for him is the phenomenon of *Anderson localization,* describing the migration of impurities within a crystal. In the 1960s Anderson concentrated particularly on superconductivity and superfluidity.*

As a Visiting Professor at Cambridge University in England, Anderson worked with one of his former professors, John H. Van Vleck, and with Nevill Francis Mott. He shared with them the Nobel Prize for physics in 1977 for his work in the electronic structure of magnetic and disordered systems. His theoretical work is, in part, an extension of Bardeen's pioneering discoveries in the field of superconductivity.

Instead of hearing Anderson on his specialty, the reader has the opportunity, in the following selection, to appreciate his insights into the

**A Biographical Encyclopedia of Scientists.* New York: Facts On File, Inc., 1981, p. 18.

broader aspects of theoretical and experimental science. Although Anderson criticizes the tendency toward an increasingly narrow and specialized focus in physics, particularly elementary particle physics, his aim is not to attack the dedicated specialists in particle physics, even though he does occasionally do so with admirable wit. Instead he concentrates on the concepts of symmetry and asymmetry in their old-fashioned, not subatomic, connotations.

By his very success in inventing labor-saving devices, modern man has manufactured an abyss of boredom that only the privileged classes in earlier civilizations have ever fathomed.

—LEWIS MUMFORD, "The Challenge to Renewal," *The Conduct of Life*

One has to look out for engineers—they begin with sewing machines and end up with the atomic bomb.

—MARCEL PAGNOL, *Critique des critiques*

Where there is the necessary technical skill to move mountains, there is no need for the faith that moves mountains.

—ERIC HOFFER, *The Passionate State of Mind*

THE REDUCTIONIST HYPOTHESIS may still be a topic for controversy among philosophers, but among the great majority of active scientists I think it is accepted without question. The workings of our minds and bodies, and of all the animate or inanimate matter of which we have any detailed knowledge, are assumed to be controlled by the same set of fundamental laws, which except under certain extreme conditions we feel we know pretty well.

It seems inevitable to go on uncritically to what appears at first sight to be an obvious corollary of reductionism: that if everything obeys the same fundamental laws, then the only scientists who are studying anything really fundamental are those who are working on those laws. In practice, that amounts to some astrophysicists, some elementary particle physicists, some logicians and other mathematicians, and few others. This point of view, which it is the main purpose of this article to oppose, is expressed in a rather well-known passage by Weisskopf:

> Looking at the development of science in the Twentieth Century one can distinguish two trends, which I will call "intensive" and "extensive" research, lacking a better terminology. In short: intensive research goes for the fundamental laws, extensive research goes for the explanation of phenomena in terms of known fundamental

---

P. W. Anderson, "More Is Different," *Science,* vol. 177, August 4, 1972, pp. 393–396. Reprinted with permission of the American Association for the Advancement of Science.

> laws. As always, distinctions of this kind are not unambiguous, but they are clear in most cases. Solid state physics, plasma physics, and perhaps also biology are extensive. High energy physics and a good part of nuclear physics are intensive. There is always much less intensive research going on than extensive. Once new fundamental laws are discovered, a large and ever increasing activity begins in order to apply the discoveries to hitherto unexplained phenomena. Thus, there are two dimensions to basic research. The frontier of science extends all along a long line from the newest and most modern intensive research, over the extensive research recently spawned by the intensive research of yesterday, to the broad and well developed web of extensive research activities based on intensive research of past decades.

The effectiveness of this message may be indicated by the fact that I heard it quoted recently by a leader in the field of materials science, who urged the participants at a meeting dedicated to "fundamental problems in condensed matter physics" to accept that there were few or no such problems and that nothing was left but extensive science, which he seemed to equate with device engineering.

The main fallacy in this kind of thinking is that the reductionist hypothesis does not by any means imply a "constructionist" one: The ability to reduce everything to simple fundamental laws does not imply the ability to start from those laws and reconstruct the universe. In fact, the more the elementary particle physicists tell us about the nature of the fundamental laws, the less relevance they seem to have to the very real problems of the rest of science, much less to those of society.

The constructionist hypothesis breaks down when confronted with the twin difficulties of scale and complexity. The behavior of large and complex aggregates of elementary particles, it turns out, is not to be understood in terms of a simple extrapolation of the properties of a few particles. Instead, at each level of complexity entirely new properties appear, and the understanding of the new behaviors requires research which I think is as fundamental in its nature as any other. That is, it seems to me that one may array the sciences roughly linearly in a hierarchy, according to the idea: The elementary entities of science X obey the laws of science Y.

X	Y
solid state or many-body physics	elementary particle physics
chemistry	many-body physics

molecular biology	chemistry
cell biology	molecular biology
•	•
•	•
•	•
psychology	physiology
social sciences	psychology

But this hierarchy does not imply that science X is "just applied Y." At each stage entirely new laws, concepts, and generalizations are necessary, requiring inspiration and creativity to just as great a degree as in the previous one. Psychology is not applied biology, nor is biology applied chemistry.

In my own field of many-body physics, we are, perhaps, closer to our fundamental, intensive underpinnings than in any other science in which nontrivial complexities occur, and as a result we have begun to formulate a general theory of just how this shift from quantitative to qualitative differentiation takes place. This formulation, called the theory of "broken symmetry," may be of help in making more generally clear the breakdown of the constructionist converse of reductionism. I will give an elementary and incomplete explanation of these ideas, and then go on to some more general speculative comments about analogies at other levels and about similar phenomena.

Before beginning this I wish to sort out two possible sources of misunderstanding. First, when I speak of scale change causing fundamental change I do not mean the rather well-understood idea that phenomena at a new scale may obey actually different fundamental laws—as, for example, general relativity is required on the cosmological scale and quantum mechanics on the atomic. I think it will be accepted that all ordinary matter obeys simple electrodynamics and quantum theory, and that really covers most of what I shall discuss. (As I said, we must all start with reductionism, which I fully accept.) A second source of confusion may be the fact that the concept of broken symmetry has been borrowed by the elementary particle physicists, but their use of the term is strictly an analogy, whether a deep or a specious one remaining to be understood.

Let me then start my discussion with an example on the simplest possible level, a natural one for me because I worked with it when I was a graduate student: the ammonia molecule. At that time everyone knew about ammonia and used it to calibrate his theory or his apparatus, and I was no exception. The chemists will tell you that ammonia "is" a triangular pyramid of one nitrogen and three hydrogen atoms, with the nitrogen negatively charged and the hydrogens positively charged, so that it has an electric dipole moment (μ), negative toward the nitrogen of the pyramid. Now this seemed very strange to me, because I was just being taught that nothing has an electric dipole moment. The professor was really proving that no nucleus has a dipole moment, because he was teaching nuclear physics, but as his arguments were based on the symmetry of space and time they should have been correct in general.

I soon learned that, in fact, they were correct (or perhaps it would be more accurate to say not incorrect) because he had been careful to say that no stationary state of a system (that is, one which does not change in time) has an electric dipole moment. If ammonia starts out from the above unsymmetrical state, it will not stay in it very long. By means of quantum mechanical tunneling, the nitrogen can leak through the triangle of hydrogens to the other side, turning the pyramid inside out, and, in fact, it can do so very rapidly. This is the so-called "inversion," which occurs at a frequency of about 3×10^{10} per second. A truly stationary state can only be an equal superposition of the unsymmetrical pyramid and its inverse. That mixture does not have a dipole moment. (I warn the reader again that I am greatly oversimplifying and refer him to the textbooks for details.)

I will not go through the proof, but the result is that the state of the system, if it is to be stationary, must always have the same symmetry as the laws of motion which govern it. A reason may be put very simply: In quantum mechanics there is always a way, unless symmetry forbids, to get from one state to another. Thus, if we start from any one unsymmetrical state, the system will make transitions to others, so only by adding up all the possible unsymmetrical states in a symmetrical way can we get a stationary state. The symmetry involved in the case of ammonia is parity, the equivalence of left- and right-handed ways of looking at things. (The elementary particle experimentalists' discovery of certain violations of parity is not relevant to this question; those effects are too weak to affect ordinary matter.)

Having seen how the ammonia molecule satisfies our theorem that there is no dipole moment, we may look into other cases and, in particular, study progressively bigger systems to see whether the state and the symmetry are always related. There are other similar pyramidal molecules, made of heavier atoms. Hydrogen phosphide, PH_3, which is twice as heavy as ammonia, inverts, but at one-tenth the ammonia frequency. Phosphorus trifluoride, PF_3, in which the much heavier fluorine is substituted for hydrogen, is not observed to invert at a measurable rate, although theoretically one can be sure that a state prepared in one orientation would invert in a reasonable time.

We may then go on to more complicated molecules, such as sugar, with about 40 atoms. For these it no longer makes any sense to expect the molecule to invert itself. Every sugar molecule made by a living organism is spiral in the same sense, and they never invert, either by quantum mechanical tunneling or even under thermal agitation at normal temperatures. At this point we must forget about the possibility of inversion and ignore the parity symmetry: the symmetry laws have been, not repealed, but broken.

If, on the other hand, we synthesize our sugar molecules by a chemical reaction more or less in thermal equilibrium, we will find that there are not, on the average, more left- than right-handed ones or vice versa. In the absence of anything more complicated than a collection of free molecules, the symmetry laws are never broken, on the average. We needed living matter to produce an actual unsymmetry in the populations.

In really large, but still inanimate, aggregates of atoms, quite a different kind of broken symmetry can occur, again leading to a net dipole moment or to a net optical rotating power, or both. Many crystals have a net dipole moment in each elementary unit cell (pyroelectricity), and in some this moment can be reversed by an electric field (ferroelectricity). This asymmetry is a spontaneous effect of the crystal's seeking its lowest energy state. Of course, the state with the opposite moment also exists and has, by symmetry, just the same energy, but the system is so large that no thermal or quantum mechanical force can cause a conversion of one to the other in a finite time compared to, say, the age of the universe.

There are at least three inferences to be drawn from this. One is that symmetry is of great importance in physics. By symmetry we mean the existence of different viewpoints from which the system appears the same. It is only slightly overstating the case to say that physics is the study of symmetry. The first demonstration of the power of this idea may have been by Newton, who may have asked himself the question: What if the matter here in my hand obeys the same laws as that up in the sky—that is, what if space and matter are homogeneous and isotropic?

The second inference is that the internal structure of a piece of matter need not be symmetrical even if the total state of it is. I would challenge you to start from the fundamental laws of quantum mechanics and predict the ammonia inversion and its easily observable properties without going through the stage of using the unsymmetrical pyramidal structure, even though no "state" ever has that structure. It is fascinating that it was not until a couple of decades ago that nuclear physicists stopped thinking of the nucleus as a featureless, symmetrical little ball and realized that while it really never has a dipole moment, it can become football-shaped or plate-shaped. This has observable consequences in the reactions and excitation spectra that are studied in nuclear physics, even though it is much more difficult to demonstrate directly than the ammonia inversion. In my opinion, whether or not one calls this intensive research, it is as fundamental in nature as many things one might so label. But it needed no new knowledge of fundamental laws and would have been extremely difficult to derive synthetically from those laws; it was simply an inspiration, based, to be sure, on everyday intuition, which suddenly fitted everything together.

The basic reason why this result would have been difficult to derive is an important one for our further thinking. If the nucleus is sufficiently small there is no real way to define its shape rigorously: Three or four or ten particles whirling about each other do not define a rotating "plate" or "football." It is only as the nucleus is considered to be a many-body system—in what is often called the $N \to \infty$ limit—that such behavior is rigorously definable. We say to ourselves: A macroscopic body of that shape would have such-and-such a spectrum of rotational and vibrational excitations, completely different in nature from those which would characterize a featureless system. When we see such a spectrum, even not so separated, and somewhat imperfect, we recognize that the nucleus is, after all, not macroscopic; it is merely approaching macroscopic behavior. Starting with the fundamental laws and a computer, we would have to do two impossible things—solve a problem with infinitely many bodies, and then apply the result to a finite system—before we synthesized this behavior.

A third insight is that the state of a really big system does not at all have to have the symmetry of the laws which govern it; in fact, it usually has less symmetry. The outstanding example of this is the crystal: Built from a substrate of atoms and space according to laws which express the perfect homogeneity of space, the crystal suddenly and unpredictably displays an entirely new and very beautiful symmetry. The general rule, however, even in the case of the crystal, is that the large system is less symmetrical than the underlying structure would suggest: Symmetrical as it is, a crystal is less symmetrical than perfect homogeneity.

Perhaps in the case of crystals this appears to be merely an exercise in confusion. The regularity of crystals could be deduced semiempirically in the mid-19th century without any complicated reasoning at all. But sometimes, as in the case of superconductivity, the new symmetry—now called broken symmetry because the original symmetry is no longer evident—may be of an entirely unexpected kind and extremely difficult to visualize. In the case of superconductivity, 30 years elapsed between the time when physicists were in possession of every fundamental law necessary for explaining it and the time when it was actually done.

The phenomenon of superconductivity is the most spectacular example of the broken symmetries which ordinary macroscopic bodies undergo, but it is of course not the only one. Antiferromagnets, ferroelectrics, liquid crystals, and matter in many other states obey a certain rather general scheme of rules and ideas, which some many-body theorists refer to under the general heading of broken symmetry. . . .

The essential idea is that in the so-called $N \rightarrow \infty$ limit of large systems (on our own, macroscopic scale) it is not only convenient but essential to realize that matter will undergo mathematically sharp, singular "phase transitions" to states in which the microscopic symmetries, and even the microscopic equations of motion, are in a sense violated. The symmetry leaves behind as its expression only certain characteristic behaviors, for instance, long-wavelength vibrations, of which the familiar example is sound waves; or the unusual macroscopic conduction phenomena of the superconductor; or, in a very deep analogy, the very rigidity of crystal lattices, and thus of most solid matter. There is, of course, no question of the system's really violating, as opposed to breaking, the symmetry of space and time, but because its parts find it energetically more favorable to maintain certain fixed relationships with each other, the symmetry allows only the body as a whole to respond to external forces.

This leads to a "rigidity," which is also an apt description of superconductivity and superfluidity in spite of their apparent "fluid" behavior. [In the former case, London noted this aspect very early.] Actually, for a hypothetical gaseous but intelligent citizen of Jupiter or of a hydrogen cloud somewhere in the galactic center, the properties of ordinary crystals might well be a more baffling and intriguing puzzle than those of superfluid helium.

I do not mean to give the impression that all is settled. For instance, I think there are still fascinating questions of principle about glasses and other amorphous phases, which may reveal even more complex types of behavior. Nevertheless, the role of this type of broken symmetry in the properties of inert but macroscopic material bodies is now understood, at least in principle. In this case we can see how the whole becomes not only more than but very different from the sum of its parts.

The next order of business logically is to ask whether an even more complete destruction of the fundamental symmetries of space and time is possible and whether new phenomena then arise, intrinsically different from the "simple" phase transition representing a condensation into a less symmetric state.

We have already excluded the apparently unsymmetric cases of liquids, gases, and glasses. (In any real sense they are more symmetric.) It seems to me that the next stage is to consider the system which is regular but contains information. That is, it is regular in space in some sense so that it can be "read out," but it contains elements which can be varied from one "cell" to the next. An obvious example is DNA; in everyday life, a line of type or a movie film has the same structure. This type of "information-bearing crystallinity" seems to be essential to life. Whether the development of life requires any further breaking of symmetry is by no means clear.

Keeping on with the attempt to characterize types of broken symmetry which occur in living things, I find that at least one further phenomenon seems to be identifiable and either universal or remarkably common, namely, ordering (regularity or periodicity) in the time dimension. A number of theories of life processes have appeared in which regular pulsing in time plays an important role: theories of development, of growth and growth limitation, and of the memory. Temporal regularity is very commonly observed in living objects. It plays at least two kinds of roles. First, most methods of extracting energy from the environment in order to set up a continuing, quasi-stable process involve time-periodic machines, such as oscillators and generators, and the processes of life work in the same way. Second, temporal regularity is a means of handling information, similar to information-bearing spatial regularity. Human spoken language is an example, and it is noteworthy that all computing machines use temporal pulsing. A possible third role is suggested in some of the theories mentioned above: the use of phase relationships of temporal pulses to handle information and control the growth and development of cells and organisms.

In some sense, structure—functional structure in a teleological sense, as opposed to mere crystalline shape—must also be considered a stage, possibly intermediate between crystallinity and information strings, in the hierarchy of broken symmetries.

To pile speculation on speculation, I would say that the next stage could be hierarchy or specialization of function, or both. At some point we have to stop talking about decreasing symmetry and start calling it increasing complication. Thus, with increasing complication at each stage, we go on up the hierarchy of the sciences. We expect to encounter fascinating and, I believe, very fundamental questions at each stage in fitting together less complicated pieces into the more complicated system and understanding the basically new types of behavior which can result.

There may well be no useful parallel to be drawn between the way in which complexity appears in the simplest cases of many-body theory and chemistry and the way it appears in the truly complex cultural and biological ones, except perhaps to say that, in general, the relationship between the system and its parts is intellectually a one-way street. Synthesis is expected to be all but impossible; analysis, on the other hand, may be not only possible but fruitful in all kinds of ways: Without an understanding of the broken symmetry in superconductivity, for instance, Josephson would probably not have discovered his effect. [Another name for the Josephson effect is "macroscopic quantum-interference phenomena": interference effects observed between macroscopic wave functions of electrons in superconductors, or of helium atoms in superfluid liquid helium. These phenomena have already enormously extended the accuracy of electromagnetic measurements, and can be expected to play a great role in future computers, among other possibilities, so that in the long run they may lead to some of the major technological achievements of this decade.] For another example, biology has certainly taken on a whole new aspect from the reduction of genetics to biochemistry and biophysics, which will have untold consequences. So it is not true, as a recent article would have it, that we each should "cultivate our own valley, and not attempt to build roads over the mountain ranges . . . between the sciences." Rather, we should recognize that such roads, while often the quickest shortcut to another part of our own science, are not visible from the viewpoint of one science alone.

The arrogance of the particle physicist and his intensive research may be behind us (the discoverer of the positron said "the rest is chemistry"), but we have yet to recover from that of some molecular biologists, who seem determined to try to reduce everything about the human organism to "only" chemistry, from the common cold and all mental disease to the religious instinct. Surely there are more levels of organization between human ethology and DNA than there are between DNA and quantum electrodynamics, and each level can require a whole new conceptual structure.

In closing, I offer two examples from economics of what I hope to have said. Marx said that quantitative differences become qualitative ones, but a dialogue in Paris in the 1920's sums it up even more clearly:

FITZGERALD: The rich are different from us.

HEMINGWAY: Yes, they have more money.

Q

Life

INTRODUCTION

AT FIRST GLANCE, the combination of themes in this section might seem contradictory, covering the work of such pioneers as von Neumann and Wiener in the field of cybernetics (which, since the early 1940s, has increasingly become the science of "artificial intelligence") and the new science known as molecular biology, which was given such professional and public impetus by the "cracking of the genetic code" by Watson, Crick, and Wilkins. (Sadly, their splendid and tragic colleague, Rosalind Franklin, who died in 1958 at the age of thirty-eight, has seldom received full credit for her vital role in the unraveling of the DNA puzzle.)

A vital bridge from the inorganic world of physics to biology was supplied by Erwin Schrödinger's little book, *What Is Life?*, published in 1944 and eagerly devoured by an entire generation of researchers who eventually came to be known as "molecular biologists." Within the life of a single generation of scientists, the sharp dividing line that once separated life and nonlife in the sciences had disappeared. Shortly after the war, a number of physicists and chemists—Delbrück, Szilard, Pauling, Wilkins—moved into biology, and their ranks were reinforced by the arrival of crystallographers, radiologists, and other technicians. Another group, centered around the brilliant French biologist-philosopher Jacques Monod at the Institut Pasteur, joined forces with California Institute of Technology specialists and with a group that had begun to organize at the Cavendish Laboratories in Cambridge, England.

Horace Freeland Judson, a one-time European correspondent for *Time* and the author of *The Eighth Day of Creation,* an exhaustive, sometimes overly detailed running history of the development of molecular biology and a remarkable account of the Watson-Crick discovery of the structure of DNA, tells us:

> John Kendrew . . . [of Cambridge University, founder of the *Journal of Molecular Biology*] proposed first in 1965, that molecular biology was the confluence of two currents, which he labeled "information and conformation." Kendrew's play with words, a memorable one, introduces an unfortunate stroke of hindsight. Information theory was a program of the 1940s and after, by which one would expect mathematicians and physicists to have influenced molecular biology; yet, in point of fact, it was almost entirely ineffectual. Except, perhaps, for Szilard's intervention in the development of the repressor model for control of enzyme biosynthesis—an intervention, assuredly dramatic, whose practical importance is easy to exaggerate—information theory has been useful to molecular biology mostly as a way to talk about discoveries and models that were reached without appeal to its principles. Kendrew used "information" in this retrospective way. He meant by it, first, the moiety of molecular biology rooted in genetics, particularly of microorganisms and including the genetical elucidation of biochemical pathways and controls, as by Jacques Monod and François Jacob [Monod's collaborator]; second, the moiety springing from studies of the structures of the large biological molecules, as by Pauling and by the British crystallographers, culminating in the elucidations of structure to function in DNA and proteins. These currents in the early years of molecular biology can be more accurately labelled the one-dimensional (thus the genetic and sequential) and the three-dimensional (thus the stereochemical or structural). Put that way the distinction moves beyond the limits of the standard view well towards the crux of the matter. . . .*

---

*Horace Freeland Judson, *The Eighth Day of Creation.* New York: Simon & Schuster, 1979, pp. 606–607.

Judson admits that the abandonment of the protein hypothesis in identifying the nature of the gene, in favor of the realization that the components of the gene are nucleic acids, is an important turning point in the development of molecular biology. But he presents a far more complex picture of the crucial developments, which he ascribes to the union of five distinct disciplines: (1) genetics and (2) x-ray crystallography, combined by Watson and Crick in their research on the structure of DNA; (3) physical chemistry, notably advanced by the work of Linus Pauling; (4) microbiology, which was undergoing a major transformation in the 1940s; and (5) biochemistry, a discipline that was also being revolutionized in those years.

Our selections include John von Neumann and Norbert Wiener on cybernetics; Max Delbrück's review of two vital decades in biology; an extract from Erwin Schrödinger's seminal book; contributions from Maurice Wilkins and Francis Crick, who shared the Nobel Prize with James Watson for the discovery of the double helix; Eugene Wigner's philosophical musings on the relations of physics and biology; and Ilya Prigogine's whirlwind survey of culture and life.

Self-Reproducing Automata

JOHN VON NEUMANN

JOHN VON NEUMANN was born in Budapest in 1903 and died in Washington in 1957 at the height of his career. Although his father was a Hungarian banker, von Neumann left his troubled homeland as a youth to study at various universities in Switzerland and Austria. While at the University of Göttingen he befriended the American physicist J. Robert Oppenheimer. Von Neumann earned his doctorate in mathematics at the University of Hamburg in 1926 and emigrated to the United States in 1930, where he later became a naturalized citizen. He taught mathematical physics at Princeton during the 1930s and, when the Institute for Advanced Study was founded at Princeton in 1933, he joined its staff. He remained at the Institute throughout his professional life.

Von Neumann's bibliography covers nearly all branches of modern mathematics. In quantum mechanics, he was one of the first to demonstrate that the wave mechanics of Schrödinger and the matrix mechanics of Heisenberg were different mathematical formulations of the same principles. His most famous work in theoretical physics, *Mathematical Physics,* contained his axiomizations of quantum mechanics. These axiomizations have survived subsequent innovations in quantum mechanics and remain vital to understanding nonrelativistic quantum theory. A later work published in 1932, *Mathematische Grundlagen der Quantenmechanik,* concerned itself with specific problems of quantum mechanics including the problem of measurement of subatomic systems, quantum statistics, and quantum thermodynamics. Von Neumann also discussed the current debate over "causality" and "indeterminancy," concluding that quantum mechanics could not be structured as a strictly causal system. He also collaborated with Subrahmanyan Chandrasekhar in producing papers about the fluctuations of gravitational fields generated by stars and wrote extensively on the physics of shock waves.

His name is perhaps most closely linked with the development of game theory, a new field that he virtually created with his book *The Theory of Games and Economic Behavior,* published in 1944. In his book, von Neumann advanced a proof applicable to all games of strategy, which states that there is at least one optimum line of play that over the long run minimizes the losses of a player. His accomplishments are all the more impressive because he gave a quantitative mathematical model for games of chance. His theory was widely accepted and soon found its way into the mainstream of the social sciences, war strategy, and economics.

With the development of cybernetics in the 1940s, von Neumann lent his talents to developing the computer technology necessary for the production of the hydrogen bomb. He introduced the idea of arranging the wires in such a way that a great variety of problems could be solved using a series of electronic "codes." His supervision of the construction of a computer at the Institute for Advanced Study guaranteed that his theories would be translated into practical applications; the designs of even the most powerful modern computers still reflect the basic concepts outlined by von Neumann.

He also received a great deal of unwanted publicity during the heyday of Senator Joseph McCarthy in the first Eisenhower administration. His old friend Oppenheimer had opposed the development of the hydrogen bomb and became the center of a great controversy in 1954. In Congressional hearings, von Neumann testified to the patriotism of Oppenheimer when Edward Teller, another Hungarian-American scientist popularly known as the "father of the H-bomb," testified against Oppenheimer. A year before his untimely death, von Neumann was honored by his scientific colleagues with the distinguished Fermi Award.

In our selection, von Neumann discusses the relationship of logic systems to animal neural networks, taking as his point of departure the theorems developed by McCulloch and Pitts, who had concluded that the functions of what they term "formal neural networks," if definable logically and unambiguously in a finite number of words, can also be realized by a formal neural network. von Neumann also cites the theories of Alan Mathison Turing, the brilliant British cyberneticist whose career was cut short by his tragic suicide, and concludes that automata capable of producing or designing other, better, automata will have to be taken into consideration in future theories.

We ought then to regard the present state of the universe as the effect of its anterior state and as the cause of the one which is to follow. Given for one instance an intelligence which could comprehend all the forces of nature and the respective situation of the things that compose it . . . for it nothing would be uncertain and the future, as the past, would be present to its eyes.

—PIERRE SIMON DE LAPLACE, *Philosophical Essay on Probabilities*

When the mind is disturbed, the multiplicity of things is produced, but when the mind is quieted, the multiplicity of things disappears.

—ASHVAGHOSHA, *The Awakening of Faith*

FORMAL NEURAL NETWORKS

I SHALL FIRST discuss here the remarkable theorems of McCulloch and Pitts on the relationship of logics and neural networks.

In this discussion I shall . . . take the strictly axiomatic point of view. I shall, therefore, view a neuron as a "black box" with a certain number of inputs that receive stimuli and an output that emits stimuli. To be specific, I shall assume that the input connections of each one of these can be of two types, excitatory and inhibitory. The boxes themselves are also of two types, threshold 1 and threshold 2. These concepts are linked and circumscribed by the following definitions. In order to stimulate such an organ it is necessary that it should receive simultaneously at least as many stimuli on its excitatory inputs as correspond to its threshold, and not a single stimulus on any one of its inhibitory inputs. If it has been thus stimulated, it will after a definite time delay (which is assumed to be always the same, and may be used to define the unit of time) emit an output pulse. This pulse can be taken by appropriate connections to any number of inputs of other neurons (also to any of its own inputs) and will produce at each of these the same type of input stimulus as the ones described above.

John von Neumann, "The General and Logical Theory of Automata." Presented at the Hixon Symposium on September 20, 1948, at the California Institute of Technology. Reprinted with permission of the California Institute of Technology.

It is, of course, understood that this is an oversimplification of the actual functioning of a neuron. . . . the character, the limitations, and the advantages of the axiomatic method . . . all apply here, and the discussion which follows is to be taken in this sense.

McCulloch and Pitts have used these units to build up complicated networks which may be called "formal neural networks." Such a system is built up of any number of these units, with their inputs and outputs suitably interconnected with arbitrary complexity. The "functioning" of such a network may be defined by singling out some of the inputs of the entire system and some of its outputs, and then describing what original stimuli on the former are to cause what ultimate stimuli on the latter.

The Main Result of the McCulloch-Pitts Theory. McCulloch and Pitts' important result is that any functioning in this sense which can be defined at all logically, strictly, and unambiguously in a finite number of words can also be realized by such a formal neural network.

It is well to pause at this point and to consider what the implications are. It has often been claimed that the activities and functions of the human nervous system are so complicated that no ordinary mechanism could possibly perform them. It has been attempted to name specific functions which by their nature exhibit this limitation. It has been attempted to show that such specific functions, logically, completely described, are per se unable of mechanical, neural realization. The McCulloch-Pitts result puts an end to this. It proves that anything that can be exhaustively and unambiguously described, anything that can be completely and unambiguously put into words, is ipso facto realizable by a suitable finite neural network. Since the converse statement is obvious, we can therefore say that there is no difference between the possibility of describing a real or imagined mode of behavior completely and unambiguously in words, and the possibility of realizing it by a finite formal neural network. The two concepts are co-extensive. A difficulty of principle embodying any mode of behavior in such a network can exist only if we are also unable to describe that behavior completely.

Thus the remaining problems are these two. First, if a certain mode of behavior can be effected by a finite neural network, the question still remains whether that network can be realized within a practical size, specifically, whether it will fit into the physical limitations of the organism in question. Second, the question arises whether every existing mode of behavior can really be put completely and unambiguously into words.

The first problem is, of course, the ultimate problem of nerve physiology, and I shall not attempt to go into it any further here. The second question is of a different character, and it has interesting logical connotations.

Interpretations of This Result. There is no doubt that any special phase of any conceivable form of behavior can be described "completely and unambiguously" in words. This description may be lengthy, but it is always possible. To deny it would amount to adhering to a form of logical mysticism which is surely far from most of us. It is, however, an important limitation, that this applies only to every element separately, and it is far from clear how it will apply to the entire syndrome of behavior. To be more specific, there is no difficulty in describing how an organism might be able to identify any two rectilinear triangles, which appear on the retina, as belonging to the same category "triangle." There is also no difficulty in adding to this, that numerous other objects, besides regularly drawn rectilinear triangles, will also be classified and identified as triangles—triangles whose sides are curved, triangles whose sides are not fully drawn, triangles that are indicated merely by a more or less homogeneous shading of their interior, etc. The more completely we attempt to describe everything that may conceivably fall under this heading, the longer the description becomes. We may have a vague and uncomfortable feeling that a complete catalogue along such lines would not only be exceedingly long, but also unavoidably indefinite at its boundaries. Nevertheless, this may be a possible operation.

All of this, however, constitutes only a small fragment of the more general concept of identification of analogous geometrical entities. This, in turn, is only a microscopic piece of the general concept of analogy. Nobody would attempt to describe and define within any practical amount of space the general concept of analogy which dominates our interpretation of vision. There is no basis for saying whether such an enterprise would require thousands or millions or altogether impractical numbers of volumes. Now it is perfectly possible that the simplest and only practical way actually to say what constitutes a visual analogy consists in giving a description of the connections of the visual brain. We are dealing here with parts of logics with which we have practically no past experience. The order of complexity is out of all proportion to anything we have ever known. We have no right to assume that the logical notations and procedures used in the past are suited to this part of the subject. It is not at all certain that in this domain a real object might not constitute the simplest description of itself, that is, any attempt to describe it by the usual literary or formal-logical method may lead to something less manageable and more involved. In fact, some results in modern logic would tend to indicate that phenomena like this have to be expected when we come to really complicated entities. It is, therefore, not at all unlikely that it is futile to look for a precise logical concept, that is, for a precise verbal description, of "visual analogy." It is possible that the connection pattern of the visual brain itself is the simplest logical expression or definition of this principle.

Obviously, there is on this level no more profit in the McCulloch-Pitts result. At this point it only furnishes another illustration of the situation outlined earlier. There is an equivalence between logical principles and their embodiment in a neural network, and while in the simpler cases the principles might furnish a simplified expression of the network, it is quite possible that in cases of extreme complexity the reverse is true.

All of this does not alter my belief that a new, essentially logical, theory is called for in order to understand high-complication automata and, in particular, the central nervous system. It may be, however, that in this process logic will have to undergo a pseudomorphosis to neurology to a much greater extent than the reverse. The foregoing analysis shows that one of the relevant things we can do at this moment with respect to the theory of the central nervous system is to point out the directions in which the real problem does not lie.

THE CONCEPT OF COMPLICATION, SELF-REPRODUCTION

The Concept of Complication. The discussions so far have shown that high complexity plays an important role in any theoretical effort relating to automata, and that this concept, in spite of its prima facie quantitative character, may in fact stand for something qualitative—for a matter of principle. For the remainder of my discussion I will consider a remoter implication of this concept, one which makes one of the qualitative aspects of its nature even more explicit.

There is a very obvious trait, of the "vicious circle" type, in nature, the simplest expression of which is the fact that very complicated organisms can reproduce themselves.

We are all inclined to suspect in a vague way the existence of a concept of "complication." This concept and its putative properties have never been clearly formulated. We are, however, always tempted to assume that they will work in this way. When an automaton performs certain operations, they must be expected to be of a lower degree of complication than the automaton itself. In particular, if an automaton has the ability to construct another one, there must be a decrease in complication as we go from the parent to the construct. That is, if *A* can produce *B*, then *A* in some way must have contained a complete description of *B*. In order to make it effective, there must be, furthermore, various arrangements in *A* that see to it that this description is interpreted and that the constructive operations that it calls for are carried out. In this sense, it would therefore seem that a certain degenerating tendency must be expected, some decrease in complexity as one automaton makes another automaton.

Although this has some indefinite plausibility to it, it is in clear contradiction with the most obvious things that go on in nature. Organisms reproduce themselves, that is, they produce new organisms with no decrease in complexity. In addition, there are long periods of evolution during which the complexity is even increasing. Organisms are indirectly derived from others which had lower complexity.

Thus there exists an apparent conflict of plausibility and evidence, if nothing worse. In view of this, it seems worth while to try to see whether there is anything involved here which can be formulated rigorously.

So far I have been rather vague and confusing, and not unintentionally at that. It seems to me that it is otherwise impossible to give a fair impression of the situation that exists here. Let me now try to become specific.

Turing's Theory of Computing Automata. The English logician, Turing, about twelve years ago attacked the following problem.

He wanted to give a general definition of what is meant by a computing automaton. The formal definition came out as follows:

An automaton is a "black box," which will not be described in detail but is expected to have the following attributes. It possesses a finite number of states, which need be prima facie characterized only by stating their number, say *n*, and by enumerating them accordingly: 1, 2, . . ., *n*. The essential operating characteristic of the automaton consists of describing how it is caused to change its state, that is, to go over from a state *i* into a state *j*. This change requires some interaction with the outside world, which will be standardized in the following manner. As far as the machine is concerned, let the whole outside world consist of a long paper tape. Let this tape be, say, 1 inch wide, and let it be subdivided into fields (squares) 1 inch long. On each field of this strip we may or may not put a sign, say, a dot, and it is assumed that it is possible to erase as well as to write in such a dot. A field marked with a dot will be called a "1," a field unmarked with a dot will be called a "0." (We might permit more ways of marking, but Turing showed that this is irrelevant and does not lead to any essential gain in generality.) In describing the position of the tape relative to the automaton it is assumed that one particular field of the tape is under direct inspection by the automaton, and that the automaton has the ability to move the tape forward and backward, say, by one field at a time. In specifying this, let the automaton be in the state i (=1 . . ., n), and let it see on the tape an e (=0, 1). It will then go over into the state j (=0, 1, . . ., n), move the tape by p fields (p=0, +1, −1; +1 is a move forward, −1 is a move backward), and inscribe into the new field that it sees f (=0, 1; inscribing 0 means erasing; inscribing 1 means putting in a dot). Specifying j, p, f as functions of i, e is then the complete definition of the functioning of such an automaton.

Turing carried out a careful analysis of what mathematical processes can be effected by automata of this type. In this connection he proved various theorems concerning the classical "decision problem" of logic, but I shall not go into these matters here. He did, however, also introduce and analyze the concept of a "universal automaton," and this is part of the subject that is relevant in the present context.

An infinite sequence of digits e (=0, 1) is one of the basic entities in mathematics. Viewed as a binary expansion, it is essentially equivalent to the concept of a real number. Turing, therefore, based his consideration on these sequences.

He investigated the question as to which automata were able to construct which sequences. That is, given a definite law for the formation of such a sequence, he inquired as to which automata can be used to form the sequence based on that law. The process of "forming" a sequence is interpreted in this manner. An automaton is able to "form" a certain sequence if it is possible to specify a finite length of tape, appropriately marked, so that, if this tape is fed to the automaton in question, the automaton will thereupon write the sequence on the remaining (infinite) free portion of the tape. This process of writing the infinite sequence is, of course, an indefinitely continuing one. What is meant is that the automaton will keep running indefinitely and, given a sufficiently long time, will have inscribed any desired (but of course finite) part of the (infinite) sequence. The finite, premarked, piece of tape constitutes the "instruction" of the automaton for this problem.

An automaton is "universal" if any sequence that can be produced by any automaton at all can also be solved by this particular automaton. It will, of course, require in general a different instruction for this purpose.

The Main Result of the Turing Theory. We might expect a priori that this is impossible. How can there be an automaton which is at least as effective as any conceivable automaton, including, for example, one of twice its size and complexity?

Turing, nevertheless, proved that this is possible. While his construction is rather involved, the underlying principle is nevertheless quite simple. Turing observed that a completely general description of any conceivable automaton can be (in the sense of the foregoing definition) given in a finite number of words. This description will contain certain empty passages—those referring to the functions mentioned earlier (*j, p, f* in terms of *i, e*), which specify the actual functioning of the automaton. When these empty passages are filled in, we deal with a specific automaton. As long as they are left empty, this schema represents the general definition of the general automaton. Now it becomes possible to describe an automaton which has the ability to interpret such a definition. In other words, which, when fed the functions that in the sense described above define a specific automaton, will thereupon function like the object described. The ability to do this is no more mysterious than the ability to read a dictionary and a grammar and to follow their instructions about the uses and principles of combinations of words. This automaton, which is constructed to read a description and to imitate the object described, is then the universal automaton in the sense of Turing. To make it duplicate any operation that any other automaton can perform, it suffices to furnish it with a description of the automaton in question and, in addition, with the instructions which that device would have required for the operation under consideration.

Broadening of the Program to Deal with Automata That Produce Automata. For the question which concerns me here, that of "self-reproduction" of automata, Turing's procedure is too narrow in one respect only. His automata are purely computing machines. Their output is a piece of tape with zeros and ones on it. What is needed for the construction to which I referred is an automaton whose output is other automata. There is, however, no difficulty in principle in dealing with this broader concept and in deriving from it the equivalent of Turing's result.

The Basic Definitions. As in the previous instance, it is again of primary importance to give a rigorous definition of what constitutes an automaton for the purpose of the investigation. First of all, we have to draw up a complete list of the elementary parts to be used. This list must contain not only a complete enumeration but also a complete operational definition of each elementary part. It is relatively easy to draw up such a list, that is, to write a catalogue of "machine parts" which is sufficiently inclusive to permit the construction of the wide variety of mechanisms here required, and which has the axiomatic rigor that is needed for this kind of consideration. The list need not be very long either. It can, of course, be made either arbitrarily long or arbitrarily short. It may be lengthened by including in it, as elementary parts, things which could be achieved by combinations of others. It can be made short—in fact, it can be made to consist of a single unit—by endowing each elementary part with a multiplicity of attributes and functions. Any statement on the number of elementary parts required will therefore represent a common-sense compromise, in which nothing too complicated is expected from any one elementary part, and no elementary part is made to perform several, obviously separate, functions. In this sense, it can be shown that about a dozen elementary parts suffice. The problem of self-reproduction can then be stated like this: Can one build an aggregate out of such elements in such a manner that if it is put into a reservoir, in which there float all these elements in large numbers, it will then begin to construct other aggregates, each of which will at the end turn out to be another automaton exactly like the original one? This is feasible, and the principle on which it can be based is closely related to Turing's principle outlined earlier.

Outline of the Derivation of the Theorem Regarding Self-reproduction. First of all, it is possible to give a complete description of everything that is an automaton in the sense considered here. This description is to be conceived as a general one, that is, it will again contain empty spaces. These empty spaces have to be filled in with the functions which describe the actual structure of an automaton. As before, the difference between these spaces filled and unfilled is the difference between the description of a specific automaton and the general description of a general automaton. There is no difficulty of principle in describing the following automata.

(*a*) Automaton *A,* which when furnished the description of any other automaton in terms of appropriate functions, will construct that entity. The description should in this case not be given in the form of a marked tape, as in Turing's case, because we will not normally choose a tape as a structural element. It is quite easy, however, to describe combinations of structural elements which have all the notational properties of a tape with fields that can be marked. A description in this sense will be called an instruction and denoted by a letter *I.*

"Constructing" is to be understood in the same sense as before. The constructing automaton is supposed to be placed in a reservoir in which all elementary components in large numbers are floating, and it will effect its construction in that milieu. One need not worry about how a fixed automaton of this sort can produce others which are larger and more complex than itself. In this case the greater size and the higher complexity of the object to be constructed will be reflected in a presumably still greater size of the instructions I that have to be furnished. These instructions, as pointed out, will have to be aggregates of elementary parts. In this sense, certainly, an entity will enter the process whose size and complexity is determined by the size and complexity of the object to be constructed.

In what follows, all automata for whose construction the facility A will be used are going to share with A this property. All of them will have a place for an instruction I, that is, a place where such an instruction can be inserted. When such an automaton is being described (as, for example, by an appropriate instruction), the specification of the location for the insertion of an instruction I in the foregoing sense is understood to form a part of the description. We may, therefore, talk of "inserting a given instruction I into a given automaton," without any further explanation.

(*b*) Automaton B, which can make a copy of any instruction I that is furnished to it. I is an aggregate of elementary parts in the sense outlined in (*a*), replacing a tape.This facility will be used when I furnishes a description of another automaton. In other words, this automaton is nothing more subtle than a "reproducer"—the machine which can read a punched tape and produce a second punched tape that is identical with the first. Note that this automaton, too, can produce objects which are larger and more complicated than itself. Note again that there is nothing surprising about it. Since it can only copy, an object of the exact size and complexity of the output will have to be furnished to it as input.

After these preliminaries, we can proceed to the decisive step.

(*c*) Combine the automata A and B with each other, and with a control mechanism C which does the following. Let A be furnished with an instruction I (again in the sense of [*a*] and [*b*]).Then C will first cause A to construct the automaton which is described by this instruction I. Next C will cause B to copy the instruction I referred to above, and insert the copy into the automaton referred to above, which has just been constructed by A. Finally, C will separate this construction from the system $A + B + C$ and "turn it loose" as an independent entity.

(*d*) Denote the total aggregate $A + B + C$ by D.

(*e*) In order to function, the aggregate $D = A + B + C$ must be furnished with an instruction I, as described above. This instruction, as pointed out above, has to be inserted into A. Now form an instruction I_D, which describes this automaton D, and insert I_D into A within D. Call the aggregate which now results E.

E is clearly self-reproductive. Note that no vicious circle is involved. The decisive step occurs in E, when the instruction I_D, describing D, is constructed and attached to D. When the construction (the copying) of I_D

is called for, D exists already, and it is in no wise modified by the construction of I_D. I_D is simply added to form E. Thus there is a definite chronological and logical order in which D and I_D have to be formed, and the process is legitimate and proper according to the rules of logic.

Interpretations of This Result and of Its Immediate Extensions. The description of this automaton E has some further attractive sides, into which I shall not go at this time at any length. For instance, it is quite clear that the instruction I_D is roughly effecting the functions of a gene. It is also clear that the copying mechanism B performs the fundamental act of reproduction, the duplication of the genetic material, which is clearly the fundamental operation in the multiplication of living cells. It is also easy to see how arbitrary alterations of the system E, and in particular of I_D, can exhibit certain typical traits which appear in connection with mutation, lethally as a rule, but with a possibility of continuing reproduction with a modification of traits. It is, of course, equally clear at which point the analogy ceases to be valid. The natural gene probably does not contain a complete description of the object whose construction its presence stimulates. It probably contains only general pointers, general cues. In the generality in which the foregoing consideration is moving, this simplification is not attempted. It is, nevertheless, clear that this simplification, and others similar to it, are in themselves of great and qualitative importance. We are very far from any real understanding of the natural processes if we do not attempt to penetrate such simplifying principles.

Small variations of the foregoing scheme also permit us to construct automata which can reproduce themselves and, in addition, construct others. (Such an automaton performs more specifically what is probably a—if not the—typical gene function, self-reproduction plus production—or stimulation of production—of certain specific enzymes.) Indeed, it suffices to replace the I_D by an instruction I_{D+F}, which describes the automaton D plus another given automaton F. Let D, with I_{D+F} inserted into A within it, be designated by E_F. This E_F clearly has the property already described. It will reproduce itself, and, besides, construct F.

Note that a "mutation" of E_F, which takes place within the F-part of I_{D+F} in E_F, is not lethal. If it replaces F by F', it changes E_F into $E_{F'}$, that is, the "mutant" is still self-reproductive; but its by-product is changed—F' instead of F. This is, of course, the typical non-lethal mutant.

All these are very crude steps in the direction of a systematic theory of automata. They represent, in addition, only one particular direction. This is, as I indicated before, the direction towards forming a rigorous concept of what constitutes "complication." They illustrate that "complication" on its lower levels is probably degenerative, that is, that every automaton that can produce other automata will only be able to produce less complicated ones. There is, however, a certain minimum level where this degenerative characteristic ceases to be universal. At this point automata which can reproduce themselves, or even construct higher entities, become possible. This fact, that complication, as well as organization, below a certain minimum level is degenerative, and beyond that level can become self-supporting and even increasing, will clearly play an important role in any future theory of the subject.

Q.2

Computing Machines and the Nervous System

Norbert Wiener

NORBERT WIENER WAS the brilliant mathematician and logician who not only coined the word *cybernetics* but who was responsible for much of the theory of that field. Born in 1894 in Missouri, he had been a child prodigy who read at three, matriculated at Tufts University at eleven, and took his Ph.D. at Harvard at eighteen. His father, a scholar of literature and linguistics, was an immigrant from Russia, who pushed his son intellectually almost to the breaking point. After Wiener got his doctorate, he went to Europe, where he studied with Bertrand Russell at Cambridge and David Hilbert at Göttingen. He tried to join the army when America entered World War I, but was turned down because of poor eyesight.

After several failed attempts to secure an academic position and a brief stint as a journalist, Wiener obtained a post in the department of mathematics at the Massachusetts Institute of Technology. Wiener then formulated a mathematical theory of Brownian motion and made several discoveries that contributed to the study of harmonic analysis. He also made important contributions to operational calculus and developed analytical methods that assumed the variability of initial conditions and used continuous feedback principles to improve analytical conclusions.

During World War II, Wiener worked in the field of antiaircraft defense. The great number of independent parameters involved in aiming an antiaircraft cannon at high-speed bombers and fighters challenged his genius, and after the war he published his findings on the processing and communication of data in his celebrated book, *Cybernetics.* In it he argued that his theories were applicable to both man-made computers and to the human nervous system, a revolutionary doctrine at the time, and one based largely on Wiener's conviction that no essential difference exists between animate and inanimate entities. His book gives an extensive mathematical analysis of feedback theory and a statistical analysis of the flow of communication. It was also the first serious treatment of such ideas as computer-controlled assembly lines and robot workers. Wiener's bold predictions captured the public imagination and brought him enduring fame outside the professional community. However, his book was also criticized by some mathematicians as being unnecessarily embellished with sophisticated mathematical terms and formulas. But the breadth of its ideas tended to fascinate even its most vocal critics. By the time the book was published, Wiener had decided to renounce all forms of collaboration in military affairs. He spent the remaining years of his life prior to his death in 1964 trying to awaken people to the full implications of the "computer age," to which he had been so vital a contributor.

From *Cybernetics* we have selected a chapter entitled "Computing Machines and the Nervous System," in which Wiener draws a comparison between the neuron, with its two states of activity—activity and repose, or firing and nonfiring—with the ON/OFF binary system of switches that guides the calculating processes of computers. In his other observations, Wiener has in a sense fallen victim to his own ingenuity: what he originated has proved so fertile a new ground in science that only a few decades after his death his convictions and expectations seem almost primitive—but brilliantly primitive.

Consciousness may be associated with all quantum mechanical processes. The uniqueness of our consciousness lies in the fact that it is a part of a logic machine, which in turn is the brain of a particular kind of physical system, a living organism. That is, the terms life, thought, and consciousness, properly defined, are separable. An organism does not have to be conscious or capable of thinking in order to be alive. A brain does not have to be conscious to be capable of "thought" (we are using the term in the restricted sense of "data processing"). Only the higher organisms have brains for data processing, and only under very special conditions, when a large part of the data processing functions of the brain is handled by an irreducible quantum mechanical process, does the organism become a conscious, thinking being.

—EVAN HARRIS WALKER, "The Nature of Consciousness," *Mathematical Biosciences*

COMPUTING MACHINES ARE essentially machines for recording numbers, operating with numbers, and giving the result in numerical form. A very considerable part of their cost, both in money and in the effort of construction, goes to the simple problem of recording numbers clearly and accurately. The simplest mode of doing this seems to be on a uniform scale, with a pointer of some sort moving over this. If we wish to record a number with an accuracy of one part in *n,* we have to assure that in each region of the scale, the pointer assumes the desired position within this accuracy. That is, for an amount of information $\log_2 n$, we must finish each part of the movement of the pointer with this degree of accuracy, and the cost will be of the form *An,* and A is not too far from a constant. More precisely, since if *n-j* regions are accurately established, the remaining region will also be determined accurately, the cost of recording an amount of information I will be about.

$$(1^1 - 1)\,A. \qquad (01)$$

Norbert Wiener, "Computing Machines and the Nervous System," in *Cybernetics.* New York: John Wiley & Sons, Inc., 1948, pp. 137–155. Reprinted with permission of the M.I.T. Press.

Now let us divide this information over two scales, each marked less accurately. The cost of recording this information will be about:

$$2\left(2^{\frac{1}{2}I}-1\right)A. \tag{02}$$

If the information be divided among N scales, the approximate cost will be:

$$N\left(2^{\frac{I}{N}}-1\right)A. \tag{03}$$

This will be a minimum when:

$$2^{\frac{I}{N}}-1=\frac{I}{N}2^{\frac{I}{N}}\log 2; \tag{04}$$

or if we put:

$$\frac{I}{N}\log 2 = x, \tag{05}$$

when

$$x=\frac{e^x-1}{e^x}=1-e^{-x}. \tag{06}$$

This will occur when, and only when, $x = 0$, or $N = \infty$. That is, N should be as large as possible to give the lowest cost for the storage of information. Let us remember that $2^{I/N}$ must be an integer, and that 1 is not a significant value, as we then have an infinite number of scales each containing no information. The best significant value for $2^{I/N}$ is 2, in which case we record our number on a number of independent scales, each divided into two equal parts. In other words, we represent our numbers in the binary system on a number of scales in which all that we know is that a certain quantity lies in one or the other of two equal portions of the scale, and in which the probability of an imperfect knowledge as to which half of the scale contains the observation is made vanishingly small. In other words, we represent a number v in the form:

$$v=v_0+\frac{1}{2}v_1+\frac{1}{2^2}v_2+\ldots+\frac{1}{2_n}v_n+\ldots, \tag{07}$$

where every v_n is either 1 or 0.

There exist at present two great types of computing machines: those, like the Bush differential analyzer, which are known as *analogy machines,* where the data are represented by measurements on some continuous scale, so that the accuracy of the machine is determined by the accuracy of construction of the scale; and those, like the ordinary desk adding and multiplying machine, which we call *numerical machines,* where the data are presented by a set of choices among a number of contingencies; and the accuracy is determined by the sharpness with which the contingencies are distinguished, the number of alternative contingencies presented at every choice, and the number of choices given. We see that for highly accurate work, at any rate, the numerical machines are preferable, and above all, those numerical machines constructed on the binary scale, in which the number of alternatives presented at each choice is two. Our use of machines on the decimal scale is conditioned merely by the historical accident that the scale of ten, based on our fingers and thumbs, was already in use when the Hindus made the great discovery of the importance of the zero and the advantage of a positional system of notation. It is worth retaining when a large part of the work done with the aid of the machine consists in transcribing onto the machine numbers in the conventional decimal form, and in taking off the machine numbers which must be written in the same conventional form.

This is in fact the use of the ordinary desk computing machine, as employed in banks, in business offices, and in many statistical laboratories. It is not the way that the larger and more automatic machines are best to be employed; in general, any computing machine is used because machine methods are faster than hand methods. In any combined use of means of computation, as in any combination of chemical reactions, it is the slowest which gives the order of magnitude of the time constants of the entire system. It is thus advantageous, as far as possible, to remove the human element from any elaborate chain of computation, and only to introduce it where it is absolutely unavoidable, at the very beginning and the very end. Under these conditions, it pays to have an instrument for the change of the scale of notation, to be used initially and finally in the chain of computations; and to perform all intermediate processes on the binary scale.

The ideal computing machine must then have all its data inserted at the beginning, and must be as free as possible from human interference to the very end. This means that not only must the numerical data be inserted at the beginning, but also all the rules of combining them, in the form of instructions covering every situation which may arise in the course of the computation. Thus the computing machine must be a logical machine as well as an arithmetic machine, and must combine contingencies in accordance with a systematic algorithm. While there are many algorithms which *might* be used for combining contingencies, the simplest of these is known as the algebra of logic *par excellence,* or the Boolean algebra. This algorithm, like the binary arithmetic, is based on the dichotomy, the choice between *yes* and *no,* the choice between being in a class and outside it. The reasons for its superiority to other systems are of the same nature as the reasons for the superiority of the binary arithmetic over other arithmetics.

Thus all the data, numerical or logical, put into the machine, are in the form of a set of choices between two alternatives; and all the operations on the data take the form of making a set of new choices depend on a set of old choices. When I add two one-digit numbers, A and B, I obtain a two-digit number commencing with 1, if A and B are both 1, and otherwise with 0. The second digit is 1 if $A \neq B$, and is otherwise 0. The addition of numbers of more than one digit follows similar but more complicated rules. Multiplication in the binary system, as in the decimal, may be reduced to the multiplication table and the addition of numbers; and the rules for multiplication for binary numbers take on the peculiarly simple form given by the table:

(08)

×		
0	0	0
1	0	1

Thus multiplication is simply a method to determine a set of new digits when old digits are given.

On the logical side, if O is a negative and I a positive decision, every operator can be derived from three : *negation,* which transforms I into O and O into I ; *logical addition,* with the table

(09)

⊕	O	I
O	O	I
I	I	I

and *logical multiplication,* with the same table as the numerical multiplication of the (1,0) system, namely:

(10)

⊙	O	I
O	O	O
I	O	I

That is, every contingency which may arise in the operation of the machine simply demands a new set of choices of contingencies I and O, depending according to a fixed set of rules on the decisions already made. In other words, the structure of the machine is that of a bank of relays, capable each of two conditions, say "on," and "off"; while at each stage, the relays assume each a position dictated by the positions of some or all the relays of the bank at a previous stage of operation. These stages of operation may be definitely "clocked" from some central clock or clocks; or the action of each relay may be held up until all the relays which should have acted earlier in the process have gone through all the steps called for.

The relays used in a computing machine may be of very varied character. They may be purely mechanical; or they may be electro-mechanical, as in the case of a solenoidal relay, in which the armature will remain in one of two possible positions of equilibrium until an appropriate impulse pulls it to the other side. They may be purely electrical systems with two alternative positions of equilibrium, either in the form of gas-filled tubes, or what is much more rapid, in the form of high-vacuum tubes. The two possible states of a relay system may both be stable in the absence of outside interference, or only one may be stable, while the other is transitory. Always in the second case, and generally in the first case, it will be desirable to have special apparatus to retain an impulse which is to act at some future time, and to avoid the clogging up of the system which will ensue if one of the relays does nothing but repeat itself indefinitely. However, we shall have more to say concerning this question of memory later.

It is a noteworthy fact that the human and animal nervous systems, which are known to be capable of the work of a computation system, contain elements which are ideally suited to act as relays. These elements are the so-called *neurons* or nerve cells. While they show rather complicated properties under the influence of electrical currents, in their ordinary physiological action they conform very nearly to the "all-or-none" principle: that is, they are either at rest; or when they "fire," they go through a series of changes almost independent of the nature and intensity of the stimulus. There is first an active phase, transmitted from one end to the other of the neuron with a definite velocity, to which there succeeds a refractory period, during which the neuron is either incapable of being stimulated, or at any rate is not capable of being stimulated by any normal, physiological process. At the end of this effective refractory period, the nerve remains inactive, but may be stimulated again into activity.

Thus the nerve may be taken to be a relay with essentially two states of activity: firing and repose. Leaving aside those neurons which accept their messages from free endings or sensory end-organs, each neuron has its messages fed into it by other neurons at points of contact known as *synapses.* For a given outgoing neuron, these vary in number from a very few to many hundred. It is the state of the incoming impulses at the various synapses, combined with the antecedent state of the outgoing neuron itself, which determines whether it will fire or not. If it is neither firing nor refractory, and the number of incoming synapses which "fire" within a certain very short fusion interval of time exceeds a certain threshold, then the neuron will fire after a known, fairly constant synaptic delay.

This is perhaps an over-simplification of the picture: the "threshold" may not depend simply on the number of synapses, but on their "weight" and their geometrical relations to one another with respect to the neuron into which they feed; and there is very convincing evidence that there exist synapses of a different nature, the so-called "inhibitory synapses," which either completely prevent the firing of the outgoing neuron, or at any rate raise its threshold with respect to stimulation at the ordinary synapses. What is pretty clear, however, is that some definite combinations of impulses on the incoming neurons having synaptic connections with a given neuron will cause it to fire, while others will not cause it to fire. This is not to say that there may not be other, non-neuronic influences, perhaps of a humoral nature, which produce slow, secular changes tending to vary that pattern of incoming impulses which is adequate for firing.

A very important function of the nervous system, and, as we have said, a function equally in demand for computing machines, is that of *memory,* the ability to preserve the results of past operations for use in the future. It will be seen that the uses of the memory are highly varied, and it is improbable that any single mechanism can satisfy the demands of all of them. There is first the memory which is necessary for the carrying out of a current process, such as a multiplication, in which the intermediate results are of no value when once the process is completed, and in which the operating apparatus should then be released for further use. Such a memory should record quickly, be read quickly, and be erased quickly. On the other hand, there is the memory which is intended to be part of the files, the permanent record, of the machine or the brain, and to contribute to the basis of all its future behavior, at least during a single run of the machine. Let it be remarked parenthetically that an important difference between the way in which we use the brain and the machine is that the machine is intended for many successive runs, either with no reference to each other, or with a minimal, limited reference; and that it can be cleared between such runs; while the brain in the course of nature, never even approximately clears out its past records. Thus the brain, under normal circumstances, is not the complete analogue of the computing machine, but rather the analogue of a single run on such a machine. We shall see later that this remark has a deep significance in psychopathology and in psychiatry.

To return to the problem of memory, a very satisfactory method for constructing a short-time memory is to keep a sequence of impulses travelling around a closed circuit, until this circuit is cleared by intervention from outside. There is much reason to believe that this happens in our brains during the retention of impulses which occurs over what is known as the specious present. This method has been imitated in several devices which have been used in computing machines, or at least suggested for such a use. There are two conditions which are desirable in such a retentive apparatus: the impulse should be transmitted in a medium in which it is not too difficult to achieve a considerable time-lag; and before the errors inherent in the instrument have blurred it too much, the impulse should be reconstructed in a form as sharp as possible. The first condition tends to rule out delays produced by the transmission of light, or even, in many cases, by electric circuits, while it favors the use of one form or another of elastic vibrations; and such vibrations have actually been employed for this purpose in computing machines. If electric circuits are used for delay purposes, the delay produced at every stage is relatively short; or as in all pieces of linear apparatus, the deformation of the message is cumulative, and very soon becomes intolerable. To avoid this, a second consideration comes into play; we must insert somewhere in the cycle a relay which does not serve to repeat the form of the incoming message, but rather to trigger off a new message of prescribed form. This is done very easily in the nervous system, where indeed all transmission is more or less of a trigger phenomenon. In the electrical industry, pieces of apparatus for this purpose have long been known, and have been used in connection with telegraph circuits. They are known as *telegraph-type repeaters.* The great difficulty of using them for memories of long duration is that they have to function without a flaw over an enormous number of consecutive cycles of operation. Their success is all the more remarkable: in a piece of apparatus designed by Mr. Williams of the University of Manchester, a device of this sort with a unit delay of the order of a hundredth of a second has continued in successful operation for several hours. What makes this more remarkable is that this apparatus was not used merely to preserve a single decision, a single "yes" or "no," but a matter of thousands of decisions.

Like other forms of apparatus intended to retain a large number of decisions, this works on the scanning principle. One of the simplest modes of storing information for a relatively short time is as the charge on a condenser; and when this is supplemented by a telegraph-type repeater, it becomes an adequate method of storage. To use to the best advantage the circuit facilities attached to such a storage system, it is desirable to be able to switch successively and very rapidly from one condenser to another. The ordinary means of doing this involve mechanical inertia, and this is never consistent with very high speeds. A much better way is the use of a large number of condensers, in which one plate is either a small piece of metal sputtered in to a dielectric, or the imperfectly insulating surface of the dielectric itself, while one of the connectors to these condensers is a pencil of cathode rays moved by the condensers and magnets of a sweep-circuit over a course like that of a plough in a ploughed field. There are various elaborations of this method, which indeed was employed in a somewhat different way by the Radio Corporation of America before it was used by Mr. Williams.

These last-named methods for storing information can hold a message for quite an appreciable time, if not for a period comparable with a human life-time. For more permanent records, there is a wide variety of alternatives among which we can choose. Leaving out such bulky, slow, and unerasable methods as the use of punched cards and punched tape, we have magnetic tape, together with its modern refinements, which have largely eliminated the tendency of messages on this material to spread; phosphorescent substances; and above all, photography. Photography is indeed ideal for the permanence and detail of its records, ideal again from the point of view of the shortness of exposure needed to record an observation. It suffers from two grave disadvantages: the time needed for development, which has been reduced to a few seconds, but is still not small enough to make photography available for a short-time memory; and (at present) the fact that a photographic record is not subject to rapid erasure and the rapid implanting of a new record. The Eastman people have been working on just these problems, which do not seem to be necessarily insoluble, and it is possible that by this time they have found the answer.

Very many of the methods of storage of information already considered have an important physical element in common. They seem to depend on systems with a high degree of quantum degeneracy; or in other words, with a large number of modes of vibration of the same frequency. This is certainly true in the case of ferromagnetism, and is also true in the case of materials with an exceptionally high dielectric constant, which are thus especially valuable for use in condensers for the storage of information. Phosphorescence as well is a phenomenon associated with a high quantum degeneracy; and the same sort of effect makes its appearance in the photographic process, where many of the substances which act as developers seem to have a great deal of internal resonance. Quantum degeneracy appears to be associated with the ability to make small causes produce appreciable and stable effects. . . . Substances with high quantum degeneracy appear to be associated with many of the problems of metabolism and reproduction. It is probably not an accident that here, in a non-living environment, we find them associated with a third fundamental property of living matter: the ability to receive and organize impulses, and to make them effective in the outer world.

We have seen in the case of photography and similar processes that it is possible to store a message in the form of a permanent alteration of certain storage elements. In reinserting this information into the system, it is necessary to cause these changes to affect the messages going through the system. One of the simplest ways to do this is to have as the storage elements which are changed parts which normally assist in the transmission of messages, and of such a nature that the change in their character due to storage affects the manner in which they will transport messages for the entire future. In the nervous system, the neurons and the synapses are elements of this sort, and it is quite plausible that information is stored over long periods by changes in the thresholds of neurons; or, what may be regarded as another way of saying the same thing, by changes in the permeability of each synapse to messages. Many of us think, in the absence of a better explanation of the phenomenon, that the storage of information in the brain can actually occur in this way. It is conceivable for such a storage to take place either by the opening of new paths or by the closure of old ones. Apparently it is adequately established that no neurons are formed in the brain after birth. It is possible, though not certain, that no new synapses are formed; and it is a plausible conjecture that the chief changes of thresholds in the memory process are increases. If this is the case, our whole life is on the pattern of Balzac's *La Peau de Chagrin,* and the very process of learning and remembering exhausts our powers of learning and remembering, until life itself squanders our capital stock of power to live. It may well be that this phenomenon does occur. This is a possible explanation for a sort of senescence. The real phenomenon of senescence, however, is much too complicated to be explained in this way alone.

We have already spoken of the computing machine, and consequently the brain, as a logical machine. It is by no means trivial to consider the light cast on logic by such machines, both natural and artificial. Here the chief work is that of Turing. We have said before that the *machina ratiocinatrix* is nothing but the *calculus ratiocinator* of Leibniz with an engine in it; and just as modern mathematical logic begins with this calculus, so it is inevitable that its present engineering development should cast a new light on logic. The science of today is operational: that is, it considers every statement as essentially concerned with possible experiments or observable processes. According to this, the study of logic must reduce to the study of the logical machine, whether nervous or mechanical, with all its non-removable limitations and imperfections.

It may be said by some readers that this reduces logic to psychology, and that the two sciences are observably and demonstrably different. This is true in the sense that many psychological states and sequences of thought do not conform to the canons of logic. Psychology contains much that is foreign to logic, but—and this is the important fact—any logic which means anything to us can contain nothing which the human mind—and hence the human nervous system—is unable to encompass. *All logic is limited by the limitations of the human mind, when it is engaged in that activity known as logical thinking.*

For example, we devote much of mathematics to discussions involving the infinite, but these discussions and their accompanying proofs are not infinite in fact. No admissible proof involves more than a finite number of stages. It is true, a proof by mathematical induction *seems* to involve an infinity of stages, but this is only apparent. In fact, it involves just the following stages:

(1) P_n is a proposition involving the number n;
(2) P_n has been proved for $n = 1$;
(3) If P_n is true, P_{n+1} is true;
(4) Therefore P_n is true for every positive integer n.

It is true that somewhere in our logical assumptions there must be one which validates this argument. However, this mathematical induction is a far different thing from complete induction over an infinite set. The same thing is true of the more refined forms of mathematical induction, such as transfinite induction, which occur in certain mathematical disciplines.

Thus some very interesting situations arise, in which we may be able—with enough time and enough computational aids—to prove every single case of a theorem P_n; but if there is no systematic way of subsuming these proofs under a single argument independent of *n,* such as we find in mathematical induction, it may be impossible to prove P_n *for all n.* This contingency is recognized in what is known as metamathematics, the discipline so brilliantly developed by Gödel and his school.

A proof represents a logical process which has come to a definitive conclusion in a finite number of stages. However, a logical machine following definite rules need never come to a conclusion. It may go on grinding through different stages without ever coming to a stop, either by describing a pattern of activity of continually increasing complexity, or by going into a repetitive process like the end of a chess game in which there is a continuing cycle of perpetual check. This occurs in the case of some of the paradoxes of Cantor and Russell. Let us consider the class of all classes which are not members of themselves. Is this class a member of itself? If it is, it is certainly not a member of itself; and if it is not, it is equally certainly a member of itself. A machine to answer this question would give the successive temporary answers "yes," "no," "yes," "no," and so on; and would never come to equilibrium.

Bertrand Russell's solution of his own paradoxes was to affix to every statement a quantity, the so-called type, which serves to distinguish between what seems to be formally the same statement, according to the character of the objects with which it concerns itself;—whether these are "things," in the simplest sense, classes of "things," classes of classes of "things," etc. The method by which we resolve the paradoxes is also to attach a parameter to each statement, this parameter being the time at which it is asserted. In both cases, we introduce what we may call a parameter of uniformization, to resolve an ambiguity which is simply due to its neglect.

We thus see that the logic of the machine resembles human logic, and following Turing, we may employ it to throw light on human logic. Has the machine a more eminently human characteristic as well—the ability to learn? To see that it may well have even this property, let us consider two closely related notions: that of the association of ideas and that of the conditioned reflex.

In the British empirical school of philosophy, from Locke to Hume, the content of the mind was considered to be made up of certain entities known to Locke as ideas, and to the later authors as ideas and impressions. The simple ideas or impressions were supposed to exist in a purely passive mind, as free from influence on the ideas it contained as a clean blackboard is on the symbols which may be written on it. By some sort of inner activity, hardly worthy to be called a force, these ideas were supposed to unite themselves into bundles, according to the principles of similarity, contiguity, and cause and effect. Of these principles, perhaps the most significant was contiguity: ideas or impressions which had often occurred together in time or in space were supposed to have acquired the ability of evoking one another, so that the presence of any one of them would produce the entire bundle.

In all this there is a dynamics implied, but the idea of a dynamics had not yet filtered through from physics to the biological and psychological sciences. The typical biologist of the eighteenth century was Linnaeus, the collector and classifier, with a point of view quite opposed to that of the evolutionists, the physiologists, the geneticists, the experimental embryologists of the present day. Indeed, with so much of the world to explore, the state of mind of the biologists could hardly have been different. Similarly, in psychology, the notion of mental content dominated that of mental process. This may well have been a survival of the scholastic emphasis on substances, in a world in which the noun was hypostasised and the verb carried little or no weight. Nevertheless, the step from these static ideas to the more dynamic point of view of the present day, as exemplified in the work of Pavlov, is perfectly clear.

Pavlov worked much more with animals than with men, and he reported visible actions rather than introspective states of mind. He found in dogs that the presence of food causes the increased secretion of saliva and of gastric juice. If then a certain visual object is shown to dogs in the presence of food, and only in the presence of food, the sight of this object in the absence of food will acquire the property of being by itself able to stimulate the flow of saliva or of gastric juice. The union by contiguity which Locke had observed introspectively in the case of ideas now becomes a similar union of patterns of behavior.

There is one important difference, however, between the point of view of Pavlov and that of Locke, and it is precisely due to this fact that Locke considers ideas and Pavlov patterns of action. The responses observed by Pavlov tend to carry a process to a successful conclusion, or to avoid a catastrophe. Salivation is important for deglutition and for digestion, while the avoidance of what we should consider a painful stimulus tends to protect the animal from bodily injury. Thus there enters into the conditioned reflex something that we may call *affective tone.* We need not associate this with our own sensations of pleasure and pain, nor need we in the abstract associate it with the advantage of the animal. The essential thing is this: that affective tone is arranged on some sort of scale from negative "pain" to positive "pleasure"; that for a considerable time, or permanently, an increase in affective tone favors all processes in the nervous system that are under way at the time, and gives them a secondary power to increase affective tone; and that a decrease in affective tone tends to inhibit all processes under way at the time, and gives them a secondary ability to decrease affective tone.

Biologically speaking, of course, a greater affective tone must occur predominantly in situations favorable for the perpetuation of the race, if not the individual, and a smaller affective tone in situations which are unfavorable for this perpetuation, if not disastrous. Any race not conforming to this requirement will go the way of Lewis Carroll's Bread-and-Butter Fly, and always die. Nevertheless, even a doomed race may show a mechanism valid so long as the race lasts. In other words, even the most suicidal apportioning of affective tone will produce a definite pattern of conduct.

Note that the mechanism of affective tone is itself a feed-back mechanism. It may even be given a diagram such as the following.

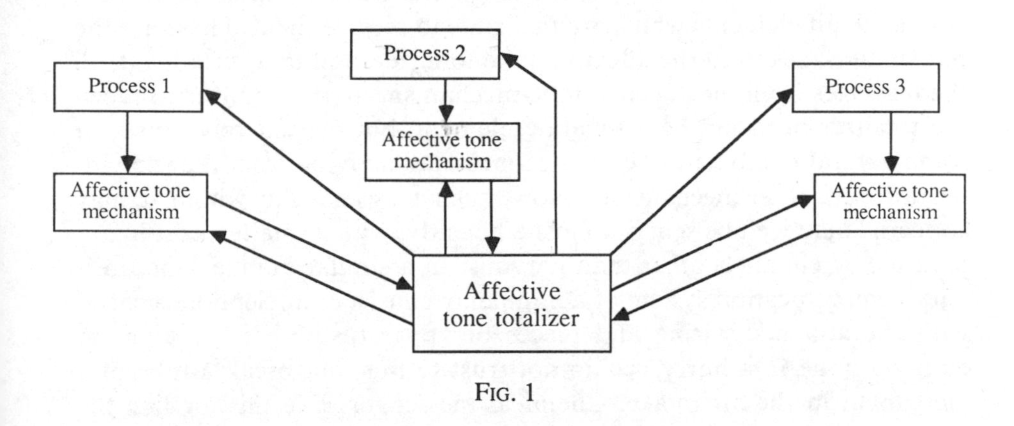

FIG. 1

Here the totalizer for affective tone combines the affective tones given by the separate affective-tone mechanisms over a short interval in the past, according to some rule which we need not specify now. The leads back to the individual affective tone mechanisms serve to modify the intrinsic affective tone of each process in the direction of the output of the totalizer, and this modification stands until it is modified by later messages from the totalizer. The leads back from the totalizer to the process mechanisms serve to lower thresholds if the total affective tone is increasing, and to raise them if the total affective tone is decreasing. They likewise have a long-time effect, which endures until it is modified by another impulse from the totalizer. This lasting effect, however, is confined to those processes actually in being at the time the return message arrives, and a similar limitation also applies to the effects on the individual affective-tone mechanisms.

I wish to emphasize that I do not say that the process of the conditioned reflex operates according to the mechanism I have given; I merely say that it *could* so operate. If, however, we assume this or any similar mechanism, there are a good many things we can say concerning it. One is that this mechanism is capable of learning. It has already been recognized that the conditioned reflex is a learning mechanism, and this idea has been used in the behaviorist studies of the learning of rats in a maze. All that is needed is that the inducements or punishments used have respectively a positive and a negative affective tone. This is certainly the case, and the experimenter learns the nature of this affective tone by experience, not simply by *a priori* considerations.

Another point of considerable interest is that such a mechanism involves a certain set of messages which go out generally into the nervous system, to all elements which are in a state to receive them. These are the return messages from the affective-tone totalizer, and to a certain extent, the messages from the affective-tone mechanisms to the totalizers. Indeed the totalizer need not be a separate element, but may merely represent some natural combinatory effects of messages arriving from the individual affective-tone mechanisms. Now, such messages "to whom it may concern" may well be sent out most efficiently, with a smallest cost in apparatus, by channels other than nervous. In a similar manner, the ordinary communication system of a mine may consist of a telephone central with the attached wiring and pieces of apparatus. When we want to empty a mine in a hurry, we do not trust to this, but break a tube of a mercaptan in the air intake. Chemical messengers like this, or like the hormones, are the simplest and most effective for a message not addressed to a specific recipient. For the moment let me break into what I know to be pure fancy. The high emotional and consequently affective content of hormonal activity is most suggestive. This does not mean that a purely nervous mechanism is not capable of affective tone and of learning, but it does mean that in the study of this aspect of our mental activity, we cannot afford to be blind to the possibilities of hormonal transmission. It may be excessively fanciful to attach this notion to the fact that in the theories of Freud, the memory—the storage function of the nervous system—and the activities of sex are both involved. Sex, on the one hand, and all affective content, on the other, contain a very strong hormonal element. This suggestion of the importance of sex and hormones has been made to me by Dr. J. Lettvin and Mr. Oliver Selfridge. While at present there is no adequate evidence to prove its validity, it is not manifestly absurd in principle.

There is nothing in the nature of the computing machine which forbids it to show conditioned reflexes. Let us remember that a computing machine in action is more than the concatenation of relays and storage mechanisms which the designer has built into it. It also contains the content of its storage mechanisms; and this content is never completely cleared in the course of a single run. We have already seen that it is the run rather than the entire existence of the mechanical structure of the computing machine which corresponds to the life of the individual. We have also seen that in the nervous computing machine, it is highly probable that information is stored largely as changes in the permeability of the synapses; and it is perfectly possible to construct artificial machines where information is stored in that way. It is perfectly possible, for example, to cause any message going into storage to change in a permanent or semipermanent way the grid bias of one or of a number of vacuum tubes, and thus to alter the numerical value of the summation of impulses which will make the tube or tubes fire.

A more detailed account of learning apparatus in computing and control machines, and the uses to which it may be put, may well be left to the engineer rather than to a preliminary book like this one. It is perhaps better to devote the rest of this chapter to the more developed, normal uses of modern computing machines. One of the chief of these is in the solution of partial differential equations. Even linear partial differential equations require the recording of an enormous mass of data to set them up, as the data involve the accurate description of functions of two or more variables. With equations of the hyperbolic type, like the wave equation, the typical problem is that of solving the equation when the initial data are given, and this can be done in a progressive manner from the initial data to the results at any given later time. This is largely true of equations of the parabolic type as well. When it comes to equations of the elliptic type, where the natural data are boundary values rather than initial values, the natural methods of solution involve an iterative process of successive approximation. This process is repeated a very large number of times, so that very fast methods such as those of the modern computing machine are almost indispensable.

In non-linear partial differential equations, we miss what we have in the case of the linear equations—a reasonably adequate purely mathematical theory. Here computational methods are not only important for the handling of particular numerical cases, but, as von Neumann has pointed out, we need them in order to form that acquaintance with a large number of particular cases without which we can scarcely formulate a general theory. To some extent this has been done with the aid of very expensive experimental apparatus, such as wind tunnels. It is in this way that we have become acquainted with the more complicated properties of shock waves, slip surfaces, turbulence, and the like, for which we are scarcely in a position to give an adequate mathematical theory. How many undiscovered phenomena of similar nature there may be, we do not know. The analogy machines are so much less accurate, and in many cases so much slower, than the digital machines, that the latter give us much more promise for the future.

It is already becoming clear in the use of these new machines that they demand purely mathematical techniques of their own, quite different from those in use in manual computation or in the use of machines of smaller capacity. For example, even the use of machines for computing determinants of moderately high order, or for the simultaneous solution of twenty or thirty simultaneous linear equations, shows difficulties which do not arise when we study analogous problems of small order. Unless care is exercised in setting a problem up these may completely deprive the solution of any significant figures whatever. It is a commonplace to say that fine, effective tools like the ultra-rapid computing machine are out of place in the hands of those not possessing a sufficient degree of technical skill to take full advantage of them. The ultra-rapid computing machine will certainly not decrease the need for mathematicians with a high level of understanding and technical training.

In the mechanical or electrical construction of computing machines, there are a few maxims which deserve consideration. One is that mechanisms which are relatively frequently used, such as multiplying or adding mechanisms, should be in the form of relatively standardized assemblages adapted for one particular use, and no other; while those of more occasional use should be assembled for the moment of use out of elements also available for other purposes. Closely related to this consideration is the one that in these more general mechanisms, the component parts should be available in accordance with their general properties, and should not be allotted permanently to a specific association with other pieces of apparatus. There should be some part of the apparatus, like an automatic telephone-switching exchange, which will search for free components and connectors of the various sorts, and allot them as they are needed. This will eliminate much of the very large expense which is due to having a great number of unused elements, which cannot be used unless their entire large assembly is used. . . .

As a final remark, let me point out that a large computing machine, whether in the form of mechanical or electric apparatus or in the form of the brain itself, uses up a considerable amount of power, all of which is wasted and dissipated in heat. The blood leaving the brain is a fraction of a degree warmer than that entering it. No other computing machine approaches the economy of energy of the brain. In a large apparatus like the Eniac or Edvac, the filaments of the tubes consume a quantity of energy which may well be measured in kilowatts, and unless adequate ventilating and cooling apparatus is provided, the system will suffer from what is the mechanical equivalent of pyrexia, until the constants of the machine are radically changed by the heat, and its performance breaks down. Nevertheless, the energy spent per individual operation is almost vanishingly small, and does not even begin to form an adequate measure of the performance of the apparatus. The mechanical brain does not secrete thought "as the liver does bile," as the earlier materialists claimed, nor does it put it out in the form of energy, as the muscle puts out its activity. Information is information, not matter or energy. No materialism which does not admit this can survive at the present day.

Q.3

A Physicist's Renewed Look at Biology: Twenty Years Later

MAX DELBRÜCK

A CURIOUS THING about Max Delbrück is that, having begun his career as a particle physicist, he later became a microbiologist of such skill and insight that he and two collaborators, Alfred Hershey and Salvador Luria, shared the 1969 Nobel Prize for physiology and medicine. He made a wide variety of contributions to the study of the propagation of viruses and, as a result of his background as a quantum physicist he was able to lay much of the groundwork for the fledgling field of molecular biology.

In physics, he had been a promising young member of the generation influenced by Niels Bohr. He was born in 1906 in Berlin and died in 1981. Delbrück took his Ph.D. in theoretical physics at Göttingen in 1930, and then worked with such subatomic specialists as Niels Bohr in Copenhagen and Otto Hahn and Lise Meitner at the Kaiser Wilhelm Institut für Chemie in Berlin until 1937. Those years were also marked by the great exodus of intelligent men and woman from Hitler's Germany, and Delbrück, like many other European scholars, brought to American science his considerable talents. He arrived in the United States in 1937, and spent two years at the California Institute of Technology on a Rockefeller scholarship in biology before becoming a physics professor at Vanderbilt University in 1939. He became a naturalized citizen in 1945, two years before returning to Cal Tech as a biology professor.

While at Cal Tech in 1937, Delbrück plunged into the biological studies that had already begun to interest him in his Berlin days. His special interest was bacteriophages, known simply as phages in the jargon of the specialists, those relatively bulky viruses that attack the cells of bacteria. Drawing upon his background in theoretical physics, and the ideas about the molecular aspects of genetics he had absorbed while working in Berlin, Delbrück refined current methods of culturing phages and discovered that a single virus can infect a bacterial cell and disintegrate it in a half hour.

In 1946, Delbrück and Hershey, working independently, discovered that different viruses can exchange genetic material and can also join to create a new virus quite different from the originals. Their experiments were the basis for the development of a whole new branch of microbiology, bacterial genetics, which in turn led to the development of the techniques applied to recombinant DNA. A collaboration between Delbrück and N. Visconti in 1953 led to a theory explaining the mating process of the bacteriophages in the host bacterium, in which genetic material is not simply exchanged but gradually transferred by repeated, random matings. Although the Visconti-Delbrück theory is somewhat dated, it still accurately describes the genetic variations that are assumed by bacterial viruses.

In his Nobel lecture, Delbrück begins by sketching a brief history of the animate/inanimate bivalence, a theory that quite understandably prevailed from the beginning of science to the relatively recent realization that "the break between the nonliving world and the living world might not be absolute." Delbrück's recollections of the "informal private meetings" held by his colleagues in theoretical physics at the time when he was an assistant to Lise Meitner in Berlin, along with his tribute to the geneticist Timofeeff-Ressovsky, who lured them into the field of molecular biology, has today the quality of folklore. In a similar way, his story of the meeting sponsored by the Connecticut Academy of the Arts and Sciences, featuring a poet, a composer, and two scientists, as "creators" and "performers," is almost wistful, a far cry from the proud boast of a particle physicist that "particle physics is the epic poetry of the twentieth century."

As in a dream it is our own will that unconsciously appears as inexorable objective destiny, everything in it proceeding out of ourselves and each of us being the secret theatre manager of our dreams with us all, our fate may be the product of our inmost selves, or our wills, and we are actually bringing about what seems to be happening to us.

—THOMAS MANN

The universe begins to look more like a great thought than a machine.

—JAMES JEANS

The pattern of organization of any biological system is established by a complex electrodynamic field which is in part determined by its atomic physio-chemical components and which in part determines the behavior and orientation of those components. This field is electrical in the physical sense and by its properties relates the entities of the biological system in a characteristic pattern and is itself, in part, a result of the existence of those entities. It determines and is determined by the components.

—HAROLD SAXTON BURR, *The Fields of Life*

PHYSICS AND BIOLOGY

AT THE VERY beginnings of science the striking dissimilarities between the behavior of living and nonliving things became obvious. Two tendencies can be discerned in the attempts to arrive at a unified view of our world. One tendency is to use the living organism as the model system. This tendency is exemplified by Aristotle. For him, the son of a physician and the keen observer of many forms of life, it was obvious that things develop according to plans. Every animal and plant is generated in some definite way, runs through a cycle of development in which it unfolds its inherent plan, and succumbs to death and decay. For Aristotle, this very obvious feature of the world which surrounds us is *the* model for under-

Max Delbrück, Nobel Prize in Physiology or Medicine Award Address, 1969. Reprinted with permission of Elsevier Publishing Company, and the Nobel Foundation.

standing our (sublunar) world. Astronomy is the exception and offers the contrast of an eternal periodic system subject to neither generation nor decay.

With the ascendance in the Renaissance of the science of physics in our modern sense of the word there seemed to develop at first a peculiar break between the living and the nonliving parts of the world. Life seemed to have unique properties quite irreducible to the world of physics and chemistry: motion generated from within, chemistry of a very distinct kind, replication, development, consciousness—each of these aspects of life turned into elements that became more and more foreign to the physicist to the extent that many physicists even today look upon biology as something outside their domain.

A partial reversal of this bizarre partition of the world into the living and the nonliving came with the many proofs that living forms are not, in fact, constant, but over the long range have evolved and that the family tree of this evolution can be traced. The interpretation of evolution in terms of natural selection, especially after the latter had been put into clearer perspective with the establishment of the science of genetics, suggested a unified view of life, but still left uncertain the connection of life with the nonliving world. The insights of chemistry and its first inroads into biochemistry made it clear that the break between the nonliving world and the living world might not be absolute.

Molecular genetics, our latest wonder, has taught us to spell out the connectivity of the tree of life in such palpable detail that we may say in plain words, "This riddle of life has been solved." The ideas of information storage, of the replication of the stored information and of its programmed readout have become commonplace and have filtered down into the popular magazines and grade school textbooks. The marvel that the mechanical and chemical machinery underlying all these affairs can in fact be worked out is keeping a host of scientists very happy and very busy. . . .

During the years 1932–1937, while I was assistant to Professor Lise Meitner in Berlin, a small group of theoretical physicists held informal private meetings, at first devoted to theoretical physics but soon turning to biology. Our principal teacher in the latter area was the geneticist Timofeeff-Ressovsky, who together with the physicist K. G. Zimmer, at that time was doing by far the best work in the area of quantitative mutation research. A few years earlier H. J. Muller had discovered that ionizing radiations produce mutations and the work of the Berlin group showed very clearly that these mutations were caused either by single pairs of ions or by small clusters of them. Discussions of these findings within our little group strengthened the notion that genes had a kind of stability similar to that of the molecules of chemistry. From the hindsight of our present knowledge one might consider this a trivial statement: what else could genes be but molecules? However, in the mid-thirties, this was not a trivial statement. Genes at that time were algebraic units of the combinatorial science of genetics and it was anything but clear that these units were molecules analyzable in terms of structural chemistry. They could have turned out to be submicroscopic steady state systems, or they could have turned out to be something unanalyzable in terms of chemistry, as first suggested by Bohr and discussed by me in a lecture twenty years ago. . . . It is true that our hope at that time to get at the chemical nature of the gene by means of radiation genetics never materialized. The road to success effectively bypassed radiation genetics. Nevertheless, radiation genetics has been through all these decades and is now more than ever a field of great importance, most recently and depressingly so because of the possibilities of large-scale military applications entailing exposure to ionizing radiations.

To illustrate our state of mind at that time I will append to this lecture a memorandum on the "Riddle of Life," written to clarify my own thinking in the fall of 1937, just before leaving Germany to go to the United States. I found this note a few years ago among my papers. This memorandum would appear to be a summary of discussions at a little meeting in Copenhagen, arranged by Niels Bohr, to which Timofeeff-Ressovsky, H. J. Muller and I had travelled from Berlin. These discussions occurred very much under the impact of the W. M. Stanley findings reporting the crystallization of tobacco mosaic virus.

NEUROBIOLOGY

While molecular genetics has taught us the proper way to reconcile the characteristics of the living world, generation, development towards a goal, and decay, with the contrasting incorruptibility and planlessness of the physical world, it has not resolved our uncertainty about the proper way to relate this language to the notions of consciousness, mind, cognition, logical thought, truth—all these notions, too, elements of our world.

What is language? How does a child come to associate meaning with a word? The ability to form abstractions is undoubtedly inherent in our brain, this marvel of a computer. The study of the brain's connectivity, the study of the development of this network in the growing animal, the study of its function and potencies—all of these studies are aspects of the neurobiology of the next decade and they are very appealing ones to many of my colleagues and to many of the new generation of graduate students.

TRANSDUCER PHYSIOLOGY

I have two reservations concerning neurobiology. The first reservation is that we are not yet ready to tackle it in a decisive way. I believe that there is a widespread underestimation of the things we do not know and do not understand about cell biology and cell-cell interaction. It simply is not enough to know that nerve fibers conduct, that synapses are inhibitory or excitatory, chemical or electrical, that sensory inputs can be transduced, that they result in trains of spikes which measure intensities of stimuli or the time derivatives of these intensities, that all kinds of accommodations occur, etc., etc. I believe that we need a much more basic and detailed understanding of these stimulus response systems, be the stimulus an outside one or a presynaptic signal.

Sensory physiology in a broad sense contains hidden at its kernel an as yet totally undeveloped but absolutely central science: transducer physiology, the study of the conversion of the outside signal to its first "interesting" output. I use the word "interesting" advisedly because I wish to exclude from the area of study I intend to delimit, for instance, the primary photochemical reactions of the visual systems. I look upon the primary photochemical processes as something "uninteresting" because they concern the conversion of a light stimulus into what might be called an olfactory stimulus. A light quantum, in order to be effective as a sensory stimulant, naturally must, in the first instance, create within the cell a primary photoproduct which carries the business further. In thus excluding the photochemistry of the visual process from transducer physiology proper I am excluding the beautiful work on the photochemistry of rhodopsin for which George Wald received the Nobel Prize two years ago. Transducer physiology proper comes after this first step, where we are dealing with devices of the cell unparalleled in anything the physicists have produced so far with respect to sensitivity, adaptability and miniaturization. Which biological material will turn out to be the most suitable for bringing us decisive insights in this field? For a number of years I have studied an organelle of the fungus *Phycomyces,* the sporangiophore, in the belief that in the field of transducer physiology, as in genetics, essential progress will require the use of a suitable microorganism. I need not detail this work here since it has very recently been critically reviewed by a group effort of those involved in this work. Let me say here only that this organelle is exquisitely sensitive to light, to gravity, to stretch, and to a stimulus which we believe to be olfactory. . . . Others have proposed and demonstrated the suitability of other systems: chemotaxis of bacteria; olfaction in insects; mechanosensitivity of motoric cilia. We may hope that each of these systems, as well as the lipid-bilayer systems, which can be made to simulate most of the astounding feats of living membranes, will contribute to the great discoveries in cell physiology which, in my opinion, are prerequisites for a truly successful venture into neurobiology.

BODY AND SOUL

My second reservation regarding the hopes of neurobiology is more disturbing to me and also more nebulous; the eagerness with which we plunge into neurobiology overlooks an essential limitation—the *a priori* aspect of the concept of truth. It is well understood that a computer can be constructed so as to operate with certain axioms and formalized rules of logic, deriving in this way any number of "proved declarative sentences." We may call these sentences true if we have faith in the axioms and the rules of logic and we may be tempted to consider the logical sum of provable sentences as the computer's definition of truth. However, our friends, the logicians, have made it clear to us long ago that in any but the simplest languages we must distinguish between an "object language" and a "metalanguage." The word "truth," and thus all discussion of truth, must be excluded from the object language if the language is to be kept free of antinomies. There then follows the strange result that there must be sentences that are true but not provable. Thus the notion of truth, if it is to be meaningful at all, must be distinct and prior to the system of provable sentences, and thus distinct from and prior to the computer which should be looked upon as the embodiment of the system of provable sentences.

Thus, even if we learn to speak about consciousness as an emergent property of nerve nets, even if we learn to understand the processes that lead to abstraction, reasoning and language, still any such development presupposes a notion of truth that is prior to all these efforts and that cannot be conceived as an emergent property . . . of a biological evolution. Our conviction of the truth of the sentence, "The number of prime numbers is infinite," must be independent of nerve nets and of evolution, if truth is to be a meaningful word at all.

ARTIST VERSUS SCIENTIST

Twenty years ago the Connecticut Academy of the Arts and Sciences had a jubilee meeting and on that occasion invited a poet, a composer, and two scientists to create and to perform. It was a very fine affair. Hindemith, conducting a composition for trumpet and percussion, and Wallace Stevens, reading a set of poems entitled "An Ordinary Evening in New Haven," were enjoyed by everybody, perhaps most by the scientists. In contrast, the scientists' performances were attended by scientists only. To my feeling this irreciprocity was fitting, although perhaps not intended by the organizers. It is quite rare that scientists are asked to meet with artists and are challenged to match the others' creativeness. Such an experience may well humble the scientist. The medium in which he works does not lend itself to the delight of the listener's ear. When he designs his experiments or executes them with devoted attention to the details he may say to himself, "This is my composition; the pipette is my clarinet." And the orchestra may include instruments of the most subtle design. To others, however, his music is as silent as the music of the spheres. He may say to himself, "My story is an everlasting possession, not a prize composition which is heard and forgotten," but he fools only himself. The books of the great scientists are gathering dust on the shelves of learned libraries. And rightly so. The scientist addresses an infinitesimal audience of fellow composers. His message is not devoid of universality but its universality is disembodied and anonymous. While the artist's communication is linked forever with its original form, that of the scientist is modified, amplified, fused with the ideas and results of others and melts into the stream of knowledge and ideas which forms our culture. The scientist has in common with the artist only this: that he can find no better retreat from the world than his work and also no stronger link with the world than his work.

The Nobel ceremonies are of a nature similar to the one I referred to. Here too scientists are brought together with a writer. Again the scientists can look back on a life during which their work addressed a diminutive audience while the writer, in the present instance Samuel Beckett, has had the deepest impact on men in all walks of life. We find, however, a strange inversion when we come to talking about our work. While the scientists seem elated to the point of garrulousness at the chance of talking about themselves and their work, Samuel Beckett, for good and valid reasons, finds it necessary to maintain a total silence with respect to himself, his work, and his critics. Even though I was more thrilled by the award of the Nobel Prize to him than about the award to me and momentarily looked forward with intense anticipation to hearing his lecture, I now realize that he is acting in accordance with the rules laid down by the old witch at the end of a marionette play entitled "The Revenge of Truth."

"The truth, my children, is that we are all of us acting in a marionette comedy. What is more important than anything else in a marionette comedy is keeping the ideas of the author clear. This is the real happiness in life and now that I have at last come into a marionette play, I will never go out of it again. But you, my fellow actors, keep the ideas of the author clear. Aye, drive them to the utmost consequences."

APPENDIX

PRELIMINARY WRITE-UP ON THE TOPIC "RIDDLE OF LIFE" (BERLIN, AUGUST 1937)

We inquire into the relevance of the recent results of virus research for a general assessment of the phenomena peculiar to life.

These recent results all agree in showing a remarkable uniformity in the behavior of individuals belonging to one species of virus in preparations employing physical or chemical treatments mild enough not to impair infective specificity. Such a collection of individuals migrates with uniform velocity in the electrophoresis apparatus. It crystallizes uniformly from solutions such that the specific infectivity is not altered by recrystallization, not even under conditions of extremely fractionated recrystallization. Elementary analysis gives reproducible results, such as might be expected for proteins, with perhaps the peculiarity that the phosphorus and sulfur contents appear to be abnormally small.

These results force us to the view that the viruses are things whose atomic constitution is as well defined as that of the large molecules of organic chemistry. True, with these latter we also cannot speak of unique spatial configurations, since most of the chemical bonds involve free rotation around the bond. We cannot even decide unambiguously which atoms do or do not belong to the molecule, since the degree of hydration and of dissociation depends not only on external conditions, but even when these are fixed, fluctuates statistically from molecule to molecule. Nevertheless, there can be no doubt that such large molecules constitute a legitimate generalization of the standard concept of the chemical molecule. The similarity between virus and molecule is particularly apparent from the fact that virus crystals can be stored indefinitely without losing either their physico-chemical or infectious properties.

Therefore we will view viruses as molecules.

If we now turn to that property of a virus which defines it as a living organism, namely, its ability to multiply within living plants, then we will ask ourselves first whether this accomplishment is that of the host, as a living organism, or whether the host is merely the provider and protector of the virus, offering it suitable nutrients under suitable physical and chemical conditions. In other words, we are asking whether we should view the injection of a virus as a stimulus which modifies the metabolism of the host in such a way as to produce the foreign virus protein instead of its own normal protein, or whether we should view the replication as an essentially autonomous accomplishment of the virus and the host as a nutrient medium which might be replaced by a suitably offered synthetic medium.

Now it appears to me, that upon close analysis the first view can be completely excluded. If we consider that the replication of the virus requires the accurate synthesis of an enormously complicated molecule which is unknown to the host, though not as to general type, yet in all the details of its pattern and therefore of the synthetic steps involved, and if we consider further what extraordinary production an organism puts on to perform in an orderly way the most minute oxidation or synthesis in all those cases that do not involve the copying of a particular pattern—setting aside serology, which is a thing by itself—then it seems impossible to assume that the enzyme system for the host could be modified in such a far-reaching way by the injection of a virus. There can be no doubt that the replication of a virus must take place with the most direct participation of the original pattern and even without the participation of any enzymes specifically produced for this purpose.

Therefore we will look on virus replication as an autonomous accomplishment of the virus, for the general discussion of which we can ignore the host.

We next ask whether we should view virus replication as a particularly pure case of replication or whether it is, from the point of view of genetics, a complex phenomenon. Here we must first point out that with higher animals and plants which reproduce bisexually replication is certainly a very complex phenomenon. This has been shown in a thousand details by genetics, based on Mendel's laws and on modern cytology, and must be so, in order to arrive at any kind of order for the infinitely varied details of inheritance. Specifically, the close cytological analysis of the details of meiosis (reduction division) has shown that it is a specialization of the simpler mitotic division. It can easily be shown that the teleological point of this specialization lies in the possibility of trying out new hereditary factors in ever-new combinations with genes already present, and thus to increase enormously the diversity of the genotypes present at any one time, in spite of low mutation rates.

However, even the simpler mitotic cell division cannot be viewed as a pure case. If we look first at somatic divisions of higher animals and plants, then we find here that an originally simple process has been modified in the most various ways to adapt it to diverse purposes of form and function, such that one cannot speak of an undifferentiated replication. The ability to differentiate is certainly a highly important step in the transition from the protists to the multicellular organisms, but it can probably be related in a natural way to the general property of protists that they can adapt themselves to their environment and change phenotypically without changing genotypically. This phenotypic variability implies that with simple algae like Chlorella we can speak of simple replication only so long as the physical conditions are kept constant. If they are not kept constant, then, strictly speaking, we can only talk of a replication of the genomes which are embedded in a more or less well-nourished, more or less mistreated, specific protoplasma, and which, in extreme cases, may even replicate without cell division.

There can be no doubt, further, that the replication of the genome in its turn is a highly complex affair, susceptible to perturbation in its details without impairing the replication of pieces of chromosomes or of genes. Certainly the crucial element in cell replication lies in the coordination of the replication of a whole set of genes with the division of the cell. With equal certainty this coordination is not a primitive phenomenon. Rather, it requires that particular modification of a simple replication system which accomplishes constancy of supply of its own nutrient. By this modification it initiates the chain of development which until now has been subsumed under the title "life."

In view of what has been said, we want to look upon the replication of viruses as a particular form of a primitive replication of genes, the segregation of which from the nourishment supplied by the host should in principle be possible. In this sense, one should view replication not as complementary to atomic physics but as a particular trick of organic chemistry.

Such a view would mean a great simplification of the question of the origin of the many highly complicated and specific molecules found in every organism in varying quantities and indispensable for carrying out its most elementary metabolism. One would assume that these, too, can replicate autonomously and that their replication is tied only loosely to the replication of the cell. It is clear that such a view in connection with the usual arguments of the theory of natural selection would let us understand the enormous variety and complexity of these molecules, which from a purely chemical point of view appears so exaggerated.

Q.4

What Is Life?

ERWIN SCHRÖDINGER

ERWIN SCHRÖDINGER, one of the most important figures in quantum mechanics because of his work on wave mechanics, possessed a wide-ranging intelligence, with a particular disposition toward philosophy. A confirmed opponent of Nazism, Schrödinger left his job as Professor of Theoretical Physics at the University of Berlin and returned to his native Austria after Hitler came to power. He was again uprooted after the *Anschluss* of 1938, in which Germany absorbed Austria. He moved first to England and then to Dublin, where he was joined for a year by Paul Dirac, his distinguished colleague in the field of quantum theory.

In February 1943 Schrödinger delivered a series of lectures under the auspices of the Dublin Institute for Advanced Studies, which were published two years later by the Cambridge University Press in Great Britain and Macmillan in the United States under the title *What Is Life: The Physical Aspect of the Living Cell.* It is a brilliant work on the physical aspects of the living cell, drawing upon experimental data as well as mathematical models of atomic phenomena to suggest that not only is a gene or an entire chromosome a giant molecule, but also that it has its own complex structure that contains a "code," which determines the future development of that organism. Schrödinger relied on quantum mechanics to suggest that rearrangements of these structures can occur in a cell to alter or mutate its pattern of development. Although such mutations or "quantum jumps" are rare, Schrödinger examined the physics and mathematics of these shifts in structural configuration to determine when, how and why these quantum jumps occur.

Schrödinger's book was received by British reviewers with praise. One reviewer writing for the *Manchester Guardian* lauded Schrödinger's work:

"Recognizing the dilemma we are all faced with owing to the fact that the sum-total of our knowledge today exceeds the capacity of any single mind 'fully to command more than a small specialised portion of it,' he [Schrödinger] has here presented a synthesis of human knowledge as regards life. . . . In order to make this synthesis he has been at pains to try to assimilate all the material made available by specialized biological research up to date." Schrödinger's little book influenced many physicists who had been disillusioned by the atomic bombings of Japan, convincing more than a few of them to focus their attention on the largely unexplored field of molecular biology. *What Is Life?* reflects to some extent Schrödinger's own reverence for life and remains a delightful attempt to reconcile physics and biology, even though Schrödinger's effort to explain mutations using quantum theory is now outdated.

Our selections from *What Is Life?* include Schrödinger's discussions of Delbrück's model [see preceding article, Q.3]; the phenomenon of order, disorder and entropy; and the ancient problem of determinism and free will. These excerpts are illustrative of the broad cultural background against which Schrödinger wrote and the dense, fact-filled, but nevertheless singularly lucid style that characterized his thinking in the many fields that interested him.

Intelligence is characterized by a natural lack of comprehension of life.

—HENRI BERGSON

Man masters nature not by force but by understanding.

—JACOB BRONOWSKI, "The Creative Mind," *Science and Human Values*

DELBRÜCK'S MODEL DISCUSSED AND TESTED

THE GENERAL PICTURE OF THE HEREDITARY SUBSTANCE

. . . ARE THESE STRUCTURES, composed of comparatively few atoms, capable of withstanding for long periods the disturbing influence of heat motion to which the hereditary substance is continually exposed? We shall assume the structure of a gene to be that of a huge molecule, capable only of discontinuous change, which consists in a rearrangement of the atoms and leads to an isomeric molecule. The rearrangement may affect only a small region of the gene, and a vast number of different rearrangements may be possible. The energy thresholds, separating the actual configuration from any possible isomeric ones, have to be high enough (compared with the average heat energy of an atom) to make the change-over a rare event. These rare events we shall identify with spontaneous mutations. . . .

THE UNIQUENESS OF THE PICTURE

Was it absolutely essential for the biological question to dig up the deepest roots and found the picture on quantum mechanics? The conjecture that a gene is a molecule is to-day, I dare say, a commonplace. Few biologists, whether familiar with quantum theory or not, would disagree with it. . . . The subsequent considerations about isomerism, threshold energy, the paramount role of the ratio $W : kT$ in determining the probability of an isomeric transition—all that could very well be introduced on a purely empirical basis, at any rate without drawing explicitly on quantum theory. Why do I so strongly insist on the quantum-mechanical point

---

Erwin Schrödinger, *What Is Life?* New York: The Macmillan Company, 1945, pp. 56–61, 68–75, 87–91. Reprinted with permission of Cambridge University Press.

of view, though I could not really make it clear in this little work and may well have bored many a reader?

Quantum mechanics is the first theoretical aspect which accounts from first principles for all kinds of aggregates of atoms actually encountered in Nature. The Heitler-London bonding is a unique, singular feature of the theory, not invented for the purpose of explaining the chemical bond. It comes in quite by itself, in a highly interesting and puzzling manner, being forced upon us by entirely different considerations. It proves to correspond exactly with the observed chemical facts, and, as I said, it is a unique feature, well enough understood to tell with reasonable certainty that "such a thing could not happen again" in the further development of quantum theory. . . .

That is the first point I wish to make.

Some Traditional Misconceptions

But it may be asked: Are there really no other endurable structures composed of atoms except molecules? Does not a gold coin, for example, buried in a tomb for a couple of thousand years, preserve the traits of the portrait stamped on it? It is true that the coin consists of an enormous number of atoms, but surely we are in this case not inclined to attribute the mere preservation of shape to the statistics of large numbers. The same remark applies to a neatly developed batch of crystals we find embedded in a rock, where it must have been for geological periods without changing.

That leads us to the second point I want to elucidate. The cases of a molecule, a solid, a crystal are not really different. In the light of present knowledge they are virtually the same. Unfortunately, school teaching keeps up certain traditional views, which have been out of date for many years and which obscure the understanding of the actual state of affairs.

Indeed, what we have learnt at school about molecules does not give the idea that they are more closely akin to the solid state than to the liquid or gaseous state. On the contrary, we have been taught to distinguish carefully between a physical change, such as melting or evaporation in which the molecules are preserved (so that, for example, alcohol, whether solid, liquid or a gas, always consists of the same molecules, C_2H_6O), and a chemical change, as, for example, the burning of alcohol,

$$C_2H_6O+3O_2=2CO_2+3H_2O,$$

where an alcohol molecule and three oxygen molecules undergo a rearrangement to form two molecules of carbon dioxide and three molecules of water.

About crystals, we have been taught that they form three-fold periodic lattices, in which the structure of the single molecule is sometimes recognizable, as in the case of alcohol and most organic compounds, while in other crystals, e.g. rock-salt (NaCl), NaCl molecules cannot be unequivocally delimited, because every Na atom is symmetrically surrounded by six Cl atoms, and vice versa, so that it is largely arbitrary what pairs, if any, are regarded as molecular partners.

Finally, we have been told that a solid can be crystalline or not, and in the latter case we call it amorphous.

DIFFERENT "STATES" OF MATTER

Now I would not go so far as to say that all these statements and distinctions are quite wrong. For practical purposes they are sometimes useful. But in the true aspect of the structure of matter the limits must be drawn in an entirely different way. The fundamental distinction is between the two lines of the following scheme of "equations":

$$\begin{aligned} &\text{molecule} = \text{solid} = \text{crystal.} \\ &\text{gas} \qquad\; = \text{liquid} = \text{amorphous.} \end{aligned}$$

We must explain these statements briefly. The so-called amorphous solids are either not really amorphous or not really solid. In "amorphous" charcoal fibre the rudimentary structure of the graphite crystal has been disclosed by X-rays. So charcoal is a solid, but also crystalline. Where we find no crystalline structure we have to regard the thing as a liquid with very high "viscosity" (internal friction). Such a substance discloses by the absence of a well-defined melting temperature and of a latent heat of melting that it is not a true solid. When heated it softens gradually and eventually liquefies without discontinuity. (I remember that at the end of the first Great War we were given in Vienna an asphalt-like substance as a substitute for coffee. It was so hard that one had to use a chisel or a hatchet to break the little brick into pieces, when it would show a smooth, shell-like cleavage. Yet, given time, it would behave as a liquid, closely packing the lower part of a vessel in which you were unwise enough to leave it for a couple of days.)

The continuity of the gaseous and liquid state is a well-known story. You can liquefy any gas without discontinuity by taking your way "around" the so-called critical point. But we shall not enter on this here.

The Distinction That Really Matters

We have thus justified everything in the above scheme, except the main point, namely, that we wish a molecule to be regarded as solid=crystal.

The reason for this is that the atoms forming a molecule, whether there be few or many of them, are united by forces of exactly the same nature as the numerous atoms which build up a true solid, a crystal. The molecule presents the same solidity of structure as a crystal. Remember that it is precisely this solidity on which we draw to account for the permanence of the gene!

The distinction that is really important in the structure of matter is whether atoms are bound together by those "solidifying" Heitler-London forces or whether they are not. In a solid and in a molecule they all are. In a gas of single atoms (as e.g. mercury vapour) they are not. In a gas composed of molecules, only the atoms within every molecule are linked in this way.

The Aperiodic Solid

A small molecule might be called "the germ of a solid." Starting from such a small solid germ, there seem to be two different ways of building up larger and larger associations. One is the comparatively dull way of repeating the same structure in three directions again and again. That is the way followed in a growing crystal. Once the periodicity is established, there is no definite limit to the size of the aggregate. The other way is that of building up a more and more extended aggregate without the dull device of repetition. That is the case of the more and more complicated organic molecule in which every atom, and every group of atoms, plays an individual role, not entirely equivalent to that of many others (as is the case in a periodic structure). We might quite properly call that an aperiodic crystal or solid and express our hypothesis by saying: We believe a gene—or perhaps the whole chromosome fibre (that it is highly flexible is no objection; so is a thin copper wire)—to be an aperiodic solid. . . .

ORDER, DISORDER AND ENTROPY

A Remarkable General Conclusion From the Model

The molecular picture of the gene made it at least conceivable that the miniature code should be in one-to-one correspondence with a highly complicated and specified plan of development and should somehow contain the means of putting it into operation. Very well then, but how does it do this? How are we going to turn "conceivability" into true understanding?

Delbrück's molecular model, in its complete generality, seems to contain no hint as to how the hereditary substance works. Indeed, I do not expect that any detailed information on this question is likely to come from physics in the near future. The advance is proceeding and will, I am sure, continue to do so, from biochemistry under the guidance of physiology and genetics.

No detailed information about the functioning of the genetical mechanism can emerge from a description of its structure so general as has been given above. That is obvious. But, strangely enough, there is just one general conclusion to be obtained from it, and that, I confess, was my only motive for this writing.

From Delbrück's general picture of the hereditary substance it emerges that living matter, while not eluding the "laws of physics" as established up to date, is likely to involve "other laws of physics" hitherto unknown, which, however, once they have been revealed, will form just as integral a part of this science as the former.

ORDER BASED ON ORDER

This is a rather subtle line of thought, open to misconception in more than one respect. All the remaining pages are concerned with making it clear. A preliminary insight, rough but not altogether erroneous, may be found in the following considerations:

. . . [T]he laws of physics, as we know them, are statistical laws. They have a lot to do with the natural tendency of things to go over into disorder.

But to reconcile the high durability of the hereditary substance with its minute size, we had to evade the tendency to disorder by "inventing the molecule," in fact, an unusually large molecule which has to be a masterpiece of highly differentiated order, safeguarded by the conjuring rod of quantum theory. The laws of chance are not invalidated by this "invention," but their outcome is modified. The physicist is familiar with the fact that the classical laws of physics are modified by quantum theory, especially at low temperature. There are many instances of this. Life seems to be one of them, a particularly striking one. Life seems to be orderly and lawful behaviour of matter, not based exclusively on its tendency to go over from order to disorder, but based partly on existing order that is kept up.

To the physicist—but only to him—I could hope to make my view clearer by saying: The living organism seems to be a macroscopic system which in part of its behaviour approaches to that purely mechanical (as contrasted with thermodynamical) conduct to which all systems tend, as the temperature approaches the absolute zero and the molecular disorder is removed.

The non-physicist finds it hard to believe that really the ordinary laws of physics, which he regards as the prototype of inviolable precision, should be based on the statistical tendency of matter to go over into disorder. . . . The general principle involved is the famous Second Law of Thermodynamics (entropy principle) and its equally famous statistical foundation. . . . I will try to sketch the bearing of the entropy principle on the large-scale behaviour of a living organism—forgetting at the moment all that is known about chromosomes, inheritance, and so on.

Living Matter Evades the Decay to Equilibrium

What is the characteristic feature of life? When is a piece of matter said to be alive? When it goes on "doing something," moving, exchanging material with its environment, and so forth, and that for a much longer period than we would expect an inanimate piece of matter to "keep going" under similar circumstances. When a system that is not alive is isolated or placed in a uniform environment, all motion usually comes to a standstill very soon as a result of various kinds of friction; differences of electric or chemical potential are equalized, substances which tend to form a chemical compound do so, temperature becomes uniform by heat conduction. After that the whole system fades away into a dead, inert lump of matter. A permanent state is reached, in which no observable events occur. The physicist calls this the state of thermodynamical equilibrium, or of "maximum entropy."

Practically, a state of this kind is usually reached very rapidly. Theoretically, it is very often not yet an absolute equilibrium, not yet the true maximum of entropy. But then the final approach to equilibrium is very slow. It could take anything between hours, years, centuries. . . . To give an example—one in which the approach is still fairly rapid: if a glass filled with pure water and a second one filled with sugared water are placed together in a hermetically closed case at constant temperature, it appears at first that nothing happens, and the impression of complete equilibrium is created. But after a day or so it is noticed that the pure water, owing to its higher vapour pressure, slowly evaporates and condenses on the solution. The latter overflows. Only after the pure water has totally evaporated has the sugar reached its aim of being equally distributed among all the liquid water available.

These ultimate slow approaches to equilibrium could never be mistaken for life, and we may disregard them here. I have referred to them in order to clear myself of a charge of inaccuracy.

It Feeds on "Negative Entropy"

It is by avoiding the rapid decay into the inert state of "equilibrium," that an organism appears so enigmatic; so much so, that from the earliest times of human thought some special non-physical or supernatural force (*vis viva,* entelechy) was claimed to be operative in the organism, and in some quarters is still claimed.

How does the living organism avoid decay? The obvious answer is: By eating, drinking, breathing and (in the case of plants) assimilating. The technical term is *metabolism.* The Greek word (μεταβαλλειν) means change or exchange. Exchange of what? Originally the underlying idea is, no doubt, exchange of material. (E.g. the German for metabolism is Stoffwechsel.) That the exchange of material should be the essential thing is absurd. Any atom of nitrogen, oxygen, sulphur, etc., is as good as any other of its kind; what could be gained by exchanging them? For a while in the past our curiosity was silenced by being told that we feed upon energy. In some very advanced country (I don't remember whether it was Germany or the U.S.A. or both) you could find menu cards in restaurants indicating, in addition to the price, the energy content of every dish. Needless to say, taken literally, this is just as absurd. For an adult organism the energy content is as stationary as the material content. Since, surely, any calorie is worth as much as any other calorie, one cannot see how a mere exchange could help.

What then is that precious something contained in our food which keeps us from death? This is easily answered. Every process, event, happening—call it what you will; in a word, everything that is going on in Nature means an increase of the entropy of the part of the world where it is going on. Thus a living organism continually increases its entropy—or, as you may say, produces positive entropy—and thus tends to approach the dangerous state of maximum entropy, which is death. It can only keep aloof from it, i.e. alive, by continually drawing from its environment negative entropy—which is something very positive as we shall immediately see. What an organism feeds upon is negative entropy. Or, to put it less paradoxically, the essential thing in metabolism is that the organism succeeds in freeing itself from all the entropy it cannot help producing while alive.

What Is Entropy?

What is entropy? Let me first emphasize that it is not a hazy concept or idea, but a measurable physical quantity just like the length of a rod, the temperature at any point of a body, the heat of fusion of a given crystal or the specific heat of any given substance. At the absolute zero point of temperature (roughly −273 ° C.) the entropy of any substance is zero. When you bring the substance into any other state by slow, reversible little steps (even if thereby the substance changes its physical or chemical nature or splits up into two or more parts of different physical or chemical nature) the entropy increases by an amount which is computed by dividing every little portion of heat you had to supply in that procedure by the absolute temperature at which it was supplied—and by summing up all these small contributions. To give an example, when you melt a solid, its entropy increases by the amount of the heat of fusion divided by the temperature at the melting-point. You see from this, that the unit in which entropy is measured is cal/ ° C (just as the calorie is the unit of heat or the centimetre the unit of length).

The Statistical Meaning of Entropy

I have mentioned this technical definition simply in order to remove entropy from the atmosphere of hazy mystery that frequently veils it. Much more important for us here is the bearing on the statistical concept of order and disorder, a connection that was revealed by the investigation of Boltzmann and Gibbs in statistical physics. This too is an exact quantitative connection, and is expressed by

$$\text{entropy} = k \log D,$$

where k is the so-called Boltzmann constant ($=3.2983 \cdot 10^{-24}$ cal/°C), and D a quantitative measure of the atomistic disorder of the body in question. To give an exact explanation of this quantity D in brief non-technical terms is well-nigh impossible. The disorder it indicates is partly that of heat motion, partly that which consists in different kinds of atoms or molecules being mixed at random, instead of being neatly separated, e.g. the sugar and water molecules in the example quoted above. Boltzmann's equation is well illustrated by that example. The gradual "spreading out" of the sugar over all the water available increases the disorder *D,* and hence (since the logarithm of *D* increases with *D*) the entropy. It is also pretty clear that any supply of heat increases the turmoil of heat motion, that is to say increases *D* and thus increases the entropy; it is particularly clear that this should be so when you melt a crystal, since you thereby destroy the neat and permanent arrangement of the atoms or molecules and turn the crystal lattice into a continually changing random distribution.

An isolated system or a system in a uniform environment (which for the present consideration we do best to include as a part of the system we contemplate) increases its entropy and more or less rapidly approaches the inert state of maximum entropy. We now recognize this fundamental law of physics to be just the natural tendency of things to approach the chaotic state (the same tendency that the books of a library or the piles of papers and manuscripts on a writing desk display) unless we obviate it. (The analogue of irregular heat motion, in this case, is our handling those objects now and again without troubling to put them back in their proper places.)

ORGANIZATION MAINTAINED BY EXTRACTING "ORDER" FROM THE ENVIRONMENT

How would we express in terms of the statistical theory the marvellous faculty of a living organism, by which it delays the decay into thermodynamical equilibrium (death)? We said before: "It feeds upon negative entropy," attracting, as it were, a stream of negative entropy upon itself, to compensate the entropy increase it produces by living and thus to maintain itself on a stationary and fairly low entropy level.

If *D* is a measure of disorder, its reciprocal, $1/D$, can be regarded as a direct measure of order. Since the logarithm of $1/D$ is just minus the logarithm of *D,* we can write Boltzmann's equation thus:

$$-(\text{entropy}) = k \log (1/D).$$

Hence the awkward expression "negative entropy" can be replaced by a better one: entropy, taken with the negative sign, is itself a measure of order. Thus the device by which an organism maintains itself stationary at a fairly high level of orderliness (=fairly low level of entropy) really consists in continually sucking orderliness from its environment. This conclusion is less paradoxical than it appears at first sight. Rather could it be blamed for triviality. Indeed, in the case of higher animals we know the kind of orderliness they feed upon well enough, viz. the extremely well-ordered state of matter in more or less complicated organic compounds, which serve them as foodstuffs. After utilizing it they return it in a very much degraded form—not entirely degraded, however, for plants can still make use of it. (These, of course, have their most powerful supply of "negative entropy" in the sunlight.)

ON DETERMINISM AND FREE WILL

As a reward for the serious trouble I have taken to expound the purely scientific aspect of our problem *sine ira et studio,* I beg leave to add my own, necessarily subjective, view of its philosophical implications.

According to the evidence put forward in the preceding pages the space-time events in the body of a living being which correspond to the activity of its mind, to its self-conscious or any other actions, are (considering also their complex structure and the accepted statistical explanation of physico-chemistry) if not strictly deterministic at any rate statistico-deterministic. To the physicist I wish to emphasize that in my opinion, and contrary to the opinion upheld in some quarters, *quantum indeterminacy* plays no biologically relevant role in them, except perhaps by enhancing their purely accidental character in such events as meiosis, natural and X-ray-induced mutation and so on—and this is in any case obvious and well recognized.

For the sake of argument, let me regard this as a fact, as I believe every unbiased biologist would, if there were not the well-known, unpleasant feeling about "declaring oneself to be a pure mechanism." For it is deemed to contradict Free Will as warranted by direct introspection.

But immediate experiences in themselves, however various and disparate they be, are logically incapable of contradicting each other. So let us see whether we cannot draw the correct, non-contradictory conclusion from the following two premises:

(i) My body functions as a pure mechanism according to the Laws of Nature.

(ii) Yet I know, by incontrovertible direct experience, that I am directing its motions, of which I foresee the effects, that may be fateful and all-important, in which case I feel and take full responsibility for them.

The only possible inference from these two facts is, I think, that I—I in the widest meaning of the word, that is to say, every conscious mind that has ever said or felt "I"—am the person, if any, who controls the "motion of the atoms" according to the Laws of Nature.

Within a cultural milieu (*Kulturkreis*) where certain conceptions (which once had or still have a wider meaning amongst other peoples) have been limited and specialized, it is daring to give to this conclusion the simple wording that it requires. In Christian terminology to say: "Hence I am God Almighty" sounds both blasphemous and lunatic. But please disregard these connotations for the moment and consider whether the above inference is not the closest a biologist can get to proving God and immortality at one stroke.

In itself, the insight is not new. The earliest records to my knowledge date back some 2500 years or more. From the early great Upanishads the recognition ATHMAN=BRAHMAN (the personal self equals the omnipresent, all-comprehending eternal self) was in Indian thought considered, far from being blasphemous, to represent the quintessence of deepest insight into the happenings of the world. The striving of all the scholars of Vedanta was, after having learnt to pronounce with their lips, really to assimilate in their minds this grandest of all thoughts.

Again, the mystics of many centuries, independently, yet in perfect harmony with each other (somewhat like the particles in an ideal gas) have described, each of them, the unique experience of his or her life in terms that can be condensed in the phrase: DEUS FACTUS SUM (I have become God).

To western ideology the thought has remained a stranger, in spite of Schopenhauer and others who stood for it and in spite of those true lovers who, as they look into each other's eyes, become aware that their thought and their joy are *numerically* one—not merely similar or identical; but they, as a rule, are emotionally too busy to indulge in clear thinking, in which respect they very much resemble the mystic.

Allow me a few further comments. Consciousness is never experienced in the plural, only in the singular. Even in the pathological cases of split consciousness or double personality the two persons alternate, they are never manifest simultaneously. In a dream we do perform several characters at the same time, but not indiscriminately: we *are* one of them; in him we act and speak directly, while we often eagerly await the answer or response of another person, unaware of the fact that it is we who control his movements and his speech just as much as our own.

How does the idea of plurality (so emphatically opposed by the Upanishad writers) arise at all? Consciousness finds itself intimately connected with, and dependent on, the physical state of a limited region of matter, the body. (Consider the changes of mind during the development of the body, as puberty, ageing, dotage, etc., or consider the effects of fever, intoxication, narcosis, lesion of the brain and so on.) Now, there is a great plurality of similar bodies. Hence the pluralization of consciousnesses or minds seems a very suggestive hypothesis. Probably all simple ingenuous people, as well as the great majority of western philosophers, have accepted it.

It leads almost immediately to the invention of souls, as many as there are bodies, and to the question whether they are mortal as the body is or whether they are immortal and capable of existing by themselves. The former alternative is distasteful, while the latter frankly forgets, ignores or disowns the facts upon which the plurality hypothesis rests. Much sillier questions have been asked: Do animals also have souls? It has even been questioned whether women, or only men, have souls.

Such consequences, even if only tentative, must make us suspicious of the plurality hypothesis, which is common to all official Western creeds. Are we not inclining to much greater nonsense, if in discarding their gross superstitions we retain their naïve idea of plurality of souls, but "remedy" it by declaring the souls to be perishable, to be annihilated with the respective bodies?

The only possible alternative is simply to keep to the immediate experience that consciousness is a singular of which the plural is unknown; that there *is* only one thing and that, what seems to be a plurality, is merely a series of different aspects of this one thing, produced by a deception (the Indian MAJA); the same illusion is produced in a gallery of mirrors, and in the same way Gaurisankar and Mt. Everest turned out to be the same peak seen from different valleys.

There are, of course, elaborate ghost-stories fixed in our minds to hamper our acceptance of such simple recognition. E.g. it has been said that there is a tree there outside my window, but I do not really see the tree. By some cunning device of which only the initial, relatively simple steps are explored, the real tree throws an image of itself into my consciousness, and that is what I perceive. If you stand by my side and look at the same tree, the latter manages to throw an image into your soul as well. I see my tree and you see yours (remarkably like mine), and what the tree in itself is we do not know. For this extravagance Kant is responsible. In the order of ideas which regards consciousness as a *singulare tantum* it is conveniently replaced by the statement that there is obviously only *one* tree and all the image business is a ghost-story.

Yet each of us has the undisputable impression that the sum total of his own experience and memory forms a unit, quite distinct from that of any other person. He refers to it as "I." *What is this "I"?*

If you analyse it closely you will, I think, find that it is just a little bit more than a collection of single data (experiences and memories), namely the canvas *upon which* they are collected. And you will, on close introspection, find that, what you really mean by "I," is that ground-stuff upon which they are collected. You may come to a distant country, lose sight of all your friends, may all but forget them; you acquire new friends, you share life with them as intensely as you ever did with your old ones. Less and less important will become the fact that, while living your new life, you still recollect the old one. "The youth that was I," you may come to speak of him in the third person, indeed the protagonist of the novel you are reading is probably nearer to your heart, certainly more intensely alive and better known to you. Yet there has been no intermediate break, no death. And even if a skilled hypnotist succeeded in blotting out entirely all your earlier reminiscences, you would not find that he had killed *you.* In no case is there a loss of personal existence to deplore.

Nor will there ever be.

Q.5

The Molecular Configuration of Nucleic Acids

MAURICE WILKINS

MAURICE HUGH FREDERICK WILKINS was born in Pongaroa, New Zealand in 1916. His father was a physician, and when the future Nobel Laureate was six years old, the family moved to England where Wilkins received his education. He did his graduate study in Birmingham, and earned his Ph.D. in solid-state physics in 1940. His early interests were in astrophysics and related matters, but during the war he came to the United States to work at the University of California at Berkeley on the Manhattan Project. His assignment involved the separation of uranium isotopes.

After the war, Wilkins abandoned nuclear physics because, like many of his colleagues who had worked on the development of the atomic bomb (most notably Harold Urey, the brilliant American chemist), he was horrified at the destruction wrought on Hiroshima and Nagasaki. Schrödinger's book *What Is Life?* [Q.4] attracted Wilkins to the concept of applying the principles of physics to biological research. Wilkins began his career in biophysics at St. Andrews University in 1945 and then joined the Biophysics Research Unit at King's College, London the following year, where he has worked in varying capacities. Basing his thinking on the examples of Max von Laue, the discoverer of x-ray diffraction, and William Lawrence Bragg, the developer of x-ray crystallography (who had shown, a generation earlier, that X rays could be diffracted by the regular spacing of atoms in a crystal and that from the manner of the diffraction the positioning of the atoms within a crystal could be deduced*), Wilkins applied their techniques to the study of DNA, working with the British biologist Rosalind Franklin. The diffraction patterns showed the DNA molecule to be very regular and to have a double-helical structure. The contributions of Wilkins's colleague, Rosalind Franklin, were important especially in showing that the phosphate groups are to the outside of the helix, so disproving Linus Pauling's theory of DNA structure.† Their data were the pathway to the famous model of the DNA molecule discovered by Francis Crick, and James Watson, for which Wilkins, Crick, and Watson shared the Nobel Prize in 1962.

Wilkins's Nobel lecture, delivered on 11 December 1962, deals with the molecular configuration of nucleic acids. In it, Wilkins recounts the background to his work in biochemistry, paying tribute to Schrödinger's seminal volume and describing the remarkable physics department assembled by J. T. Randall in England and Scotland, where biologists, biochemists, and other specialists worked together with physicists. Wilkins's x-ray diffraction work with DNA fibers led to the realization that genetic material in animals is a pure chemical substance and to the hypothesis that its molecular structure might be surprisingly simple. Soon Wilkins joined forces with Crick and Watson, all of them aided by Franklin (who died of cancer in 1958 at the age of thirty-seven), and the result of their collaboration—the discovery of the double helical structure of DNA—became one of the great landmarks in the history of science.

*Isaac Asimov, *Asimov's Biographical Encyclopedia of Science and Technology* (2nd rev'd ed.) New York: Doubleday & Co., 1982, p. 863.

†*A Biographical Encyclopedia of Scientists.* New York: Facts On File, Inc., 1981, p. 848.

Structure with life is dead. But life without structure is unseen.
—John Cage, "Lecture on Nothing," *Silence*

The movement of life has its rest in its own music.
—Rabindranath Tagore, *Stray Birds*

Nucleic acids are basically simple. They are at the root of very fundamental biological processes, growth and inheritance. The simplicity of nucleic acid molecular structure and of its relation to function expresses the underlying simplicity of the biological phenomena, clarifies their nature, and has given rise to the first extensive interpretation of living processes in terms of macromolecular structure. These matters have only become clear by an unprecedented combination of biological, chemical and physical studies, ranging from genetics to hydrogen-bond stereochemistry. I shall not discuss all this here but concentrate on the field in which I have worked, and show how X-ray diffraction analysis has made its contribution. I shall describe some of the background of my own researches, for I suspect I am not alone in finding such accounts often more interesting than general reviews.

EARLY BACKGROUND

I took a physics degree at Cambridge in 1938, with some training in X-ray crystallography. This X-ray background was influenced by J. D. Bernal, then at the Cavendish. I began research at Birmingham, under J. T. Randall, studying luminescence and how electrons move in crystals. My contemporaries at Cambridge had mainly been interested in elementary particles, but the organization of the solid state and the special properties which depended on this organization interested me more. This may have been a forerunner of my interest in biological macromolecules and how their structure related to their highly specific properties which so largely determine the processes of life.

Maurice H. F. Wilkins, Nobel Prize in Physiology or Medicine Award Address, 1962. Reprinted with permission of Elsevier Publishing Company, and the Nobel Foundation.

During the war I took part in making the atomic bomb. When the war was ending, I, like many others, cast around for a new field of research. Partly on account of the bomb, I had lost some interest in physics. I was therefore very interested when I read Schrödinger's book "What Is Life?" and was struck by the concept of a highly complex molecular structure which controlled living processes. Research on such matters seemed more ambitious than solid-state physics. At that time many leading physicists such as Massey, Oliphant, and Randall (and later I learned that Bohr shared their view) believed that physics would contribute significantly to biology; their advice encouraged me to move into biology.

I went to work in the Physics Department at St. Andrews, Scotland, where Randall had invited me to join a biophysics project he had begun. Stimulated by Muller's experimental modification, by means of X-radiation, of genetic substance, I thought it might be interesting to investigate the effects of ultrasonics; but the results were not very encouraging.

The biophysics work then moved to King's College, London, where Randall took the Wheatstone Chair of Physics and built up, with the help of the Medical Research Council, an unusual laboratory for a Physics Department, where biologists, biochemists and others worked with the physicists. He suggested I might take over some ultraviolet microscope studies of the quantities of nucleic acids in cells. This work followed that of Caspersson, but made use of the achromatism of reflecting microscopes. By this time, the work of Caspersson and Brachet had made the scientific world generally aware that nucleic acids had important biological roles which were connected with protein synthesis. The idea that DNA might itself be the genetic substance was, however, barely hinted at. Its function in chromosomes was supposed to be associated with replication of the protein chromosome thread. The work of Avery, MacLeod, and McCarty, showing that bacteria could be genetically transformed by DNA, was published in 1944, but even in 1946 seemed almost unknown, or if known, its significance was often belittled.

It was fascinating to look through microscopes at chromosomes in cells, but I began to feel that as a physicist I might contribute more to biology by studying macromolecules isolated from cells. I was encouraged in this by Gerald Oster who came from Stanley's virus laboratory and interested me in particles of tobacco mosaic virus. As Caspersson had shown, ultraviolet microscopes could be used to find the orientation of ultraviolet absorbing groups in molecules as well as to measure quantities of nucleic acids in cells. Bill Seeds and I studied DNA, proteins, tobacco mosaic virus, vitamin B_{12}, etc. While examining oriented films of DNA prepared for ultraviolet dichroism studies, I saw in the polarizing microscope extremely uniform fibres giving clear extinction between crossed nicols. I found that fibres had been produced unwittingly while I was manipulating DNA gel. Each time that I touched the gel with a glass rod and removed the rod, a thin and almost invisible fibre of DNA was drawn out like a filament of spider's web. The perfection and uniformity of the fibres suggested that the molecules in them were regularly arranged. I immediately thought the fibres might be excellent objects to study by X-ray diffraction analysis. I took them to Raymond Gosling, who had our only X-ray equipment (made from war-surplus radiography parts) and who was using it to obtain diffraction photographs from heads of ram spermatozoa. This research was directed by Randall, who had been trained under W. L. Bragg and had worked with X-ray diffraction. Almost immediately, Gosling obtained very encouraging diffraction patterns. . . . One reason for this success was that we kept the fibres moist. We remembered that, to obtain detailed X-ray patterns from proteins. Bernal had kept protein crystals in their mother liquor. It seemed likely that the configuration of all kinds of water-soluble biological macromolecules would depend on their aqueous environment. We obtained good diffraction patterns with DNA made by Signer and Schwander which Signer brought to London to a Faraday Society meeting on nucleic acids and which he generously distributed so that all workers, using their various techniques, could study it.

REALIZATION THAT THE GENETIC MATERIAL WAS A PURE CHEMICAL SUBSTANCE AND SIGNS THAT ITS MOLECULAR STRUCTURE WAS SINGULARLY SIMPLE

Between 1946 and 1950 many lines of evidence were uncovered indicating that the genetic substance was DNA, not protein or nucleoprotein. For instance, it was found that the DNA content of a set of chromosomes was constant, and that DNA from a given species had a constant composition although the nucleotide sequence in DNA molecules was complex. It was suggested that genetic information was carried in the polynucleotide chain in a complicated sequence of the four nucleotides. The great significance of bacterial transformation now became generally recognized, and the demonstration by Hershey and Chase that bacteriophage DNA carried the viral genetic information from parent to progeny helped to complete what was a fairly considerable revolution in thought. . . .

The remarkable conclusion that a pure chemical substance was invested with a deeply significant biological activity coincided with a considerable growth of many-sided knowledge of the nature of the substance. Meanwhile we began to obtain detailed X-ray diffraction from DNA. This was the only type of data that could provide an adequate description of the 3-dimensional configuration of the molecule.

THE NEED FOR COMBINING X-RAY DIFFRACTION STUDIES OF DNA WITH MOLECULAR MODEL-BUILDING

As soon as good diffraction patterns were obtained from fibres of DNA, great interest was aroused. In our laboratory, Alex Stokes provided a theory of diffraction from helical DNA. Rosalind Franklin (who died some years later at the peak of her career) made very valuable contributions to the X-ray analysis. In Cambridge, at the Medical Research Council laboratory where structures of biological macromolecules were studied, my friends Francis Crick and Jim Watson were deeply interested in DNA structure. Watson was a biologist who had gone to Cambridge to study molecular structure. He had worked on bacteriophage reproduction and was keenly aware of the great possibilities that might be opened up by finding the molecular structure of DNA. Crick was working on helical protein structure and was interested in what controlled protein synthesis. Pauling and Corey, by their discovery of the protein α-helix, had shown that precise molecular model-building was a powerful analytical tool in its own right. The X-ray data from DNA were not so complete that a detailed picture of DNA structure could be derived without considerable aid from stereochemistry. It was clear that the X-ray studies of DNA needed to be complemented by precise molecular model-building. In our laboratory we concentrated on amplifying the X-ray data. In Cambridge, Watson and Crick built molecular models.

THE PARADOX OF THE REGULARITY OF THE DNA MOLECULE

The sharpness of the X-ray diffraction patterns of DNA showed that DNA molecules were highly regular—so regular that DNA could crystallize. The form of the patterns gave clear indications that the molecule was helical, the polynucleotide chains in the molecular thread being regularly twisted. It was known, however, that the purines and pyrimidines of various dimensions were arranged in irregular sequence along the polynucleotide chains. How could such an irregular arrangement give a highly regular structure? This paradox pointed to the solution of the DNA structure problem and was resolved by the structural hypothesis of Watson and Crick.

THE HELICAL STRUCTURE OF THE DNA MOLECULE

The key to DNA molecular structure was the discovery by Watson and Crick that, if the bases in DNA were joined in pairs by hydrogen-bonding, the overall dimensions of the pairs of adenine and thymine and of guanine and cytosine were identical. This meant that a DNA molecule containing these pairs could be highly regular in spite of the sequence of bases being irregular. Watson and Crick proposed that the DNA molecule consisted of two polynucleotide chains joined together by base-pairs. . . . If two polynucleotide chains are joined by the base-pairs, the distance between the two chains is the same for both base-pairs and, because the angle between the bonds and the C_1–C_1 line is the same for all bases, the geometry of the deoxyribose and phosphate parts of the molecule can be exactly regular.

Watson and Crick built a two-chain molecular model of this kind, the chains being helical and the main dimensions being as indicated by the X-ray data. In the model, one polynucleotide chain is twisted round the other and the sequence of atoms in one chain runs in opposite direction to that in the other. As a result, one chain is identical with the other if turned upside down, and every nucleotide in the molecule has identical structure and environment. The only irregularities are in the base sequences. The sequence along one chain can vary without restriction, but base-pairing requires that adenine in one chain be linked to thymine in the other, and similarly guanine to cytosine. The sequence in one chain is, therefore, determined by the sequence in the other, and is said to be complementary to it.

The structure of the DNA molecule in the *B* configuration is as follows: The bases are stacked on each other 3.4 Å apart and their planes are almost perpendicular to the helix axis. The flat sides of the bases cannot bind water molecules; as a result there is attraction between the bases when DNA is in an aqueous medium. This hydrophobic bonding, together with the base-pair hydrogen-bonding, stabilizes the structure.

THE WATSON-CRICK MODEL OF DNA REPLICATION, AND TRANSFER OF INFORMATION FROM ONE POLYNUCLEOTIDE CHAIN TO ANOTHER

It is essential for genetic material to be able to make exact copies of itself; otherwise growth would produce disorder, life could not originate, and favourable forms would not be perpetuated by natural selection. Base-pairing provides the means of self-replication (Watson and Crick). It also appears to be the basis of information transfer during various stages in protein synthesis.

Genetic information is written in a four-letter code in the sequence of the four bases along a polynucleotide chain. This information may be transferred from one polynucleotide chain to another. A polynucleotide chain acts as a template on which nucleotides are arranged to build a new chain. Provided that the two-chain molecule so formed is exactly regular, base-pairing ensures that the sequence in the new chain is exactly complementary to that in the parent chain. If the two chains then separate, the new chain can act as a template, and a further chain is formed; this is identical with the original chain. Most DNA molecules consist of two chains; clearly the copying process can be used to replicate such a molecule. It can also be used to transfer information from a DNA chain to an RNA chain (as is believed to be the case in the formation of messenger RNA).

Base-pairing also enables specific attachments to be made between part of one polynucleotide chain and a complementary sequence in another. Such specific interaction may be the means by which amino acids are attached to the requisite portions of a polynucleotide chain that has encoded in it the sequence of amino acids that specifies a protein. In this case the amino acid is attached to a transfer RNA molecule and part of the polynucleotide chain in this RNA pairs with the coding chain. . . .

The remarkable precision of the base-pairs reflects the exactness of DNA replication. One wonders, however, why the precision is so great, for the energy required to distort the base-pairs so that their perfection is appreciably less, is probably no greater than one quantum of thermal energy. The explanation may be that replication is a co-operative phenomenon involving many base-pairs. In any case, it must be emphasized that the specificity of the base-pairing depends on the bonds joining the bases to the deoxyribose groups being correctly placed in relation to each other. This placing is probably determined by the DNA polymerizing enzyme. Whatever the mechanics of the process are, the exact equivalence of geometry and environment of every nucleotide in the double-helix should be conducive to precise replication. Mistakes in the copying process will be produced if there are tautomeric shifts of protons involved in the hydrogen-bonding or chemical alterations of the bases. These mistakes can correspond to mutations.

THE UNIVERSAL NATURE AND CONSTANCY OF THE HELICAL STRUCTURE OF DNA

After our preliminary X-ray studies had been made, my friend Leonard Hamilton sent me human DNA he and Ralph Barclay had isolated from human leukocytes of a patient with chronic myeloid leukemia. He was studying nucleic acid metabolism in man in relation to cancer and had prepared the DNA in order to compare the DNA of normal and leukemic leukocytes. The DNA gave a very well-defined X-ray pattern. Thus began a collaboration that has lasted over many years and in which we have used Hamilton's DNA, in the form of many salts, to establish the correctness of the double-helix structure. Hamilton prepared DNA from a very wide range of species and diverse tissues. Thus it has been shown that the DNA double-helix is present in inert genetic material in sperm and bacteriophage, and in cells slowly or rapidly dividing or secreting protein. No difference of structure has been found between DNA from normal and from cancerous tissues, or in calf thymus DNA separated into fractions of different base composition by my colleague Geoffrey Brown. . . .

THE X-RAY DIFFRACTION PATTERNS OF DNA AND THE VARIOUS CONFIGURATIONS OF THE MOLECULE

X-ray diffraction analysis is the only technique that can give very detailed information about the configuration of the DNA molecule. Optical techniques, though valuable as being complementary to X-ray analysis, provide much more limited information—mainly about orientation of bonds and groups. X-ray data contributed to the deriving of the structure of DNA at two stages. First, in providing information that helped in building the Watson-Crick model; and second, in showing that the Watson-Crick proposal was correct in its essentials, which involved readjusting and refining the model.

The X-ray studies show that DNA molecules are remarkable in that they adopt a large number of different conformations, most of which can exist in several crystal forms. The main factors determining the molecular conformation and crystal form are the water and salt contents of the fibre and the cation used to neutralize the phosphate groups. . . .

In all cases the diffraction data are satisfactorily accounted for in terms of the same basic Watson-Crick structure. This is a much more convincing demonstration of the correctness of the structure than if one configuration alone were studied. The basic procedure is to adjust the molecular model until the calculated intensities of diffraction from the model correspond to those observed. . . .

RELATION OF THE MOLECULAR STRUCTURE OF RNA TO FUNCTION

Molecular model-building shows that the number of nucleotides forming the fold at the end of a transfer RNA molecule must be three or more. In our model, the fold consists of three nucleotides, each with an unpaired base. It might be that this base-triplet is the part of the molecule that attaches to the requisite part of the coding RNA polynucleotide chain that determines the sequence of amino acids in the polypeptide chain of a protein. It is believed that a base-triplet in the coding RNA corresponds to each amino acid. The triplet in the transfer RNA could attach itself specifically to the coding triplet by hydrogen-bonding and formation of base-pairs. It must be emphasized, however, that these ideas are speculative.

We suppose that part of the transfer RNA molecule interacts specifically with the enzyme that is involved in attaching the amino acid to the RNA; but we do not know how this takes place. Similarly, we know little of the way in which the enzyme involved in DNA replication interacts with DNA, or of other aspects of the mechanics of DNA replication. The presence of complementary base sequences in the transfer RNA molecule, suggests that it might be self-replicating like DNA; but there is at present little evidence to support this idea. The diffraction patterns of virus and ribosome RNA show that these molecules also contain helical regions; the function of these are uncertain too.

In the case of DNA, the discovery of its molecular structure led immediately to the replication hypothesis. This was due to the simplicity of the structure of DNA. It seems that molecular structure and function are in most cases less directly related. Derivation of the helical configuration of RNA molecules is a step towards interpreting RNA function; but more complete structural information, e.g. determination of base sequences, and more knowledge about how the various kinds of RNA interact in the ribosome, will probably be required before an adequate picture of RNA function emerges.

THE POSSIBILITY OF DETERMINING THE BASE SEQUENCE OF TRANSFER RNA BY X-RAY DIFFRACTION ANALYSIS

Since the biological specificity of nucleic acids appears to be entirely determined by their base sequences in them, determination of these sequences is probably the most fundamental problem in nucleic acid research today. The number of bases in a DNA molecule is too large for determination of base sequence by X-ray diffraction to be feasible. However, in transfer RNA the number of bases is not too large. The possibility of complete structure analysis of transfer RNA by means of X-rays is indicated by two observations. First, we have observed, in X-ray patterns of transfer RNA, separate spots each corresponding to a single crystal of RNA. We estimated their size to be about 10μ and have confirmed this estimate by observing, in the polarizing microscope, birefringent regions that probably are the crystals. It should not be too difficult to grow crystals several times larger, which is large enough for single-crystal X-ray analysis.

The second encouraging observation is that the X-ray data from DNA have restricted resolution almost entirely on account of disorientation of the microcrystals in DNA fibres. The DNA intensity data indicate that the temperature factor (B=4Å) is the same for DNA as for simple compounds. It thus appears that DNA crystals have fairly perfect crystallinity and that, if single crystals of DNA could be obtained, the intensity data would be adequate for precise determination of all atomic positions in DNA (apart from the non-periodic base sequence).

We are investigating the possibility of obtaining single crystals of DNA, but the more exciting problem is to obtain single crystals of transfer RNA with crystalline perfection equal to that of DNA, and thereby analyse base sequence. At present, the RNA crystals are much less perfect than those of DNA. However, most of our experiments have been made with RNA that is a mixture of RNA's specific for different amino acids. We have seldom used RNA that is very largely specific for one amino acid only. We hope that good preparations of such RNA may be obtained consisting of one type of molecule only. We might expect such RNA to form crystals as perfect as those of DNA. If so, there should be no obstacle to the direct analysis of the whole structure of the molecule, including the sequence of the bases and the fold at the end of the helix. We may be over-optimistic, but the recent and somewhat unexpected successes of X-ray diffraction analysis in the nucleic acid and protein fields are cause for optimism. . . .

Q.6

On the Genetic Code

FRANCIS CRICK

AS ONE OF the protagonists in the American biologist James Watson's acclaimed bestseller, *The Double Helix,* Francis Crick is well-known as a public figure, a rare phenomenon in the higher reaches of science. Horace Freeland Judson, author of *The Eighth Day of Creation: The Makers of the Revolution in Biology,* has written:

> Talk is Crick's life in science as it is to few others, for he has deliberately taken on a singular role in molecular biology: Francis Crick is the theorist. When I have heard him, over the lunch table, explain a new idea to his peers, his reminding them of details they should already know has seemed an almost absent-minded orderliness of exposition, just to save later backtracking.*

However annoying this oddity may be to some of his fellow scientists, Crick always knows what he is talking about. The late Jacques Monod, one of the greatest minds in modern biology, once remarked, "No one man discovered or created molecular biology. But one man dominates intellectually the whole field, because he knows the most and understands the most. Francis Crick."†

Crick was born in England in 1916 and was educated at University College at London and at Cambridge University, where he obtained his Ph.D. in 1953. During World War II, he worked on radar and magnetic mines. When he left his wartime post in 1947 (where he had lingered for a

---

*Horace Freeland Judson, *The Eighth Day of Creation.* New York: Simon & Schuster, 1979, p. 108.
†*Ibid.*, p. 109.

time after peace came), Crick started to teach himself biology, apparently reading the relevant literature with a voraciousness that astounded even his well-read colleagues. Crick confided to Judson that he had turned to biology in part because "he is an atheist and was impatient to throw light into the remaining shadowy sanctuaries of vitalistic illusions."*

"I had read Schrödinger's little book, too," Judson quotes Crick as saying. "Essentially if you read that book fairly critically, the main import is very peculiar; for one thing, it's a book written by a physicist who doesn't know any chemistry! But the impact—there's no doubt that Schrödinger wrote it in a compelling style, not like the junk that most people write, and it was imaginative. It suggested that biological problems could be *thought* about, in physical terms—and thus it gave the impression that exciting things in this field were not far off."†

In 1953, Crick earned lasting fame when he and Watson proposed a double-helical chain model for the deoxyribonucleic acid (DNA) molecule. Their hypothesis drew upon the x-ray diffraction photographs of Maurice Wilkins, which had indicated that DNA was helical, although it fell to Crick and Watson to perceive that DNA consists of a double helix, with two strands wound around each other, each strand having the opposite sequence of atoms. The double helix model offers an ingenious explanation for the replication of hereditary material, suggesting that the two coiled strands of the DNA molecule separate and that each strand forms a copy of its opposing DNA strand from nearby molecules. Crick and Watson's discovery of the structure of DNA (which owed a heavy debt to Wilkins' x-ray diffraction photography) was rightly hailed as one of the greatest discoveries in biology of the century and provided the impetus for a wide range of investigations into how enzymes are formed, and the ways in which they promote the metabolism, growth, and differentiation of cells.

People who talk as much and as well as Crick usually write well if they are relaxed, and the British molecular biologist confirms this generalization here. His Nobel lecture, "On the Genetic Code," is a masterpiece of clear scientific writing. Crick never takes refuge in scientific jargon, the traditional refuge of academics in moments of confusion.

---

* *Ibid.*

† *Ibid.*

Nothing becomes real till it is experienced—even a proverb is no proverb to you till your life has illustrated it.

—JOHN KEATS, *Letters*

May the universe in some strange sense be "brought into being" by the participation of those who participate? . . . the vital act is the act of participation. "Participator" is the incontrovertible new concept given by quantum mechanics. It strikes down the term "observer" of classical theory, the man who stands safely behind the thick glass wall and watches what goes on without taking part. It can't be done, quantum mechanics says.

—JOHN ARCHIBALD WHEELER, *Gravitation*

. . . I SHALL DISCUSS here the present state of a . . . problem in information transfer in living material—that of the genetic code—which has long interested me, and on which my colleagues and I, among many others, have recently been doing some experimental work.

It now seems certain that the amino acid sequence of any protein is determined by the sequence of bases in some region of a particular nucleic acid molecule. Twenty different kinds of amino acid are commonly found in protein, and four main kinds of base occur in nucleic acid.The genetic code describes the way in which a sequence of twenty or more things is determined by a sequence of four things of a different type.

It is hardly necessary to stress the biological importance of the problem. It seems likely that most if not all the genetic information in any organism is carried by nucleic acid—usually by DNA, although certain small viruses use RNA as their genetic material. It is probable that much of this information is used to determine the amino acid sequence of the proteins of that organism. (Whether the genetic information has any other major function we do not yet know.) This idea is expressed by the classic slogan of Beadle: "one gene–one enzyme," or in the more sophisticated but cumbersome terminology of today: "one cistron–one polypeptide chain."

It is one of the more striking generalizations of biochemistry—which

---

Francis H. C. Crick, Nobel Prize in Physiology and Medicine Award Address, 1962. Reprinted with permission of Elsevier Publishing Company, and the Nobel Foundation.

surprisingly is hardly ever mentioned in the biochemical text-books—that the twenty amino acids and the four bases, are, with minor reservations, the same throughout Nature. As far as I am aware the presently accepted set of twenty amino acids was first drawn up by Watson and myself in the summer of 1953 in response to a letter of Gamow's.

I shall not deal with the intimate technical details of the problem. . . . Rather I shall ask certain general questions about the genetic code and ask how far we can now answer them.

Let us assume that the genetic code is a simple one and ask how many bases code for one amino acid? This can hardly be done by a pair of bases, as from four different things we can only form $4 \times 4 = 16$ different pairs, whereas we need at least twenty and probably one or two more to act as spaces or for other purposes. However, triplets of bases would give us 64 possibilities. It is convenient to have a word for a set of bases which codes one amino acid and I shall use the word "codon" for this.

This brings us to our first question. Do codons overlap? In other words, as we read along the genetic message do we find a base which is a member of two or more codons? It now seems fairly certain that codons do *not* overlap. If they did, the change of a single base, due to mutation, should alter two or more (adjacent) amino acids, whereas the typical change is to a single amino acid, both in the case of the "spontaneous" mutations, such as occur in the abnormal human haemoglobin or in chemically induced mutations, such as those produced by the action of nitrous acid and other chemicals on tobacco mosaic virus. In all probability, therefore, codons do not overlap.

This leads us to the next problem. How is the base sequence divided into codons? There is nothing in the backbone of the nucleic acid, which is perfectly regular, to show us how to group the bases into codons. If, for example, all the codons are triplets, then in addition to the correct reading of the message, there are two *in*correct readings which we shall obtain if we do not start the grouping into sets of three at the right place. My colleagues and I have recently obtained experimental evidence that each section of the genetic message is indeed read from a fixed point, probably from one end. This fits in very well with the experimental evidence, most clearly shown in the work of Dintzis that the amino acids are assembled into the polypeptide chain in a linear order, starting at the amino end of the chain.

This leads us to the next general question: the size of the codon. How many bases are there in any one codon? The same experiments to which I have just referred strongly suggest that all (or almost all) codons consist of a triplet of bases, though a small multiple of three, such as six or nine, is not completely ruled out by our data. We were led to this conclusion by the study of mutations in the A and B cistrons of the r_{11} locus of bacteriophage T4. These mutations are believed to be due to the addition or subtraction of one or more bases from the genetic message. They are typically produced by acridines, and cannot be reversed by mutagens which merely change one base into another. Moreover these mutations almost always render the gene completely inactive, rather than partly so.

By testing such mutants in pairs we can assign them all without exception to one of two classes which we call + and –. For simplicity one can think of the + class as having one extra base at some point or other in the genetic message and the – class as having one too few. The crucial experiment is to put together, by genetic recombination, three mutants of the same type into one gene. That is, either (+with+with+) or (–with–with–). Whereas a single + or a pair of them (+with+) makes the gene completely inactive, a set of three, suitably chosen, has some activity. Detailed examination of these results show that they are exactly what we should expect if the message were read in triplets starting from one end.

We are sometimes asked what the result would be if we put four +'s in one gene. To answer this my colleagues have recently put together not merely four but six +'s. Such a combination is active as expected on our theory, although sets of four or five of them are not. We have also gone a long way to explaining the production of "minutes" as they are called. That is, combinations in which the gene is working at very low efficiency. Our detailed results fit the hypothesis that in some cases when the mechanism comes to a triplet which does not stand for an amino acid (called a "nonsense" triplet) it very occasionally makes a slip and reads, say, only two bases instead of the usual three. These results also enable us to tie down the direction of reading of the genetic message, which in this case is from left to right, as the r_{11} region is conventionally drawn. We plan to write up a detailed technical account of all this work shortly. A final proof of our ideas can only be obtained by detailed studies on the alterations produced in the amino acid sequence of a protein by mutations of the type discussed here.

One further conclusion of a general nature is suggested by our results. It would appear that the number of nonsense triplets is rather low, since we only occasionally come across them. However this conclusion is less secure than our other deductions about the general nature of the genetic code.

It has not yet been shown directly that the genetic message is co-linear with its product. That is, that one end of the gene codes for the amino end of the polypeptide chain and the other for the carboxyl end, and that as one proceeds along the gene one comes in turn to the codons in between in the linear order in which the amino acids are found in the polypeptide chain. This seems highly likely, especially as it has been shown that in several systems mutations affecting the same amino acid are extremely near together on the genetic map. The experimental proof of the co-linearity of a gene and the polypeptide chain it produces may be confidently expected within the next year or so.

There is one further general question about the genetic code which we can ask at this point. Is the code universal, that is, the same in all organisms. Preliminary evidence suggests that it may well be. For example something very like rabbit haemoglobin can be synthesized using a cell-free system, part of which comes from rabbit reticulocytes and part from *Escherichia coli.* This would not be very probable if the code were very different in these two organisms. However as we shall see it is now possible to test the universality of the code by more direct experiments.

In a cell in which DNA is the genetic material it is not believed that DNA itself controls protein synthesis directly. As Watson has described, it is believed that the base sequence of the DNA—probably of only one of its chains—is copied onto RNA, and that this special RNA then acts as the genetic messenger and directs the actual process of joining up the amino acids into polypeptide chains. The breakthrough in the coding problem has come from the discovery, made by Nirenberg and Matthaei, that one can use synthetic RNA for this purpose. In particular they found that polyuridylic acid—an RNA in which every base is uracil—will promote the synthesis of polyphenylalanine when added to a cell-free system which was already known to synthesize polypeptide chains. Thus one codon for phenylalanine appears to be the sequence UUU (where U stands for uracil: in the same way we shall use A, G, and C for adenine, guanine, and cytosine respectively). This discovery has opened the way to a rapid although somewhat confused attack on the genetic code.

It would not be appropriate to review this work in detail here. . . . Such is the pace of work in this field that more recent experiments have already made it out of date to some extent. However, some general conclusions can safely be drawn.

The technique mainly used so far, both by Nirenberg and his colleagues and by Ochoa and his group, has been to synthesize enzymatically "random" polymers of two or three of the four bases. For example, a polynucleotide, which I shall call poly (U,C), having about equal amounts of uracil and cytosine in (presumably) random order will increase the incorporation of the amino acids phenylalanine, serine, leucine, and proline, and possibly threonine. By using polymers of different composition and assuming a triplet code one can deduce limited information about the composition of certain triplets.

From such work it appears that, with minor reservations, each polynucleotide incorporates a characteristic set of amino acids. Moreover the four bases appear quite distinct in their effects. A comparison between the triplets tentatively deduced by these methods with the *changes* in amino acid sequence produced by mutation shows a fair measure of agreement. Moreover the incorporation requires the same components needed for protein synthesis, and is inhibited by the same inhibitors. Thus the system is most unlikely to be a complete artefact and is very probably closely related to genuine protein synthesis. . . .

It now seems very likely that many of the 64 triplets, possibly most of them, may code one amino acid or another, and that in general several distinct triplets may code one amino acid. In particular a very elegant experiment suggests that both (UUC) and (UUG) code leucine (the brackets imply that the order within the triplets is not yet known). This general idea is supported by several indirect lines of evidence which cannot be detailed here. Unfortunately it makes the unambiguous determination of triplets by these methods much more difficult than would be the case if there were only one triplet for each amino acid. Moreover, it is not possible by using polynucleotides of "random" sequence to determine the *order* of bases in a triplet. A start has been made to construct polynucleotides whose exact sequence is known at one end, but the results obtained so far are suggestive rather than conclusive. It seems likely however from this and other unpublished evidence that the amino end of the polypeptide chain corresponds to the "right-hand" end of the polynucleotide chain—that is, the one with the 2′, 3′ hydroxyls on the sugar.

It seems virtually certain that a single chain of RNA can act as messenger RNA, since poly U is a single chain without secondary structure. If poly A is added to poly U, to form a double or triple helix, the combination is inactive. Moreover there is preliminary evidence which suggests that secondary structure within a polynucleotide inhibits the power to stimulate protein synthesis.

It has yet to be shown by direct biochemical methods, as opposed to the indirect genetic evidence mentioned earlier, that the code is indeed a triplet code.

Attempts have been made from a study of the changes produced by mutation to obtain the relative order of the bases within various triplets, but my own view is that these are premature until there is more extensive and more reliable data on the composition of the triplets.

Evidence presented by several groups suggests that poly U stimulates both the incorporation of phenylalanine and also a lesser amount of leucine. The meaning of this observation is unclear, but it raises the unfortunate possibility of ambiguous triplets; that is, triplets which may code more than one amino acid. However, one would certainly expect such triplets to be in a minority.

It would seem likely, then, that most of the sixty-four possible triplets will be grouped into twenty groups. The balance of evidence both from the cell-free system and from the study of mutation, suggests that this does not occur at random, and that triplets coding the same amino acid may well be rather similar. This raises the main theoretical problem now outstanding. Can this grouping be deduced from theoretical postulates? Unfortunately, it is not difficult to see how it might have arisen at an extremely early stage in evolution by random mutations, so that the particular code we have may perhaps be the result of a series of historical accidents. This point is of more than abstract interest. If the code does indeed have some logical foundation then it is legitimate to consider all the evidence, both good and bad, in any attempt to deduce it. The same is not true if the codons have no simple logical connection. In that case, it makes little sense to guess a codon. The important thing is to provide enough evidence to prove each codon independently. It is not yet clear what evidence can safely be accepted as establishing a codon. What is clear is that most of the experimental evidence so far presented falls short of proof in almost all cases.

In spite of the uncertainty of much of the experimental data there are certain codes which have been suggested in the past which we can now reject with some degree of confidence.

COMMA-LESS TRIPLET CODES

All such codes are unlikely, not only because of the genetic evidence but also because of the detailed results from the cell-free system.

TWO-LETTER OR THREE-LETTER CODES

For example a code in which A is equivalent to O, and G to U. As already stated, the results from the cell-free system rule out all such codes.

THE COMBINATION TRIPLET CODE

In this code all permutations of a given combination code the same amino acid. The experimental results can only be made to fit such a code by very special pleading.

COMPLEMENTARY CODES

There are several classes of these. Consider a certain triplet in relation to the triplet which is complementary to it on the other chain of the double helix. The second triplet may be considered either as being read in the same direction as the first, or in the opposite direction. Thus if the first triplet is UCC, we consider it in relation to either AGG or (reading in the opposite direction) GGA.

It has been suggested that if a triplet stands for an amino acid its complement necessarily stands for the same amino acids, or, alternatively in another class of codes, that its complement will stand for no amino acid, i.e. be nonsense.

It has recently been shown by Ochoa's group that poly A stimulates the incorporation of lysine. Thus presumably AAA codes lysine. However since UUU codes phenylalanine these facts rule out all the above codes. It is also found that poly (U,G) incorporates quite different amino acids from poly (A,C). Similarly poly (U,C) differs from poly (A,G). Thus there is little chance that any of this class of theories will prove correct. Moreover they are all, in my opinion, unlikely for general theoretical reasons.

A start has already been made, using the same polynucleotides in cell-free systems from different species, to see if the code is the same in all organisms. Eventually it should be relatively easy to discover in this way if the code is universal, and, if not, how it differs from organism to organism. The preliminary results presented so far disclose no clear difference between *E. coli* and mammals, which is encouraging.

At the present time, therefore, the genetic code appears to have the following general properties:

(1) Most if not all codons consist of three (adjacent) bases.
(2) Adjacent codons do not overlap.
(3) The message is read in the correct groups of three by starting at some fixed point.
(4) The code sequence in the gene is co-linear with the amino acid se-

quence, the polypeptide chain being synthesized sequentially from the amino end.

(5) In general more than one triplet codes each amino acid.

(6) It is not certain that some triplets may not code more than one amino acid, i.e. they may be ambiguous.

(7) Triplets which code for the same amino acid are probably rather similar.

(8) It is not known whether there is any general rule which groups such codons together, or whether the grouping is mainly the result of historical accident.

(9) The number of triplets which do not code an amino acid is probably small.

(10) Certain codes proposed earlier, such as comma-less codes, two- or three-letter codes, the combination code, and various transposible codes are all unlikely to be correct.

(11) The code in different organisms is probably similar. It may be the same in all organisms but this is not yet known.

Finally one should add that in spite of the great complexity of protein synthesis and in spite of the considerable technical difficulties in synthesizing polynucleotides with defined sequences it is not unreasonable to hope that all these points will be clarified in the near future, and that the genetic code will be completely established on a sound experimental basis within a few years.

Q.7

Physics and the Explanation of Life

EUGENE P. WIGNER

IN 1969, THE Hungarian-American physicist Eugene P. Wigner, whose long career ranges from his early studies in mathematics under the direction of David Hilbert in Göttingen to the winning of the U.S. Atomic Energy Commission's Enrico Fermi Prize in 1958 and the Atoms for Peace Award in 1960, turned his attention to the burgeoning science of molecular biology in an address delivered in Boston at the annual meeting of the American Association for the Advancement of Science. His argument is simple but fetchingly presented: In the past, Wigner says, physical theory has often dealt selectively with unusual situations (and he regards the application of physical laws to life processes as unusual); the application of theories to phenomena on one level often requires that phenomena on other levels be ignored.

Planetary theory, for example, neglects all but gravitational forces; macroscopic—that is, nonmolecular or nonatomic—physics neglects fluctuations due to the atomic structure of matter; nuclear physics disregards weak and gravitational interactions. In some of these cases, Wigner claims, scientists were quite aware they were dealing with special situations, or limiting cases, as he calls them; in other cases they were not so aware.

Even if present-day physics could accurately describe the motion of the physical constituents of living bodies, the full story about how physical laws govern biological processes would not be forthcoming. And Wigner's final point seems to be that the laws of physics, in their various developmental versions so splendidly fitted to inanimate matter, will have to be modified when dealing with the more general situation in which life and consciousness play significant roles.

That Wigner lectures on the relationship between physics and biology is not surprising given the variety of areas in which he has made important contributions: He has worked on the spectra of chemical compounds, introduced the notion of parity in nuclear reactions based on his observations of the transformation of Schrödinger wave functions, showed that the nuclear force which binds protons and neutrons is short-ranged and charge-independent, helped to formulate the Breit-Wigner formula which explains the absorption of neutrons by compound nuclei and done much of the early work (along with Enrico Fermi) which led to the first controlled fission reaction. The breadth of his work in physics was illustrated by the vagueness of the Nobel Foundation's presentation to him of the 1963 Nobel Prize in physics for "systematically improving and extending the methods of quantum mechanics and applying them widely."

Wigner is not adamant; he simply questions. It may be, he seems to say, that if science reaches a point where life scientists and quantum physicists meet to work on mutual problems, the ingenious craft of the latter will have to be modified (as has often happened when the subatomic world of particles has revealed itself in its varied mystery) to cope with the new challenges.

The being with a consciousness must have a different role in quantum mechanics than the inanimate measuring device.

In other words, the impression which one gains at an interaction, called also the result of an observation, modifies the wave function of the system. The modified wave function is, furthermore, in general unpredictable before the impression gained at the interaction has entered our consciousness: it is the entering of an impression into our consciousness which alters the wave function because it modifies our appraisal of the probabilities for different impressions which we expect to receive in the future. It is at this point that the consciousness enters the theory unavoidably and unalterably.

—EUGENE P. WIGNER, *Symmetries and Reflections*

A PHYSICIST'S VIEW ON THE MIND-BODY PROBLEM: THE FIRST ALTERNATIVE

PHYSICS HAS, at present, no perfect, well-rounded theory which could be applied to all phenomena involving inanimate matter. Nevertheless, most of us physicists feel that such a theory is attainable without delving deeper into the problems of life and consciousness, that the basic principles of quantum mechanics need not be modified in order to arrive at a theory which correctly accounts for all regularities in the behavior of inanimate matter. Even though the present writer is well aware of the cavalier nature of the picture which present quantum-mechanical theory presupposes about consciousness, as long as this theory deals with inanimate matter he would not seek for its improvement by means of a closer analysis of the content of consciousness. To put it in a somewhat vulgar fashion, even most physicists, if unexpectedly presented with the question of the validity of the laws of physics for organic matter, would affirm that validity. On the other side of the chasm, many, if not most, microbiolo-

---

Eugene P. Wigner, "Physics and the Explanation of Life," in *Foundations of Physics,* vol. 1, 1970, pp. 35–45. Reprinted with permission of Plenum Publishing Company.

gists would concur in this view. The view does not lead automatically to Huxley's view that we are automata, since the present laws of physics are not deterministic but have a probabilistic character. Furthermore, if we are honest about it,we cannot now formulate laws of physics valid for inanimate objects under all conditions. Hence, the statement which we are considering should be formulated somewhat more cautiously: that laws of nature, for the formulation of which observations on inanimate matter suffice, are valid also for living beings. In other words, physical laws, obtained by studying the traditional subjects of physics, and perhaps not very different from those that physicists are trying to formulate now, will form the basis from which the behavior of living matter can be derived—derived perhaps with a great deal of effort and computing, but still correctly derived.

The assumption just formulated is surely logically possible. It is very close to Huxley's views which were mentioned before. Would it mean that, eventually, the whole science of the mind will become applied physics? In my opinion this would not be the case even if the assumption which we are discussing were correct. What we are interested in is not only, and not principally, the motion of the molecules in a brain but, to use Descartes' terminology, the sensations which are experienced by the soul which is linked to that brain, whether pain or pleasure, stimulation or anxiety, whether it thinks of love or prime numbers. In order to obtain an answer to these questions, the physical characterization of the state of the brain would have to be translated into psychological-emotional terms.

It may be useful to give an example from purely physical theory for the need for such a translation. The example which I most like to present derives from the classical theory of the electromagnetic field in vacuum, that is, the simplest form of Maxwell's equations. These give the time derivative of the electric field E in terms of the magnetic field H, and the time derivative of the magnetic field in terms of the electric field. Both fields are free of sources. Although the actual form of the equations is not very relevant for our discussion, writing out these equations may render the discussion more concrete:

$$\partial H/\partial t = -c \text{ curl } E, \qquad \partial E/\partial t = c \text{ curl } H \tag{1}$$

These equations will be called, briefly, Maxwell's equations; actually, they are Maxwell's equations for empty space; c is the velocity of light. If E and H are given at one instant of time, these equations permit their calculation for all later times, and for all earlier times. They will serve as model equations for the discussion which follows—they give both sides of the picture, the electric and the magnetic side, and do not prefer one over the other.

It is possible, however, to formulate an equation for the magnetic field alone. This is again, and should remain, free of sources and its time dependence is regulated by the equation

$$\frac{\partial^2 H}{\partial t^2} = c^2 \left(\frac{\partial^2 H}{\partial x^2} + \frac{\partial^2 H}{\partial y^2} + \frac{\partial^2 H}{\partial z^2} \right) \tag{2}$$

One can observe now that, if H and $\partial H/\partial t$ are given at one instant of time, (2) permits the calculation of H for all later times—and for all earlier times. Is now this equation, which is just as valid as Maxwell's original equations, a full substitute for the latter? The answer is no. If we want to obtain the force on a small charge at rest, the original form of the equations furnishes this directly: it is the electric field at the place where the charge is, multiplied by the magnitude of the charge. In order to obtain the force from (2), referring only to the magnetic field H, one has to calculate first the electric field in terms of H. This can be done, though the formula is not quite simple:

$$E(r) = -\operatorname{curl} \int\int\int \frac{\partial H(r')}{\partial t} \frac{d^3 r'}{4\pi c|r - r'|} \tag{3}$$

Equation (3) gives the translation of the magnetic field into the electric one and this is, in the case considered, more relevant than the magnetic field itself. We have, therefore, an example before us in which a theory—Eq. (2) for H alone—is completely valid but is not very useful without the translation which should go with it.

The preceding example also shows that the translation into the more relevant quantity can be quite complicated—more complicated than the underlying theory, that is, Eq. (2). The translation equation is also more complicated than the set of equations, in this case Maxwell's equations, which uses both concepts: the one which turns out to be the more relevant one, that is, E, along with the other, H, which does suffice for the formulation of the time dependence. It is unnecessary to remark that, in the preceding illustration of a future theory of life, H plays the role of the purely physical variables, E plays the role of the psychological variables. In this illustration, the use of both types of variables in the basic equations is much preferable to the use of only one of them—the problem of translation does not arise in that case.

The example just given illustrates also the observation on the meaningless nature of the concept of causation in a deterministic theory—such as Maxwell's theory of the electromagnetic field in vacuum. Looking only at Eq. (2), and the translation Eq. (3) thereof, one will conclude that the magnetic field is the prime quantity, its development is determined by its magnitude in the past. The electric field will appear as a derived quantity, caused and generated by the magnetic field. Maxwell's original form of the equations shows, on the other hand, the possibility (and, in the opinion of the physicists, the desirability) to consider the two to have equal rank and primitivity. As a matter of fact, relativity theory teaches us that the three components of E and the three components of H are all components of a single tensor, the electromagnetic field tensor. The fact that they appear in Maxwell's equations as equals was helpful when the special theory of relativity was first formulated.

I believe I have discussed the assumption that the laws of physics, in the sense described, are valid also for living matter. We also saw that this assumption need not imply, as is often postulated, that the mind and the consciousness are only unimportant derived concepts which need not enter the theory at all. It may even be possible to give them the privileged status. Let us now turn to the assumption opposite to the "first alternative" considered so far: that the laws of physics will have to be modified drastically if they are to account for the phenomena of life. Actually, I believe that this second assumption is the correct one.

THE SECOND ALTERNATIVE: LIFE MODIFIES THE LAWS VALID FOR INANIMATE NATURE

I wish to begin this discussion by recalling how wonderfully actual situations in the world have helped us to discover laws of nature. The story may well begin with Newton and his law of gravitation. It is hard to imagine how he could have discovered this, had he not had the solar system before him in which only gravitational forces play a significant role. The discovery, also due to Newton, that these forces also determine the motion of the Moon around the Earth, and the motion of freely falling bodies too heavy to be much affected by air resistance, was a wonderful example for science's power to create a unified point of view for phenomena which had, originally, widely differing characters. Newton, of course, recognized that there must be other forces in addition to the gravitational ones—forces which, however, remain of negligible importance as far as the motion of the planets is concerned.

Newton's discovery was followed by the discovery of most laws of macroscopic physics. Maxwell's laws of electromagnetism—the ones which were just considered in the special case of absence of matter—are perhaps the most remarkable among these. Again, these laws—those of macroscopic physics—could not have been discovered were not all the common objects which surround us of macroscopic nature, containing many millions of atoms, so that quantum effects, for instance, play no role in their gross behavior. Again, the unifying power of science manifested itself in a spectacular way: it turned out that Maxwell's equations also describe light and, as we now know, all electromagnetic radiation from radio waves to X-rays.

The next step of comparable, perhaps even greater, importance was the development of microscopic physics, starting with the theory of heat and soon leading to quantum theory. Most of this development took place in the first half of our century but, in some regards, the development is still incomplete. If we assume that it can and will be completed—most of us believe this—the question which we should face is whether our present microscopic theories also presuppose some special situation, the absence of certain forces or circumstances. The point of view which we are discussing maintains that this is the case. Just as gravitational theory can describe only the situation in which no other but gravitational forces play a role, and macroscopic physics describes only situations in which all bodies present consist of many millions of atoms, present microscopic theory describes only situations in which life and consciousness play no active role.

Similarly, just as macroscopic physics contains gravitational theory as a special case, applicable whenever only gravitational forces play a significant role, and just as microscopic physics contains macroscopic physics as a special case, valid for bodies which contain millions of atoms, in the same way, the theory which is here anticipated should contain present microscopic physics as a special case, valid for inanimate objects. Thus, each successive theory is expected to be a generalization of the preceding one, to recognize the regularities which its antecedent postulated, but to recognize them as valid only under special conditions. This should apply also to the theory postulated here, in the form of the "second alternative."

Naturally, the preceding story does not *prove* that the present, microscopic physics will also have to be generalized, that the laws of nature as we now know them, or try to establish them, are only limiting cases, just as the planetary system, macroscopic physics, were limiting cases. In other words, it does not prove that our second alternative, rather than the first one, is correct. Can arguments be adduced to show the need for modification? There seem to be two such arguments.

The first is that, if one entity is influenced by another entity, in all known cases the latter one is also influenced by the former. The most striking and originally least expected example for this is the influence of light on matter, most obviously in the form of light pressure. That matter influences light is an obvious fact—if it were not so, we could not see objects. The influence of light on matter is, however, a more subtle effect and is virtually unobservable under the conditions which surround us. Light pressure is, however, by now a well demonstrated phenomenon and it plays a decisive role in the interior of stars. More generally, we do not know any case in which the influence is entirely one-sided. Since matter clearly influences the content of our consciousness, it is natural to assume that the opposite influence also exists, thus demanding a modification of the presently accepted laws of nature which disregard this influence.

The second argument which I like to put forward is that all extensions of physics to new sets of phenomena were accompanied by drastic changes in the theory. In fact, most were accompanied by drastic changes of the entities for which the laws of physics were supposed to establish regularities. These were the positions of bodies in Newton's theory and the developments which soon followed his theory. They were the intensities of fields as functions of position and time in Maxwell's theory. These were replaced then by the outcomes of observations (the perceptions referred to before) in modern microscopic physics, that is, quantum mechanics. In the development which we are trying to envisage, leading to the incorporation of life, consciousness, and mind into physical theory, the change of the basic entities indeed appears unavoidable: the observation, being the entity which plays the primitive role in the theory, cannot be further analyzed within that theory. Similarly, Newtonian theory did not further analyze the meaning of the position of an object, field theory did not analyze further the concept of the field. If the concept of observation is to be further analyzed, it cannot play the primitive role it now plays in the theory and this will have to establish regularities between entities different from the outcomes of observations. An alteration of the basic concepts of the theory is necessary.

These are the two arguments in favor of the second alternative, that the laws of physics which result from the study of inanimate objects only are not adequate for formulating the laws for situations in which life and consciousness are relevant parts of the picture.

CONCLUSION AND SUMMARY

Clearly, the hope expressed in the last two sections, that man shall acquire deeper insights into mental processes, into the character of our consciousness, is only a hope. The intellectual capabilities of man may have their limits just as the capabilities of other animals have. The hope does imply, though, that the mental and emotional processes of men and animals will be the subjects of scrutiny just as processes in inanimate matter are subjects of scrutiny now. The knowledge of mind and consciousness may be less sharp and detailed than is the knowledge given by present-day physics on the behavior of inanimate objects. The expectation is, nevertheless, that we can view mind and consciousness—at least those of other living beings—from the outside so that their perceptions will not be the primitive concepts in terms of which all laws and correlations are formulated. As to the loss in the sharpness and detail of the laws, this is probably unavoidable. It has taken place throughout the history of physics. Newton could determine all the initial conditions of the system of his interest and could foresee its behavior into the indefinite future. Maxwell's and his contemporaries' theories can be verified only by creating conditions artificially under which a verification is possible. Even then, it is possible only for limited periods of time. The laws of quantum mechanics, finally, neither make definite predictions under all conditions, nor have its equations of motion been verified in any detail similar to those of macroscopic theories. A further retrenchment of our demands for detail of verification is probably unavoidable whenever we extend our interest to a wider variety of phenomena.

As to the usefulness of the preceding considerations, I must admit that I do not see much of it. This may well be the reason for the lack of a more acute interest on the part of physicists in the questions discussed. Even if not useful, I would like to summarize them in concluding this article.

I believe that the present laws of physics are at least incomplete without a translation into terms of mental phenomena. More likely, they are inaccurate, the inaccuracy increasing with the increase of the role which life plays in the phenomena considered. The example of the wasp which does not seem to have sensations may indicate that even animals of considerable complexity are not far from being automata, largely subject to the present ideas of physics. On the other hand, the fact that the laws of microscopic physics are formulated in terms of observations is strong evidence that these laws become invalid for the description of observations whenever consciousness plays a decisive role. This also constitutes the difference between the view here represented and the views of traditional philosophers. They considered body and soul as two different and separate entities, though interacting with each other. The view given here considers inanimate matter as a limiting case in which the phenomena of life and consciousness play as little a role as the nongravitational forces play in planetary motion, as fluctuations play in macroscopic physics. It is argued that, as we consider situations in which consciousness is more and more relevant, the necessity for modifications of the regularities obtained for inanimate objects will be more and more apparent. . . .

Q.8

Only an Illusion

Ilya Prigogine

Ilya Prigogine, a Belgian physical chemist, who was born in Russia in 1917 but was taken by his parents to Belgium while still a child, is a scientist whose interests have ranged far afield from his specialty (he won the Nobel Prize for chemistry in 1977) to touch on deeper, more universal facets of his science. Prigogine studied at the Free University of Brussels (Université Libre de Bruxelles), where he soon became a professor, and today occupies the chair at his alma mater, in addition to teaching at the Center for Statistical Mechanics and Thermodynamics in Austin, Texas. He has specialized, among other subjects, in the second law of thermodynamics, questioning its implication of entropy and arguing that, in life, order can be maintained and even, at times, increased.

> The second law, first enunciated in its full form by Clausius nearly a century before, asserts that spontaneous change is in the direction of increasing disorder. And yet there are phenomena that seem to move spontaneously toward increasing order—the phenomenon of life particularly. Prigogine produced mathematical models to show how life maintains order within the requirements of the second law.*

It was for his mathematical models illustrating such increasing order that he was made a Nobel Laureate. In 1979, he coauthored and published a volume entitled *La Nouvelle Alliance* (*The New Alliance*), released in English in 1983 under the title *Order and Chaos.*

---

*Isaac Asimov, *Asimov's Biographical Encyclopedia of Science and Technology* (2nd rev'd ed.) New York: Doubleday & Co., 1982, p. 863.

Our selection is a lecture that was given by Prigogine at the Jawaharlal Nehru University in New Delhi, in December 1982. It opens with an astounding quotation from Werner Heisenberg concerning Kronberg Castle, where Hamlet may once have lived. This leads Prigogine immediately to consider what reality actually is, and then to cite another quotation from Albert Einstein, from which the title of Prigogine's lecture ("Only an Illusion") has been taken. Prigogine seems to have every aspect of Western culture at his fingertips. Einstein's letter reminds him of the doctrines of Giordano Bruno whose opinions brought him to the stake in the Campo dei Fiori in Rome in 1600. Prigogine then darts off to visit Homer's *Iliad* and *Odyssey,* reconsidering and returning to the modern age, quoting Paul Valéry, and discussing Proust, Bergson, Teilhard de Chardin, Freud, Peirce, and Whitehead before coming to rest for a moment in the seventeenth century of Newton and Leibniz. Despite his whirlwind approach to science and cultural history, Prigogine manages to weave a colorful tapestry that includes threads of physics, biology, and philosophy.

From an inner center the psyche seems to move outward, in the sense of an extraversion, into the physical world.

—WOLFGANG PAULI, *The Interpretation of Nature and the Psyche*

Be it clearly understood that space is nothing but a mode of particularization and that it has no real existence of its own.... Space exists only in relation to our particularizing consciousness.

—ASHVAGHOSHA, *The Awakening of Faith*

1.

LET ME START with a recollection of Werner Heisenberg when, as a young man, he took a walking tour with Niels Bohr. This is Heisenberg's account of what Bohr said when they came to Kronberg Castle.

> Isn't it strange how this castle changes as soon as one imagines that Hamlet lived here? As scientists, we believe that a castle consists only of stones, and admire the way the architect put them together. The stones, the green roof with its patina, the wood carvings in the church, constitute the whole castle. None of this should be changed by the fact that Hamlet lived here, and yet it is changed completely. Suddenly the walls and the ramparts speak a different language.... Yet all we really know about Hamlet is that his name appears in a thirteenth-century chronicle.... But everyone knows the questions Shakespeare had him ask, the human depths he was made to reveal, and so he, too, had to be found a place on earth, here in Kronberg.

Obviously this story brings us a question which is as old as humanity itself: *the meaning of reality.*

This question cannot be dissociated from another one, the meaning of time. To us time and human existence, and therefore also reality, are concepts which are indissociable. But is this necessarily so? I like to quote the correspondence between Einstein and his old friend Besso. In the lat-

---

Ilya Prigogine, "Only an Illusion." This lecture was delivered as a Tanner Lecture on Human Values in 1982 at Jawaharlal Nehru University, New Delhi, India, and is reprinted with permission from Volume V of *The Tanner Lectures on Human Values,* University of Utah Press, Salt Lake City, Utah, and Cambridge University Press, Cambridge, England, 1984.

ter years, Besso comes back again and again to the question of time. What is time? What is irreversibility? Patiently, Einstein answers again and again, irreversibility is an *illusion,* a subjective impression, coming from exceptional initial conditions.

Besso's death only a few months before Einstein's was to interrupt this correspondence. At Besso's death, in a moving letter to Besso's sister and son, Einstein wrote:

> Michele has preceded me a little in leaving this strange world. This is not important. For us who are convinced physicists, the distinction between past, present, and future is only an illusion, however persistent.

"Only an illusion." I must confess that this sentence has greatly impressed me. It seems to me that it expresses in an exceptionally striking way the symbolic power of the mind.

In fact, in his letter Einstein was reiterating what Giordano Bruno had written in the sixteenth century and what had become for centuries the credo of science.

> *The universe is, therefore, one, infinite, immobile.* One, I say, is the absolute possibility, one the act, one the form or soul, one the matter or body, one the thing, one the being, one the maximum and optimum; which is not capable of being comprehended; and yet is without end and interminable, and to that extent infinite and interminate, and consequently immobile. *It does not move itself locally,* because it has nothing outside itself to which it might be transported, it being understood that it is all. *It does not generate itself* since there is no other thing into which it could desire or look for, it being understood that it has all the beings. *It is not corruptible,* since there is no other thing into which it could change itself, it being understood that it is everything. It cannot diminish or increase, it being understood that it is infinite, thus being that to which nothing can be added, and that from which nothing can be subtracted, for the reason that the universe does not have proportional parts. *It is not alterable* into any other disposition because it does not have anything external through which it could suffer and through which it could be affected.

For a long time, Bruno's vision was to dominate the scientific view of the western world. It was to lead to the mechanical world view with its two basic elements:

a) changeless substances such as atoms, molecules or elementary particles

b) locomotion

Of course many changes came through quantum theory, to which I shall come back, but still some basic features of this conception remain even now. But how to understand this timeless nature which puts man outside the reality he describes? As Carl Rubino has emphasized, Homer's *Iliad* centers around the problem of time. Achilles *embarks in a search for something permanent and immutable.* "But the wisdom of the *Iliad,* a bitter lesson that Achilles, its hero, learns too late, is that such perfection can be gained only at the cost of one's humanity: he must lose his life in order to gain this new degree of glory. For human men and women, *for us,* immutability, freedom from change, total security, immunity from life's maddening ups and downs will come only when we depart this life, by dying, or becoming gods: the gods, Horace tells us, are the only living beings who lead secure lives, free from anxiety and change."

Homer's Odyssey appears as the dialectical counterpart to the Iliad. Odysseus has the choice—he is fortunate enough to be able to choose between agelessness, immortality, remaining forever the lover of Calypso, or the return to humanity, and ultimately to old age and death: Still he chooses time over eternity, human's fate over god's fate.

Let us still stay in literature, but come closer to our time. In his *Cimetière marin* Paul Valéry describes man's struggle to come to terms with time as duration, with its limited span open to us. In his *Cahiers*—those numerous volumes of notes he used to write in the early mornings—he comes back again and again to the problem of time: Time, science to be constructed. There is a deep feeling for the unexpected in Valéry, why things are happening as they do. Obviously, Valéry could not be satisfied with simple explanations such as schemes implying a universal determinism which supposes that in some sense *all is given.* Valéry writes:

> Le déterminisme—subtil anthropomorphisme—dit que tout se passe comme dans une machine telle qu'elle est comprise par moi. Mais toute loi mécanique est au fond irrationnelle—expérimentale. . . . *Le sens du mot déterminisme est du même degré de vague que celui du mot liberté. . . . Le déterminisme rigoureux est profondément déiste. Car il faudrait un dieu pour apercevoir cet enchaînement infini complet.* Il faut imaginer un dieu, un front de dieu pour imaginer cette logique. C'est un point de vue divin. De sorte que le dieu retranché de la création et de l'invention de l'univers est restitué pour la compréhension de cet univers. Qu'on le veuille ou non, un dieu

> est posé nécessairement dans la pensée du déterminisme—et c'est une rigoureuse ironie.

Valéry is making an important remark to which I shall come back. Determinism is only possible for an observer outside his world, while we describe the world from within.

In Valéry, this preoccupation with time is not an isolated phenomenon in the early part of this century. We may quote in disorder Proust, Bergson, Teilhard de Chardin, Freud, Peirce or Whitehead.

As we have mentioned, the verdict of science seemed final. Time *is* an illusion. Still again and again the question was asked: How is this possible? Do we really have to make a tragic choice between a timeless reality which leads to human alienation or an affirmation of time which seems to break with scientific rationality?

Most of European philosophy, from Kant to Whitehead, appears as an attempt to overcome in one way or another the necessity of this choice. We cannot go into details, but obviously Kant's distinction between a noumenal world and a phenomenal one was a step in this direction, as is the idea of the process philosophy by Whitehead. None of these attempts has met with more than a mitigated success. As a result, we have seen a progressive decay of "philosophy of nature." I agree completely with Leclerc when he writes:

> In the present century we are suffering the consequences of the separation of science and philosophy which followed upon the triumph of Newtonian physics in the eighteenth century. It is not only the dialogue between science and philosophy which has suffered.

Here is one of the roots of the dichotomy into "two cultures." There is an irreducible opposition between classical *reason* with its atemporal vision and our *existence* with its vision of time as this *twirl* which Nabokov describes in "Look at the Harlequins." But something very dramatic is happening in science, something as unexpected as the birth of geometry, or the grandiose vision of the cosmos as expressed in Newton's work. We become progressively more and more conscious of the fact that on all levels from elementary particles up to cosmology, science is rediscovering time.

We are still embedded in this process of reconceptualization of physics—we still don't know where it will lead. But certainly it opens a new chapter in the dialogue between men and nature. In this perspective, the problem of the relation between science and human values, the central subject of the Tanner lectures, can be seen in a new perspective. A dialogue between natural sciences, human sciences, including arts and literature, may take a new start and perhaps develop into something as fruitful as it was during the classic period of Greece or during the seventeenth century at the time of Newton and Leibniz.

2.

To understand the changes which are going on in our time, it may be useful to start with our scientific heritage from the nineteenth century. I believe that this heritage included *two basic contradictions* or at least *two basic questions* to which no answer was provided.

As you know, the nineteenth century was essentially the century of evolution. Think about the work of Darwin in biology, of Hegel in philosophy, or of the formulation of the famous entropy law in physics.

Let us start with *Darwin.* The present year is the centenary of the death of Darwin. Beyond the importance of *The Origin of Species,* published in 1859, for biological evolution proper, there is a general element involved in Darwin's approach, which I want to emphasize. His approach combines *two* elements: on one side, the assumption of spontaneous *fluctuations* in biological species, which then through selection from the medium lead to *irreversible* biological evolution. Therefore, his model combines two elements to which we shall very often come back in this lecture: the idea of *fluctuations,* or randomness, of stochastic processes, and the idea of evolution, of *irreversibility.* Let us emphasize that on the level of biology this association leads to evolution corresponding to increasing complexity, to self-organization.

This is in complete contrast to the meaning that is generally associated with the law of *entropy* increase as formulated by Clausius in 1865. The basic element in this law is the distinction between reversible and irreversible processes. Reversible processes do not know any privileged direction of time. You may think about a spring oscillating in a frictionless medium or about planetary motion. On the other hand, irreversible processes involve an arrow of time. If you bring together two liquids, they tend to mix, but the unmixing is not observed as a spontaneous process. All of chemistry corresponds to irreversible processes. This distinction is taken up in the formulation of the second law, which postulates the existence of a function, the entropy (entropy means evolution in Greek), which in an isolated system, can only increase because of the presence of irreversible processes, while it remains unchanged through reversible processes. Therefore, in an isolated system, entropy will finally reach its maximum whenever the system has come to equilibrium and the irreversible processes come to a final halt.

It is the lifework of one of the greatest theoretical physicists of all times, Ludwig Boltzmann, to have given a first microscopic interpretation to this increase of entropy. He turned to kinetic theory of gases with the idea that the mechanism of change, of "evolution," is then described in terms of collisions between molecules. His main conclusion was that entropy S is closely related to *probability* P. Everybody has heard quoted the famous formula $S = k \ln P$ which was engraved on the tombstone of Boltzmann after his tragic suicide in 1906. Here k is a universal constant named after Boltzmann by Planck. Again, as with Darwin, evolution and probability, randomness, are closely related. However, Boltzmann's result is different and even contradictory to that of Darwin. Probability will reach its maximum when *uniformity will be achieved.* Think about a system formed by two boxes which may communicate through a small hole. Equilibrium will obviously be achieved when the number of particles in the two boxes will be the same. Therefore, the approach to equilibrium corresponds to destruction of privileged initial conditions, to forgetting of initial structures, in contrast to Darwin where evolution means creation of new structures.

We come, therefore, to the first question, to the first contradiction which we have inherited from the nineteenth century: how can *Boltzmann and Darwin both be right?* How can we describe both the destruction of structures and processes involving self-organization? Still, as I have already emphasized, both approaches use common elements: the idea of probability (expressed in Boltzmann's theory in terms of the collisions between particles) and irreversibility emerging as a result of this probabilistic description. Before I shall explain how both Boltzmann and Darwin can be right, let us describe the second contradiction which we had to face.

3.

The problems to which we come now lie much deeper than the opposition between Boltzmann and Darwin. The prototype of classical physics is classical mechanics, the study of motion, the *description of trajectories* leading a point from position A to position B. Some of the basic characteristics of the dynamical description are its deterministic and *reversible character.* Once appropriate initial conditions are given, we can predict rigorously the trajectory. Moreover, the direction of time does not play any role. Prediction and retrodiction are identical. Therefore, on the fundamental dynamic level, there seems to be no place for randomness nor for irreversibility. To some extent, the situation remains the same in quantum theory. Here we do not speak any more about trajectories but about wave functions. Again the wave function evolves according to a reversible deterministic law.

As a consequence, the universe appears as a vast automaton. We have already mentioned that, for Einstein, time in the sense of directed time, of irreversibility, was an illusion. Quite generally, the classical attitude in respect to time was some form of distrust, as it appears in innumerable books and publications. For example, in his monograph *The Ambidextrous Universe,* Martin Gardner writes that the second law makes only certain processes *improbable,* but never impossible. In other words the law of increase of entropy would refer only to a *practical* difficulty without any deep foundation. Similarly, in his classic book *Chance and Necessity,* Jacques Monod expresses the view that life is only an *accident* in the history of nature. It would be a kind of fluctuation, which, for some not very clearly understood reasons, is able to maintain itself.

It is certain that, whatever our final appreciation of these complex problems, our universe has a pluralistic, complex character. Structures may disappear, like in a diffusion process, but structures may appear, as in biology and, even more clearly, in social processes. Some phenomena are, so far as we know, well described by deterministic equations, as is the case for planetary motions, but some, like biological evolution, likely involve stochastic processes. Even a scientist convinced of the validity of deterministic descriptions would probably hesitate to imply that at the very moment of the "Big Bang," the date of this lecture was already inscribed in the laws of nature.

How then to overcome the apparent contradiction between these concepts? We are living in a *single universe.* As we shall see, we begin to appreciate the meaning of these problems; we begin to see that irreversibility, life, are inserted in the basic laws, even on the microscopic level. Moreover, the importance that we give to the various phenomena we may observe and describe is quite different, I would even say opposite, to what was suggested by classical physics. There the basic processes, as I mentioned, were considered as *deterministic and reversible.* Processes involving randomness or irreversibility were considered to be exceptions, mere artifacts. Today, *we see everywhere the role of irreversible processes, of fluctuations.* The models considered by classical physics appear to us now to correspond only to limiting situations which we can create artificially, for example, by putting matter into a box and waiting for it to reach equilibrium.

The *artificial* may be deterministic and reversible. The *natural* contains essential elements of randomness and irreversibility. This leads to a new view of matter, in which matter is no longer passive as described in the mechanical world view, but is associated with spontaneous activity. This change is so deep that I believe we can really speak about a *new dialogue of man with nature.*

4.

Of course, it has taken many unexpected developments, both in theoretical concepts and experimental discoveries, to go from the classic description of nature to the new one which is emerging. In brief, we were looking for all-embracing schemes, for symmetries, for immutable general laws, and we have discovered the mutable, the temporal, the complex. Examples abound. As you know, quantum theory predicts a *remarkable symmetry,* the one which should exist between matter and antimatter, but our world does not have this symmetry. Matter dominates largely over antimatter. This is a quite happy circumstance as otherwise the annihilation between matter and antimatter would mean the end of all massive particles. The discovery of a large number of *unstable* particles is another example; it may even be that all particles are unstable. Anyway, the idea of an unchanging, permanent substrate for matter has been shattered.

Who could expect that (in contrast with the views of Giordano Bruno) the concept of evolution would be applicable to the world as a whole? As a matter of fact, astrophysical discoveries, and especially the famous residual black-body radiation, leave little doubt that the world as a whole has undergone a remarkable evolution.

How to speak then about immutable, eternal laws? We can certainly not speak about laws of life at a moment when there was no life. The very concept of law which emerged at the time of Descartes and Newton, a time of absolute monarchies, has to be revised.

Of special importance in the context of this lecture are experiments dealing with macroscopic physics, with chemistry, in other words, with nature on our own scale. The classical view (remember our discussion of Boltzmann's interpretation of the second law of thermodynamics) focused its interest on the transition from order to disorder. Now we find everywhere transitions from disorder to order, processes involving *self-organization* of matter. If you would have asked a physicist a few years ago what exactly physics permits to explain and what remains open, he may have answered that we obviously do not understand sufficiently elementary particles or cosmological features of the universe as a whole, but in between our knowledge is pretty satisfactory. Today, a growing minority (to which I belong) would not share this optimistic view. I am, on the contrary, convinced that we are only at the beginning of a deeper understanding of the nature around us, and this seems to me of outstanding importance for the embedding of *life in matter as well as of man in life.*

5.

We shall now briefly review the way in which the two contradictions that we have mentioned can be approached. First of all, how can we describe the origin of structures, of self-organization? This problem has been the object of many publications, and I may be quite brief. Once we may attach an entropy to a physical system, we may distinguish between equilibrium or near equilibrium on one hand, and situations corresponding to *far from equilibrium* on the other. What has been shown is that near equilibrium matter indeed conforms to Boltzmann's paradigm; structures are destroyed. If we perturb such a system, the system responds by restoring its initial condition; such systems are therefore stable. In a sense, such systems are always able to develop mechanisms which make them *immune to perturbation.* However, these properties do not extend to far-from-equilibrium conditions. The key words there are *"nonlinearity," "instability," "bifurcations."* In brief, this means that if we drive a system sufficiently far from equilibrium, its state may become unstable in respect to perturbation. The exact point at which this may happen is called the *bifurcation point.* At this point, the old solution becoming unstable, new solutions emerge which may correspond to a quite different behavior of matter. A spectacular example is the appearance of chemical clocks in far-from-equilibrium conditions. The experimental demonstration of the existence of chemical clocks is today a routine experiment which is performed in most courses in chemistry at colleges and universities. It is, therefore, a very simple experiment, and, still, I believe it is perhaps one of the most important experiments of the century. Let me briefly explain why I think so.

In this experiment we have basically two types of molecules. Let us call one species A (the red molecules), the other B (the blue molecules). When we think about them interacting, we think of some chaotic collisions going on at random. Therefore, we would expect that the interchange between A and B would lead to a uniform color with occasional flashes of red or blue. This is not what happens with appropriate chemicals in far-from-equilibrium conditions. The whole system becomes red, then blue, and again red. This shows that molecules may *communicate* over large, macroscopic distances and over macroscopic times. They have means to signal each other their state in order to react together. This is very unexpected, indeed. We always thought that molecules interact only through short-range forces; each molecule would only know its direct neighbors. Here, on the contrary, the system acts as a whole. Such behavior was traditionally associated with biological systems, and here we see it already arising in relatively simple nonliving systems.

A second aspect I want to emphasize is the idea of *symmetry breaking* associated with some of the bifurcations. The equations of reaction and diffusion are highly symmetric; if we replace the geometric coordinates x, y, z by -x, -y, -z, which corresponds to space inversion, these equations would not change. Still, after bifurcation we may have different solutions each of which has a broken symmetry. Of course, if we would have, say, a "left" solution, we would have also a "right" solution, but it may happen that in nature we observe for some reason only one of the solutions. Everybody has observed that shells often have a preferential chirality. Pasteur went so far as to see in the breaking of symmetry the very characteristic of life. Again we see in non-life a precursor of this property. Here I want to emphasize that solutions of symmetrical equations may have less symmetry than the equations themselves. This will be an essential point when we discuss the roots of time in nature.

Finally, the appearance of bifurcations in far-from-equilibrium conditions leads to an irreducible stochastic random element on the macroscopic level. Deterministic theories are of no help to permit us to predict which of the branches that arise at the bifurcation point will be followed at the bifurcation point. We have here an example of the essential role of *probabilities.* You may remember that in quantum mechanics probabilities already play an essential role; this is the essence of the famous Heisenberg uncertainty relation. There one could object by saying that we living beings are made by so many elementary particles, that quantum effects are being washed out by the laws of large numbers. However, this is no longer possible when we speak about bifurcation of chemical systems far from equilibrium. Here irreducible probabilistic effects appear on our *own level.* Clearly, there is a relation with the role of fluctuations and the Darwinian theory of the origin of species. Again you see why I mentioned before that, in the present perspective, life appears much less isolated, as having much deeper roots in the basic laws of nature.

6.

We come now to the second problem, which, I have to tell you immediately, is vastly more difficult. The second law of thermodynamics belongs traditionally to macroscopic physics, but, curiously, its meaning has some elements in common with microscopic theories like quantum theory and relativity. Indeed, all these theoretical constructs have one element in common: they indicate some limit to our manipulation of nature. For example, the existence of the velocity of light as a universal constant indicates that we cannot transmit signals with a speed larger than that of light in a vacuum. Similarly, the existence of the quantum mechanical constant *h*, Planck's constant, indicates that we cannot measure simultaneously momentum and position of an elementary particle. In the same spirit, the second law of thermodynamics indicates that we cannot realize certain types of experiments in spite of the fact that they are compatible with all other known laws of physics. For example, we cannot drive a thermal engine using the heat of a single heat source, such as the ocean. That is the meaning of the impossibility of a "perpetuum mobile of the second kind."

I believe that this does not mean, however, that physics now becomes a subjectivistic physics, some result of our preferences or convictions, but it is indeed a physics subjected to intrinsic constraints that identify us as a part of the physical world we are describing. It is this physics, which presupposes an observer situated in the world that is confirmed by experiment. Our dialogue with nature is successful only if carried on from *within* nature.

But how to understand irreversibility, no longer in terms of macroscopic physics, but in terms of the basic laws, be they classical or quantum? In the introduction I have already mentioned the bold essay of Boltzmann to relate irreversibility to probability theory. But, in turn, what can probability mean in a world in which particles or wave functions evolve according to deterministic laws? In his beautiful book *Unended Quest,* Popper has described the tragic struggle of Boltzmann, and the way in which Boltzmann was finally obliged to retreat and to admit that there would be no intrinsic arrow of time in nature. Again we come back to Einstein's stony conclusion: Time is an illusion.

We can now take up Boltzmann's quest because we have a much better understanding of dynamics, as a result of the work of great mathematicians such as Poincaré, Lyapounov, and, more recently, Kolmogoroff. Without their work, this problem would still be a question of conjecture. Let us first observe that irreversibility *is not a universal.* I have already mentioned that there are systems, like an isolated spring, for which entropy has no relevance, its motion being entirely reversible. Therefore, we cannot hope that irreversibility may be a property of *all* dynamical systems. What we have to do is to identify dynamical systems of the right complexity for which a formulation of the second law on a microscopic basis becomes possible.

We can, of course, not go here into technical details. However, the main point is the recent discovery of *highly unstable dynamical systems.* In such systems, the trajectories starting with two points as near as we want diverge exponentially in time. But then the concept of trajectory ceases to be meaningful. We can only reach finite accuracy. In spite of the fact that we start with deterministic equations, the solutions appear as "chaotic," and some authors speak of "deterministic chaos." Curiously strong probabilities elements appear in the core of dynamics.

Such systems can be called *"intrinsically random";* we can only speak about their average behaviour. Indeed, their behavior is so stochastic that they can be mapped into a probabilistic process called a Markov process, reaching equilibrium either for $t \rightarrow +\infty$ in the distant future or $t \rightarrow -\infty$ in the distant past.

So we have already justified one of Boltzmann's basic intuitions. It is indeed meaningful to speak about probabilities even in the frame of classical mechanics, but *not for all systems; only for highly unstable systems for which the concept of a trajectory loses its meaning.* Now, how can we go further and go from intrinsically random to *intrinsically irreversible* systems?

This requires supplementary conditions. We need representations of dynamics, which have less symmetry than the full time-inversion symmetry of the basic equations.

For example in a system of colliding hard spheres, a possible situation is one in which, for distant past, the velocities of a group of particles were really parallel and, for distant future, the distribution becomes random as required by equilibrium. The time inversion symmetry requires that there would also exist a situation in which, in the distant past, velocities were random and, in a distant future, they would tend to be parallel. One situation is obtained through the velocity inversion of the other. In fact, only the first situation is observed, while the second is not. The second law of thermodynamics on the macroscopic level expresses precisely the exclusion of one of the two situations, which are velocity inverses, one of the other.

Irreversibility can have a microscopic meaning only if there are representations of dynamics which are not invariant in respect to time inversion, in spite of the fact that the initial equations are.

Let us emphasize the remarkable analogy between such situations and symmetry-breaking bifurcations we mentioned before. There also we may derive in some cases from a symmetrical equation two solutions, one "left," one "right," each of which when taken separately breaks the space symmetry of the equation. We may now make precise what the second law may mean on the microscopic level. It states that only situations that go to equilibrium in the future may be prepared or observed in nature. This means that the second law is an *exclusion principle,* which excludes situations in which, in the distant past, the velocities of colliding spheres would be distributed uniformly, while, in the distant future, they would tend to parallel velocities. On the contrary, the situation in which we start in the distant past, particles with nearly parallel velocities that are then randomized by collisions is an experiment which we can perform easily.

I have used here *physical* images. But the important point is that the existence of these representations of dynamics with broken-time symmetry can be proved rigorously for highly unstable systems.

For such systems, we may associate to each initial condition expressed by a distribution function in phase space a number measuring the information necessary to prepare this state. The initial conditions that are excluded are those for which this information would be infinite.

Note also that the entropy principle cannot be derived from dynamics; it appears as a supplementary condition which has to be tested experimentally as does any other law of physics. The basic point, however, is that this exclusion principle is not contradictory with dynamics; once admitted at a given time it is *propagated* by dynamics.

The probability interpretation of Boltzmann is only possible because there exists this exclusion principle which provides us with an arrow of time.

Irreversibility, as in the theory of Darwin, or also as in the theory of Boltzmann, is an even stronger property than randomness. I find this quite natural. Indeed, what could irreversibility mean in a deterministic concept of the universe in which tomorrow is already potentially present today? Irreversibility presupposes a universe in which there are limitations in the prediction of the future. I want again to emphasize that, in the spirit of this explanation, *irreversibility is not a universal property.* However, the world as a whole seems to belong to these complex systems, intrinsically random, for which irreversibility is meaningful, and it is to this category of systems with broken-time symmetries that all phenomena of life belong and, as a consequence, all human existence.

You may be astonished that I have spoken little about cosmological theories. Certainly, the global state of our universe plays an essential role. It provides the nonequilibrium environment which makes the formation of structures possible. However, I do not believe that the existence of the expanding universe and of the initial Big Bang can provide by themselves an explanation of irreversibility. We observe, as already indicated, both reversible and irreversible processes in spite of the fact that all processes, reversible or not, are embedded in the expanding universe.

7.

The microscopic interpretation of the second law is very recent. I am personally convinced that it will lead to deep changes in our conception of matter. Some preliminary results have been worked out by my colleagues and myself, but what I shall say now is to some extent an anticipation that may or may not be confirmed by later developments.

If we take the second law together with its probabilistic interpretation seriously, we have to associate equilibrium with maximum probability. Now maximum probability in terms of particles means chaotic, uncorrelated motion similar to the way the Greek atomists imagined the physical world. Inversely, we may define particles as units that are uncorrelated and behave in a chaotic way in thermodynamic equilibrium. What is, then, the effect of nonequilibrium? It is to create *correlations* between these units, to create order out of the chaotic motions arising in the equilibrium state. This description of nature, in which order is generated out of chaos *through nonequilibrium conditions* provided by our cosmological environment, leads to a physics that is quite similar in its spirit to the world of "processes" imagined by Whitehead. It leads us to conceive matter as active, as in a continuous state of becoming. This picture deviates significantly from the classical description of physics, of change in terms of forces or fields. It is a momentous step to go away from the royal road opened by Newton, Maxwell and Einstein. But I believe that the unification of dynamics and thermodynamics paves the way to a radically new description of temporal evolution of physical systems, a description which again, to my mind, is much closer to what we see on the macroscopic level, be it in the nonliving or the living world.

We may quote as examples the highly correlated distribution of nucleotides in the fundamental biological molecules, and perhaps even of the distribution of letters which are assembled in words to form our language.

8.

Over all my scientific career, the attitude I have taken was to consider the law of entropy increase, the second law of thermodynamics, as a basic law of nature. I was following the views Planck exposed in the following text:

> The impracticability of perpetual motion of the second kind is granted, yet its absolute impossibility is contested, since our limited experimental appliances, supposing it were possible, would be insufficient for the realization of the ideal processes which the line of proof presupposes. This position, however, proves untenable. It would be absurd to assume that the validity of the second law depends in any way on the skill of the physicist or chemist in observing or experimenting. The gist of the second law has nothing to do with experiment; the law asserts briefly that *there exists in nature a quantity which changes always in the same sense in all natural processes.* The proposition stated in this general form may be correct or incorrect; but whichever it may be, it will remain so, irrespective of whether thinking and measuring beings exist on the earth or not, and whether or not, assuming they do exist, they are able to measure the details of physical or chemical processes more accurately by one, two, or a hundred decimal places than we can. The limitations to the law, if any, must lie in the same province as its essential idea, in the observed Nature, and not in the Observer. That man's experience is called upon in the deduction of the law is of no consequence; for that is, in fact, our only way of arriving at a knowledge of natural law.

However, Planck's views remained isolated. As we noticed, most scientists considered the second law as the result of approximation, or as the intrusion of subjective views into the exact laws of physics. Our attitude is the opposite: we have looked for the limits, which the second law brings into the world of dynamics.

In other words, our goal is to unify dynamics and thermodynamics. It is clear that in such a view randomness, fluctuations and irreversibility will play an essential role at the microscopic level, quite different from the marginal role they played in the traditional descriptions of nature.

This goal is far from being realized, but on the road we have been led to a series of surprising findings, some of which I have summarized in this lecture.

I remain stunned by the variety of non-equilibrium structures that have been discovered experimentally, some of which we may now describe theoretically. Still, we are only at the level of "taxonomy."

We have already mentioned the work of great mathematicians, such as Poincaré, or Kolmogroff in classical mechanics. As a result, we know that classical dynamics may lead to situations where the concept of trajectories loses its meaning, and where we can only make probabilistic statements. Curiously, chemistry is now also going through a comparable reconceptualization. In many instances, we have to go beyond the deterministic approach of chemical kinetics and take into account fluctuation and randomness, even in systems formed of a large number of molecules. At the microscopic level, irreversibility emerges as symmetry-breaking in systems reaching a sufficient degree of randomness.

The second law limits what is observable. It appears as an exclusion principle propagated by classical or quantum mechanics.

Perhaps the most unexpected aspect is that, at all levels, order, coherence emerge from chaos for non-equilibrium conditions: An equilibrium world would be chaotic, the non-equilibrium world achieves a degree of coherence which, at least for me, is a source of surprise.

9.

In this lecture I have discussed some steps in the rediscovery of time in the physical sciences. We have seen that time in the sense of duration, of irreversibility, is basically related to the role of randomness, in full accord with the brilliant intuition of Boltzmann.

Since the discovery of quantum mechanics, in which probability plays an essential role, the meaning of randomness has led to many controversies. It appears today that deterministic schemes, which make predictions valid in each individual case, are unaccessible to us in a wide range of phenomena from microscopic physics up to the level of molecules and of life. Of course this situation may change, but we see no sign for such a change to occur over the next ten years.

In this context, let us emphasize that we don't know how to describe the reality as it would appear to an observer who, in some sense, would be situated outside this world. We can only deal with the problem of determinism or randomness as they are included in the schemes we formulate to describe our experience with the world and us.

One is reminded of the dialogue between Einstein and Tagore. In this most interesting dialogue on the nature of reality, Einstein was emphasizing that science has to be independent of the existence of any observer. As I mentioned at the beginning of this lecture, his realism led him to paradoxes. Time and, therefore, human existence become illusions. On the contrary, Tagore emphasized that even if absolute truth could have a meaning, it would be inaccessible to human mind. . . .

The controversy between Einstein and Tagore is only meaningful if man is supposed to be separated from nature. If the imbedding of man in nature is taken into account, human truths become truths of nature. Curiously enough, the present evolution of science goes in the direction stated by the great Indian poet. Whatever we call reality, it is only open to us through constructs of our mind. This has been concisely expressed by D. S. Kothari:

> The simple fact is that no measurement, no experiment or observation is possible without a relevant theoretical framework.

In a more sophisticated form this appears in quantum theory through the intervention of "operators" which are associated with physical quantities.

The problems of the limits of determinism, randomness, irreversibility and the notion of reality are closely connected, and we begin to see their relations.

As we are able to find the roots of time in nature, time ceases to be the concept which separates men from nature. It now expresses our belonging to nature, not our alienation.

The vision of the world around us and of the world in us converge. As I deliver this lecture in Delhi, why not stress that type of convergence, of synthesis of the external world around us and the internal world inside us, that is one of the recurrent themes of Indian philosophy.

We now overcome the temptation to reject time as an illusion. Far from that, we are back to Valéry's anticipation: "Durée est construction, vie est construction." In a universe in which tomorrow is not continued in today, time is to be constructed. Valéry's sentence expresses our responsibility to this construction of the future, not only *our* future, but the future of mankind. With this conclusion the problem of human values, of ethics, even of art takes a new form. We may now see music with its elements of expectation, of improvisation, with its arrow of time as an allegory of becoming, of physics in its etymological Greek sense.

The dialectics between what is *in* time and what is *out* of time, between external truths and time oriented existence will probably continue for ever. But perhaps we are now in a privileged moment where we begin to perceive a little better the junction, the passage between stillness and motion, time arrested and time passing. It is this moment with its incertitudes, its open questions, but also its hopes for a more integrated human world, which I have tried to describe in this lecture.

R
Particles

INTRODUCTION

THE HISTORY OF particle physics covers the period from about 430 B.C. to the present—some twenty-four centuries. Democritus, the student of Leucippus, was one of the earliest Greek philosophers to suggest that, if a unit of matter is divided in successive operations, one will eventually arrive at an uncuttable unit, which Democritus called *atomos.*

From the time of Democritus until 1897, when J. J. Thomson discovered the electron and thus showed that the atom was indeed divisible into smaller components, little happened in particle physics. But from Thomson's day to the present, particle physics has developed along with the two principal intellectual movements in modern physics—relativity theory and quantum mechanics—into one of the most fruitful areas of science.

The formulation of the relativistic wave equation by the British physicist Paul Dirac in 1930 led many physicists to believe that the basic problems of matter had been solved. Since the neutron had not yet been discovered, the electron and the proton were accepted as the basic elementary particles of nature. But although physicists had achieved great success, certain aspects of the atomic and subatomic world were disturbing and could not be incorporated into the relatively simple world of protons and electrons. Thus the neutrino, which was postulated by Wolfgang Pauli to conserve energy and balance spins in the beta decay of heavy nuclei, seemed to have no place in the then-prevailing theory, and it remained one of the great mysteries in particle physics until relatively recently. Further, whereas the Dirac relativistic wave equation describes the electromagnetic dynamics of the electron quite accurately, it gives the wrong results for the magnetic properties of the proton, indicating a puzzling asymmetry between the proton and electron which was more disturbing to physicists in the 1930s than the proton-electron mass asymmetry. All attempts to construct a correct electron-proton model of the atomic nucleus failed because all the models conflicted violently with the Heisenberg uncertainty principle and give wrong values for the nuclear spins. Something new was needed, and it came with the discovery of the neutron by James Chadwick in 1932. In the meantime, the positron (positively charged electron) had also been discovered by Carl Anderson in 1932 but that breakthrough just added to the confusion.

The neutron was just what was needed to solve the nuclear problem and it soon became clear that all atomic nuclei consist of protons and neutrons, in about equal numbers in the light nuclei like helium, carbon and oxygen, but with the neutrons increasingly outnumbering the protons in the heavy nuclei so that the uranium-238 nucleus, for example, contains 142 neutrons but only 96 protons. Since the neutron's mass exceeds the proton's mass by a few electron masses, the neutron is unstable when it is not in a nucleus, and it decays into a proton, an electron, and an antineutrino in about twelve minutes. For that reason Werner Heisenberg suggested that the proton and neutron are two different states of the same basic particle, the nucleon, and he called the neutron and proton an "isotopic double."

Since the neutron is uncharged, it can penetrate into nuclei unopposed and alter their structures drastically. This discovery triggered the tremendous development of theoretical and experimental nuclear physics, which ultimately led to nuclear bombs and nuclear reactors. Experimental nuclear physics quickly outstripped theory. New discoveries about nuclear forces led the Japanese physicist Hideki Yukawa to propose a new massive particle—the meson—and to deduce the meson's mass as about 250 times that of the electron. The muon, which Anderson discovered in cosmic rays in 1932, was first thought to be Yukawa's meson, but that assumption was discarded by the time the meson was discovered by Cecil Powell in the late 1940s.

With the discovery of mesons (later, more massive mesons called kaons were discovered) and muons, the known particles were divided into three groups: baryons (protons, neutrons and some other massive particles), mesons, and light particles called leptons (electrons, muons, neutrinos and the so-called tau particles, which, though more massive than many baryons, are still listed as leptons). The difference between leptons, baryons and mesons is that the baryons and mesons interact via the strong force, whereas leptons react only to the weak interaction and do not respond to the strong force. All charged particles, of course, respond to the electromagnetic force in the same way (via protons).

Through the electron high energy scattering experiments of Richard Hofstadter it became clear that nucleons, unlike electrons, have no internal structure, which led to Murray Gell-Mann's quark hypothesis; namely that nucleons consist of three fractionally charged point-like particles of unknown mass. A slightly different form of the hypothesis was introduced at about the same time by George Zweig, who called these constituent particles "aces" because he assigned integral charges to them; today the name "quark" is universally accepted.

The composition of the proton and neutron can be understood in terms of two kinds of quarks (called quark "flavors" by particle physicists); an "up" quark *u*, and a "down" quark *d*. The proton consists of a *d* and two *u* quarks and the neutron of a *u* and two *d* quarks. The present theory of quark dynamics, based on the interaction of quarks via the strong force, called quantum electrodynamics, can say nothing about the arrangement or dynamics of the three quarks in a nucleon and predicts very little about the internal properties of nucleons.

The discovery of baryons increasingly more massive than the nucleon led to the hypothesis of a third quark, for it soon became clear that these new "strange" baryons, as they are called, cannot be constructed from just two quarks *u* and *d*, and so a third one, called a strange quark *s*, is needed.

As things stood in the later 1960s and early 1970s, the eighteen known baryons could be organized into groups (charge multiplets) by properly arranging the three assumed quarks into all possible triplet combinations.

But this led to an immediate difficulty because certain arrangements conflict with the Pauli exclusion principle, which states that no more than two identical particles can be in the same spatial configuration. This difficulty was brushed away by fiat; Gell-Mann and his adherents arbitrarily, with no experimental or observational justification, proposed that each of the quarks *u, d,* and *s* comes in three varieties called "color": "red," "green," and "blue," for example. This assumed property is not to be related to the usual concept of color but rather to a kind of charge, similar to electric charge. Just as the force between the electric charges is carried by a photon, so the strong force between quarks is carried by a particle called the gluon. But the "color" or gluon field is more complicated than the electromagnetic field because whereas photons, the quanta of the electromagnetic field, do not interact with each other, gluons do because the gluons themselves are colored. Eight different colored gluons are assumed in this theory of the strong force. Since quarks in this theory interact by emitting and absorbing colored gluons, the quarks can change color as they interact with each other, but in all such interactions, the triplet quark combination must always have zero "color," that is, "color" must always be an internal property of a quark triplet and never apparent from the ouside; the three quarks in a baryon must always be of three different "colors."

Scarcely a week goes by that a publisher does not bring out a new book, usually written by an expert, purporting to explain particle physics in terms readily understandable to the layman. Many physics professors complain that these books, almost without exception, are inaccurate when understandable, and laymen complain that when accurate to the satisfaction of the professors, they are not understandable. The articles in this selection, first, serve as a basis for a layman's understanding of the current state of particle physics; second, they supply historical accounts of its development by several of the leading participants.

R.1

Carriers of Negative Electricity

JOSEPH JOHN THOMSON

SIR JOSEPH JOHN THOMSON, the precocious son of a Manchester bookseller, was born in Cheetham Hill, a suburb of Manchester, in 1856. He matriculated at Owens College in Manchester at the age of fourteen intending to become an engineer. His major interests were mathematics and physics but the untimely death of his father caused him to focus on engineering. Thomson's mathematics professor, Thomas Barker, encouraged him to delay beginning his apprenticeship so he could compete for a scholarship to Trinity College, Cambridge; Thomson was successful on his second try and entered Cambridge at the age of twenty. Four years later, he became a Fellow of Trinity College, then Lecturer in 1883, and Master in 1918. His advancement in the academic world continued at a breathless pace: At twenty-seven, Thomson succeeded Lord Rayleigh, who was retiring, as Professor of Physics. A year later, he became director of the famed Cavendish Laboratory, a post he held until 1919. Almost single-handedly, he assured Great Britain's dominance in atomic studies for the first three decades of the century.

Toward the end of the nineteenth century, interest in cathode rays was widespread among scientists, and in 1897 Thomson demonstrated the particulate nature of cathode rays, the particles eventually identified as basic electrical units to which the name *electron* was given. He discussed this radiation in a paper in the *Philosophical Magazine* in 1897, where he described the experiments by which Jean-Baptiste Perrin (a friend and neighbor of the Curies, and a winner of the Nobel Prize in 1926 for his work on the discontinuous structure of matter) proved that cathode rays carry a negative charge; he then described his own experiments on the electromagnetic deflection of the rays under various conditions. He concluded: "I can see no escape from the conclusion that they are charges of negative electricity carried by particles of matter," that is, that the electrons are particles.

Thomson's paper is interesting as it is one of the earliest descriptions of the identification of a subatomic particle, a feat that was to bring Thomson the Nobel Prize for physics in 1906 and a knighthood in 1908. He was also an outstanding teacher; seven of his students were eventually awarded Nobel Prizes. He visited the United States in 1904 to deliver a series of lectures on electricity and matter at Yale University, where he lectured about the structure of the atom. Soon afterward, he discovered a method for separating certain types of atoms and molecules by using positive rays, an idea that directed other physicists, such as Aston and Dempster, toward the discovery of many types of isotopes.

After a long and singularly productive career, Thomson died just before the "Battle of Britain" began in 1940 and was honored by burial in Westminster Abbey close to the tomb of Isaac Newton.

We have selected Thomson's Nobel address, which recalls some of the experiments that led him to conclude that "carriers of negative electricity" are corpuscular in nature and that marked the discovery of the electron.

The way in which science arrives at its beliefs is quite different from that of mediaeval theology. Experience has shown that it is dangerous to start from general principles and proceed deductively, both because the principles may be untrue and because the reasoning based upon them may be fallacious. Science starts, not from large assumptions, but from particular facts discovered by observation or experiment. From a number of such facts a general rule is arrived at, of which, if it is true, the facts in question are instances. This rule is not positively asserted, but is accepted, to begin with, as a working hypothesis. . . . But whereas, in mediaeval thinking, the most general principles were the starting point, in science they are the final conclusion—final, that is to say, at a given moment, though liable to become instances of some still wider law at a later stage.

—BERTRAND RUSSELL, *Religion and Science*

INTRODUCTORY

. . . I WISH TO give an account of some investigations which have led to the conclusion that the carriers of negative electricity are bodies, which I have called corpuscles, having a mass very much smaller than that of the atom of any known element, and are of the same character from whatever source the negative electricity may be derived.

The first place in which corpuscles were detected was a highly exhausted tube through which an electric discharge was passing. When an electric discharge is sent through a highly exhausted tube, the sides of the tube glow with a vivid green phosphorescence. That this is due to something proceeding in straight lines from the cathode—the electrode where the negative electricity enters the tube—can be shown in the following way (the experiment is one made many years ago by Sir William Crookes): A Maltese cross made of thin mica is placed between the cathode and the walls of the tube. When the discharge is past, the green phosphorescence no longer extends all over the end of the tube, as it did when

Joseph John Thomson, Nobel Prize in Physics Award Address, 1906. Reprinted with permission of Elsevier Publishing Company, and the Nobel Foundation.

the cross was absent. There is now a well-defined cross in the phosphorescence at the end of the tube; the mica cross has thrown a shadow and the shape of the shadow proves that the phosphorescence is due to something travelling from the cathode in straight lines, which is stopped by a thin plate of mica. The green phosphorescence is caused by cathode rays and at one time there was a keen controversy as to the nature of these rays. Two views were prevalent: one, which was chiefly supported by English physicists, was that the rays are negatively electrified bodies shot off from the cathode with great velocity; the other view, which was held by the great majority of German physicists, was that the rays are some kind of ethereal vibration or waves.

The arguments in favour of the rays being negatively charged particles are primarily that they are deflected by a magnet in just the same way as moving, negatively electrified particles. We know that such particles, when a magnet is placed near them, are acted upon by a force whose direction is at right angles to the magnetic force, and also at right angles to the direction in which the particles are moving.

Thus, if the particles are moving horizontally from east to west, and the magnetic force is horizontal from north to south, the force acting on the negatively electrified particles will be vertical and downwards.

When the magnet is placed so that the magnetic force is along the direction in which the particle is moving, the latter will not be affected by the magnet.

The next step in the proof that cathode rays are negatively charged particles was to show that when they are caught in a metal vessel they give up to it a charge of negative electricity. This was first done by Perrin. This experiment was made conclusive by placing the catching vessel out of the path of the rays, and bending them into it by means of a magnet, when the vessel became negatively charged.

ELECTRIC DEFLECTION OF THE RAYS

If the rays are charged with negative electricity they ought to be deflected by an electrified body as well as by a magnet. In the earlier experiments made on this point no such deflection was observed. The reason of this has been shown to be that when cathode rays pass through a gas they make it a conductor of electricity, so that if there is any appreciable quantity of gas in the vessel through which the rays are passing, this gas will become a conductor of electricity and the rays will be surrounded by a conductor which will screen them from the effect of electric force, just as the metal covering of an electroscope screens off all external electric effects.

By exhausting the vacuum tube until there was only an exceedingly small quantity of air left in to be made a conductor, I was able to get rid of this effect and to obtain the electric deflection of the cathode rays. This deflection had a direction which indicated a negative charge on the rays.

Thus, cathode rays are deflected by both magnetic and electric forces, just as negatively electrified particles would be.

Hertz showed, however, that cathode particles possess another property which seemed inconsistent with the idea that they are particles of matter, for he found that they were able to penetrate very thin sheets of metal, e.g. pieces of gold leaf, and produce appreciable luminosity on glass behind them. The idea of particles as large as the molecules of a gas passing through a solid plate was a somewhat startling one, and this led me to investigate more closely the nature of the particles which form the cathode rays.

The principle of the method used is as follows: When a particle carrying a charge e is moving with velocity v across the lines of force in a magnetic field, placed so that the lines of magnetic force are at right angles to the motion of the particle, then, if H is the magnetic force, the moving particle will be acted on by a force equal to Hev. This force acts in the direction which is at right angles to the magnetic force and to the direction of motion of the particle. If also we have an electric field of force X, the cathode ray will be acted upon by a force Xe. If the electric and magnetic fields are arranged so that they oppose each other, then, when the force Hev due to the magnetic field is adjusted to balance the force due to the electric field Xe, the green patch of phosphorescence due to the cathode rays striking the end of the tube will be undisturbed, and we have

$$Hev = Xe$$

or

$$v = \frac{X}{H}$$

Thus if we measure, as we can do without difficulty, the values of X and H when the rays are not deflected, we can determine the value of v, the velocity of the particles. In a very highly exhausted tube this may be ⅓ the velocity of light, or about 60,000 miles per second; in tubes not so highly exhausted it may not be more than 5,000 miles per second, but in all cases when the cathode rays are produced in tubes their velocity is much greater than the velocity of any other moving body with which we are acquainted. It is, for example, many thousand times the average velocity with which the molecules of hydrogen are moving at ordinary temperatures, or indeed at any temperature yet realized.

DETERMINATION OF E/M

Having found the velocity of the rays, let us now subject them to the action of the electric field alone. Then the particles forming the rays are acted upon by a constant force and the problem is like that of a bullet projected horizontally with a velocity v and falling under gravity. We know that in time t, the bullet will fall a depth equal to $\frac{1}{2}gt^2$, where g is the acceleration due to gravity. In our case the acceleration due to the electric field is equal to Xe/m, where m is the mass of the particle. The time $t = l/v$, where l is the length of path, and v the velocity of projection.

Thus the displacement of the patch of phosphorescence where the rays strike the glass is equal to

$$\frac{1}{2}\frac{Xe}{m}\cdot\frac{l^2}{v^2}$$

We can easily measure this displacement d, and we can thus find e/m from the equation

$$\frac{e}{m} = \frac{2d}{X}\cdot\frac{v^2}{l^2}$$

The results of the determinations of the values of e/m made by this method are very interesting, for it is found that, however the cathode rays are produced, we always get the same value of e/m for all the particles in the rays. We may, for example, by altering the shape of the discharge tube and the pressure of the gas in the tube, produce great changes in the velocity of the particles, but unless the velocity of the particles becomes so great that they are moving nearly as fast as light, when other considerations have to be taken into account, the value of e/m is constant. The value of e/m is not merely independent of the velocity. What is even more remarkable is that it is independent of the kind of electrodes we use and also of the kind of gas in the tube. The particles which form the cathode rays must come either from the gas in the tube or from the electrodes; we may, however, use any kind of substance we please for the electrodes and fill the tube with gas of any kind and yet the value of *e/m* will remain unaltered.

This constant value, when we measure *e/m* in the c.g.s. system of magnetic units, is equal to about 1.7 x 10^7. If we compare this with the value of the ratio of the mass to the charge of electricity carried by any system previously known, we find that it is of quite a different order of magnitude. Before the cathode rays were investigated, the charged atom of hydrogen met with in the electrolysis of liquids was the system which had the greatest known value of *e/m,* and in this case the value is only 10^4, hence for the corpuscle in the cathode rays the value of *e/m* is 1,700 times the value of the corresponding quantity for the charged hydrogen atom. This discrepancy must arise in one or other of two ways: either the mass of the corpuscle must be very small compared with that of the atom of hydrogen, which until quite recently was the smallest mass recognized in physics, or else the charge on the corpuscle must be very much greater than that on the hydrogen atom. Now it has been shown by a method which I shall shortly describe, that the electric charge is practically the same in the two cases; hence we are driven to the conclusion that the mass of the corpuscle is only about 1/1,700 of that of the hydrogen atom. Thus the atom is not the ultimate limit to the subdivision of matter; we may go further and get to the corpuscle, and at this stage the corpuscle is the same from whatever source it may be derived.

CORPUSCLES VERY WIDELY DISTRIBUTED

It is not only from what may be regarded as a somewhat artificial and sophisticated source, viz. cathode rays, that we can obtain corpuscles. When once they had been discovered, it was found that they are of very general occurrence. They are given out by metals when raised to a red heat; indeed any substance when heated gives out corpuscles to some extent. We can detect the emission of them from some substances, such as rubidium and the alloy of sodium and potassium, even when they are cold; and it is perhaps allowable to suppose that there is some emission by all substances, though our instruments are not at present sufficiently delicate to detect it unless it is unusually large.

Corpuscles are also given out by metals and other bodies, but especially by the alkali metals, when these are exposed to light.

They are being continually given out in large quantities and with very great velocities by radioactive substances such as uranium and radium;

they are produced in large quantities when salts are put into flames, and there is good reason to suppose that corpuscles reach us from the sun.

The corpuscle is thus very widely distributed, but wherever it is found, it preserves its individuality, e/m being always equal to a certain constant value.

The corpuscle appears to form a part of all kinds of matter under the most diverse conditions; it seems natural therefore to regard it as one of the bricks of which atoms are built up.

MAGNITUDE OF THE ELECTRIC CHARGE CARRIED BY THE CORPUSCLE

I shall now return to the proof that the very large value of e/m for the corpuscle, as compared with that for the atom of hydrogen, is due to the smallness of m the mass, and not to the greatness of e the charge. We can do this by actually measuring the value of e, availing ourselves for this purpose of a discovery by C.T.R. Wilson, that a charged particle acts as a nucleus round which water vapour condenses and forms drops of water. If we have air saturated with water vapour and cool it, so that it would be supersaturated if there were no deposition of moisture, we know that if any dust is present, the particles of dust act as nuclei round which the water condenses and we get the familiar phenomena of fog and rain. If the air is quite dust-free, we can, however, cool it very considerably without any deposition of moisture taking place. If there is no dust, C.T.R. Wilson has shown that the cloud does not form until the temperature has been lowered to such a point that the supersaturation is about eightfold. When however this temperature is reached, a thick fog forms even in dust-free air.

When charged particles are present in the gas, Wilson showed that a much smaller amount of cooling is sufficient to produce the fog, a fourfold supersaturation being all that is required when the charged particles are those which occur in a gas when it is in a state in which it conducts electricity. Each of the charged particles becomes the centre round which a drop of water forms; the drops form a cloud, and thus the charged particles, however small to begin with, now become visible and can be observed.

The effect of the charged particles on the formation of a cloud can be shown very distinctly by the following experiment:

A vessel which is in contact with water is saturated with moisture at the temperature of the room. This vessel is in communication with a cylinder in which a large piston slides up and down. The piston to begin with is at the top of its travel; by suddenly exhausting the air from below the piston, the pressure of the air above it will force it down with great rapidity, and the air in the vessel will expand very quickly. When, however, air expands, it gets cool; thus the air in the vessel previously saturated is now supersaturated. If there is no dust present, no deposition of moisture will take place, unless the air is cooled to such a low temperature that the amount of moisture required to saturate it is only about ⅛ of that actually present.

Now the amount of cooling, and therefore of supersaturation, depends upon the travel of the piston; the greater the travel the greater the cooling. Suppose the travel is regulated so that the supersaturation is less than eightfold and greater than fourfold. We now free the air from dust by forming cloud after cloud in the dusty air; as the clouds fall they carry the dust down with them, just as in nature the air is cleared by showers. We find at last that when we make the expansion no cloud is visible.

The gas is now made in a conducting state by bringing a little radium near the vessel; this fills the gas with large quantities of both positively and negatively electrified particles. On making the expansion now an exceedingly dense cloud is formed. That this is due to the electrification in the gas can be shown by the following experiment:

Along the inside walls of the vessel we have two vertical insulated plates which can be electrified. If these plates are charged, they will drag the electrified particles out of the gas as fast as they are formed, so that in this way we can get rid of, or at any rate largely reduce, the number of electrified particles in the gas. If the expansion is now made with the plates charged before bringing up the radium, there is only a very small cloud formed.

We can use the drops to find the charge on the particles, for when we know the travel of the piston, we can deduce the amount of supersaturation, and hence the amount of water deposited when the cloud forms. The water is deposited in the form of a number of small drops all of the same size; thus the number of drops will be the volume of the water deposited divided by the volume of one of the drops. Hence, if we find the volume of one of the drops, we can find the number of drops which are formed round the charged particles. If the particles are not too numerous, each will have a drop round it, and we can thus find the number of electrified particles.

From the rate at which the drops slowly fall we can determine their size. In consequence of the viscosity or friction of the air small bodies do not fall with a constantly accelerated velocity, but soon reach a speed which remains uniform for the rest of the fall; the smaller the body the slower this speed. Sir George Stokes has shown that v, the speed at which a drop of rain falls, is given by the formula

$$v = \frac{2}{9} \times \frac{g a^2}{\mu}$$

where a is the radius of the drop, g the acceleration due to gravity, and u the coefficient of viscosity of the air.

If we substitute the values of g and u, we get

$$v = 1.28 \times 10^6 \cdot a^2$$

Hence if we measure v we can determine a, the radius of the drop.

We can in this way find the volume of a drop, and may therefore, as explained above, calculate the number of drops and therefore the number of electrified particles.

It is a simple matter to find by electrical methods the total quantity of electricity on these particles; and hence, as we know the number of particles, we can deduce at once the charge on each particle.

This was the method by which I first determined the charge on the particle; H.A. Wilson has since used a simpler method founded on the following principles: C.T.R. Wilson has shown that the drops of water condense more easily on negatively electrified particles than on positively electrified ones. Thus, by adjusting the expansion, it is possible to get drops of water round the negative particles and not round the positive; with this expansion, therefore, all the drops are negatively electrified. The size of these drops and therefore their weight can, as before, be determined by measuring the speed at which they fall under gravity. Suppose now, that we hold above the drops a positively electrified body: then, since the drops are negatively electrified, they will be attracted towards the positive electricity, and thus the downward force on the drops will be diminished and they will not fall so rapidly as they did when free from electrical attraction. If we adjust the electrical attraction so that the upward force on each drop is equal to the weight of the drop, the drops will not fall at all, but will, like Mahomet's coffin, remain suspended between heaven and earth. If then we adjust the electrical force until the drops are in equilibrium and neither fall nor rise, we know that the upward force on each drop is equal to the weight of the drop, which we have already determined by measuring the rate of fall when the drop was not exposed to any electrical force. If X is the electrical force, e the charge on the drop, and w its weight, we have, when there is equilibrium,

$$Xe = w$$

Since X can easily be measured and w is known, we can use this relation to determine e, the charge on the drop. The value of e, found by these methods, is 3.1×10^{-10} electrostatic units, or 10^{-20} electromagnetic units. This value is the same as that of the charge carried by a hydrogen atom in the electrolysis of dilute solutions, an approximate value of which has been long known.

It might be objected that the charge measured in the preceding experiments is the charge on a molecule or collection of molecules of the gas, and not the charge on a corpuscle.

This objection does not, however, apply to another form in which I tried the experiment, where the charges on the corpuscles were got, not by exposing the gas to the effects of radium, but by allowing ultraviolet light to fall on a metal plate in contact with the gas. In this case, as experiments made in a very high vacuum show, the electrification, which is entirely negative, escapes from the metal in the form of corpuscles. When a gas is present, the corpuscles strike against the molecules of the gas and stick to them.

Thus, though it is the molecules which are charged, the charge on a molecule is equal to the charge on a corpuscle, and when we determine the charge on the molecules by the methods I have just described, we determine the charge carried by the corpuscle.

The value of the charge when the electrification is produced by ultraviolet light is the same as when the electrification is produced by radium.

We have just seen that e, the charge on the corpuscle, is in electromagnetic units equal to 10^{-20}, and we have previously found that e/m, m being the mass of a corpuscle, is equal to 1.7×10^7, hence $m = 6 \times 10^{-28}$ grammes.

We can realize more easily what this means if we express the mass of the corpuscle in terms of the mass of the atom of hydrogen.

We have seen that for the corpuscle $e/m = 1.7 \times 10^7$. If E is the charge carried by an atom of hydrogen in the electrolysis of dilute solutions, and M is the mass of the hydrogen atom, $E/M = 10^4$; hence $e/m = 1{,}700\ E/M$.

We have already stated that the value of e found by the preceding methods agrees well with the value of E which has long been approximately known. Townsend has used a method in which the value of e/E is directly measured, and has shown in this way also that e equal to E. Hence, since $e/m = 1{,}700\ E/M$, we have $M = 1{,}700\ m$, i.e. the mass of a corpuscle is only about 1/1,700 part of the mass of the hydrogen atom.

In all known cases in which negative electricity occurs in gases at very low pressures, it occurs in the form of corpuscles, small bodies with an invariable charge and mass. The case is entirely different with positive electricity.

The Neutron and Its Properties

JAMES CHADWICK

THIRTY-FIVE YEARS passed between the discovery of the electron and that of the neutron. James Chadwick was a child when Thomson made his great discovery. Chadwick was born in Cheshire, England, in 1891, and graduated from the University of Manchester in 1911. He spent the next two years working on radioactivity problems under Ernest Rutherford in the Physical Laboratory at Manchester; Chadwick received his Master of Science degree in 1913 and, with the aid of a scholarship, traveled to Berlin to work for Hans Geiger at the Physikalisch-Technische Reichsanstalt at Charlottenburg. Unluckily, World War I broke out while Chadwick was in Germany, and he was locked up as an enemy alien at Ruhleben for its duration. After the war, he rejoined Rutherford at Cambridge.

Chadwick began his career at a time when only two subatomic particles were known: Thomson's electron and the proton, which had been identified by Rutherford. Protons form part of the nucleus of the atom, but experiments showed that the mass of the helium nucleus was equal to that of four protons, while its charge was that of only two protons. This fact vaguely suggested that a chargeless particle might exist, but the idea seemed too radical in the 1920s, and the efforts of Rutherford and Chadwick to identify such a particle were in vain. Contributions to nuclear theory by the Joliot-Curies and others experimenting with the effect of alpha particles on beryllium might have led to the proper conclusion, but such an interpretation was not given to their results at the time. When Chadwick repeated their experiments in 1932, his evidence was sufficient for him to announce the discovery of the neutral particle that was named the neutron.

The discovery of the neutron was extremely important since, through the work of Enrico Fermi and his colleagues in Rome and that of the Joliot-Curies and of Otto Hahn and Lise Meitner in Germany, the neutron (slowed down by appropriate means) became the key to atomic fission. Unlike the helium nuclei which are charged and must overcome the repulsion of the electrical forces in the nuclei of heavy atoms, the chargeless neutron is not repelled by this energy barrier and can penetrate and split the nuclei of even the heaviest elements. For this work, Chadwick was awarded the Nobel Prize in 1935. In that same year, he assumed the Chair in Physics at the University of Liverpool. During World War II, he spent three years in the United States directing Great Britain's work on the atomic bomb. He became Sir James in 1945, retired from his position at Liverpool in 1948, and served as Master of Gonville and Caius College at Cambridge until 1958. Chadwick worked for the United Kingdom Atomic Energy Authority from 1957 until 1962, then retired to his home in North Wales until his death in 1974.

The following selection is excerpted from Chadwick's Nobel lecture, which is essentially the story of Chadwick's discovery as outlined above, but has the added virtue of being a fine, no-nonsense treatment of his work.

How do we discover the individual laws of Physics, and what is their nature? It should be remarked, to begin with, that we have no right to assume that any physical laws exist, or if they have existed up to now, that they will continue to exist in a similar manner in the future. It is perfectly conceivable that one fine day Nature should cause an unexpected event to occur which would baffle us all; and if this were to happen we would be powerless to make any objection, even if the result would be that, in spite of our endeavours, we should fail to introduce order into the resulting confusion. In such an event, the only course open to science would be to declare itself bankrupt. For this reason, science is compelled to begin by the general assumption that a general rule of law dominates throughout Nature.

—MAX PLANCK, *Universe in the Light of Modern Physics*

THE IDEA THAT there might exist small particles with no electrical charge has been put forward several times. Nernst, for example, suggested that a neutral particle might be formed by a negative electron and an equal positive charge, and that these "neutrons" might possess many of the properties of the ether; while Bragg at one time suggested that the γ-rays emitted by radioactive substances consisted of small neutral particles, which, on breaking up, released a negative electron.

The first suggestion of a neutral particle with the properties of the neutron we now know, was made by Rutherford in 1920. He thought that a proton and an electron might unite in a much more intimate way than they do in the hydrogen atom, and so form a particle of no net charge and with a mass nearly the same as that of the hydrogen atom. His view was that with such a particle as the first step in the formation of atomic nuclei from the two elementary units in the structure of matter—the proton and the electron—it would be much easier to picture how heavy complex nuclei can be gradually built up from the simpler ones. He pointed out that this neutral particle would have peculiar and interesting properties. It may be of interest to quote his remarks:

"Under some conditions, however, it may be possible for an electron to

---

James Chadwick, Nobel Prize in Physics Award Address, 1935. Reprinted with permission of Elsevier Publishing Company, and the Nobel Foundation.

combine much more closely with the H nucleus, forming a kind of neutral doublet. Such an atom would have very novel properties. Its external field would be practically zero, except very close to the nucleus, and in consequence it should be able to move freely through matter. Its presence would probably be difficult to detect by the spectroscope, and it may be impossible to contain it in a sealed vessel. On the other hand, it should enter readily the structure of atoms, and may either unite with the nucleus or be disintegrated by its intense field.

"The existence of such atoms seems almost necessary to explain the building up of the nuclei of heavy elements; for unless we suppose the production of charged particles of very high velocities it is difficult to see how any positively charged particle can reach the nucleus of a heavy atom against its intense repulsive field."

Rutherford's conception of closely combined proton and electron was adopted in pictures of nuclear structure developed by Ono (1926), by Fournier and others, but nothing essentially new was added to it.

No experimental evidence for the existence of neutral particles could be obtained for years. Some experiments were made in the Cavendish Laboratory in 1921 by Glasson and by Roberts, hoping to detect the formation of such particles when an electric discharge was passed through hydrogen. Their results were negative.

The possibility that neutral particles might exist was, nevertheless, not lost sight of. I myself made several attempts to detect them—in discharge tubes actuated in different ways, in the disintegration of radioactive substances, and in artificial disintegrations produced by α-particles.* No doubt similar experiments were made in other laboratories, with the same result.

Later, Bothe and Becker showed that γ-radiations were excited in some light elements when bombarded by α-particles. Mr. H. C. Webster, in the Cavendish Laboratory, had also been making similar experiments, and he proceeded to examine closely the production of these radiations. The radiation emitted by beryllium showed some rather peculiar features, which were very difficult to explain. I suggested therefore that the radiation might consist of neutral particles and that a test of this hypothesis might be made by passing the radiation into an expansion chamber. Several photographs were taken: some β-particle tracks—presumably recoil electrons—were observed, but nothing unexpected.†

*Cf. Rutherford and Chadwick, *Proc. Cambridge Phil. Soc., 25* (1929), 186.
†The failure was partly due to the weakness of the polonium source.

The first real step towards the discovery of the neutron was given by a very beautiful experiment of Mme. and M. Joliot-Curie, who were also investigating the properties of this beryllium radiation. They passed the radiation through a very thin window into an ionization vessel containing air. When paraffin wax or any other matter containing hydrogen was placed in front of the window the ionization in the vessel increased. They showed that this increase was due to the ejection from the wax of protons moving with very high velocities.

This behaviour of the beryllium radiation was very difficult to explain if it were a quantum radiation. I therefore began immediately the study of this new effect using different methods—the counter, the expansion chamber, and the high-pressure ionization chamber.

It appeared at once that the beryllium radiation could eject particles not only from paraffin wax but also from other light substances, such as lithium, beryllium, boron, etc., though in these cases the particles had a range of only a few millimetres in air. The experiments showed that the particles are recoil atoms of the element through which the radiation passes, set in motion by the impact of the radiation. . . .

The beryllium radiation thus behaved very differently from a quantum radiation. This property of setting in motion the atoms of matter in its path suggests that the radiation consists of particles.

Let us suppose that the radiation consists of particles of mass M moving with velocities up to a maximum velocity V. Then the maximum velocity which can be imparted to a hydrogen atom, mass 1, by the impart of such a particle will be

$$U_p = \frac{2M}{M+1} V$$

and the maximum velocity imparted to a nitrogen atom will be

$$U_n = \frac{2M}{M+14} V$$

Then

$$\frac{M+14}{M+1} = \frac{U_p}{U_n}$$

The velocities U_p and U_n were found by experiment. The maximum range of the protons ejected from paraffin wax was measured and also the ranges of the recoil atoms produced in an expansion chamber filled with nitrogen. From these ranges the velocities U_p and U_n can be deduced approximately: U_p = ca. 3.7×10^9 cm/sec, U_n = ca. 4.7×10^8 cm/sec. Thus we find $M = 0.9$.

We must conclude that the beryllium radiation does in fact consist of particles, and that these particles have a mass about the same as that of a proton. Now the experiments further showed that these particles can pass easily through thicknesses of matter, e.g. 10 or even 20 cm lead. But a proton of the same velocity as this particle is stopped by a thickness of ¼ mm of lead. Since the penetrating power of particles of the same mass and speed depends only on the charge carried by the particle, it was clear that the particle of the beryllium radiation must have a very small charge compared with that of the proton. It was simplest to assume that it has no charge at all. All the properties of the beryllium radiation could be readily explained on this assumption, that the radiation consists of particles of mass 1 and charge 0, or neutrons.

THE NATURE OF THE NEUTRON

I have already mentioned Rutherford's suggestion that there might exist a neutral particle formed by the close combination of a proton and an electron, and it was at first natural to suppose that the neutron might be such a complex particle. On the other hand, a structure of this kind cannot be fitted into the scheme of the quantum mechanics, in which the hydrogen atom represents the only possible combination of a proton and an electron. Moreover, an argument derived from the spins of the particles is against this view. The statistics and spins of the lighter elements can only be given a consistent description if we assume that the neutron is an elementary particle.

Similar arguments make it difficult to suppose that the proton is a combination of neutron and positive electron. It seems at present useless to discuss whether the neutron and proton are elementary particles or not; it may be that they are two different states of the fundamental heavy particle.

In the present view of the β-transformations of radioactive bodies the hypothesis is made that a neutron in the nucleus may transform into a proton and a negative electron with the emission of the electron, or conversely a proton in the nucleus may transform into a neutron and a positive electron with the emission of the positron. Thus

$$\mathrm{n} \rightarrow \mathrm{p} + \mathrm{e}^-$$
$$\mathrm{p} \rightarrow \mathrm{n} + \mathrm{e}^+$$

If spin is to be conserved in this process we must invoke the aid of another particle—Pauli's neutrino; we then write

$$n \rightarrow n + e^- + \text{neutrino}$$
$$p \rightarrow n + e^+ + \text{antineutrino}$$

where the neutrino is a particle of very small mass, no charge, and spin ½.

If we knew the masses of the neutron and proton accurately, these considerations would give the mass of the hypothetical neutrino.

As I have shown, observations of the momenta transferred in collisions of a neutron with atomic nuclei lead to a value of the mass of the neutron but the measurements cannot be made with precision. To obtain an accurate estimate of the neutron mass we must use the energy relations in a disintegration process in which a neutron is liberated from an atomic nucleus. The best estimate at present is obtained from the disintegration of the deuteron by the photoelectric effect of a γ-ray

$$^2_1\text{D} + h\nu \rightarrow {}^1_1\text{p} + {}^1_0\text{n}$$

The energy of the protons liberated by a γ-ray quantum of $h\nu = 2.62 \times 10^6$ eV has been measured recently by Feather, Bretscher, and myself. It is 180,000 eV. Thus the total kinetic energy set free is 360,000 eV, giving a binding energy of the deuteron of 2.26×10^6 eV. Using the value of the deuteron mass given by Oliphant, Kempton, and Rutherford, we then obtain a value for the mass of the neutron of 1.0085. (Recent measurements of the mass of deuterium lead to a value of 1.0090 for the mass of the neutron.) The mass of the hydrogen atom is 1.0081. It would seem therefore that a free neutron should be unstable, i.e. it can change spontaneously into a proton + electron + neutrino, unless the neutrino has a mass of the order of the mass of an electron. On the other hand, an argument from the shape of the β-ray spectra suggests that the mass of the neutrino is zero. One must await more exact measurements of the masses of hydrogen and deuterium before speculating further on this matter.

PASSAGE OF NEUTRONS THROUGH MATTER

The neutron in its passage through matter loses its energy in collisions with the atomic nuclei and not with the electrons. The experiments of Dee showed that the primary ionization along the track of a neutron in air could not be as much as 1 ion pair in 3 metres' path, while Massey has calculated that it may be as low as 1 ion pair in 10^5 km. This behaviour is very different from that of a charged particle, such as a proton, which dissipates its energy almost entirely in electron collisions. The collision of a neutron with an atomic nucleus, although much more frequent than with an electron, is also a rare event, for the forces between a neutron and a nucleus are very small except at distances of the order of 10^{-12} cm. In a close collision the neutron may be deflected from its path and the struck nucleus may acquire sufficient energy to produce ions. The recoiling nucleus can then be detected either in an ionization chamber or by its track in an expansion chamber. In some of these collisions, however, the neutron enters the nucleus and a disintegration is produced. Such disintegrations were first observed by Feather in his observations on the passage of neutrons through an expansion chamber filled with nitrogen. . . . The disintegration process is

$$^{14}_{7}N + ^{1}_{0}n \rightarrow ^{11}_{5}B + ^{4}_{2}He$$

Since these early experiments many examples of this type of disintegration have been observed by different workers.

Fermi and his collaborators have also shown that the phenomenon of artificial radioactivity can be provoked in the great majority of all elements, even in those of large atomic number, by the bombardment of neutrons. They have also shown that neutrons of very small kinetic energy are peculiarly effective in many cases.

In some cases an α-particle is emitted in the disintegration process; in others a proton is emitted; while in others an unstable species of nucleus is formed by the simple capture of the neutron.

Examples of these types are:

$$^{31}_{15}P + ^{1}_{0}n \rightarrow ^{28}_{13}Al + ^{4}_{2}He$$

$$^{28}_{14}Si + ^{1}_{0}n \rightarrow ^{28}_{13}Al + ^{1}_{1}H$$

$$^{27}_{13}Al + ^{1}_{0}n \rightarrow ^{28}_{13}Al$$

$$^{127}_{53}I + ^{1}_{0}n \rightarrow ^{128}_{53}I$$

In the cases just cited the nuclei formed in the reaction are unstable, showing the phenomenon of induced activity discovered by Mme. and M. Joliot-Curie, and return to a stable form with the emission of negative electrons.

In the transformations produced in heavy elements by neutrons, the process is, with very few exceptions, one of simple capture. The nucleus so formed, an isotope of the original nucleus, is often unstable but not invariably so. For example, in the reaction

$$_{48}Cd + {}_{0}n \rightarrow {}_{48}Cd + h\nu,$$

the cadmium isotope formed is stable, but a γ-ray quantum is emitted of energy corresponding to the binding energy of the neutron.

Other cases of this type of transformation are known.

The great effectiveness of the neutron in producing nuclear transmutations is not difficult to explain. In the collisions of a charged particle with a nucleus, the chance of entry is limited by the Coulomb forces between the particle and the nucleus; these impose a minimum distance of approach which increases with the atomic number of the nucleus and soon becomes so large that the chance of the particle entering the nucleus is very small. In the case of collisions of a neutron with a nucleus there is no limitation of this kind. The force between a neutron and a nucleus is inappreciable except at very small distances, when it increases very rapidly and is attractive. Instead of the potential wall in the case of the charged particle, the neutron encounters a potential hole. Thus even neutrons of very small energy can penetrate into a nucleus. Indeed slow neutrons may be enormously more effective than fast neutrons, for they spend a longer time in the nucleus. The calculations of Hans Bethe show that the chance of capture of a neutron may be inversely proportional to its velocity. The possibility of capture will depend on whether the nucleus possesses an unoccupied p-level or a level with azimuthal quantum number $l = 1$.

In cases where a particle (α-particle or proton) is ejected from the nucleus, the possibility of disintegration will depend on whether the particle can escape through the potential barrier. This will be easier the greater the energy set free in the disintegration process. As a rule disintegration by neutrons will take place with absorption of kinetic energy if a proton is released in the transformation, and may take place with release of kinetic energy if at least one of the products is an α-particle. Thus processes in which a proton is emitted can only occur with fast neutrons, even in collisions with elements of low atomic number; while processes in which α-particles are emitted can occur with slow neutrons in elements of low atomic number, but again only with fast neutrons in elements of higher atomic number. If the atomic number is sufficiently high, the neutrons at present at our disposal have insufficient energy and the particles cannot escape through the potential barrier. Thus with elements of high atomic number, only capture processes are observed, although there may be a few exceptions. There may be, however, special cases in which the particles escape through a resonance level. These would be characterized by the phenomenon that the energy of the escaping particle would be independent of the energy of the incident neutron. These special cases may explain the exceptional disintegrations in which a particle is emitted from a heavy nucleus. They may be of particular interest in giving information about the resonance levels of atomic nuclei.

There is also the possibility of resonance capture of the neutrons, more particularly with very slow neutrons. The capture of neutrons of a certain energy may take place with very great frequency in one species of nucleus while for another neighbouring nucleus the same neutrons may have a long free path. These resonance regions may perhaps be rather broad and therefore comparatively easy to observe experimentally.

THE STRUCTURE OF THE NUCLEUS

Before the discovery of the neutron we had to assume that the fundamental particles from which an atomic nucleus was built up were the proton and the electron, with the α-particle as a secondary unit. The behaviour of an electron in a space of nuclear dimensions cannot be described on present theory; and other difficulties, e.g. the statistics of the nitrogen nucleus, the peculiarities in the mass defect curve in the region of the heavy elements, also arose. These difficulties are removed if we suppose that the nuclei are built up from protons and neutrons. The forces which determine the stability of a nucleus will then be of three types, the interactions between proton and proton, between proton and neutron, and between neutron and neutron. It is assumed, with Heisenberg and Majorana, that the interaction between neutron and proton is of the exchange type—similar to that between the hydrogen atom and the hydrogen ion—and that the interaction between neutron and neutron is small.

For a nucleus of mass number A and charge Ze we shall have

$$N_n + N_p = A \qquad N_p = Z$$
$$N_n/N_p = (A - Z)/Z$$

The value of N_n/N_p for the most stable nucleus of a given mass number will be determined by the condition that the binding energy is a maximum. The repulsive Coulomb force between the protons tends to diminish the number of protons in a nucleus, while the neutron-proton interaction tends to make $N_n = N_p$, $Z = A/2$; the neutron-neutron interaction is probably very small. Now in existing nuclei $N_p \sim N_n$, and therefore the neutron-proton interaction must be the predominating force in the nucleus. In heavy elements $N_n > N_p$. This relative increase in the number of neutrons may be due either to an attractive force between neutron-neutron, or more probably to the Coulomb forces between proton–proton.

Thus it appears that the interaction between proton and neutron is of the highest significance in nuclear structure and governs the stability of a nucleus. It is most important to obtain all experimental evidence about the nature of this interaction. The information we have at present is very meagre, but I think that it does to some degree support the view that the interaction is of the exchange type. Dr. Feather and I hope to obtain more definite information on this subject by an extensive study of the collisions of neutrons and protons.

Heisenberg's considerations of nuclear structure point very strongly to this exchange interaction. Such an interaction provides an attractive force at large distances between the particles and a repulsive force at very small distances, thus giving the effect of a more or less definite radius of the particles. A system of particles interacting with exchange forces will keep together due to the attraction, but there will be a minimum distance of approach of the particles; thus the system will not collapse together but will have a more or less definite "radius."

The exchange forces between a hydrogen atom and a hydrogen ion are large compared with the forces between neutral atoms; by analogy we explain why the neutron-proton interaction is so much stronger than the proton–proton or neutron–neutron interactions.

By a suitable choice of the exchange forces it is possible to obtain a saturation effect, analogous to the saturation of valency bindings between two atoms, when each neutron is bound to two protons and each proton to two neutrons. Thus two neutrons and two protons form a closed system—the α-particle.

These ideas thus explain the general features of the structure of atomic nuclei and it can be confidently expected that further work on these lines may reveal the elementary laws which govern the structure of matter.

Theory of Electrons and Positrons

Paul A.M. Dirac

The history of particle physics is marked by a half dozen or so towering figures, men and women who stand like mighty monuments on the road to understanding the invisible or the undetectable. One of the greatest of these particle physicists is Paul Dirac, a British physicist whose parents were Swiss immigrants. Like many other physicists, Dirac wanted originally to be an engineer, but turned to pure science when the scarcity of jobs led—or forced—many into the less lucrative teaching profession. Dirac took his graduate degree in mathematics, but while studying for his Ph.D. changed his field to mathematical physics. He earned his Ph.D. in 1926, and showed such promise so early in his career that by 1932 he was firmly ensconced in Newton's old Chair as Lucasian Professor of Mathematics at Cambridge University.

Dirac established his primacy in the field of particle physics in the mid-1920s, when he devised a sound mathematical model of the wave-particle duality of electrons, and showed that the Heisenberg matrix formulation and the Schrödinger wave formulation are two different aspects of a single model. Dirac's greatest achievement was his discovery of the relativistic wave equations of the electron (the Dirac equation), which predicted the spin of the electron and the existence of the positron (the anti-electron) discovered by Carl Anderson in 1932.

For his discoveries in quantum mechanics and of antiparticles, Dirac shared the 1933 Nobel Prize in physics with Erwin Schrödinger. From 1940 until his retirement, he served as Lucasian Professor at Cambridge and as a Fellow of the Dublin Institute for Advanced Studies. Following his retirement, he moved to Florida and taught physics at Florida State University until his death in 1983.

Dirac's Nobel lecture, which encapsulates his theory of electrons and positrons, is a straightforward, nonmathematical presentation of his original theory, written as he says, "to discuss the simpler kinds of particles and to consider *what can be inferred about them from purely theoretical arguments.*" The italics are Dirac's and constitute a crystalline definition of how the problems of subatomic physics were approached before the advent of the "atom smashers," as particle accelerators were called in the 1930s. Dirac limited his lecture to a treatment of electrons and positrons, "not because they are the most interesting [particles], but because in their case the theory has been developed further."

The study of mathematics is apt to commence in disappointment. The important applications of the science, the theoretical interest of its ideas, and the logical rigour of its methods, all generate the expectation of a speedy introduction to processes of interest. We are told that by its aid the stars are weighed and the billions of molecules in a drop of water are counted. Yet, like the ghost of Hamlet's father, this great science eludes the efforts of our mental weapons to grasp it—" 'Tis here, 'tis there, 'tis gone"—and what we do see does not suggest the same excuse for illusiveness as sufficed for the ghost, that it is too noble for our gross methods. "A show of violence," if ever excusable, may surely be "offered" to the trivial results which occupy the pages of some elementary mathematical treatises.

—ALFRED NORTH WHITEHEAD, *Introduction to Mathematics*

MATTER HAS BEEN found by experimental physicists to be made up of small particles of various kinds, the particles of each kind being all exactly alike. Some of these kinds have definitely been shown to be composite, that is, to be composed of other particles of a simpler nature. But there are other kinds which have not been shown to be composite and which one expects will never be shown to be composite, so that one considers them as elementary and fundamental.

From general philosophical grounds one would at first sight like to have as few kinds of elementary particles as possible, say only one kind, or at most two, and to have all matter built up of these elementary kinds. It appears from the experimental results, though, that there must be more than this. In fact the number of kinds of elementary particles has shown a rather alarming tendency to increase during recent years.

The situation is perhaps not so bad, though, because on closer investigation it appears that the distinction between elementary and composite particles cannot be made rigorous. To get an interpretation of some modern experimental results one must suppose that particles can be created and annihilated. Thus if a particle is observed to come out from another

---

Paul A. M. Dirac, Nobel Prize in Physics Award Address, 1933. Reprinted with permission of Elsevier Publishing Company, and the Nobel Foundation.

particle, one can no longer be sure that the latter is composite. The former may have been created. The distinction between elementary particles and composite particles now becomes a matter of convenience. This reason alone is sufficient to compel one to give up the attractive philosophical idea that all matter is made up of one kind, or perhaps two kinds of bricks.

I should like here to discuss the simpler kinds of particles and to consider *what can be inferred about them from purely theoretical arguments.* The simpler kinds of particles are:

(*i*) the photons or light-quanta, of which light is composed;
(*ii*) the electrons, and the recently discovered positrons (which appear to be a sort of mirror image of the electrons, differing from them only in the sign of their electric charge);
(*iii*) the heavier particles—protons and neutrons.

Of these, I shall deal almost entirely with the electrons and the positrons—not because they are the most interesting ones, but because in their case the theory has been developed further. There is, in fact, hardly anything that can be inferred theoretically about the properties of the others. The photons, on the one hand, are so simple that they can easily be fitted into any theoretical scheme, and the theory therefore does not put any restrictions on their properties. The protons and neutrons, on the other hand, seem to be too complicated and no reliable basis for a theory of them has yet been discovered.

The question that we must first consider is how theory can give any information at all about the properties of elementary particles. There exists at the present time a general quantum mechanics which can be used to describe the motion of any kind of particle, no matter what its properties are. The general quantum mechanics, however, is valid only when the particles have small velocities and fails for velocities comparable with the velocity of light, when effects of relativity come in. There exists no relativistic quantum mechanics (that is, one valid for large velocities) which can be applied to particles with arbitrary properties. Thus when one subjects quantum mechanics to relativistic requirements, one imposes restrictions on the properties of the particle. In this way one can deduce information about the particles from purely theoretical considerations, based on general physical principles.

This procedure is successful in the case of electrons and positrons. It is to be hoped that in the future some such procedure will be found for the case of the other particles. I should like here to outline the method for electrons and positrons, showing how one can deduce the spin properties of the electron, and then how one can infer the existence of positrons with similar spin properties and with the possibility of being annihilated in collisions with electrons.

We begin with the equation connecting the kinetic energy W and momentum p_r, ($r = 1, 2, 3$), of a particle in relativistic classical mechanics

$$\frac{W^2}{c^2} - p_r^{\,2} - m^2c^2 = 0 \tag{1}$$

From this we can get a wave equation of quantum mechanics, by letting the left-hand side operate on the wave function ψ and understanding W and p_r to be the operators $ih\partial/\partial t$ and—$ih\partial/\partial x_r$. With this understanding, the wave equation reads

$$\left[\frac{W^2}{c^2} - p_r^{\,2} - m^2c^2\right]\psi = 0 \tag{2}$$

Now it is a general requirement of quantum mechanics that its wave equations shall be linear in the operator W or $\partial/\partial t$, so this equation will not do. We must replace it by some equation linear in W, and in order that this equation may have relativistic invariance it must also be linear in the p's.

We are thus led to consider an equation of the type

$$\left[\frac{W}{c} - \alpha_r p_r - \alpha_o mc\right]\psi = 0 \tag{3}$$

This involves four new variables α_r, and α_v, which are operators that can operate on ψ. We assume they satisfy the following conditions,

$$\alpha_\mu^{\,2} = 1 \qquad \alpha_\mu\alpha_v + \alpha_v\alpha_\mu = 0$$

for

$$\mu \neq \nu \text{ and } \mu, \nu = 0, 1, 2, 3$$

and also the α's commute with the p's and W. These special properties for the α's make Eq. (3) to a certain extent equivalent to Eq. (2), since if we then multiply (3) on the left-hand side by $W/c + \alpha_r p_r + \alpha_o mc$ we get exactly (2).

The new variables α, which we have to introduce to get a relativistic wave equation linear in W, give rise to the spin of the electron. From the general principles of quantum mechanics one can easily deduce that these variables α give the electron a spin angular momentum of half a quantum and a magnetic moment of one Bohr magneton in the reverse direction to the angular momentum. These results are in agreement with experiment. They were, in fact, first obtained from the experimental evidence provided by spectroscopy and afterwards confirmed by the theory.

The variables α also give rise to some rather unexpected phenomena concerning the motion of the electron. These have been fully worked out by Schrödinger. It is found that an electron which seems to us to be moving slowly, must actually have a very high frequency oscillatory motion of small amplitude superposed on the regular motion which appears to us. As a result of this oscillatory motion, the velocity of the electron at any time equals the velocity of light. This is a prediction which cannot be directly verified by experiment, since the frequency of the oscillatory motion is so high and its amplitude is so small. But one must believe in this consequence of the theory, since other consequences of the theory which are inseparably bound up with this one, such as the law of scattering of light by an electron, are confirmed by experiment.

There is one other feature of these equations which I should now like to discuss, a feature which led to the prediction of the positron. If one looks at Eq. (1), one sees that it allows the kinetic energy W to be either a positive quantity greater than mc^2 or a negative quantity less than $-mc^2$. This result is preserved when one passes over to the quantum equation (2) or (3). These quantum equations are such that, when interpreted according to the general scheme of quantum dynamics, they allow as the possible results of a measurement of W either something greater than mc^2 or something less than $-mc^2$.

Now in practice the kinetic energy of a particle is always positive. We thus see that our equations allow of two kinds of motion for an electron, only one of which corresponds to what we are familiar with. The other corresponds to electrons with a very peculiar motion such that the faster they move, the less energy they have, and one must put energy into them to bring them to rest.

One would thus be inclined to introduce, as a new assumption of the theory, that only one of the two kinds of motion occurs in practice. But this gives rise to a difficulty, since we find from the theory that if we disturb the electron, we may cause a transition from a positive-energy state of motion to a negative-energy one, so that, even if we suppose all the electrons in the world to be started off in positive-energy states, after a time some of them would be in negative-energy states.

Thus in allowing negative-energy states, the theory gives something which appears not to correspond to anything known experimentally, but which we cannot simply reject by a new assumption. We must find some meaning for these states.

An examination of the behaviour of these states in an electromagnetic field shows that they correspond to the motion of an electron with a positive charge instead of the usual negative one—what the experimenters now call a positron. One might, therefore, be inclined to assume that electrons in negative-energy states are just positrons, but this will not do, because the observed positrons certainly do not have negative energies. We can, however, establish a connection between electrons in negative-energy states and positrons, in a rather more indirect way.

We make use of the exclusion principle of Pauli, according to which there can be only one electron in any state of motion. We now make the assumptions that in the world as we know it, nearly all the states of negative energy for the electrons are occupied, with just one electron in each state, and that a uniform filling of all the negative-energy states is completely unobservable to us. Further, *any unoccupied negative-energy state, being a departure from uniformity, is observable and is just a positron.*

An unoccupied negative-energy state, or *hole,* as we may call it for brevity, will have a positive energy, since it is a place where there is a shortage of negative energy. A hole is, in fact, just like an ordinary particle, and its identification with the positron seems the most reasonable way of getting over the difficulty of the appearance of negative energies in our equations. On this view the positron is just a mirror-image of the electron, having exactly the same mass and opposite charge. This has already been roughly confirmed by experiment. The positron should also have similar spin properties to the electron, but this has not yet been confirmed by experiment.

From our theoretical picture, we should expect an ordinary electron, with positive energy, to be able to drop into a hole and fill up this hole, the energy being liberated in the form of electromagnetic radiation. This would mean a process in which an electron and a positron annihilate one another. The converse process, namely the creation of an electron and a positron from electromagnetic radiation, should also be able to take place. Such processes appear to have been found experimentally, and are at present being more closely investigated by experimenters.

The theory of electrons and positrons which I have just outlined is a self-consistent theory which fits the experimental facts so far as is yet known. One would like to have an equally satisfactory theory for protons.

One might perhaps think that the same theory could be applied to protons. This would require the possibility of existence of negatively charged protons forming a mirror-image of the usual positively charged ones. There is, however, some recent experimental evidence obtained by Stern about the spin magnetic moment of the proton, which conflicts with this theory for the proton. As the proton is so much heavier than the electron, it is quite likely that it requires some more complicated theory, though one cannot at the present time say what this theory is.

In any case I think it is probable that negative protons can exist, since as far as the theory is yet definite, there is a complete and perfect symmetry between positive and negative electric charge, and if this symmetry is really fundamental in nature, it must be possible to reverse the charge on any kind of particle. The negative protons would of course be much harder to produce experimentally, since a much larger energy would be required, corresponding to the larger mass.

If we accept the view of complete symmetry between positive and negative electric charge so far as concerns the fundamental laws of Nature, we must regard it rather as an accident that the Earth (and presumably the whole solar system), contains a preponderance of negative electrons and positive protons. It is quite possible that for some of the stars it is the other way about, these stars being built up mainly of positrons and negative protons. In fact, there may be half the stars of each kind. The two kinds of stars would both show exactly the same spectra, and there would be no way of distinguishing them by present astronomical methods.

R.4

The Production and Properties of Positrons

CARL ANDERSON

CARL ANDERSON was born of Swedish parents in New York City in 1905. He is a product of the California Institute of Technology, where he obtained both his undergraduate degree in physics and engineering in 1927 and his Ph.D. in 1930. He did postdoctoral research on cosmic rays under Robert Millikan, which eventually led to the discovery of the positron. Anderson was appointed to the Chair of Physics at Cal Tech in 1939 and became the chairman of the department in 1962.

In connection with their cosmic-ray research, Anderson and Millikan devised a cloud chamber containing a lead-plate divider. Such a plate slows the incredibly fast cosmic rays so that their paths in the cloud chamber are curved. In the summer of 1932, Anderson observed that some photographs of these curves were exactly like the curves described by electrons, except that they curved in the opposite direction. Their behavior, in other words, was that of an electron carrying a positive charge. Dirac had predicted the positive counterpart of the electron several years before, and Anderson, confident that he had photographed these particles, named them positrons. (He also came up with the term negatron for the electron, but the name did not catch on; Millikan, by the way, liked to call positrons "positive electrons.")

In the mid-1930s, while doing cosmic-ray research in the Colorado mountains, Anderson spotted a new curve on some photographic plates, which proved to be that of a particle proposed by the Japanese physicist Hideki Yukawa. It was 130 times as massive as the electron, which is itself about one quarter the mass of a proton. Anderson's original name for the particle was mesotron, which was soon shortened to meson. Anderson shared the 1936 Nobel Prize in physics with Victor Hess, whose original discovery of cosmic rays had laid the groundwork for Anderson's own professional career.

Later, Cecil Powell found a slightly heavier meson, the pi-meson or pion, that was identified as Yukawa's predicted particle, and in 1961 Anderson's meson was found to be identical with the electron in every respect except for mass; it is known as the muon and its mass is 208 times that of the electron.

Anderson's Nobel lecture of 12 December 1936 is an historical account of his work, followed by a theoretical interpretation that is both clear and free of overly technical complication.

In the light of quantum theory . . . elementary particles are no longer real in the same sense as objects of daily life, trees or stones . . . .

—WERNER HEISENBERG, *On Modern Physics*

INFORMATION OF FUNDAMENTAL importance to the general problem of atomic structure has resulted from systematic studies of the cosmic radiation carried out by the Wilson cloud-chamber method.

After Skobelzyn in 1927 had first shown photographs of tracks of cosmic-ray particles, Professor R. A. Millikan and the writer in the spring of 1930 planned a cloud-chamber apparatus suitable for cosmic-ray studies, in particular to measure the energies of cosmic-ray particles by means of their curvatures in a strong magnetic field. The chamber, of dimensions $17 \times 17 \times 3$ cm, was arranged with its long dimension vertical, and incorporated into a powerful electromagnet capable of maintaining a uniform magnetic field up to 24,000 gauss strength.

In the summer of 1931 the first results were obtained with this technique. The direct measurement of the energies of atomic particles was extended from about 15 million electron-volts, the highest energy measured before that time, to 5 billion electron-volts. In the spring of 1932 a preliminary paper on the energies of cosmic-ray particles was published in which energies over 1 billion electron-volts were reported. It was here shown that particles of positive charge occurred about as abundantly as did those of negative charge, and in many cases several positive and negative particles were found to be projected simultaneously from a single center. The presence of positively charged particles and the occurrence of "showers" of several particles showed clearly that the absorption of cosmic rays in material substances is due primarily to a nuclear phenomenon of a new type.

Measurements of the specific ionization of both the positive and negative particles, by counting the number of droplets per unit length along the tracks, showed the great majority of both the positive and negative particles to possess unit electric charge. The particles of negative charge were readily interpreted as electrons, and those of positive charge were at

---

Carl D. Anderson, Nobel Prize in Physics Award Address, 1936. Reprinted with permission of Elsevier Publishing Company, and the Nobel Foundation.

first tentatively interpreted as protons, at that time the only known particle of unit positive charge.

If the particles of positive charge were to be ascribed to protons then those of low energy and sharp curvature in the magnetic field (e.g. a curvature greater than that corresponding to an electron having an energy of about 500 million electron-volts), should be expected to exhibit an appreciably greater ionization than the negatively charged electrons. In general, however, the positive particles seemed to differ in specific ionization only inappreciably from the negative ones. To avoid the assumption, which appeared very radical at that time, that the positive particles had electronic mass, serious consideration was given to the possibility that the particles which appeared to be positively charged and directed downward into the earth were in reality negatively charged electrons which through scattering had suffered a reversal of direction and were projected upwards away from the earth. Although such a reversal of direction through scattering might be expected to occur occasionally it seemed inadequate to account for the large number of particle tracks which showed a specific ionization anomalously small if they were to be ascribed to protons.

To differentiate with certainty between the particles of positive and negative charge it was necessary only to determine without ambiguity their direction of motion. To accomplish this purpose a plate of lead was inserted across a horizontal diameter of the chamber. The direction of motion of the particles could then be readily ascertained due to the lower energy and therefore the smaller radius of curvature of the particles in the magnetic field after they had traversed the plate and suffered a loss in energy.

Results were then obtained which could logically be interpreted only in terms of particles of a positive charge and a mass of the same order of magnitude as that normally possessed by the free negative electron. In particular one photograph . . . shows a particle of positive charge traversing a 6 mm plate of lead. If electronic mass is assigned to this particle its energy before it traverses the plate is 63 million electron-volts and after it emerges its energy is 23 million electron-volts. The possibility that this particle of positive charge could represent a proton is ruled out on the basis of range and curvature. A proton of the curvature [described] after it emerges from the plate would have an energy of 200,000 electron-volts, and according to previously well-established experimental data would have a range of only 5 mm whereas the observed range was greater than 50 mm. The only possible conclusion seemed to be that this track, indeed, was the track of a positively charged electron. Examples similar to this and others in which two or more particles were found to be produced at one center gave additional evidence for the existence of particles of positive charge and mass, small compared with that of the proton. These results formed the basis of the paper published in September 1932 announcing the existence of free positive electrons.

Measurements by the droplet counting method of the magnitude of the specific ionization of the positive and negative electrons which occur with energies low enough to be appreciably curved in the magnetic field have shown that the mass and charge of the positive electron cannot differ by more than 20 percent and 10 percent, respectively, from the mass and charge of the negative electron.

Blackett and Occhialini, using an apparatus similar to ours but with the added advantage that through the use of control by Geiger-Müller tube counters their apparatus was made to respond automatically to the passage of a cosmic-ray particle, in the spring of 1933 confirmed the existence of positive electrons, or positrons, and obtained many beautiful photographs of complex electron showers.

That positrons could be produced by an agent other than cosmic rays was first shown by Chadwick, Blackett and Occhialini when they observed that positrons were produced by the radiation generated in the impact of alpha particles upon beryllium. The radiation produced in the beryllium is complex in character, consisting both of neutrons and gamma rays. In their experiment it was not possible to determine which of these rays was responsible for the production of positrons. Curie and Joliot by a similar experiment, in which they interposed blocks of lead and paraffin into the path of the rays from beryllium and measured the yield of positrons as a function of the thickness and material of the absorber concluded that the positrons arose more likely as a result of the gamma rays than of the neutrons. . . .

In addition to the methods of producing positrons already mentioned, i.e. by absorption of cosmic-ray photons and electrons, and by the absorption of sufficiently high energy gamma rays from terrestrial sources, positrons have also been observed among the disintegration products of certain radioactive substances. The artificially produced radioactive elements first discovered by Curie and Joliot in 1934 are found to disintegrate either by the ejection of a positive or negative electron. Those elements whose atomic number is greater than that of the stable elements of the same mass number in general disintegrate by the ejection of a positron. . . .

THEORETICAL INTERPRETATION

The present electron theory of Dirac provides a means of describing many of the phenomena governing the production and annihilation of positrons. Blackett and Occhialini first suggested that the appearance of pairs of positive and negative electrons could be understood in terms of this theory as the "creation" of a positive-negative electron pair in the neighborhood of an atomic nucleus. The energy corresponding to the proper mass of both of the particles, as well as to their kinetic energies, is, according to this view, supplied by the incident radiation. Since the energy corresponding to the proper mass of a pair of electrons is approximately one million electron-volts one should expect gamma rays of energy greater than this amount to produce positrons in their passage through matter, and further that the sum of the kinetic energies of the positive and negative electrons should be equal to the energy of the radiation producing them diminished by approximately one million electron-volts.

Experiments by Neddermeyer and the writer, and by Chadwick, Blackett and Occhialini, and others, have shown this relation to obtain in the production of positrons by . . . gamma rays, providing evidence for the correctness of this view of the origin of positive-negative electron pairs.

The theory of Dirac requires further that a positron, when it finds itself in a very ordinary environment, as, for example, in passing through common substances, will, on the average, have only a very short life, of the order of one billionth of a second or less. The positrons and negative electrons will mutually annihilate one another in pairs, and in their stead will appear a pair of photons, each of approximately one-half million electron-volts energy. Although the lifetime of positrons has not been actually measured, it has been shown to be very short, and the radiation which results from their annihilation has been observed. The first to do this were Joliot and Thibaud. The annihilation radiation is of the proper intensity and the energy of its individual corpuscles is approximately the required amount of one-half million electron-volts, corresponding to the complete annihilation of the positrons.

POSITRONS OF HIGH ENERGY

The experimental results on the production of positrons out of radiation have been shown to be in approximate agreement with the theory for those processes where the quantum energies are not too high. Gamma radiations of quantum energy extending up to some 15 million electron-volts arise in certain nuclear transformations produced in the laboratory. Measurements of the absorption of these radiations and of the numbers and distribution in energy of the positive and negative electrons produced by these radiations are in sufficiently good agreement with the calculations of Oppenheimer, Heitler, and Bethe based on the Dirac theory to provide evidence for the essential correctness of the theory of absorption of gamma radiations in the range of quantum energy up to some 15 million electron-volts.

In the broad range of energies, however, which lies above 15 million electron-volts and extends up to at least 20,000 million electron-volts, such as the energies with which the cosmic-ray particles are endowed, the experiments have only very recently provided strong evidence leading to a detailed understanding of the absorption of photons and electrons in this range of energies and to an explanation of the cosmic-ray showers.

Closely related to the process of the production of positive and negative electrons out of radiation, is the one which may be considered its inverse, namely, the production of radiation through nuclear impacts by a positive or negative electron in its passage through matter. Direct measurements on the energy loss of electrons, in the energy range up to about 400 million electron-volts, in their traversals through thin plates of lead, have shown that the loss in energy due to direct ionization by the electrons is but a small fraction of the total energy loss, and that the loss in energy over that due to ionization is in good accord with that to be expected theoretically through the production of radiation by nuclear impact. Furthermore a small number of measurements at energies up to 1,000 million electron-volts has shown no significant deviation from the theoretical loss. These data on energy loss of high-energy electrons afford strong evidence that, at least in part, the origin of the cosmic-ray showers of photons and positive and negative electrons can be understood in terms of a chain of successive processes of photon production by radiative impacts with nuclei on the part of the high-energy positive and negative electrons, and the subsequent absorption of these photons in nuclear collisions resulting in the production of numerous positive-negative electron pairs which appear as the cosmic-ray showers. After more detailed theoretical computations have been carried out on the rate of building-up of positive and negative electron secondaries resulting from these multiple processes, and their subsequent removal through absorption, a more adequate test of the theory will be possible. At present, however, it is very difficult to doubt that the highly absorbable component of the primary cosmic-ray beam consists largely of electrons absorbed principally through the mechanisms discussed above, which give rise to the electron showers.

Until quite recently it was not clear that the high-energy positive and negative electrons which have now been shown to exhibit a high absorbability, behaved in a manner essentially different from the cosmic-ray particles of highly penetrating character. These highly penetrating particles, although not free positive and negative electrons, appear to consist of both positive and negative particles of unit electric charge, and will provide interesting material for future study. . . .

R.5

Meson Theory in Its Developments

Hideki Yukawa

Hideki Yukawa was born in Tokyo, Japan, in 1907. His father was a professor of geology at Kyoto University and Yukawa spent his childhood in Kyoto, graduating from the local university in 1929. He earned his Ph.D. at Osaka University in 1938 while employed as a lecturer there. One year later he became, and continues to be, Professor of Theoretical Physics at Kyoto University. While working as an instructor, Yukawa had published in 1935 a paper in which he proposed a new nuclear force-field theory, and hypothesized the existence of the meson as the mediator of that field. The theoretical and experimental work then being done in the United States encouraged Yukawa to develop his meson theory further. Later, he worked on a "non-local field" theory of elementary particles.

The view of the atom that prevailed during the 1930s was that the nucleus consisted solely of protons (with a positive charge) and neutrons (with no charge). Protons alone could not form a stable nucleus since their mutual repulsions would prohibit such a structure; the neutron provided the needed counterbalance to the proton. But what is the nature of the attractive force between protons and neutrons? Yukawa suggested that inside the nucleus there is some extraordinarily powerful "nuclear force" operating over an extremely short range (a trillionth of a millimeter, the diameter of the nucleus), dying away to zero within a distance of about the same magnitude. He developed the theory that the strong nuclear force is mediated by massive particles that are tossed back and forth between nucleons (neutrons and protons). He showed that the masses of such particles must be about 250 times the mass of an electron to account for such a short range and strong force as the nuclear force. These particles, called *mesons,* were discovered by Cecil Powell.

Yukawa's Nobel lecture, which we have selected, is a straightforward account of the development of his meson theory, with mathematics held well in check. It documents one of the monumental discoveries in quantum mechanics and particle physics.

The field exists always and everywhere; it can never be removed. It is the carrier of all material phenomena. It is the "void" out of which the proton creates the pi-mesons. Being and fading of particles are merely forms of motion of the field.

—W. THIRRING

THE MESON THEORY started from the extension of the concept of the field of force so as to include the nuclear forces in addition to the gravitational and electromagnetic forces. The necessity of introduction of specific nuclear forces, which could not be reduced to electromagnetic interactions between charged particles, was realized soon after the discovery of the neutron, which was to be bound strongly to the protons and other neutrons in the atomic nucleus. As pointed out by Wigner, specific nuclear forces between two nucleons, each of which can be either in the neutron state or the proton state, must have a very short range of the order of 10^{-13} cm, in order to account for the rapid increase of the binding energy from the deuteron to the alpha-particle. The binding energies of nuclei heavier than the alpha-particle do not increase as rapidly as if they were proportional to the square of the mass number A, i.e. the number of nucleons in each nucleus, but they are in fact approximately proportional to A. This indicates that nuclear forces are saturated for some reason. Heisenberg suggested that this could be accounted for, if we assumed a force between a neutron and a proton, for instance, due to the exchange of the electron or, more generally, due to the exchange of the electric charge, as in the case of the chemical bond between a hydrogen atom and a proton. Soon afterwards, Fermi developed a theory of beta-decay based on the hypothesis by Pauli, according to which a neutron, for instance, could decay into a proton, an electron, and a neutrino, which was supposed to be a very penetrating neutral particle with a very small mass.

This gave rise, in turn, to the expectation that nuclear forces could be reduced to the exchange of a pair of an electron and a neutrino between two nucleons, just as electromagnetic forces were regarded as due to the exchange of photons between charged particles. It turned out, however,

---

Hideki Yukawa, Nobel Prize in Physics Award Address, 1949.

that the nuclear forces thus obtained were much too small because the beta-decay was a very slow process compared with the supposed rapid exchange of the electric charge responsible for the actual nuclear forces. The idea of the meson field was introduced in 1935 in order to make up this gap. Original assumptions of the meson theory were as follows:
I. The nuclear forces are described by a scalar field U, which satisfies the wave equation

$$\left(\frac{\partial^2}{\partial x^2}+\frac{\partial^2}{\partial y^2}+\frac{\partial^2}{\partial z^2}-\frac{1}{c^2}\frac{\partial^2}{\partial t^2}-\varkappa^2\right)U=0 \tag{1}$$

in vacuum, where $\varkappa$ is a constant with the dimension of reciprocal length. Thus, the static potential between two nucleons at a distance r is proportional to $\exp(-\varkappa r)/r$, the range of forces being given by $1/\varkappa$.
II. According to the general principle of quantum theory, the field U is inevitably accompanied by new particles or quanta, which have the mass

$$\mu=\frac{\varkappa h}{c} \tag{2}$$

and the spin 0, obeying Bose-Einstein statistics. The mass of these particles can be inferred from the range of nuclear forces. If we assume, for instance, $\varkappa = 5 \times 10^{12}\ \mathrm{cm}^{-1}$, we obtain $\mu \cong 200 m_e$, where m_e is the mass of the electron.
III. In order to obtain exchange forces, we must assume that these mesons have the electric charge $+e$ or $-e$, and that a positive (negative) meson is emitted (absorbed) when the nucleon jumps from the proton state to the neutron state, whereas a negative (positive) meson is emitted (absorbed) when the nucleon jumps from the neutron to the proton. Thus a neutron and a proton can interact with each other by exchanging mesons just as two charged particles interact by exchanging photons. In fact, we obtain an exchange force of Heisenberg type between the neutron and the proton of the correct magnitude, if we assume that the coupling constant g between the nucleon and the meson field, which has the same dimension as the elementary charge e, is a few times larger than e.

However, the above simple theory was incomplete in various respects. For one thing, the exchange force thus obtained was repulsive for the triplet S-state of the deuteron in contradiction to the experiment, and moreover we could not deduce the exchange force of Majorana type, which was necessary in order to account for the saturation of nuclear forces just at the alpha-particle. In order to remove these defects, more general types of meson fields including vector, pseudoscalar and pseudovector fields in addition to the scalar fields, were considered by various authors. In particular, the vector field was investigated in detail, because it could give a combination of exchange forces of Heisenberg and Majorana types with correct signs and could further account for the anomalous magnetic moments of the neutron and the proton qualitatively. Furthermore, the vector theory predicted the existence of non-central forces between a neutron and a proton, so that the deuteron might have the electric quadripole moment. However, the actual electric quadripole moment turned out to be positive in sign, whereas the vector theory anticipated the sign to be negative. The only meson field which gave the correct signs both for nuclear forces and for the electric quadripole moment of the deuteron was the pseudoscalar field. There was, however, another feature of nuclear forces, which was to be accounted for as a consequence of the meson theory. Namely, the results of experiments on the scattering of protons by protons indicated that the type and magnitude of interaction between two protons was, at least approximately, the same as that between a neutron and a proton, apart from the Coulomb force. Now the interaction between two protons or two neutrons was obtained only if we took into account the terms proportional to g^4, whereas that between a neutron and a proton was proportional to g^2, as long as we were considering charged mesons alone. Thus it seemed necessary to assume further: IV. In addition to charged mesons, there are neutral mesons with the mass either exactly or approximately equal to that of charged mesons. They must also have the integer spin, obey Bose-Einstein statistics and interact with nucleons as strongly as charged mesons.

This assumption obviously increased the number of arbitrary constants in meson theory, which could be so adjusted as to agree with a variety of experimental facts. These experimental facts could not be restricted to those of nuclear physics in the narrow sense, but was to include those related to cosmic rays, because we expected that mesons could be created and annihilated due to the interaction of cosmic ray particles with energies much larger than μc^2 with matter. In fact, the discovery of particles of intermediate mass in cosmic rays in 1937 was a great encouragement to further developments of meson theory. At that time, we came naturally to the conclusion that the mesons which constituted the main part of the hard component of cosmic rays at sea level were to be identified with the mesons which were responsible for nuclear forces. Indeed, cosmic ray

mesons had the mass around 200 m_e as predicted and moreover, there was the definite evidence for the spontaneous decay, which was the consequence of the following assumption of the original meson theory:
V. Mesons interact also with light particles, i.e. electrons and neutrinos, just as they interact with nucleons, the only difference being the smallness of the coupling constant g' in this case compared with g. Thus a positive (negative) meson can change spontaneously into a positive (negative) electron and a neutrino, as pointed out first by Bhabha. The proper lifetime, i.e. the mean lifetime at rest, of the charged scalar meson, for example, is given by

$$\tau_0 = 2\left(\frac{\hbar c}{(g')^2}\right)\left(\frac{\hbar}{\mu c^2}\right) \quad (3)$$

For the meson moving with velocity v, the lifetime increases by a factor $1/\sqrt{1-(v/c)^2}$ due to the well-known relativistic delay of the moving clock. Although the spontaneous decay and the velocity dependence of the lifetime of cosmic ray mesons were remarkably confirmed by various experiments, there was an undeniable discrepancy between theoretical and experimental values for the lifetime. The original intention of meson theory was to account for the beta-decay by combining the assumptions III and V together. However, the coupling constant g', which was so adjusted as to give the correct result for the beta-decay, turned out to be too large in that it gave the lifetime τ_0 of mesons of the order of 10^{-8} sec, which was much smaller than the observed lifetime 2×10^{-6} sec. Moreover, there were indications, which were by no means in favour of the expectation that cosmic-ray mesons interacted strongly with nucleons. For example, the observed cross-section of scattering of cosmic-ray mesons by nuclei was much smaller than that obtained theoretically. Thus, already in 1941, the identification of the cosmic-ray meson with the meson, which was supposed to be responsible for nuclear forces, became doubtful. In fact, Tanikawa and Sakata proposed in 1942 a new hypothesis as follows: The mesons which constitute the hard component of cosmic rays at sea level are not directly connected with nuclear forces, but are produced by the decay of heavier mesons which interacted strongly with nucleons.

However, we had to wait for a few years before this two-meson hypothesis was confirmed, until 1947, when two very important facts were discovered. First, it was discovered by Italian physicists that the negative mesons in cosmic rays, which were captured by lighter atoms, did not

disappear instantly, but very often decayed into electrons in a mean time interval of the order of 10^{-6} sec. This could be understood only if we supposed that ordinary mesons in cosmic rays interacted very weakly with nucleons. Soon afterwards, Powell and others discovered two types of mesons in cosmic rays, the heavier mesons decaying in a very short time into lighter mesons. Just before the latter discovery, the two-meson hypothesis was proposed by Marshak and Bethe independent of the Japanese physicists above mentioned. In 1948, mesons were created artificially in Berkeley and subsequent experiments confirmed the general picture of two-meson theory. The fundamental assumptions are now:
(i) The heavier mesons, i.e. π-mesons with the mass m_π about 280 m_e interact strongly with nucleons and can decay into lighter mesons, i.e. μ-mesons and neutrinos with a lifetime of the order of 10^{-8} sec; π-mesons have integer spin (very probably spin 0) and obey Bose-Einstein statistics. They are responsible for, at least, a part of nuclear forces. In fact, the shape of nuclear potential at a distance of the order of $\hbar/m_\pi c$ or larger could be accounted for as due to the exchange of π-mesons between nucleons.
(ii) The lighter mesons, i.e. μ-mesons with the mass about 210 m_e are the main constituent of the hard component of cosmic rays at sea level and can decay into electrons and neutrinos with the lifetime 2×10^{-6} sec. They have very probably spin ½ and obey Fermi-Dirac statistics. As they interact only weakly with nucleons, they have nothing to do with nuclear forces.

Now, if we accept the view that π-mesons are the mesons that have been anticipated from the beginning, then we may expect the existence of neutral π-mesons in addition to charged π-mesons. Such neutral mesons, which have integer spin and interact as strongly as charged mesons with nucleons, must be very unstable, because each of them can decay into two or three photons. In particular, a neutral meson with spin 0 can decay into two photons and the lifetime is of the order of 10^{-14} sec or even less than that. Very recently, it became clear that some of the experimental results obtained in Berkeley could be accounted for consistently by considering that, in addition to charged π-mesons, neutral π-mesons with the mass approximately equal to that of charged π-mesons were created by collisions of high-energy protons with atomic nuclei and that each of these neutral mesons decayed into two mesons with the lifetime of the order of 10^{-13} sec or less. Thus, the neutral mesons must have spin 0.

In this way, meson theory has changed a great deal during these fifteen years. Nevertheless, there remain still many questions unanswered.

Among other things, we know very little about mesons heavier than π-mesons. We do not know yet whether some of the heavier mesons are responsible for nuclear forces at very short distances. The present form of meson theory is not free from the divergence difficulties, although recent development of relativistic field theory has succeeded in removing some of them. We do not yet know whether the remaining divergence difficulties are due to our ignorance of the structure of elementary particles themselves. We shall probably have to go through another change of the theory, before we shall be able to arrive at the complete understanding of the nuclear structure and of various phenomena, which will occur in high energy regions.

R.6

Conservation Laws

Kenneth Ford

CONSERVATION PRINCIPLES ARE essential for explaining the characteristics of all the newly discovered particles. For example, the conservation of momentum principle holds that in the absence of any external force acting on a system, the total momentum of the system remains constant regardless of the movements and interactions of the individual particles. Momentum conservation is also maintained when a particle decays, since the amount of momentum before the decay began is equal to the sum total of the momenta of all the products (particles) of decay. A second major principle is the conservation of energy, which asserts that no physical process can take place unless enough energy is available. This is not to say that the ratio of energy to mass in a system remains constant; Einstein's mass-energy equivalence equation ($E = mc^2$) tells us that matter can be converted into energy and vice versa. The total amount of energy of a system undergoing radioactive decay is conserved, but some of the mass may be converted into energy during the process.

A third important conservation principle asserts that the angular momentum, or total rotational motion, of an isolated system before and after a process is unchanged. The total angular momentum is computed by summing the orbital and spinning motions for all the particles in that system. Thus, regardless of how a system of elementary particles is transformed during a process, the summed spins of all the particles must remain the same. Another conservation law states that electric charge can be neither created or destroyed; the total charge in the Universe is constant. As a result, if a new positive charge appears during a process, an additional negative charge must also appear to offset it.

Yet another important guide for particle physics is the conservation of the total number of nucleons (protons and neutrons) in the Universe, even though this principle has recently been challenged by advocates of the Weinberg-Salam "electroweak theory," which asserts that protons decay after 10^{32} years. Although light particles called *leptons* are not conserved individually, the total number of leptons (consisting of neutrinos, electrons, and muons, as well as their antiparticle counterparts) is conserved. The conservation of parity, which was originally assumed to be valid among all four forces, was shown by Yang and Lee to be violated in weak interactions. Parity itself simply refers to the mirror-reflected states or configurations of particles. Finally, there is the partial conservation of strangeness, a concept that was introduced to better understand the decay processes of heavy particles, and which is explored more fully elsewhere in this section.

Physicists have always been guided by the conservation laws and accepted them as absolute truths, because, as with checkbooks, when debits and credits do not balance, something is awry. Bertrand Russell, a fine writer and one of the great minds of the nineteenth and twentieth centuries, did not think much of conservation laws, regarding them as "truisms," or cheap profundities. But they are here to stay, and Kenneth Ford has provided us with a fine summary of conservation principles.

Untwisting all the chains that tie/ The hidden soul of harmony.

—JOHN MILTON, "L'Allegro"

Every why hath a wherefore.

—WILLIAM SHAKESPEARE, *The Comedy of Errors*

IN A SLOW and subtle, yet inexorable, way conservation laws have moved in the past few centuries from the role of interesting sidelight in physics to the most central position. What little we now understand about the interactions and transformations of particles comes in large part through certain conservation laws that govern elementary-particle behavior.

A conservation law is a statement of constancy in nature. If there is a room full of people, say, at a cocktail party, and no one comes in or leaves, we can say that there is a law of conservation of the number of people; that number is a constant. This would be a rather uninteresting law. But suppose the conservation law remained valid as guests came and went. This would be more interesting, for it would imply that the rate of arrival of guests was exactly equal to the rate of departure. During a process of change, something is remaining constant. The significant conservation laws in nature are of this type, laws of constancy during change. It is not surprising that scientists, in their search for simplicity, fasten on conservation laws with particular enthusiasm, for what could be simpler than a quantity that remains absolutely constant during complicated processes of change. To cite an example from the world of particles, the total electric charge remains precisely constant in every collision, regardless of how many particles may be created or annihilated in the process.

The classical laws of physics are expressed primarily as laws of change, rather than as laws of constancy. Newton's law of motion describes how the motion of objects responds to forces that act upon them. Maxwell's equations of electromagnetism connect the rate of change of electric and magnetic fields in space and time. The early emphasis in fundamental science was rather naturally on discovering those laws which successfully describe the changes actually occurring in nature. Briefly, the "classical" philosophy concerning nature's laws is this. One can imagine countless

---

Kenneth Ford, "Conservation Laws," in *The World of Elementary Particles.* Waltham, Massachusetts: Blaisdell Publishing Company, 1963, pp. 81-112. Reprinted with permission of the author.

possible laws, indeed infinitely many, that might describe a particular phenomenon. Of these, nature has chosen only one simple law, and the job of science is to find it. Having successfully found laws of change, scientists may derive from them certain conservation laws, such as the conservation of energy in mechanics. These appear as particularly interesting and useful consequences of the theory, but are not themselves taken as fundamental statements of the theory.

Gradually conservation laws have percolated to the top in the hierarchy of natural laws. This is not merely because of their simplicity, although this has been an important factor. It comes about also for two other reasons. One is the connection between conservation laws and principles of invariance and symmetry in nature—surely, one of the most beautiful aspects of modern science. The meaning of this connection will be discussed near the end of this chapter. The other reason we want to discuss here might best be described simply as a new view of the world, in which conservation laws appear naturally as the most fundamental statements of natural law. This new view is a view of order upon chaos—the order of conservation laws imposed upon the chaos of continual annihilation and creation taking place in the submicroscopic world. The strong hint emerging from recent studies of elementary particles is that the only inhibition imposed upon the chaotic flux of events in the world of the very small is that imposed by the conservation laws. Everything that *can* happen without violating a conservation law *does* happen.

This new view of democracy in nature—freedom under law—represents a revolutionary change in our view of natural law. The older view of a fundamental law of nature was that it must be a law of *permission.* It defined what *can* (and must) happen in natural phenomena. According to the new view, the more fundamental law is a law of *prohibition.* It defines what *cannot* happen. A conservation law is, in effect, a law of prohibition. It prohibits any phenomenon that would change the conserved quantity, but otherwise allows any events. Consider, for example, the production of pions in a proton-proton collision,

$$p + p \rightarrow p + p + \pi + \pi + \pi + \cdots .$$

If a law of permission were operative, one might expect that, for protons colliding in a particular way, the law would specify the number and the type of pions produced. A conservation law is less restrictive. The conservation of energy limits the number of pions that can be produced, because the mass of each one uses up some of the available energy. It might say, for example, that not more than six pions can be produced. In the actual collision there might be none, or one, or any number up to six. The law of charge conservation says that the total charge of the pions must be zero, but places no restriction on the charge of any particular pion; this could be positive, negative, or neutral.

To make more clear the distinction between laws of permission and laws of prohibition, let us return to the cocktail party. A law of change, which is a law of permission, might describe the rate of arrival and the rate of departure of guests as functions of time. In simplest form, it might say that three guests per minute arrive at 6:00, two guests per minute at 6:15, and so on. Or it might say, without changing its essential character as a law of permission, that the rate of arrival of guests is given by the formula:

$$R = \frac{A}{\pi D} \frac{1}{1 + \left(T - 5 - 2\frac{A}{D}\right)^2},$$

where R is the number of guests arriving per minute. A is the annual income of the host in thousands of dollars, D is the distance in miles from the nearest metropolitan center, and T is the time in hours after noon. This law resembles a classical law of physics. It covers many situations, but for any particular situation it predicts exactly what will happen.

A conservation law is simpler and less restrictive. Suppose it is observed that between 7 and 10 o'clock the number of guests is conserved at all parties. This is a grand general statement, appealing for its breadth of application and its simplicity. It would, were it true, be regarded as a deep truth, a very profound law of human behavior. But it gives much less detailed information than the formula for R above. The conservation law allows the guests to arrive at any rate whatever, so long as guests depart at the same rate. To push the analogy with natural law a bit further, we should say that according to the old view, since the cocktail-party attendance is a fundamental aspect of human behavior, we seek and expect to find simple explicit laws governing the flow of guests. According to the new view, we expect to find the flux of arriving and departing guests limited only by certain conservation principles. Any behavior not prohibited by the conservation laws will, sooner or later, at some cocktail party, actually occur.

It should be clear that there is a close connection between this view of nature and the fundamental role of probability in nature. If the conservation law does not prohibit various possible results of an experiment, as in the proton-proton collision cited above, then these various possibilities will occur, each with some definite probability. The very fact that we can use the word chaos to describe the creation and annihilation events occurring continually among the particles rests on the existence of laws of probability. At best the probability, never the certainty, of these endless changes in the particle world can be known.

Are the laws of probability themselves derivable from conservation laws? The answer to this question is not yet known, but the trend of recent history is enough to make this author and many other physicists willing to bet on the affirmative. It appears possible, at least, that the conservation laws may not only be the most fundamental laws, but may be *all* the laws. They may be sufficient to characterize the elementary-particle world completely, specifying not only which events may occur and which are forbidden, but giving also the relative probabilities of those events that do occur.

We have so far emphasized that a conservation law is less restrictive than an explicit law of change, or law of permission. However, there are a number of different conservation laws and, taken all together, they may be very strongly restrictive, far more so than any one taken alone. In the ideal case, they may leave open only one possibility. The laws of prohibition, all taken together, then imply a unique law of permission. The most beautiful example of this kind of power of conservation laws concerns the nature of the photon. From conservation principles alone, it has been possible to show that the photon must be a massless particle of unit spin and no charge, emitted and absorbed by charged particles in a particular characteristic way. This truly amazing result has been expressed vividly by J. J. Sakurai who wrote, "The Creator was supremely imaginative when he declared, 'Let there be light.' "* In the world of human law, a man so hemmed in by restrictions that there is only one course of action open to him is not very happy. In the world of natural law it is remarkable and satisfying to learn that a few simple statements about constant properties in nature can have locked within them such latent power that they determine uniquely the nature of light and its interaction with matter.

There are conservation laws and conservation laws. Some things in nature are constant, but others are even more constant. To convert this jar-

**Annals of Physics,* Volume 11, page 5 (1960).

gon into sense, some quantities in nature seem to be absolutely conserved, remaining unchanged in all events whatever; other quantities seem to be conserved in some kinds of processes and not in others. The rules governing the latter are still called conservation laws, but nature is permitted to violate them under certain circumstances. Here we consider seven of the conservation laws that *may* be absolute.

What does it mean to call a law "absolute"? The ones we focus on have been tested innumerable times and have never been seen to fail. Yet there is reason to believe that some of them may, in fact, be less than absolute. If any are found to fail, no matter how rarely, it will be of extraordinary interest.

We begin by listing by name the seven quantities that are conserved:

1. Energy (including mass)
2. Momentum
3. Angular momentum, including spin
4. Charge
5. Electron-family number
6. Muon-family number
7. Baryon-family number.

There are two different kinds of quantities here, which can be called properties of motion and intrinsic properties, but the two are not clearly separated. The intrinsic particle properties that enter into the conservation laws are mass, spin, charge, and the several "family numbers." The properties of motion are kinetic energy, momentum, and angular momentum, the last frequently being called orbital angular momentum to avoid possible confusion with intrinsic spin, which is a form of angular momentum. In the laws of energy conservation and angular-momentum conservation, the intrinsic properties and properties of motion become mixed.

The interactions and transformations of the elementary particles serve admirably to illustrate the conservation laws and we shall focus attention on the particles for illustrative purposes. It is through studies of the particles that all of these conservation laws have been verified, although the first four were already known in the macroscopic world. The particles provide the best possible testing ground for conservation laws, for any law satisfied by small numbers of particles is necessarily satisfied for all larger collections of particles, including the macroscopic objects of our everyday world. Whether the extrapolation of the submicroscopic conservation laws on into the cosmological domain is justified is uncertain. Gravity, whose effects in the particle world appear to be entirely negligible, becomes of dominant importance in the astronomical realm. And events so rare as to escape detection in the laboratory may nevertheless have cosmological significance.

We shall examine first the conservation laws that have to do with the intrinsic properties of the particles.

Singly charged particles are most common, carrying the same electric charge as the electron (negative), or the equal and opposite charge of the proton (positive). The charge is a measure of the strength of electric force which the particle can exert and, correspondingly, a measure of the strength of electric force which the particle experiences. A neutral particle, of course, neither exerts nor responds to an electric force. A charged particle does both.

In terms of the proton charge as a unit, most particles have charge +1, −1, or 0. The conservation of charge requires that the total charge remain unchanged during every process of interaction or transformation. For any event involving particles, then, the total charge before the event must add up to the same value as the total charge after it. In the decay of a lambda into a neutron and a pion,

$$\Lambda^{\circ} \rightarrow n + \pi^{\circ},$$

the charge is zero both before and after. In the positive pion decay,

$$\pi^{+} \rightarrow \mu^{+} + \nu_{\mu},$$

the products are a positive muon and a neutral neutrino. A possible high-energy nuclear collision might proceed as follows:

$$p + p \rightarrow n + \Lambda^{\circ} + K^{+} + \pi^{+}.$$

Neither positively charged proton survives the collision, but the net charge +2 appears on the particles created.

Notice that the law of charge conservation provides a partial explanation for the fact that particle charges come in only one size. If the charge on a pion were, say, 0.73 electron charges, it would be quite difficult to balance the books in transformation processes and maintain charge conservation. Actually, according to the present picture of elementary processes, the charge is conserved not only from "before" to "after," but at every intermediate stage of the process. One can visualize a single charge as an indivisible unit which, like a baton in a relay race, can be handed off from one particle to another, but never dropped or divided.

Perhaps the most salutary effect of the law of charge conservation in human affairs is the stabilization of the electron. The electron is the lightest charged particle and, for this reason alone, it cannot decay. The only lighter particles, the photon and neutrinos (and graviton) are neutral, and a decay of the electron would therefore necessarily violate the law of charge conservation. The stability of the electron is one of the simplest, yet one of the most stringent tests of the law of charge conservation. Nothing else prevents electron decay. If the law were almost, but not quite, valid, the electron should have a finite lifetime. A 1985 experiment places the electron lifetime beyond 10^{25} years; this means that charge conservation must be regarded as at least a very good approximation to an absolute law.

Unlike the other four laws, which were already known in the macroscopic world, the laws of family-number conservation were discovered through studies of particle transformation. We can best explain their meaning through examples. The proton and all heavier particles are called baryons, that is, they belong to the baryon family. In the decay of the unstable Λ particle,

$$\Lambda^{\circ} \rightarrow p + \pi^{-},$$

one baryon, the Λ, disappears, but another, the proton, appears. Similarly, in the decay of the Σ^{0},

$$\Sigma^{\circ} \rightarrow \Lambda^{\circ} + \gamma,$$

the number of baryons is conserved. Notice that, in one of these examples, a pion is created; in the other, a photon. Pions and photons belong to none of the special family groups and can come and go in any number. In a typical proton-proton collision the number of baryons (2) remains unchanged, as in the example,

$$p + p \rightarrow p + \Sigma^{+} + K^{\circ}.$$

These and numerous other examples have made it appear that the number of baryons remains forever constant—in every single event, and therefore, of course, in any larger structure.

Each of the Ω, Ξ, Σ, and Λ particles, and the neutron, undergoes spontaneous decay into a lighter baryon. But the lightest baryon, the proton, has nowhere to go. The law of baryon conservation stabilizes the proton and makes possible the structure of nuclei and atoms and, therefore, of our world. From the particle physicist's point of view, this is a truly miraculous phenomenon, for the proton stands perched at a mass nearly 2,000 times the electron mass, having an intrinsic energy of about one billion electron volts, while beneath it lie the lighter unstable kaon, pion, and muon. Only the law of baryon conservation holds this enormous energy locked within the proton and makes it a suitable building block for the universe. The proton appears, so far, to be absolutely stable. If it is unstable it has, according to a 1985 experimental result, a half life greater than 4×10^{31} years, unimaginably greater than the age of the earth, a mere 4×10^{9} years. Yet a half life in excess of 10^{31} years is in fact forecast for the proton.

Our statement of the law of baryon conservation needs some amplification, for we have not yet taken into account antibaryons. A typical antiproton-production event might go as follows:

$$p + p \rightarrow p + p + p + \bar{p}.$$

(The bar over the letter designates the antiparticle. Since the antiproton has negative charge, the total charge of plus 2 is conserved.) It appears that we have transformed two baryons into four. Similarly, in the antiproton annihilation event,

$$p + \bar{p} \rightarrow \pi^{+} + \pi^{-} + \pi^{0},$$

two baryons have apparently vanished. The obvious way to patch up the law of baryon conservation is to assign to the antiparticles baryon number -1, and to the particles baryon number $+1$. Then the law would read: In every event the total number of baryons *minus* the total number of antibaryons is conserved; or, equivalently, the total baryon number remains unchanged.

The cynic might say that with so many arbitrary definitions—which particles should be called baryons and which not, and the use of negative baryon numbers—it is no wonder that a conservation law can be constructed. To this objection, two excellent answers can be given. The first is that it is not so easy to find an absolute conservation law. To find any quantity absolutely conserved in nature is so important that it easily justifies a few arbitrary definitions. The arbitrariness at this stage of history only reflects our lack of any deep understanding of the reason for baryon conservation, but it does not detract from the obvious significance of baryon conservation as a law of nature. The other answer, based on the mathematics of the quantum theory, is that the use of negative baryon number for antiparticles is perfectly natural, in fact, is demanded by the theory. This comes about because the description of the appearance of an antiparticle is "equivalent" (in a mathematical sense we cannot delve into*) to the description of the disappearance of a particle; and conversely antiparticle annihilation is "equivalent" to particle creation.

The "electron family" contains only the electron and its neutrino, the "muon family" only the muon and its neutrino. For each of these small groups, there is a conservation of family members exactly like the conservation of baryons. The antiparticles must be considered negative members of the families, the particles positive members. These light-particle conservation laws are not nearly as well tested as the other absolute conservation laws because of the difficulties of studying neutrons, but there are no known exceptions to them.

The beta decay of the neutron,

$$n \rightarrow p + e^- + \overline{\nu_e},$$

illustrates nicely the conservation laws we have discussed. Initially, the single neutron has charge zero, baryon number 1, and electron-family number zero. The oppositely charged proton and electron preserve zero charge; the single proton preserves the baryon number; and the electron with its antineutrino ($\overline{\nu_e}$) together preserve zero electron-family number. In the pion decay processes,

$$\pi^+ \rightarrow \mu^+ + \nu_\mu \qquad \text{and} \qquad \pi^- \rightarrow \mu^- + \overline{\nu_\mu},$$

muon-family conservation demands that a neutrino accompany the μ^+ antimuon, and an antineutrino accompany the μ^- muon. The muon, in turn, decays into three particles, for example,

$$\mu^- \rightarrow e^- + \nu_\mu + \overline{\nu_e},$$

which conserves the members of the muon family and of the electron family.

* Antiparticles can be described mathematically as "particles moving backwards in time."

The general rule enunciated earlier was that whatever *can* happen without violating a conservation law *does* happen. Until 1962, there was a notable exception to this rule; its resolution has beautifully strengthened the idea that conservation laws play a central role in the world of elementary particles. The decay of a muon into an electron and a photon,

$$\mu^- \rightarrow e^- + \gamma,$$

has never been seen, a circumstance which had come to be known as the μ-e-γ puzzle. Before the discovery of the muon's neutrino it was believed that electron, muon, and one neutrino formed a single family (called the lepton family) with a single family-conservation law. If this were the case, no conservation law prohibited the decay of muon into electron and photon, since the lost muon was replaced with an electron, and charge and all other quantities were conserved as well. According to the classical view of physical law, the absence of this process should have caused no concern. There was, after all, no law of permission which said that it should occur. There was only the double negative: No conservation law was known to prohibit the decay.

However, the view of the fundamental role of conservation laws in nature as the only inhibition on physical processes had become so ingrained in the thinking of physicists that the absence of this particular decay mode of the muon was regarded as a significant mystery. It was largely this mystery that stimulated the search for a second neutrino belonging exclusively to the muon. The discovery of the muon's neutrino established as a near certainty that the electron and muon belong to two different small families which are separately conserved. With the electron and muon governed by two separate laws of conservation, the prohibition of the μ-e-γ decay became immediately explicable, and the faith that what can happen does happen was further bolstered.

We turn now to the conservation laws that involve properties of motion (the first three in the list on page 766).

In the world of particles there are only two kinds of energy: energy of motion, or kinetic energy, and energy of being, which is equivalent to mass. Whenever particles are created or annihilated (except the massless particles) energy is transformed from one form to the other, but the total energy in every process always remains conserved. The simplest consequence of energy conservation for the spontaneous decay of unstable particles is that the total mass of the products must be less than the mass of the parent. For each of the following decay processes the masses on the right add up to less than the mass on the left:

$$K^+ \rightarrow \pi^+ + \pi^+ + \pi^-,$$
$$\Xi^- \rightarrow \Lambda^o + \pi^-,$$
$$\mu^+ \rightarrow e^+ + \nu_e + \overline{\nu_\mu}.$$

Energy conservation prohibits every "uphill" decay in which the products are heavier than the parent. An unstable particle at rest has only its energy of being, no energy of motion. The difference between this parent mass and the mass of the product particles is transformed into kinetic energy, which the product particles carry away as they rapidly leave the scene.

One might suppose that if the parent particle is moving when it decays it has some energy of motion of its own which might be transformed to mass. The conservation of momentum prohibits this. The extra energy of motion is in fact "unavailable" for conversion into mass. If a particle loses kinetic energy, it also loses momentum. It turns out that momentum and energy conservation taken together forbid uphill decays into heavier particles no matter how fast the initial particle might be moving.

If two particles collide, on the other hand, some—but usually not all—of their energy of motion is available to create mass. It is in this way that the various unstable particles are manufactured in the laboratory. In an actual typical collision in the vicinity of an accelerator, one of the two particles, the projectile, is moving rapidly, and the other, the target, is at rest. Under these conditions, the requirement that the final particles should have as much momentum as the initial projectile severely restricts the amount of energy that can be converted into mass. This is too bad, for the projectile has been given a great energy at a great expense. To make a proton-antiproton pair, for example, by the projectile-hitting-fixed-target method, the projectile must have a kinetic energy of GeV (billion electron volts), of which only 2 GeV goes into making the mass. The 6 GeV Berkeley Bevatron was designed with this fact in mind in order to be able to make antiprotons and antineutrons. Typical processes for protons striking protons are:

$$p + p \rightarrow p + p + p + \overline{p},$$
$$p + p \rightarrow p + p + n + \overline{n}.$$

The unfortunate waste of 4 GeV in these processes could be avoided if the target proton were not quiescent, but flew at the projectile with equal and opposite speed. It is hard enough to produce one high-energy beam, and far more difficult to produce two at once. Nevertheless, the gain in available energy makes it worth the trouble, and "colliding beams" are now in use in several laboratories in Europe and America.

Momentum is purely a property of motion—that is, if there is no motion, there is no momentum. It is somewhat trickier than energy, for momentum is what is called a vector quantity. It has direction as well as magnitude. Vectors are actually familiar in everyday life, whether or not we know them by that name. The velocity of an automobile is a vector, with a magnitude (50 miles per hour, for example) and a direction (northbound, for example). Force is a vector, a push or pull of some strength in some direction. Mass, on the other hand, is not a vector. It points in no particular direction. Energy also has no direction. The momentum of a rolling freight car, however, is directed along the tracks, and the momentum of an elementary particle is directed along its course through space.

In order to appreciate the law of momentum conservation, one must know how to add vectors. Two people pushing on a stalled car are engaged in adding vectors. If they push with equal strength *and* in the same direction, the total force exerted is twice the force each one exerts and, of course, in the direction they are pushing [Figure 1(a)]. If they push with equal strength but at opposite ends of the car, their effort comes to naught, for the sum of two vector quantities that are equal in strength but opposite in direction is zero [Figure 1(b)]. If they get on opposite sides of the car and push partly inward, partly forward, the net force exerted will be forward, but less than twice the force of each [Figure 1(c)]. Depending on their degree of co-operation, the two people may achieve a strength of force from zero up to twice the force each can exert.

This is a general characteristic of the sum of two vectors. It may have a wide range of values depending on the orientation of the two vectors.

Consider the law of momentum conservation applied to the decay of a kaon into muon and neutrino,

$$K^+ \rightarrow \mu^+ + \nu_\mu.$$

Before the decay, suppose the kaon is at rest [Figure 2(a)]. After the decay, momentum conservation requires that muon and neutrino fly off with equal magnitudes of momenta *and* that the momenta be oppositely directed [Figure 2(b)]. Only in this way can the vector sum of the two final momenta be equal to the original momentum, namely zero. This

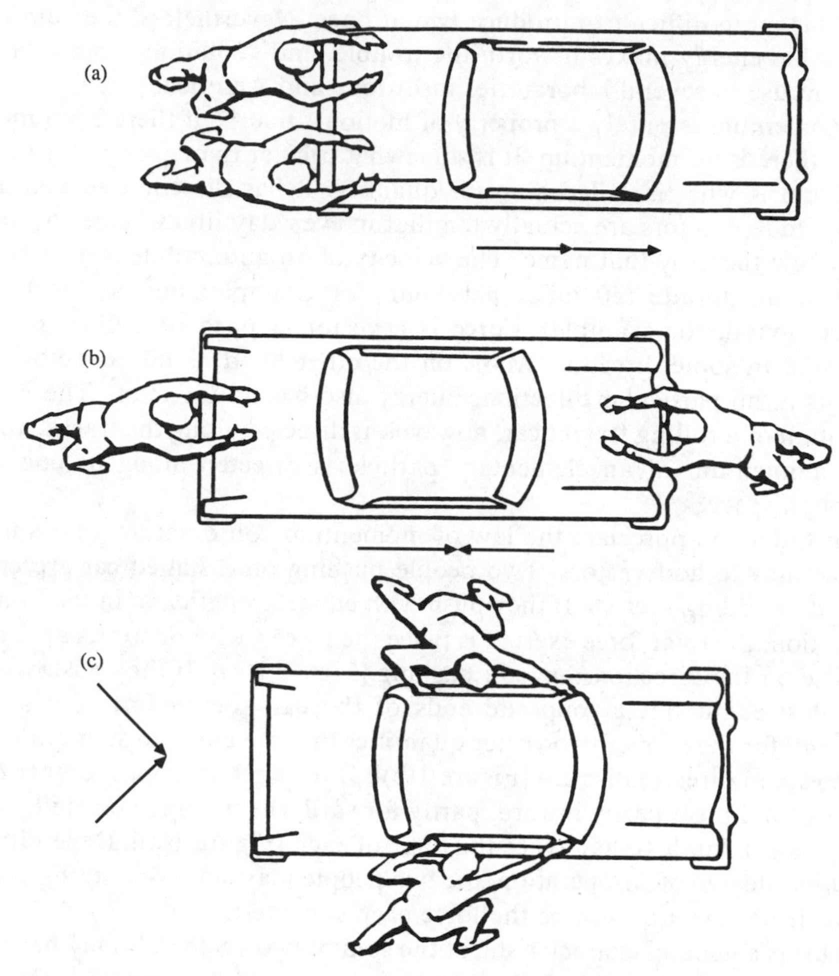

FIG. 1. *The addition of vectors.* The forces exerted by two people pushing equally hard may be "added," that is, combined, to give any total from zero up to twice the force of each.

type of decay, called a two-body decay, is rather common, and is always characterized by particles emerging in exactly opposite directions.

In a three-body decay, the emerging particles have more freedom. They can fan out in a variety of different directions. Recalling the analogy between momentum and force, one can visualize a situation in which three different forces are acting and producing no net effect—two fighters

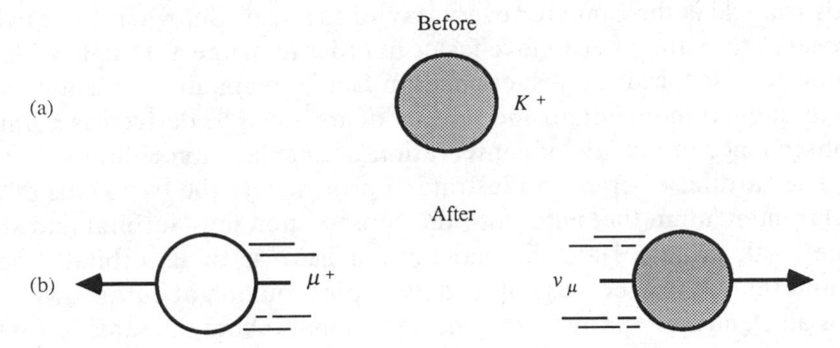

FIG. 2. *Momentum conservation in kaon decay.* The total momentum is zero both before and after the decay.

and a referee all pushing in different directions in a clinch. Similarly, the momentum vectors must adjust themselves to produce no net effect; that is, they must add up to give zero. . . .

One vital prohibition of the law of momentum conservation is that against one-body decays. Consider, for example, this possibility,

$$K^+ \rightarrow \pi^+,$$

the transformation of kaon to pion. It satisfies the laws of charge and family-number conservation. It is consistent with energy conservation, for it is downhill in mass, and it also satisfies spin conservation. But the kaon-pion mass difference must get converted to energy of motion, so that if the kaon was at rest, the pion will fly away. In whatever direction it moves, it has some momentum and therefore violates momentum conservation, since the kaon had none. On the other hand, if we enforce the law of momentum conservation, and keep the pion at rest, we shall have violated energy conservation, for in this case the extra energy arising from the mass difference will be unaccounted for.

Angular momentum, a measure of the strength of rotational motion, has been a key concept in physics since the time of Kepler. Actually, Kepler did not recognize it as such, but the second of his three laws of planetary motion—the so-called law of areas—is equivalent to a law of conservation of angular momentum. According to this law, an imaginary straight line drawn from the earth to the sun sweeps out area in space at a constant rate. During a single day this line sweeps across a thin triangular region with apex at the sun and base along the earth's orbit. The area of this triangle is the same for every day of the year. So, when the earth is closer to the sun, it must move faster in order to define a triangle with the same area. It speeds up just enough, in fact, to maintain a constant value of its angular momentum, and the law of areas can be derived as a simple consequence of the law of conservation of angular momentum.

The earth also serves to illustrate approximately the two kinds of angular momentum that enter into the conservation law—orbital and spin. The earth possesses angular momentum because of its orbital motion round the sun and because of its daily (spin) rotation about its own axis. For an elementary particle, the notion of spin is about the same as for the earth—rotational motion about an axis.

If a photographer in space took a time exposure of the earth and sun, his photograph would contain a short blur for the sun and a longer blur for the earth. He would notice that the blurs were not directed toward each other, and from this fact alone could conclude that earth and sun possess relative angular momentum. He would not need to know whether the earth swings around the sun or whether it proceeds into interstellar space. The key fact defining orbital angular momentum is some transverse motion of two objects. Any two moving objects, not aimed directly at each other, possess relative angular momentum. Two trains passing on the great plains have relative angular momentum, even though each is proceeding straight as an arrow. But if, through some mischance, both were on the same track on a collision course, they would have zero angular momentum. In particle collisions and decays, orbital angular momentum is usually of this trains-in-the-plains type, not involving actual orbiting of one particle round another. Figure 3 illustrates several examples of motion with angular momentum.

Angular momentum is a vector quantity. Its direction is taken to be the axis of rotation. The axis is well defined for spin, but what about orbital motion? For the passing trains, imagine again the blurred photograph indicating their direction of motion. Then ask: What would the axis be if the trains rotated about each other, instead of proceeding onward? The answer is a vertical axis; the angular momentum is directed upward. One more fact about orbital angular momentum needs to be known. Unlike spin, which comes in units of $\frac{1}{2}\hbar$, it comes only in units of $\hbar$.*

The spinless pion decays into muon and neutrino, each with spin ½. In Figure 4 we use artistic license and represent the particles by little spheres with arrows to indicate their direction of spin. Muon and neutrino spin

*The symbol $\hbar$ (pronounced "h-bar") stands for Planck's constant h divided by 2π.

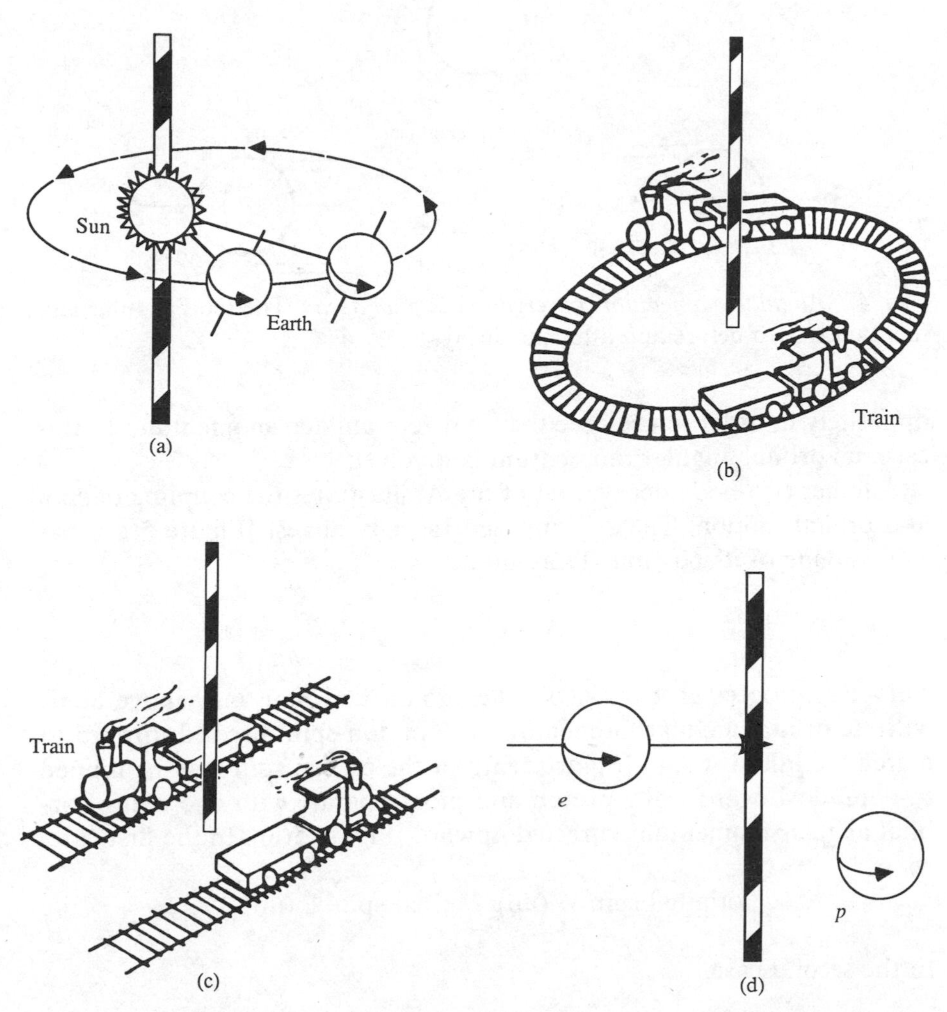

FIG. 3. *Examples of motion with angular momentum.* (a) The earth possesses spin angular momentum about its axis as well as orbital angular momentum about an axis designated by the giant barber pole. The constancy of the earth's orbital angular momentum means that the shaded area swept out in one day is the same for every day of the year. (b) Trains on a circular track possess angular momentum about a vertical axis. (c) Even on straight tracks, a similar relative motion of trains represents angular momentum. (d) An electron flies past a proton. Both particles possess spin angular momentum and, because they are not on a collision course, they also have orbital angular momentum.

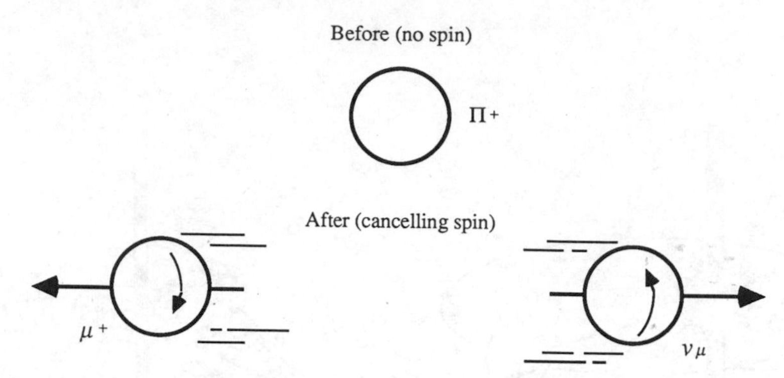

FIG. 4. *Angular-momentum conservation in pion decay.* The total angular momentum is zero before and after the decay.

oppositely in order to preserve the total zero angular momentum. In this case, no orbital angular momentum is involved.

Another two-body decay, that of the Λ, illustrates the coupling of spin and orbital motion. The Λ, supposed initially at rest [Figure 5(a)], has spin ½. One of its possible decay modes is

$$\Lambda^{o} \rightarrow p + \pi^{-}.$$

This may proceed in two ways. The proton and pion may move apart with no orbital angular momentum, the proton spin directed upward to match the initial Λ spin [Figure 5(b)]; or the proton spin may be flipped to point downward while proton and pion separate with one unit of orbital angular momentum, directed upward [Figure 5(c)]. In the first case,

original spin ½ (up) → final spin ½ (up).

In the second case,

original spin ½ (up) → final spin ½ (down) + orbital angular momentum 1 (up).

Beta decay, the earliest known particle decay process, serves nicely to illustrate all of the absolute conservation laws discussed. The beta decay of the neutron, indicated symbolically by

$$n \rightarrow p + e^{-} + \overline{\nu_e},$$

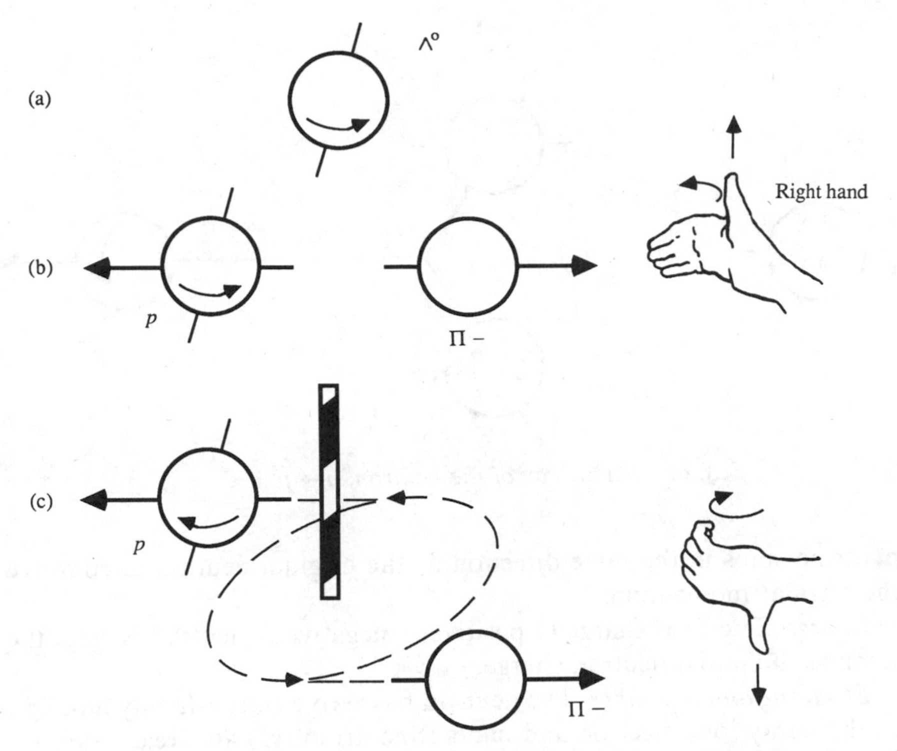

FIG. 5. *Angular-momentum conservation in lambda decay.* The direction of angular momentum is defined by the right-hand rule. If the curved fingers of the right hand point in the direction of rotational motion, the right thumb defines the direction assigned to the angular momentum. Thus the particle spin is up in diagrams (a) and (b) and down in diagram (c); the orbital angular momentum is up in diagram (c).

is pictured in Figure 6. Consider now the conservation laws applied to this decay.

Energy. The sum of the masses of the proton (1836.10), the electron (1.0), and the electron's neutrino (0), add up to less than the neutron mass (1838.63). The decay is therefore an allowed downhill decay, the slight excess mass going into kinetic energy of the products.

Momentum. The three particles must fan off in different directions with the available excess energy so distributed among them that the sum of the three momentum vectors is zero.

Angular momentum. One possibility, illustrated in Figure 6, is that the departing electron and proton have opposite cancelling spins, while the

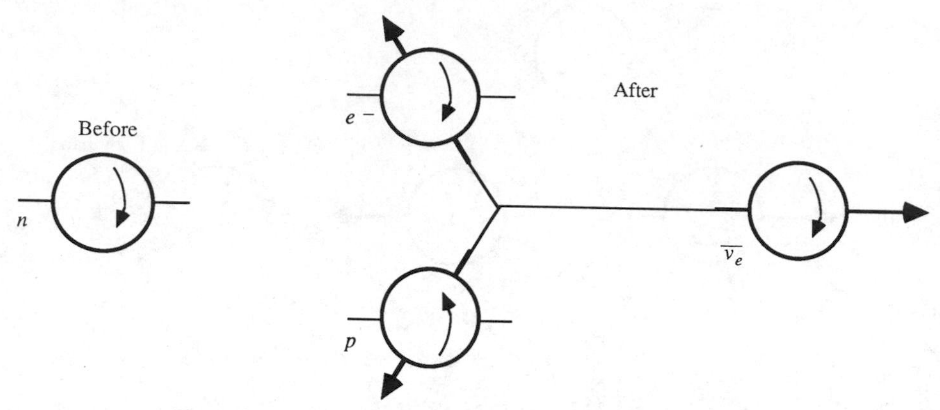

FIG. 6. *Beta decay of the neutron,* $n \rightarrow p + e^- + \overline{\nu_e}$.

neutrino spins in the same direction as the original neutron to conserve the angular momentum.

Charge. The final charge (1 positive, 1 negative, 1 neutral) is zero, the same as the initial neutron charge.

Electron-family number. The neutron has zero electron-family number. In the decay, one electron and one antineutrino ($\overline{\nu_e}$) are created to preserve zero electron-family number.

Muon-family number. No members of the muon family are created or destroyed.

Baryon number. The proton is the single baryon among the final three particles, matching the single original baryon.

Now we propose an exercise for the reader. Below are listed a few decays and transformations which do *not* occur in nature. If only one particle stands on the left, a decay process is understood. If two particles stand on the left, a collision process is understood. At least one conservation law prohibits each of these processes. Find at least one conservation law violated by each process. Several violate more than one law and one of those listed violates five of the seven conservation laws. (Answers are given on page 789.)

a. $\mu^+ \rightarrow \pi^+ \nu_\mu$
b. $e^- \rightarrow \nu_e + \gamma$
c. $p + p \rightarrow p + \Lambda^\circ + \Sigma^+$
d. $\mu^+ \rightarrow \Lambda^\circ$
e. $n \rightarrow \mu^+ + e^- + \gamma$

f. $\Lambda^o \rightarrow p + e^-$
g. $\pi^- + p \rightarrow \pi^- + n + \Lambda^o + K^+$
h. $e^+ + e^- \rightarrow \mu^+ + \pi^-$
i. $\mu^- \rightarrow e^- + e^+ + \nu_\mu$

The aspect of conservation laws that makes them appear to the theorist and the philosopher to be the most beautiful and profound statements of natural law is their connection with principles of symmetry in nature. Baldly stated, energy, momentum, and angular momentum are all conserved because space and time are isotropic (the same in every direction) and homogeneous (the same at every place). This is a breath-taking statement when one reflects upon it, for it says that three of the conservation laws arise solely because empty space has no distinguishing characteristics, and is everywhere equally empty and equally undistinguished. (Because of the relativistic link between space and time, we really mean space-time.) It seems, in the truest sense, that we are getting something from nothing.

Yet there can be no doubt about the connection between the properties of empty space and the fundamental conservation laws that govern elementary-particle behavior. This connection raises philosophical questions which we will mention but not pursue at any length. On the one hand, it may be interpreted to mean that conservation laws, being based on the most elementary and intuitive ideas, are the most profound statements of natural law. On the other hand, one may argue, as Bertrand Russell* has done, that it only demonstrates the hollowness of conservation laws ("truisms," according to Russell), energy, momentum, and angular momentum all being defined in just such a way that they must be conserved. Now, in fact, it is not inconsistent to hold both views at once. If the aim of science is the self-consistent description of natural phenomena based upon the simplest set of basic assumptions, what could be more satisfying than to have basic assumptions so completely elementary and self-evident (the uniformity of space-time) that the laws derived from them can be called truisms? Since the scientist generally is inclined to call most profound that which is most simple and most general, he is not above calling a truism profound. Speaking more pragmatically, we must recognize the discovery of *anything* that is absolutely conserved as something of an achievement, regardless of the arbitrariness of definition involved. Looking at those conservation laws whose basis we do not

* Bertrand Russell, *The ABC of Relativity*. New York: New American Library, 1959.

understand (the three family-number-conservation laws) also brings home the fact that it is easier to call a conservation law a truism after it is understood than before. It seems quite likely that we shall gain a deeper understanding of nature and of natural laws before the conservation of baryon number appears to anyone to be a self-evident truth.

Before trying to clarify through simple examples the connection between conservation laws and the uniformity of space, we consider the question, "What is symmetry?" In most general terms, symmetry means that when one thing (A) is changed in some particular way, something else (B) remains unchanged. A symmetrical face is one whose appearance (B) would remain the same if its two sides (A) were interchanged. If a square figure (A) is rotated through 90 degrees, its appearance (B) is not changed. Among plane figures, the circle is the most symmetrical, for if it is rotated about its center through any angle whatever, it remains indistinguishable from the original circle—or, in the language of modern physics, its form remains invariant. In the language of ancient Greece, the circle is the most perfect and most beautiful of plane figures.

Aristotle regarded the motion of the celestial bodies as necessarily circular because of the perfection (the symmetry) of the circle. Now, from a still deeper symmetry of space-time, we can derive the ellipses of Kepler. Modern science, which could begin only after breaking loose from the centuries-old hold of Aristotelian physics, now finds itself with an unexpected Aristotelian flavor, coming both from the increasingly dominant role of symmetry principles and from the increasingly geometrical basis of physics.

We are accustomed to think of symmetry in spatial terms. The symmetry of the circle, the square, and the face are associated with rotations or inversions in space. Symmetry in time is an obvious extension of spatial symmetry; the fact that nature's laws appear to remain unchanged as time passes is a fundamental symmetry of nature. However, there exist some subtler symmetries, and it is reasonable to guess that the understanding of baryon conservation, for example, will come through the discovery of new symmetries not directly connected with space and time.*

In the symmetry of interest to the scientist, the unchanging thing—the invariant element—is the form of natural laws. The thing changed may be orientation in space, or position in space or time, or some more abstract change (not necessarily realizable in practice) such as the inter-

* If baryon number is not absolutely conserved, these symmetries will be slightly imperfect—like an otherwise symmetrical face with a mole on one cheek.

change of two particles. The inversion of space and the reversal of the direction of flow of time are other examples of changes not realizable in practice, but nonetheless of interest for the symmetries of natural law. . . .

If scientists in Stanford, Chicago, and Geneva perform the same experiment and get the same answer (within experimental error) they are demonstrating one of the symmetries of nature, the homogeneity of space. If the experiment is repeated later with the same result, no one is surprised, for we have to come to accept the homogeneity of time. The laws of nature are the same, so far as we know, at all points in space, and for all times. This invariance is important and is related to the laws of conservation of energy and momentum, but ordinary experience conditions us to expect such invariance so that it seems at first to be trivial or self-evident. It might seem hard to visualize any science at all if natural law changed from place to place and time to time, but, in fact, quantitative science would be perfectly possible without the homogeneity of space-time. Imagine yourself, for example, on a merry-go-round that speeded up and slowed down according to a regular schedule. If you carried out experiments to deduce the laws of mechanics and had no way of knowing that you were on a rotating system, you would conclude that falling balls were governed by laws that varied with time and with position (distance from central axis), but you would be quite able to work out the laws in detail and predict accurately the results of future experiments, provided you knew where and when the experiment was to be carried out. Thanks to the actual homogeneity of space and time, the results of future experiments can in fact be predicted without any knowledge of the where or when.

A slightly less obvious kind of invariance, although one also familiar from ordinary experience, is the invariance of the laws of nature for systems in uniform motion. Passengers on an ideally smooth train or in an ideally smooth elevator are unaware of motion. If the laws of mechanics were significantly altered, the riders would be aware of it through unusual bodily sensations. Such a qualitative guide is, of course, not entirely reliable, but careful experiments performed inside the ideal uniformly moving train would reveal the same laws of nature revealed by corresponding experiments conducted in a stationary laboratory. This particular invariance underlies the theory of relativity, and is a manifestation of the isotropy of four-dimensional space-time, a point we can regrettably not discuss in detail. What, to our limited three-dimensional vision, appears to be uniform motion is, to a more enlightened brain capable of encompassing four dimensions, merely a rotation. Instead of turning, say, from north to east, the experimenter who climbs aboard the train is, from the more general view, turning from space partly toward the time direction. According to relativity, which joins space and time together in a four-dimensional space-time, the laws of nature should no more be changed by "turning" experimental apparatus toward the time direction (that is, loading it aboard the train) than by turning it through 90 degrees in the laboratory.

The chain of connection we have been discussing is: Symmetry → invariance → conservation. The symmetry of space and time, or possibly some subtler symmetry of nature, implies the invariance of physical laws under certain changes associated with the symmetry. In the simplest case, for example, the symmetry of space which we call its homogeneity implies the invariance of experimental results when the apparatus is moved from one place to another. This invariance, in turn, implies the existence of certain conservation laws. The relation between conservation laws and symmetry principles is what we now wish to illuminate through two examples. . . .

Suppose we imagine a single isolated hydrogen atom alone and at rest in empty space. If we could draw up a chair and observe it without influencing it, what should we expect to see? (For this discussion, we ignore quantum mechanics and the wave nature of particles, pretending that electron and proton may be separately seen as particles, and be uninfluenced by the observer. The reader will have to accept the fact that these false assumptions are permissible and irrelevant for the present discussion.) We should see an electron in rapid motion circling about a proton, and the proton itself moving more slowly in a smaller circle. Were we to back off until the whole atom could only be discerned as a single spot, that spot, if initially motionless, would remain at rest forever. We now must ask whether this circumstance is significant or insignificant, important or dull. It certainly does not seem surprising. Why should the atom move, we may ask. It is isolated from the rest of the universe; no forces act upon it from outside; therefore there is nothing to set it into motion. If we leave a book on a table and come back later, we expect to find it there. Everyday experience conditions us to expect that an object on which no external forces act will not spontaneously set itself into motion. There is no more reason for the atom to begin to move than for the book to migrate across the table and fly into a corner. The trouble with this argument is that it makes use of the common sense of ordinary experience, without offering any explanation for the ordinary experience.

If we put aside "common sense" and ask what the atom might do, it is by no means obvious that it should remain at rest. In spite of the fact that no external forces are acting, strong internal forces are at work. The proton exerts a force on the electron which constantly alters its motion; the electron, in turn, exerts a force on the proton. Both atomic constituents are experiencing force. Why should these forces not combine to set the atom as a whole into motion? Having put the question in this way, we may consider the book on the table again. It consists of countless billions of atoms, each one exerting forces on its neighboring atoms. Through what miracle do these forces so precisely cancel out that no net force acts upon the book as a whole and it remains quiescent on the table?

The classical approach to this problem is to look for a positive, or permissive, law, a law that tells what *does* happen. Newton first enunciated this law which (except for some modification made necessary by the theory of relativity) has withstood the test of time to the present day. It is called Newton's third law, and says that all forces in nature occur in equal and opposite balanced pairs. The proton's force on the electron is exactly equal and opposite to the electron's force on the proton. The sum of these two forces (the *vector* sum) is zero, so that there is no tendency for the structure as a whole to move in any direction. The balancing of forces, moreover, can be related to a balancing of momenta. By making use of Newton's second law,* which relates the motion to the force, one can discover that, in a hydrogen atom initially at rest, the balanced forces will cause the momenta of electron and proton to be equal and opposite. At a given instant, the two particles are moving in opposite directions. The heavier proton moves more slowly, but has the same momentum as the electron. As the electron swings to a new direction and a new speed in its track, the proton swings too in just such a way that its momentum remains equal and opposite to that of the electron. In spite of the continuously changing momenta of the two particles, the total momentum of the atom remains zero; the atom does not move. In this way—by "discovering" and applying two laws, Newton's second and third laws of motion—one derives the law of momentum conservation and finds an explanation of the fact that an isolated atom does not move.

Without difficulty, the same arguments may be applied to the book on the table. Since all forces come in equal and opposite pairs, the forces be-

---

* Newton's second law, usually written $F = ma$, says that the acceleration a experienced by a particle multiplied by its mass m is equal to the force F acting upon it. The law may also be stated in this way: The rate at which the momentum of a particle is changing is equal to the force applied.

tween every pair of atoms cancel, so that the total force is zero, no matter how many billions of billions of atoms and individual forces there might be.

It is worth reviewing the steps in the argument above. Two laws of permission were discovered, telling what does happen. One law relates the motion to the force; the other says that the forces between pairs of particles are always equal and opposite. From these laws, the conservation of momentum was derived as an interesting consequence, and this conservation law in turn explained the fact that an isolated atom at rest remains at rest.

The modern approach to the problem starts in quite a different way, by seeking a law of prohibition, a principle explaining why the atom does *not* move. This principle is the invariance of laws of nature to a change of position. Recall the chain of key ideas referred to above: symmetry → invariance → conservation. In the example of the isolated hydrogen atom, the symmetry of interest is the homogeneity of space. Founded upon this symmetry is the invariance principle just cited. Finally, the conservation law resting on this invariance principle is the conservation of momentum.

In order to clarify, through the example of the hydrogen atom, the connecting links between the assumed homogeneity of space and the conservation of momentum, we must begin with an exact statement of the invariance principle as applied to our isolated atom. The principle is this: No aspect of the motion of an isolated atom depends upon the location of the center of mass of the atom. The center of mass of any object is the average position of all of the mass in the object. In a hydrogen atom, the center of mass is a point in space between the electron and the proton, close to the more massive proton.

Let us visualize our hydrogen atom isolated in empty space with its center of mass at rest. Suppose now that its center of mass starts to move. In which direction should it move? We confront at once the question of the homogeneity of space. Investing our atom with human qualities for a moment, we can say that it has no basis upon which to "decide" how to move. To the atom surveying the possibilities, every direction is precisely as good or bad as every other direction. It is therefore frustrated in its "desire" to move and simply remains at rest.

This anthropomorphic description of the situation can be replaced by sound mathematics. What the mathematics shows is that an acceleration of the center of mass—for example, changing from a state of rest to a state of motion—is not consistent with the assumption that the laws of motion of the atom are independent of the location of the center of mass. If the center of mass of the atom is initially at rest at point *A* and it then begins to move, it will later pass through another point *B*. At point *A*, the center of mass had no velocity. At point *B* it does have a velocity. Therefore, the state of motion of the atom depends on the location of the center of mass, contrary to the invariance principle. Only if the center of mass remains at rest can the atom satisfy the invariance principle.* The immobility of the center of mass requires, in turn, that the two particles composing the atom have equal and opposite momenta. A continual balancing of the two momenta means that their sum, the total momentum, is a constant.

The argument thus proceeds directly from the symmetry principle to the conservation law without making use of Newton's laws of motion. That this is a deeper approach to conservation laws, as well as a more esthetically pleasing one, has been verified by history. Although Newton's laws of motion have been altered by relativity and by quantum mechanics, the direct connection between the symmetry of space and the conservation of momentum has been unaffected—or even strengthened—by these modern theories, and momentum conservation remains one of the pillars of modern physics. We must recognize that a violation of the law of momentum conservation would imply an inhomogeneity of space; this is not an impossibility, but it would have far-reaching consequences for our view of the universe.

Returning finally to the book on the table, we want to emphasize that the quiescence of the undisturbed book—a macroscopic object—at least strongly suggests that momentum conservation must be a valid law in the microscopic world. Viewed microscopically, the book is a collection of an enormous number of atoms, each one in motion. That this continuous microscopic motion never makes itself felt as spontaneous bulk motion of the whole book is true only because of the conservation of momentum which requires that every time an atom changes its momentum (as it is constantly doing) one or more other atoms must undergo exactly compensating changes of their momentum.

Through similar examples it is possible to relate the law of conservation of angular momentum to the isotropy of space. A compass needle that is held pointing east and is then released will swing toward the north because of the action of the earth's magnetic field upon it. But if the same compass needle is taken to the depths of empty space, far removed from all external influences, and set to point in some direction, it will remain

* If the center of mass of the atom had been moving initially, the invariance principle requires that it continue moving with constant velocity.

pointing in that direction. A swing in one direction or the other would imply a nonuniformity* of space. If the uniformity of space is adopted as a fundamental symmetry principle, it can be concluded that the total angular momentum of all the atomic constituents of the needle must be a constant. Otherwise, the internal motions within the needle could set the whole needle into spontaneous rotation and its motion would violate the symmetry principle.

Energy conservation, in a way that is not so easy to see, is related to the homogeneity of time. Thus all three conservation laws—of energy, momentum, and angular momentum—are "understood" in terms of the symmetry of space-time, and indeed the theory of relativity has shown that these three laws are all parts of a single general conservation law in the four-dimensional world.

Only one of the three conservation laws governing the intrinsic properties of the particles has so far been understood in terms of a symmetry principle. This is the law of charge conservation. (Recall, however, that the *quantization* of charge is not yet understood.) The symmetry principle underlying charge conservation is considerably more subtle than the space-time symmetry underlying the conservation laws of properties of motion. The modern version of this symmetry principle rests upon technical aspects of the theory of quantum mechanics (it may be based also on equally technical aspects of the theory of electromagnetism). Nevertheless, it is such a stunning victory for the power of a symmetry principle that we must try, however crudely, to indicate the modern view of this symmetry.

In the main, the classical theories of physics deal directly with quantities that are measurable, usually called observables. Force, mass, velocity, and almost all the other concepts described by the classical laws are themselves observables. The equations of quantum mechanics, however, contain quantities that are not themselves observables. From these quantities—one step removed from reality—the observables are derived. The "wave function" is one of the unobservable quantities; it determines the probability, say, that the electron is at any particular point in the hydrogen atom, but is itself not that probability nor any other measurable thing. Now enters the idea of symmetry. Any change that can be made in

* Strictly, momentum conservation rests on the *homogeneity* of space (uniformity of place), and angular momentum conservation rests on the *isotropy* of space (uniformity of direction). The distinction is not important for our purposes, and it is satisfactory to think of space simply as everywhere the same, homogeneity and isotropy being summarized by the word uniformity.

the unobservable quantity without resulting in a change of the observables ought to leave all the laws of nature unchanged. After careful scrutiny, this statement seems so obviously true that it is hard to understand how it could have any important consequences. Of course one ought to be able to do anything whatever to unobservable quantities so long as observables are not changed. But remember how important were the properties of empty space. Equally important are the properties of unobservables such as wave functions.

Space itself may be regarded as an unobservable. The uniformity of space means that it is impossible, by any experimental means, to ascertain one's absolute position in space. An experiment carried out at one place will yield results identical to the results of the same experiment carried out at another place. Any change in the unobservable space (for instance, moving the apparatus from one place to another) must leave unchanged the laws of nature and the observable results of experiment. As we have just seen, this symmetry principle or invariance requirement underlies the law of momentum conservation.

When an analogous symmetry principle is applied to the unobservable wave function of the electron a conservation law results, the conservation of charge. Expressed negatively, if charge were not conserved, the form of the equations of quantum mechanics would depend upon unobservable quantities, a situation at variance with our symmetry principle. The analogous statement for spatial homogeneity would be: If momentum were not conserved, the laws of mechanics would depend upon the absolute location in space and such dependence is at variance with the assumed symmetry of space.

Regrettably, we can not explain the law of charge conservation more fully without mathematics. It is expected, but not yet verified, that some undiscovered subtle symmetries of nature underlie the laws of electron-family conservation, muon-family conservation, and baryon conservation. The fact that the proton keeps its enormous intrinsic energy locked forever in the form of mass can be no accident, but the reason still remains hidden. And, if these subtle symmetries turn out to be slightly flawed (or "broken"), we shall grasp the limits of the "absolute" conservation laws, limits likely to tie us to the beginning of time.

Answers

The particle transformations listed on page 780 *violate* the following conservation laws:

a. Energy (an "uphill" decay); muon-family number (since μ^+ is an antiparticle).
b. Charge.
c. Angular momentum; baryon number.
d. Energy; momentum (a one-particle decay); charge; muon-family number; baryon number.
e. Angular momentum; baryon number; muon-family number; electron-family number.
f. Angular momentum; electron-family number.
g. Angular momentum; baryon number.
h. Angular momentum; muon-family number.
i. Charge. (Why is angular momentum conservation satisfied?)

Strongly Interacting Particles

Murray Gell-Mann, Geoffrey Chew, Arthur Rosenfeld

Murray Gell-Mann was born in New York in 1929 into a family of Austrian immigrants. He showed an early talent for mathematics and physics and entered Yale as a freshman at the age of fifteen. He earned his degree at Yale in 1948 and his doctorate at M.I.T. in 1951. He worked at the Institute for Advanced Study for one year before moving on to the University of Chicago, where he became a professor in 1956. It was at this early period in his career that he began to develop his strangeness theory and his eightfold way theory.

Gell-Mann's first major contribution to the field of theoretical physics was made at the age of twenty-four, when he introduced his concept of strangeness to account for the unexpectedly long lifetimes of certain massive particles (hadrons). He defined "strangeness" as a quantum property (a quantum number) that appears in the strong interaction among particles. Although conserved in strong and electromagnetic interactions, it is not conserved in weak interactions. Strangeness and its conservation in strong interactions is important in Gell-Mann's symmetry schemes for classifying strongly interacting particles.

Gell-Mann was one of the first physicists to detect patterns in the seemingly chaotic particle zoo that had proliferated during the 1950s and 1960s. He used a symmetry scheme called SU(3) to classify hadrons (mesons and baryons) into so-called "isotopic multiplets"; all particles in the same isotopic multiplet are regarded as different manifestations of the same particle. In 1961, he extended this classification scheme, which he called the "eightfold way," to include all known mesons and baryons, that is hadrons (all strongly interacting particles). Like Mendeleev with his periodic chart, Gell-Mann, using this SU(3) symmetry, predicted the existence of certain particles having particular properties. On the basis of the scattering of electrons from baryons, he and G. Zweig independently proposed that all hadrons consist of triplet or doublet combinations of three kinds of fundamental objects known as "up," "down," and "strange" *quarks,* a name taken from a line in James Joyce's *Finnegans Wake.* The electric charge on each of these quarks is ⅓ or −⅔ the charge on the electron, which, although arbitrary, has always been accepted as the unit electric charge. Quarks appear to behave like points of matter and have no internal structure. Since electrons, muons, and neutrinos—known collectively as *leptons*—also appear to be point-like and to have no internal structure, particle physicists assume that all matter in the Universe consists of leptons and various combinations of quarks. The symmetries observed in these two major families of particles have prompted more than one physicist to suggest that the search for the fundamental building blocks of nature is over.

The present selection, which Gell-Mann cowrote with Geoffrey Chew and Arthur Rosenfeld, was one of the earliest articles to be directed to a nonprofessional audience that highlighted the symmetries, which were beginning to appear at that time in particle physics.

Syllogism is of necessary use, even to the lovers of truth, to show them the fallacies that are often concealed in florid, witty or involved discourses.

—JOHN LOCKE

It was a saying of the ancients, that "truth lies in a well"; and to carry on the metaphor, we may justly say, that logic supplies us with steps whereby we may go down to reach the water.

—ISAAC WATTS

ONLY FIVE YEARS ago it was possible to draw up a tidy list of 30 subatomic particles that could be called, without too many misgivings, elementary. Since then another 60 or 70 subatomic objects have been discovered, and it has become obvious that the adjective "elementary" cannot be applied to all of them. For this reason the adjective has been carefully avoided in the title of this article. There is now a widespread belief among physicists that none of the particles with which this article is mainly concerned deserves to be singled out as elementary.

What is happening has happened before in physics: the old way of looking at things, which was adequate for perceiving order in a limited number of observations, finally proved cumbersome and inadequate when the accuracy and range of observation increased. This happened with the Ptolemaic scheme of epicycles for describing the motions of the planets. Much the same thing occurred early in this century when spectroscopists, studying the light emitted by excited atoms, found a profusion of discrete wavelengths that were at total variance with the wavelengths predicted by classical electrodynamics. The spectroscopists accumulated so much empirical information, including sets of "selection rules" governing the permissible states of excited atoms, that it finally became possible in 1926 for Werner Heisenberg, Erwin Schrödinger and others to formulate a new mechanics—quantum mechanics—capable of predicting most of the states of matter on the atomic and molecular scale.

A similar situation may exist today in particle physics. The great unifying invention analogous to quantum mechanics is still not clearly in

---

Murray Gell-Mann, Geoffrey Chew, Arthur Rosenfeld. "Strongly Interacting Particles," *Scientific American* (February 1964), pp. 74–93.

sight, but the experimental data are beginning to fall into striking and partly predictable patterns. What can be said to summarize the vast amount of particle information now available?

First of all, there is a clear distinction between strongly interacting particles, such as the neutron and proton, and other particles. The neutron and proton are known to interact through the strong, short-range nuclear force, which is responsible for the binding of these particles in atomic nuclei. All particles discovered to date participate in this strong interaction except the photon (the particle of light and other electromagnetic radiation) and the four particles called leptons: the electron, the muon (or mu particle) and the two kinds of neutrino.

Another striking property of the strongly interacting particles is that none of them has a small rest mass. Rest mass is the mass that a particle would have if it were motionless; this is the minimum mass the particle can have. It is now common to express this mass as its equivalent in energy, rather than in units of the electron's mass, as was often done in the past. The lightest strongly interacting particle is the pion (or pi meson), which has a mass with an energy equivalent of some 137 million electron volts (Mev). In contrast, the mass of the electron is about .5 Mev and that of the photon and the neutrinos is believed to be zero.

A third general observation is that the recent proliferation of particles has so far occurred almost exclusively among the strongly interacting particles. Although this proliferation came as a surprise to physicists, a precedent for this state of affairs can be found in ordinary atomic nuclei. It is well known that all compound nuclei, from the nucleus of deuterium (heavy hydrogen) to those of the heaviest elements, can exist at a variety of energy levels, comprising a "ground" state and many excited states. These levels, which can be detected in several ways, indicate different degrees of binding energy among the component nucleons (neutrons and protons) in the nucleus. The binding energy, of course, is an expression of the strong nuclear force.

It is now clear that the nuclear force can similarly give rise to numerous states among those strongly interacting particles sometimes designated elementary. The lower states are "bound," or stable; the higher states are only partly bound, or unstable, decaying in a tiny fraction of a second. The result is that all strongly interacting particles exhibit a spectrum of energy levels with no sharp upper limit.

Since the leptons do not participate in strong interactions, it is not surprising that their spectrum of states, beginning with the massless neutrino and apparently terminating sharply at the muon, with a mass of 106 Mev, bears no resemblance to any known dynamical spectrum. In recent years physicists have learned much about the simplicity and regularity in the properties of leptons, but they have learned nothing of why these particles exist.

In the following discussion we shall begin by considering the place of the strong force in the hierarchy of four forces that seem to underlie all the operations of the physical universe. Next we shall describe a new nomenclature that assigns each of the strongly interacting particles to one of a small number of families, each characterized by a distinctive set of properties. One group of these families embraces the baryons, which in general are the heaviest particles; a second group consists of mesons, the first members of which to be discovered were lighter than the baryons. The new naming system will require a brief review of the seven quantum numbers, or physical quantities, that are conserved in strong interactions.

We shall next describe how these quantities are conserved when particles decay into different "channels," representing different modes of decay. This will lead to a description of "resonances," or unstable particles, which account for most of the proliferation among strongly interacting particles.

We shall then be ready to discuss two classification systems, or rules for the formation of groups, that have brought to light deep-seated family relations among strongly interacting particles. These rules have made it possible to predict the existence of still undiscovered particles, their approximate masses and certain other properties. One system is based on the concept of the "Regge trajectory"; the other is the "eightfold way."

Next we shall explain why the term "elementary" has fallen into disrepute for describing strongly interacting particles. There is growing evidence that all such particles can be regarded as composite structures. Finally we shall describe the "bootstrap" hypothesis, which may make it possible to explain mathematically the existence and properties of the strongly interacting particles. According to this hypothesis all these particles are dynamical structures in the sense that they represent a delicate balance of forces; indeed, they owe their existence to the same forces through which they mutually interact. In this view the neutron and proton would not be in any sense fundamental, as was formerly thought, but would merely be two low-lying states of strongly interacting matter, enjoying a status no different from that of the more recently discovered baryons and mesons and the nuclei of atoms heavier than those of hydrogen.

FORCES AND REACTION TIMES

In present-day physics the concepts of force and interaction are used interchangeably. The strong, or nuclear, force is the most powerful of the four basic interactions that, together with cosmology, account for all known natural phenomena. (Cosmology provides the stage on which the forces play their roles.) The strong interaction is limited to a short range: about 10^{-13} centimeter, which is about the diameter of a strongly interacting particle.

The next force, in order of strength, is the electromagnetic force, which is about 1 per cent as powerful as the strong force. Its strength decreases as the square of the distance between interacting particles, and its range is in principle unlimited. This force acts on all particles with an electric charge and involves the uncharged photon, which is the carrier of the electromagnetic-force field. The electromagnetic force binds electrons to the positively charged nucleus to form atoms, binds the atoms together to form molecules and thus in its manifold workings is responsible for all chemistry and biology.

Next in order, with only about a one-hundred-trillionth (10^{-14}) of the strength of the strong interaction, is the weak interaction. It also has a short range and cannot, as far as anyone knows, bind anything, but it governs the decay of many strongly interacting particles and is responsible for the decay of certain radioactive nuclei. It is most easily studied in the behavior of the four leptons, which do not respond to the strong force.

The fourth and weakest force is gravity, which has only about 10^{-39} of the strength of the strong interaction. It produces large-scale effects because it is always attractive and operates at long range. On the scale of atomic nuclei, however, its effects are undetectable.

Many particles are "coupled" to all four of these interactions. Take, for example, the proton. It is a strongly interacting particle, and since it is electrically charged it must also "feel" the electromagnetic force. It can be created by the beta decay of a neutron, a decay in which the neutron emits a negative electron and an antineutrino by a weak-interaction process; hence it must be involved in weak interactions. And like all other matter the proton is attracted by gravity. The least reactive particle is the neutrino, which is directly coupled only to the weak interaction and to gravitation. The neutrino shares with the other leptons a total immunity to the strong force.

An important idea, not self-evident in the foregoing, is that the basic forces can do more than bind particles together. For instance, when two particles collide and go off in different directions (the phenomenon called scattering), an interaction is involved. If a particle is moving with enough energy before striking a particle at rest, a new particle can be created in the collision. The collision of a proton and a neutron can yield a proton, a neutron and a neutral pion, or it can yield two neutrons and a positive pion. The collision can also yield strongly interacting particles that are more massive than either of the colliding particles. This, in fact, is the process by which particle-accelerating machines have created the scores of new particles heavier than protons and neutrons. Thus the basic forces are interactions that can scatter, create, annihilate and transform particles.

The interactions of principal interest in high-energy physics take place when one of the particles in the interaction is traveling at nearly the speed of light, or more than 10^{10} centimeters per second. Since the size of a particle is typically about 10^{-13} centimeter, the minimum reaction time is less than 10^{-23} second for a particle moving at the speed of light. What we mean when we call the strong interaction "strong" is that even in that brief time the strong force is powerful enough to cause a reaction to take place. Electromagnetic reactions, being 100 times weaker than strong reactions, take around 100 times longer, or typically 10^{-21} second. Processes involving the weak interaction, which is 10^{-14} times weaker than the strong interaction, commonly take about 10^{-9} second.

CONSERVATION LAWS

When one of the present authors (Gell-Mann) and E. P. Rosenbaum discussed particles in these pages not quite seven years ago . . . 30 well-established particles and antiparticles were singled out for attention. In this collection there were 16 baryons and antibaryons, seven mesons, six leptons and antileptons and the photon. . . . At that time it was customary to classify as elementary not only the photon and the leptons but also all the baryons and mesons—the strongly interacting particles.

A distinction, which we now regard as unjustified, was drawn between these strongly interacting particles and the states of ordinary atomic nuclei containing two or more nucleons, which, of course, are also strongly interacting. These nuclei, such as the deuteron (the nucleus of heavy hydrogen) and the alpha particle (the nucleus of helium), had been classified as composite structures, made essentially of protons and neutrons, almost from the beginning of nuclear physics because of the small binding energies involved in them. This article will place little emphasis on such nuclei. We shall concentrate on the lighter particles, not because we believe them to be more elementary than their heavy brothers but because their status is still in doubt. If they are in fact composite dynamical structures, their binding energies are often enormous. Furthermore, if elementary particles do exist, they will certainly not include the obviously composite nuclei.

We shall introduce the reader to a new system of nomenclature that has developed quite recently and that provides a great deal of information about each particle. Although it may look forbidding at first sight, it is really no harder to master than the telephone company's all-digit dialing system.

The new classification scheme takes advantage of the fact that nature conserves many quantities (in addition to energy and momentum) and shows various symmetries (such as that between left and right). As a result groups of particles have similar properties, which, as we shall see, can be indicated by a common notation. There is a close relation between symmetries and conservation laws, and in a particular case one can refer either to a symmetry or to the associated conservation law, whichever is more convenient. The conserved quantity appears in quantum mechanics as a quantum number, which is often either an integer (such as 0, 1, 2, 3 and so on) or a half-integer (such as ½, 3/2, 5/2 and so on).

Some conservation laws appear to be universal: they are obeyed by all four basic interactions. This inviolable group includes the conservation of energy, of momentum, of angular momentum (the momentum associated with rotation) and of electric charge. Another exact conservation law is best described as a kind of mirror-image symmetry. It is the symmetry between particles and antiparticles, in which whatever is left-handed for one is right-handed for the other. For each particle there is an antiparticle with the same mass and lifetime but with some properties, such as electric charge, reversed. Some neutral particles, such as the photon and the neutral pion, are their own antiparticles.

In the new system for naming strongly interacting particles we shall make use of five quantities, each indicated by a letter symbol, that are conserved by the strong interactions but not necessarily by the electromagnetic or weak interactions. These five quantities are: atomic mass number (A), hypercharge (Y), isotopic spin (I), spin angular momentum (J) and parity (P). The chart on the next page should help the reader to keep these five quantum numbers in mind as we discuss them in more detail. Also included in the chart are two other quantities that are conserved by strong interactions but that are not essential to the naming

Conserved Quality	Symbol	Observed Values	Description	Examples: proton	Examples: negative pion
ELECTRIC CHARGE	Q	0, ± 1, ± 2, ± 3 . . .	Represents the number of electric-charge units carried by a particle, or atomic nucleus, in units of the positive charge on the proton. Charge multiplets, such as the neutron-proton doublet or the pion triplet, can be assigned an average charge, $\bar{Q}$.	Q = + 1 $\bar{Q}$ = + ½	Q = −1 $\bar{Q}$ = O
ATOMIC MASS NUMBER, OR BARYON NUMBER	A	0, ± 1, ± 2, ± 3 . . .	Represents the familiar atomic mass number long used for nuclei. For uranium 235, A = 235. For baryons, A = +1; for antibaryons, A = −1; for mesons, A = O.	A = +1	A = O
HYPERCHARGE (Related to average charge, $\bar{Q}$, and to strangeness, S)	Y	− 2, −1,0, + 1	Defined as twice the average charge, $\bar{Q}$, of a multiplet. Strangeness, S, is hypercharge minus the atomic mass number (S = Y − A).	Y = + 1 S = O	Y = Q S = O
ISOTOPIC SPIN (related to multiplicity, M)	I	0, ½, 1, 3⁄2	Groups nuclear states into multiplets whose members differ only in electric charge. The number of charge states, or multiplicity, M, is related to I by the equation M = 2I + 1.	I = ½ M = 2	I = 1 M = 3
SPIN ANGULAR MOMENTUM	J	½, 3⁄2, 5⁄2 . . . 0, 1, 2, 3 . . .	Indicates how fast a particle rotates about its axis, expressed in units of Planck's constant, h.	J = ½	J = O
PARITY	P	−1, +1	An intrinsic property related to left-right symmetry.	P = +1	P = −1
G	G	−1, +1	An intrinsic property found only in mesons with zero hypercharge.	not defined	G = −1

CHART OF QUANTUM NUMBERS shows seven quantities conserved by the strong interaction but not necessarily by the electromagnetic or weak interaction. The three quantities in color (*A, Y, I*) are easily established by experiment and provide the basis for assigning a family name to each particle. Only 10 combinations of *A, Y* and *I* are now known, each represented by a Greek letter.

system: electric charge (Q) and a quantity called G that has only two values, $+1$ and -1, and can be assigned only to mesons that have a hypercharge of 0.

The first three quantum numbers—A, Y and I—provide the basis of the naming system. . . . Mesons and baryons occur in "charge multiplets," or families of states differing only in their electric charge. The number of particles and their charges occur in different patterns: singlets, doublets, triplets and quadruplets. Only 10 different patterns are known or predicted at present, and each pattern represents a different set of values for A, Y and I. As we shall explain, each of the 10 patterns is identified by a different Greek letter.

Now we shall describe the physical significance of A, Y, I, J and P, but for convenience we shall discuss them in a slightly different order to emphasize certain relations among them. A is simply the long-familiar atomic mass number used to describe atomic nuclei. It is also known as baryon number. Like electric charge, A can be 0, ± 1, ± 2, ± 3 and so on. For uranium 235, A is 235, indicating that the nucleus of this isotope contains 235 neutrons and protons, for each of which A equals 1. Neutrons and protons are baryons and, by definition, so are all other particles with an A of 1. Particles with an A of -1 are antibaryons. Mesons have an A of 0 (as do the leptons and the photon). The law of baryon conservation states that the total value of A, like electric charge, can never change in a reaction. Baryons cannot be created or destroyed, except when a baryon-antibaryon pair annihilate each other or are created together.

The second conserved quantity is J, or spin angular momentum, which measures how fast a particle rotates about its axis. It is a fundamental feature of quantum theory that a particle can have a spin of only integral or half-integral multiples of Planck's constant. (This constant, $\hbar$, relates the energy of a quantum of radiation to its wavelength: energy equals 2π times frequency times $\hbar$.) For baryons J is always half-integral (that is, half an odd integer, such as ½, 3/2, 5/2 and so on) and for mesons J is always integral (that is, 0, 1, 2 and so on).

The third conserved quantity, closely associated with J, is P, or intrinsic parity. Parity is conserved when nature does not distinguish between left and right. Because such symmetry is observed in strong interactions, quantum mechanics tells us that an intrinsic parity value of $+1$ or -1 can be assigned to each strongly interacting particle. In the case of weak interactions, however, nature does distinguish between left and right, and the symmetry is violated.

The bookkeeping on parity is not quite so simple as that for electric charge and baryon number; the intrinsic parity values on each side of an equation are not necessarily the same. The reason is that the total parity is affected by spin angular momentum as well as by intrinsic parity. The close connection between parity and spin angular momentum makes it convenient to write the spin angular momentum quantum number *J* and the intrinsic parity *P* next to each other in describing each particle. For the proton, for example, *J* equals ½ and *P* equals +1, which is shortened to J^P equals $\frac{1}{2}^+$. For the pion *J* equals 0 and *P* is −1, so that one writes J^P equals 0^-. (The system of bookkeeping for *J* is actually quite complicated in quantum mechanics, but the details need not concern us here.)

The fourth quantity conserved in strong interactions is *I*, or isotopic spin. This quantum number has nothing to do with spin or angular momentum, except that its peculiar quantum-mechanical bookkeeping is similar to that for *J*. The concept of isotopic spin was originally introduced into quantum mechanics to accommodate the fact that the nucleon exists in two charge states: one positively charged (the proton) and the other neutral (the neutron). As far as strong interactions are concerned, these two states behave alike; they are related to each other by the symmetry of isotopic spin. Moreover, if the symmetry were observed by the electromagnetic interaction, the proton and the neutron would have the same mass. Precisely because isotopic-spin symmetry is violated by the electromagnetic interaction the neutron is 1.3 Mev (or .14 per cent) more massive than the proton.

A set of particles or particle states (we use the terms interchangeably) related to each other by isotopic-spin symmetry is a charge multiplet and is given a single name. Thus the nucleon doublet consists of the two charge states, positive and negative. The pion triplet consists of negative, neutral and positive charge states. The number of different charge states in a multiplet, or its "multiplicity" (*M*), is directly related to the isotopic-spin quantum number *I* by the equation $M = 2I + 1$. For the nucleon *M* equals 2 and *I* equals ½; for the pion *M* equals 3 and *I* equals 1.

The fifth conserved quantity goes by any of three names: average charge ($\bar{Q}$), hypercharge (*Y*) or strangeness (*S*), which are related to each other in a simple way. Average charge is just what its name implies: the average of the electric charges in a multiplet. Hence for the nucleon it is ½ (0 plus 1 divided by 2); for the pion it is 0. Hypercharge is defined as twice the average charge (*Y* equals $2\bar{Q}$) merely to make it an integral number. And strangeness is hypercharge minus the baryon number (*S* equals *Y* minus *A*). It is clear that the three quantities are in effect interchangeable.

The concept of strangeness and its conservation is only 11 years old. In the early 1950's certain particles such as the K, the sigma and the xi were being observed for the first time, and because of their unusual behavior they were referred to as "strange particles." Most of them have relatively long lifetimes, which indicates that they decay by the weak interaction rather than by the electromagnetic or strong interaction. On the other hand, they are readily produced in high-energy collisions of "ordinary" particles (pions and nucleons), which proves that the strange particles too are strongly interacting. When invariable behavior patterns of this sort are observed, the physicist suspects that a conservation law (or symmetry) is at work. One of the authors of this article (Gell-Mann) and the Japanese physicist Kazuhiko Nishijima independently proposed that a previously unsuspected quantity (strangeness, or hypercharge) is conserved in strong and electromagnetic interactions but violated by weak interactions. The hypothesis made it possible to predict the existence and general properties of several strange particles before they were discovered.

THE NEW NOMENCLATURE

We are now ready to describe how the five quantum numbers can provide the basis for a new naming system. By appropriate selection of three of the five quantum numbers we can indicate immediately whether a strongly interacting particle is a baryon or a meson, how many members it has in its immediate family (that is, its multiplicity) and what its degree of strangeness is. The three quantum numbers that provide this information are the atomic mass, or baryon, number A, the hypercharge Y and the isotopic spin I. (It will be recalled that Y is directly related to strangeness and I to multiplicity.)

Now, partly as a mnemonic aid and partly out of respect for the old pet names, we shall employ a letter symbol to indicate various combinations of A, Y and I. To designate the known mesons, particles for which A is 0, it is sufficient to use four lower-case Greek letters: η (eta), π (pi), κ (kappa) and $\bar{\kappa}$ (antikappa, or kappa bar). The chart at the botton of the next page shows the Y and I values for each symbol. Even though the multiplicity M can be found simply by doubling I and adding 1, it is shown separately for easy reference.

To designate baryons, for which A is 1, we will use the following upper-case Greek letters: Λ (lambda), Σ (sigma), N (which stands for the nucleon and is pronounced "en," not "nu"), Ξ (xi), Ω (omega) and Δ (delta). The values of Y, I and M for each symbol are also shown at the bottom of the next page.

PARTICLE STABILITY

We mentioned earlier that particles can decay in one of three ways: via the strong, the electromagnetic or the weak interaction. A few particles (the photon, the two neutrinos, the electron and the proton) are absolutely stable provided that they do not come into contact with their antiparticles, whereupon they are annihilated. Particles that decay through the electromagnetic or weak interaction are said to be metastable. Those that decay through the strong interaction are called unstable and have very short lifetimes, typically a few times 10^{-23} second. This is still a considerable length of time, however, compared with less than 10^{-23} second, which is the characteristic time required for a collision between high-speed particles.

Unstable particles are those with enough energy to decay into two or more strongly interacting particles without violating any of the conservation laws respected by the strong interaction. Some unstable particles have only one possible decay mode; others have more than one. . . .

How can the existence of several decay modes he explained? This can be answered by introducing the concept of "communicating states." A nuclear state can be either a single particle or a combination of two or more particles. We have seen that each particle has definite values of the conserved quantum numbers *A, Y, I, J, P* and, where applicable, *G*. The strong interaction allows transitions, or communication, only among nuclear states with the same values for all the conserved quantum numbers.

Now, one can write down many nuclear states, consisting of two or more particles, that have in the aggregate the same set of quantum numbers as any particular unstable particle. For decay to take place, however, the unstable particle must have a rest mass at least equal to the threshold

Mesons	*Y*	*I*	*M*	Baryons	*Y*	*I*	*M*
η	0	0	*1*	Λ	0	0	*1*
π	0	1	*3*	Σ	0	1	*3*
κ	+1	½	*2*	N	+1	½	*2*
$\bar{\kappa}$	−1	½	*2*	Ξ	−1	½	*2*
				Ω	−2	0	*1*
				Δ	+1	3⁄2	*4*

SYMBOLS FOR MESONS AND BARYONS are Greek letters. For mesons atomic mass number *A* is 0; for baryons it is 1. The 10 letters identify the 10 known combinations of mass number (*A*), hypercharge (*Y*) and isotopic spin (*I*). Multiplicity (*M*) is related to *I*.

energy (that is, the sum of the rest masses) of the particles into which it might conceivably decay. In other words, energy must be conserved. The various states into which an unstable particle has sufficient energy to decay are called open channels. Communicating states with threshold energies greater than that available to the unstable particle are called closed channels; decay into them is allowed by everything *but* conservation of energy. . . .

To explain how an unstable particle can communicate with several open channels we have found it helpful to draw an analogy between the behavior of unstable particles and the behavior of resonant cavities such as organ pipes and electromagnetic cavities. Cavities of the latter sort (such as the magnetron tube employed in radar) are used in electronics to create intense electromagnetic waves of a desired frequency, which is a resonant frequency of the cavity. Each cavity has a characteristic "lifetime": the time required for the electromagnetic radiation to leak out.

In quantum mechanics, particles and waves are complementary concepts, and the amount of energy associated with a particle, or nuclear state, can be expressed as an equivalent frequency. In other words, energy is proportional to frequency. The fact that the Δ particle appears when a pion is scattered by a proton at or near a certain energy—the resonance energy—is equivalent to saying that the particle appears at a certain frequency. Thus a resonance energy in particle physics can be compared to the resonance frequency of an acoustic or electromagnetic cavity. What is the "cavity" in particle physics? It is an imaginary structure: one cavity, each with its own special properties, for each set of values of the quantum numbers conserved in strong interactions.

The analogy between unstable particles and the resonant modes of electromagnetic cavities can be carried further. To the electromagnetic cavity one can attach the long pipes known as wave guides, which have the property of efficiently transmitting electromagnetic waves of high frequency but not those of low frequency. When the electromagnetic wavelength is slightly larger than the dimensions of the wave guide, the guide refuses to transmit. In this sense the wave guide acts like a particle channel that is open only above its characteristic threshold energy. If a cavity has attached to it several wave guides of different sizes, high-frequency radiation can flow into the cavity through one guide and flow out through the same or different guides.

By analogy energy can flow into a nuclear interaction through one channel and pass out through one or more open channels. As the energy (frequency) is increased from low values, the channels open up one by one and new nuclear reactions become possible, with energy going out through any of the open channels. Now, as the frequency is increased, suppose it passes through a resonance frequency of the nuclear cavity. At this point it becomes easier for the cavity to absorb and reradiate energy. The resonance appears as a peak in the scattering cross section of a nuclear reaction.

One can use the wave guide analogy to describe not only unstable particles but also stable ones. A stable particle is merely one that has such a low mass that all the communicating channels are closed. Therefore it is a "bound" state rather than a scattering resonance. For an electromagnetic cavity this condition would correspond to a resonant mode whose frequency is below the threshold frequency of all the wave guide outlets. If radiation could be put into the cavity in such a mode, it could not leak out. Of course, an actual cavity would eventually lose radiation by leakage into and through its walls. Such leakage corresponds to the decay of metastable particles via the weak and electromagnetic reactions. An absolutely stable particle really does live forever.

The reader who is unfamiliar with the phenomenon of resonance in electromagnetic cavities may be wondering if we have simplified his task by introducing the electromagnetic analogy. Would it not be just as easy to explain resonances in particle physics directly? Possibly so. But by drawing attention to similar behavior in two apparently different fields we hope we have illustrated a unity in physics that may make particle behavior seem less esoteric. The more basic value of the analogy, however, is that it has helped theorists to understand some deeper points in particle resonances than we have been able to talk about here.

REGGE TRAJECTORIES

As the strongly interacting particles proliferated, physicists sought to find patterns that would show relations among them. In particular they tried to find classification systems that might predict new particles on the basis of those already known. The first concept to prove useful in this respect developed from an idea introduced into particle physics in 1959 by the Italian physicist Tullio Regge. The concept already had counterparts throughout quantum physics, notably in the study of atomic- and nuclear-energy levels.

It had been observed that as particles increase in mass they frequently (but not invariably) exhibit a higher value of spin angular momentum J. Regge pointed out the existence, in many important cases, of a mathematical relation between the value of J and particle mass. He showed that certain properties of particles can be regarded as "smooth" mathematical functions of *J,* that is, functions that vary continuously as *J* varies. But since in quantum mechanics *J* can have only integral and half-integral values, the functions have direct physical meaning only for those permitted values. The smooth mathematical curve that gives the physical mass for different values of *J* is called a "Regge trajectory."

A spaceship analogy may help to clarify the concept of the Regge trajectory. Suppose in the nearly circular orbit occupied by each of the sun's nine planets one were to place a one-ton spacecraft. These craft circle the sun as if they were miniature planets. The nearer a craft is to the sun, the more strongly it "feels" the sun's gravitational force and the more strongly it is bound. This binding energy, therefore, is highest for the craft in Mercury's orbit and least for that in Pluto's orbit. (The binding energy is just that amount needed to release the craft from the sun's attractive force.)

Each craft can be assigned another quantity that also varies with its distance from the sun: angular momentum. It frequently happens in physics that, other things being equal, the greater the binding energy, the smaller the angular momentum. In our example this means that angular momentum increases with the distance from the sun. One can now draw a graph for the nine spacecraft in which angular momentum is plotted on the vertical axis and binding energy on the horizontal axis. . . . The curve drawn through the plotted points is analogous to a Regge trajectory.

Now, suppose quantum mechanics were to have a controlling effect on the macroscopic scale of spacecraft and solar orbits. Suppose, that is, the angular momentum of the spacecraft in Mercury's orbit represented an elementary quantum of spin. If such were the case, a one-ton craft would be allowed to occupy only those orbits in which the angular momenta (expressed in Mercury units) assumed integral values. This is equivalent to saying that a one-ton craft in a circular orbit could exist only at certain energy levels. The Regge trajectory for the spaceship would then be physical only at these points. Another Regge trajectory could be drawn for a two-ton spacecraft. In any given circular orbit its binding energy and angular momentum would be twice that of a one-ton spacecraft.

Although it was not used by physicists until recently, the notion of a Regge trajectory applies to problems long familiar in atomic physics. It is well known, for example, that the electron-proton system constituting the hydrogen atom can exist in various states of excitation. The electron can occupy various orbits around the proton, just as the spaceship could occupy many orbits around the sun. In the case of the electron, of course, the quantization of the orbits is very conspicuous. When the value of a certain quantum number, n_r (a number characterizing the energy of motion in a radical direction), is held fixed, the binding energies of the various hydrogen states decrease with increasing values of the angular momentum *J*. If a smooth curve is drawn through the permitted values of *J*, one obtains a Regge trajectory similar to that connecting the spacecraft in different orbits. . . . For each value of n_r in the hydrogen atom there is a different trajectory, just as there is for each spacecraft of different mass.

In the case of the hydrogen atom the intersection of a Regge trajectory with a permitted value of *J* (0, 1, 2 and so on) corresponds to the occurrence of a bound state. From an experimental standpoint each of these occurrences is a different "particle" with a different mass. The series of occurrences is brought to an end when the excitation energy becomes so great that the electron is dissociated from the proton. This energy limit divides stable states from unstable states.

Just as Regge trajectories can be drawn for the gravitational force (spaceship example) and the electromagnetic force (hydrogen-atom example), trajectories can be drawn for the strong force governing strongly interacting particles. In this case the trajectories do not terminate at the boundary between stable and unstable states but continue across, cutting further integer values of *J*. . . . An intersection in the stable region indicates the existence of a bound state, meaning a particle that is either stable or meta-stable. Intersections in the unstable region indicate the existence of resonances, or unstable particles. It can be shown that for strongly interacting particles a particular trajectory joins up real states for either odd or even values of *J* but not for both. This means that an interval of two units of *J* must intervene between states on the same trajectory. The lowest state is called an occurrence; higher states can be referred to as Regge recurrences, or as a series of excited rotational states.

How can the existence of a trajectory be demonstrated? In analogy with the spaceship or hydrogen-atom example, one plots the angular momentum *J* against mass (in Mev) for all the particles that share all the same quantum numbers except *J*. One can then quickly see whether or not they fall into groups that lie on rising curves. If they do, one has an indication of Regge trajectories. . . .

THE EIGHTFOLD WAY

Now we shall turn to another classification scheme that has proved valuable in predicting the existence of previously undiscovered particles. We have seen how the notion of Regge trajectories makes it possible to perceive family connections between particles with different values of J but the same values of all other quantum numbers. Now we shall describe a relation that seems to exist among particles with the same values of J and of parity P but different values of mass, hypercharge Y and isotopic spin I.

We mentioned earlier that the difference in mass between charge multiplets such as the nucleon doublet (neutron and proton) can be regarded as a "splitting" caused by the fact that isotopic spin is not conserved by the electromagnetic interaction, which underlies the electric charge. This violation produces a maximum mass difference of about 12 Mev in the case of the Σ triplet.

Now, it is a remarkable fact that the four best-known members of the baryon family, N, Λ, Σ and Ξ, are separated by average mass differences only about a factor of 10 greater than that separating members within each multiplet. The gaps in average mass separating the four baryon states are only 77, 75 and 130 Mev respectively. Moreover, these four baryons all seem to have the same J^P; it is $\frac{1}{2}^+$. (Actually the J of Ξ is not firmly established and its parity is still unmeasured.)

If the difference in mass within a multiplet is caused by a violation of the isotopic spin I, is it conceivable that the somewhat greater difference in mass between neighboring multiplets is caused by the violation of the conservation of some other quantum numbers? The kind of solution needed is one in which Y and I are exactly conserved by the strong interaction but certain other conservation laws are broken by some aspect, or some part, of this same interaction. If such a partial violation of new symmetry principles were permitted, one might be able to group baryon multiplets into "supermultiplets" with various values of Y and I but the same J and P. This new system of symmetries would connect different Y and I values in the same way that isotopic spin connects different values of electric charge. The aspect of the strong interaction that violates the new symmetries—represented by new quantum numbers—would split each supermultiplet into charge multiplets of different mass, much as the electromagnetic interaction causes splitting of mass among the members of a charge multiplet by violating isotopic-spin symmetry. The scale of the mass splitting within the supermultiplet, however, would be much greater than that observed within a multiplet, since it is an appreciable fraction of the strong force that is at work rather than the electromagnetic force, which is much weaker.

Early in 1961 an Israeli army colonel and engineer-turned-physicist, Y. Ne'eman, and one of the authors (Gell-Mann), working independently, suggested a particular unified system of symmetries and a particular pattern of violations that made plausible the existence of supermultiplets. The new system of symmetries has been referred to as the "eightfold way" because it involves the operation of eight quantum numbers and also because it recalls an aphorism attributed to Buddha: "Now this, O monks, is noble truth that leads to the cessation of pain: this is the noble *Eightfold Way:* namely, right views, right intention, right speech, right action, right living, right effort, right mindfulness, right concentration."

The mathematical basis of the eightfold way is to be found in what are called Lie groups and Lie algebras, which are algebraic systems developed in the 19th century by the Norwegian mathematician Sophus Lie. The simplest Lie algebra involves the relation of three components, each of which is a symmetry operation of the kind used in quantum mechanics. Isotopic spin consists of three such components (I_+, I_- and I_z) related by the rules of this simplest algebra. The algebra is that of the Lie group called SU(2), which stands for special unitary group for arrays of size 2×2; there is one condition in the 2×2 arrays that reduces the number of independent components from four to three (hence the term "special").

The component operations of the eightfold way satisfy the mathematical relations of the next higher Lie algebra, which has eight independent components. Here the Lie group is called SU(3), which stands for special unitary group for arrays of size 3×3; again a special condition reduces the number of components from nine to eight. The eight conserved quantities of the eightfold way consist of the three components of isotopic spin, the hypercharge Y and four new symmetries not yet formally named. Two of the new symmetries change Y up or down by one unit without changing electric charge; the other two symmetries change both Y and electric charge by one unit. . . . The violation of all four new symmetries by part of the strong interaction changes the masses of the multiplets forming a supermultiplet. An example of a supermultiplet (an octet) is provided by N, Λ, Σ and Ξ, if indeed they all have an angular momentum J of ½ and positive parity, that is, a J^P of $\frac{1}{2}^+$.

We close this section with the remark that the symmetry game may not yet be finished for strongly interacting particles. For example, there might exist some undiscovered quantum number that is conserved by the strong interactions and that has the value 0 for all known particles. Before strange particles were discovered, the strangeness quantum number (equivalent to Y) was of this kind. Experiments at very high energies with the next generation of accelerators might produce a similar situation with respect to an entirely new quantum number.

COMPOSITE PARTICLES

The meaning of the term "elementary particle" has varied enormously as man's view of the physical universe has become more detailed. In the past few years it has become increasingly awkward to consider several scores of particles as elementary. Evidently a reappraisal of the entire elementary-particle concept is in order.

Let us begin by asking why we feel sure that certain particles such as the hydrogen atom are *not* elementary. The answer is that even though these particles have properties qualitatively similar to those of neutrons, protons and electrons, it has been possible theoretically to explain their properties by assuming that they are composites of other particles.

The hydrogen atom itself provides an outstanding example of what is meant by composite, because its properties have been theoretically predicted with enormous accuracy. It is important to realize that the hydrogen atom is not exactly composed of one proton and one electron. It is more accurate to say that it is so composed *most* of the time. The ground state of the hydrogen atom is a stable "particle" that communicates (via strong, electromagnetic and weak interactions) with a great variety of closed channels, of which electron plus proton is the most important. According to quantum mechanics any state consists part of the time of each of the channels that communicate with it. As an example, for a certain small fraction of the time the ground state of the hydrogen atom consists of the expected electron and proton plus an electron-positron pair. The effect of this channel on the energy of the atom is tiny, but it has been calculated and measured; the agreement is excellent. There are infinitely many other closed channels that contribute to the structure of the hydrogen atom, but fortunately their effect is negligibly small.

In strongly interacting systems complicated channels are more important. For example, the properties of the deuteron (A equals 2) have been predicted, assuming that this particle is a composite of neutron and proton, but here the accuracy of the predictions is much poorer than it is for the hydrogen atom because the effect of additional channels (involving pions, say) is substantial. Nevertheless, there is a general belief that, since the simplest channel accounts for the bulk of the observed properties of the deuteron, it should eventually be possible to improve the predictions systematically by inclusion of more channels. The same kind of statement can be made for all nuclei heavier than the deuteron, and there is no disposition to regard any of these compound nuclei as "elementary."

Confusion about the distinction between composite and elementary particles has arisen for particles with an *A* of 0 and 1 (mesons and baryons) because here one rarely has a single dominant channel nearby in energy. . . .

We may nevertheless employ the operational definition: a particle is nonelementary if all its properties can be calculated *in principle* by treating it as a composite. Such a calculation must yield various probabilities for the various closed channels; the binding forces in these channels must yield the right mass for the particle.

The problem of including all the significant channels is in most cases still too difficult, but suppose the calculation could be carried out. Would we then get a correct description of each particle? Would the quantum numbers and the mass come out right? Until recently there was an almost universal belief that a few strongly interacting particles, including the nucleon, could not be calculated on such a basis. In the present theory of electrons and photons, which gives such an excellent description of electromagnetic phenomena, the properties of the electron and photon cannot be dynamically predicted. The reason is that the known forces are not powerful enough to form bound states with masses as small as those of the electron and the photon.

Reasoning by analogy, theorists tended to give the nucleon a special status parallel to that of the electron. Thus they were inhibited from trying to treat the nucleon as a composite particle. Gradually, however, this select status seemed increasingly dubious. . . .

It seems, furthermore, on the basis of recent developments in which the concept of the Regge trajectory plays an important role, that in all such dynamical calculations no distinction need be made on the basis of the angular momentum or other quantum numbers of the particle involved. If there is no need for an aristocracy among strongly interacting particles, may there not be democracy?

"BOOTSTRAP" DYNAMICS

Composite particles owe their existence to the forces acting in channels with which the particles communicate. How do these forces arise and how can they be calculated?

The key concept behind the calculation is "crossing." Consider the following reaction involving four particles:

$$a + b \longleftrightarrow c + d$$

This says that the channel *a,b* is coupled to the channel *c,d.* The probability that this reaction will take place (in either direction) is expressed mathematically as the absolute value squared of the "reaction amplitude," which depends on the energies of the four particles involved. The principle of crossing states that the same reaction amplitude also applies to the two "crossed" reactions in which *ingoing* particles are replaced by *outgoing* antiparticles (indicated by a bar over a letter) thus:

$$a + \bar{c} \longleftrightarrow \bar{b} + d$$
$$a + \bar{d} \longleftrightarrow \bar{b} + c$$

These different reactions are distinguished by the signs of the energy variables, which are positive or negative according to whether ingoing or outgoing particles are involved, but if the reaction amplitude is known for any one of the three reactions, it can be obtained for the other two by inserting the proper signs for energy.

An example of crossing is the following pair of reactions involving neutrons and protons:

$$\text{(a) } n + p \longleftrightarrow n + p$$
$$\text{(b) } n + \bar{n} \longleftrightarrow p + \bar{p}$$

Both reactions are described by the same reaction amplitude, an important aspect of which can be depicted diagrammatically. The first way of drawing the arrows in this diagram is appropriate to reaction *a* and the second to reaction *b.* The two figures differ, of course, only in the direction one reads them, either from bottom to top or left to right, as indicated by the arrowheads.

One interprets the first figure by saying that in a scattering collision between a neutron and a proton a meson is "exchanged," and it can be shown that this exchange is a way of representing the force acting between those two baryons. The interpretation of reaction *b* is that a meson that communicates with both the *n,n* and $p,\bar{p}$ channels provides a means for connecting the two channels of the reaction. Thus a single diagram corresponds both to a force in one reaction and to an intermediate particle in the crossed reaction. It follows that forces in a given channel may be said to arise, in general, from the exchange of intermediate particles that communicate with crossed channels.

With this as background we return to the idea mentioned in the introduction to this article, that the strongly interacting particles are all dynamical structures that owe their existence to the same forces through which they mutually interact. In short, the strongly interacting particles are the creatures of the strong interaction. We refer to this as the "bootstrap" hypothesis. . . .

According to the bootstrap hypothesis each strongly interacting particle is assumed to be a bound state of those channels with which it communicates, owing its existence entirely to forces associated with the exchange of strongly interacting particles that communicate with crossed channels. Each of these latter particles in turn owes its existence to a set of forces to which the first particle makes a contribution. In other words, each particle helps to generate other particles, which in turn generate it. In this circular and violently nonlinear situation it is possible to imagine that no free, or arbitrary, variables appear (except for something to establish the energy scale) and that the only self-consistent set of particles is the one found in nature.

We remind the reader that in electromagnetic theory a few special particles (leptons and photon) are *not* treated as bound (or composite) states, the masses and coupling characteristics of each particle being freely adjustable. Conventional electrodynamics, as far as anyone knows, is not a bootstrap regime.

It is too soon to be sure that free variables are absent for strong interactions, but we shall close in an optimistic spirit by mentioning a fascinating possibility that would represent the ultimate contribution of the bootstrap hypothesis. If the system of strongly interacting particles is in fact self-determining through a dynamical mechanism, perhaps the special strong-interaction symmetries are not arbitrarily imposed from the outside, so to speak, but will emerge as necessary components of self-consistency. It is remarkable, and puzzling, that isotopic-spin symmetry, strangeness and now the broader eightfold-way symmetry have never been related to other physical symmetries. Perhaps their origin is destined to be understood at the same moment we understand the pattern of masses and spins for strongly interacting particles. Both this pattern and the puzzling symmetries may emerge together from the bootstrap dynamics.

R.8

Quarks with Color and Flavor

SHELDON GLASHOW

SHELDON GLASHOW WAS born in New York City in 1932 and graduated from the Bronx High School of Science. He earned his undergraduate degree at Cornell in 1954 and received a doctorate in physics from Harvard in 1959. Glashow did postdoctoral research at the Bohr Institute, CERN in Geneva and the California Institute of Technology. He began his faculty career at Berkeley, but left in 1966 to become a physics professor at Harvard.

Glashow's work, which ultimately earned him one-third of the 1979 Nobel Prize in physics, was for helping to develop the theory of the so-called "electroweak force," a theory that is said to unify the weak and electromagnetic interactions between elementary particles, and which predicts the neutrino neutral current. Howard Georgi and Glashow have extended the electroweak unification to try to include the strong interaction in it and thus to obtain a grand unification theory (GUT); this effort has not been successful. Glashow also introduced a fourth quark called "charm" to account for a certain discrepancy between the theoretical prediction of the rate of decay of certain strange mesons, called kaons, and the observed rate of decay. The introduction of a charmed quark replaced the Gell-Mann SU(3) symmetry group by the SU(4) group. The evidence for the charmed quark is assumed to be the discovery in 1974 of the J/psi particle by Burton Richter and Samuel C.C. Ting.

Glashow's article provides a concise historical overview of particle physics developments that date back to the time of Rutherford. He shows how a multitude of elementary particles can be explained by combinations of up, down, and strange quarks. The charm quark is used to explain in a systematic fashion the proliferation of baryons and mesons.

Glashow also explains the functions of gluon particles, "gluons," which bind the quarks inside hadrons and produce the exchange of "colors" among quarks—names given to three new kinds of charge on each quark and designated as "red," "green," and "blue." Although gluons must also have colors, quark theory states that no colored particle can exist in isolation. Particles exist only when their constituent quarks combine so that the resulting color is "white." Any attempt to extract a single "colored" quark fails, because the introduction of the energy into that subatomic system to extract a quark creates only combinations of quarks and antiquarks which are also color-neutral.

The tendency of modern science is to reduce proof to absurdity by continually reducing absurdity to proof.

—SAMUEL BUTLER

Plurality which is not reduced to unity is confusion; unity which does not depend on plurality is tyranny.

—BLAISE PASCAL, *Pensées*

ATOMOS, THE GREEK root of "atom," means indivisible, and it was once thought that atoms were the ultimate, indivisible constituents of matter, that is, they were regarded as elementary particles. One of the principal achievements of physics in the 20th century has been the revelation that the atom is not indivisible or elementary at all but has a complex structure. In 1911 Ernest Rutherford showed that the atom consists of a small, dense nucleus surrounded by a cloud of electrons. It was subsequently revealed that the nucleus itself can be broken down into discrete particles, the protons and neutrons, and since then a great many related particles have been identified. During the past decade it has become apparent that those particles too are complex rather than elementary. They are now thought to be made up of the simpler things called quarks. A solitary quark has never been observed, in spite of many attempts to isolate one. Nonetheless, there are excellent grounds for believing they do exist. More important, quarks may be the last in the long series of progressively finer structures. They seem to be truly elementary.

When the quark hypothesis was first proposed more than 10 years ago, there were supposed to be three kinds of quark. The revised version of the theory I shall describe here requires 12 kinds. In the whimsical terminology that has evolved for the discussion of quarks they are said to come in four flavors, and each flavor is said to come in three colors. ("Flavor" and "color" are, of course, arbitrary labels; they have no relation to the usual meanings of those words.) One of the quark flavors is distinguished by the property called charm (another arbitrary term). The concept of charm was suggested in 1964, but until last year it had remained an untested conjecture. Several recent experimental findings, including the discovery

---

Sheldon Glashow, "Quarks with Color and Flavor," *Scientific American* (October 1975), pp. 38-50.

last fall of the particles called *J* or psi, can be interpreted as supporting the charm hypothesis.

The basic notion that some subatomic particles are made of quarks has gained widespread acceptance, even in the absence of direct observational evidence. The more elaborate theory incorporating color and charm remains much more speculative. The views presented here are my own, and they are far from being accepted dogma. On the other hand, a growing body of evidence argues that these novel concepts must play some part in the description of nature. They help to bring together many seemingly unrelated theoretical developments of the past 15 years to form an elegant picture of the structure of matter. Indeed, quarks are at once the most rewarding and the most mystifying creation of modern particle physics. They are remarkably successful in explaining the structure of subatomic particles, but we cannot yet understand why they should be so successful.

The particles thought to be made up of quarks form the class called the hadrons. They are the only particles that interact through the "strong" force. Included are the protons and neutrons, and indeed it is the strong force that binds protons and neutrons together to form atomic nuclei. The strong force is also responsible for the rapid decay of many hadrons.

Another class of particles, defined in distinction to the hadrons, are the leptons. There are just four of them: the electron and the electron neutrino and the muon and the muon neutrino (and their four antiparticles). The leptons are not subject to the strong force. Because the electron and the muon bear an electric charge, they "feel" the electromagnetic force, which is roughly 100 times weaker than the strong force. The two kinds of neutrino, which have no electric charge, feel neither the strong force nor the electromagnetic force, but interact solely through a third kind of force, weaker by several orders of magnitude and called the weak force. The strong force, the electromagnetic force and the weak force, together with gravitation, are believed to account for all interactions of matter.

The leptons give every indication of being elementary particles. The electron, for example, behaves as a point charge, and even when it is probed at the energies of the largest particle accelerators, no internal structure can be detected. The hadrons, on the other hand, seem complex. They have a measurable size: about 10^{-13} centimeter. Moreover, there are hundreds of them, all but a handful discovered in the past 25 years. Finally, all the hadrons, with the significant exception of the proton and the antiproton, are unstable in isolation. They decay into stable particles such as protons, electrons, neutrinos or photons. (The photon, which is the carrier of the electromagnetic force, is in a category apart; it is neither a lepton nor a hadron.)

The hadrons are subdivided into three families: baryons, antibaryons and mesons. The baryons include the proton and the neutron; the mesons include such particles as the pion. Baryons can be neither created nor destroyed except as pairs of baryons and antibaryons. This principle defines a conservation law, and it can be treated most conveniently in the system of bookkeeping that assigns simple numerical values, called quantum numbers, to conserved properties. In this case the quantum number is called baryon number. For baryons it is +1, for antibaryons −1 and for mesons 0. The conservation of baryon number then reduces to the rule that in any interaction the sum of the baryon numbers cannot change.

Baryon number provides a means of distinguishing baryons from mesons, but it is an artificial means, and it tells us nothing about the properties of the two kinds of particle. A more meaningful distinction can be established by examining another quantum number, spin angular momentum.

Under the rules of quantum mechanics a particle or a system of particles can assume only certain specified states of rotation, and hence can have only discrete values of angular momentum. The angular momentum is measured in units of $h/2\pi$, where h is Planck's constant, equal to about 6.6×10^{-27} erg-second. Baryons are particles with a spin angular momentum measured in half-integral units, that is, with values of half an odd integer, such as ½ or 3/2. Mesons have integral values of spin angular momentum, such as 0 or 1.

The difference in spin angular momentum has important consequences for the behavior of the two kinds of hadron. Particles with integral spin are said to obey Bose-Einstein statistics (and are therefore called bosons). Those with half-integral spin obey Fermi-Dirac statistics (and are called fermions). In this context "statistics" refers to the behavior of a population of identical particles. Those that obey Bose-Einstein statistics can be brought together without restriction; an unlimited number of pions, for example, can occupy the same state. The Fermi-Dirac statistics, on the other hand, require that no two particles within a given system have the same energy and be identical in all their quantum numbers. This statement is equivalent to the exclusion principle formulated in 1925 by Wolfgang Pauli. He applied it in particular to electrons, which have a spin of ½ and are therefore fermions. It requires that each energy level in an atom contain only two electrons, with their spins aligned in opposite directions.

One of the clues to the complex nature of the hadrons is that there are so many of them. Much of the endeavor to understand them has consisted of a search for some ordering principle that would make sense of the multitude.

The hadrons were first organized into small families of particles called charge multiplets or isotopic-spin multiplets; each multiplet consists of particles that have approximately the same mass and are identical in all their other properties except electric charge. The multiplets have one, two, three or four members. The proton and the neutron compose a multiplet of two (a doublet); both are considered to be manifestations of a single state of matter, the nucleon, with an average mass equivalent to an energy of .939 GeV (billion electron volts). The pion is a triplet with an average mass of .137 GeV and three charge states: +1, 0 and −1. In the strong interactions the members of a multiplet are all equivalent, since electric charge plays no role in the strong interactions.

In 1962 a grander order was revealed when the charge multiplets were organized into "supermultiplets" that revealed relations between particles that differ in other properties in addition to charge. The creation of the supermultiplets was proposed independently by Murray Gell-Mann of the California Institute of Technology and by Yuval Ne'eman of Tel-Aviv University. . . . The introduction of the new system led directly to the quark hypothesis.

QUARKS					LEPTONS			
Symbol	**Mass (GeV)**	**Electric Charge**	**Strangeness**	**Charm**	**Name**	**Symbol**	**Mass (GeV)**	**Electric Charge**
d	.338	−½	0	0	ELECTRON	e−	.0005	−1
u	.336	+⅔	0	0	ELECTRON NEUTRINO	ν_e	0	0
s	.540	−⅓	−1	0	MUON NEUTRINO	ν_μ	0	0
c	1.5	+⅔	0	+1	MUON	μ^-	.105	−1

QUARKS AND LEPTONS, the two kinds of particles that seem to be elementary, exhibit an apparent symmetry. The quarks are much more massive than the leptons, and they have fractional charges instead of integral ones, but both groups consist of two pairs of particles (indicated by colored rectangles). Either member of a pair is readily transformed into the other by the weak interaction. All ordinary matter can be constructed of just the *d* and *u* quarks and the electron and electron neutrino; the muon, the muon neutrino and the *c* and *s* quarks, which display the properties of strangeness and charm respectively, are important only in high-energy physics. Each kind, or flavor, of quark comes in three colors.

	Name	Symbol	Mass (GeV)	Charge States	Baryon Number	Spin	Isotopic Spin	Strangeness
BARYONS	NUCLEON	N	.939	0, +1	+1	½	½	0
	LAMBDA	Λ	1.115	0	+1	½	0	−1
	OMEGA	Ω	1.672	−1	+1	3/2	0	−2
MESONS	PION	π	.139	−1, 0, +1	0	0	1	0
	K	K	.496	0, +1	0	0	½	+1
	PHI	ϕ	1.019	0	0	1	0	0
	J	J	3.095	0	0	1	0	0

HADRONS form the class of particles thought to be constructed of quarks. They are divided into baryons, made of three quarks, and mesons, made of a quark and an antiquark. (Antibaryons consist of three antiquarks.) The groups are distinguished by baryon number and by spin angular momentum, which has half-integral values for baryons and integral values for mesons. Each line in the table represents a multiplet of particles identical in all properties except electric charge, provided that small differences in mass are ignored. Isotopic spin is a function of the number of particles in a multiplet, and strangeness measures distribution of electric charge among them. Only a few representative hadrons are shown.

The grouping of the hadrons into supermultiplets involves eight quantum numbers and has been referred to as the "eightfold way." Its mathematical basis is a branch of group theory invented in the 19th century by the Norwegian mathematician Sophus Lie. The Lie group that generates the eightfold way is called SU(3), which stands for the special unitary group of matrices of size 3 × 3. The theory requires that all hadrons belong to families corresponding to representations of the group SU(3). The families can have one, three, six, eight, 10 or more members. If the eightfold way were an exact theory, all the members of a given family would have the same mass. The eightfold way is only an approximation, however, and within the families there are significant differences in mass.

The construction of the eightfold way begins with the classification of the hadrons into broad families sharing a common value of spin angular momentum. Each family of particles with identical spin is then depicted by plotting the distribution of two more quantum numbers: isotopic spin and strangeness.

Isotopic spin has nothing to do with the spin of a particle; it was given its name because it shares certain algebraic properties with the spin quantum number. It is a measure of the number of particles in a multiplet, and it is calculated according to the formula that the number of particles in the multiplet is one more than twice the isotopic spin. Thus the nucleon (a doublet) has an isotopic spin of ½; for the pion triplet the isotopic spin is 1.

Strangeness is a quantum number introduced to describe certain hadrons first observed in the 1950's, and called strange particles because of their anomalously long lifetimes. They generally decay in from 10^{-10} to 10^{-7} second. Although that is a brief interval by everyday standards, it is much longer than the lifetime of 10^{-23} second characteristic of many other hadrons.

Like isotopic spin, strangeness depends on the properties of the multiplet, but it measures the distribution of charge among the particles rather than their number. The strangeness quantum number is equal to twice the average charge (the sum of the charges divided by the number of particles in a multiplet) minus the baryon number. By this contrivance it is made to vanish for all hadrons except the strange ones. The triplet of pions, for example, has an average charge of 0 and a baryon number of 0; its strangeness is therefore also 0. The nucleon doublet has an average charge of + ½ and a baryon number of +1, so that those particles too have a strangeness of 0. On the other hand, the lambda particle is a neutral baryon that forms a family of one (a singlet). Its average charge of 0 and its baryon number of +1 give it a strangeness of −1.

On a graph that plots electric charge against strangeness the hadrons form orderly arrays. The mesons with a spin angular momentum of 0 compose an octet and a singlet; the octet is represented graphically as a hexagon with a particle at each vertex and two particles in the center, and the singlet is represented as a point at the origin. The mesons with a spin of 1 form an identical representation, and so do the baryons with a spin of ½. Finally, the baryons with a spin of 3/2 form a decimet (a group of 10) that can be graphed as a large triangle made up of a singlet, a doublet, a triplet and a quartet. The eightfold way was initially greeted with some skepticism, but the discovery in 1964 of the negatively charged omega particle, the predicted singlet in the baryon decimet, made converts of us all.

The regularity and economy of the supermultiplets are aesthetically satisfying, but they are also somewhat mystifying. The known hadrons do fit into such families, without exception. Mesons come only in families of one and eight, and baryons come only in families of one, eight and 10. The singlet, octet and decimet, however, are only a few of many possible representations of SU(3). Families of three particles or six particles are entirely plausible, but they are not observed. Indeed, the variety of possible families is in principle infinite. Why, then, do only three representations appear in nature? It early became apparent that the eightfold way is in some approximate sense true, but it was also plain from the start that there is more to the story.

In 1963 an explanation was proposed independently by Gell-Mann and by George Zweig, also of Cal Tech. They perceived that the unexpected regularities could be understood if all hadrons were constructed from more fundamental constituents, which Gell-Mann named quarks. The quarks were to belong to the simplest nontrivial family of the eightfold way: a family of three. (There is also, of course, another family of three antiquarks.)

The quarks are required to have rather peculiar properties. Principal among these is their electric charge. All observed particles, without exception, bear integer multiples of the electron's charge; quarks, however, must have charges that are fractions of the electron's charge. Gell-Mann designated the three quarks *u, d* and *s,* for the arbitrary labels "up," "down" and "sideways."

The mechanics of the original quark model are completely specified by three simple rules. Mesons are invariably made of one quark and one antiquark. Baryons are invariably made of three quarks and antibaryons of three antiquarks. No other assemblage of quarks can exist as a hadron. The combinations of the three quarks under these rules are sufficient to account for all the hadrons that had been observed or predicted at the time. Furthermore, every allowed combination of quarks yields a known particle.

Many of the necessary properties of the quarks can be deduced from these rules. It is mandatory, for example, that each of the quarks be assigned a baryon number of $+1/3$ and each of the antiquarks a baryon number of $-1/3$. In that way any aggregate of three quarks has a total baryon number of $+1$ and hence defines a baryon; three antiquarks yield a particle with a baryon number of -1, an antibaryon. For mesons the baryon numbers of the quarks ($+1/3$ and $-1/3$) cancel, so that the meson, as required, has a baryon number of 0.

In a similar way the angular momentum of the hadrons is described by giving the quarks half-integral units of spin. A particle made of an odd number of quarks, such as a baryon, must therefore also have half-integral spin, conforming to the known characteristics of baryons. A particle made of an even number of quarks, such as a meson, must have integral spin.

The *u* quark and the *s* quark compose an isotopic-spin doublet: they have nearly the same mass and they are identical in all other properties except electric charge. The *u* quark is assigned a charge of +⅔ and the *d* quark is assigned a charge of −⅓. The average charge of the doublet is therefore +⅙ and twice the average charge is +⅓; since the baryon number of all quarks is +⅓, the definition of strangeness gives both the *u* and the *d* quarks a strangeness of 0. The *s* quark has a larger mass than either the *u* or the *d* and makes up an isotopic-spin singlet. It is given an electric charge of −⅓ and consequently has a strangeness of −1. The antiquarks, denoted by writing the quark symbol with a bar over it, have opposite properties. The $\bar{u}$ has a charge of −⅔ and the $\bar{d}$ +⅓; both have zero strangeness. The $\bar{s}$ antiquark has a charge of +⅓ and a strangeness of +1.

Just two of the quarks, the *u* and the *d,* suffice to explain the structure of all the hadrons encountered in ordinary matter. The proton, for example, can be described by assembling two *u* quarks and a *d* quark; its composition is written *uud.* A quick accounting will show that all the properties of the proton determined by its quark constitution are in accord with the measured values. Its charge is equal to ⅔ + ⅔ − ⅓, or +1. Similarly, its baryon number can be shown to be +1 and its spin ½. A positive pion is composed of a *u* quark and a $\bar{d}$ antiquark (written $u\bar{d}$). Its charge is ⅔ + ⅓, or +1; its spin and baryon number are both 0.

The third quark, *s,* is needed only to construct strange particles, and indeed it provides an explicit definition of strangeness: A strange particle is one that contains at least one *s* quark or $\bar{s}$ antiquark. The lambda baryon, for example, can be shown from the charge distribution of its multiplet to have a strangeness of −1; that result is confirmed by its quark constitution of *uds.* Similarly, the neutral *K* meson, a strange particle, has a strangeness of + 1, as confirmed by its composition of $d\bar{s}$.

Until quite recently these three kinds of quark were sufficient to describe all the known hadrons. As we shall see, experiments conducted during the past year seem to have created hadrons whose properties cannot be explained in terms of the original three quarks. The experiments can be interpreted as implying the existence of a fourth kind of quark, called the charmed quark and designated *c.*

The statement that the *u, d* and *s* quarks are sufficient to construct all the observed hadrons can be made more precisely in the mathematical formalism of the eightfold way. Since a meson is made up of one quark and one antiquark, and since there are three kinds, or flavors, of quark, there are nine possible combinations of quarks and antiquarks that can form a meson. It can be shown that one of these combinations represents a singlet and the remaining eight form an octet. Similarly, since a baryon is made up of three quarks, there are 27 possible combinations of quarks that can make up a baryon. They can be broken up into a singlet, two octets and a decimet. Those groupings correspond exactly to the observed families of hadrons. The quark theory thus explains why only a few of the possible representations of SU(3) are realized in nature as hadron supermultiplets.

The quark rules provide a remarkably economical explanation for the formation of the observed hadron families. What principles, however, can explain the quark rules, which seem quite arbitrary? Why is it possible to bind together three quarks but not two or four? Why can we not create a single quark in isolation? A line of thought that leads to possible answers to these questions appeared at first as a defect in the quark theory.

As we have seen, it is necessary that the quarks have half-integral values of spin angular momentum; otherwise the known spins of the baryons and mesons would be predicted wrongly. Particles with half-integral spin are expected to obey Fermi-Dirac statistics and are therefore subject to the Pauli exclusion principle: No two particles within a particular system can have exactly the same quantum numbers. Quarks, however, seem to violate the principle. In making up a baryon it is often necessary that two identical quarks occupy the same state. The omega particle, for example, is made up of three *s* quarks, and all three must be in precisely the same state. That is possible only for particles that obey Bose-Einstein statistics. We are at an impasse: quarks must have half-integral spin but they must satisfy the statistics appropriate to particles having integral spin.

The connection between spin and statistics is an unshakable tenet of relativistic quantum mechanics. It can be deduced directly from the theory, and a violation has never been discovered. Since it holds for all other known particles, quarks could not reasonably be excluded from its domain.

The concept that has proved essential to the solution of the quark statistics problem was proposed in 1964 by Oscar W. Greenberg of the University of Maryland. He suggested that each flavor of quark comes in three varieties, identical in mass, spin, electric charge and all other measurable quantities but different in an additional property, which has come to be known as color. The exclusion principle could then be satisfied, and quarks could remain fermions, because the quarks in a baryon would not all occupy the same state. The quarks could differ in color even if they were the same in all other respects.

The color hypothesis requires two additional quark rules. The first simply restates the condition that color was introduced to satisfy: Baryons must be made up of three quarks, all of which have different colors. The second describes the application of color to mesons: Mesons are made of a quark and an antiquark of the same color, but with equal representation of each of the three colors. The effect of these rules is that no hadron can exhibit net color. A baryon invariably contains quarks of each of the three colors, say red, yellow and blue. In the meson one can imagine the quark and antiquark as being a single color at any given moment, but continually and simultaneously changing color, so that over any measurable interval they will both spend equal amounts of time as red, blue and yellow quarks.

The price of the color hypothesis is a tripling of the number of quarks; there must be nine instead of three (with charm yet to be considered). At first it may also appear that we have greatly increased the number of hadrons, but that is an illusion. With color there seem to be nine times as many mesons and 27 times as many baryons, but the rules for assembling hadrons from colored quarks ensure that none of the additional particles are observable.

Although the quark rules imply that we will never see a colored particle, the color hypothesis is not merely a formal construct without predictive value. The increase it requires in the number of quarks can be detected in at least two ways. One is through the effect of color on the lifetime of the neutral pion, which almost always decays into two photons. Stephen L. Adler of the Institute for Advanced Study has shown that its rate of decay depends on the square of the number of quark colors. Just the observed lifetime is obtained by assuming that there are three colors.

Another effect of color can be detected in experiments in which electrons and their antiparticles, the positrons, annihilate each other at high energy. The outcome of such an event is sometimes a group of hadrons and sometimes a muon and an antimuon. At sufficiently high energy the ratio of the number of hadrons to the number of muon-antimuon pairs is expected to approach a constant value, equal to the sum of the squares of the charges of the quarks. Tripling the number of quarks also triples the expected value of the ratio. The experimental result at energies of from 2 GeV to 3 GeV is in reasonable agreement with the color hypothesis (which predicts a value of 2) and is quite incompatible with the original theory of quarks without color.

The introduction of the color quantum number solves the problem of quark statistics, but it once again requires a set of rules that seem arbitrary. The rules can be accounted for, however, by establishing another hypothetical symmetry group analogous to the SU(3) symmetry proposed by Gell-Mann and by Ne'eman. The earlier SU(3) is concerned entirely with combinations of the three quark flavors; the new one deals exlusively with the three quark colors. Moreover, unlike the earlier theory, which is only approximate, color SU(3) is supposed to be an exact symmetry, so that quarks of the same flavor but different color will have identical masses.

In the color SU(3) theory all the quark rules can be explained if we accept one postulate: All hadrons must be represented by color singlets; no larger multiplets can be allowed. A color singlet can be constructed in two ways: by combining an identically colored quark and antiquark with all three colors equally represented, or by combining three quarks or three antiquarks in such a way that the three colors are all included. These conditions, of course, are equivalent to the rules for building mesons, baryons and antibaryons, and they ensure that all hadrons will be colorless. There are no other ways to make a singlet in color SU(3); a particle made any other way would be a member of a larger multiplet, and it would display a particular color.

Although the color SU(3) theory of the hadrons can explain the quark rules, it cannot entirely eliminate the arbitrary element in their nature. We can ask a still more fundamental question: What explains the postulate that all hadrons must be color singlets? One approach to an answer, admittedly a speculative one, has been suggested recently by many investigators; it incorporates the color SU(3) model of the hadrons into one of the class of theories called gauge theories.

The color gauge theory postulates the existence of eight massless particles, sometimes called gluons, that are the carriers of the strong force, just as the photon is the carrier of the electromagnetic force. Like the photon, they are electrically neutral, and they have a spin of 1; they are therefore called vector bosons (bosons because they have integer spin and obey Bose-Einstein statistics, vector because a particle with a spin of 1 is described by a wave function that takes the form of a four-dimensional vector). Gluons, like quarks, have not been detected.

When a quark emits or absorbs a gluon, the quark changes its color but not its flavor. For example, the emission of a gluon might transform a red *u* quark into a blue or a yellow *u* quark, but it could not change it into a *d* or an *s* quark of any color. Since the color gluons are the quanta of the strong force, it follows that color is the aspect of quarks that is most important in the strong interactions. In fact, when describing interactions that involve only the strong force, one can virtually ignore the flavors of quarks.

The color gauge theory proposes that the force that binds together colored quarks represents the true character of the strong interaction. The more familiar strong interactions of hadrons (such as the binding of protons and neutrons in a nucleus) are manifestations of the same fundamental force, but the interactions of colorless hadrons are no more than a pale remnant of the underlying interaction between colored quarks. Just as the van der Waals force between molecules is only a feeble vestige of the electromagnetic force that binds electrons to nuclei, the strong force observed between hadrons is only a vestige of that operating within the individual hadron.

From these theoretical arguments one can derive an intriguing, if speculative, explanation of the confinement of quarks. It has been formulated by John Kogut and Kenneth Wilson of Cornell University and by Leonard Susskind of Yeshiva University. If it should be proved correct, it would show that the failure to observe colored particles (such as isolated quarks and gluons) is not the result of any experimental deficiency but is a direct consequence of the nature of the strong force.

The electromagnetic force between two charged particles is described by Coulomb's law: The force decreases as the square of the distance between the charges. Gravitation obeys a fundamentally similar law. At large distances both forces dwindle to insignificance. Kogut, Wilson and Susskind argue that the strong force between two colored quarks behaves quite differently: it does not diminish with distance but remains constant, independent of the separation of the quarks. If their argument is sound, an enormous amount of energy would be required to isolate a quark.

Separating an electron from the valence shell of an atom requires a few electron volts. Splitting an atomic nucleus requires a few million electron volts. In contrast to these values, the separation of a single quark by just an inch from the proton of which it is a constituent would require the investment of 10^{13} GeV, enough energy to separate the author from the earth by some 30 feet. Long before such an energy level could be attained another process would intervene. From the energy supplied in the effort to extract a single quark, a new quark and antiquark would materialize. The new quark would replace the one removed from the proton, and would reconstitute that particle. The new antiquark would adhere to the dislodged quark, making a meson. Instead of isolating a colored quark, all that is accomplished is the creation of a colorless meson. . . . By this mechanism we are prohibited from ever seeing a solitary quark or a gluon or any combination of quarks or gluons that exhibits color.

If this interpretation of quark confinement is correct, it suggests an ingenious way to terminate the apparently infinite regression of finer structures in matter. Atoms can be analyzed into electrons and nuclei, nuclei into protons and neutrons, and protons and neutrons into quarks, but the theory of quark confinement suggests that the series stops there. It is difficult to imagine how a particle could have an internal structure if the particle cannot even be created.

Quarks of the same flavor but different color are expected to be identical in all properties except color; indeed, that is why the concept of color was introduced. Quarks that differ in flavor, however, have quite different properties. It is because the *u* quark and the *d* quark differ in electric charge that the proton is charged and the neutron is not. Similarly, it is because the *s* quark is considerably more massive than either the *u* or the *d* quark that strange particles are generally the heaviest members of their families. The charmed quark, *c,* must be heavier still, and charmed particles as a rule should therefore be heavier than all others. It is the flavor of quarks that brings variety to the world of hadrons, not their color.

As we have seen, the flavors of quarks are unaffected by the strong interactions. In a weak interaction, on the other hand, a quark can change its flavor (but not its color). The weak interactions also couple quarks to the leptons. The classical example of this coupling is nuclear beta decay, in which a neutron is converted into a proton with the emission of an electron and an antineutrino. In terms of quarks the reaction represents the conversion of a *d* quark to a *u* quark, accompanied by the emission of the two leptons.

The weak interactions are thought to be mediated by vector bosons, just as the strong and the electromagnetic interactions are. The principal one, labeled W and long called the intermediate vector boson, was predicted in 1938 by Hideki Yukawa. It has an electric charge of −1, and it differs from the photon and the color gluons in that it has mass, indeed a quite large mass. Quarks can change their flavor by emitting or absorbing a *W* particle. Beta decay, for example, is interpreted as the emission of a *W* by a *d* quark, which converts the quark into a *u;* the *W* then decays to yield the electron and antineutrino. From this process it follows that the *W* can also interact with leptons, and it thus provides a link between the two groups of apparently elementary particles.

The realization that the strong, weak and electromagnetic forces are all carried by the same kind of particle—bosons with a spin of 1—invites speculation that all three might have a common basis in some simple unified theory. A step toward such a unification would be the reconciliation of the weak interactions and electromagnetism. Julian Schwinger of Harvard University attempted such a unification in the mid-1950's (when I was one of his doctoral students, working on these very questions). His theory had serious flaws. One was eliminated in 1961, when I introduced a second, neutral vector boson, now called *Z*, to complement the electrically charged *W*. Other difficulties persisted for 10 years, until in 1967 Steven Weinberg of Harvard and Abdus Salam of the International Center for Theoretical Physics in Trieste independently suggested a resolution. By 1971 it was generally agreed, largely because of the work of Gerard 't Hooft of the University of Utrecht, that the Weinberg-Salam conjecture is successful. . . .

Through the unified weak and electromagnetic interactions, quarks and leptons are intimately related. These interactions "see" the four leptons and distinguish between the three quark flavors. The *W* particle can induce one kind of neutrino to become an electron and the other kind of neutrino to become a muon. Similarly, the *W* can convert a *u* quark into a *d* quark; it can also influence the *u* quark to become an *s quark, although much less readily.*

*There is an obvious lack of sym*metry in these relations. The leptons consist of two couples, married to each other by the weak interaction: the electron with the electron neutrino and the muon with the muon neutrino. The quarks, on the other hand, come in only three flavors, and so one must remain unwed. The scheme could be made much tidier if there were a fourth quark flavor, in order to provide a partner for the unwed quark. Both the quarks and the leptons would then consist of two pairs of particles, and each member of a pair could change into the other member of a same pair simply by emitting a *W*. The desirability of such lepton-quark symmetry led James Bjorken and me, among others, to postulate the existence of a fourth quark in 1964. Bjorken and I called it the charmed quark. When provisions are made for quark colors, charm becomes a fourth quark flavor, and a new triplet of colored quarks is required. There are thus a total of 12 quarks.

Since 1964 several additional arguments for charm have developed. To me the most compelling of them is the need to explain the suppression of certain interactions called strangeness-changing neutral currents. An explanation that relies on the properties of the charmed quark was presented in 1967 by John Iliopoulos, Luciano Maiani and me.

Strangeness-changing neutral currents are weak interactions in which the net electric charge of the hadrons does not change but the strangeness does; typically an *s* quark is transformed into a *d* quark, and two leptons are emitted. An example is the decay of the neutral *K* meson (a strange particle) into two oppositely charged muons. Such processes are found by experiment to be extremely rare. The three-quark theory cannot account for their suppression, and in fact the unified theory of weak and electromagnetic interactions predicts rates more than a million times greater than those observed.

The addition of a fourth quark flavor with the same electric charge as the *u* quark neatly accounts for the suppression, although the mechanism by which it does so may seem bizarre. With two pairs of quarks there are two possible paths for the strangeness-changing interactions, instead of just one when there are only three quarks. In the macroscopic world the addition of a second path, or channel, would be expected always to bring an increase in the reaction rate. In a world governed by quantum mechanics, however, it is possible to subtract as well as to add. As it happens, a sign in the equation that defines one of the reactions is negative, and the two interactions cancel each other.

The addition of a fourth quark flavor must obviously increase the number of hadrons. In order to accommodate the newly predicted particles in supermultiplets the eightfold way must be expanded. In particular another dimension must be added to the graphs employed to represent the families, so that the plane figures of the earlier symmetry become Platonic and Archimedean solids.

To the meson octet are added six charmed particles and one uncharmed particle to make up a new family of 15. It is represented as a cuboctahedron, in which one plane contains the hexagon of the original uncharmed meson octet. The baryon octets and decimet are expected to form two families having 20 members each. They are represented as a tetrahedron truncated at each vertex and as a regular tetrahedron. In addition there is a smaller regular tetrahedron consisting of just four baryons. Again, each figure contains one plane of uncharmed particles. . . .

It now appears that the first of the new particles to be discovered is a meson that is not itself charmed. That conclusion is based on the assumption that the predicted meson is the same particle as the *J* or psi particle discovered last November. The announcement of the discovery was made simultaneously by Samuel C. C. Ting and his colleagues at the Brookhaven National Laboratory and by Burton Richter, Jr., and a group of other physicists at the Stanford Linear Accelerator Center (SLAC). At Brookhaven it was named *J,* at Stanford psi. Here I shall adopt the name *J*. For two excited states of the same particle, however, the names psi′ and psi″ will be employed, since they were seen only in the SLAC experiments.

The *J* particle was found as a resonance, an enhancement at a particular energy in the probability of an interaction between other particles. At Brookhaven the resonance was detected in the number of electron-positron pairs produced in collisions between protons and atomic nuclei. At SLAC it was observed in the products of annihilations of electrons and positrons. The energy at which the resonances were observed—and thus the energy or mass of the *J* particle— is about 3.1 GeV. . . .

The *J* particle decays in about 10^{-20} second, certainly a brief interval, but nevertheless 1,000 times longer than the expected lifetime of a particle having the *J*'s mass. The considerable excitement generated by the discovery of the *J* was largely a result of its long lifetime.

A great many explanations of the particle were proposed; for example, it was suggested that it might be the Z. I believe there is good reason to interpret the *J* as being a meson made up of a charmed quark and a charmed antiquark, that is, a meson with the quark constitution $c\bar{c}$. Thomas Appelquist and H. David Politzer of Harvard have named such a meson "charmonium," by analogy to positronium, a bound state of an electron and a positron. Charmonium is without charm because the charm quantum numbers of its quarks (+1 and −1) add up to zero.

The charmonium hypothesis can account for the anomalous lifetime of the *J* if one considers the ultimate fate of the decaying particle's quarks. There are three possibilities: they can be split up to become constituents of two daughter hadrons, they can both become part of a single daughter particle or they can be annihilated. An empirical rule, first noted by Zweig, states that decays of the first kind are allowed but the other two are suppressed. For the *J* particle to decay in the allowed manner it must create two charmed particles, that is, two hadrons, one containing a charmed quark and the other a charmed antiquark. That decay is possible only if the mass of the *J* is greater than the combined masses of the charmed daughter particles. There is reason to believe the lightest charmed particle has a mass greater than half of the mass of the *J*, and it therefore appears that the *J* cannot decay in the allowed mode. The *J* cannot decay in the second way, either, keeping both its quarks in a single particle, because the *J* is the least massive state containing a charmed quark and a charmed antiquark. It must therefore decay by the annihilation of its quarks, a decay suppressed by Zweig's rule. The suppression offers a partial explanation for the particle's extended lifetime.

Zweig's rule was formulated to explain the decay of the phi meson, which is made up of a strange quark and a strange antiquark and has a mass of about 1 GeV. The two particles are closely analogous, but the decay of the *J* is appreciably slower than that of the phi. Why should Zweig's rule be more effective for *J* than it is for phi? Furthermore, what explains Zweig's arbitrary rule?

A possible answer is provided by the theoretical concept called asymptotic freedom, which holds that the strong interactions become less strong at high energy. At sufficiently high energy the proton behaves as if it were made up of three freely moving quarks instead of three tightly bound ones. The concept takes its name from the fact that the quarks approach the state of free motion asymptotically as the energy is increased. Asymptotic freedom offers an explanation for the discrepancy between the phi and the *J* particles in the application of Zweig's rule. Because the *J* is so massive, or alternatively so energetic, the strong interaction is of diminished strength, and it is particularly difficult for the quark and the antiquark to annihilate each other.

Like postitronium, charmonium should appear in many energy states. Two were discovered at SLAC soon after the first state was found; they are psi′, with a mass of about 3.7 GeV, and psi″, with a mass of about 4.1 GeV. They appear to be simple excited states of the lowest-lying state of charmonium, the *J* particle. Psi′ decays only a little more quickly than *J*, and half the time its decay products are the *J* particle itself and two pions. Thus it sometimes decays by the second suppressed process described by Zweig's rule, that is, by contributing both of its quarks to a single daughter particle. The extended lifetime implies that psi′ also lies below the energy threshold for the creation of a pair of charmed particles.

Psi″ decays much more quickly and therefore must be decaying in some mode permitted by Zweig's rule. Its decay products have not yet been determined, but it is possible they include charmed hadrons.

Numerous other excited states of charmonium follow inevitably from the theory of quark interactions. . . . One, called *p*-wave charmonium, is formed when the particle takes on an additional unit of angular momentum. Some fraction of the time psi′ should decay into *p*-wave charmonium, which should subsequently decay predominantly to the ground state, *J*. At each transition a photon of characteristic energy must be emitted. Recent experiments at the DORIS particle-storage rings of the German Electron Synchrotron in Hamburg have apparently detected the decays associated with the *p*-wave particle. In a few percent of its decays psi′ yields the *J* particle and two photons, with energies of .2 GeV and .4 GeV. At SLAC psi′ has been found to decay into an intermediate state and a single photon with an energy of .2 GeV. The intermediate state, which is presumably the same particle as the one observed at DORIS, then decays directly into hadrons.

The correspondence of theory and experiment revealed by the discovery of the *p*-wave transitions inspires considerable confidence that the charmonium interpretation of the *J* particle is correct. There is at least one more predicted state, called paracharmonium, that must be found if this explanation of the particle is to be confirmed. It differs from the observed states in the orientation of the quark spins: in *J*, psi′ and psi″ (collectively called orthocharmonium) they are parallel; in paracharmonium they are antiparallel. Paracharmonium has so far evaded detection, but if the theoretical description is to make sense, paracharmonium must exist.

In addition to the various states of (uncharmed) charmonium, all the predicted charmed particles must also exist. If the *J* is in fact a state of charmonium, we can deduce from its mass the masses of all the hadrons containing charmed quarks.

An important initial constraint on the range of possible masses was provided by the interpretation of the suppression of strangeness-changing neutral currents. If the suppression mechanism is to work, the charmed quark cannot be too much heavier than its siblings. On the other hand, it cannot be very light or charmed hadrons would already have been observed. An estimate from these conditions suggested that charmed particles would be found to have masses of about 2 or 3 GeV.

After the discovery of the *J*, I performed a more formal analysis with my colleagues at Harvard, Alvaro De Rújula and Howard Georgi. So did many others. Our estimates indicate that the least massive charmed states are mesons made up of a c quark and a $\bar{u}$ or $\bar{d}$ antiquark; their mass should fall between 1.8 GeV and 2.0 GeV. A value within that range could be in agreement with the supposition that psi′ lies below the threshold for the creation of a pair of charmed mesons, but psi″ lies above it.

The least massive charmed baryon has a quark composition of *udc;* we predict that its mass is near 2.2 GeV. As might be expected, since the c quark is the heaviest of the four, the most massive predicted charmed hadron is the *ccc* baryon. We estimate its mass at about 5 GeV.

An important principle guiding experimental searches for charmed hadrons is the requirement that in most kinds of interaction charmed particles can be created only in pairs. Two hadrons must be produced, one containing a charmed quark, the other a charmed antiquark; the obvious consequence is a doubling of the energy required to create a charmed particle. An important exception to this rule is the interaction of neutrinos with other kinds of particles, such as protons. Neutrino events are exempt because neutrinos have only weak interactions and quark flavor can be changed in weak processes. Many experimental techniques have been tried in the search for charm during the past 10 years, yet no charmed particle has been unambiguously identified. Nevertheless, two recent experiments, both involving neutrino interactions, are encouraging. In both charm may at last have appeared, but even if that proves to be an illusion, the experiments suggest promising lines of research.

One of the experiments was conducted at the Fermi National Accelerator Laboratory in Batavia, Ill., by a group of physicists headed by David B. Cline of the University of Wisconsin, Alfred K. Mann of the University of Pennsylvania and Carlo Rubbia of Harvard. In examining the interactions of high-energy neutrinos they found that in several percent of the events the products included two oppositely charged muons. One of the muons could be created directly from the incident neutrino, but the other is difficult to account for with only the ensemble of known, uncharmed particles. The most likely interpretation is that a heavy particle created in the reaction decays by the weak force to emit the muon. The particle would have a mass of between 2 and 4 GeV, and if it is a hadron, some explanation must be found for its weak decay. Most particles with masses that large decay by the strong force. The presence of a charmed quark in the particle might provide the required explanation.

The second experiment was performed at Brookhaven by a group of investigators under Nicholas P. Samios. They photographed the tracks resulting from the interaction of neutrinos with protons in a bubble chamber. In a sample of several hundred observed collisions one photograph seemed to have no conventional interpretation. . . . The final state can be construed as the decay products of a charmed baryon. The process would provide convincing evidence for the existence of charm if it were not attested to by only one event. A few more observations of the same reaction would settle the matter.

It would be misleading to give the impression that the description of hadrons in terms of quarks of three colors and four flavors had solved all the outstanding problems in the physics of elementary particles. For example, continuing measurements of the ratio of hadrons to muon pairs produced in electron-positron annihilations have confounded prediction. The ratio discriminates between various quark models, and an argument in support of the color hypothesis was that at energies of from 2 to 3 GeV the ratio is about 2. At higher energy, high enough for charmed hadrons to be created in pairs, the ratio was expected to rise from 2 to about 3.3. The ratio does increase, but it overshoots the mark and appears to stabilize at a value of about 5. Perhaps charmed particles are being formed, but it seems that something else is happening as well: some particle is being made that does not appear in the theory I have described. One of my colleagues at Harvard, Michael Barnett, believes we have not been ambitious enough. He invokes six quark flavors rather than four, so that there are three flavors of charmed quark. It is also possible there are heavier leptons we know nothing about.

Finally, even if a completely consistent and verifiable quark model could be devised, many fundamental questions would remain. One such perplexity is implicit in the quark-lepton symmetry that led to the charm hypothesis. Both the quarks and the leptons, all of them apparently elementary, can be divided into two subgroups. In one group are the *u* and *d* quarks and the electron and electron neutrino. These four particles are the only ones needed to construct the world; they are sufficient to build all atoms and molecules, and even to keep the sun and other stars shining. The other subgroup consists of the strange and charmed quarks and the muon and muon neutrino. Some of them are seen occasionally in cosmic rays, but mainly they are made in high-energy particle accelerators. It would appear that nature could have made do with half as many fundamental things. Surely the second group was not created simply for the entertainment or edification of physicists, but what is the purpose of this grand doubling? At this point we have no answer.

R.9

The Nature of Elementary Particles

WERNER HEISENBERG

THE LIFE OF Werner Heisenberg, one of the truly great names in modern physics, has been described elsewhere in this volume. The paper selected here was adapted from a translation of his opening lecture to the German Physical Society's spring meeting of 1975, and the original version was published in the German journal *Naturwissenschaften* for February 1976.

This selection, like much of Heisenberg's writings, incorporates considerations about physics and philosophy, viewing the latter as necessary to a better understanding of the former. Heisenberg views philosophy as a guide for posing the right questions, which then enable physics to be defined better and developed more efficiently. While favoring the quark hypothesis and the search for constituent particles of matter, Heisenberg concludes that there is little difference between elementary particles and compound systems since the latter are merely various combinations of the former. Heisenberg also discusses what he refers to as the "spectrum of matter," an exhaustive representation of the various states that can be assumed by combinations of elementary particles. He concludes with a discussion of the dynamics of matter and reviews the history of particle physics, which has centered around deciding what happens when a particular object is divided into smaller parts. Yet he suggests that we may have reached the ultimate limits of particle physics, since further attempts at division on the subatomic level have become essentially meaningless. Instead, he regards it as more useful to search for patterns of symmetry in elementary particle groups and to focus on the dynamics of matter.

If the world were good for nothing else, it is a fine subject for speculation.

—WILLIAM HAZLITT, *Characteristics*

The mathematically formulated laws of quantum theory show clearly that our ordinary intuitive concepts cannot be unambiguously applied to the smallest particles. All the words or concepts we use to describe ordinary physical objects, such as position, velocity, color, size, and so on, become indefinite and problematic if we try to use them of elementary particles.

—WERNER HEISENBERG, *Across the Frontiers*

THE QUESTION, "What is an elementary particle?" must find its answer primarily in experiment, although it must also be confronted with philosophical considerations. I will therefore begin by giving a short survey of the important experimental results of the last fifty years. This survey will show that a critical unbiased study of these results already gives an answer to the question; theory, as we shall see, cannot add much to this answer.

Next I will deal with the philosophical problems that arise in connection with the concept of an elementary particle. It may be objected that in this question we should concentrate on physics rather than on philosophy. But this separation is not so simple. In fact, I believe that certain erroneous developments in particle theory—and I am afraid that such developments do exist—are caused by a misconception by some physicists that it is possible to avoid philosophical arguments altogether. Starting with poor philosophy, they pose the wrong questions. It is only a slight exaggeration to say that good physics has at times been spoiled by poor philosophy.

Finally I will discuss these problematic developments. Having witnessed similar mistakes in the development of quantum mechanics fifty years ago, I am in a position to make some suggestions to avoid such errors in the future. This will lead us to conclude on an optimistic note.

Werner Heisenberg, "The Nature of Elementary Particles." Reprinted with permission from *Physics Today.* Vol. 29, No. 3, pp. 32-39 (1976).

PARTICLE NUMBER NOT CONSERVED

Let us start with the experimental facts. Nearly fifty years have passed since P. A. M. Dirac predicted, on the basis of his theory of the electron, the existence of its antiparticle, the positron: A few years later the existence of these was demonstrated experimentally by Carl Anderson and P. M. S. Blackett. They produced them in pair creation, forming what we now call "antimatter" artificially.

This was a discovery of prime importance. Before this time it was assumed that there were two fundamental kinds of particles, electrons and protons, which, unlike most other particles, were immutable. Therefore their number was fixed and they were referred to as "elementary" particles. Matter was seen as being ultimately constructed of electrons and protons. The experiments of Anderson and Blackett provided definite proof that this hypothesis was wrong. Electrons can be created and annihilated; their number is not constant; they are not "elementary" in the original meaning of the word.

The next important step was the discovery of artificial radioactivity by Frédéric Joliot and Irène Curie. From many experiments it became clear that an atomic nucleus can be transmuted into another nucleus by emitting particles, provided the laws of conservation of energy, angular momentum, electric charge and so on permit this transmutation. The transformation of energy into matter, predicted as a possibility very early in the theory of special relativity, has become recognized as a rather common phenomenon. Contrary to the earlier views, there was no conservation of particle number. However, there are physical properties that can be characterized by quantum numbers, for instance angular momentum and electric charge; these quantum numbers may assume positive or negative values and are subject to laws of conservation.

The 1930's brought a few other important experimental discoveries. In cosmic radiation very energetic particles are observed. A cosmic-ray particle, when colliding with another particle such as an atomic nucleus in a photographic emulsion, can produce a shower of many secondary particles. For some time many physicists believed that such showers can be produced only by cascades in the interior of a heavy nucleus. Later it became clear, however, that even the collision of two protons can lead to the production of many secondaries in one step. In the late 1940's Cecil Powell discovered the pions, which play the main role in this process of multiple production of particles. His results emphasized again that the transmutation of energy into matter is the decisive process, and that it would be meaningless to speak about the "division" of the original particles. Experimentally, the concept of "dividing" had lost its meaning.

This new situation was confirmed again and again in the experiments of the 1950's and 1960's; many new particles of various lifetimes were discovered and there was no answer to the question, "What do these particles consist of?" A proton could be obtained from a neutron and a pion, or from a Λ hyperon and a kaon, or from two nucleons and one antinucleon, and so on. Could we therefore simply say a proton consists of continuous matter? Such a statement would be neither right nor wrong: There is no difference in principle between elementary particles and compound systems. This is probably the most important experimental result of the last fifty years.

This development convincingly suggests the following analogy: Let us compare the so-called "elementary" particles with the stationary states of an atom or a molecule. We may think of these as various states of one single molecule or as the many different molecules of chemistry. One may therefore speak simply of the "spectrum of matter." Experiments in the 1960's and 1970's with large accelerators have demonstrated that this picture fits elementary particles as well. Like the stationary states of atoms, the elementary particles can be characterized by quantum numbers—that is, by their behavior under certain transformations. The corresponding laws of conservation determine what transmutations are possible. For an excited hydrogen atom it is its behavior under rotation that determines whether it can fall into a lower state with the emission of a light quantum. In the same way for a φ boson, it is its symmetry properties that determine whether it can disintegrate with the emission of a pion into a ρ boson.

The stationary states of atoms have very different lifetimes, and the same is true of particles. The ground state of an atom has an infinite lifetime, and there are many particles with this property, including the electron, the proton and the deuteron. But these stable particles are no more elementary than the unstable ones—the ground state of hydrogen is a solution of the same Schrödinger equation as any of the excited states. Similarly, the electron and the proton are no more elementary than the Λ hyperon.

During recent years experimental particle physics has applied itself to tasks similar to those of spectroscopy in the early 1920's. Just as at that time all stationary states of atoms were collected in large tables, the "Paschen-Götze" volumes, so nowadays the "reviews of particle properties" every year collect new or improved data on the masses and quantum numbers of particles. This kind of work corresponds to the astronomical surveys, and obviously every observer hopes occasionally to find an especially interesting object in his sector.

TWO TYPES OF BROKEN SYMMETRY

There are nevertheless certain characteristic differences between the physics of atomic shells and particle physics. In the shells the relevant energies are small, so the typical features of the theory of relativity play no role, and their behavior can be described at least approximately by non-relativistic quantum mechanics. Hence in shell physics and in particle physics the underlying symmetry groups are different. The Galilean group in shell physics is replaced by the Lorentz group in particle physics. Other groups have to be added, such as the isospin group, which is isomorphic to SU_2; then SU_3, the scale group and others. It is an important experimental proposition to determine all groups relevant to particle physics, and this problem has been solved to a large extent during the past twenty years.

From the physics of atomic shells we can learn that among those groups that describe only approximate symmetries in nature, two essentially different types can be distinguished. In the optical spectra of atoms, for example, the groups O_3 and $O_3 \times O_3$ play very different roles. The fundamental equations of quantum mechanics of atoms are strictly invariant under O_3. Therefore the stationary states with higher angular momenta are strictly degenerate, so that there are always several states with exactly the same energy. Only when external electromagnetic fields are applied do these states split and the well known fine structures of the Zeeman effect or the Stark effect appear.

A similar effect can be produced in systems in which the ground state is not invariant under rotations, such as a crystal or a ferromagnet: The two directions of electron spin in a ferromagnet do not belong to precisely the same energy. In this case, according to a well known theorem of Goldstone, bosons must also exist, with energy that tends to zero with increasing wavelength; in the case of ferromagnetism the spin waves of Bloch, "magnons," take the place of the Goldstone waves.

An entirely different situation is met with in the group $O_3 \times O_3$, which produces the well known multiplets in the optical spectra. The group $O_3 \times O_3$ is only an approximate symmetry, which comes about if the spin-orbit interactions become small in a certain part of the spectrum, so that the orbits and the spins of the particles can be rotated almost independently. The symmetry $O_3 \times O_3$ results from the dynamics of the system and is useful only in certain parts of the spectrum. Empirically the two types of broken symmetries can be distinguished by the existence or nonexistence of the Goldstone modes. If they are found, it is plausible to assume that the degeneracy of the ground state plays an important role.

If we apply the experience of shell physics to particle physics, the experiments suggest that the Lorentz group and the group SU_2 should be interpreted as fundamental symmetries of the underlying natural law. Electromagnetism and gravitation then appear as those long-range effects that, according to Goldstone, are connected with the broken symmetry of the ground state. The more complicated groups such as SU_3, SU_4, SU_6, $SU_2 \times SU_2$ or $SU_3 \times SU_3$, should be taken as dynamical symmetries, like $O_3 \times O_3$ in shell physics. We may doubt whether the dilatation group or scale group should be counted among the fundamental symmetries. They are disturbed by the existence of particles of finite mass and by the gravitational influence of the big masses in the universe. Because of their close mathematical relation to the Lorentz group they probably belong to the fundamental group. This particular coordination of the broken empirical symmetries with the two proposed types of broken symmetries is suggested by the existing experimental evidence, but it may not yet be finally settled. It is important to emphasize that, for any symmetry group arising out of the phenomenological analysis of the spectrum, the question must be asked—and if possible answered—to which of these two types it belongs.

Another special feature of shell physics should be mentioned. Among the optical spectra there are non-combining or weakly combining term systems such as the parahelium and the orthohelium spectrum. In particle physics one could perhaps compare the division of the fermion spectrum into baryons and leptons with this feature.

It is evident that the analogy between the stationary states of an atom or molecule on one hand and the particles of high-energy physics on the other is nearly complete, qualitatively answering the question about the nature of the "elementary particle." But only qualitatively. For the theoretician the further question arises whether this interpretation can be based on quantitative calculations. Here we must answer a preliminary question: What does it mean to "understand a spectrum quantitatively"?

DYNAMICS AND CONTINGENT CONDITIONS

A number of examples, both from classical physics and from quantum mechanics, teach the general procedure by which we acquire physical understanding. Let us think of the spectrum of elastic vibrations of a steel plate. If a qualitative theoretical interpretation is not enough, we have to start with the elastic properties of the steel plate, which should be represented mathematically. When this has been done, we must add boundary conditions that tell whether the plate is a circle or a square, and whether it is stretched in a frame or free. This knowledge then should be sufficient to calculate, at least in principle, the spectrum of the acoustic vibrations. It is true that, because of the high degree of complication, we are frequently unable to calculate all the vibrational frequencies precisely, but only the lowest ones—those with the smallest number of nodes.

Hence two elements are prerequisite to a quantitative understanding of the spectrum: a precise knowledge of the elastic behavior of the plate and the boundary conditions. The latter may be called "contingent" because they depend on the particular circumstances—the plate could have been cut differently. A similar case would be the electro-dynamical vibrations of a cavity. Maxwell's equations determine the dynamical behavior, and the shape of the cavity defines the boundary conditions. Another comparable situation is met with in the optical spectrum of the iron atom. The Schrödinger equation for a system of one nucleus and 26 electrons determines the dynamical behavior, while the boundary condition determines that the wave function vanishes at infinity. If the atom were enclosed in a small box the spectrum would be different.

To relate these results to particle physics, our first problem must be to determine experimentally the dynamical properties of the system "matter" and to formulate them mathematically. Then the contingent element, the boundary conditions, has to be added. These contain in this case statements about the so-called "empty space": the cosmos and its symmetry properties. In other words: The first step must be the attempt to formulate mathematically a natural law that defines the dynamics of matter. The second step is to determine the boundary conditions. Without such conditions the spectrum can not be defined. We might, for example, conjecture that inside a black hole the spectrum of elementary particles would be quite different from the spectrum in normal space—but unfortunately we can not check this by experiments!

Let me add a word about that decisive first step, the formulation of the governing dynamical law. There are pessimists among the particle physicists who believe that no such natural law defining the dynamics of matter exists. This view appears quite absurd to me. There must be some clearly defined dynamics of matter or there could be no spectrum; therefore a mathematical description should be possible. The pessimistic view implies that particle physics aims at no other goal than presenting an immense volume of particle data, a "super review of particle properties." In this super volume nothing could be understood since there is no dynamics of matter and so the volume would scarcely be read.

But I would like to emphasize strongly that I can not see any reason for such pessimism. A spectrum with sharp lines is observed, and this should imply a well defined dynamics of matter. The experimental results mentioned in the beginning give definite hints as to the fundamental invariances of the underlying natural law, and from the dispersion relations a lot is known about the degree of causality this law contains. Essential parts of the natural law therefore already belong to our definite knowledge. Because so many other spectra in physics have finally come to be understood quantitatively, this should be also possible here—in spite of the high degree of complication. It is on account of this intricacy that I will not discuss here the special proposal that Wolfgang Pauli and I made many years ago for the mathematical formulation of the underlying law, one I still regard as having the best chance of being the correct one. It is more important to emphasize that the formulation of such a law is the unavoidable precondition for a quantitative understanding of the spectrum. Anything else can not be called understanding; it would be scarcely more than looking up the table of data, and theoreticians at least should not be content with that.

PHILOSOPHICAL PROBLEMS

Taking this point of view, I am now going to discuss the philosophy that, whether consciously or unintentionally, has determined the direction of particle physics. For 2500 years philosophers and scientists have pondered the questions: "What happens if one tries to divide matter again and again? What are the smallest particles of matter?" Different philosophers have given very different answers, all of which have influenced the history of natural science. The best known answer is that of the philosopher Democritos: In the attempt to divide again and again one finally ends up with indivisible, unchangeable units, called atoms, of which all matter is composed. The positions and motions of the atoms determine the qualities of matter.

For Aristotle and his medieval successors, on the other hand, the concept of the smallest particle is not so well defined. It is true that for every kind of matter smallest particles are assumed—further division would change the characteristic qualities of the substance—but these smallest particles can be changed continuously like the substances themselves. Mathematically the substances can be divided *ad infinitum.* Matter is taken as continuous.

The clearest position against Democritos was taken by Plato. In his opinion the attempt to divide again and again results in mathematical forms: the regular bodies of stereometry, defined by their symmetry, and the triangles of which they are composed. These forms are not matter themselves, but they make up matter. For the element earth, for example, the characteristic body is the cube; for fire, the tetrahedron. What all these philosophies have in common is the attempt to deal with the antinomy of the infinitely small, which was discussed extensively by Immanuel Kant.

There have been even more naive attempts to rationalize this paradox. Some biologists have proposed the idea that within the seed of an apple an invisibly small apple tree is concealed, which again bears flowers and fruits; that the fruits again contain a still much smaller apple tree and so on *ad infinitum.* It was an amusing game, in the early days of the Bohr-Rutherford theory of the atom as small planetary system, to develop a similar idea: The electrons, the planets of the system, are inhabited by very small living bodies who build houses, cultivate their soil and study atomic physics—and find that their atoms are again small planetary systems, and so on in an unending progression.

In the background of every such fiction is Kant's antinomy: It is difficult to imagine that matter can be divided again and again, but it is equally difficult to imagine that this division must necessarily come to an end. As we now know, the paradox is caused by the erroneous assumption that our intuition can be applied to the smallest dimensions.

The strongest influence on the physics and chemistry of the last century undoubtedly came from the atomism of Democritos. This view allows an intuitive description of chemical processes on a small scale. Atoms can be compared with the mass points of Newtonian mechanics, and from this a satisfactory statistical theory of heat was developed. It is true that the atoms of the chemists turned out to be not mass points but small planetary systems, and the atomic nucleus likewise was a compound system formed of protons and neutrons. Nevertheless the electron, the proton and possibly the neutron could, it seemed, be considered as the genuine atoms, the indivisible building blocks, of matter. In this way the atomism of Democritos became an essential part of the materialistic interpretation of the world during the last century: Easily understood and intuitively plausible, it determined the way of thinking of even those physicists who insisted on not dealing with philosophy. At this point let me substantiate my earlier statement that, in the physics of elementary particles of our time, good physics has sometimes been unconsciously spoiled by poor philosophy.

We can not avoid using a language bound up with the traditional philosophy. We ask: "What does a proton consist of? Can an electron be divided or is it indivisible? Is a photon simple or compound?" But all these questions are wrongly put, because words such as "divide" or "consist of" have to a large extent lost their meaning. It must be our task to adapt our thinking and speaking—indeed our scientific philosophy—to the new situation created by the experimental evidence. Unfortunately this is very difficult. Wrong questions and wrong pictures creep automatically into particle physics and lead to developments that do not fit the real situation in nature. We will discuss these fallacies below.

But first a word should be added concerning the postulate that understanding requires a visual picture of the phenomena. Some philosophers have claimed that such pictures are the preconditions for understanding. For example, the philosopher Hugo Dingler of Munich held the view, in regard to the theory of relativity, that Euclidean geometry was the only possible correct geometry because we assume its correctness in building our measuring apparatus; on this latter point Dingler was right, of course. Therefore he argued that the experimental facts that are at the foundation of general relativity can not be described by a Riemannian, non-Euclidean geometry, because this would lead to contradictions. Here the postulate seems to be overdrawn: It is enough to know that, within the dimensions of our apparatus, Euclidean geometry applies with sufficient accuracy.

We will have to accept the fact that experimental data on a very large or a very small scale do not necessarily produce pictures, and we must learn to do without them. We then come to recognize that the antinomy of the smallest dimensions is solved in particle physics in a very subtle manner, of which neither Kant nor the ancient philosophers could have thought: The word "dividing" loses its meaning.

If we wish to compare the results of present-day particle physics with any of the old philosophies, the philosophy of Plato appears to be the most adequate: The particles of modern physics are representations of symmetry groups and to that extent they resemble the symmetrical bodies of Plato's philosophy.

WRONG QUESTIONS

My intention, however, is not to deal with philosophy but with physics. Therefore I will now discuss that development of theoretical particle physics that, I believe, begins with the wrong questions. First of all there is the thesis that the observed particles such as the proton, the pion, the hyperon consist of smaller particles: quarks, partons, gluons, charmed particles or whatever else, none of which have been observed. Apparently here the question was asked: "What does a proton consist of?" But the questioners appear to have forgotten that the phrase "consist of" has a tolerably clear meaning only if the particle can be divided into pieces with a small amount of energy, much smaller than the rest mass of the particle itself.

To demonstrate how a word that seems to be well defined can lose its meaning in special situations, I can not resist repeating a story that Niels Bohr loved to quote: A small boy enters a shop with two pennies in his hand and asks the grocer for two pennies' worth of mixed sweets. The grocer hands him two sweets and adds: "You can do the mixing yourself." The concept "consist of" for a proton has just as much meaning as the concept of "mixing" in the story of the boy.

At this point many readers may object that the quark hypothesis was derived from experimental material, namely from the empirical relevance of the SU_3 group, and that it has been successfully applied in the interpretation of many experiments even beyond the use of the SU_3 group. This can not be denied. But I would like to mention a counter-example from the history of quantum mechanics, of which I have been a witness; a counter-example that shows clearly the weakness of such arguments.

Before Bohr's theory of the atom many physicists believed that an atom must consist of harmonic oscillators. The optical spectrum contains sharp lines, and sharp lines can only be emitted by harmonic oscillators. The charges in these oscillators, however, would correspond to e/m values different from those of the electron, and it would be necessary to assume very many different oscillators, since there are very many lines in the spectrum.

Notwithstanding these difficulties Woldemar Voigt in Göttingen in 1912 developed a theory of the anomalous Zeeman effect of the D lines in the optical spectrum of sodium, on the following basis: He assumed two coupled oscillators which, without an external magnetic field, emitted the frequencies of the two D lines. The interaction between the two oscillators, and their coupling with an outer magnetic field could be arranged such that for weak fields the anomalous Zeeman effect, and for strong fields the Paschen-Back effect, could be reproduced correctly. For the intermediate range of medium fields the frequencies and intensities were described by long and complicated square roots, which, however, seemed to fit the experiments well.

Fifteen years later Pascual Jordan and I took the trouble to treat the same problem on the basis of the perturbational calculus of quantum mechanics. It came as a great surprise to us that we got exactly the old formulas of Voigt both for the frequencies and for the intensities, even for intermediate fields. The reason, as we understood later, was of purely formal mathematical nature. The perturbational calculus leads to a system of coupled linear equations and the frequencies are determined by the eigenvalues of the system. A system of coupled harmonic oscillators in classical theory also leads to a system of coupled linear equations. Since in Voigt's theory the important parameters had been adjusted to the empirical situation, it was not strange that the results ended up the same in the two cases. But Voigt's theory has not contributed to our understanding of atomic structure.

Why was this attempt by Voigt so successful on the one hand, and so useless on the other hand? Because Voigt intended to discuss only the D lines, without taking notice of the complete spectrum. Voigt phenomenologically used a special aspect of the oscillator hypothesis but ignored the other problems and difficulties of the model; at least he consciously left them undefined. He did not really take the oscillator hypothesis seriously. In the same way I am afraid that the quark hypothesis is not really taken seriously today by its proponents. Questions dealing with the statistics of quarks, the forces that keep them together, the reason why the quarks are never seen as free particles, the creation of pairs of quarks inside an elementary particle, are all left more or less undefined. If the quark hypothesis is really to be taken seriously it is necessary to formulate precise mathematical assumptions for the quarks and for the forces that keep them together and to show, at least qualitatively, that all these assumptions reproduce the known features of particle physics.

There should be no problem in particle physics to which these assumptions can not be applied. I do not know of any such attempt, and I am afraid that every attempt written down in precise mathematical language would easily lead to contradictions. My objections to the quark hypothesis can therefore be put in the form of questions:

▶Does the quark hypothesis contribute more to the understanding of the particle spectrum than Voigt's hypothesis contributed in an earlier period to the understanding of the structure of the atomic shells?

▶Do we not find behind the quark hypothesis the old idea—refuted long ago by experiments—that simple and compound particles can be distinguished?

NO MORE SURPRISES?

Let us now discuss a few special questions. If the SU_3 group plays an important role in the structure of the spectrum—and this appears from the experimental evidence to be the case—then it is important to decide whether SU_3 is a fundamental symmetry of the underlying natural law or a dynamical symmetry, which by its very nature can be only approximately valid. If this decision is left open, the other assumptions concerning the dynamics of the system are left open as well, and no understanding can be gained. The higher symmetries such as SU_4, SU_6, SU_{12}, $SU_2 \times SU_2$ probably belong to the dynamical symmetries that may be useful for a phenomenological description; but their heuristic value could probably be compared with the heuristic values of cycles and epicycles in Ptolemy's astronomy. They give only very indirect information about the underlying natural law.

In the most important experimental results of recent years, bosons with relatively high mass (3–4 GeV) and long life have been discovered. Such states were to be expected, as has been especially emphasized by Hans-Peter Dürr. Whether their long life allows us to interpret them as being "composed of" other particles of long life is a difficult dynamical question, in which all the complications of many-body physics play a role. I would, however, consider it as useless speculation once again to introduce *ad hoc* new particles, of which these objects are assumed to "consist." This would again be the wrong question, which can not contribute to the understanding of the spectrum.

Recently, in the storage rings in Geneva and in the Batavia machine, total cross sections of collisions between protons at extremely high energies were measured. The result was that at very high energies the cross sections increase roughly as the square of the logarithm of the energy. This behavior had long ago been conjectured in the theory for the asymptotic region, independent of the nature of the particles. In the meantime the collisions between other particles have led to similar results, and this outcome strongly suggests that in the big accelerators the asymptotic region has already been reached—that even at highest energies no more surprises are to be expected.

New experiments generally can not be expected to yield a *deus ex machina* that suddenly leads to an understanding of the spectrum. The experiments of the last fifty years have already given qualitatively a quite satisfactory, consistent and complete answer to the question of the nature of elementary particles. The quantitative details can—as in quantum chemistry—be analyzed only in the course of years by much detailed work in physics and mathematics, far from being solved in a single step.

Therefore this article can be concluded with a more optimistic view of those developments in particle physics that promise success. New experimental results are always valuable, even if they only enlarge the data table; but they are especially interesting if they answer critical questions of the theory. In the theory one should try to make precise assumptions concerning the dynamics of matter, without any philosophical prejudices. The dynamics must be taken seriously, and we should not be content with vaguely defined hypotheses that leave essential points open. Everything outside of the dynamics is just a verbal description of the table of data, and even then the data table probably yields more information than the verbal description can. The particle spectrum can be understood only if the underlying dynamics of matter is known; dynamics is the central problem.

R.10

When Is a Particle?

SIDNEY D. DRELL

AS ONE OF the directors of the Stanford Linear Accelerator Center, Sidney D. Drell has probably spent as much time dealing with accelerated particles as any physicist around. We have chosen an article he wrote in 1978 as a particularly representative account in the chronicle of the hunting of the quark. At a point when Gell-Mann and Zweig were hypothesizing the existence of only three quarks, Drell, who was then a professor at Stanford, took his cue from Archibald MacLeish's "A poem should not mean but be" to ask a poignant question: When is a particle?

He discusses the evolution and utility of the quark theory and highlights the tension inherent in relying on a still unobserved particle to provide an orderly explanation for the structure of matter, especially when physics is rooted in a philosophy of verification by observation. Drell is unsure whether quarks will continue to be inferred by the properties of larger structures of matter such as protons, or if they may one day be detected on their own. Drell also examines the basic features of quantum chromodynamics and suggests that quarks and leptons may provide the key to formulating a unified understanding of all the basic interactions in nature. Yet Drell is troubled by the proliferation of quark degrees of freedom—five flavors, each of which appears in three colors—and suggests that they may simply be too numerous to be simple enough to provide a unified explanation of nature.

Drell discusses the concept of confined quarks, which has a distinguished history, going back to the days of Leucippus, Democritus, Epicurus and Lucretius, the last of whom talked about *minimae partes,* a fair Latin name for the elusive quark.

The hidden harmony is better than the obvious.

—HERACLITUS, *Fragments*

How much finer things are in composition than alone.

—RALPH WALDO EMERSON, *Journals*

MAN'S EFFORT TO understand what we and our world are made of is one of the greatest adventure stories of the human race. It dates from the beginning of recorded history. Science first flourished 2500 years ago with the quest of the early Greek philosophers for an underlying unity to the rich diversity observed in the world around them. They realized that the search for an understanding of Nature at a fundamental level in terms of basic processes and constituents necessarily carried them beyond the sensory world of appearance.

In his essay on Lucretius, George Santayana described the emergence of this idea that "all we observe about us, and ourselves also, may be but passing forms of a permanent substance" as one of mankind's greatest thoughts. We now recognize it as the original search, in its most primitive form, for Nature's conservation laws and elementary particles. However, the early Greek "metaphysicists" predated by two millennia the modern scientific method—with its insistence on experimental observation. In their inquiry they relied purely on rational analysis, free from the discipline of direct observational content. Not surprisingly, therefore, they went off in widely differing directions: Leucippus and Democritus to the concept of indivisible atoms; Anaxagoras to the original bootstrap model of infinitely divisible seeds within seeds, each as complex as the whole, and Anaximander, Pythagoras and Plato to more abstract mathematical concepts of numbers and symmetries.

On occasion since then, scientists have arrogantly alleged that the end of this search for Nature's basic building blocks is in sight. However, such delusions have been short lived, especially in modern times, crumbling midst the debris that emerges from increasingly powerful atom smashers. We have come to appreciate how much richer Nature's imagination is than our own vision of what lies beyond the next frontier, as we explore

---

Sidney D. Drell, "When Is a Particle?" Reprinted with permission from *Physics Today*. Vol. 31, No. 6, pp. 23-32 (1978).

with ever more powerful and sensitive instruments on ever-shrinking space-time scales. This is particularly true now, after three explosive years of remarkable discoveries of new particles, starting with J/ψ in November 1974. In the wake of these momentous discoveries it is timely to explore what has happened in recent years to our concept of the elementary constituents—the building blocks—of Nature.

QUARKS—FIT FOR THE SOCIAL REGISTER?

During the past two decades we have come to the point of accepting into the exclusive family of elementary particles a guest who would not have made the Social Register of an earlier generation. I am, of course, referring to the *quark,* which was introduced in 1964 independently by Murray Gell-Mann and George Zweig in their efforts to summarize and systematize the great proliferation of nuclear particles that were being produced by accelerators on the high-energy frontiers of the 1950's. Regularities had been perceived in the masses of these particles as well as in the characteristics of their creation, their interactions and their decays; Gell-Mann and Zweig showed that these regularities, as well as new ones found later, could be accounted for in terms of the simple motions and interactions of just three different kinds of fractionally charged spin-½ quarks. The quantum numbers of these quarks are shown in table 1. In this scheme, known as SU(3), several hundred hadronic resonances are successfully interpreted as excited states of just two simple quark configurations:

►three quarks (qqq) form a baryon and
►a quark-antiquark pair ($q\bar{q}$) forms a meson.

Table 1 lists the quantum numbers and "flavor" labels—up, down and strange—of the three original quarks. In it Q is their electric charge in units of the proton charge, I_3 is the third component of their isotopic spin and S is their strangeness number.

TABLE 1. QUARKS AND THEIR QUANTUM NUMBERS IN SU(3)

Quark flavor	Q	I_3	S
u (up)	⅔	½	0
d (down)	−⅓	−½	0
s (strange)	−⅓	0	−1

Because the quark hypothesis made correct predictions, provided a systematic organization of a large mass of data and brought simplicity along with a unifying harmony to our view of Nature, it was a crucial step forward, similar in many ways to the discovery of the nuclear atom by Ernest Rutherford in 1911. The role of quarks in subnuclear spectra is similar to that of electrons in atomic spectra, and that of neutrons and protons in nuclear spectra. What is new, however, is that in contrast to both our atomic and nuclear experience we do not "see" the individual quarks isolated from one another. When we break apart—ionize—an atom, its electrons and its nucleus are clearly evident and are detected in the debris. The same is true when we identify individual protons and neutrons in the debris stripped away in nuclear collisions. How then do we account for our failure to see quarks in the debris of a shattered proton or of any other subnuclear particle? What does this unprecedented experimental situation do to our very concept of an elementary particle?

Indeed, when we now ask what we mean by an elementary particle, we can feel a certain kinship with the modern poet who asks, "When is a poem?" Modern poetry clearly no longer lives by rigid rules of meter and rhyme, and many times even—or should I say especially—the meaning is obscure. As Archibald MacLeish wrote in his poem "Ars Poetica": "A poem should not mean / But be."

How do we scientists now respond to the question, "When is a particle?"

A TALE OF TWO PARTICLES

To appreciate how far we have moved in our modern concept of elementary particles we need look back only to 1930 and the birth of the neutrino. On the historical time scale this is less than 2% of the way back to the earliest Greek metaphysicists of 2500 years ago, but our concepts of "When is a particle?" have evolved considerably during this relatively brief period. Before the neutrino could fully establish its credentials as a socially respectable elementary particle in the 1930's and 1940's, many physicists insisted on seeing it carry away both energy and momentum in proper proportion during beta decay; more conservatively, some insisted on being able to observe its arrival. In today's world of quarks we have apparently discarded such requirements as a standard against which to test the concept of particle. Indeed, according to current dogma we never can nor shall observe isolated quarks or record the emission and absorption of individual quarks. How and why have we come to such a revolutionary new perception of elementary particles? By way of contrast, recall first the story of the neutrino.

The neutrino idea was born in 1930, following the accumulation of experimental data from the microcalorimetric measurements of the average energy of beta particles from radium E. This evidence showed that the average energy of disintegration in the decay equaled the mean energy of the continuous beta spectrum, rather than its upper limit, and that furthermore there was no appreciable accompanying gamma radiation.

Wolfgang Pauli was convinced that these results were very significant. In an open letter to Hans Geiger and Lise Meitner, who were attending a meeting at Tübingen in December of 1930, Pauli pointed out that in beta decay not only the energy but also the spin and the statistics were apparently not conserved. A half integer of angular momentum is missing in beta decay if the beta particle is the only particle emitted.

Incidentally, at this same time an understanding of the spins and statistics of nuclei based on the fledgling quantum theory provided a strong argument against the hypothesis then current that the fundamental constituents of nuclei were protons and electrons. For example, in those days prior to the discovery of the neutron, how could one explain Bose statistics for the N^{14} nucleus if its basic building blocks were in fact 14 protons and 7 electrons!

Emphasizing the importance of spin and statistics, Pauli went on to propose the outlandish idea of introducing a very penetrating, new neutral particle of vanishingly small mass in beta decay to save the situation. However, as Chien-Shiung Wu relates in the fascinating account she prepared as a memorial tribute to Pauli, ". . . he was most modest and conciliatory in his pleading for a hearing." Admitting that his remedy might appear an unlikely one, Pauli commented in his letter.

> Nothing venture, nothing win. And the gravity of the situation with regard to the continuous beta spectrum is illuminated by a pronouncement of my respected predecessor in office, Herr [Peter] Debye, who recently said to me in Brussels, "Oh, it is best not to think about it at all . . . like the new taxes." One ought therefore to discuss seriously every avenue of rescue. So, dear radioactive folks, put it to the test and judge.

Pauli made public his proposal of this strange new particle at the American Physical Society meeting in Pasadena in June 1931. He insisted that the sum of the energies of the beta particle and of the very penetrating neutral particle (or particles) emitted by the nucleus in one process should equal the energy that corresponds to the upper limit of the beta spectrum. In addition to energy, Pauli assumed that linear momentum, angular momentum and statistics are conserved in all elementary processes.

TABLE 2. **FOUR-QUARK SCHEME, SU(4)**

Quark flavor	Q	I_3	S	C
u	⅔	½	0	0
d	−⅓	−½	0	0
s	−⅓	0	−1	0
c (charm)	⅔	0	0	1

In his reverence for conservation laws Pauli adopted a very different approach from Niels Bohr, who had suggested some years earlier that energy and momentum may be conserved only statistically and not in individual nuclear processes. Let me quote here from Bohr's 1930 Faraday Lecture:

> At the present stage of atomic theory . . . we have no argument, either empirical or theoretical, for upholding the energy principle in the case of beta-ray disintegrations, and are even led to complications and difficulties in trying to do so.

After conceding some difficulties in so radical a departure from the principle of energy conservation, particularly with regard to time reversal, Bohr remarked:

> Just as the account of those aspects of atomic constitution essential for the explanation of the ordinary physical and chemical properties of matter implies a renunciation of the classical ideal of causality, the features of atomic stability, still deeper-lying, responsible for the existence and the properties of atomic nuclei, may force us to renounce the very idea of energy balance.

PAULI'S "RADICAL" CONSERVATISM

Pauli's hypothesis of a new unobserved particle met with skepticism, as it was too radical for most physicists to accept with ease. From our present perspective, the "radical" Pauli, with his insistence on maintaining conservation laws—even at the expense of introducing a new and invisible particle—was really the conservative in his approach. Indeed, Pauli was so conservative that he did not publish his proposal. Only after James Chadwick's discovery of neutrons in 1932 and the consequent collapse of the proton-electron picture of nuclei did Pauli put aside the reservations he had about his neutrino hypothesis.

At the Solvay Congress in 1933 Pauli (correctly) conjectured the neutrino to be a massless spin-½ fermion with a penetrating power far greater than that of photons of the same energy. He did this because of conservation laws and in spite of the fact that, as he noted, "experiments do not provide us with any direct proof of this hypothesis"; furthermore, "we don't know anything about the interaction of neutrinos with other material particles and with photons." Pauli argued forcefully against Bohr:

> The interpretation supported by Bohr admits that the laws of conservation of energy and momentum do not hold when one deals with a nuclear process where light particles play an essential part. This hypothesis does not seem to me either satisfying or even plausible. In the first place the electric charge is conserved in the process, and I don't see why conservation of charge would be more fundamental than conservation of energy and momentum.

Pauli also emphasized the crucial importance of investigating the relation between the energy and momentum carried away by the neutrino by means of sensitive measurements of the amount of nuclear recoil.

Shortly after the close of the discussions at the Solvay Congress, Enrico Fermi gave a quantitative formulation of the neutrino hypothesis, and from 1934 on this hypothesis gained enormous strength from its successful predictions of the energy spectrum and of the angular momentum selection rules in beta processes. However, the evidence for the neutrino was still indirect and would remain so for more than 20 years until it was directly observed as a particle—until the advent of the nuclear reactor there were no intense neutrino sources. As a result, even though all observations and predictions were as if the neutrino were emitted in beta decay, one could still adopt the agnostic stance of weak faith and worry that Nature was simply fooling us. We could invent the neutrino, but did we actually *need* it? Could we do without it?

I remember how real such questions appeared when I entered graduate school at the University of Illinois in Urbana. There I was witness to the first of the series of delicate and ingenious measurements of the nuclear recoils in beta decay that Chalmers Sherwin initiated in 1947. Pauli had judged these measurements to be insurmountably difficult at the Solvay Congress in 1933, but Sherwin accomplished them using a thin-film source of radioactive P^{32} and time-of-flight techniques with then-fast electronics. He clearly demonstrated that the ratio of the missing energy to momentum was given approximately by the velocity of light. Hence within the errors of his measurements the neutrino was, if it indeed existed at all, behaving kinematically as a massless particle.

By then there was in fact little if any doubt about the neutrino, and skepticism on this score was hardly stylish. But I remember vividly a physics-colloquium debate in 1948 between Maurice Goldhaber and Sid Dancoff on whether the neutrino really existed. With a conservative logic that, in these days of confined quarks, seems downright reactionary, Goldhaber advised caution, even in the light of Sherwin's results, and emphasized the importance of looking for evidence of neutrino absorption. After all, we may see it disappear but before all doubts can be removed he advised that we should see the neutrino arrive and hit us over the head—the ultimate litmus test for the full respectability of an elementary particle. To which Dancoff replied, in the spirit of the logical positivist, that we had a respectable wave function, a dignified Dirac wave equation and the unambiguous principles of quantum theory for describing, predicting and analyzing neutrinos in beta processes. What more did we need? The neutrino was in fact no less respectable than, say, the proton!

No skeptics whatsoever remained eight years later, in 1956, when Clyde Cowan and Frederick Reines used a powerful nuclear reactor as an intense neutrino source and observed its effects. The importance attached by some to being able to detect the arrival as well as the departure of neutrinos is reflected in the 1963 edition of the *Encyclopaedia Britannica:*

> Were it not for the quite convincing experimental evidence of their existence . . . one might regard neutrinos as the necessary but undetectable scapegoats whose subtle function was to permit the application of conservation of energy and momentum to atomic reactions.

I have sketched this history of the neutrino and its development from a radical idea to a respectable particle, in order to contrast and compare it with the current quark theory and dogma. Where do we stand now with the quarks? On one hand, we still lack conclusive evidence of ever having "seen" isolated quarks in the laboratory, in spite of many efforts to find them. On the other hand, should we not insist on seeing them if they are indeed the building blocks of the proton? Is such observation not required in the spirit of the modern scientific method, with its primary aim and its central goal, as Herman von Helmholtz characterized it in paying tribute to Faraday and his work ". . . to purify science from the last remnants of metaphysics"?

Early in this century the remarkable art of experimentation developed to the extent that it was possible to study the properties of individual atoms. As a result of this fantastic sensitivity of measurement, our entire conception of "seeing" underwent revolutionary changes. On the atomic frontier the most profound and radical change was the realization that it is necessary to take into account the effect of the observation itself on the physical system being observed. This fundamental limitation of the measurement process led to a major revolution in our concept of the elementary particle, driving us beyond classical ideas alone to a quantum description. But there is no uncertainty in what we mean when we say that we *observe* an electron as an elementary particle.

Now that we have come upon the quarks, the situation is very different. Are they objects whose existence can be inferred *only* from the properties of larger, complex structures, such as a proton, in which they are the constituents confined to one another by unbreakable bonds? If quarks are indeed not observed singly or in isolation and if they never get beyond being the "undetectable scapegoats" of the *Encyclopaedia Britannica* phrase, will we still attach so central and fundamental an importance to them, or even to the elementary-particle concept itself?

QUARKS: THE EVIDENCE

For the quarks to survive as fundamental there are two possibilities:

►They will be discovered, that is, observed singly. In this case they will constitute the atoms of yet another layer of matter, presumably with an internal structure of their own to be studied by another generation to come.

►They will not be discovered in the same sense as the neutrino was, but they will persist in fulfilling the goal that motivated their being introduced in the first place, of providing a simple basis for the explanation of the observed multiplet structure and properties of subnuclear particles.

We know that there exist simple general laws that explain the rich diversity of Nature; this is our fundamental faith as scientists. If the quarks are indeed not observed directly, their survival as vital ingredients in the structure of matter will depend on how successful the quark idea is in unifying, simplifying and correctly predicting diverse observations, and thereby leading us to such general laws. At least from today's perspective, the quarks seem to have done enough for particle physics that there is little danger of their fading away with the ether. In brief, what is the evidence most strongly supporting quarks?

The original quark idea was put forward to explain why baryons occur with the observed multiplet structure of an octet of spin ½ and a decuplet of spin $3/2$, and why mesons form nonets of spin zero or one. General features of the hadronic mass spectra, their transition matrix elements and such static properties as baryon magnetic moments, could be understood in terms of their quark content, three quarks for baryons and a quark-antiquark pair for mesons. What emerged was an intuitively simple picture of relatively light point-like quark constituents moving approximately as independent particles within a hadron.

We had to pay a price for these successes. At the very outset it was realized that a successful classification scheme for the three-quark baryonic spectra required that the quarks be assigned to *symmetric* configurations. This was in apparent violation of the heretofore sacred relation between spin and statistics that requires half-integral spin particles such as quarks to be in *antisymmetric* configurations. The way out of this dilemma was to assign to the quarks a new quantum number, dubbed color, which could take any one of three values, and to require the baryon wave functions to be antisymmetric in color. This effectively triples the number of quarks, and is therefore reminiscent of Pauli's original proposal for the neutrino. He also introduced a new particle, in part to satisfy the requirements of statistics in beta decay. The added quantum number of color introduces the possibility of many additional but unobserved states, corresponding to hadrons of different colors.

To remove this difficulty we must insist that the three quarks forming a baryon are in an antisymmetric color-singlet state. Similarly, the quark-antiquark pair composing a meson must form an anticolor pair, with each color occurring in equal parts. All hadronic states that fail to hide their color are ruled out. An explanation of why Nature is color-blind is fundamental to a complete theory of quarks. Although a theoretical derivation of color blindness still remains to be given, we can correctly and simply describe the observed spectra with no unwanted states by insisting that color remain Nature's secret.

The existence of point-like constituents within the hadron also provided a basis for understanding the observed character of hard, very inelastic high-energy collisions between two hadrons, or between the hadron and an electron, a neutrino or a muon. The nature of the observed scattering patterns between an electron and a proton, for example, required the existence of strong local electromagnetic charges and currents due to point-like constituents within the proton, which were presumably the quarks. (They play the same role as the nucleus in Rutherford scattering.) Specifically, the constituents of the proton scatter the incoming high-energy electrons in high-momentum-transfer collisions as if they themselves had no inner structure, and as if they were relatively light and essentially unbound.

The high-energy inelastic scattering measurements further emphasize the enigma of unobserved quarks. We resort to quark constituents for the most direct and simple interpretation of the observed scattering pattern. However, the proton or neutron, smashed hard and shattered into bits and pieces in the collision, does not spill out quarks in its debris—just other normal hadronic states of mesons and baryons.

THE CHARM OF J/ψ

Unquestionably the most important recent evidence that decisively supports the quark picture was provided by the new discoveries of charmed matter, beginning with the J/ψ close to four years ago. It sent an explosive shock through the scientific community because, on the subnuclear scale of times, it was an almost stable, very narrow, resonance that could not be accommodated in the existing quark scheme. These properties differentiated it from the hundreds of other hadronic resonances with typical decay widths for 10–100 MeV unless they are suppressed by selection rules. The J/ψ, however, was found to have a total decay width of only 70 keV and a mass of 3095 MeV.

Evidently there was a selection rule operating to account for the narrowness of this state. Because the new particle is heavy—weighing more than three times as much as the proton—there is no inhibition in its decay due to threshold effects or lack of phase space. Hence the suppression of its decay can not be explained on kinematic grounds alone. Moreover, the measured quantum numbers of the J/ψ are quite conventional: zero charge, one unit of angular momentum and zero strangeness—just like the photon, which is the source of the J/ψ in electron-positron annihilation. Furthermore, its decay products are familiar particles—predominantly electrons, muons and pions. The narrowness of the J/ψ therefore can not be explained in terms of a selection rule corresponding to the conservation of known quantum numbers.

What then was holding it together for such a long time—about 1,000–10,000 times longer than expected? A new quantum number was required above and beyond what could be accommodated in the three-quark scheme, which was now found to be too restrictive. This new quantum number already had been dubbed "charm" by James Bjorken and Sheldon Glashow. Indeed, the existence of a fourth quark with charm had been anticipated by Glashow and his colleagues several years before, as a simple and natural way of theoretically suppressing unobserved weak decays involving a change of strangeness but not of electrical charge between the interacting particles. The easiest way to account for the J/ψ and its long lifetime was to assume it to be a meson made up of a new, massive charmed quark bound to its antiparticle; this is illustrated in table 2.

The value and beauty of such a simple model lies in its predictive power. The successes for this model have been extensive—and although the new discoveries were a shocking surprise, they are now recognized as contributing importantly to the impressive successes of the quark hypothesis. In the decade preceding these new discoveries it was established that every known hadron could be explained as a combination of a quark and an antiquark for the mesons and of three quarks for the baryons. Moreover, all possible combinations of the three "ordinary" quarks correspond to a known hadron, without fail. With the discovery of the J/ψ and its interpretation in terms of the theoretically anticipated fourth, charmed, quark a whole new set of spectroscopic levels had to be hunted for and interpreted.

In particular, we should observe a complete spectrum of charmonium, the bound states of a charmed quark-antiquark pair. This is analogous to positronium, with its spectrum of excited states. The energy spacings and branching ratios in charmonium can be understood qualitatively in terms of the binding of a heavy quark-antiquark pair. Detailed analyses of the fine structure reveal information of the shape of their interaction potential.

Furthermore, new mesonic states will be formed when a charmed quark binds with an ordinary uncharmed antiquark. These, the D mesons, have also been observed and fit into the four-quark classification, SU(4). Already there are starts toward the spectroscopy of a charmed quark bound to a strange antiquark, the so-called "F meson," and toward the spectroscopy of charmed baryons.

NATURAL FLAVORS

A very important parameter for the quark hypothesis is the ratio, *R*, of the cross sections for an electron-positron pair to annihilate to all possible configu-quarks below the region in which charm is excited, as well as for the four flavors (including the charmed quarks) in the higher-energy region, provided each flavor occurs in three colors. Otherwise—without color—there would be a sharp discrepancy of a factor of three.

These results thus represent a triumph for the hypothesis of color triplets of quarks. On a descriptive level the quark hypothesis evidently accounts very well for a broad set of observations. Furthermore—does anyone seriously doubt that *R* will increase onto a higher plateau when the total energy of the colliding electrons and positrons exceeds 9.4 GeV, the threshold for producing the upsilon meson recently discovered at Fermilab? The upsilon is believed to be a bound quark-antiquark pair like charmonium, but built of yet another new flavor of quark pairs.

By now we have had such a proliferation of quark degrees of freedom—presumably at least fifteen, five flavors times three colors—that it can hardly be said they are entering a very exclusive Social Register!

If we look back once more to the 1930's, we recall that the neutrino became a strong, and to most a persuasive, candidate for the Social Register of elementary particles long before it was seen, when Fermi provided it with effective and indisputable theoretical credentials. There is optimism, at least among many theorists, that quarks have also been gaining the dignity of a pedigree during the past few years. The reason for this optimism is recent theoretical progress that has identified important features of successful quark dynamics in a well defined class of quantum field theories known as non-abelian gauge theories. These are generalizations, pioneered in 1954 by Chen Ning Yang and Robert Mills, of the precisely tested and unfailingly successful theory of quantum electrodynamics (QED). The photons, which are the vector quanta of QED, are themselves electrically neutral; this is characteristic of an abelian gauge theory. In non-abelian gauge theory the vector quanta, dubbed "gluons," are themselves also charged. The gluons can exchange this charge between sources, or between one another. In the theory of quarks and gluons known as "quantum chromodynamics" (QCD), the color quantum number plays the same role as the electric charge in QED. In QCD an octet family of colored gluons replaces the single photon of QED as the messenger of the color electric and magnetic fields.

The case for QCD, pioneered in 1973 by H. David Politzer, and by

David Gross and Frank Wilczek, is based on the crucial observation that such theories can lead to forces between the quarks, mediated by gluons, that grow weaker at short distances. This behavior is known as "asymptotic freedom." It contrasts with the familiar forces of electromagnetism, which grow even stronger than the $1/r^2$ of the Coulomb law at short distances when quantum effects—in particular vacuum polarization—are included. Asymptotically free forces between quarks provide a basis for explaining the observed behavior of the hard collisions, such as Bjorken scaling, which look like scattering from almost free, point-like light quarks within the hadrons. These forces must remain in effect for large separations, however, so that the quarks that behave as almost free at short distances on the scale of hadronic sizes will be confined and can not be pulled apart. Further, the theory must allow only the formation of color-singlet states, to account for the observed spectra and quark structure of hadrons. In QCD the simplest quark configurations that can form color-singlet states are just the observed ones with three quarks or a quark-antiquark pair.

COLOR CONFINEMENT

In the framework of quantum chromodynamics, quark confinement becomes synonymous with color confinement. This is the other basic ingredient in addition to asymptotic freedom that we want to find in QCD if it is to form the basis of a fundamental dynamical theory of hadrons in terms of gluons and of quark constituents. The theoretical challenge to prove whether or not QCD actually confines is formidable because of the difficulty in solving—or even attempting to solve—quantum field theory when the forces are strong and we can not resort to weak-coupling perturbative treatments.

With an appropriate dash of the theorists' optimism, let us suppose for a moment that these technical challenges will be surmounted and a convincing case made for color confinement in QCD. We may even further imagine that approximately correct mass spectra will be calculated for the hadrons; this includes understanding why the pion is much lighter than all other hadrons. How compelling will this make the case for quarks as elementary fundamental constituents of the hadron?

The case for quarks might appear more compelling than that of the neutrino back in 1934, because we now see all the elements of a grand synthesis in place. The quantum chromodynamics to which we have turned for a theory of hadrons is a non-abelian local gauge theory of the same formal structure as the one introduced in unifying the weak and electromagnetic interactions. . . . The strong, the electromagnetic and the weak interactions have very different characteristics, such as their ranges and strengths, as studied at present laboratory energies. It has been conjectured, starting with the pioneering work in 1967 of Weinberg and Abdus Salam, that the difference between the weak and electromagnetic interactions is a consequence of a partially broken symmetry for the weak processes. The intermediate vector mesons of the weak interactions acquire large masses due to the symmetry breaking; only at truly high energies exceeding these masses—thought to be in the range of 70 GeV or higher—will the common characteristics of the weak and electromagnetic interactions be apparent. Some predictions of this approach to a unified theory of weak and electromagnetic processes have already been verified, in particular, neutral-current effects in neutrino scattering.

A further extension of the symmetry considerations underlying the gauge theories to the strong interactions puts the quarks on the same basis as the leptons. The difference between these derives from the fact that quarks carry the color quantum number—the charge of the strong interactions—whereas leptons do not and are immune to the strong forces. Hence leptons, in contrast to quarks, are not confined by the requirement that color remain Nature's secret. Such a picture is very attractive. It provides a giant step toward one of the principal goals of modern physics: a unified understanding of all the basic interactions in Nature.

Were we to achieve such a theoretical synthesis, the case for admitting quarks to the Social Register as the hadron's basic constituents would clearly be very strong. However, recalling our earlier discussion of the neutrino, do we no longer care that we are now identifying as fundamental constituents of hadronic matter things that, in contrast to all our prior experience, can not even in principle be isolated and observed? When *is* a particle? Are we now willing to say, as a variation of Archibald MacLeish's lines, that in contrast to a poem a particle need not be, but should mean?

The burden of proof for quarks is very different from the original argument for the neutrino. Whereas the neutrino was postulated to protect energy, momentum and spin conservation laws, which involve observable space-time properties, no conservation laws *require* quarks. However, the quarks do present the very strong operational credentials I have described—both experimental and theoretical.

What troubles me most about accepting quarks as the fundamental hadronic constituents is quite simple and has nothing to do with their confinement and Helmholtz's dictum "to purify physics from the last remnants of metaphysics." It is that we already have so many quark degrees of freedom—at least five flavors in each of three colors. The Social Register of particles surely must be more exclusive than that if it is to be valued and honored as it was in those good old days!

BREAK A MESON, MAKE TWO QUARKS

Having said this, I have a sneaky suspicion that quarks may turn out to be somewhat like magnetic poles, and nothing more. When broken in two, a bar magnet becomes, not isolated north and south poles separated from one another, but two magnets, each with its own north and south poles. As many have noted, this is very similar to what happens when a meson made of a quark and an antiquark is smashed apart. . . . The debris of the shattered meson consists not of isolated quarks but of more mesons, each with its own quark and antiquark. This is not a literal analogy, of course, because the non-abelian color gauge theory also allows baryons made of three quarks to be formed in color-singlet states. Nevertheless, it is sufficiently close and accurate to be a useful guide.

Our curiosity and present plight with quarks may not be very different from that of an inquisitive mariner at sea some ten centuries or so ago. In a moment of calm on a passage he might have viewed a compass needle—a spare one, I hope!—with idle bafflement or scientific curiosity, and tried to break it apart to separate the two ends with opposite properties, that is, the north pole from the south pole. But to no avail, for with each breaking of the compass needle he ended up with an additional one having both a north and south pole. The understanding of this impossibility to isolate single magnetic poles came only when, in 1820, André Ampere explained magnetism in terms of electric currents. In fact, a fundamental theory of magnetism at the atomic level in terms of the currents of circulating and spinning electrons was achieved only in this century on the basis of modern quantum theory.

Perhaps the quarks are not the fundamental particles of hadron dynamics, just as magnetic poles are but phenomenological manifestations of amperian currents. Presumably they will remain no less important for the description and understanding of subnuclear processes than bar magnets are for understanding a lot of the physics of magnetism. The quarks have done too much already to be forgotten and discarded. If, however, there are underlying dynamics, the whole question of the meaning of confined constituents as elementary particles will disappear. Whether or not the analogy with magnetism has any merit, the notion of a new "elementary structure" underlying the quarks destroys the very attractive idea of a quark-lepton parallel unless we similarly modify and elaborate our picture of leptons. This hardly appears to be an attractive prospect now, particularly because no one has either seen or theorized creatively as to what these new "elementary structures" might be. On the other hand, the lepton degrees of freedom, along with those of the quark, have also begun to proliferate.

Some physicists were discouraged already 30 years ago with the discovery of the muon, about which I. I. Rabi is quoted as remarking, "Who ordered that?" More recently we have apparently encountered a new, third strain of leptons in the tau, which is also thought to be accompanied by its own neutrino, as are the electron and muon. As we continue to raise our energy frontier, are we fated to meet proliferating families of leptons as well as quarks? Facing this dilemma, Werner Heisenberg proposed a different viewpoint in a lecture on the nature of elementary particles, which he delivered in 1975 shortly before his death. He raised the possibility that we are asking the wrong question in particle physics when we ask what a proton "consists of":

> I will now discuss that development of theoretical particle physics that, I believe, begins with the wrong questions. First of all there is the thesis that the observed particles such as the proton . . . consist of smaller particles: quarks . . . or whatever else, none of which have been observed. Apparently here the question was asked: What does a proton consist of? But the questioners appear to have forgotten that the phrase "consist of" has a tolerably clear meaning only if the particle can be divided into pieces with a small amount of energy, much smaller than the rest mass of the particle itself.

Heisenberg is referring here to the fact that in the realm of quarks, in contrast to atomic or even nuclear physics, we are no longer dealing with energies that are but small fractions of the rest masses of the particles themselves. The strong subnuclear forces confining the three valence quarks in the baryon also create many virtual quark pairs and gluons. Retardation effects, as well as the energy-momentum content of the gluon fields that bind the quarks together, will be important. All of these effects and virtual particles—the gluons as well as the fluctuating numbers of quark pairs—must be included in a dynamical description of "life" within the hadrons. When we apply quantum mechanics to a relativistic strong-interaction problem, our basic elements are no longer simply a fixed small number of particles, but field amplitudes that create eigenstates of definite quantum numbers. In view of this, Heisenberg suggested that our quest for simplicity and an underlying level of unification should be formulated in terms of the fundamental currents or symmetries of the theory.

Heisenberg's emphasis on symmetries is reminiscent of the ideas of Pythagoras and Plato. Pythagoras first explictly emphasized the importance of symmetry 25 centuries ago and insisted that ultimately all order is capable of being understood and expressed in terms of number. Plato provided a specific form for this idea by identifying the fundamental symmetries as the basic atoms in his scheme. Dressing his idea in modern garb, we should not so much focus on the problem of the many quark degrees of freedom as seek simplicity in the underlying group structure of the fundamental equations. Perhaps in our quest for simplicity we should follow the lead of Albert Einstein, by incorporating the sources and forces of a unified field theory in the physical geometry of space-time.

THE HIDDEN ONE

The fate of the idea of hidden building blocks can be settled only by experiments, including the very fundamental and difficult quark searches already in progress. Whatever their future fate, the concept of confined quarks already has a distinguished history, which dates back to the classical Greek and Roman times and includes the writings of Leucippus, Democritus, Epicurus and Lucretius. Modern historians of science still debate intensely whether the atoms of Democritus and Leucippus are physically indivisible (because they are solid and impenetrable) or whether they are logically and mathematically indivisible (because they have no parts). Some suggest that both kinds of atoms are to be found in their writings. Apparently a full development of the idea of indivisible elementary atoms consisting of minimal parts that are permanently confined and can not be pulled apart dates to Epicurus, around 300 B.C. In a charming article, Julia and Thomas Gaisser refer to the elaboration of this idea by Lucretius, who refers to the *minimae partes* of the indivisible atoms in his great poem, *De Rerum Natura.* Clearly these were the early versions of quarks—or partons, as we sometimes call the minimal parts of the hadrons.

I was so intrigued by these references to confined quarks, or *minimae partes,* in classical philosophy that I found myself also wondering whether Greek mythology could not provide some roots—perhaps even a name—for these hidden basic things. What I came up with after anything but a scholarly, thorough search was the nymph goddess Calypso, referred to as the hidden one, who kept Odysseus confined to her island of Ogygia for seven years after his shipwreck en route home from the Trojan War. Calypso offered Odysseus immortality if he would share eternal, blissful confinement with her. However, when the gods called on Odysseus to resume his human destiny and his hazardous journey home, he rejected her offer.

Whether quarks will remain mysterious as hidden elementary "calypsons," be understood as phenomenological manifestations of an underlying dynamics or reveal themselves directly to experiment remains for the future. So does the fate of those theorists among us who concern ourselves with quantum chromodynamics, asymptotic freedom and quark confinement. Those theorists who have said that it is impossible to liberate quarks should perhaps look again to Epicurus, the metaphysical father of confined quarks, for a second message that is frequently—and erroneously—attributed to him: Eat, drink and be merry, for tomorrow an isolated quark may actually be found.

R.11

Elementary Particles and Forces

CHRIS QUIGG

THE STATUS OF particle physics changes frequently; this volume contains a number of reports concerning subatomic physics written over the years by specialists. One of the most recent accounts is an article written by Chris Quigg, head of the department of theoretical physics at the Fermi National Acceleratory Laboratory at Batavia, Illinois, and a professor of physics at the University of Chicago, in which the author explains that a coherent view of the fundamental constitution of matter and the interactions governing matter has emerged—a view embracing a number of disparate theories, which physicists, however, hope may soon be united into a single comprehensive description of what takes place within subatomic particles. An especially useful part of Quigg's article, for the amateur interested in subatomic phenomena, is the wonderfully succinct charts of leptons, quarks, interactions and carriers, together with the author's explanations of them.

Because particle physics is so complex, so different from the common-sense world familiar to the rest of humanity, it is almost impossible for those who are not physicists to understand it. At times it seems that physicists, in bursts of professional whimsy, are inventing jargon that conceals rather than reveals meaning. They use *up* and *down* in ways that have nothing to do with above and below, *top* and *bottom* in ways that have nothing to do with sides of something to keep up or down, *color* and *flavor* in ways that have nothing to do with hue and taste, all to describe *quarks,* which may be the ultimate, truly indivisible points of matter that make up the subatomic world and the entire Universe.

A little over fifty years ago, physicists and interested laymen were happy with the fourfold division of the atom into protons, neutrons, electrons and neutrinos. By the 1950s, a kind of panic had set in, with the discovery of hundreds of subatomic particles, some of them born to exist for only fractions of a trillionth of a second, others to endure eternally (or at least for an estimated 10^{31} years). As order emerged from the chaos of this particle menagerie, a basic scheme of nature was painstakingly developed. As of this writing it seems to include twelve particles (six leptons, and six quarks described by the quirky adjectives mentioned above), each with its antiparticle for a total of an even two dozen particles. These particles are subject to the manipulations of the gravitational, electromagnetic, weak, and strong forces. These four forces respectively exert themselves through "carrier" particles known as gravitons (hypothetical); photons; the "bosons" W^+, W^-, Z°; and gluons.

The errors of definitions multiply themselves according as the reckoning proceeds; and lead men into absurdities, which at last they see but cannot avoid, without reckoning anew from the beginning.

—THOMAS HOBBES

There can never be surprises in logic.

—LUDWIG WITTGENSTEIN

"Cheshire-Puss," she began, rather timidly . . .
"Would you tell me please, which way I ought to go from here?"
"That depends a good deal on where you want to get to," said the Cat.
"I don't much care where—" said Alice.
"Then it doesn't matter which way you go," said the Cat.

—LEWIS CARROLL

THE NOTION THAT a fundamental simplicity lies below the observed diversity of the universe has carried physics far. Historically the list of particles and forces considered to be elementary has changed continually as closer scrutiny of matter and its interactions revealed microcosms within microcosms: atoms within molecules, nuclei and electrons within atoms, and successively deeper levels of structure within the nucleus. Over the past decade, however, experimental results and the convergence of theoretical ideas have brought new coherence to the subject of particle physics, raising hopes that an enduring understanding of the laws of nature is within reach.

Higher accelerator energies have made it possible to collide particles with greater violence, revealing the subatomic realm in correspondingly finer detail; the limit of experimental resolution now stands at about 10^{-16} centimeter, about a thousandth the diameter of a proton. A decade ago physics recognized hundreds of apparently elementary particles; at today's resolution that diversity has been shown to represent combina-

---

Chris Quigg, "Elementary Particles and Forces," *Scientific American* (April 1985), pp. 84–96.

tions of a much smaller number of fundamental entities. Meanwhile the forces through which these constituents interact have begun to display underlying similarities. A deep connection between two of the forces, electromagnetism and the weak force that is familiar in nuclear decay, has been established, and prospects are good for a description of fundamental forces that also encompasses the strong force that binds atomic nuclei.

Of the particles that now appear to be structureless and indivisible, and therefore fundamental, those that are not affected by the strong force are known as leptons. Six distinct types, fancifully called flavors, of lepton have been identified. Three of the leptons, the electron, the muon and the tau, carry an identical electric charge of −1; they differ, however, in mass. The electron is the lightest and the tau the heaviest of the three. The other three, the neutrinos, are, as their name suggests, electrically neutral. Two of them, the electron neutrino and the muon neutrino, have been shown to be nearly massless. In spite of their varied masses all six leptons carry precisely the same amount of spin angular momentum. They are designated spin ½ because each particle can spin in one of two directions. A lepton is said to be right-handed if the curled fingers of a right hand indicate its rotation when the thumb points in its direction of travel and left-handed when the fingers and thumb of the left hand indicate its spin and direction.

For each lepton there is a corresponding antilepton, a variety of antiparticle. Antiparticles have the same mass and spin as their respective particles but carry opposite values for other properties, such as electric charge. The antileptons, for example, include the antielectron, or positron, the antimuon and the antitau, all of which are positively charged, and three electrically neutral antineutrinos.

In their interactions the leptons seem to observe boundaries that define three families, each composed of a charged lepton and its neutrino. The families are distinguished mathematically by lepton numbers; for example, the electron and the electron neutrino are assigned electron number 1, muon number 0 and tau number 0. Antileptons are assigned lepton numbers of the opposite sign. Although some of the leptons decay into other leptons, the total lepton number of the decay products is equal to that of the original particle; consequently the family lines are preserved.

The muon, for example, is unstable. It decays after a mean lifetime of 2.2 microseconds into an electron, an electron antineutrino and a muon neutrino through a process mediated by the weak force. Total lepton number is unaltered in the transformation. The muon number of the muon neutrino is 1, the electron number of the electron is 1 and that of the electron antineutrino is -1. The electron numbers cancel, leaving the initial muon number of 1 unchanged. Lepton number is also conserved in the decay of the tau, which endures for a mean lifetime of 3×10^{-13} second.

The electron, however, is absolutely stable. Electric charge must be conserved in all interactions, and there is no less massive charged particle into which an electron could decay. The decay of neutrinos has not been observed. Because neutrinos are the less massive members of their respective families, their decay would necessarily cross family lines.

Where are leptons observed? The electron is familiar as the carrier of electric charge in metals and semiconductors. Electron antineutrinos are emitted in the beta decay of neutrons into protons. Nuclear reactors, which produce large numbers of unstable free neutrons, are abundant sources of antineutrinos. The remaining species of lepton are produced mainly in high-energy collisions of subnuclear particles, which occur naturally as cosmic rays interact with the atmosphere and under controlled conditions in particle accelerators. Only the tau neutrino has not been observed directly, but the indirect evidence for its existence is convincing.

QUARKS

Subnuclear particles that experience the strong force make up the second great class of particles studied in the laboratory. These are the hadrons; among them are the protons, the neutrons and the mesons. A host of other less familiar hadrons exist only ephemerally as the products of high-energy collisions, from which extremely massive and very unstable particles can materialize. Hundreds of species of hadron have been catalogued, varying in mass, spin, charge and other properties.

Hadrons are not elementary particles, however, since they have internal structure. In 1964 Murray Gell-Mann of the California Institute of Technology and George Zweig, then working at CERN, the European laboratory for particle physics in Geneva, independently attempted to account for the bewildering variety of hadrons by suggesting they are composite particles, each a different combination of a small number of fundamental constituents. Gell-Mann called them quarks. Studies at the Stanford Linear Accelerator Center (SLAC) in the late 1960s in which high-energy electrons were fired at protons and neutrons bolstered the hypothesis. The distribution in energy and angle of the scattered electrons indicated that some were colliding with pointlike, electrically charged objects within the protons and neutrons.

Particle physics now attributes all known hadron species to combinations of these fundamental entities. Five kinds, also termed flavors, of quark have been identified—the up (*u*), down (*d*), charm (*c*), strange (*s*) and bottom (*b*) quarks—and a sixth flavor, the top (*t*) quark, is believed to exist. Like the leptons, quarks have half a unit of spin and can therefore exist in left- and right-handed states. They also carry electric charge equal to a precise fraction of an electron's charge: the *d, s* and *b* quarks have a charge of $-1/3$, and the *u, c* and the conjectured *t* quark have a charge of $+2/3$. The corresponding antiquarks have electric charges of the same magnitude but opposite sign.

Such fractional charges are never observed in hadrons, because quarks form combinations in which the sum of their charges is integral. Mesons, for example, consist of a quark and an antiquark, whose charges add up to -1, 0 or $+1$. Protons and neutrons consist respectively of two *u* quarks and a *d* quark, for a total charge of $+1$, and of a *u* quark and two *d* quarks, for a total charge of 0.

Like leptons, the quarks experience weak interactions that change one species, or flavor, into another. For example, in the beta decay of a neutron into a proton one of the neutron's *d* quarks metamorphoses into a *u* quark, emitting an electron and an antineutrino in the process. Similar transformations of *c* quarks into *s* quarks have been observed. The pattern of decays suggests two family groupings, one of them thought to contain the *u* and the *d* quarks and the second the *c* and the *s* quarks. In apparent contrast to the behavior of leptons, some quark decays do cross family lines, however; transformations of *u* quarks into *s* quarks and of *c* quarks into *d* quarks have been observed. It is the similarity of the two known quark families to the families of leptons that first suggested the existence of a *t* quark, to serve as the partner of the *b* quark in a third family.

In contrast to the leptons, free quarks have never been observed. Yet circumstantial evidence for their existence has mounted steadily. One indication of the soundness of the quark model is its success in predicting the outcome of high-energy collisions of an electron and a positron. Because they represent matter and antimatter, the two particles annihilate each other, releasing energy in the form of a photon. The quark model predicts that the energy of the photon can materialize into a quark and an antiquark. Because the colliding electron-positron pair had a net momentum of 0, the quark-antiquark pair must diverge in opposite directions at equal velocities so that their net momentum is also 0. The quarks themselves go unobserved because their energy is converted into additional quarks and antiquarks, which materialize and combine with the original pair, giving rise to two jets of hadrons (most of them pions, a species of meson). Such jets are indeed observed, and their focused nature confirms that the hadrons did not arise directly from the collision but from single, indivisible particles whose trajectories the jets preserve.

The case for the reality of quarks is also supported by the variety of energy levels, or masses, at which certain species of hadron, notably the psi and the upsilon particles, can be observed in accelerator experiments. Such energy spectra appear analogous to atomic spectra: they seem to represent the quantum states of a bound system of two smaller components. Each of its quantum states would represent a different degree of excitation and a different combination of the components' spins and orbital motion. To most physicists the conclusion that such particles are made up of quarks is irresistible. The psi particle is held to consist of a *c* quark and its antiquark, and the upsilon particle is believed to comprise a *b* quark and its antiquark.

What rules govern the combinations of quarks that form hadrons? Mesons are composed of a quark and an antiquark. Because each quark has a spin of ½, the net spin of a meson is 0 if its constituents spin in opposite directions and 1 if they spin in the same direction, although in their excited states mesons may have larger values of spin owing to the quarks' orbital motion. The other class of hadrons, the baryons, consist of three quarks each. Summing the constituent quarks' possible spins and directions yields two possible values for the spin of the least energetic baryons: ½ and 3⁄2. No other combinations of quarks have been observed; hadrons that consist of two or four quarks seem to be ruled out.

The reason is linked with the answer to another puzzle. According to the exclusion principle of Wolfgang Pauli, no two particles occupying a minute region of space and possessing half-integral spins can have the same quantum number—the same values of momentum, charge and spin. The Pauli exclusion principle accounts elegantly for the configurations of electrons that determine an element's place in the periodic table. We should expect it to be a reliable guide to the panoply of hadrons as well. The principle would seem to suggest, however, that exotic hadrons such as the delta plus plus and the omega minus particles, which materialize

Leptons				Quarks			
Particle Name	Symbol	Mass at Rest (MeV/c^2)	Electric Charge	Particle Name	Symbol	Mass at Rest (MeV/c^2)	Electric Charge
ELECTRON NEUTRINO	$^{\nu}e$	ABOUT 0	0	UP	u	310	⅔
ELECTRON	e or e^-	0.511	−1	DOWN	d	310	−⅓
MUON NEUTRINO	$^{\nu}\mu$	ABOUT 0	0	CHARM	c	1,500	⅔
MUON	μ or μ^-	106.6	−1	STRANGE	s	505	−⅓
TAU NEUTRINO	$^{\nu}\tau$	LESS THAN 164	0	TOP/TRUTH	t	>22,500; HYPOTHETICAL PARTICLE	⅔
TAU	τ or τ^-	1,784	−1	BOTTOM/ BEAUTY	b	ABOUT 5,000	−⅓

Force	Range	Strength at 10^{-13} Centimeter in Comparison with Strong Force	Carrier	Mass at Rest (GeV/c^2)	Spin	Electric Charge	Remarks
GRAVITY	INFINITE	10^{-38}	GRAVITON	0	2	0	CONJECTURED
ELECTROMAGNETISM	INFINITE	10^{-2}	PHOTON	0	1	0	OBSERVED DIRECTLY
WEAK	LESS THAN 10^{-16} CENTIMETER	10^{-13}	INTERMEDIATE BOSONS: W^+	81	1	+1	OBSERVED DIRECTLY
			W^-	81	1	−1	OBSERVED DIRECTLY
			Z^0	93	1	0	OBSERVED DIRECTLY
STRONG	LESS THAN 10^{-13} CENTIMETER	1	GLUONS	0	1	0	PERMANENTLY CONFINED

FUNDAMENTAL SCHEME OF NATURE, according to current theory, embraces 12 elementary particles (*top*) and four forces (*bottom*). All the particles listed are thought to be structureless and indivisible; among their properties are an identical amount of spin, given by convention as ½, and differing values of electric charge, color charge and mass, given as energy in millions of electron volts (MeV) divided by the square of the speed of light (c). Only the pairs of leptons and quarks at the top of each column are found in ordinary matter; the other particles are observed briefly in the aftermath of high-energy collisions. The four forces thought to govern matter vary in range and strength; although the strong force is the most powerful, it acts only over a distance of less than 10^{-13} centimeter, the diameter of a proton. All the forces are conveyed by force particles, whose masses are given in billions of electron volts (GeV) divided by the square of the speed of light. Because of its weakness, gravity has not been studied experimentally by particle physicists.

briefly following high-energy collisions, cannot exist. They consist respectively of three *u* and three *s* quarks and possess a spin of 3/2; all three quarks in each of the hadrons must be identical in spin as well as in other properties and hence must occupy the same quantum state.

COLORS

To explain such observed combinations it is necessary to suppose the three otherwise identical quarks are distinguished by another trait: a new kind of charge, whimsically termed color, on which the strong force acts. Each flavor of quark can carry one of three kinds of color charge: red, green or blue. To a red quark the force between colored quarks must be extraordinarily powerful, perhaps powerful enough to confine quarks permanently within colorless, or color-neutral, hadrons. The description of violent electron-positron collisions according to the quark model, however, assumes the quarks that give rise to the observed jets of hadrons diverge freely during the first instant following the collision. The apparent independence of quarks at very short distances is known as asymptotic freedom. . . .

Analogy yields an operational understanding of this paradoxical state of affairs, in which quarks interact only weakly when they are close together and yet cannot be separated. We may think of a hadron as a bubble within which quarks are imprisoned. Within the bubble the quarks move freely, but they cannot escape from it. The bubbles, of course, are only a metaphor for the dynamical behavior of the force between quarks, and a fuller explanation for what is known as quark confinement can come only from an examination of the forces through which particles interact.

THE FUNDAMENTAL INTERACTIONS

Nature contrives enormous complexity of structure and dynamics from the six leptons and six quarks now thought to be the fundamental constituents of matter. Four forces govern their relations: electromagnetism, gravity and the strong and weak forces. In the larger world we experience directly, a force can be defined as an agent that alters the velocity of a body by changing its speed or direction. In the realm of elementary particles, where quantum mechanics and relativity replace the Newtonian mechanics of the larger world, a more comprehensive notion of force is in order, and with it a more general term, interaction. An interaction can cause changes of energy, momentum or kind to occur among several colliding particles; an interaction can also affect a particle in isolation, in a spontaneous decay process.

Only gravity has not been studied at the scale on which elementary particles exist; its effects on such minute masses are so small that they can safely be ignored. Physicists have attempted with considerable success to predict the behavior of the other three interactions through mathematical descriptions known as gauge theories.

The notion of symmetry is central to gauge theories. A symmetry, in the mathematical sense, arises when the solutions to a set of equations remain the same even though a characteristic of the system they describe is altered. If a mathematical theory remains valid when a characteristic of the system is changed by an identical amount at every point in space, it can be said that the equations display a global symmetry with respect to that characteristic. If the characteristic can be altered independently at every point in space and the theory is still valid, its equations display local symmetry with respect to the characteristic.

Each of the four fundamental forces is now thought to arise from the invariance of a law of nature, such as the conservation of charge or energy, under a local symmetry operation, in which a certain parameter is altered independently at every point in space. An analogy with an ideal rubber disk may help to visualize the effect of the mathematics. If the shape of the rubber disk is likened to a natural principle and the displacement of a point within the disk is regarded as a local symmetry operation, the disk must keep its shape even as each point within it is displaced independently. The displacements stretch the disk and introduce forces between points. Similarly, in gauge theories the fundamental forces are the inevitable consequences of local symmetry operations; they are required in order to preserve symmetry.

Of the three interactions studied in the realm of elementary particles, only electromagnetism is the stuff of everyday experience, familiar in the form of sunlight, the spark of a static discharge and the gentle swing of a compass needle. On the subatomic level it takes on an unfamiliar aspect. According to relativistic quantum theory, which links matter and energy, electromagnetic interactions are mediated by photons: massless "force particles" that embody precise quantities of energy. The quantum theory of electromagnetism, which describes the photon-mediated interactions of electrically charged particles, is known as quantum electrodynamics (QED).

In common with other theories of the fundamental interactions, QED is a gauge theory. In QED the electromagnetic force can be derived by requiring that the equations describing the motion of a charged particle remain unchanged in the course of local symmetry operations. Specifically, if the phase of the wave function by which a charged particle is described in quantum theory is altered independently at every point in space, QED requires that the electromagnetic interaction and its mediating particle, the photon, exist in order to maintain symmetry.

QED is the most successful of physical theories. Using calculation methods developed in the 1940's by Richard P. Feynman and others, it has achieved predictions of enormous accuracy, such as the infinitesimal effect of the photons radiated and absorbed by an electron on the magnetic moment generated by the electron's innate spin. Moreover, QED's descriptions of the electromagnetic interaction have been verified over an extraordinary range of distances, varying from less than 10^{-18} meter to more than 10^{8} meters.

SCREENING

In particular QED has explained the effective weakening of the electromagnetic charge with distance. The electric charge carried by an object is a fixed and definite quantity. When a charge is surrounded by the other freely moving charges, however, its effects may be modified. If an electron enters a medium composed of molecules that have positively and negatively charged ends, for example, it will polarize the molecules. The electron will repel their negative ends and attract their positive ends, in effect screening itself in positive charge. The result of the polarization is to reduce the electron's effective charge by an amount that increases with distance. Only when the electron is inspected at very close range—on a submolecular scale, within the screen of positive charges—is its full charge apparent.

Such a screening effect seemingly should not arise in a vacuum, in which there are no molecules to become polarized. The uncertainty principle of Werner Heisenberg suggests, however, that the vacuum is not empty. According to the principle, uncertainty about the energy of a system increases as it is examined on progressively shorter time scales. Particles may violate the law of the conservation of energy for unobservably brief instants; in effect, they may materialize from nothingness. In QED the vacuum is seen as a complicated and seething medium in which pairs of charged "virtual" particles, particularly electrons and positrons, have a fleeting existence. These ephemeral vacuum fluctuations are polarizable just as are the molecules of a gas or a liquid. Accordingly QED predicts that in a vacuum too electric charge will be screened and effectively reduced at large distances.

The strong interaction affecting quarks that is based on the color charge also varies with distance, although in a contrary manner: instead of weakening with distance the color charge appears to grow stronger. Only at distances of less than about 10^{-13} centimeter, the diameter of a proton, does it diminish enough to allow mutually bound quarks a degree of independence. Yet the explanation for this peculiar behavior is found in a theory that is closely modeled on QED. It is a theory called quantum chromodynamics (QCD), the gauge theory of the strong interactions.

Like QED, QCD postulates force particles, which mediate interactions. Colored quarks interact through the exchange of entities called gluons, just as charged particles trade photons. Whereas QED recognizes only one kind of photon, however, QCD admits eight kinds of gluon. In contrast to the photons of QED, which do not alter the charge of interacting particles, the emission or absorption of a gluon can change a quark's color; each of the eight gluons mediates a different transformation. The mediating gluon is itself colored, bearing both a color and an anticolor.

The fact that the gluons are color-charged, in contrast to the electrically neutral photons of QED, accounts for the differing behaviors over distance of the electromagnetic and strong interactions. In QCD two competing effects govern the effective charge: screening, analogous to the screening of QED, and a new effect known as camouflage. The screening, or vacuum polarization, resembles that in electromagnetic interactions. The vacuum of QCD is populated by pairs of virtual quarks and antiquarks, winking into and out of existence. If a quark is introduced into the vacuum, virtual particles bearing contrasting color charges will be attracted to the quark; those bearing a like charge will be repelled. Hence the quark's color charge will be hidden within a cloud of unlike colors, which serves to reduce the effective charge of the quark at greater distances.

CAMOUFLAGE

Within this polarized vacuum, however, the quark itself continuously emits and reabsorbs gluons, thereby changing its color. The color-charged gluons propagate to appreciable distances. In effect they spread the color charge throughout space, thus camouflaging the quark that is the source of the charge. The smaller an arbitrary region of space centered on the quark is, the smaller will be the proportion of the quark's color charge contained in it. Thus the color charge felt by a quark of another color will diminish as it approaches the first quark. Only at a large distance will the full magnitude of the color charge be apparent.

In QCD the behavior of the strong force represents the net effect of screening and camouflage. The equations of QCD yield a behavior that is consistent with the observed paradox of quarks: they are both permanently confined and asymptotically free. The strong interaction is calculated to become extraordinarily strong at appreciable distances, resulting in quark confinement, but to weaken and free quarks at very close range.

In the regime of short distances that is probed in high-energy collisions, strong interactions are so enfeebled that they can be described using the methods developed in the context of QED for the much weaker electromagnetic interaction. Hence some of the same precision that characterizes QED can be imparted to QCD. The evolution of jets of hadrons from a quark and an antiquark generated in electron-positron annihilation, for example, is a strong interaction. QCD predicts that if the energy of the collision is high enough, the quark and the antiquark moving off in opposite directions may generate not two but three jets of hadrons. One of the particles will radiate a gluon, moving in a third direction. It will also evolve into hadrons, giving rise to a third distinct jet—a feature that indeed is commonly seen in high-energy collisions.

The three jets continue along paths set by quarks and gluons moving within an extremely confined space, less than 10^{-13} centimeter. The quark-antiquark pair cannot proceed as isolated particles beyond that distance, the limit of asymptotic freedom. Yet the confinement of quarks and of their interactions is not absolute. Although a hadron as a whole is color-neutral, its quarks do respond to the individual color charges of quarks in neighboring hadrons. The interaction, feeble compared with the color forces within hadrons, generates the binding force that holds the protons and neutrons together in nuclei.

Moreover, it seems likely that when hadronic matter is compressed and heated to extreme temperatures, the hadrons lose their individual identities. The hadronic bubbles of the image used above overlap and merge, possibly freeing their constituent quarks and gluons to migrate over great distances. The resulting state of matter, called quark-gluon plasma, may exist in the cores of collapsing supernovas and in neutron stars. Workers are now studying the possibility of creating quark-gluon plasma in the laboratory through collisions of heavy nuclei at very high energy. . . .

ELECTROWEAK SYMMETRY

Understanding of the third interaction that elementary-particle physics must reckon with, the weak interaction, also has advanced by analogy with QED. In 1933 Enrico Fermi constructed the first mathematical description of the weak interaction, as manifested in beta radioactivity, by direct analogy with QED. Subsequent work revealed several important differences between the weak and the electromagnetic interactions. The weak force acts only over distances of less than 10^{-16} centimeter (in contrast to the long range of electromagnetism), and it is intimately associated with the spin of the interacting particles. Only particles with a left-handed spin are affected by weak interactions in which electric charge is changed, as in the beta decay of a neutron, whereas right-handed ones are unaffected.

In spite of these distinctions theorists extended the analogy and proposed that the weak interaction, like electromagnetism, is carried by a force particle, which came to be known as the intermediate boson, also called the *W* (for weak) particle. In order to mediate decays in which charge is changed, the *W* boson would need to carry electric charge. The range of a force is inversely proportional to the mass of the particle that transmits it; because the photon is massless, the electromagnetic interaction can act over infinite distances. The very short range of the weak force suggests an extremely massive boson.

A number of apparent connections between electromagnetism and the weak interaction, including the fact that the mediating particle of weak interactions is electrically charged, encouraged some workers to propose a synthesis. One immediate result of the proposal that the two interactions are only different manifestations of a single underlying phenomenon was an estimate for the mass of the *W* boson. The proposed unification implied that at very short distances and therefore at very high energies the weak force is equal to the electromagnetic force. Its apparent weakness in experiments done at lower energies merely reflects its short range. Therefore the whole of the difference in the apparent strengths of the two interactions must be due to the mass of the *W* boson. Under that assumption the *W* boson's mass can be estimated at about 100 times the mass of the proton.

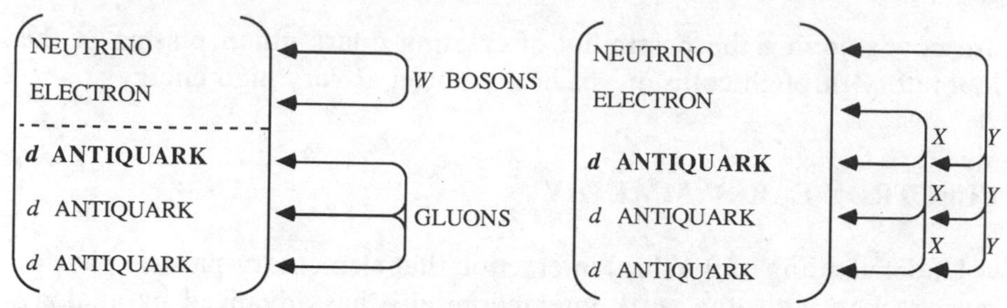

KINSHIP OF ALL MATTER is implied by unified theories of the fundamental forces; one branch of a unified family of elementary particles is shown here. Particles that are equivalent within a theory can metamorphose into one another. Because leptons, such as the electron and the neutrino, respond to the electroweak force alone whereas quarks also respond to the strong force, the two kinds of particles are not equivalent in current theory, and transformations of one into the other have not been observed (*left*). If the simplest unified theories are correct and the fundamental forces are ultimately identical, then at some very high energy quarks and leptons are interconvertible (*right*). Known transformations are mediated by force particles such as the *W* bosons and the gluons; transitions between the quark and lepton groups would be mediated by new force particles, here given as *X* and *Y*.

To advance from the notion of a synthesis to a viable theory unifying the weak and the electromagnetic interactions has required half a century of experiments and theoretical insight, culminating in the work for which Sheldon Lee Glashow and Steven Weinberg, then at Harvard University, and Abdus Salam of the Imperial College of Science and Technology in London and the International Center for Theoretical Physics in Trieste won the 1979 Nobel prize in physics. Like QED itself, the unified, or electroweak, theory is a gauge theory derived from a symmetry principle, one that is manifested in the family groupings of quarks and leptons.

Not one but three intermediate bosons, along with the photon, serve as force particles in electroweak theory. They are the positively charged W^+ and negatively charged W^- bosons, which respectively mediate the exchange of positive and negative charge in weak interactions, and the Z^0 particle, which mediates a class of weak interactions known as neutral current processes. Neutral current processes such as the elastic scattering of a neutrino from a proton, a weak interaction in which no charge is exchanged, were predicted by the electroweak theory and first observed at CERN in 1973. They represent a further point of convergence between electromagnetism and the weak interaction in that electromagnetic inter-

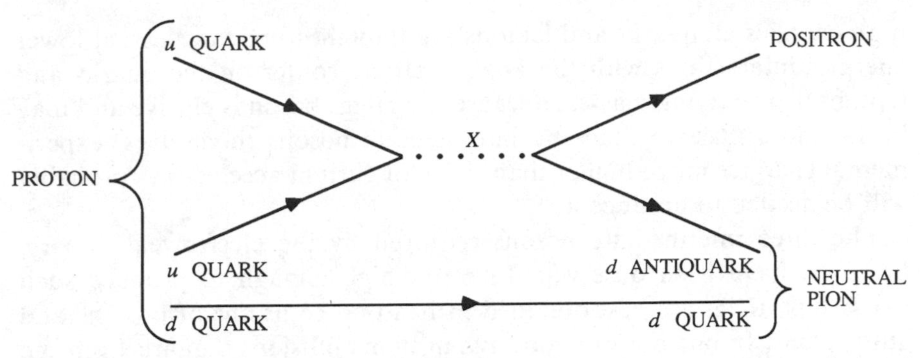

DECAY OF THE PROTON is a possible consequence of transformations of quarks into leptons, a phenomenon unified theories would allow. The diagram shows one of several proposed decay routes. The proton's constituent *u* quarks combine to form an *X* particle, which disintegrates into a *d* antiquark and a positron (a lepton). The *d* antiquark combines with the remaining quark of the proton, a *d* quark, to form a neutral pion. Because pions are composed of matter and antimatter, they are short-lived; the mutual annihilation of their constituents will release energy in the form of two photons. The positron is also ephemeral: an encounter with a stray electron, its antiparticle, will convert it into energy as well.

actions do not change the charge of participating particles either.

To account for the fact that the electromagnetic and weak interactions, although they are intimately related, take different guises, the electroweak theory holds that the symmetry uniting them is apparent only at high energies. At lower energies it is concealed. An analogy can be drawn to the magnetic behavior of iron. When iron is warm, its molecules, which can be regarded as a set of infinitesimal magnets, are in hectic thermal motion and therefore randomly oriented. Viewed in the large the magnetic behavior of the iron is the same from all directions, reflecting the rotational symmetry of the laws of electromagnetism. When the iron cools below a critical temperature, however, its molecules line up in an arbitrary direction, leaving the metal magnetized along one axis. The symmetry of the underlying laws is now concealed.

The principal actor in the breaking of the symmetry that unites electromagnetism and the weak interaction at high energies is a postulated particle called the Higgs boson. It is through interactions with the Higgs boson that the symmetry-hiding masses of the intermediate bosons are generated. The Higgs boson is also held to be responsible for the fact that quarks and leptons within the same family have different masses. At very high energies all quarks and leptons are thought to be massless; at lower energies interactions with the Higgs particle confer on the quarks and leptons their varying masses. Because the Higgs boson is elusive and may be far more massive than the intermediate bosons themselves, experimental energies much higher than those of current accelerators probably will be needed to produce it.

The three intermediate bosons required by the electroweak theory, however, have been observed. Energies high enough to produce such massive particles are best obtained in head-on collisions of protons and antiprotons. In one out of about five million collisions a quark from the proton and an antiquark from the antiproton fuse, yielding an intermediate boson. The boson disintegrates less than 10^{-24} second after its formation. Its brief existence, however, can be detected from its decay products.

In a triumph of accelerator art, experimental technique and theoretical reasoning, international teams at CERN led by Carlo Rubbia of Harvard and Pierre Darriulat devised experiments that in 1983 detected the W bosons and the Z^0 particle. An elaborate detector identified and recorded in the debris of violent proton-antiproton collisions single electrons whose trajectory matched the one expected in a W^- particle's decay; the detector also recorded electrons and positrons traveling in precisely opposite directions, unmistakable evidence of the Z^0 particle. For their part in the experiments and in the design and construction of the proton-antiproton collider and the detector Rubbia and Simon van der Meer of CERN were awarded the 1984 Nobel prize in physics.

UNIFICATION

With QCD and the electroweak theory in hand, what remains to be understood? If both theories are correct, can they also be complete? Many observations are explained only in part, if at all, by the separate theories of the strong and the electroweak interactions. Some of them seem to invite a further unification of the strong, weak and electromagnetic interactions.

Among the hints of deeper patterns is the striking resemblance of quarks and leptons. Particles in both groups are structureless at current experimental resolution. Quarks possess color charges whereas leptons do not, but both carry a half unit of spin and take part in electromagnetic and weak interactions. Moreover, the electroweak theory itself suggests a relation between quarks and leptons. Unless each of the three lepton families (the electron and its neutrino, for example) can be linked with the corresponding family of quarks (the *u* and *d* quarks, in their three colors), the electroweak theory will be beset with mathematical inconsistencies.

What is known about the fundamental forces also points to a unification. All three can be described by gauge theories, which are similar in their mathematical structure. Moreover, the strengths of the three forces appear likely to converge at very short distances, a phenomenon that would be apparent only at extremely large energies. We have seen that the electromagnetic charge grows strong at short distances, whereas the strong, or color, charge becomes increasingly feeble. Might all the interactions become comparable at some gigantic energy?

If the interactions are fundamentally the same, the distinction between quarks, which respond to the strong force, and leptons, which do not, begins to dissolve. In the simplest example of a unified theory, put forward by Glashow and Howard Georgi of Harvard in 1974, each matched set of quarks and leptons gives rise to an extended family containing all the various states of charge and spin of each of the particles.

The mathematical consistency of the proposed organization of matter is impressive. Moreover, regularities in the scheme require that electric charge be apportioned among elementary particles in multiples of exactly ⅓, thereby accounting for the electrical neutrality of stable matter. The atom is neutral only because when quarks are grouped in threes, as they are in the nucleus, their individual charges combine to give a charge that is a precise integer, equal and opposite to the charge of an integral number of electrons. If quarks were unrelated to leptons, the precise relation of their electric charges could only be a remarkable coincidence.

In such a unification only one gauge theory is required to describe all the interactions of matter. In a gauge theory each particle in a set can be transformed into any other particle. Transformations of quarks into other quarks and of leptons into other leptons, mediated by gluons and intermediate bosons, are familiar. A unified theory suggests that quarks can change into leptons and vice versa. As in any gauge theory, such an interaction would be mediated by a force particle: a postulated *X* or *Y* boson. Like other gauge theories, the unified theory describes the variation over distance of interaction strengths. According to the simplest of the unified theories, the separate strong and electroweak interactions converge and become a single interaction at a distance of 10^{-29} centimeter, corresponding to an energy of 10^{24} electron volts.

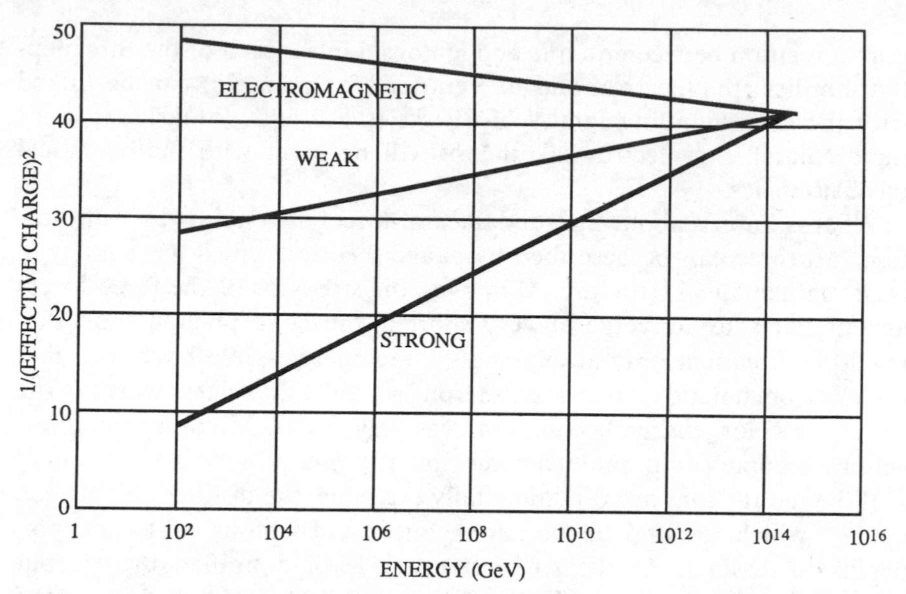

CONVERGENCE OF FORCES at extreme high energies, which are equivalent to very small scales of distance, is expected in unified theories. The graph gives an inverse measure of the forces' intrinsic strength; that of the strong and weak forces diminishes with energy, whereas that of electromagnetism increases. The simplest unified theory predicts that the fundamental identity of the three forces is revealed in interactions taking place at an energy of more that 10^{15} GeV, which corresponds to a distance of less than 10^{-29} centimeter.

Such an energy is far higher than may ever be attained in an accelerator, but certain consequences of unification might be apparent even in the low-energy world we inhabit. The supposition that transformations can cross the boundary between quarks and leptons implies that matter, much of whose mass consists of quarks, can decay. If, for example, the two *u* quarks in a proton were to approach each other closer than 10^{-29} centimeter, they might combine to form an *X* boson, which would disintegrate into a positron and a *d* antiquark. The antiquark would then combine with the one remaining quark of the proton, a *d* quark, to form a neutral pion, which itself would quickly decay into two photons. In the course of the process much of the proton's mass would be converted into energy.

The observation of proton decay would lend considerable support to a unified theory. It would also have interesting cosmological consequences. The universe contains far more matter than it does antimatter. Since matter and antimatter are equivalent in almost every respect, it is appealing to speculate that the universe was formed with equal amounts of both. If the number of baryons—three-quark particles such as the proton and the neutron, which constitute the bulk of ordinary matter—can change, as the decay of the proton would imply, then the current excess of matter need not represent the initial state of the universe. Originally matter and antimatter may indeed have been present in equal quantities, but during the first instants after the big bang, while the universe remained in a state of extremely high energy, processes that alter baryon number may have upset the balance.

A number of experiments have been mounted to search for proton decay. The large unification energy implies that the mean lifetime of the proton must be extraordinarily long—10^{30} years or more. To have a reasonable chance of observing a single decay it is necessary to monitor an extremely large number of protons; a key feature of proton-decay experiments has therefore been large scale. The most ambitious experiment mounted to date is an instrumented tank of purified water 21 meters on a side in the Morton salt mine near Cleveland. During almost three years of monitoring none of the water's more than 10^{33} protons has been observed to decay, suggesting that the proton's lifetime is even longer than the simplest unified theory predicts. In some rival theories, however, the lifetime of the proton is considerably longer, and there are other theories in which protons decay in ways that would be difficult to detect in existing experiments. Furthermore, results from other experiments hint that protons can indeed decay.

OPEN QUESTIONS

Besides pointing the way to a possible unification, the standard model, consisting of QCD and the electroweak theory, has suggested numerous sharp questions for present and future accelerators. Among the many goals for current facilities is an effort to test the predictions of QCD in greater detail. Over the next decade accelerators with the higher energies needed to produce the massive W and Z^0 bosons in adequate numbers will also add detail to electroweak theory. It would be presumptuous to say these investigations will turn up no surprises. The consistency and experimental successes of the standard model at familiar energies strongly suggest, however, that to resolve fundamental issues we need to take a large step up in interaction energy from the several hundred GeV (billion electron volts) attainable in the most powerful accelerators now being built.

Although the standard model is remarkably free of inconsistencies, it is incomplete; one is left hungry for further explanation. The model does not account for the pattern of quark and lepton masses or for the fact that although weak transitions usually observe family lines, they occasionally cross them. The family pattern itself remains to be explained. Why should there be three matched sets of quarks and leptons? Might there be more?

Twenty or more parameters, constants not accounted for by theory, are required to specify the standard model completely. These include the coupling strengths of the strong, weak and electromagnetic interactions, the masses of the quarks and leptons, and parameters specifying the interactions of the Higgs boson. Furthermore, the apparently fundamental constituents and force particles number at least 34: 15 quarks (five flavors, each in three colors), six leptons, the photon, eight gluons, three intermediate bosons and the postulated Higgs boson. By the criterion of simplicity the standard model does not seem to represent progress over the ancient view of matter as made up of earth, air, fire and water, interacting through love and strife. Encouraged by historical precedent, many physicists account for the diversity by proposing that these seemingly fundamental particles are made up of still smaller particles in varying combinations.

There are two other crucial points at which the standard model seems to falter. Neither the separate theories of the strong and the electroweak interactions nor the conjectured unification of the two takes any account of gravity. Whether gravity can be described in a quantum theory and unified with the other fundamental forces remains an open question. Another basic deficiency of the standard model concerns the Higgs boson. The electroweak theory requires that the Higgs boson exist but does not specify precisely how the particle must interact with other particles or even what its mass must be, except in the broadest terms.

THE SUPERCONDUCTING SUPERCOLLIDER

What energy must we reach, and what new instruments do we need, to shed light on such fundamental problems? The questions surrounding the Higgs boson, although they are by no means the only challenges we face, are particularly well defined, and their answers will bear on the entire strategy of unification. They set a useful target for the next generation of machines.

It has been proposed that the Higgs boson is not an elementary particle at all but rather a composite object made up of elementary constituents analogous to quarks and leptons but subject to a new kind of strong interaction, often called technicolor, which would confine them within about 10^{-17} centimeter. The phenomena that would reveal such an interaction would become apparent at energies of about 1 TeV (trillion electron volts). A second approach to the question of the Higgs boson's mass and behavior employs a postulated principle known as supersymmetry, which relates particles that differ in spin. Supersymmetry entails the existence of an entirely new set of elusive, extremely massive particles. The new particles would correspond to known quarks, leptons and bosons but would differ in their spins. Because of their mass, such particles would reveal themselves fully only in interactions taking place at very high energy, probably about 1 TeV.

Our best hope for producing interactions of fundamental particles at energies of 1 TeV is an accelerator known as the Superconducting Supercollider (SSC). Formally recommended to the Department of Energy in 1983 by the High Energy Physics Advisory Panel, it would incorporate proved technology on an unprecedented scale. A number of designs have been put forward, but all envision a proton-proton or proton-antiproton collider. High-energy beams of protons are produced more readily with current technology than beams of electrons and positrons, although electron-positron collisions are generally simpler to analyze; because protons are composite particles, their collisions yield a larger variety of interactions than collisions of electrons and positrons. Another common feature of the designs is the use of superconducting magnets, first employed on a large scale in the Tevatron Collider at the Fermi National Accelerator Laboratory (Fermilab) in Batavia, Ill. The technology increases the field strength and lowers the power consumption of the magnets that bend and confine the beam.

One of the more compact designs incorporates niobium-titanium alloy magnets cooled to 4.4 degrees Celsius above absolute zero. If the magnets generated fields of five tesla (100,000 times the strength of the earth's magnetic field), two counterrotating beams of protons accelerated to energies of 20 TeV (needed to produce 1-TeV interactions of the quarks and gluons within the protons) could be confined within a loop about 30 kilometers in diameter. In other designs magnetic fields are lower and the proposed facility is correspondingly larger.

It is believed such a device could be operational in 1994, at a cost of $3 billion. The Department of Energy has encouraged the establishment of a Central Design Group to formulate a specific construction proposal within three years and is currently funding the development of magnets for the SSC at several laboratories.

The SSC represents basic research at unprecedented cost on an unmatched scale. Yet the rewards will be proportionate. The advances of the past decade have brought us tantalizingly close to a profound new understanding of the fundamental constituents of nature and their interactions. Current theory suggests that the frontier of our ignorance falls at energies of about 1 TeV. Whatever clues about the unification of the forces of nature and the constituents of matter wait beyond that frontier, the SSC is likely to reveal them.

Index